Ionic Liquids as Green Solvents

ACS SYMPOSIUM SERIES **856**

Ionic Liquids as Green Solvents

Progress and Prospects

Robin D. Rogers, Editor
The University of Alabama

Kenneth R. Seddon, Editor
The Queen's University of Belfast

American Chemical Society, Washington, DC

Library of Congress Cataloging-in-Publication Data

Ionic liquids as green solvents : progress and prospects / Robin D. Rogers, editor [and] Kenneth R. Seddon, editor.

p. cm.—(ACS symposium series ; 856)

"Sponsored by the ACS Division of Industrial and Engineering Chemistry, Inc.

"Developed from a symposium ... at the 244th National Meeting of the American Chemical Society, Boston, Massachusetts, August 18–22, 2002"--

Includes bibliographical references and index.

ISBN 0–8412–3856–1

1. Ionic solutions—Congresses. 2. Solution (Chemistry)—Congresses.

I. Rogers, Robin D. II. Seddon, Kenneth R., 1950- III. American Chemical Society. Division of Industrial and Engineering Chemistry, Inc. IV. American Chemical Society. Meeting (224th : 2002 : Boston, Mass.) V. Series.

QD543.I58 2003
541.3′482—dc21 2003048138

The paper used in this publication meets the minimum requirements of American National Standard for Information Sciences—Permanence of Paper for Printed Library Materials, ANSI Z39.48–1984.

Distributed by Oxford University Press

PRINTED IN THE UNITED STATES OF AMERICA

Foreword

The ACS Symposium Series was first published in 1974 to provide a mechanism for publishing symposia quickly in book form. The purpose of the series is to publish timely, comprehensive books developed from ACS sponsored symposia based on current scientific research. Occasionally, books are developed from symposia sponsored by other organizations when the topic is of keen interest to the chemistry audience.

Before agreeing to publish a book, the proposed table of contents is reviewed for appropriate and comprehensive coverage and for interest to the audience. Some papers may be excluded to better focus the book; others may be added to provide comprehensiveness. When appropriate, overview or introductory chapters are added. Drafts of chapters are peer-reviewed prior to final acceptance or rejection, and manuscripts are prepared in camera-ready format.

As a rule, only original research papers and original review papers are included in the volumes. Verbatim reproductions of previously published papers are not accepted.

ACS Books Department

Contents

Preface..........xiii

Overview

1. **Selection of Ionic Liquids for Green Chemical Applications..........2**
   John D. Holbrey, Megan B. Turner, and Robin D. Rogers

Ionic Liquids: Manufacture and Synthesis

2. **Ionic Liquids: Improved Syntheses and New Products..........14**
   Adrian J. Carmichael, Maggel Deetlefs, Martyn J. Earle, Ute Fröhlich, and Kenneth R. Seddon

3. **Challenges to the Commercial Production of Ionic Liquids..........32**
   Philip E. Rakita

4. **Industrial Preparation of Phosphonium Ionic Liquids..........41**
   Christine J. Bradaric, Andrew Downard, Christine Kennedy, Allan J. Robertson, and Yuehui Zhou

5. **New Ionic Liquids Based on Alkylsulfate and Alkyl Oligoether Sulfate Anions: Synthesis and Applications..........57**
   Peter Wasserscheid, Roy van Hal, Andreas Bösmann, Jochen Eßer, and Andreas Jess

6. **Green Synthesis of Ionic Liquids for Green Chemistry..........70**
   Rex X. Ren

7. **Expeditious Synthesis of Ionic Liquids Using Ultrasound and Microwave Irradiation..........82**
   Rajender S. Varma

8. **Unprecedented Synthesis of 1,3-dialkylimidazolium-2-carboxylate: Applications in the Synthesis of Halogen-Free Ionic Liquids and Reactivity as Carbon Dioxide Transfer Agent to Active C–H Bonds....................................93**
M. Aresta, I. Tkatchenko, and I. Tommasi

9. **Commercially Available Salts as Building Blocks for New Ionic Liquids..100**
James H. Davis, Jr. and Phillip A. Fox

Characterization and Engineering

10. **Phase Equilibria of Gases and Liquids with 1-*n*-butyl-3-Methylimidazolium Tetrafluoroborate..110**
Jennifer L. Anthony, Jacob M. Crosthwaite, Daniel G. Hert, Sudhir N. V. K. Aki, Edward J. Maginn, and Joan F. Brennecke

11. **Heat Capacities of Ionic Liquids and Their Applications as Thermal Fluids...121**
John D. Holbrey, W. Matthew Reichert, Ramana G. Reddy, and Robin D. Rogers

12. **Thermodynamic Properties of Liquid Mixtures Containing Ionic Liquids...134**
Andreas Heintz, Jochen K. Lehmann, and Sergey P. Verevkin

13. **Liquid and Solid-State Structures of 1,3-Dimethylimidazolium Salts...151**
D. T. Bowron, C. Hardacre, J. D. Holbrey, S. E. J. McMath, M. Nieuwenhuyzen, and A. K. Soper

14. **Molecular Structure of Various Ionic Liquids from Gas Phase Ab Initio Calculations...162**
Timothy I. Morrow and Edward J. Maginn

15. **Transition Structure Models of Organic Reactions in Chloroaluminate Ionic Liquids: Cyclopentadiene and Methyl Acrylate Diels–Alder Reaction in Acidic and Basic Melts of 1-Ethyl-3-methylimidazolium Chloride with Aluminum(III) Chloride..174**
Orlando Acevedo and Jeffrey D. Evanseck

Biotechnology in Ionic Liquids

16. **Biotransformations in Ionic Liquids: An Overview............................192**
Roger A. Sheldon, F. van Rantwijk, and R. Madeira Lau

17. **Enzymatic Condensation Reactions in Ionic Liquids..........................206**
Nicole Kaftzik, Sebastian Neumann, Maria-Regina Kula, and Udo Kragl

18. **Aspects of Chemical Recognition and Biosolvation within Room Temperature Ionic Liquids...212**
Gary A. Baker, Sheila N. Baker, T. Mark McCleskey, and James H. Werner

19. **Ionic Liquids Create New Opportunities for Nonaqueous Biocatalysis with Polar Substrates: Acylation of Glucose and Ascorbic Acid..225**
Seongsoon Park, Fredrik Viklund, Karl Hult, and Romas J. Kazlauskas

20. **Enzymatic Catalysis in Ionic Liquids and Supercritical Carbon Dioxide..239**
Pedro Lozano, Teresa De Diego, Daniel Carrié, Michel Vaultier, and José L. Iborra

21. **Efficient Lipase-Catalyzed Enantioselective Acylation in an Ionic Liquid Solvent System...251**
Toshiyuki Itoh, Yoshihito Nishimura, Masaya Kashiwagi, and Makoto Onaka

Non-Catalytic and Calalytic Chemistry

22. **Acids and Bases in Ionic Liquids..264**
Douglas R. MacFarlane and Stewart A. Forsyth

23. **Catalytic Olefin Epoxidation and Dihydroxylation with Hydrogen Peroxide in Common Ionic Liquids: Comparative Kinetics and Mechanistic Study...................................277**
Mahdi M. Abu-Omar, Gregory S. Owens, and Armando Durazo

24. **Polarity Variation of Room Temperature Ionic Liquids and Its Influence on a Diels–Alder Reaction....................................289**
Richard A. Bartsch and Sergei V. Dzyuba

25. **Polar, Non-Coordinating Ionic Liquids as Solvents for Coordination Polymerization of Olefins.....................................300**
Kevin H. Shaughnessy, Marc A. Klingshirn, Steven J. P'Pool, John D. Holbrey, and Robin D. Rogers

26. **The Importance of Hydrogen Bonding to Catalysis in Ionic Liquids: Inhibition of Allylic Substitution and Isomerization by [bmim][BF_4]...314**
James Ross and Jianliang Xiao

27. **Recent Developments in the Use of *N*-Heterocyclic Carbenes: Applications in Catalysis...323**
Rebecca M. Kissling, Mihai S. Viciu, Gabriela A. Grasa, Romain F. Germaneau, Tatyana Güveli, Marie-Christiane Pasareanu, Oscar Navarro-Fernandez, and Steven P. Nolan

Photochemistry

28. **An Overview of Photochemistry in Ionic Liquids..............................344**
Richard M. Pagni

29. **Diffusion-Controlled Reactions in Room Temperature Ionic Liquids...357**
Charles M. Gordon, Andrew J. McLean, Mark J. Muldoon, and Ian R. Dunkin

30. **Amine Mediated Photoreduction of Aryl Ketones in *N*-Heterocyclic Ionic Liquids: Novel Solvent Effects Leading to Altered Product Distribution..370**
Paul B. Jones, John L. Reynolds, Robert G. Brinson, and Ryan L. Butke

31. **Radiation Chemistry of Ionic Liquids: Reactivity of Primary Species...381**
James F. Wishart

32. **Pulse Radiolysis Studies of Reaction Kinetics in Ionic Liquids..397**
P. Neta, D. Behar, and J. Grodkowski

33. **Organic Electrochemistry in Ionic Liquids......................................410**
Andrew P. Doherty and Claudine A. Brooks

34. **Solvent–Solute Interactions in Ionic Liquid Media: Electrochemical Studies of the Ferricenium–Ferrocene Couple...421**
M. Cristina Lagunas, William R. Pitner, Jan-Albert van den Berg, and Kenneth R. Seddon

35. **Electrochemical Studies of Ambient Temperature Ionic Liquids Based on Choline Chloride...439**
Andrew P. Abbott, Glen Capper, David L. Davies, Helen Munro, Raymond K. Rasheed, and Vasuki Tambyrajah

36. **Electrodeposition of Nanoscale Metals and Semiconductors from Ionic Liquids...453**
Sherif Zein El Abedin and Frank Endres

Novel Applications

37. **Plasticizing Effects of Imidazolium Salts in PMMA: High-Temperature Stable Flexible Engineering Materials...............468**
Mark P. Scott, Michael G. Benton, Mustafizur Rahman, and Christopher S. Brazel

38. **The Use of Ionic Liquids in Polymer Gel Electrolytes........................478**
Hugh C. De Long, Paul C. Trulove, and Thomas E. Sutto

39. **Electrochemistry: Ionic Liquid Electroprocessing of Reactive Metals...495**
Jianming Lu and David Dreisinger

40. **Electrochemistry: Electrochemically Generated Superoxide Ion in Ionic Liquids: Applications to Green Chemistry.....................509**
I. M. AlNashef, M. A. Matthews, and J. W. Weidner

41. **Conventional Aspects of Unconventional Solvents: Room Temperature Ionic Liquids as Ion-Exchangers and Ionic Surfactants...526**
Mark L. Dietz, Julie A. Dzielawa, Mark P. Jensen, and Millicent A. Firestone

42. Extraction of Chlorophenols from Water Using Room Temperature Ionic Liquids..544
Evangelia Bekou, Dionysios D. Dionysiou, Ru-Ying Qian, and Gregory D. Botsaris

Indices

Author Index...563

Subject Index...565

Preface

The chapters in this book are based on papers that were presented at the symposium *Ionic Liquids as Green Solvents: Progress and Prospects* at the 224th American Chemical Society (ACS) National Meeting held August 18–22, 2002, in Boston, Massachusetts. It followed, by eighteen months, the first successful ionic liquids symposium at the ACS meeting in April 2001 in San Diego, California. Judging by the presentations and by the symposium attendance (more than 300 at one session), the field is rapidly moving forward with high-quality work, and interest in ionic liquids remains high from both academia and industry. The talks showed the depth of research currently being undertaken, the broad and diverse base for activities, and the excitement and potential opportunities that exist and are continuing to emerge, in the field.

Ionic Liquids are now defined as salts that melt below about 100 °C. It should be noted that these salts are not 'defined' as 'green,' and care must be taken when examining materials with unknown toxicity, Biological Oxygen Demand, etc. Ionic liquids define a *class* of fluids rather than a small group of individual examples: Ionic liquids can be designed to be flammable, unstable, or even toxic. Although this may be obvious to many in the field, it is clear that confusion can occur leading to an implicit assumption that all ionic liquids are always (1) green, (2) non-toxic, and (3) environmentally friendly, or (indeed) the converse. As scientists and engineers, we must continue to present a balanced view of ionic liquids and strive to answer questions regarding the environmental sustainability of their use.

The Boston meeting comprised ten half-day sessions that broadly reflected the areas of development and interest in ionic liquids as green solvents for chemistry. We are indebted to the session organizers who each planned and developed a half-day session, invited the speakers, and presided over the session. The featured topics and the presiding organizers for each session were Ionic Liquid Tutorial (K. R. Seddon and R. D. Rogers), Manufacture and Synthesis of Ionic Liquids: Industrial (A. J. Robertson) and Academic (K. R. Seddon), Characterization and Engineering (J. F. Brennecke), Novel Applications of Ionic Liquids (J. D. Holbrey), Separations (R. D. Rogers), Biotechnology (R. A. Sheldon), Catalytic Chemistry (T. Welton), Non-Catalytic Chemistry (M. J. Earle), Electrochemistry (W. R. Pitner), and Photochemistry and Reaction Intermediates (C. M. Gordon). Starting with an experience-based series of tutorials on using ionic liquids, which gave the attendees an extended opportunity to ask questions of experienced practitioners in the field, the symposium moved through the manufacture and production of ionic liquids, on both industrial and lab scale, and into the use of ionic liquids as solvents for chemical applications. Page restrictions prevented the publication of all the presentations, but we have tried to select a representative subset of the papers.

It is remarkable to think that, since the publication of the first book dedicated to room temperature ionic liquids (*Ionic Liquids: Industrial Applications for Green Chemistry*, ACS Symposium Series 818; the volume based on the San Diego meeting) in Autumn 2002, two other books have already appeared. The first of these is a multi-authored volume (edited by Peter Wasserscheid and Tom Welton) entitled *Ionic Liquids in Synthesis* (Wiley, 2003), the second is the proceedings volume (edited by ourselves and Sergei Volkov) of the NATO ARW held in Crete, the first international meeting on ionic liquids, entitled *Green Industrial Applications of Ionic Liquids* (Kluwer, 2002). This current volume, therefore, is the fourth book to appear on the subject of ionic liquids within nine months. What is truly remarkable is that very little overlap occurs in content, and that all four books are complementary; if you need one, you need them all! If this were not enough to reflect the astonishing intensification of interest in the field, Figure 1 illustrates the growth in publications on ionic liquids. If this trend is continued into the future (and the 2002 data were not complete at the time we went to proof), then at least another 500 papers will have appeared by the time this book is published.

It can be seen, both by the range and quality of contributions to this symposium volume, and also by the vastly increased volume of publications in the open and patent literature, that significant efforts are being

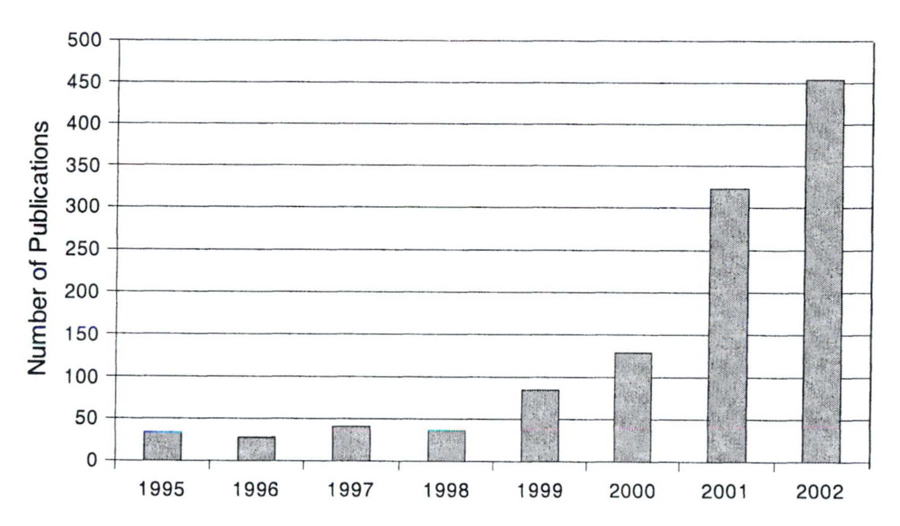

Figure 1: Publications containing the phrase 'ionic liquid' in the title, abstract, or keywords, determined by ISI Web of Science, as a function of time. (Courtesy of Dr. Annegret Stark.)

made in utilizing ionic liquids in green chemistry. It is clear that *progress* is indeed being made, and that the *prospects* for the future are good, based on the continuing commitment to excellent and innovative research from both industry and academia that are reflected in the contributions to this volume. For these reasons, we are planning a third symposium in New York, at the Autumn 2003 ACS National Meeting, September 7–11, 2003. This will be the final meeting of the ACS trilogy, by which time we will have (we hope) achieved our aim of communicating the excitement, breadth, and depth of the applications of ionic liquids to an audience with strong representation from academia, industry, and government, with particular emphasis on green industrial applications. We will then be doffing our editorial clothes, hanging up our hats, and passing the gauntlet!

This symposium was successful because of the invaluable support it received from industry, academia, government, and our professional society. Industrial support was received from Cytec Industries Inc., Fluka, Merck KGaA, Ozark Fluorine Specialties, SACHEM Inc., Solvent Innovation, and Strem Chemicals Inc. Academic contributions were received from The University of Alabama Center for Green Manufacturing and The Queen's University Ionic Liquid Laboratory (QUILL).

The U.S. Environmental Protection Agency's Green Chemistry Program also supported the meeting. Of course, we are (as always) indebted to the ACS and its many programs for their help, encouragement, and support. We especially thank the Industrial and Engineering Chemistry Division, the I&EC Separations Science as well as the Technology and Green Chemistry Subdivisions, and the Green Chemistry Institute. Another measure of success was the impressive strength of the student contributions, in the tutorial, oral, and poster sessions; on this basis, our future is in safe hands. Thank you all for helping maintain a strong program!

Robin D. Rogers
Center for Green Manufacturing
Box 870336
The University of Alabama
Tuscaloosa, AL 35487
Telephone: +1 205–348–4323
Fax: +1 205–348–0823
Email: RDRogers@bama.ua.edu
URL: http://bama.ua.edu/~rdrogers

Kenneth R. Seddon
QUILL Research Centre
The Queen's University of Belfast
Stranmillis Road
Belfast, Northern Ireland BT9 5AG
United Kingdom
Telephone: +44 28 90335420
Fax: +44 28 90665297
Email: k.seddon@qub.ac.uk
URL: http://quill.qub.ac.uk/

Overview

Chapter 1

Selection of Ionic Liquids for Green Chemical Applications

John D. Holbrey, Megan B. Turner, and Robin D. Rogers

Center for Green Manufacturing and Department of Chemistry, The University of Alabama, Tuscaloosa, AL 35487

Ionic liquids (ILs) are proving to be increasingly promising as viable media for not only potentially 'green' synthesis and separations operations, but also for novel applications, where the unique property set of the IL materials provides new options based upon different chemical and physical properties. The range and variability in the properties between individual examples within the class of solvents that are known as ILs, however, are both challenges, and opportunities for developing new and improved processes. Some of the challenges in understanding IL behavior and in selecting specific IL media for applications is presented in the context of research from The University of Alabama.

Introduction

The 'greening' of chemical technology is based on opportunities, both real and perceived, to improve processes by using 'green principles' (*1*) that translate to improvements in production processes. These can be achieved by reducing or eliminating waste, improving chemical syntheses, extractions and/or separations, and reusability of reaction solvent. One of the core aspects of green chemistry that is attractive for both industrial and academic research teams to tackle is the redesign of chemical processes to reduce, or eliminate, losses of solvents, particularly in situations where these are volatile organic compounds (VOCs). A number of approaches can be made, including incremental modification to reaction design (process optimization), increased efforts in recovery and recycling, and implementing a switch to solvents that are more environmentally benign. This can include both examples that are intrinsically biodegradable, or 'naturally occurring', or by design, utilizing properties of the solvents to ensure that they are not inadvertently released to the environment.

The properties of ILs, and their uses as solvents for chemical reactions, have been reviewed extensively in recent years (*2,3*). "Ideal" solvent requirements may include low toxicity, low cost, high solute selectivity, inertness to materials, non-flammability, high capacity for solutes, low carrier selectivity, and moderate interfacial tension. Ionic liquids (ILs) can meet some of these requirements now, but it remains necessary to address issues of cost, availability, toxicity, and recycling.

Ionic Liquids – An Endless Trail of Possibilities

The term 'Ionic Liquid' is used loosely to describe organic salts that melt below about 100 °C (*4*) and have an appreciable liquid range. ILs define a *class* of fluids rather than a small group of individual examples – the implications of this, with respect to the choice of ILs for particular, specific, processes will be developed later. The most commonly studied systems contain ammonium, phosphonium, pyridinium, or imidazolium cations, with varying heteroatom functionality.

Seddon (*5*) has remarked that over 10^{18} simple organic salts that might be potential ionic liquids could be prepared by varying the substitution patterns and anion choices, even just within imidazolium and pyridinium systems. In fact, over 30,000 1,3-functionalized imidazolium entries are recorded in the CAS database. Further scope for derivatization beyond ramification of linear alkyl-substituents, for example with branched, chiral, fluorinated, or an active-functionality, can yield further useful materials. The degree and type of substitution renders the salts low melting, largely by reducing cation-anion Coulombic interactions and disrupting ion-ion packing. This results in low melting salts with reduced lattice energy and a marked tendency to form glasses on cooling, rather than crystalline solids (*6,7*).

Common anions that yield useful ILs include hexafluorophosphate $[PF_6]^-$, tetrafluoroborate, $[BF_4]^-$, bis(trifyl)imide, $[NTf_2]^-$, and chloride, Cl^-. Although high symmetry *pseudo*-spherical, non-coordinating anions are commonly regarded as optimal for formation of ILs, the existence of low melting ILs containing anions such as methylsulfate (*8*), dicyanamide (*9*), and bis(trifyl)imide (*10*), show that shape and ion-interaction factors are far from clear-cut.

Anions can control the solvent's reactivity with water, coordinating ability, and hydrophobicity. The $[PF_6]^-$ and $[NTf_2]^-$ anions produce hydrophobic solvents due to their lack of hydrogen-bond accepting ability (*11*), though not all ILs containing these anions are hydrophobic. Control over hydrophobicity and other physical properties is governed by cation-anion pair interactions. The range of functionality and resulting properties of ILs appear to suggest that chemometric design and factorial approaches to developing new ILs and studying the properties and characteristics would be advantageous: however, as yet, this does not appear to have been exploited.

ILs are curious materials to be posited as solvents for 'green chemistry'. ILs are advanced, technological solvents that can be designed to fit a particular application. One regularly suggested advantage of ILs over VOCs as solvents, for both synthetic chemistry and for electrochemistry, is the intrinsic lack of vapor pressure. However, it is important, when discussing potential benefits from green chemical approaches, to remember that the aim is for improvements in the overall process. Although replacement of a VOC solvent in a process with an IL solvent will, necessarily, reduce VOC emissions from that reaction, overall efficiency (atom efficiency or *E*-factor (*12*)) depends on consideration of the overall process, not just the solvent used.

In particular, it is important to emphasize that, although ILs are chemicals that can be applied as solvents and catalysts in green chemistry processes, they are not necessarily green chemicals. ILs can be designed to be flammable, unstable, or even toxic.

A general absence of data on toxicity, environmental fate (decomposition, BOD, bioinhibition data, *etc.*) and, it appears, a reluctance to collect some of these data, does not help to support the green platform. While acute toxicity determinations take both time and money, environmental fate and cellular inhibition measurements are relatively quick and easy to perform and should be made.

Choice of Ionic Liquid – The Case Against 1-Butyl-3-methylimidazolium Hexafluorophosphate

In terms of investigating the applications of ILs as media for chemical reactions and separations, as distinct from electrochemistry, the most widely studied IL is 1-butyl-3-methylimidazolium hexafluorophosphate ($[C_4mim][PF_6]$) (*13*). The desirable features of an ionic, yet essentially water-immiscible, solvent as a replacement of VOCs in processes is obvious and does not need restating here.

Hexafluorophosphate-containing ILs have been used for a combination reasons, including historically wide use within the peer group, its hydrophobic and non-coordinating nature, and its ease of preparation (*11*) as shown in Figure 1. This is despite the well established instability towards hydrolysis in contact with moisture forming volatiles, including HF, POF_3, *etc.*, which can dissolve glassware and damage steel autoclaves and reactors. Proper care should be exercised when using the $[PF_6]^-$-containing ILs, as with all compounds containing possibly harmful decomposition products.

Figure 1. Ready synthesis and separation of $[C_4mim][PF_6]$. Metathesis of the halide salts in aqueous solution results in separation of the $[PF_6]^-$-containing IL as a dense, separate phase allowing simple, rapid isolation. If a neutral salt (i.e. $Na[PF_6]$ or $[NH_4][PF_6]$) is used for the metathesis step, extensive washing and neutralization of acid from the IL product is not required.

One situation where this is common is in the published procedures for preparing these ILs from aqueous solution, followed by extended drying under vacuum at high temperature to remove the residual water. Many novice researchers, preparing these ILs, may not be aware of the white fumes rising from the IL. This white fume may contain HF, among other species, which is very toxic and corrosive. It is important to emphasize that considerable care should be taken when handling possible HF-containing compounds.

The use of the $[PF_6]^-$ anion is probably not entirely acceptable within the 'green paradigm'. It is worth noting that hydrolysis followed by phosphate analysis by ion chromatography is a standard method for the assay of $[PF_6]^-$ anions. Although these ILs do not conform to the green principals discussed earlier, they will continue to have a use mostly in primary research rather than commercial applications. $[C_4mim][PF_6]$ has been widely studied, is well characterized, and readily available either commercially or *via* the synthesis indicated in Figure 1. Choosing the best IL, however, will still depend on many factors including reactivity, chemical reaction or separation process, physical properties, and non-chemical factors like cost.

We would suggest that greater consideration of end-of-life factors be made in the development of 'green' solvents, as potentially full-scale commercial applications of ILs come to fruition. From this perspective, environmentally acceptable disposal of ILs will require the ILs to contain only elements and functionalities that are amenable to either biodegradation or incineration. From both these perspectives, the hexafluorophosphate anion is undesirable, despite the attractive properties that are introduced into ILs with fluorine-containing anion. Thus, there are plenty of opportunities still available for fundamental development of new IL types. Of interest are both new anions and cations, for example, hydrophobic ILs containing new hydrolytically stable $[PF_3(R_f)_3]^-$ and npn-toxic anions including octylsulfate and docusate (dioctylsulfosuccinate).

Within the pharmaceutical and food-additive industries, the concepts of *non-toxic pharmaceutically acceptable ions* and *GRAS* (generally regarded as safe) materials are well understood, in terms of providing guidelines to chemical (and ion) types for which the toxicological and environmental hazards are established and considered to be acceptably low. The list of *non-toxic pharmaceutically acceptable anions* includes inorganic anions such as chloride, bromide, sulfate, phosphate, nitrate, and organic anions such as acetate, propionate, succinate, glycolate, stearate, lactate, malate, tartrate, citrate, ascorbate, glutamate, benzoate, salicylate, methanesulfonate, and toluenesulfonate. It will immediately be recognized that many of the anions from this (incomplete) list support the formation of ILs with many organic cations. But it may also be obvious that many of these anions are used for preparation of *crystalline* organic salts. Consideration of both structural and chemical properties of both the anion and cation must be made in order to (i) prepare ionic liquids and (ii) obtain the desired properties of the IL as a solvent.

Linear Solvent Energy Relationships and Property Contributions

Because of the large variability within the class of solvents known as ILs, the ability to develop IL mixtures, and the need to select an IL for a particular application, it will be increasingly important to be able to characterize the solvent properties of many different types of ILs. In contrast to many molecular solvents, that have relatively simple solvation interactions, ILs are complex solvents that can support many types of solvent-solute interactions (hydrogen-bond donation and accepting, π-π, dipolar, ionic, *etc.*). In any IL, many different interaction types will be simultaneously present, and the resulting properties of the IL will depend on which interactions are dominant for the particular cation-anion combination(s) *and* solute present. Effective use of the solvent characteristics of ILs can only come about with increased understanding of the properties and behavior of these fluids. Developing methods to characterize IL properties in terms of the different contributions to their solvent characteristics would greatly increase this understanding.

In our research, we have looked at methods of developing core fundamental understanding of IL properties. One of these approaches is modeling of the solvent properties of ILs through a Linear Solvent Energy Relationship (LSER) in which the partitioning of organic molecules between ILs and a second phase such as water (*14*) is related to physio-chemical properties of the solvents through a multiple linear regression using Abraham's generalized solvent equation (*15*) shown in eq. 1:

$$\text{Log}\, D = c + rR_2 + s\pi_2^H + a\Sigma\alpha_2^H + b\Sigma\beta_2^H + vV_x \qquad (1)$$

where R_2 is the excess molar refraction, π_2^H is the dipolarity/polarizability, $\Sigma\alpha_2^H$ is the overall and effective hydrogen bonding acidity, $\Sigma\beta_2^H$ is the overall and effective hydrogen bonding basicity, and V_x is the McGowan characteristic volume of the probe solutes. The corresponding terms *r*, *s*, *a*, *b*, and *v* which relate the solute descriptors to the properties of the solvent system are extracted from the multiple regression. Distribution ratios (D = ([solute]$_{IL}$/[solute]$_{org}$)) for a representative set of 20-30 organic solutes with varying parameters are determined using radiotracers. Since the tracers are introduced at extremely low concentrations, the data obtained can be related to partitioning at infinite dilution.

In a recent paper, Armstrong and co-workers (*16*) have also applied an LSER approach using Abraham's generalized solvent equation to characterize the properties of ILs from their interactions with organic solutes when used as stationary phases for gas-liquid chromatography. This provided the authors the ability to rank ionic liquids according to their usefulness in specific applications.

Other approaches include using solvatochromatic probes such as Reichardt's dye (*17*) or Nile Red (*18*), and fluorescent probes (*10,19*) in order to determine the 'polarity' of ILs using empirical polarity scales (*20*). In all cases, the 'polarity' values fall within a relatively narrow range, similar to short chain alcohols (*10,16-18*), even though ILs with similar 'polarities' can have remarkably different properties to each other, and to traditional molecular solvents. Studies using a range of different dyes, which respond to differing molecular interactions have been used to examine different polarity contributions. Another approach is the QSPR/CODESSA work of Katritzky and co-workers (*21*) which has been used to model the melting-points of imidazolium and pyridinium bromide salts.

Partitioning of organic molecules in IL/water systems have been shown to follow traditional octanol/water distributions (*22*), which has useful implications for applying hydrophobic ILs as direct replacements for solvents such as benzene, toluene, dichloromethane, or chloroform in two-phase system separation schemes (*21-24*). The LSER analysis shows that the volume parameter, followed by the hydrogen-bond donating ability of the IL is the largest contributors to partitioning. The overall analysis indicates that the ILs studied are less polar than water (*i.e.* in terms of HBD and HBA) which is a reasonable conclusion, but has important ramifications for separations. Ions and ionic compounds will not, in the main, partition to an ionic liquid from water.

To observe true solvent properties of ILs, partitioning studies of various solutes in IL/organic solvent systems have been conducted. Conclusions drawn from these studies are relevant, and effective, in predicting IL solvent properties due to the fact hexane has no polarizability or hydrogen bonding acidity or basicity leaving all effects as the result of the IL.

The distribution ratios for the set of 21 probe solutes between toluene and $[C_4mim][NTf_2]$ are shown in Figure 2, plotted against their regression line. Table I lists the interaction parameters obtained for partitioning of organic solutes in a range of IL/organic biphasic systems and are compared to the descriptors calculated for IL/aqueous partitioning where available. These data indicate that the dominant interactions for transfer of a solute from an organic phase to the IL are hydrogen bond basicity (a) and dipolarity (s) of the IL. In contrast, for solute partitioning between ILs and aqueous phases, the volume (v) and hydrogen bond acidity (b) terms were most important. This approach will allow ILs to be classified on the basis of their interactions with solute probes.

Compared to conventional molecular organic solvents, ILs are much more complex, both chemically, and in terms of the wide range of interactions that can take place. Characterization of ILs using simple 'polarity' scales fails to differentiate between the different types of interactions present, whereas LSER

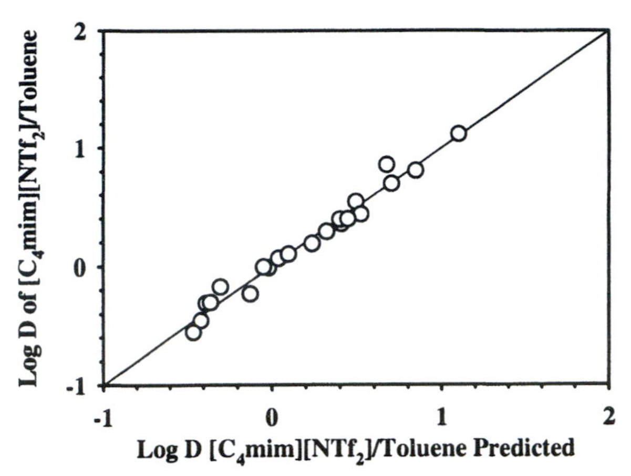

Figure 2. Distribution ratios for organic solutes from the standard screening set between toluene and [C_4mim][NTf_2] plotted against the predicted partition values from the LSER regression parameters shown in Table I.

Table I. Interaction Parameters Obtained from the LSER Model for IL/Organic and IL/Water Partitioning of Organic Solutes (*14, 25*).

	Parameters				
System	*a*	*b*	*s*	*v*	*r*
[C_4mim][PF_6]/hexane	1.82	1.27	2.25	-1.78	-0.09
[C_4mim][PF_6]/water	-1.82	-1.63	-0.004	2.14	0.63
[C_4mim][PF_6]/toluene	1.3	0.88	0.90	-1.23	0.20
[C_6mim][PF_6]/hexane	2.11	1.06	2.02	-1.36	0.03
[C_6mim][PF_6]/water	-1.48	-2.15	0.27	2.31	0.14
[C_6mim][PF_6]/toluene	1.26	0.53	0.93	-0.74	-0.08
[C_4mim][NTf_2]/hexane	2.56	0.80	2.13	-1.11	-0.36
[C_4mim][NTf_2]/toluene	1.13	0.89	0.49	-0.60	0.11

analysis using a range of solute probes provides a route to obtain direct determination of the magnitude and importance of the different interaction contributions. However, partitioning between ILs and aromatic solvents presents a second consideration that needs to be understood.

Liquid Clathrate Formation with IL/Aromatic Mixtures

The remarkable solubility, but rarely complete miscibility, of benzene with ILs (*26*) may be a result of liquid clathrate formation, first described by Atwood

(*27*) for highly reactive air-sensitive alkylaluminum salts and benzene or toluene. Since this discovery, an expanded range of organic salts, for example $[AlCl_4]^-$, $[HX_n]X^-$, $[X_3]^-$, and also $[BF_4]^-$ anions (*28*) have been shown to support liquid clathrate formation (*29*). It has been recognized that ILs can support liquid clathrate formation (*30*) and Zaworotko and co-workers (*31*) have suggested that an approach to developing further liquid clathrate sustaining systems would be to investigate organic salts with low melting points, that is, *ionic liquids*.

Liquid clathrate formation between conventional 1-alkyl-3-methylimidazolium ILs and aromatic hydrocarbons (benzene, toluene, and xylenes) have been investigated (*32*). Liquid clathrate phases form spontaneously under ambient conditions on mixing of the aromatic solvent with the IL. The liquid clathrates phases obtained exhibited typical behavior characteristics, namely, low viscosity (especially relative to the initial neat ILs), immiscibility with excess aromatic solvents, and non-stoichiometric, but reproducible, compositions. The molar composition of the lower liquid clathrate phase formed on contacting ILs with a range of aromatic hydrocarbons are shown in Figure 3, as determined by proton NMR. IL concentration in the upper, aromatic phase was below the NMR detection limit in all the systems examined.

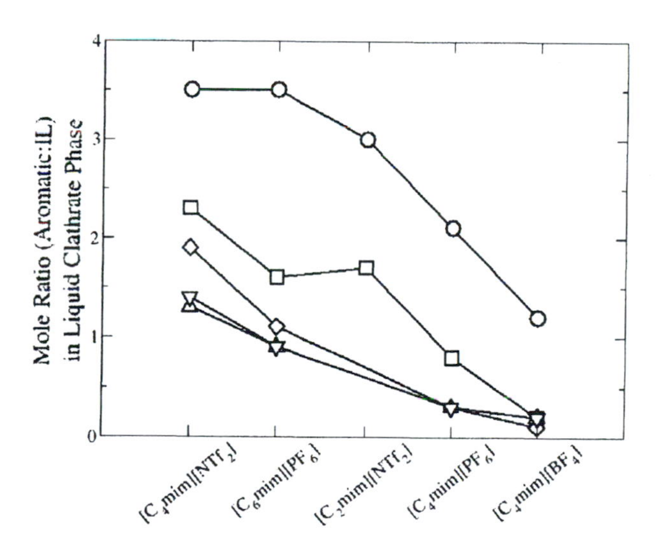

Figure 3. Ratio of aromatic to ionic liquid in the lower phase of the liquid clathrate biphasic systems, data from ref 32. Connecting lines are a visual guide to changes in the liquid clathrate phase composition. Organic components are benzene (circle), toluene (square), and o-xylene(diamond) , m-xylene (down triangle), and p-xylene (up triangle) repectively.

In all the examples studied, miscibility with benzene was greatest, and the maximum aromatic content in the lower phase of the liquid clathrate biphase decreases following the order benzene > toluene > xylenes. The aromatic content of the liquid clathrate phase is somewhat lower than that found in $[Al_2R_6I]^-$ systems (28), but largely comparable with the values observed in $[(HX)_nX]^-$-containing liquid clathrates (28), and with the 0.66 mole fraction solubility of benzene in $[C_4mim][PF_6]$ reported by Blanchard and Brennecke (*33*). Note that the examples described here all contain between 1.5-3.5 moles of benzene for every mole of IL in the lower phase.

This has important consequences for interpreting IL/aromatic biphasic reactions and separations, in that both phases may be 'aromatic-rich', leading to significant differences in solubilities, reaction kinetics, or extraction behavior, when compared to non-clathrate forming IL-biphases such as with alkanes or ethers. Much of the chemistry developed in ILs utilizes IL-organic biphasic systems as reaction and extraction media. Recent examples include hydrogenation (*34*), Friedel-Crafts alkylation (*35*), ring-closing metathesis (*36*), and ring-opening metathesis polymerization (*37*) reactions, and comparisons of results using different solvents may need to be interpreted in the context of liquid clathrate formation.

Summary

As is reflected in the contributions to this Symposium Series volume, significant efforts are being made in utilizing ILs in green chemistry. It is clear that *progress* is indeed being made, and that the *prospects* for the future are good, based on the continuing commitment to excellent and innovative research from both industry and academia. Choosing the right ionic liquid for a given task can be based upon a number of factors: performance, availability, or cost, and correct selection can be vital to the overall success of the process. Models to understand contributions to the solvent properties of ILs are being developed and will act as yet another tool to aid in choosing the appropriate IL(s) for a given task. Both fundamental and applied research into the properties of a broad range of ILs continue to be needed to better understand their potential in replacing traditional organic solvents and operating in the 'green' paradigm.

Acknowledgements

The ionic liquids research at The University of Alabama is supported by the U.S. Environmental Protection Agency STAR program through grant number R-82825701-0 (Although the research described in this article has been funded in part by EPA, it has not been subjected to the Agency's required peer and policy review and therefore does not necessarily reflect the views of the Agency and no official endorsement should be inferred).

References

1 Anastas, P. T.; Warner, J. C. *Green Chemistry: Theory and Practice*; Oxford University Press: New York, 1998.

2 Welton, T. *Chem. Rev.* **1999**, *99*, 2071; Holbrey, J. D.; Seddon, K. R. *Clean Prod. Proc.* **1999**, *1*, 223; Wasserscheid P.; Keim, W. *Angew. Chem. Int. Ed.* **2000**, *39*, 3772; Sheldon, R. *Chem. Commun.* **2001**, 2399; Gordon, C. M. *Appl. Catal. A* **2001**, *222*, 101; Olivier-Bourbigou, H.; Magna, L. *J. Mol. Catal. A.: Chem.* **2002**, *182-183*, 419; Dupont, J.; de Souza, R. F.; Suarez, P. A. Z. *Chem. Rev.* **2002**, *102*, 3667.

3 *Ionic Liquids: Industrial Applications for Green Chemistry*; Rogers, R. D.; Seddon, K.R., Eds.; ACS Symposium Series 818, American Chemical Society: Washington, DC, 2002.

4 Holbrey, J. D.; Rogers, R. D. In *Ionic Liquids in Synthesis*; Wasserscheid, P.; Welton, T., Eds.; VCH-Wiley: Weinheim, 2002; p 41.

5 Seddon, K. R. In *The International George Papatheodorou Symposium: Proceedings*; Boghosian, S., Dracopoulos, V., Kontoyannis, C. G., Voyiatzis, G. A., Eds.; Institute of Chemical Engineering and High Temperature Chemical Processes: Patras, 1999; pp. 131-135.

6 Easteal, E. J.; Angell, C. A. *J. Phys. Chem.* **1970**, *74*, 3987.

7 Golding, J.; Forsyth, S.; MacFarlane, D. R.; Forsyth, M.; Deacon, G. B. *Green Chem.* **2002**, *4*, 223.

8 Holbrey, J. D.; Reichert, W. M.; Swatloski, R. P.; Broker, G. A.; Pitner, W. R.; Seddon, K. R.; Rogers, R. D. *Green Chem.* **2002**, *4*, 407.

9 MacFarlane, D. R.; Golding, J.; Forsyth, S.; Forsyth, M; Deacon, G. B. *Chem. Commun.* **2001**, 1430.

10 Bonhôte, P.; Das, A.; Papageorgiou, N.; Kalanasundram, K.; Grätzel, M. *Inorg. Chem.* **1996**, *35*, 1168.

11 Wilkes, J. S.; Zaworotko, M. J. *Chem. Commun.* **1992**, 965.

12 Sheldon, R. A. In *Precision Process Technology: Perspectives for Pollution Prevention*; Weijnen, M. P. C.; Drinkenburg, A. A. H., Eds.; Kluwer: Dordrecht, 1993; p 125.

13 Chauvin, Y.; Mussmann, L.; Olivier, H. *Angew. Chem. Int. Ed. Engl.* **1995**, *34*, 2698; Suarez, P. A. Z. Dullius, J. E. L.; Einloft, S.; de Souza, R. F.; Dupont, J. *Polyhedron* **1996**, *15*, 1217.

14 Huddleston, J. G.; Visser, A. E.; Reichert, W. R.; Willauer, H. D.; Broker, G. A.; Rogers, R. D. *Green Chem.* **2001**, *3*, 156.

15 Abraham, M. H.; Andonian-Haftvan, J.; Whiting, G. S.; Leo, A.; Taft, R. S. *J. Chem. Soc., Perkin Trans. 2* **1994**, 1777.

16 Anderson, J. L.; Ding, J.; Welton, T.; Armstrong, D. W. *J. Am. Chem. Soc.* **2002**, *124*, 14247.

17 Muldoon, M. J.; Gordon, C. M.; Dunkin, I. R. *J. Chem. Soc., Perkin Trans. 2* **2001**, 433.

18 Carmichael, A. J.; Seddon, K. R. *J. Phys. Org. Chem.* **2000**, *13*, 591.

19 Aki, S. N.; Brennecke, J. F.; Samanta, A. *Chem. Commun.* **2001**, 413.

20 Reichardt, C. *Angew. Chem. Int. Ed. Engl.* **1965**, *4*, 29.

21 Katritzky, A. R.; Lomaka, A.; Petrukhin, R.; Jain, R.; Karelson, M.; Visser, A. E.; Rogers, R. D. *J. Chem. Inf. Comp. Sci.* **2002**, *42*, 71. (b) Katritzky, A. R.; Jain, R.; Lomaka, A.; Petrukhin, R.; Karelson, M.; Visser, A. E.; Rogers, R. D. *J. Chem. Inf. Comp. Sci.* **2002**, *42*, 225.

22 Huddleston, J. G.; Willauer, H. D.; Swatloski, R. P.; Visser, A. E.; Rogers, R. D. *Chem. Commun.* **1998**, 1765.

23 Visser, A. E.; Swatloski, R. P.; Griffin, S. T.; Hartman, D. H.; Rogers, R. D. *Sep. Sci. Technol.* **2001**, *36*, 785.

24 Cull, S. G.; Holbrey, J. D.; Vargas-Mora, V.; Seddon, K. R.; Lye, G. J. *Biotech. Bioeng.* **2000**, *69*, 227.

25 Huddleston, J. G.; Broker, G. A.; Willauer, H. D.; Rogers, R. D. In *Ionic Liquids: Industrial Applications for Green Chemistry*; Rogers, R. D.; Seddon, K. R., Eds.; ACS Symposium Series 818, American Chemical Society: Washington, DC, 2002; p 270.

26 Chauvin, Y.; Olivier-Bourbigou, H. *CHEMTECH* **1995**, *25*, 26.

27 Atwood, J. L. In *Inclusion Compounds*; Atwood, J. L.; Davies, J. E. D.; MacNicol, D. D., Eds.; Academic Press: London, 1984; Vol. 1; Atwood, J. L.; Atwood, J. D. *Inorganic Compounds with Unusual Properties*; Advances in Chemistry Series 150; American Chemical Society: Washington, DC, 1976; p 112; Atwood, J. L. *Rec. Develop. Sep. Sci.*, **1977**, *3*, 195.

28 Christie, S.; Dubois, R. H.; Rogers, R. D.; White, P. S.; Zaworotko, M. J. *J. Incl. Phenomen.* **1991**, *11*, 103; Coleman, A. W.; Means, C. M.; Bott S.G.; Atwood. J. L. *J. Cryst. Spect. Res.* **1990**, *20*, 199; Pickett, C. J. *J. Chem. Soc., Chem. Commun.* **1985**, 323.

29 Steed, J. W.; Atwood, J. L. In *Supramolecular Chemistry*; Wiley: Chichester, 2000; p 707.

30 Surette, J. K. D.; Green, L.; Singer R. D. *Chem. Commun.* **1996**, 2753.

31 Gaudet, M. V.; Peterson, D. C.; Zaworotko, M. J. *J. Incl. Phenomen.* **1988**, *6*, 425.

32 Holbrey, J. D.; Reichert, W. M. Nieuwenhuyzen, M.; Sheppard, O.; Hardacre, C.; Rogers, R. D. *Chem. Commun* **2003**, *in press*.

33 Blanchard, L. A.; Brennecke, J. F. *Ind. Eng. Chem. Res.*, **2001** *40*, 287.

34 Boxwell, C. J.; Dyson, P. J.; Ellis, D. J.; Welton, T. *J. Am. Chem. Soc.* **2002**, *124*, 9334; Dyson, P. J.; Ellis, D. J.; Parker, D. G.; Welton, T. *Chem. Commun.* **1999**, 25.

35 Abdul-Sada, A.-K.; Seddon, K. R.; Stewart, N. J., World Patent WO 95 21872 (1995); Boon, J. A.; Levisky, J. A.; Pflug, J. L.; Wilkes, J. S. *J. Org. Chem.* **1985**, *51*, 480; DeCastro, C.; Sauvage, E.; Valkenberg, M. H.; Holderich, W. F. *J. Catal.* **2000**, *196*, 86.

36 Sémeril, D.; Olivier-Bourbigou, H.; Bruneau, C.; Dixneuf. P. H. *Chem. Commun.* **2002**, 143.

37 Csihony, S.; Fischmeister, C.; Bruneau, C.; Horváth, I.; Dixneuf, P. H. *New. J. Chem.* **2002**, *26*, 1667.

Ionic Liquids: Manufacture and Synthesis

Chapter 2

Ionic Liquids: Improved Syntheses and New Products

Adrian J. Carmichael, Maggel Deetlefs, Martyn J. Earle, Ute Fröhlich, and Kenneth R. Seddon

The QUILL Centre, The Queen's University of Belfast, Stranmillis Road, Belfast BT9 5AG, United Kingdom

We report here the improved syntheses of 1-alkyl-3-methylimidazolium ionic liquids. Microwave irradiation drastically reduces the preparation time of 1-alkyl-3-methylimidazolium and *N*-alkylpyridinium halide salts and, in addition, three halide-free routes to ionic liquids have been developed. New, chiral, imidazolium-based ionic liquids were prepared using both conventional and halide-free procedures. Chirality was introduced in the new compounds at either the cation or the anion, or both.

Introduction

At present, the use of ionic liquids as solvents and/or catalysts for chemical reactions is well past infancy, with many excellent review articles available summarising their preparation, use, and advantages compared to traditional solvents (*1,2,3,4,5,6,7*). While much research emphasis is still placed on the syntheses of new ionic liquids, systematic physico-chemical property studies of these neoteric solvents are rare (*8*). Both these topics are vital in gaining a better understanding of the factors governing chemistry in an ionic liquid environment, but the impact of purity on the latter remains a neglected issue. This can be illustrated by the various melting points reported for 1-ethyl-3-

methylimidazolium tetrafluoroborate, $[C_2mim][BF_4]$; 15 °C (*9*), 5.8 °C (*10*), 12.0-12.5 °C (*11*), 11 °C (*12*) and 14.6 °C (*13*). Further examples of the effects of contaminants in ionic liquids are the varying reaction efficiencies and/or specificities reported for reactions in the same ionic liquid (*3,6,7*). If ionic liquid purity issues are not addressed, the reasons for "ionic liquid effects" will continue as a barrier for the predictive preparation of ionic liquids for particular applications.

One might ask whether the increased cost of stringently purified ionic liquids is justified. In some instances the answer may be negative if no clear advantages are presented by using an uncontaminated ionic liquid. However, by disregarding the impact of impurities on reaction rates, physico-chemical properties and toxicological data, the reliability and reproducibility of reported results becomes problematic. Furthermore, when imidazolium ionic liquids reach fruition and find industrial application, their involatile nature and recyclability advantages will certainly outweigh their price tag. In our view, an ionic liquid should first be prepared in its purest form and used as such. Thereafter water, or chloride, or both, can be added to determine the impact, if any, on the reaction being studied. If no significant reactivity differences are found, the use of a "dirty" ionic liquid could then be justified and at the same time, the production of incorrect literature data avoided.

General Syntheses of Ionic Liquids

The general preparation of 1-alkyl-3-methylimidazolium, $[C_nmim]^+$ ionic liquids (Figure 1) involves a consecutive quaternisation-metathetic/acid-base procedure. The first step affords a 1-alkyl-3-methylimidazolium halide precursor, $[C_nmim]X$ (*1*) and has two disadvantages: (a) it is time consuming; and (b) an excess of haloalkane (10-100 %) is required to achieve good yields. The latter renders the quaternisation reaction dirty, especially when long chained derivatives are prepared, since high boiling haloalkanes are difficult to remove from the reaction mixture. Together, both (a) and (b) make the first step in ionic liquid preparation cost and atom inefficient. Our efforts to clean and speed-up the initial stage of ionic liquid synthesis using microwave (mw) radiation are presented here. Although the focus of this paper is on imidazolium-based ionic liquids, the same synthetic strategies apply to pyridinium-based ionic liquids.

The second step of ionic liquid preparation proceeds *via* metathesis with a metal salt or an acid-base neutralisation reaction, respectively, producing a stoicheiometric amount of waste MX or HX (Figure 1). Due to the excellent solvating properties of ionic liquids, these by-products become trapped and contaminate the ionic liquid. A previous study in our group has shown that another source of halide contamination is the incomplete conversion of the

$[C_n mim]X$ precursor to the target ionic liquid (*14*). This study also detailed various procedures to minimise and analyse the halide content of hydrophobic and hydrophilic ionic liquids. Although the preparation of ionic liquids derived from halide-containing starting materials is still the most widely used, the elimination of halide contamination during synthesis is attractive. This approach was first employed by Bonhôte *et al* (*15*) and we have also developed three such alternative strategies, which are described here, using respectively fluorinated esters, alkyl sulfonates and free carbenes.

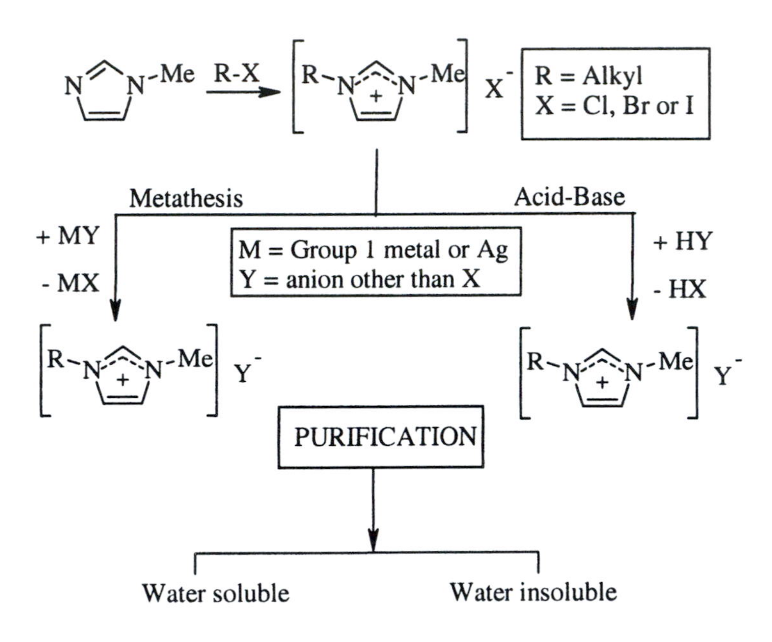

Figure 1. General preparation of 1-alkyl-3-methylimidazolium ionic liquids.

The preparation of ionic liquid halide precursors is not limited to the use of 1-methylimidazole or pyridine. Various acyclic and heterocyclic salts derived from the chiral pool (*16*) are also available for this purpose and include natural α-hydroxy acids, alkaloids and terpenes. Alternatively, chiral heterocycles can be prepared and employed as ionic liquid substrates. Chiral ionic liquids have the potential to introduce chirality and this can be achieved by using either a chiral cation or anion, or both. Surprisingly, relatively few examples of chiral ionic liquids have been published thus far (Figure 2). Howarth *et al.* were the first to describe the synthesis of a chiral 1,3-dialkylimidazolium cation (*17*) starting from imidazole and a chiral alkyl chloride and our group investigated Diels-Alder reactions in lactate ionic liquids (*18*). More recently, three patents (*16,19,20*) as well as one paper (*21*) detailing the syntheses of various chiral ionic liquids have been published.

As for the general imidazolium-based ionic liquid preparations, the majority of chiral ionic liquid syntheses involve a consecutive quaternisation-metathetic/acid-base procedure (compare Figure 1) (*22*). Using this general methodology, we have recently prepared new imidazolium-based ionic liquids where chirality is introduced either at the cation or the anion.

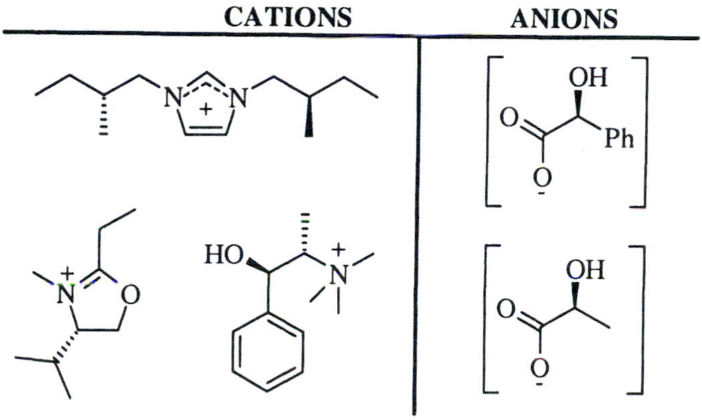

Figure 2. Some reported chiral cations and anions (17,18,21).

Preparation of Ionic Liquids using Microwave Irradiation

The use of microwave ovens as tools for synthetic chemistry is one of the fastest growing areas of research (*23*, *24*). Since the first reports of microwave-assisted synthesis in 1986 (*25,26*) the technique has been accepted as a method for drastically reducing reaction times and for increasing yields of products compared to conventional methods (*27*).

A key advantage of modern scientific microwave equipment is the ability to control reaction conditions very specifically, monitoring temperature, pressure, and reaction times (*28*). In order to take advantage of microwave heating effects, a covalent, non-conducting reaction medium or reactant needs to have a high dielectric constant. In contrast, ionic liquids are ideal candidates for the exploitation of microwave heating due to their conducting, ionic nature. When an irradiated sample is an electrical conductor, the ions move through the material under the influence of an electric field, which results in a polarisation and these induced currents cause heating due to electrical resistance.

Our interest in utilising microwave radiation is diverse, but the preparation of ionic liquid halide salts using multimode microwave radiation was selected as a preliminary study. This is the first time ionic liquids have been prepared using controlled microwave radiation on a large scale. Although previous reports have shown that multimode microwave ovens can be used to prepare imidazolium halide (*29,30*), tetrafluoroborate (*31*) and aluminate (*32*) salts (Figure 3), realistic power control cannot be achieved with a domestic microwave oven. Although the rate of heating with a domestic microwave can be moderated either by using heat dissipaters (*33*), or by increasing reactant volumes, these methods are undesirable since they respectively provide no power control, which can lead to product charring, and produce large amounts of unwanted solvent vapour. Therefore, due to power control restrictions associated with the use of domestic microwave ovens, most previous ionic liquid preparations could only be performed on a small scale (3 – 150 mmol) in open vessels, which necessitates an excess of haloalkane of up to 100 %, due to evaporative loss.

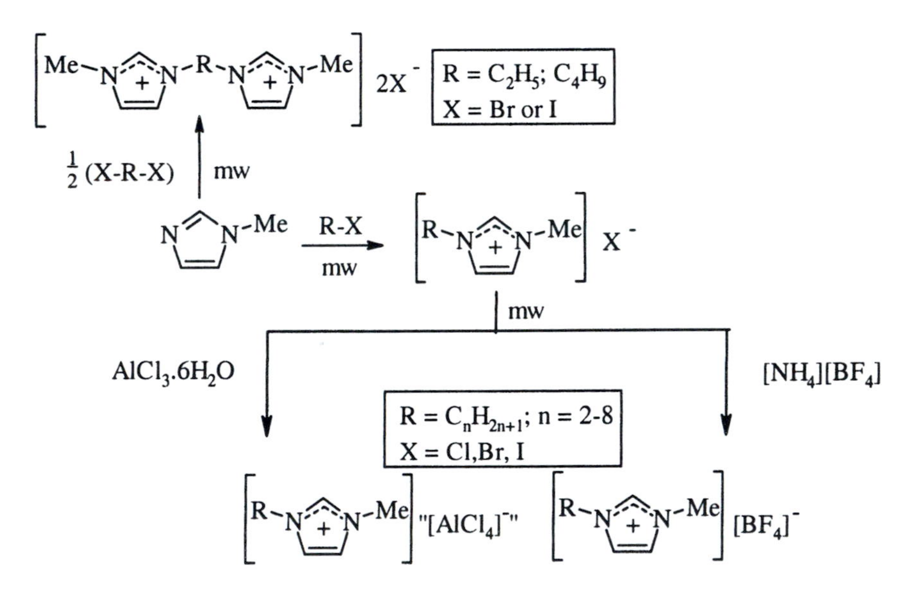

Figure 3. Previous $[Rmim]^+$ syntheses using domestic mw heating (29,31,32).

It should be noted that the reported preparation of tetrachloroaluminate compounds (*32*) <u>must</u> be incorrect, as employing $AlCl_3{\cdot}6H_2O$ would result in irreversible formation of hydroxoaluminate species *via* hydrolysis of the anion.

Our attempts to clean-up and speed-up the synthesis of ionic liquids included the preparation of a series of both [Rmim]X and [Rpy]X salts using controlled, multimode microwave radiation (Figure 4) in sealed vessels, as well as scaling up these reactions in an open vessel.

Imidazolium and pyridinium halide salts were prepared in duplicate by adding 1.1 equivalents of a haloalkane (110 mmol) to either 1-methylimidazole or pyridine (100 mmol) in a sealed quartz reaction vessel fitted with a temperature and pressure probe (Figure 4). The former regulates the selected temperature by adjusting the microwave power input and the latter monitors the autogenous pressure. The optimum reaction conditions determined for the medium-scale, sealed vessel syntheses of [Rmim]X and [Rpy]X salts are shown respectively in Tables I and II. The most effective reaction temperature was established to be the approximate average boiling point of the employed haloalkane and the heterocycle.

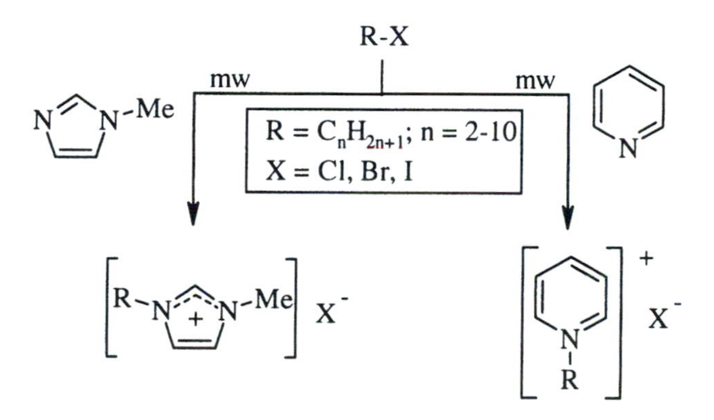

Figure 4. [Rmim]X (or [$C_n$mim]X) and [Rpy]X (or [$C_n$py]X) preparations using controlled mw heating.

Due to evaporative loss, the preparation of imidazolium and pyridinium halides in open vessels often requires a large excess of the haloalkane to obtain good yields. For example, a 100% excess of 1-chlorobutane or 2-bromobutane was used by Varma *et al.* (*29*) to respectively prepare [C_4mim]Cl (76 % yield) and [C_4mim]Br (61 % yield). In contrast, all the sealed vessel preparations in the current study required less than 10 % excess of the appropriate haloalkane to obtain ≥ 95 % conversions (based on NMR) and ≥ 87 % work-up yields, making this synthetic route more cost and reagent efficient, and hence greener.

The rate at which the quaternisation of 1-methylimidazole or pyridine proceeds follows the conventional order; R-I > R-Br > R-Cl. Using microwave, as opposed to conductive heating, this reactivity order remains the same but reaction times are significantly decreased. For example, compared to conductive heating, microwave radiation accelerates the formation of [C_4mim]Cl, [C_6mim]Cl and [C_8mim]Cl by a factor of *ca.* 70 and [C_{10}mim]Cl is generated *ca.* 110 times faster. [$C_n$mim] and [$C_n$py] bromide and iodide salts show a similar trend.

Table I. Preparation of [Rmim]X Salts on a Medium Scale: 150 – 300 mmol.

[Rmim]Cl	*R =C_4H_9*	*R = C_6H_{13}*	*R = C_8H_{17}*	*R = $C_{10}H_{21}$*
Power/ W	300	300	300	300
Temperature /°C	150	150	180	120
Microwave irradiation time / min.	20	20	20	10
Conventional heating time/ min.	1440	1440	1440	1080
Conversion / %	> 95	> 95	> 95	> 95
Work-up yield / %	95	95	90	95
[Rmim]Br	*R =C_4H_9*	*R = C_6H_{13}*	*R = C_8H_{17}*	*R = $C_{10}H_{21}$*
Power/ W	240	240	240	240
Temperature °C	80	80	110	120
Microwave irradiation time / min.	6	8	8	10
Conventional heating time/ min.	840	960	960	1080
Conversion / %	> 95	> 95	> 95	> 95
Work-up yield / %	87	91	95	95
[Rmim]I	*R =C_4H_9*	*R = C_6H_{13}*	*R = C_8H_{17}*	*R = $C_{10}H_{21}$*
Power / W	200	200	200	200
Temperature °C	165	190	210	165
Microwave irradiation time / min.	4	7	9	11
Conventional heating time/ min.	640	720	720	720
Conversion / %	> 95	> 95	> 95	> 95
Yield / %	93	90	95	94

A further result of haloalkane reactivity differences is that the chloride preparations require higher power levels (300 W) than the bromides (240 W), which in turn require greater levels than the iodides (200 W) to achieve similar yields. In addition, shorter irradiation times are required for bromide and iodide preparations than for the chlorides. It is of interest to note that despite the high temperatures employed during synthesis, no imidazolium-based disproportionation products were observed.

Table II. Preparation of [Rpy]X Salts on a Medium Scale: 150 – 300 mmol.

[Rpy]Cl	*R =* C_4H_9	R = C_6H_{13}	R = C_8H_{17}	*R =* $C_{10}H_{21}$
Power/ W	300	300	300	300
Temperature / °C	120	120	150	180
Microwave irradiation time /min.	40	60	60	60
Conventional heating time/ min.	3360	3360	3360	2880
Conversion / %	> 95	> 95	> 95	> 95
Work-up yield / %	87	94	94	93
[Rpy]Br	*R =* C_4H_9	R = C_6H_{13}	R = C_8H_{17}	*R =* $C_{10}H_{21}$
Power/ W	240	240	240	240
Temperature / °C	130	130	130	130
Microwave irradiation time /min.	30	30	30	20
Conventional heating time/ min.	2880	2880	2880	2880
Conversion / %	> 95	> 95	> 95	> 95
Work-up yield / %	98	100	98	93
[Rpy]I	*R =* C_4H_9	R = C_6H_{13}	R = C_8H_{17}	*R =* $C_{10}H_{21}$
Power/ W	200	200	200	200
Temperature / °C	165	190	210	165
Microwave irradiation time / min.	10	12	15	15
Conventional heating time/ min.	960	960	1080	1140
Conversion / %	> 95	> 95	> 95	> 95
Work-up yield / %	95	97	97	93

For the large scale [Rmim]X and [Rpy]X salt preparations (300 mmol–2 mol) a conventional reflux arrangement was employed. A one-litre round-bottomed flask equipped with a temperature probe housing was placed inside the microwave reactor cavity, with a reflux condenser protruding through an aperture in the reactor roof. All reactions were performed at the boiling point of the appropriate haloalkane and gave conversions ≥ 95 and yields ≥ 85 %. On these scales, the reaction times were between 200 – 400 times shorter compared to conductive heating. Both the sealed and open vessel reactions were optimised by determining the time required for complete conversion of 1-methylimidazole to product and purity subsequently ensured washing with either ethyl ethanoate or hexane.

To summarise, using microwave radiation, the first step in the general synthesis of ionic liquids has been transformed from a time-consuming, dirty process to a rapid, more environmentally-friendly procedure. This was achieved in two ways: (a) using 100 cm^3 sealed reaction vessels; or (b) a one-litre reflux set-up. Although the latter is a less laborious method, the former allows up to fourteen sealed vessels to be used simultaneously in a parallel approach.

Preparation of Halide-free Ionic Liquids

During the second step of the general syntheses of ionic liquids (see Figure 1) the intrinsically good solvating properties of ionic liquids become a problem. Both the metathetic and acid-base neutralisation reactions generate a stoicheiometric amount of halide waste. Many ionic liquids solvate the generated halide waste so effectively that complete removal can often not be achieved. When metathesis is carried out using a silver(I) salt, the route gives purer ionic liquid products but becomes prohibitively expensive upon scale up and residual silver must be removed electrochemically (*34*). Employing alkali metal salts reduces the cost, but not the waste. Halide contamination of ionic liquids is a problem that must be overcome for them to be used as reaction solvents on a large scale. For instance, when used as media for many processes catalysed by transition metals the presence of strongly coordinating halide ions has been shown to reduce catalyst activity (*35,36,37,38*). The opportunity also exists in many reactions for the residual halides to be oxidised to halogens, which will react with many substrates and can corrode equipment.

We have recently developed three synthetic strategies that eliminate halide contamination during ionic liquid synthesis and consequently reduce the amount of waste products. The first two methods are based on the use of fluorinated esters and alkyl sulfonates (*39*) as replacements for haloalkanes while the third makes use of free carbenes (*40*). All three methods are discussed.

Halide-free Ionic Liquids from Fluorinated Esters and Alkyl Sulfonates

Using an analogous procedure to Bonhôte's (*15*) original method, trifluoroethanoate, $[CF_3CO_2]^-$, or methanesulfonate, $[CH_3SO_3]^-$, ionic liquids were respectively obtained in excellent yields by heating equimolar amounts of 1-methylimidazole and either a fluorinated ester or an alkyl sulfonate, under reflux, (Figure 5). The quaternisation reactions are time-consuming but this could be overcome using microwave radiation in a solvent-free procedure. The $[C_n\text{mim}]Y$ (n = 4 or 6; Y = $[CF_3CO_2]$) ionic liquids are available as room temperature reaction media, but the $[C_2\text{mim}]$ analogues melt above ambient temperature and can therefore not be employed as room-temperature solvents. Nevertheless, these methods provide two new routes to novel, halide-free ionic liquids.

If ionic liquids with anions other than $[CF_3CO_2]^-$ or $[CH_3SO_3]^-$ are desired, the former can be used as metathetic substrates to produce *e.g.* hexafluorophosphate or tetrafluoroborate halide-free ionic liquids (Figure 5).

The reactions respectively release trifluoroethanoic (CF_3COOH) and methanesulfonic (CH_3SO_3H) acid as by-products (Figure 5). Trifluoroethanoic acid is too expensive a commodity to be produced as waste, but fortunately its low boiling point of 72 °C makes it recyclable (Figure 6). Recycling is simple

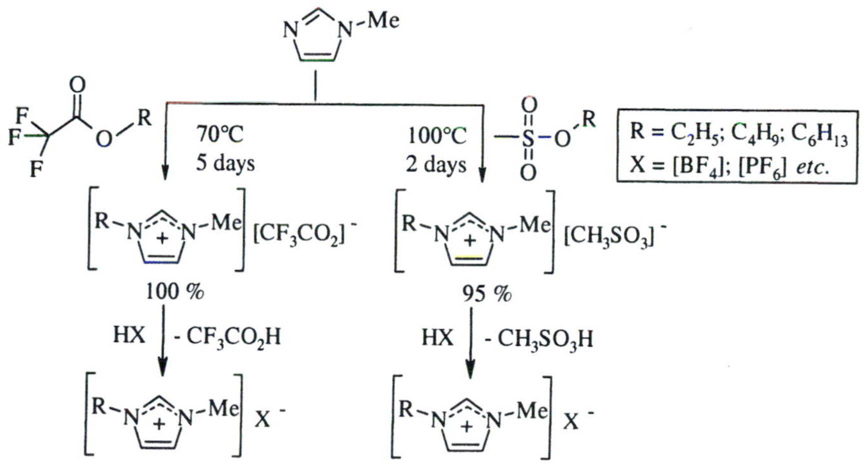

Figure 5. [Rmim][CF_3CO_2] and [Rmim][CH_3SO_3] preparations (39).

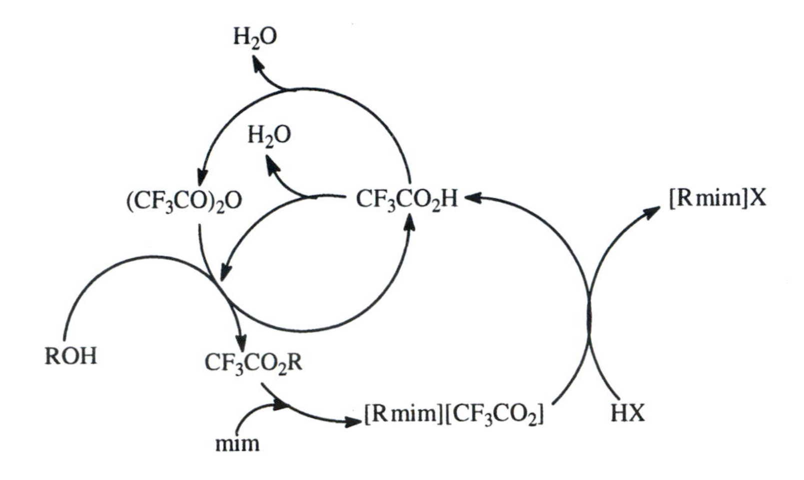

Figure 6. Recycling of trifluoroethanoic acid (39).

and involves distilling trifluoroethanoic acid out of the non-volatile ionic liquid at 100 °C and at atmospheric pressure.

Trifluoroethanoic acid is easily collected by distillation and reintroduced into the reaction cycle either by direct esterification of the acid, or by dehydration via the anhydride (Figure 6). In this way, the costly recyclable reagent is reused in a sustainable cycle.

Methanesulfonic acid can similarly be recycled although its high boiling point of *ca.* 150 °C makes its separation from the product ionic liquid more difficult. It should also be noted that due to the low pH of methanesulfonic acid, it can only normally be exchanged using an acid with a higher pH. However, this potential problem can be overcome by anion exchange with Group 1 metal salts of the desired anion in water, followed by extraction with a water immiscible solvent such as ethyl ethanoate or dichloromethane. This method bypasses the anion exchange problem since [Rmim][CH_3SO_3] and methanesulfonic acid are hydrophilic while [Rmim][anion] almost always dissolves in organic solvents. Exceptions of course exist for hydrophilic anions such as sulphate or phosphate. In addition, if the desired ionic liquid is hydrophobic, it will separate from the water layer making its isolation simple.

Halide-free Ionic Liquids from Imidazolium-based Carbenes

The third method we have recently developed to prepare halide-free ionic liquids exploits the acidity of H-2 in [Rmim]X salts *(40)*. These salts are easily deprotonated with a strong base such as K[$OCMe_3$] to afford the corresponding free carbene (Figure 7). These reactions have been extensively studied since Arduengo isolated the first stable free carbene in 1991 *(41)* and a variety of procedures to synthesise free heterocyclic (diamino)carbenes are currently available *(42,43,.44)* Our initial work was based on these methods and involved the use of alkyllithium bases such as butyllithium or lithium diisopropylamide to deprotonate [Rmim]X salts in tetrahydrofuran. Although these methods produced imidazolium-based free carbenes, product isolation was laborious and yields low. Nevertheless, due to the stability of Arduengo-type free carbenes *(44)* we were able to isolate a series of 1-alkyl-3-methylimidazol-2-ylidenes (Figure 7) by Kugelrohr distillation in a solvent free procedure.

Heating an equimolar mixture of [Rmim]X with K[$OCMe_3$] affords free carbenes in excellent yields (Table III), which were isolated by vacuum distillation. A variety of anions was subsequently introduced by equimolar

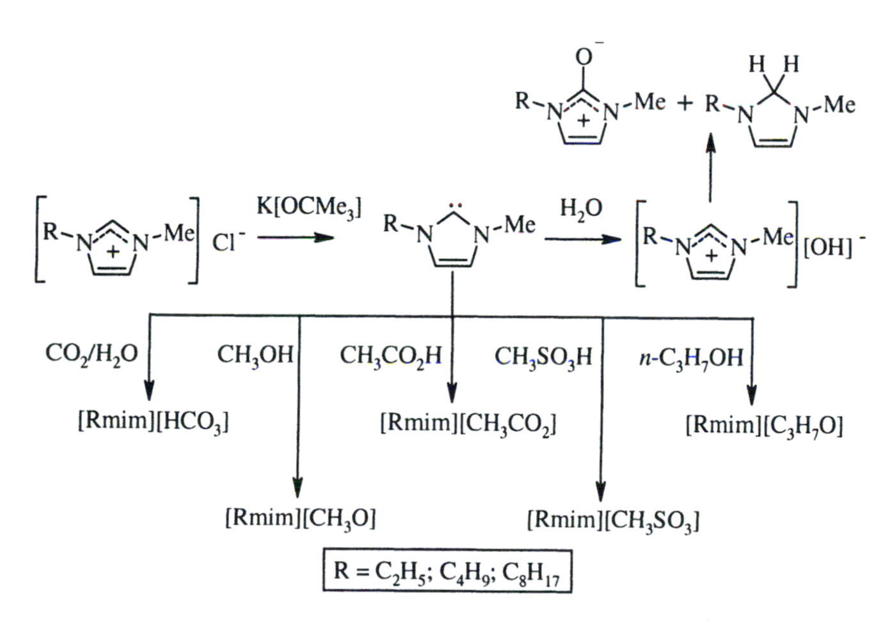

Figure 7. Imidazolium carbene routes to halide-free ionic liquids (40).

addition of a free carbene to the Brønsted acid form (Figure 7) of the desired anion. In addition to providing a straightforward route to prepare a wide range of halide-free ionic liquids, this is the only known method to prepare unstable [Rmim][OH] salts as well as [Rmim][HCO_3] and [Rmim][OR] ionic liquids. The hydroxide was isolated as the hydrated salt and characterised by NMR spectroscopy. Complete dehydration of the salt was not possible. The hydroxide (a viscous oil) is unstable and slowly disproportionates at room temperature as illustrated in Figure 7. The ^{13}C-NMR spectra of the "propoxide ionic liquid" (C-2 signal at δ 190) suggest that it is intermediate between a solution of the carbene (δC 210 p.p.m.) and an imidazolium salt with the C-2 signal at δ 135 p.p.m. This also suggests that the pK_a's of the imidazolium salt and 1-propanol are similar.

Table III. Synthesis of 1-Alkyl-3-methylimidazol-2-ylidenes using K[$OCMe_3$]

[Rmim]Cl	*Reaction Temp. / °C*	*Product bp / °C at 1 mm Hg*	*Yield/ %*
[C_2mim]Cl	120	90	95
[C_4mim]Cl	150	120	90
[C_8mim]Cl	200	180	70

At the time of our initial study, it was generally accepted that free imidazolinylidenes required bulky groups on the nitrogen atoms to provide the necessary steric and electronic stability to exist in free form. In our hands, free carbenes such as 1-ethyl-3-methylimidazol-2-ylidene proved to be stable enough to be handled at both room temperature and elevated temperatures under dry conditions. To summarise, the utilisation of imidazolium-based free carbenes provides a simple, relatively cheap method to prepare halide-free ionic liquids without the generation of noxious waste products.

Chiral Ionic Liquids

With the growing number of reactions that can be performed in ionic liquids, the possibility of chiral induction through the use of a chiral ionic liquid is intriguing. The improved synthesis of an ionic liquid, based on a chiral cation, previously reported by Herrmann *et al. (45)*, was performed as shown in Figure 8. The chiral imidazolium cation was prepared starting from the readily available chiral amine, 1-phenylethylamine, using a method based on the current

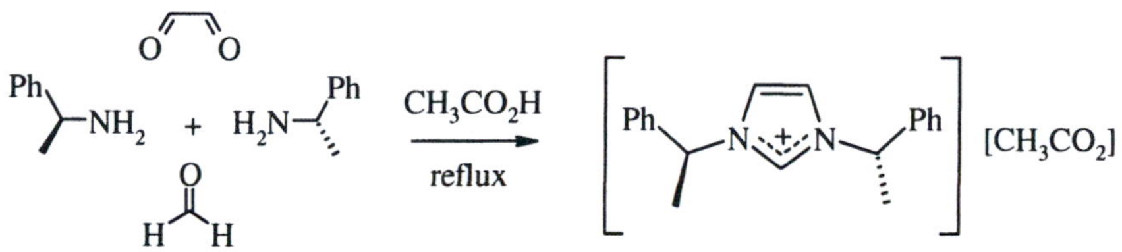

Figure 8. Synthesis of S-[dpeim][CH_3CO_2].

industrial process for the manufacture of 1-methylimidazole (*46*). The amine is heated for twelve hours under reflux with 0.6 equivalents of aqueous glyoxal and methanal in ethanoic acid. Removal of water and ethanoic acid yielded 95 % of [dpeim][CH_3CO_2].

Ethanoic acid is a comparatively weak acid, and so the anion can easily be exchanged by an acid-base procedure as already described in Figure 1. In this way a range of *R*- and *S*-[dpeim]$^+$ salts with the following anions was prepared: triflate, chloride, bromide, iodide, nitrate, trifluoroethanoate, tetrafluoroborate and hexafluorophosphate. The acid was added to [dpeim][CH_3CO_2] and the aqueous solution extracted with dichloromethane. Evaporation of the solvent and drying in high vacuum yielded the corresponding chiral imidazolium salts.

The [dpeim][$N(SO_2CF_3)_2$] ionic liquid was prepared by metathesis of [dpeim][CH_3CO_2] with Li[$N(SO_2CF_3)_2$] in aqueous solution. The dicyanamide salt was synthesised starting from the [dpeim]Cl salt with Na[$N(CN)_2$] in propanone. For some physical properties of the prepared [dpeim]$^+$ salts see Table IV.

Table IV. Physical Properties of the Chiral [dpeim]$^+$ Salts.

Anion	*Melting point / °C*	$\lvert[\alpha]_D\rvert$ */° in $CHCl_3$*
$[N(SO_2CF_3)_2]^-$	Glass	8.84
$[OTf]^-$	Glass	11.3
$[CF_3SO_3]^-$	-40 (glass transition)	16.2
$[N(CN)_2]^-$	-31 (glass transition)	11.5
$[NO_3]^-$	7	14.0
$[BF_4]^-$	96	12.5
$[PF_6]^-$	101	11.9
I^-	177	21.6
Br^-	191	34.0
Cl^-	198	36.2

In general, chiral [dpeim]$^+$ salts show the same melting point trends as the corresponding [$C_n$mim]$^+$ salts, but melting points are higher due to the presence of phenyl groups. The [dpeim]X salts with X = [$N(SO_2CF_3)_2$], [OTf], [CF_3SO_3] and [$N(CN)_2$] do not exhibit a melting point in the differential scanning analysis, instead glass transitions are detected for [dpeim][$N(SO_2CF_3)_2$] and [dpeim][OTf] at temperatures as low as –40 and –31 °C. These ionic liquids are good candidates for reaction media since they are comparatively fluid at room temperature. The [dpeim] halides, tetrafluoroborate, hexafluorophosphate and perchlorate salts could be crystallised, and crystal structures were obtained for the bromide, iodide, perchlorate and tetrafluoroborate. The partial X-ray structure of *S*-[dpeim]Br is illustrated in Figure 9.

We are currently also developing the synthesis of chiral ionic liquids where the chirality resides in the anion. Examples include 1,3,-dialkylimidazolium mandelates, lactates and camphorsulfonates.

Our present work and recent publications from other groups (16,19,21) demonstrate that these chiral ionic liquids can be synthesised from readily available starting materials, including compounds from the chiral pool. The synthesis is often no more complex than for non-chiral ionic liquids. Future work will have to investigate their influence on the stereochemical outcome of reactions, although unpublished work from our laboratory shows that enantiomeric excesses of up to 44 % can be achieved for some prochiral reactions (*47*).

Summary

Having demonstrated faster, as well as halide-free routes to a variety of ionic liquids, many new opportunities for synthesis are now available. These

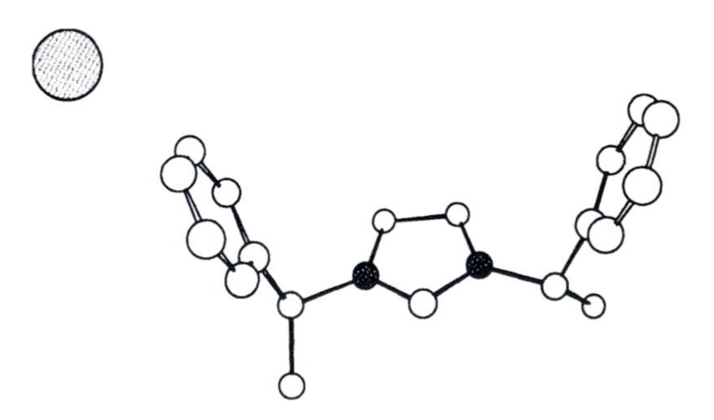

Figure 9. Structure of S-[dpeim] bromide (hydrogen atoms removed for clarity).

include the synthesis of ionic liquids that are difficult to prepare conventionally using conductive heating. For example, pyrazolium- and thiazolioum-based ionic liquids are notoriously time-consuming to prepare, but the use of microwave radiation makes these syntheses easily achievable.

Based on our work, a further opportunity that exists for ionic liquids is the introduction of chirality *via* the solvent, but also by the use of a chiral catalyst within the chiral solvent. The way in which this will impact on the stereochemical outcome of reactions represents another of our research interests. The use of chiral ionic liquids as both solvent and ligand substrates is yet another research option.

Acknowledgements

We (AJC, MD, MJE and UF) would like to thank QUILL for the funding of this study. We are also indebted to Dr. M. Nieuwenhuyzen for the crystal structure determination.

References

1. Welton, T. *Chem. Rev.* **1999**, *99*, 2071-2083.
2. Holbrey, J. D.; Seddon, K. R. *Clean Prod. Proc.* **1999**, *1*, 223-236.
3. Wasserscheid, P.; Keim, W. *Angew. Chem. Int. Ed. Engl.* **2000**, *39*, 3772-3789.
4. Sheldon, R. *Chem. Commun.* **2001**, 2399-2407.
5. Gordon, C. M. *Appl. Catal. A.* **2001**, *222*, 101-117.
6. Zhao, D.; Wu, M.; Kou, Y.; Min, E. *Catalysis Today.* **2002**, *2654*, 1-33.
7. Olivier-Bourbigou. H.; Magna, H. *J. Mol. Catal. A.* **2002**, *164*, 1-19.
8. Seddon, K. R. *Green Chem.* **2002**, *4*, G25-G26.
9. Wilkes, J. S.; Zaworotko M. J., *J. Chem. Soc, Chem. Commun.* **1992**, 965-967.
10. Holbrey, J. D.; Seddon, K. R. *J. Chem. Soc., Dalton Trans.* **1999**, 2133-2139.
11. Fuller, J.; Carlin, R. T.; Osteryoung, R. A. *J. Electrochem. Soc.* **1997**, *144*, 3881-3886.
12. McEwen, A. B.; Ngo, H. L.; LeCompte, K.; Goldman, J. L. *J. Electrochem. Soc.* **1999**, *145*, 1687-1695.
13. Noda, A.; Watanabe, M.; *Electrochim. Acta.* **2000**, *45*, 1265-1270.
14. Seddon, K. R.; Stark, A.; Torres, M-J. *Pure Appl. Chem.* **2000**, *72*, 2275-2287.

15. Bonhôte, P.; Dias, A. –P.; Papeprgeorgiou, N.; Kalyanasundaram, N.; Grätzel, M. *Inorg. Chem.* **1996**, *35*, 1168-1172.
16. Bolm, C.; Keim, W.; Bösmann, A.; Wasserscheid, P. German Patent, DE 10003708, **2001**.
17. Howarth, J.; Hanlon, K.; Fayne, D.; McCormac, P. B. *Tetrahedron Lett.* **1997**, *38*, 3097-3100.
18. Earle, M. J.; McCormac, P. B.; Seddon, K. R. *Green Chem.* **1999**, *1*, 23-25.
19. Kitazume, T. U.S. Patent, US 2001031875, **2001**.
20. Abbott, A. P.; Davies, D.L. World Patent, WO 2000056700, **1999**.
21. Wasserscheid, P.; Bösmann, A.; Bolm, C. *Chem. Commun.* **2002**, 200-201.
22. Herrmann, W. A.; Goossen, L. J.; Artus, G. R. J.; Köcher, C. *Organometallics*, **1997**, *16*, 2472-2474.
23. Lindström, P.; Tierney, J.; Wathey, B.; Westman, J. *Tetrahedron.* **2001**, *57*, 9225-9283.
24. Varma, R.S. *Green Chem.* **1999**, *1,* 43-55.
25. Gedye, R.; Smith, F.; Westaway, K.; Humera, A.; Baldisera, L.; Laberge, L.; Rousell, L. *Tetrahedron Lett.* **1986**, *27*, 279-282.
26. Giguere, R.; Bray, T. L.; Duncan, S. M.; Majetich, G. *Tetrahedron Lett.* **1986**, *27*, 4945-4948.
27. (a) Westman, J. *Org. Lett.* **2001**, *3*, 3745-3747 (b) Kuhnert, N.; Danks, T. N. *Green Chem.* **2001**, *3*, 68-70; (c) Loupy, A.; Regnier, S. *Tetrahedron Lett.* **1999**, *40*, 6221-6224.
28. Mars Synthesis System: http://www.cem.com/products/MARS5.html
29. Varma, R.S.; Namboodiri, V. V. *Chem. Comm.* **2001**, 643-645.
30. Khadilkar, B. M.; Rebeiro, G. L.; *Org. Process Res. Dev.* **2002**; ASAP Article.
31. Varma, R. S.; Namboodiri, V. V. *Tetrahedron Lett.* **2002**, *43*, 5381-5383.
32. Namboodiri, V. V.; Varma, R. S. *Chem. Comm.* **2002**, 342-345.
33. Law, M. C.; Wong, K. –Y, Chan, T. H. *Green Chem.* **2002**, *4*, 328-330.
34. Thied, R. C.; Hatter, J. E.; Seddon, K. R.; Pitner, W. R.; Rooney, D. W.; Hebditch, D. World Patent, WO 20000818, BNFL, **2000**.
35. Chauvin, Y.; Mussman, L.; Olivier, H. *Angew. Chem. Int. Ed. Engl.* **1996**, *34*, 2698-2700.
36. Suarez, P. A. Z.; Einloft, S.; de Souza, R. F.; Dupont, J. *Polyhedron* **1996**, 1217-1219.
37. Monteiro, A. L.; Zinn, F. K.; de Souza, R. F.; Dupont, J. *Tetrahedron-Asym.* **1997**, *8*, 177-179.
38. Suarez, P. A. Z.; Dullius, J. E. L., Einloft, S.; de Souza, R. F.; Dupont, J.; *Inorg. Chim. Acta.* **1997**, *255*, 207-209.

39. Seddon, K. R.; Carmichael, A. J.; Earle, M. J. World Patent, WO 0140146.
40. Earle, M. J.; Seddon, K. R. World Patent Patent, WO 0177081, **2001**.
41. Arduengo III, A. J.; Harlow, R. L.; Kline, M. *J. Am. Chem. Soc.;* **1991**, *113*, 2801-2801.
42. Enders, D.; Breuer, K.; Raabe, G.; Runsink, J.; Teles, J. H.; Melder, J. P.; Ebel, K.; Brode, S. *Angew. Chem. Int. Ed. Engl.* **1995**, *34*, 1021-1025.
43. Kuhn, N; Kratz, T. *Synthesis,* **1993**, 561-565.
44. Bourissou, D.; Guerret, O.; Gabbaï, F. P.; Bertrand, G. *Chem. Rev.* **2000**, *100*, 39-91.
45. Herrmann, W. A.; Goossen, L. J.; Köcher, C.; Artus, G. R. J.; *Angew. Chem. Int. Ed., Engl.*, **1996**, *35*, 2805-2807.
46. Graf, F.; Hupfer, L. US Patent 4450277, BASF, **1984**.
47. Anderson, K.; Earle, M. J.; Goodrich, P. W.;. Hardacre, C.; Seddon, K. R. unpublished results, **2000 - 2002**.

Chapter 3

Challenges to the Commercial Production of Ionic Liquids

Philip E. Rakita

Director of Business Development, Ozark Fluorine Specialties, 1830 Columbia Avenue, Folcroft, PA 19032

Room temperature ionic liquids – particularly imidazolium hexafluorophosphates and tetrafluoroborates -- have been widely studied as substitutes for aromatic solvents in a broad range of chemical reactions and separation processes. In the absence of a reliable commercial supplier, researchers have had to either synthesize their own RTIL's or purchase them from research chemical supply houses in small amounts at high price. Ozark Fluorine Specialties has undertaken the pilot synthesis of a representative RTIL at the 200-kg scale. This paper outlines the issues that need to be addressed to achieve efficient commercial synthesis and make these compounds available on an industrial scale.

Novel Solvents and Reagents for Diverse Applications

As a novel class of solvents, room temperature ionic liquids (RTIL's), sometimes called simply ionic liquids (IL's), have properties that make them desirable alternatives for many industrial applications (such as synthesis and extraction processes). To date, researchers have had to either synthesize their own RTIL's or purchase them from research chemical supply houses in small amounts at high price. In order for these compounds to achieve their commercial potential, it must be demonstrated that they can be produced in bulk quantities (hundreds of kilograms at a time). Ozark Fluorine Specialties has undertaken the pilot synthesis of a representative RTIL at the 200-kg scale. This paper outlines the issues that need to be addressed to achieve efficient commercial synthesis and make these compounds available on an industrial scale.

What are ionic liquids?

Different groups, depending of the focus of their particular interests, have described this large and growing class of novel solvents and reagents in various ways. Researchers have offered several definitions for this class of materials:

- "Organic salts with melting points under 100 °C, often below room temperature" --Fluka product bulletin (*1*)
- "Liquids...composed entirely of anions and cations in contrast to molecular solvents" --Covalent Associates (*2*)
- "Liquids with a wide temperature range and no vapor pressure" --Covalent Associates (*2*)
- "Fused salts are liquids containing only ions" --Tom Welton (*3*)

The key features underlying most of these definitions are that ionic liquids have (A) a large liquid range and (B) no measurable vapor pressure because they are composed of ions rather than discrete molecules.

Why are they chemically interesting?

RTIL's generally exhibit good solvent properties and can often facilitate and influence chemical reactions without being transformed in the process (*3*). Most RTIL's have negligible vapor pressure and miniscule flammability. They frequently exhibit high thermal stability and wide working temperature ranges. For example, the TGA curve for 1-butyl-3-methylimidazolium hexafluorophosphate ($BMIPF_6$) shows stability in air up to 300 °C. (*4*) Owing to the multitude of possible combinations of cation and anion, they are susceptible to numerous permutations that allow the various physical and chemical properties to be adjusted almost at will.

Why are they commercially important?

Considerable research (5,6) has shown that certain RTIL's are effective substitutes for common organic solvents (for example, benzene) where control of VOC emissions is critical. Some hydrophobic ionic liquids will form three liquid phase systems with water and hexane, offering interesting possibilities for extraction and separation technology as well as phase transfer chemistry. Other ionic liquids can act as liquid supports for re-usable catalyst systems. Ionic Liquids are being investigated as potential electrolytes in high performance energy storage systems such as batteries and double layer supercapacitors. Other researchers are investigating ionic liquids as heat transfer fluids in solar energy collectors. (5)

How are they obtained?

Until recently, researchers were obliged to synthesize the ionic liquids they wished to study. Now an increasing number can be purchased from research chemical supply houses. Welton (*3*) lists three basic routes to the synthesis of IL's. These are:

- Metathesis of a halide salt with a salt of the desired anion
- Acid-base neutralization reactions
- Direct combination of a halide salt with a metal halide

For the limited quantity needed for laboratory study, the metathesis route provides the simplest and most direct route. Typically a Group I or ammonium or silver salt of the requisite anion is combined with a quaternary halide, often in a solvent. If the intended IL is not miscible with water, the by-product metal or ammonium halide can often be removed by extraction and separation, or in the case of the silver salts, by filtration of the insoluble precipitate. However, methods that work on the bench top can prove onerous in the chemical plant where tons of products need to be made efficiently and cheaply. Questions of raw material cost, by product removal and disposal and the recovery and reuse of solvents dominate the equation.

A Specific Example—1-butyl-3-methylimidazolium hexafluorophosphate, $BMIPF_6$, is one of the most widely studied IL's to date (*7*). There have been literally hundreds of R&D publications on its preparation, properties and use. Numerous reactions have been studied using $BMIPF_6$ as solvent. Its physical properties have been extensively examined. Table I gives some representative values. It is possibly one of the best-evaluated IL's known. The compound has certain attractive attributes from an industrial perspective. The cost of the raw materials for its manufacture is modest. Several synthetic routes are known. For these reasons, Ozark Fluorine Specialties has chosen this compound as the first to scale up to production at the >100 kg. batch size. The remainder of this paper will address the specific issues and challenges of that scale-up.

Table I. Typical Properties of $BMIPF_6$

Form:	Liquid
Specific gravity:	1.36
Melting point:	~8 °C
Boiling point:	dec. >350 °C
Solubility in water:	Insoluble
Viscosity:	3.12 g/cm-sec at 30 °C

SOURCE: Ozark Fluorine Specialties, product data sheet, April 2002.

Selecting the Synthetic Route

Raw Material Issues

The industrial preparation of $BMIPF_6$, involves first the synthesis of the quaternary ammonium cation, BMI^+, as a halide salt, followed by replacement of the halide with the PF_6 anion, either through metathesis or via an acid-base neutralization. Either route requires a quaternization using butyl chloride and 1-methylimidazole, both of which are commercially available. The anion is also commercially available, either as the potassium salt or as the free acid, HPF6. Since the potassium (or other cationic salt) is typically made from the free acid, the latter route (neutralization) is favored from a cost perspective as one less step is required. There are a number of other factors, including process parameters, which must be considered. These are summarized in Table II.

Table II. Process Parameters to be considered in the synthesis of BMIPF6: Comparison of the metathesis and neutralization routes

	Metathesis route	*Neutralization route*
Materials of construction	Glass OK	Monel or other metal resistant to HF
Thermodynamic considerations	Negligible ΔH	ΔH of neutralization (heat removal)
Handling safety	Non-hazardous reagents	Treat like HF
By-products to dispose of	KCl (aq) or other halide salt	HCl (aqueous)

Generally speaking, when an IL containing the PF_6 or BF_4 anion is prepared, the process parameters such as materials of construction of the reactor, dealing with the enthalpy of reaction and the handling safety of strong acid starting materials favor the metathesis route over neutralization. Thus, when a few grams or a few hundred grams are needed for bench top studies, metathesis is the route of choice. By contrast, where the synthetic infrastructure is already in place for the preparation and handling of HPF_6 or HBF_4, the subsequent conversion of the acid to the "quat salt" IL involves little additional effort. Because of its extensive experience with handling HF in bulk, Ozark Fluorine Specialties has elected the inherently simpler and less costly neutralization route to $BMIPF_6$ using HPF_6.

Other Issues

As mentioned above, using HPF_6 as a reagent for synthesis is inherently more hazardous than using its salts. The other components of the cation, butyl chloride and 1-methylimidazole also require respect, owing to their own specific hazard classifications (see Table III). The byproduct of the final coupling reaction of the cation and anion leaves only an aqueous solution of hydrochloric acid, a byproduct readily handled by a modern chemical synthesis plant.

Table III. Hazard Classifications of Raw Materials for $BMIPF_6$ Synthesis

Raw Material	*Hazard Classification*
1-Butyl chloride	Flammable liquid, irritant
1-Methyl imidazole	Corrosive, hygroscopic liquid
Hexafluorophosphoric acid	Corrosive, toxic Liquid
Potassium Hexafluorophosphate	None

SOURCE: Supplier Material Safety Data Sheets

Although the product has been extensively studied as a chemical substance, both as a reagent and solvent, little is known about its toxicology. Efforts to extrapolate from the properties of "similar" substances may not be valid. Certainly a substance without a measurable vapor pressure poses minimal risks from inhalation. The lack of vapor also reduces the risk of flammability to negligible. As a result, for transportation purposes, $BMIPF_6$ does not require special regulation or packaging.

To date $BMIPF_6$ has not been submitted to the U.S. Environmental Protection Agency for inclusion on the directory of chemical substances known as the TSCA inventory. EPA regulations permit the manufacture and use of new substances not registered, provided "suitably qualified individuals" use them for "R&D purposes" and not entered into commerce. Certain exemptions apply which could allow the commercial use of limited volumes under the so-called "low volume exemption." Limited to a specific time period, a specific volume, and a specific application, this exemption could provide a suitable mechanism for commercial trials using $BMIPF_6$. Ozark Fluorine Specialties is prepared to work with commercial partners to identify and support such an application for exemption.

Cost is an issue central to any commercial use of $BMIPF_6$. On the face of it, the compound, as available today in small volumes, is not cheap. Realistically, it may never be "cheap" when compared to other common solvents. However, there are lots of ways to look at cost and purchase price is only part of the equation. With the increasing regulation of volatile organic compounds (VOC's) in the workplace and the environment, a solvent that has no "volatile" character has some advantages. The cost of more conventional solvents should also reflect the expenses and capital costs required for personal protective equipment (PPE) and emission control hardware and monitoring equipment. The example of benzene, once a common "aromatic" solvent for chemical synthesis should serve to put the potential for RTIL's in perspective. Table IV lists comparative prices for various solvents in use today. Although $BMIPF_6$ may never approach the price of $1 per pound, it is reasonable to expect, based on economies of scale, that pricing at the $20 per pound level could be achieved.

Table IV. Comparative Prices of Various Solvents

Solvent	*Cost of One Liter*	*Cost in Bulk*
	Catalog Supply House	*(per pound or gallon)*
Benzene	$27.30	$0.75/gallon
Dimethylformamid	22.90	1.20/pound
Dimethylsulfoxide	29.67	1.00/pound
Hexamethylphosphoramide	200.00	
Acetonitrile	18.04	0.65-0.75/pound
Sulfolane	44.80	
N,N-Dimethylacetamide	23.70	1.12/pound

Note: One-liter prices from Aldrich or Lancaster catalogs. Bulk prices taken from Chemical Market Reporter (April 8, 2002) or company price sheets.

And, if so, what does this mean for the commercial potential of $BMIPF_6$? First of all, it means that the compound would have to be used in applications where it was recycled. No one today chooses to dispose of solvents or by-products if they have commercial value. For a high priced solvent like $BMIPF_6$, re-use is essential to economic viability. That means either applications such as a heat transfer fluid, where re-use is built into the application, or recycle. And this poses the next big unsolved challenge—how to purify IL's.

$BMIPF_6$ can be synthesized to high levels of purity. Multiple water washings of the reaction product, followed by vacuum drying or sparging with inert gas lead to low levels of water, free acid (measured as HF), and chloride impurities. Table V gives a typical analysis of product made via Ozark's pilot process. A perhaps more important question is what does the IL look like after it has been used in a synthetic procedure. Are there contaminants left behind in the solvent? What are they? Do they matter and if so how can they be removed? These questions need to be answered in the context of specific applications, not abstractly or in an absolute sense.

Table V. Typical Analysis of $BMIPF_6$

Property	*Observed Value*	*Calculated value*
C	33.53%	33.81%
H	5.50%	5.32%
N	9.72%	9.86%
PF_6^-	51.3%	51.01%
F	39.3%	40.11%
Chloride	20 ppm	
Free HF	None detected.	
Moisture	0.42%.	

Methods are being developed to recover and reuse IL's via vacuum evaporation of volatiles and extraction. Extractions with VOC solvents obviate some of the advantages of using IL's in the first place, but supercritical CO_2 extraction may offer promise.

In summary, the manufacturing technology for $BMIPF_6$ and other IL's is substantially advanced. There are no significant impediments to the commercial production and use of these important new solvents.

References

1. Anon. *Chem Files.* **2001,** *vol.1, number 7; p 3, Fluka, Buchs Switzerland.*
2. Anon., *"Ionic Liquids: Enabling Solvents"*, Covalent Associates, Woburn, MA, April 2001.
3. Welton, T. *Chem. Rev.* **1999,** *vol.99, pp 2071-2083.*
4. Koel, M. Proc. *Estonian Acad. Sci. Chem.*, **49** (3), 145 (2000).

5. *Ionic Liquids—Industrial Applications to Green Chemistry;* Rogers, R. D.; Seddon, K. R., eds. ACS Symposium Series 818; American Chemical Society: Washington, DC, 2002.
6. See for example: Freemantle, M., *Chem. Eng. News*, 15 May p. 37 (2000) and 1 January, p. 21 (2001).
7. Wilkes, J. S.; Zaworotko, M. J., *J. Chem. Soc., Chem. Commun.* **1992**, 965.

Chapter 4

Industrial Preparation of Phosphonium Ionic Liquids

Christine J. Bradaric[1], Andrew Downard[2,*], Christine Kennedy[3], Allan J. Robertson[4], and Yuehui Zhou[5]

Cytec Canada Inc., P.O. Box 240, Niagara Falls, Ontario L2E 6T4, Canada
[1]christine_bradaric@we.cytec.com
[2]Corresponding author: phone: 905–356–9000; fax: 905–374–5819);
andrew_downard@we.cytec.com
[3]christine_kennedy@we.cytec.com
[4]al_robertson@we.cytec.com
[5]joey_zhou@we.cytec.com

Abstract

While a great deal of attention has been given to imidazolium ionic liquids in recent years, very few investigations involving phosphonium ionic liquids have been reported in the open literature. Here we present an account of our research into ionic liquids from the perspective of a future, large-scale producer of ionic liquids for industrial applications.

1 - Introduction

The field and phenomenon of room temperature ionic liquids [*1*] are now well past infancy, but much work remains to be done to fulfill the true potential of these neoteric solvents. While ionic liquids containing quaternary nitrogen-based cations have undergone extensive investigation in a myriad of applications over the last several years (see reviews by the groups of Gordon [*2*], Rogers [*3*], Seddon [*4*], Sheldon [*5*], Welton [*6*], Wasserscheid [*7*], and others [*8*]), studies involving quaternary phosphonium systems are much rarer [*9*]. As the global leader in the production of phosphine and phosphine derivatives, Cytec Industries has a great deal of experience in the manufacture of quaternary phosphonium salts that translates naturally to the manufacture of ionic liquids. For instance, we routinely produce tetraalkylphosphonium halides such as the ionic liquid trihexyl(tetradecyl)phosphonium chloride, $[(C_6H_{13})_3P(C_{14}H_{29})]Cl$ (trade names: CYPHOS® 3653 and CYPHOS IL 101) [*10*], in tonne quantities. Over the past several years our research program has developed a diverse range of phosphonium ionic liquids, pairing tetraalkylphosphonium cations with anions such as halides, tetrafluoroborate, hexafluorophosphate, dicyanamide, bis(trifluoromethanesulfonyl)amide, carboxylates, phosphinates, tosylates, alkylsulfates, and dialkylphosphates, among others (Figure 1).

The history of ionic liquids chemistry has been described in detail elsewhere [*2, 11*] but is worth reviewing briefly here. The first report of a room temperature molten salt was made by Walden in 1914, who noted the physical properties of ethylammonium nitrate (mp: 12-14°C) formed by the reaction of ethylamine with concentrated nitric acid [*12*]. This discovery evidently did not arouse great or immediate interest in the scientific community of the day. Nonetheless, the next half century saw sporadic reports of the use of ionic liquids as media for electrochemical studies and, less commonly, as solvents for organic reactions [*2*]. Much of this work involved eutectic mixtures of chloroaluminate-based salts such as $AlCl_3$–NaCl and pyridinium hydrochloride [*13*]. To our knowledge, no quaternary phosphonium cations were employed for such work during this period.

Ionic liquids didn't reach a more general audience until seminal research efforts by the groups of Osteryoung [*14*] and Wilkes [*11, 15*] in the 1970s, and Hussey [*16*] and Seddon [*17*] in the 1980s. This period also saw the first use of ionic liquids as reaction media for organic synthesis [*18*], and, in 1990, for biphasic catalysis [*19*]. In the early 1990s, a report by Wilkes and co-workers describing the first air and moisture stable imidazolium salts, based on tetrafluoroborate, $[BF_4]^-$, and hexafluorophosphate, $[PF_6]^-$ [*20*], fueled further interest in the field. This interest has seen explosive growth during the past decade [*21*], expanding to include diverse applications such as catalysis [*2, 5, 8a*], separations [*22*], electrochemistry [*23*], electrodeposition [*24*], photochemistry [*25*], liquid crystals [*26*], CO_2 capture [*27*], desulfurization of fuel [*28*], enzymatic syntheses [*5*], lubrication [*29*], rocket propulsion [*30*], and

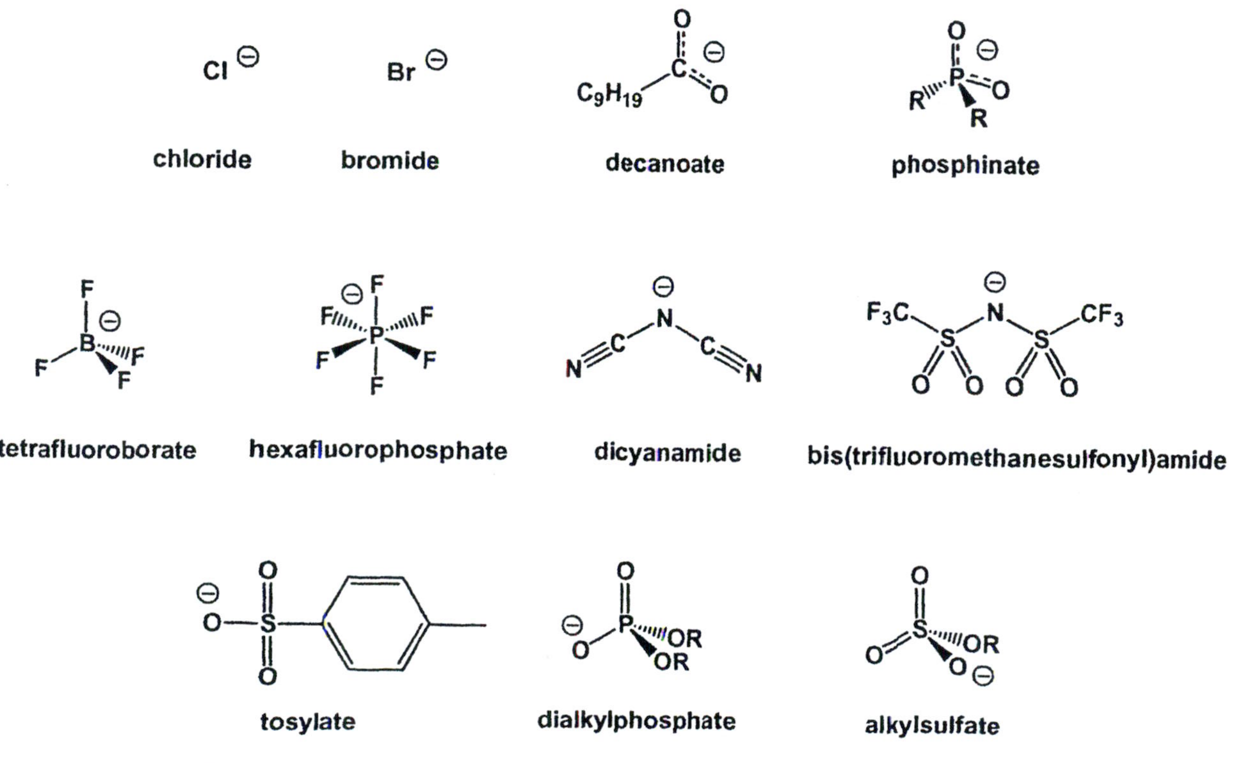

Figure 1: Examples of anions that can be paired with tetraalkylphosphonium cations to produce ionic liquids. (Adapted with permission from reference 9a. ©2003 The Royal Society of Chemistry.)

thermal storage devices [*31*] to name a just a few. Reflecting this, the number of papers published on ionic liquids has grown from approximately 40 per year in the early 1990s to multiple hundreds per year today [*21*].

Relative to their quaternary nitrogen based cousins, specific accounts of ionic liquids containing quaternary phosphorus cations are quite rare, a lacuna which is somewhat surprising given how often phosphonium ionic liquids are mentioned in review articles [*2-8*]. One recent report described some trialkylphosphonium salts, $[HPR_3]^+X^-$, that are liquid at room temperature [*32*]. Three accounts of catalysis in phosphonium ionic liquids have been reported: 1) tetraalkylphosphonium tosylates as solvents for hydroformylation [*9*]; 2) tetraalkylphosphonium halides as solvents for palladium catalyzed Heck reactions [*33*]; and 3) trihexyl(tetradecyl)phosphonium chloride as a solvent for palladium mediated Suzuki cross-coupling reactions [*34*]. We note also that many low melting tetraalkylphosphonium salts are already well known as phase transfer catalysts [*35*].

2 - Features and Advantages of Phosphonium Ionic Liquids

Asymmetrical tetraalkylphosphonium halides, $[R'PR_3^+]X^-$ are typically prepared by nucleophilic (S_N2) addition of tertiary phosphines, $[PR_3]$, to haloalkanes, [R'X (X = Cl, Br, I)] (Eq. 1) [*36*], although other methods have been reported [*37*].

$$PR_3 + R'X \longrightarrow [R'PR_3]^+X^- \quad (1)$$

While the pK_as for tertiary phosphines are typically lower than the corresponding amines, their larger radii and more polarizable lone pair make them more nucleophilic. Hence the kinetics of salt formation are, in general, much faster than for amines [*38, 39*]. The requisite tertiary phosphine starting materials can be prepared *via* free radical addition of phosphine gas (PH_3) [*40*] to alpha olefins [*41*], often in the presence of a suitable promoter such as DuPont's Vazo® series [*42*].

The large number of commercially available haloalkanes and trialkylphosphines suggests a potentially large number of possible tetraalkylphosphonium salts. In our experience, however, there are some practical restrictions that limit the number of systems than can be synthesized easily. For example, only primary haloalkanes have reasonably fast kinetics. Furthermore, alkylphosphines containing branched alkyl chains also tend to be rather slow to react for what we believe are steric reasons. These restrictions notwithstanding, there are still a very large number of phosphonium salts that can be made by quaternization and/or anion exchange reactions.

While quaternization reactions such as those discussed above are well known (particularly for the syntheses of transfer catalysts and Wittig reagents), the realization that judicious selection of anion and cation can produce true room temperature ionic liquids is comparatively recent. For example, we have found that quaternizing PR_3 (R = pentyl, hexyl, octyl) with 1-chloro- or 1-bromotetradecane produces phosphonium halides that are liquid at room temperature[*43*]. In addition, other phosphonium salts such as tetrabutylphosphonium chloride (mp: 67 °C), tetraoctylphosphonium bromide (mp: 45 °C) and tributyl(tetradecyl)phosphonium chloride (mp: 60 °C) are also low melting and fall within the generally accepted, broader definition of ionic liquids, *i.e.* salts which melt below ~100 °C. Though innumerable phosphonium cations can be imagined as constituents of phosphonium ionic liquids, we have utilized the trihexyl(tetradecyl)phosphonium cation, $[(C_6H_{13})_3P(C_{14}H_{29})]^+$, in much of our work. This is for reasons of cost and convenience, and because we have found it works well in many cases.

As is well known for imidazolium halides [*4, 6, 11*] salts containing chlorometallate anions (*e.g.* $AlCl_4^-/Al_2Cl_7^-$, $FeCl_4^-$, *etc.*) addition of metal chlorides (*e.g.* $AlCl_3$, $FeCl_3$, *etc.*) to phosphonium chlorides [*43*]. For example, we have prepared trihexyl(tetradecyl)phosphonium tetrachloropalladate, $[(C_6H_{13})_3P(C_{14}H_{29})]_2[PdCl_4]$, by simple addition of $PdCl_2$ to two equivalents of $[(C_6H_{13})_3P(C_{14}H_{29})]Cl$. This deep red ionic liquid has a melting point of –50 °C, with onset of decomposition occurring above 400 °C. Although free flowing well below room temperature, the viscosity of trihexyl(tetradecyl)phosphonium tetrachloropalladate is approximately an order of magnitude greater than that of trihexyl(tetradecyl)phosphonium chloride (*e.g.* 104 P versus 12 P at 30 °C). It's composition has been confirmed by NMR spectroscopy and elemental analysis [*44*].

Phosphonium halides can also be converted by metathesis methods (see Eq. 2-3) to other anions such as phosphinate, carboxylate, tetrafluoroborate, hexafluorophosphate, *etc.* (Figure 1) [*45*].

$$[R'PR_3]^+X^- + MA \longrightarrow [R'PR_3]^+A^- + MX \qquad (2)$$
$$[R'PR_3]^+X^- + HA + MOH \longrightarrow [R'PR_3]^+A^- + MX + H_2O \qquad (3)$$

[R, R' = alkyl; X = halogen; M = alkali metal; A = anion]

While such methods are both powerful and versatile, the ionic liquids thus produced inevitably contain residual halide ions [*2-8, 46*], making them unsuitable for many applications. Halogen free systems can be produced by direct reaction of tertiary phosphines with alkylating agents such as alkyltosylates, dialkysulfates, and trialkylphosphates, among others.

As suggested by Eq. 1, typical phosphonium cations have the general formula $[R'PR_3]^+$, in which three of the alkyl groups are identical while the fourth is different. However, this does not have to be the case. Primary and secondary alkylphosphines (RPH_2, R_2PH) are also available and can be converted to asymmetric tertiary phosphines (RR'_2P or $R_2R'P$) through free radical addition to olefins [*41*]. The resulting phosphonium cations have generic formulas of $RR'_2R''P^+$ and $R_2R'R''P^+$. This path offers another way to tune the properties of phosphonium ionic liquids.

One difference between phosphonium and ammonium salts is their stability with respect to degradation under various conditions [*9*, *47*]. For example, although both can decompose by internal displacement at higher temperatures (Eq. 4), phosphonium salts are generally more thermally stable than ammonium salts in this respect [*47*].

$$[R_4E]^+X^- \xrightarrow{\Delta} R_3E + R\text{-}X \quad (E = N, P) \qquad (4)$$

Unlike their ammonium counterparts, which can undergo facile Hoffmann- or β-elimination in the presence of base [*48*], phosphonium salts decompose to yield a tertiary phosphine oxide and alkane under alkaline conditions (Eq. 5) [*49*]. Alternatively, depending on the nature of R and R', stable phosphoranes can be formed (Eq. 6); these are well known as Wittig reagents. While the decomposition of phosphonium salts by these pathways may occur even at room temperature in some cases, contrasting examples are known where tetraalkylphosphonium halides can be combined with concentrated sodium hydroxide well above room temperature without any degradation [*47*] (*e.g.* $[(C_{16}H_{33})P(C_4H_9)]Br$ [*50*]).

$$[R_3P\text{-}CH_2\text{-}R']^+ + OH^- \longrightarrow R_3P{=}O + CH_3\text{-}R' \qquad (5)$$
$$[R_3P\text{-}CH_2\text{-}R']^+ + OH^- \longrightarrow R3P{=}CHR' + H_2O \qquad (6)$$

While the decomposition point of neat phosphonium ionic liquids on heating varies somewhat depending on the anion, thermogravimetric analyses (TGA) indicate dynamic thermal stability in excess of 300 °C for many species [*51*]. Figure 2 shows TGA data for trihexyl(tetradecyl)phosphonium tetrafluoroborate, which exhibits a profile typical of most phosphonium salts. This enhanced thermal stability relative to quaternary nitrogen based salts is an important factor when, for example, reaction products must be removed from an ionic liquid by high temperature distillation.

Viscosity is a particularly important characteristic for solvents being considered in industrial applications. Phosphonium based ionic liquids tend to have viscosities somewhat higher than their ammonium counterparts, especially at or near room temperature. However, on heating from ambient to typical industrial reaction temperatures (*e.g.* 70-100 °C) their viscosities generally decrease to < 1 P. This is shown for trihexyl(tetradecyl)phosphonium chloride in Figure 3. Ionic liquid viscosities are also very sensitive to solutes [*52*], and the addition of reactants and/or catalysts can be expected to further reduce viscosity.

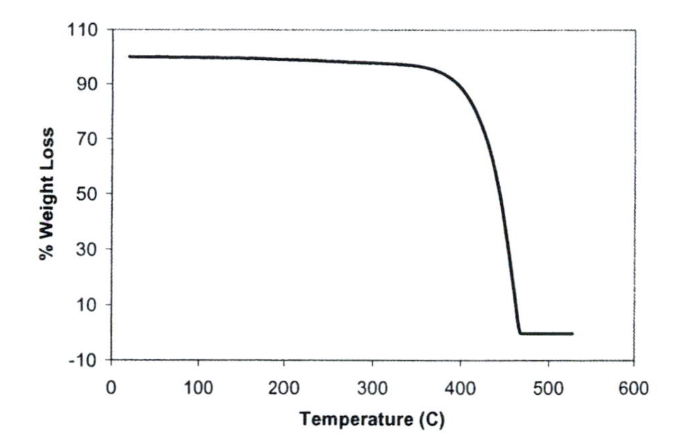

Figure 2: Stability with respect to temperature, as demonstrated by thermogravimetric analysis (TGA), for trihexyl(tetradecyl)phosphonium tetrafluoroborate, $[(C_6H_{13})_3P(C_{14}H_{29})][BF_4]$. (Reproduced with permission from reference 9a. ©2003 The Royal Society of Chemistry.)

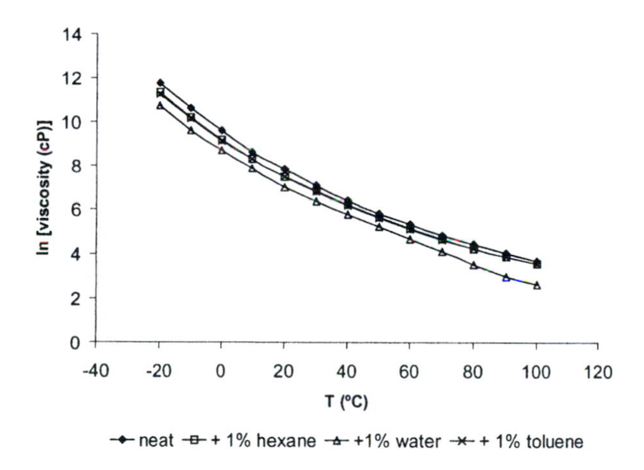

Figure 3: Viscosity with respect to temperature for trihexyl(tetradecyl)phosphonium chloride, $[(C_6H_{13})_3P(C_{14}H_{29})]Cl$. (Reproduced with permission from reference 9a. ©2003 The Royal Society of Chemistry.)

For example, mixing trihexyl(tetradecyl)phosphonium chloride with 1% (w/w) of hexane, water, or toluene decreases viscosity at all temperatures (Figure 3).

While the densities of many imidazolium ionic liquids have been reported previously [*52*], few phosphonium ionic liquids are available for comparison. In contrast to most imidazolium ionic liquids (which generally have densities > 1 g/mL) [*52*], tetraalkylphosphonium salts tend to have densities in

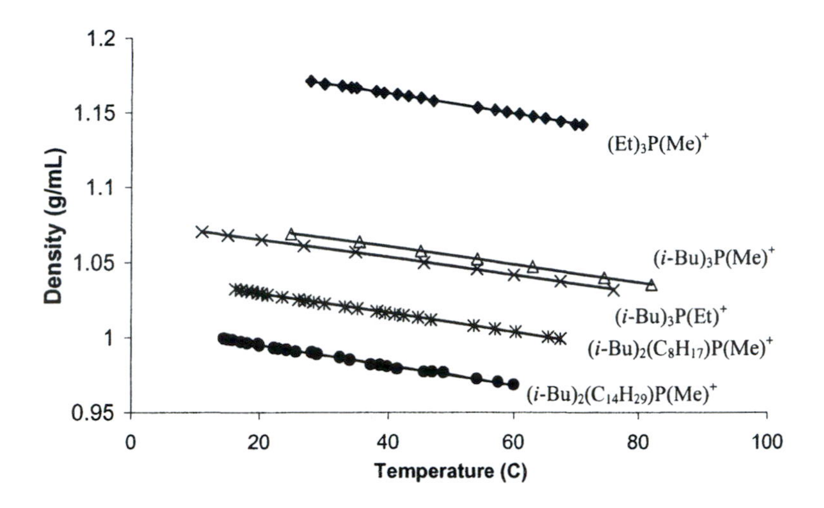

Figure 4: Density as a function of temperature for several tetraalkylphosphonium tosylates. (Reproduced with permission from reference 9a. ©2003 The Royal Society of Chemistry.)

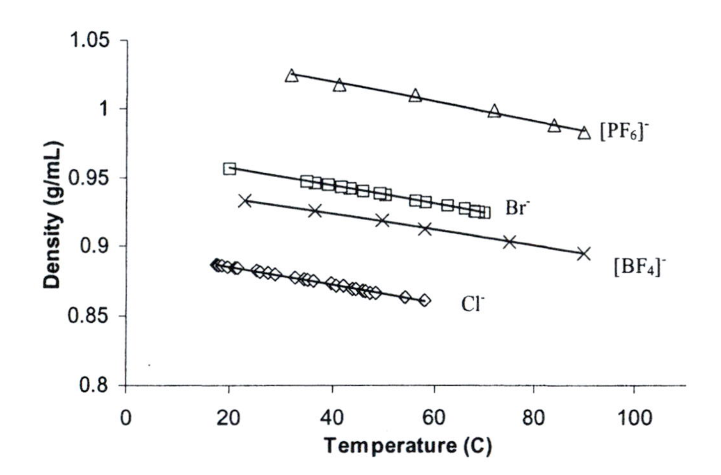

Figure 5: Density as a function of temperature for several salts of the trihexyl(tetradecyl)phosphonium cation. (Reproduced with permission from reference 9a. ©2003 The Royal Society of Chemistry.)

the range 0.8 to 1.2 g/mL range with densities < 1 g/mL being the norm. Phosphonium cations containing more carbon atoms have slightly lower densities, as illustrated in Figure 4, which shows density as a function of temperature for several tetraalkylphosphonium tosylates.

The anion employed also has an impact on density; for example, Figure 5 shows the following order with respect to density for trihexyl(tetradecyl)phosphonium cations paired with several anions:

$$Cl^- < [BF_4]^- < Br^- < [PF_6]^-$$

Another important difference between imidazolium and phosphonium salts are the acidic protons present in the former. Relative to phosphonium cations, imidazolium cations are not entirely inert and can interact with solutes either through hydrogen bonding interactions or through the aromatic nature of the ring system [*53*]. Tetraalkylphosphonium salts do not have such acidic protons or aromatic rings, and consequently there is little potential for interaction with solutes.

3. Metathesis Routes to Phosphonium Ionic Liquids

Phosphonium salts, especially halides, have been available commercially for many years. They are typically made by quaternizing an alkylphosphine with a haloalkane [*36*], and historically have had use as biocides [*54*] and phase transfer catalysts [*35*]. Burgeoning interest in ionic liquids prompted us to closely examine our own range of phosphonium salts in order to target their potential as ionic liquids. In fact, we discovered several ionic compounds that were liquid at or near room temperature. Trihexyl(tetradecyl)phosphonium chloride, long a commercial product, has subsequently seen new life as a starting material for the synthesis of numerous phosphonium-based ionic liquids by anion exchange reactions [*45*]. These generally fall into two categories (as shown in Eq. 2-3); ionic liquids containing the anions shown in Figure 1 can be synthesized by one or the other of these routes.

Phosphonium phosphinates are of particular interest, both because they couple an existing expertise in phosphonium salt and phosphinic acid manufacturing, and because many are apparently novel ionic liquids. A notable example from this series is derived from bis(2,4,4-trimethylpentyl)phosphinic acid, better known as CYANEX® 272 Extractant (see Eq. 10). This is a popular solvent for the extraction of cobalt from nickel in both sulfate and chloride media [*55*], and is currently used to produce more than half of the western world's cobalt [*56*]. Ionic liquids containing the bis(2,4,4-trimethylpentyl)phosphinate anion are thus of interest not only for the usual reasons, but particularly for solvent extraction applications.

We note that both trihexyl(tetradecyl)phosphonium bis(2,4,4-trimethylpentyl)phosphinate and trihexyl(tetradecyl)phosphonium chloride form a middle layer when combined with hexane (or petroleum ether) and water. This behaviour differs from that of the generally more dense dialkyl imidazolium salts, which typically form the bottom phase in such three-layer systems [*57*]. An example of the advantageous nature of this feature was observed recently in the Suzuki cross-coupling of an aryl boronic acid with an aryl halide, which shown to proceed under mild conditions in trihexyl(tetradecyl)phosphonium chloride [*34*]. Product separation was easily accomplished by washing with hexane and water to form a three-layer system; the palladium catalyst remained fully dissolved in the central phosphonium layer and could be easily recycled.

3.1 - Synthesis of trihexyl(tetradecyl)phosphonium chloride

The synthesis of trihexyl(tetradecyl)phosphonium chloride is described here specifically, but similar methodology can be applied to the synthesis of other tetraalkylphosphonium halides. Trihexyl(tetradecyl)phosphonium chloride was synthesized by adding trihexylphosphine to one equivalent of 1-chlorotetradecane at 140 °C under nitrogen and stirring for 12 hours. After the reaction was complete, the mixture was vacuum stripped to remove any volatile components such as hexene, tetradecene isomers, and excess 1-chlorotetradecane. This yielded a clear, pale yellow liquid in 98 % yield (by mass), containing 93.9 % trihexyl(tetradecyl)phosphonium chloride, 4.4 % trihexylphosphonium hydrochloride and 0.3 % hydrochloric acid. Other minor impurities include tetradecene isomers (< 0.3 %) and R_2PH (< 0.7 %).

3.2: Synthesis of trihexyl(tetradecyl)phosphonium bis(2,4,4-trimethylpentyl) phosphinate [*58*].

Trihexyl(tetradecyl)phosphonium bis(2,4,4-trimethylpentyl)phosphinate was prepared by combining trihexyl(tetradecyl)phosphonium chloride (0.66 kg, 1.26 mol, Cytec Industries) and bis(2,4,4-trimethylpentyl)phosphinic acid (86.5 % purity, 0.48 kg, 1.64 mol, Cytec Industries) in the presence of an excess of sodium hydroxide (40 % w/w in water, 0.24 kg, 2.43 mol) and water (Eq. 9). This mixture was heated to 55 °C with vigorous agitation for four hours, and washed three times with water to remove sodium chloride. Vacuum stripping at 135 °C to remove any residual water gave the product, an orange/brown liquid, in 95 % isolated yield (0.93 kg, 1.20 mol, chloride content 0.082% w/w by titration with $AgNO_3$).

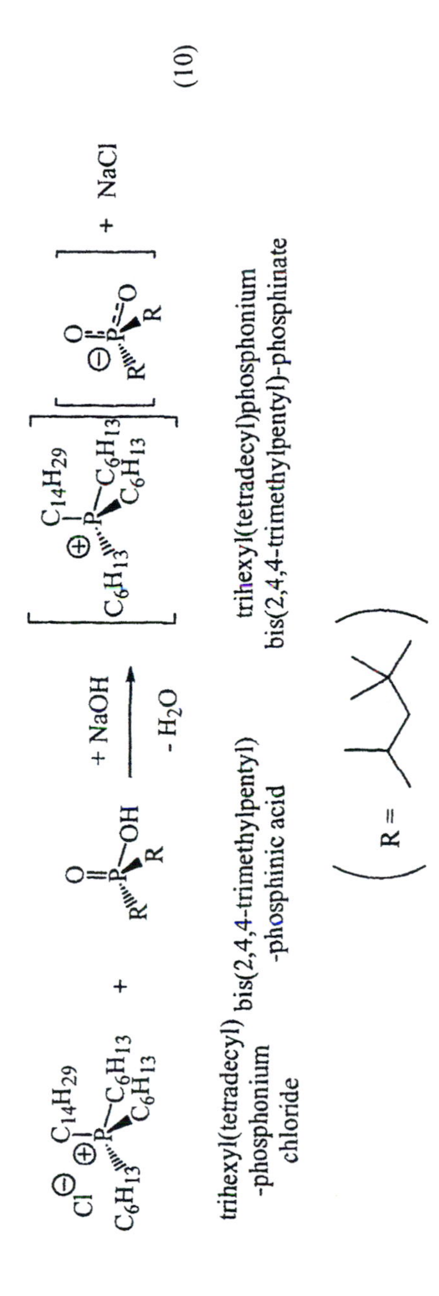
+ NaOH
- H2O
+ NaCl
trihexyl(tetradecyl)
-phosphonium
chloride
bis(2,4,4-trimethylpentyl)
-phosphinic acid
trihexyl(tetradecyl)phosphonium
bis(2,4,4-trimethylpentyl)-phosphinate
R =

(10)

4 - Halide Free Routes to Phosphonium Ionic Liquids

As indicated above, many ionic liquids are prepared through quaternization of nitrogen or phosphorus centers to form chlorides, with subsequent anion exchange if other anions are required [*2-8*]. The materials thus prepared inevitably contain residual chloride ions, which may adversely affect metal catalysts [*2-8, 46*] and/or contaminate reaction products. This is true, for example, of phosphonium ionic liquids used in the production of halogen-free epoxy resins [*59*] and polycarbonates [*60*]. In addition, anion exchange processes are typically wasteful and expensive, often involving the use of environmentally hazardous molecular solvents. Factors such as these increase the final cost of ionic liquids produced industrially, and effectively limit their application range. For these reasons and others, chloride free routes to phosphonium salts are desirable [*61*]. Ren and co-workers, for example, have developed a direct route to imidazolium hexafluorophosphate without anion exchange [*62*]. The preparation of halide free ionic liquids is a topic of ongoing investigation in our laboratories, and we have developed several systems based on direct alkylation of tertiary phosphines, *e.g.* to form phosphonium tosylates (some examples of which have been described previously [*9*]), alkylsulfates, and dialkylphosphates. Here we describe one example of this work: the synthesis and characterization of tri-*iso*-butyl(methyl)phosphonium tosylate.

4.1 – Synthesis of tri-*iso*-butyl(methyl)phosphonium tosylate [*58*]

Methyl tosylate (mp: 27.5 °C, 0.71 kg, 3.83 mol, Esprix Technologies) was heated in a water bath using a steam lance until fully liquefied. This was then charged into a jacketed, glass-lined reactor with paddle agitation as a neat liquid and heated to 80 °C. An equimolar amount of tri-*iso*-butylphosphine (0.77 kg, 3.79 mol, Cytec Industries) was added dropwise, causing the temperature of the reaction mixture to rise. The jacketed reactor was cooled as necessary to keep the internal temperature near 100 °C (+/- 5 °C) throughout the course of the reaction. Addition was complete after six hours, and the reaction mixture was allowed to stand under ambient conditions overnight. After this time, the product was heated to 160 °C under ~10 mmHg vacuum for 0.5 hours to remove any residual starting material. The product was then cooled to 50 °C before being discharged as a clear, very faintly yellow liquid, identified as tri-*iso*-butyl(methyl)phosphonium tosylate (1.43 kg, 3.67 mol, 97% yield). This product has a slight odor that can be removed, if needed, by oxidizing any residual tri-*iso*-butylphosphine with a dilute solution of hydrogen peroxide to produce tri-*iso*-butylphosphine oxide and water.

Conclusions

While the future for ionic liquids is anything but certain, the potential for the field as a whole is undeniably enormous. We believe that successful commercialization of technologies utilizing these neoteric solvents will be a key driver for their continued development and integration into the chemical industry. Large-scale, industrial manufacture of the ionic liquids themselves is clearly a necessary precursor for this process. Here we have described a small part of our continuing efforts in this area.

Ionic liquids based on quaternary nitrogen cations such as imidazolium and pyridinium have been extensively investigated in the academic literature. By comparison, phosphonium ionic liquids have previously received scant attention. Judging by patent activity, however, there is significant interest in phosphonium ionic liquids for industrial use. We believe that phosphonium salts offer a good alternative to ammonium salts for many applications; while neither family is "better" than the other, each will undoubtedly offer advantages and disadvantages for any particular function. This being the case, we anticipate that phosphonium ionic liquids will take their place alongside molecular solvents, imidazolium ionic liquids, and other modern materials in the toolboxes of chemists, chemical engineers, process developers, and inventors.

This is not to say, however, that further investigation of a more fundamental level is not required. Indeed, the physical behaviour of phosphonium ionic liquids still manages to surprise us regularly, and we look forward to continued research and discovery in this area.

Acknowledgements

We thank: Eduardo Kamenetzky, Chermeine Rivera, and William Mealmaker of Cytec Industries Research and Development for physical characterization of ionic liquids; Donato Nucciarone, John Hillhouse and Mike Humeniuk of Cytec Canada Inc. for input in the preparation of this manuscript; Ken Seddon and Alwar Ramani of the QUILL Centre for continued collaboration in this area.

Notes and References

[1] The terms "ionic liquid" and "room temperature ionic liquid" are used here, interchangeably, to described ionic salts with melting points < ~100 °C. Many systems of interest have melting points around or below room temperature.

[2] Gordon, C. M. *Appl. Cat. A* **2001,** *222*, 101.

[3] Huddleston, J. G.; Visser, A. E.; Reichert, W. M.; Willauer, H. D.; Broker, G. A.; Rogers, R. D. *Green Chem.* **2001,** *3*, 156.

[4] a) Seddon, K. R. *J. Chem. Tech. Biotech.* **1997,** *68*, 351; b) Holbrey, J. D.; Seddon, K. R. *Clean Prod. Proc.* **1999**, *1*, 223; c) Seddon, K. R.; Stark, A.; Torres, M. –J. *Pure Appl. Chem.* **2000**, *72*, 2275
[5] Sheldon, R. *Chem. Commun.* **2001**, 2399.
[6] Welton, T. *Chem. Rev.* **1999**, *99*, 2071.
[7] Wasserscheid, P.; Keim, W. *Angew. Chem. Int. Ed. Engl.* **2000**, *39*, 3772.
[8] a) Zhao, D.; Wu, M.; Kou, Y.; Min, E. *Catalysis Today C* **2002**, *2654*, 1; b) Vygodskii, Y. S.; Lozinskaya, E. I.; Shaplov, A. S. *Polymer Science Ser. C* **2001**, *34*, 236; c) Dupont, J.; Consorti, C. S.; Spencer, J. *J. Braz. Chem. Soc.* **2000**, *11*, 337.
[9] For examples, see: a) Bradaric, C. J.; Downard, A.; Kennedy, C.; Robertson, A. J.; Zhou, Y. *Green Chem.*, **2003**, in press. b) Karodia, N.; Guise, S.; Newlands, C.; Andersen, J. *Chem. Commun.* **1998**, 2341.
[10] Cytec phosphonium salts are, in general, sold under the CYPHOS® trade name. Those specifically designated as ionic liquids will heretofore also be known as CYPHOS IL phosphonium ionic liquids.
[11] Wilkes, J. S. *Green Chem.* **2002**, 73.
[12] Walden, P. *Bull. Acad. Imper. Sci.* **1914**, 1800.
[13] Pagni, R. M., in: Mamantov, G.; Mamantov, C. B.; Braunstein, J. (Eds.), *Advances in Molten Salt Chemistry, Vol. 6* **1987**, Elsevier, New York, 1987, pp. 211–346.
[14] a) Chum, H. L.; Koch, V. R.; Miller, L. L.; Osteryoung. R. A. *J. Am. Chem. Soc.* **1975**, *97*, 3264; b) Robinson, J.; Osteryoung, R. A. *J. Am. Chem. Soc.* **1979**, *101*, 323.
[15] Wilkes, J. S.; Levisky, J. A.; Wilson, R. A.; Hussey, C. L. *Inorg. Chem.* **1982**, *21*, 1263.
[16] Scheffler, T. B.; Hussey, C. L.; Seddon, K. R.; Kear, C. M.; Armitage, P. D. *Inorg. Chem.* **1983**, *22*, 2099.
[17] Appleby, D.; Hussey, C. L.; Seddon, K. R.; Turp, J. E. *Nature* **1986**, *323*, 614.
[18] Boon, J. A.; Levisky, J. A.; Pflug, J. L.; Wilkes, J. S. *J. Org. Chem.* **1986**, *51*, 480.
[19] a) Chauvin, Y.; Gilbert, B.; Guibard, I. *J. Chem. Soc., Chem. Commun.* **1990**, 1715; b) Carlin, R. T.; Osteryoung, R. A. *J. Mol. Catal.* **1990**, *63*, 125.
[20] Wilkes, J. S.; Zaworotko, M. J. *J. Chem. Soc., Chem. Commun.* **1992**, 965.
[21] Seddon, K. R. *Green Chem.* **2002**, G25.
[22] Visser, A. E.; Swatloski, R. P.; Rogers, R. D. *Green Chem.* **2001**, *2*, 1.
[23] Quinn, B. M.; Ding, Z.; Moulton, R.; Bard, A. J. *Langmuir* **2002**, *18*, 1734.
[24] Endres, F. *Chem. Phys. Chem.* **2002**, *3*, 144.
[25] Gordon, C. M.; McLean, A. J. *Chem. Commun.* **2000**, 1395.

[26] Gordon, C. M.; Holbrey, J. D.; Kennedy, A. R.; Seddon, K. R. *J. Mater. Chem.* **1998**, *8*, 2627.
[27] Bates, E. D.; Mayton, R. D.; Ntai, I.; Davis, Jr., J. H. *J. Am. Chem. Soc.* **2002**, *6*, 926.
[28] Bösmann, A.; Datsevich, L.; Jess, A.; Lauter, A.; Schmitz, C.; Wasserscheid, P. *Chem. Commun.* **2001**, 2494.
[29] a) Ye, C.; Liu, W.; Chen, Y.; Yu, L, *Chem. Commun.* **2001**, 2244; b) Liu, W.; Ye, C.; Wang, H.; Yu, L. *Tribology* **2001**, *21*, 482.
[30] Gamero-Castaño, G.; Hruby, V. *Journal of Propulsion and Power* **2001**, *5*, 977.
[31] Wu, B.; Reddy, R. G.; Rogers, R. D. *Solar Engineering* **2001**, 445
[32] Netherton, M. R.; Fu, G. C. *Org. Lett.* **2001**, *26*, 4295.
[33] Kaufmann, D. E.; Nouroozian, M.; Henze, H. *Synlett* **1996**, 1091.
[34] McNulty, J.; Capretta, A.; Wilson, J.; Dyck, J.; Adjabeng, G.; Robertson, A. *Chem. Commun.* **2002**, 1986.
[35] a) Herriott, A. W.; Picker, D. *J. Am. Chem. Soc.* **1975**, *97*, 2345; b) Starks, C.M.; Liotta, C. *Phase Transfer Catalysis*, Academic Press, Inc., New York, 1978; c) Wolff, M.O.; Alexander, K.M.; Belder, G. *Chemica Oggi*, **2000,** January/February, 29.
[36] a) Johnson, W. A. *Ylids and Imines of Phosphorus* **1993**, John Wiley and Sons, Inc., New York, USA; b) Hartley, F. R., Ed. *The Chemistry of Organophosphorus Compounds Vol. 3: Phosphonium Salts, Ylids and Phosphoranes* **1994**, John Wiley and Sons, Inc., New York, USA.
[37] For example, see: Balema, V. P.; Wiench, J. W.; Pruski, M.; Pecharsky, V. K. *Chem Commun.* **2002**, 724.
[38] Buckler, S. A.; Henderson, W. A. *J. Am. Chem. Soc.* **1960**, *82*, 5791.
[39] Henderson, W. A.; Schultz, C. J. *J. Am. Chem. Soc.* **1960**, *82*, 5759.
[40] We note that phosphine gas is highly toxic and pyrophoric, necessitating dedicated equipment and special handling techniques. The same is true, to a lesser extent, of many tertiary phosphines. Reactions involving these reagents should only be attempted by appropriately trained personnel.
[41] Rahut, M. M.; Currier, H. A.; Semsel, A. M.; Wystrach, V. P. *J. Org. Chem.* **1961**, *26*, 5138.
[42] See http://www.dupont.com/vazo/.
[43] Robertson, A. J. WO0187900 **2001** (Cytec Industries).
[44] Downard, A.; Seddon, K. R. unpublished results.
[45] Seddon, K. R.; Robertson, A. J. patent pending.
[46] a) Mørk, P. C.; Norgård, D. *8th Scandinavian Symposium on Lipids, Helsinki, Proceedings* **1975**, 339; b) Mørk P. C.; Norgård, D. *J. Am. Oil Chem. Soc.* **1976**, *53*, 506; c) Menzel, A.; Swamy, K.; Beer, R.; Hanesch, P.; Bertel, E.; Birkenheuer, U. *Surface Science* **2000**, *454*, 88.
[47] Wolff, M. O.; Alexander, K. M.; Belder, G *Chimica Oggi* **2000**, 29.

[48] a) Ingold, C. K; Schumacher, O. *J. Chem. Soc.* **1928**, 3125; b) Hanhart, W.; Ingold, C. K. *J. Chem. Soc.* **1927**, 999; c) March, J. *Advanced Organic Chemistry, 4th Ed.* **1992**, John Wiley and Sons, Inc., New York, USA.
[49] Zanger, M.; VanderWerf, C. A.; McEwen, W. E. *J. Am. Chem. Soc.* **1959**, *81*, 3806.
[50] Campbell, J. B. US3639493 **1972** (DuPont)
[51] We note that *dynamic* thermal stabilities normally reported for ionic liquids are typically much higher than *static* thermal stabilities.
[52] Seddon, K. R.; Stark, A.; Torres, M. –J. *Pure Appl. Chem.* **2000**, *72*, 2275; b) Seddon, K. R.; Stark, A.; Torres, M. –J. in *Clean Solvents: Alternative Media for Chemical Reactions and Processing* **2002**, ACS Symp. Ser. 819, M. Abraham, L. Moens, Eds., American Chemical Society: Washington D. C., Vol. 819.
[53] Avent, A. G.; Chaloner, P. A.; Day, M. P.; Seddon, K. R;. Welton, T. *J. Chem. Soc. Dalton Trans.* **1994**, 3405.
[54] a) Jerchel, D. *Chem. Ber.* **1943,** *76B*, 600; b) Jerchel D.; Kimmig, J. *Chem. Ber.* **1950**, *83*, 277; c) Davis, K.P.; Talbot, R.E. US4673509 (Albright and Wilson, Ltd.); d) Talbot R. E. EP275207A2 (Albright and Wilson, Ltd.). e) *Tolcide PS 72A Product Brochure,* Albright and Wilson Biocides, Oldbury, Worley, West Midlands, U.K. f) Wehner, W.; Grade, R. CA2082994AA (FMC Corp. (U.K.) Ltd.).
[55] a) Rickelton, W. A.; Nucciarone, D. *The Treatment of Cobalt/Nickel Solutions Using CYANEX Extractants* in: Cooper W. C.; Mihaylov, I., Ed., *Proceedings of the Nickel-Cobalt 97 International Symposium - Hydrometallurgy and Refining of Nickel and Cobalt* **1997**, Canadian Institute of Mining, Metallurgy and Petroleum, Montreal, pp. 275-292; b) Danesi, P. R.; Reichley-Yinger, L.; Mason, G.; Kaplan, L.; Horwitz, E. P.; Diamond, H. *Solvent Extraction and Ion Exchange* **1985** *4*, 435.
[56] a) Rickelton, W.A.; Robertson, A.J. US4353883 **1982** (Cytec Technology Corp.). b) Roberston, A.J. US4374780 **1983** (Cytec Technology Corp.). c) Rickelton, W. A.; Flett, D. S.; West, D.W. *Solv. Extr. Ion. Exch.* **1984**, *2*, 815.
[57] Carmichael, A.J.; Earle, M.J.; Holbrey, J.D.; McCormac, P.B.; Seddon, K.R. *Org. Lett.* **1999**, *1*, 997.
[58] For proprietary reasons, quantities of reagents in this section have been scaled.
[59] Bayer, H; Fischer, W.; Hekele, W.; Wipfelder, E. US 6160077 **2000** (Siemens AG).
[60] Yokoyama, M.; Takano, J.; Hasegawa, M.; Tatsukawa, Y. US 5578694 **1996** (Mitsubishi).
[61] Mori, S.; Ida, K.; Ue, M. US 4892944 **1990** (Mitsubishi).
[62] *Chemical and Engineering News* **2002** (April 22), 32.

Chapter 5

New Ionic Liquids Based on Alkylsulfate and Alkyl Oligoether Sulfate Anions: Synthesis and Applications

Peter Wasserscheid[1], Roy van Hal[1], Andreas Bösmann[1], Jochen Eßer[1], and Andreas Jess[2]

[1]**Institut für Technische Chemie und Makromolekulare Chemie, University of Technology at Aachen, Worringer Weg 1, D–52074 Aachen, Germany**
[2]**Lehrstuhl für Chemische Verfahrenstechnik, Universität Bayreuth, Universitätsstrasse 30, D–95440 Bayreuth, Germany**

Typical ionic liquids consist of halogen containing anions such as $[AlCl_4]^-$, $[PF_6]^-$, $[BF_4]^-$, $[CF_3SO_3]^-$ or $[(CF_3SO_2)_2N]^-$. However, for many technical applications the presence of halogen atoms in the ionic liquid's anion may cause concerns if the hydrolytic stability of the anion is poor (e. g. for chloroaluminate and hexafluorophosphate systems) or if a thermal treatment of the spent ionic liquid is desired. In this contribution, synthesis, properties and application of several new alkylsulfate and alkyloligoethersulfate ionic liquids are presented. The described systems are easily available from technical raw materials. Some candidates combine low melting points with high hydrolytic stability and acceptable viscosity. Some of the new ionic liquids have been tested as catalyst layer in the Rh-catalyzed hydroformylation of 1-octene.

Why develop new, halogen-free ionic liquids?

The historical development of ionic liquids can be structured according to the different classes of anions that were found to form low melting salts in combination with imidazolium, pyridinium, ammonium and phosphonium cations. Low melting chloroaluminate salts can be regarded as the *"first generation ionic liquids"*. They were described as early as in 1948 by Hurley and Wier at the Rice Institute in Texas as bath solutions for electroplatinating aluminum (*1*). Later in the seventies and eighties, these systems were intensively studied by the groups of Osteryoung (*2*), Wilkes (*3*), Hussey (*4*) and Seddon (*4b,5*).
In 1992, ionic liquid methodology received a substantial boost when Wilkes and Zaworotko described the synthesis of non-chloroaluminate, room temperature liquids (e. g. low melting tetrafluoroborate melts) which may be regarded as *"second generation ionic liquids"* (*6*). Nowadays, tetrafluoroborate and [the slightly later published(*7*)] hexafluorophosphate ionic liquids are among the "working horses" in ionic liquid research. However, their use in many technical applications is still limited by their relatively high sensitivity *vs.* hydrolysis. The tendency of anion hydrolysis is of course much less pronounced than for the chloroaluminate melts but still existent. The $[PF_6]^-$ anion of 1-butyl-3-methylimidazolium ([BMIM]) hexafluorophosphate – for example - has been found in our laboratories to completely hydrolyze after addition of excess water when the sample was kept for 8h at 100°C. HF (toxic and highly corrosive) and phosphoric acid was formed. Under the same conditions hydrolysis of the tetrafluoroborate ion of $[BMIM][BF_4]$ was observed as well, however to a much smaller extent (*8*). Consequently, the application of tetrafluoroborate and hexafluorophosphate ionic liquids is effectively restricted– at least under a technical scenario – to those applications where water-free conditions can be realized at acceptable costs.
In 1996, Grätzel, Bonhôte and coworkers published synthesis and properties of ionic liquids with anions containing CF_3- groups and other fluorinated alkyl groups (*9*). These do not show the same sensitivity towards hydrolysis than $[BF_4]^-$ and $[PF_6]^-$ containing systems. In fact, heating $[BMIM][(CF_3SO_2)_2N]$ with excess of water to 100°C for 24h did not reveal any hint for anion hydrolysis *(9)*. However, despite the very high stability of these salts against hydrolysis and a number of other very suitable properties (e. g. low viscosity, high conductivity, high thermal stability, easy preparation in halogen-free form due to miscibility-gap with water etc.) the high price of $[(CF_3SO_2)_2N]^-$ and of related anions may be a major problem for their practical application in larger quantities. Moreover, the

presence of fluorine in the anion may still be problematic even if hydrolysis is not an issue. Besides the elevated price of the anion (which is also related to the presence of fluorine), the relatively obvious idea to dispose technical amounts of spent ionic liquid by thermal treatment becomes complicated with these ionic liquids. Additional efforts to avoid the liberation of toxic and highly corrosive HF during the combustion of these systems is needed.

In the last two years, an interesting process can be observed in the research aiming for the development of new ionic liquids. Depending on the complexity of the combination of properties required and the amount of ionic liquid consumed for a given application, the recently developed ionic liquids can be divided in two groups:
The first group falls under the definition of "bulk ionic liquids". This means a class of ionic liquids that is designed to be produced, used and somehow consumed in larger quantities. Applications for these ionic liquids are expected to be solvents for organic reactions, homogeneous catalysis, biocatalysis and other synthetic applications with some ionic liquid consumption as well as non synthetic applications such as the application as heat carriers, lubricants, additives, surfactants, phase transfer catalysts, extraction solvents, solvents for extractive distillation, antistatics etc. Cation and anion of these "bulk ionic liquids" are chosen to make a relatively cheap (expected price on a multi-hundred litre scale: ca 30€/litre) and toxologically well-characterized liquid [a preliminary study about the acute toxicity of a non-chloroaluminate ionic liquid has been recently published(*10*)].
The second group comprises highly specialised, task-specific ionic liquids that - of course - will be used in much smaller quantities. Fields of applications for the latter are expected to be special solvents for organic synthesis, homogeneous catalysis, biocatalysis and all other synthetic applications with very low ionic liquid consumption (e. g. due to very efficient multiphasic operation). Non-synthetic applications for these materials are analytic applications (stationary or mobile phases for chromatography, matrices for MS etc.), sensors, batteries etc. These ionic liquids are designed and optimised for the best performance in high-value-adding applications.

Concerning the first group of "bulk ionic liquids" it can be expect that only a very limited number of candidates will be selected for an industrial use on larger scale. However, these candidates will become well characterised and – due to their larger production quantities – readily available. From the above mentioned considerations concerning price, hydrolysis stability and disposal options we anticipate that all future "bulk ionic liquids" will contain no halogen atoms.

A number of halogen-free ionic liquids are already known from the literature. However, none of these systems fulfils the complex combination of properties that is – according to our experiences – required for such a technically suitable "bulk ionic liquid": a) melting point or glass point below 40°C; b) thermal stability > 250 °C; c) stability *vs.* hydrolysis in neutral aqueous solution up to 80°C; d) possible disposal by combustion without formation of highly corrosive gases; e) possible biodegradation of the used anion in ordinary waste water treatment; f) synthesis from cheap, technical available raw materials e. g. alkali salts.
Imidazolium salts with nitrate (*6*), nitrite (*6*), sulfate (*6*), benzene sulfonate (*11*) and phosphonium salts with toluene sulfonate (*12*) anions are described in the literature but their reported melting points are usually higher then 40 °C. Hydrogensulfate and hydrogenphosphate ionic liquids abstract their protons in aqueous solution to form acidic solutions (*13*) Ionic liquids with methylsulfate and ethylsulfate anions show significant hydrolysis in aqueous solution at 80°C. Hydrogensulfate is formed together with the corresponding alcohol (*14*).

In this contribution we like to give a brief description of our recent research on new, halogen-free ionic liquids which fulfil the above mentioned technical criteria in a more promising manner. The synthesis and some properties of selected, new alkylsulfate and alkyloligoethersulfate ionic liquids will be presented together with preliminary results on their physico-chemical properties. Some of the described systems have been applied as catalyst solvent in the Rh-catalyzed hydroformylation of 1-octene.

Synthesis of Octylsulfate Ionic Liquids

$Na[n\text{-}C_8H_{17}O\text{-}SO_3]$ is a commercial chemical which is produced in a multi-thousand ton scale as detergent and ingredient for cosmetics [e. g. from Cognis GmbH, Düsseldorf/Germany (*15*) in ≥87% purity;main impurities are inorganic water soluble salts e. g. Na_2SO_4]. From the technical application of this salt it becomes quite clear that hydrolysis stability, biological degradation and toxicity of the octylsulfate anion are very well documented.
For the ionic liquid synthesis, this anion was combined with the chloride salt of the desired cation in a metathesis reaction either by precipitation of NaCl (in dry acetone) or by extraction with CH_2Cl_2 from an aqueous solution containing $Na[n\text{-}C_8H_{17}O\text{-}SO_3]$. The latter method is preferred because it tolerates water in the sodium salt and allows as well the removal of all ionic impurities originating from the technical quality of the used $Na[n\text{-}C_8H_{17}O\text{-}SO_3]$. The synthesis route is displayed in Scheme 1.

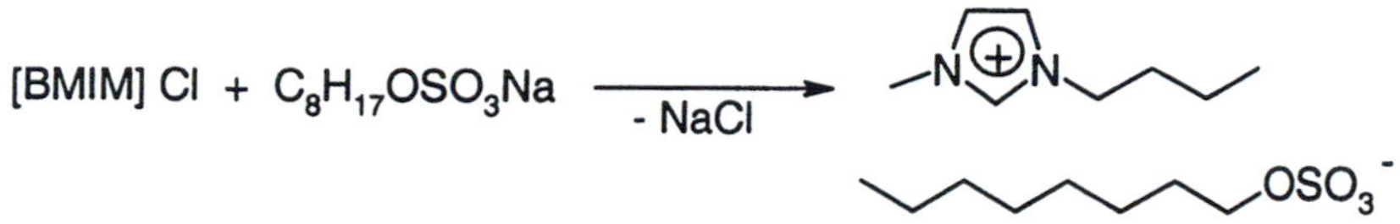

$$[BMIM]\,Cl + C_8H_{17}OSO_3Na \xrightarrow[-\,NaCl]{}$$

Scheme 1: Synthesis of 1-butyl-3-methyl ([BMIM]) octylsulfate.

It is well known that impurities in an ionic liquid can have large effects on the physico-chemical properties of the material under investigation (*16*). For the synthesis of 1-butyl-3-methylimidazolium([BMIM]) octylsulfate the following potential impurities were identified. a) organic volatiles (e. g. traces of methylimidazol from the synthesis of the chloride salt); b) halide impurities from incomplete metathesis reaction; c) other ionic impurities resulting from the technical grade of the applied Na[n-$C_8H_{17}OSO_3$] or from some solubility of Na[n-$C_8H_{17}OSO_3$] in the ionic liquid product; d) water.
In order to obtain reliable data for physico-chemical properties of [BMIM][n-$C_8H_{17}OSO_3$] we took maximum care to either eliminate the impurities completely during synthesis and purification [in case of a)-c)] or to investigate a material with a clearly defined amount of the impurity (in case of water).
To achieve this we checked for volatile impurities (e.g. methylimidazol) in the [BMIM]Cl prior to its application in synthesis, e.g. using the known methods described earlier by Holbrey, Seddon and Wareing (*17*). The amount of chloride in the final product was checked with $AgNO_3$ from an acidic aqueous solution ([BMIM][n-$C_8H_{17}OSO_3$] is water soluble). Using the above described extraction of [BMIM][n-$C_8H_{17}OSO_3$] from an aqueous solution with CH_2Cl_2 [+ some additional washing of the CH_2Cl_2 phase with small portions of water (*18*)] the product could be easily obtained in chloride-free quality. A chloride-free quality of the ionic liquid is not only important for the determination of physico-chemical data but also crucial for the application of the material in hydroformylation catalysis.
Apart from chloride impurities, one has to consider that the synthesis of [BMIM][n-$C_8H_{17}OSO_3$] from technical [BMIM]Cl and Na[n-$C_8H_{17}OSO_3$] may result in contamination of the final products with other ionic impurities. Two main sources of such a potential contamination have to be considered here namely the fact that Na[n-$C_8H_{17}OSO_3$] shows significant solubility in [BMIM][n-$C_8H_{17}OSO_3$] and the fact that the Na[n-$C_8H_{17}OSO_3$] in the applied technical quality contains inorganic salts (mainly Na_2SO_4) in considerable amounts.
While ionic impurities such as Na^+ or SO_4^{2-} may not be a problem for some catalytic applications of the ionic liquid (such as e.g. in Rh- catalyzed hydroformylation) it is of great relevance for the determination physico-chemical properties of the melt.

Fortunately, the applied synthetic method by extraction of [BMIM][*n*-$C_8H_{17}OSO_3$] from water with CH_2Cl_2 allows to remove all ionic impurities due to their high water solubility.
Concerning the amount of water in [BMIM][*n*-$C_8H_{17}OSO_3$] we found it very difficult to reduce the latter in large samples to less than 200 ppm by evaporation at 80°C under high vacuum (10^{-3} bar). This reflects the highly hygroscopic nature of the "dry" octylsulfate salt and gives rise to the question whether anybody would use such an ionic liquid under absolute dry condition. Therefore, we decided to determine physico-chemical data for [BMIM][*n*-$C_8H_{17}OSO_3$] of a defined, low water content rather than for the absolute "dry" material. We found that the water content in [BMIM][*n*-$C_8H_{17}OSO_3$] can be adjusted quite reproducibly to 1000± 100 ppm (0.1±0.01 mass%) after a defined drying procedure of the material [80°C; 3h, high vacuum (10^{-3} bar)]. This water content was checked by coulometric Karl-Fischer titration prior to the measurements and catalytic experiments.

Synthesis of alkyloligoethersulfate ionic liquids

Alternatively to the application of higher alkylsulfate anions, we were interested to synthesize and investigate ionic liquids with alkyloligoethersulfate anions as well. The motivation for this research stem from the fact that we had found in earlier research that the replacement of alkyl groups by oligoether groups at the ionic liquid's cation can decrease the ionic liquid's viscosity to a significant extent (8) [these findings have been very recently confirmed by Afonso and coworkers (*19*)]. Therefore we were interested to look for similar effects on the anion side by replacing the alkyl group in higher alkylsulfate anions by a alkyloligoether group.

A very promising access to alkyloligoethersulfate ionic liquids uses technically available oligoethylenglycol monoalkylethers of the general type R-(O-CH_2-$CH_2)_n$-OH as starting materials. Sulfation of these alcohols can be readily achieved with pyridine-SO_3 (at room temperature) or with NH_2SO_3H (80-160 °C) yielding quantitatively the corresponding pyridinium or ammonium sulfate salts respectively. For a later cation exchange by metathesis (to obtain e.g. the corresponding imidazolium salts) the synthesis of the ammonium salt is preferred since the ammonium cation allows a simple metathesis reaction with all chloride salt in dry CH_2Cl_2 under precipitation of NH_4Cl. The overall route for the synthesis of alkyloligoethersulfate ionic liquids and two examples prepared by this method are presented in Scheme 2.

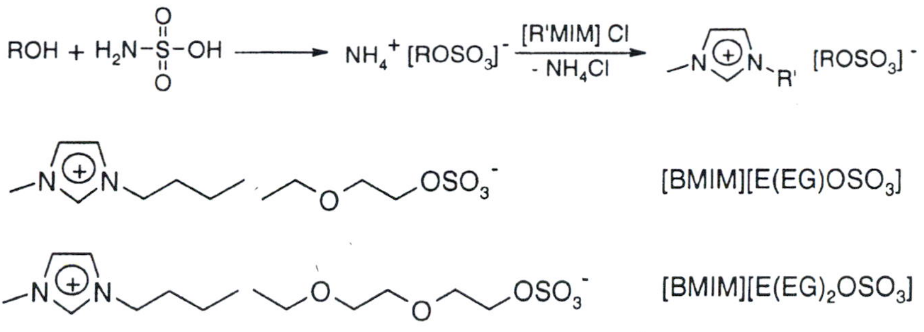

$ROH + H_2N\text{-}SO_2\text{-}OH \longrightarrow NH_4^+ [ROSO_3]^- \xrightarrow[- NH_4Cl]{[R'MIM]\ Cl} [R'MIM][ROSO_3]^-$

OSO_3^- [BMIM][E(EG)OSO_3]

OSO_3^- [BMIM][E$(EG)_2OSO_3$]

Scheme 2: General synthesis route for the preparation of alkyloligoethersulfate ionic liquids and two examples that have been prepared following this route.

It is noteworthy, that the here described synthesis method for the new alkyloligoethersulfate ionic liquids may be more generally applied to prepare a whole range of different ionic liquids with functionalized alkylsulfates from their correspondung alcohols.
Concerning the quality of the ionic liquids obtained after the here described method, the main problems occur during the metathesis step. While the sulfatation reaction proceeds smoothly to quantitative yield, it is not easy to dry the resulting, very hygroscopic ammonium salt to very low levels of water. Consequently, during the metathesis reaction it is not easy to obtain a fully halide-free product. Therefore all physico-chemical data for alkyloligoethersulfate ionic liquids discribed in the contribution will be related to the amount of chloride detected in the sample under investigation.

Selected physico-chemical properties of alkylsulfate and alkyloligoethersulfate ionic liquids

Thermal properties

[BMIM][n-$C_8H_{17}OSO_3$] is often obtained as a sub-cooled melt which slowly crystallizes only below room temperature. The crystalline material has a melting point of 34-35°C (according to DSC) and the heat of fusion was determined to be 12.7 kJ/mol. [BMIM][n-$C_8H_{17}OSO_3$] has a thermal decomposition temperature of 341°C (determined by TGA) and is stable vs. hydrolysis in neutral solution at 80°C for at least 8 hours.

In contrast to [BMIM][*n*-$C_8H_{17}OSO_3$] we never observed so far the crystallization of [BMIM][E(EG)OSO_3] or [BMIM][E$(EG)_2OSO_3$]. The determination of a complete set of thermal properties of these two new ionic liquids is actually on-going in our group.

Viscosity

As mentioned earlier, the viscosity of an ionic liquid is very dependent on the amount of impurities in the ionic liquid material. While water and other solvents present in the ionic liquid are known to reduce the viscosity drastically, chloride impurities have been found to increase the ionic liquid's viscosity with increasing concentration in the liquid (*16*).
To produce reliable viscosity date we routinely correlate all viscosity data to the specific impurities that are found in the specific sample under investigation. The remaining water was analyzed by coulometric Karl-Fischer titration using a Metrohm 756 KF Coulometer with a Hydranal® Coulomat AG reagent. The chloride content was determined as earlier described by Seddon and coworkers (*16*).
For practical reasons, we decided to determine our viscosity data for [BMIM][*n*-$C_8H_{17}OSO_3$] for a material 1000± 100 ppm water (0.1±0.01 mass%) (for more details see above). The investigated samples of [BMIM][E(EG)OSO_3] and [BMIM][E$(EG)_2OSO_3$] contained less water (166 ppm in case of [BMIM][E(EG)OSO_3] and 127 ppm in case of [BMIM][E$(EG)_2OSO_3$]).No chloride was detected in the [BMIM][n-$C_8H_{17}OSO_3$] under investigation, while the amount of chloride in [BMIM][E(EG)OSO_3] and [BMIM][E$(EG)_2OSO_3$] was about 500ppm in each sample (from incomplete metathesis).The viscosity of all samples were determined using RS 100 viscometer from Haake.

The viscosity of [BMIM][n-$C_8H_{17}OSO_3$] was found to be 874.5 cP at 20°C (as sub-cooled liquid) and 152.3 cP at 50 °C. To compare the viscosity of [BMIM][*n*-$C_8H_{17}OSO_3$] to other known solvent systems it may be of interest to note that [BMIM][*n*-$C_8H_{17}OSO_3$] reaches at about 45°C the viscosity of [BMIM][PF_6] at room temperature [which is 207 cP according to a paper by Bright et al.(*20*)] and comes at 100°C close to the room temperature viscosity of ethanediol [16.1 cP according to reference(*21*)].
The viscosities of [BMIM][E(EG)OSO_3] and [BMIM][E$(EG)_2OSO_3$] are surprisingly low. The viscosity of [BMIM][E$(EG)_2OSO_3$] allows a direct comparison with [BMIM][*n*-$C_8H_{17}OSO_3$] since the number of chain members at the substituted sulfate is identical and only two CH_2-groups are replaced by ether bridges in the case of [BMIM][E$(EG)_2OSO_3$]. For this ionic liquid in the above described quality we determined a viscosity of 568 cP (20°C) which clearly demonstrates the effect of the ether functionalities on the viscosity. In fact, the

difference is supposed to be even more significant for a comparison of very pure materials since the viscosity for dry [BMIM][n-$C_8H_{17}OSO_3$] is supposed to be higher than 873 cP (the water in the investigated sample has obviously a viscosity decreasing effect) while the viscosity of halide free [BMIM] [E(EG)$_2$$OSO_3$] can be expected to be even lower than 568 cP (due to the known viscosity increasing effect of chloride impurities *(16)*.
Finally, for [BMIM][E(EG)OSO_3] a viscosity of 92cP was determined at 20°C. This value indicates that viscosity is decreasing with shorter substituents at the sulfate ion even for alkyloligoethersulfate systems. This value represents – to our best knowledge – the lowest reported viscosity for a halogen-free ionic liquid based on alkylsulfate, alkylsulfonate or related anions, so far.

Catalytic Application in the Rh-catalyzed hydroformylation of 1-octene

Since the pioneering work of Chauvin et al. who described in 1995 the first Rh-catalyzed biphasic hydroformylation using room temperature liquid ionic liquids (*22*, *23*) the research efforts in this field were largely dominated by attempts to improve the immobilization of phosphine ligands in the ionic liquid solvent. The use of phosphine ligands with cobaltocenium(*24*), guanidinium(*25* , *26*), imidazolium and pyridinium(*27*) or sulfonate(22)were described by our group and others.

Thus coming closer to an industrial realization of hydroformylation reactions using ionic liquids, potential users of this technology have drawn our attention to the fact that the application of ionic liquids containing fluorine atoms [which have been almost exclusively used in the earlier work; for one exception see (12)] gives rise to serious concerns with regard to the ionic liquid's price, stability and disposal options (see earlier in this contribution).

Therefore it appeared highly interesting to test our new halogen-free systems as solvent for the Rh-catalyzed hydroformylation of 1-octene. So far, only the octylsulfate systems have been evaluated in the hydroformylation catalysis in more detail since the residual chloride contaminants in the alkyloligoethersulfate ionic liquids act as a catalyst poison for the Rh-catalyst.

All catalytic experiments were carried out with [BMIM][n-$C_8H_{17}OSO_3$] synthesised after the above described method. As catalyst system we used a Rh(acac)$(CO)_2$ precursor in combination with the earlier described (*25*) phenylguadinium modified triphenylphosphine ligand precursor **1**. Under reaction conditions the anion attached to **1** is readily exchanged by the anion of the ionic liquid.

The results of the hydroformylation experiments are given in Table 1.

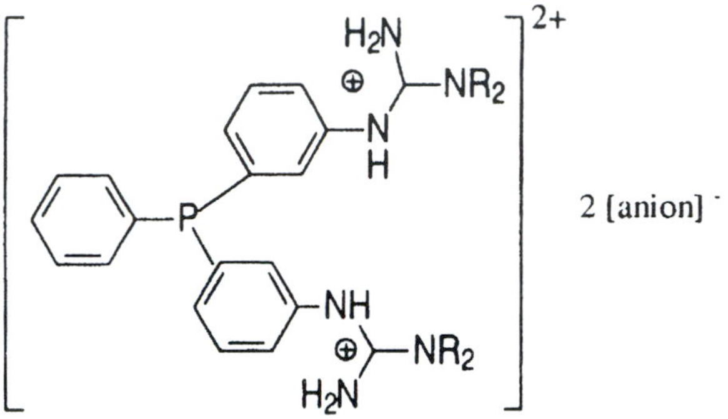

2 [anion]⁻

1

Table 1: Rh-catalyzed hydroformylation of 1-octene in different ionic liquids.

Ionic liquid	Reaction	TOF (h^{-1})	*n:i*–ratio (Sel.)
[BMIM] [Octylsulfate]	monophasic	892	2.86 / 74%
[BMIM] [Octylsulfate] + C_6H_{12}	biphasic	862	2.53 / 72%
For comparison			
[BMIM] [PF_6]	biphasic	276	2.00 / 67%
[BMIM] [BF_4]	biphasic	317	2.60 / 72%

conditions: 25-28 bar CO/H_2 (1:1), 100°C, 1h, 1-octene/Rh= 1000, 5 ml IL, Rh-precursor: Rh(acac)$(CO)_2$, 2 eq. of ligand precursor **1** as hexafluorophosphate salt.

Remarkably, the reaction mixture with [BMIM][n-$C_8H_{17}OSO_3$] does not form a biphasic system any more at high 1-octene conversion. Consequently the reaction mixture becomes monophasic during reaction. However, with cyclohexane added as extraction solvent the biphasic reaction system is maintained even at very high 1-octene conversion and the ionic catalyst solution can be easily separated from the colorless product layer by simple decantation. Interestingly, the activity of the Rh-catalyst is significantly higher with [BMIM][n-$C_8H_{17}OSO_3$] being the solvent in comparison to commonly used hexafluorophosphate and tetrafluoroborate ionic liquids with the same cation. This may be due to the higher 1-octene solubility in [BMIM][n-$C_8H_{17}OSO_3$] (600 mmol/mol IL at 25°C) vs. [BMIM][PF_6] (25 mmol/mol IL at 25°C). Another advantage may arise from the fact that the fluoride anion is a well-known catalyst poison for the Rh-catalyst and the formation of traces of fluoride during the reaction conditions can not be excluded absolutely if hexafluorophosphate ionic liquids are used.

For all examples displayed in Table 1 the ratio between linear and branched hydroformylation products is between 2 and 3 which is the expected range for triphenylphosphine derived ligands.

Conclusion

For good reasons, ionic liquids are often discussed as 'green solvents'. Besides their negligible vapor pressure which prevents solvent evaporation into the atmosphere, two additional options are of interest for transition metal catalysis. Firstly, the special solubility characteristics of the ionic reaction medium enables often a biphasic operation mode of the reaction allowing effective separation of the catalyst from the product and catalyst recycling. Secondly, the non-volatile nature of ionic liquids allows a more effective product isolation by distillation. However, the use of typical ionic liquids consisting of halogen containing anions (such as $[AlCl_4]^-$, $[PF_6]^-$, $[BF_4]^-$, $[CF_3SO_3]^-$ or $[(CF_3SO_2)_2N]^-$) restrict in some regard their 'greenness'. The presence of halogen atoms may cause serious concerns if the anion hydrolyzes under the reaction conditions or if thermal or biological treatment of a spent ionic liquids is desired. In this context, we propose halogen-free octylsulfate and alkyloligoethersulfate ionic liquids as promising candidates for many catalytic, biocatalytic and engineering applications. Although it is still much too early for a final evaluation of the usefulness of these systems (mainly due to missing physico-chemical data), their synthesis from technically available, cheap starting materals and their halogen-free character make these systems especially promising for many of the expected, future "bulk" application of ionic liquids.

References

1 a) Hurley, F. H. U.S. Patent 2,446,331, 1948 [*Chem. Abstr.* **1949**, *43*, P7645b]; b) Hurley, F. H.; Wier, T. P. Jr. *J. Electrochem. Soc.* **1951**, *98*, 207-212.

2 a) Chum, H. L.; Koch, V. R.; Miller, L. L.; Osteryoung, R. A. *J. Am. Chem. Soc.* **1975**, *97*, 3264-3265; b) Robinson, J.; Osteryoung, R. A. *J. Am. Chem. Soc.* **1979**, *101*, 323-327.

3 Wilkes, J. S.; Levisky, J. A.; Wilson, R. A.; Hussey, C. L. *Inorg. Chem.* **1982**, *21*, 1263-1264.

4 a) Laher, T. M.; Hussey, C. L. *Inorg. Chem.* **1983**, *22*, 3247-3251; b) Scheffler, T. B.; Hussey, C. L. *Inorg. Chem.* **1984**, *23*, 1926-1932; c) Appleby, D.; Hussey, C. L.; Seddon, K. R.; Turp, J. E. *Nature* **1986**, *323*, 614-616.

5 Hitchcock, P. B.; Mohammed, T. J.; Seddon, K. R.; Zora, J. A.; Hussey, C. L.; Ward, E. H. *Inorg. Chim. Acta* **1986**, *113*, L25-L26.

6 Wilkes, J. S.; Zaworotko, M. J. *J. Chem. Soc. Chem. Commun.* **1992**, 965-967.

7 Fuller, J.; Carlin, R. T.; de Long, H. C.; Haworth, D. *J. Chem. Soc. Chem. Commun.* **1994**, 299-300.

8 Wasserscheid, P.; Bösmann, A. *unpublished results*.

9 Bonhôte, P.; Dias, A.-P.; Papageorgiou, N.; Kalyanasundaram, K.; Grätzel, M. *Inorg. Chem.* **1996**, *35*, 1168-1178.

10 Pernak, J.; Czepukowicz, A.; Pozniak, R. *Ind. Eng. Chem. Res.* **2001**, *40*, 2379-2383.

11 Waffenschmidt, H. Ph.D. thesis, RWTH Aachen, Germany, 2000.

12 Karodia, N.; Guise, S.; Newlands, C.; Andersen, J.-A. *Chem. Commun.* **1998**, 2341-2342.

13 Keim, W.; Korth, W.; Wasserscheid, P. WO 0016902 to BP Chemicals Limited, UK; Akzo Nobel NV; Elementis UK Limited.

14 Wasserscheid, P.; Bösmann, A.; van Hal, R., RWTH Aachen, Germany, *unpublished results*.

15 Cognis GmbH, Düsseldorf/Germany: http://www.cognis.com.

16 Seddon, K. R.; Stark, A.; Torres, M. –J. *Pure Appl. Chem.* **2000**, *72*, 2275-2287.

17 Holbrey, J. D.; Seddon, K. R.; Wareing, R. *Green Chem.* **2001**, *3*, 33-36.

18 Wasserscheid, P.; van Hal, R.; Bösmann, A. *Green Chem.* **2002**, *4*, 400-404.

19 Branco, L. C.; Rosa, J. N; J. J., Moura Ramos; Afonso, C. A. M. *Chem. Europ. J.*, **2002**, *8*, 3671-3677.

20 Baker, S. N.; Baker, G. A.; Kane, M. A.; Bright, F. V. *J. Phys. Chem. B* **2001**, *105*, 9663-9668.

21 *Handbook of Chemistry and Physics*, 82nd Edition, Linde, D. R., Ed., CRC Press: New York, 2001, pp. 6-182.

22 Chauvin, Y.; Mußmann, L.; Olivier, H. *Angew. Chem.* **1995**, *107*, 2941-2943; *Angew. Chem. Int. Ed. Engl.* **1995**, *34*, 2698-2700.
23 Chauvin, Y.; Olivier, H.; Mußmann, L. EP 0776 880 A1, 1996, to IFP.
24 Brasse, C. C.; Englert, U.; Salzer, A.; Waffenschmidt, H.; Wasserscheid, P. *Organometallics* **2000**, *19*, 3818-3823.
25 Wasserscheid, P.; Waffenschmidt, H.; Machnitzki, P.; Kottsieper, K. W.; Stelzer, O. *Chem. Commun.* **2001**, 451-452.
26 Favre, F.; Olivier-Bourbigou, H.; Commereuc, D.; Saussine, L. *Chem. Commun.* **2001**, 1360-1361.
27 Brauer, D. J.; Kottsieper, K. W.; Liek, C.; Stelzer, O.; Waffenschmidt, H.; Wasserscheid, P. *J. Organomet. Chem.* **2001**, *630*, 177-184.

Chapter 6

Green Synthesis of Ionic Liquids for Green Chemistry

Rex X. Ren

Max Tishler Laboratory of Organic Chemistry, Department of Chemistry, Wesleyan University, Middletown, CT 06459

An environmentally friendly ("green") method for synthesis of ionic liquids has been developed. A number of useful ionic liquids have been prepared in large scale in a solvent-free and waste-free manner and in one reaction vessel, where halide-based ionic liquids were converted to non-halide-based ionic liquids in the presence of Brønsted acids and primary alcohols (ethanol, 1-propanol, 1-butanol). The alkyl halides can be recovered and reused. This novel, innovative technology eliminates the shortcomings in the previously widely used methods of making ionic liquids via the anion metathesis approaches which employ conventional organic solvents, generate aqueous and solid wastes and have a technical difficulty in industrial scale-up. The utilization of such a novel synthetic method for ionic liquids should facilitate further development of green chemistry and green chemical technologies.

Introduction

Green chemistry represents a trend in chemically related research, aiming at waste minimization and cost effectiveness (*1*). Recently, there has been a gradual but steadfast rise of general interest in environmental science, technology and practice, which aims at improving health, aesthetics and in many cases, the economics of individual chemical operations (*2*). In the chemical world, this can be achieved via designing new reactions or modifying existing chemical processes. Conventional organic solvents are used in a range of industrial applications (*3*). However, volatile organic compounds readily evaporate, which often have complex negative effects on the environment. They have been implicated as one of the sources of ozone depletion, global climate change and smog formation. The avoidance of hazardous, volatile organic solvents is essential in future industrialization of green chemistry technologies. On the other hand, reactivity and selectivity as well as product separation are important issues for efficient chemical synthesis and process chemistry (*4*).

Waste minimization and pollution prevention represent significant challenges that will require alternative chemistries and processes. In response to theses challenges, several strategies have been developed: (a) chemistry and chemical reaction-based as well as non-reaction-based engineering for pollution avoidance and prevention, (b) green design, manufacturing and industrial ecology for sustainability. Recently, solvent-free reactions (*5*), supercritical CO_2 (*6*), aqueous reactions (*7*) and fluorous phase synthesis (*8*) have staged their attack on these challenges, with the goal of improving organic transformations with facile separation of products, limiting the efflux of by-products. The past few years have also witnessed an interest in the use of *ionic liquids* for synthetic organic chemistry (*9*).

Room temperature ionic liquids have been considered to be possible environmentally friendly, recyclable media for synthetic organic chemistry, separation sciences and other chemical sciences and engineering. Ionic liquids contain organic cations (for example, 1,3-dialkylimidazolium) and anions, which melt at or close to room temperature. Some examples of ionic liquids are shown in Figure 1.

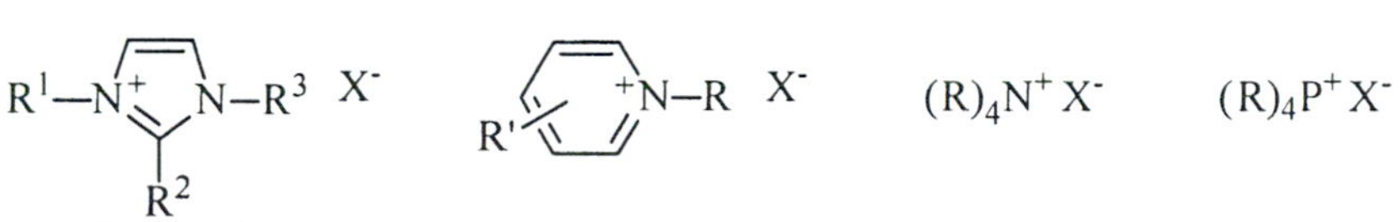

R^1–N⁺ imidazolium N–R^3 X^- (with R^2 at C-2)   R'-pyridinium ⁺N–R X^- $(R)_4N^+X^-$ $(R)_4P^+X^-$

R^1, R^2, R^3, R and R' are alkyl groups with variable chain lengths.
X=Cl, Br, I, $AlCl_4$, BF_4, PF_6, $N(SO_2CF_3)_2$, CH_3SO_3, HSO_4, H_2PO_4, etc.

Figure 1. Examples of Ionic Liquids

In contrast to the conventional organic solvents that are composed of molecular entities such as DMSO, DMF, CH_2Cl_2, $CHCl_3$ and THF, ionic liquids have no significant vapor pressure, thus allow chemical processes to be carried out with essentially zero emission of toxic organic solvents into the environment. In some cases, products can be obtained through distillation from these non-volatile reaction media. On the other hand, low miscibility of various ionic liquids in a number of organic solvents (such as ether, hexanes, ethyl acetate) as well as in supercritical carbon dioxide (a potentially green solvent) (*10*) provides potential solutions for biphasic separation of reaction products.

Ionic liquids possess high ionic conductivity, high ion concentration, and good electrochemical stability. Applications benefiting from these materials may include electrical energy storage devices (such as electrolytic capacitors, batteries and fuel cells), as well as supporting media for catalysts. They have also been found to be efficient in standard separating processes, which eliminates the need of noxious organic solvents (*11*). Therefore such systems are becoming increasingly technologically important. However, due to considerable lack of appreciation and basic research, the use of ionic liquids in chemical industry and other disciplines has not yet become widespread. Nevertheless, although it is still in its infancy, ionic liquid research is recently gathering considerable steam in chemical industry (*12*).

Challenges for Current Research and Development of Ionic Liquids

Chloroaluminate-Based Ionic Liquids

Hurley and Wier first disclosed the room-temperature melts, composed of *N*-alkylpyridinium halides and $AlCl_3$, in a series of U.S. patents (*13*). The room-temperature ionic liquids composed of varying portions of $AlCl_3$ and 1-ethyl-3-methylimidazolium chloride was discovered in 1982 (*14*). However, a disadvantage of this type of room temperature molten salts (ionic liquids) is the liberation of toxic gas HCl when exposed to moisture, thus limiting its role as practical ionic liquids in chemical processes.

$$R^1{-}N^{+}\text{(imidazolium)}N{-}R^2\ Cl^- \xrightarrow{nAlCl_3} R^1{-}N^{+}\text{(imidazolium)}N{-}R^2\ [(AlCl_3)_nCl]^-$$

Figure 2. Preparation of Chloroaluminate-based Ionic Liquids

Anion Metathesis Route to Air- and Water-Stable Ionic Liquids (Figure 3)

Air- and water-stable ionic liquids containing 1,3-dialkylimidazolium cation and tetrafluoroborate anion can be prepared using the anion metathesis method using 1,3-dialkylimidazolium halides and sodium tetrafluoroborate in organic solvents (*15*). The preparation of water-immiscible ("hydrophobic") ionic liquids containing one 1,3-dialkylimidazolium cation via anion metathesis of halides (mostly chloride) with anions such as hexafluorophosphate in organic solvents or water is also known (*16*).

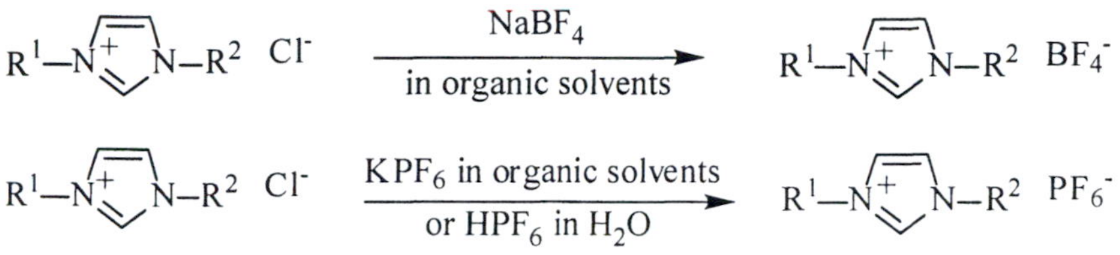

R^1–N(+)N–R^2 Cl^- —($NaBF_4$, in organic solvents)→ R^1–N(+)N–R^2 BF_4^-

R^1–N(+)N–R^2 Cl^- —(KPF_6 in organic solvents or HPF_6 in H_2O)→ R^1–N(+)N–R^2 PF_6^-

Figure 3. Preparation of Air- and Water-Stable Ionic Liquids via Anion Metathesis.

So far, several ionic liquids have been tested as novel reaction solvents in organic synthesis and transition metal catalysis, such as 1-butyl-3-methylimidazolium tetrafluoroborate ([bmi]BF_4, *water-miscible, "hydrophilic"*), and 1-butyl-3-methylimidazolium hexafluorophosphate ([bmi]PF_6, *water-immiscible, "hydrophobic"*). They have shown great potential to replace hazardous organic solvents while maintaining similar and sometimes enhanced chemical selectivity and reactivity. These ionic liquids have been prepared on a laboratory scale via anion metathesis method using halide-based ionic liquids and alkali metal salts or acids, shown in Figure 3. However, these methods have their intrinsic shortcomings.

Compared with the existing industrial technology for manufacturing of conventional organic solvents such as DMSO, DMF, CH_2Cl_2, $CHCl_3$ and THF, the synthesis and manufacturing of useful ionic liquids has been the biggest hurdle that hampers the development of this potentially important technology. The anion metathesis process unavoidably generates large quantities of organic and solid wastes, and is proven to be difficult to scale up. Excess of reagents such as $NaBF_4$, KPF_6 and HPF_6, large quantities of water and organic solvents such as CH_2Cl_2, acetone and acetonitrile have to be used. The anion exchanges readily establish equilibrium among the ions. The purification of ionic liquids generated from anion metathesis process requires extensive washing, filtration and drying. Thus these reactions often give yields in the range of 70-90%. The aqueous wastes contain halide-based ionic liquids, ionic liquid products and excess of acid (hydrogen halides and HPF_6). Solid wastes will necessarily contain ionic liquid starting materials and ionic liquid products, excess of reagents as well as the alkaline metal halide salts. Moreover, the desired ionic liquid products inevitably contain various levels of halides and other contaminants. Besides, the PF_6^- anion-based ionic liquids are found to be unstable, giving little possibility for their recycling, particularly under acidic and Lewis acid conditions. Thus there is an urgent need for development of truly useful ionic liquids with stable anions. On the other hand, optimal preparative methods for manufacturing of ionic liquids should also be sought.

Development of a Novel Waste-Free Method for Preparation of Non-halide Based Ionic Liquids via Alcohol Conversion Route

Introduction

In order to solve the problems associated with the preparation of ionic liquids mentioned above and to search for cheaper and more stable ionic liquids, we set out to investigate alternative approaches with the goal of minimizing and eventually eliminating the formation of wastes.

The readily available halide-based ionic liquids (such as 1-*n*-butyl-3-methylimidazolium chloride "[bmi]Cl", 1-butyl-3-methylimidazolium bromide "[bmi]Br") have been used as precursors in the anion metathesis for preparation of non-halide based ionic liquids. However, the utility of halide anions themselves as nucleophiles has been ignored, considering that these anions in ionic liquid media may have much enhanced nucleophilicity. Therefore, we investigated the utilization of halide-based ionic liquids in a particular organic transformation, that is, the conversion of alcohols to alkyl halides in the presence of acids. These halide-based ionic liquids may serve as both non-aqueous reaction media and nucleophiles. Some preliminary results were published earlier (*17*). We observed that when 1 eq. of 95% sulfuric acid was added to a mixture of 1-butanol in 1 eq. of [bmi]Br at room temperature, a smooth transformation of 1-butanol to 1-bromobutane occurred (Figure 4).

$$\text{1-Butanol} \xrightarrow[\text{(>95\% conversion)}]{\text{1 eq. [bmi]Br} \atop \text{1 eq. } H_2SO_4\text{, 2-3 hr}} \text{1-Bromobutane} + [\text{bmi}]HSO_4 + H_2O$$

Figure 4. Mild Conversion of 1-Butanol to 1-Bromobutane.

Simple decantation or extraction with hexanes can achieve the separation of 1-bromobutane product from the reaction mixture. The mechanism should be close to the well-established S_N2 reaction via protonation of OH group of 1-butanol followed by nucleophilic displacement by bromide. Other acids such as methanesulfonic acid also behaved similarly. This transformation is significant from the viewpoint of pollution avoidance, because the existing method of converting 1-butanol to 1-bromobutane involves heating under reflux of 1-butanol with NaBr in excess of H_2SO_4, followed by azeotropic removal of 1-bromobutane from reaction mixture. It should be pointed out that, in principle, the 1-butyl-3-methylimidazolium cation could be recycled in the form of another ionic liquid (1-butyl-3-methylimidazolium hydrogensulfate "[bmi]HSO_4" or 1-butyl-3-methylimidazolium methansulfonate "[bmi][OMs]").

Synthesis of Non-Halide Based Ionic Liquids Via Alcohol Conversion Route

Proposed methodology

Our preliminary results showed that a primary alcohol was smoothly converted to the primary halide. Although we did not examined in detail the identity of the by-product, it was presumed that it was 1-butyl-3-methylimidazolium hydrogensulfate ("[bmi]HSO_4"), an ionic liquid itself. Thus, by using primary alcohols (ROH) and suitable acids (HA), 1,3-dialkylimidazolium halides or other halides such as pyridinium halides, tetraalkylammonium halides and tetraalkylphosphonium halides (all designated as ILX) can be converted to halide-free ionic liquids (ILA) with anions being the conjugated bases of the acids used (Figure 5).

$$\text{ILX} + \text{ROH} + \text{HA} \longrightarrow \text{ILA} + \text{RX} + H_2O$$

ILX: halide-based ionic liquids (IL=imidazoliums, pyridiniums, tetraalkylammoniums and phosphoniums, X=Cl, Br, I).

ROH: alcohols.

HA: typically Brønsted acids.

ILA: ionic liquids with the conjugated anions (A^-) of HA.

RX: recovered primary alkyl halides.

Figure 5. Proposed Method for Synthesis of Non-Halide Ionic Liquids (ILA).

In the meantime, the recycled primary alkyl halides (RX) can be reused for quaternization of imidazoles, pyridines, trialkylamines and trialkylphosphines to generate more halide-based ionic liquids. The only by-product in this process will be merely one equivalent of water, as shown in Figure 6. This process will eliminate the use of organic solvents and water as required in the conventional anion metathesis methods. Since no alkali metal salts are used, no solid wastes will be generated, and it will not be necessary to carry out a technically difficult filtration step.

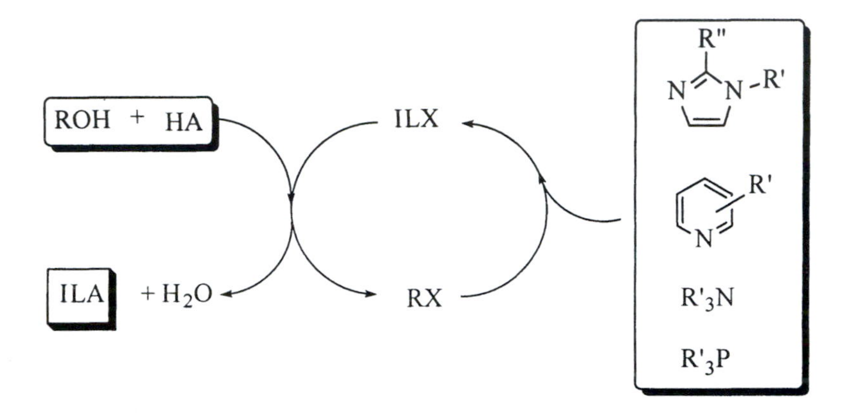

Figure 6. A Waste-Free Process for Synthesis of Ionic Liquids via the Alcohol Conversion Route.

Results and Discussion (18)

With the preliminary results above, we initiated a program in which we began systematically investigating the use of various types of halide-based ionic liquids, alcohols (particularly primary alcohols) and Brønsted acids. In our initial studies, we chose to use methanesulfonic acid because it can be readily purified via distillation. Other acids such as *p*-toluenesulfonic acid and camphorsulfonic acid (both racemic and enantiopure) were also used.

Because bromide anion (Br^-) has a higher nucleophilicity than chloride (Cl^-) and is more stable than iodide (I^-), we chose to use bromide-based ionic liquids such as 1-propyl-3-methylimidazolium bromide ([pmi]Br). Thus, to this ionic liquid (1 mole scale) was added one equivalent of methanesulfonic acid at 0°C followed by addition of one equivalent of 1-propanol with stirring. The 1-bromopropane (b.p. 71 °C; Caution: a flammable liquid) formed was distilled out and collected in a chilled receiver. Then water was also distilled. In order to complete the removal of bromide anions, more 1-propanol (usually 2-3 equivalent) was added followed by distillation until no bromide could be detected via $AgNO_3$ test method. The use of extra amount of 1-propanol also helped the removal of water via formation of an azeotrope. The resulting ionic liquid 1-propyl-3-methylimidazolium methanesulfonate ([pmi][OMs]) was an off-white solid at room temperature with a melting point of approximately 62°C. The water content is determined to be less than 0.01 wt%. 1-bromopropane and excess of 1-propanol were also recovered and reused. Thus a quantitative and waste-free conversion of 1-propyl-3-methylimidazolium bromide to 1-propyl-3-methylimidazolium methanesulfonate was realized.

We also investigated the use of ethanol in this process. The use of ethanol should be more advantageous than 1-propanol or 1-butanol, because the resulting bromoethane (Caution: a flammable liquid) has a much lower boiling point (37°C) than 1-bromopropane or 1-bromobutane (b. p. 100 °C). Methanol can also be used; however, it is less attractive because the methyl chloride and methyl bromide generated are gases at room temperature, which are difficult to contain. Thus, to 1-butyl-3-ethylimidazolium bromide ([bei]Br), which was synthesized from 1-butylimidazole and bromoethane, was added one equivalent of ethanol and one equivalent of methanesulfonic acid. Bromoethane was distilled and recovered in a chilled receiver, followed by removal of water. Again in order to drive the reaction to completion, 2-3 equivalents of ethanol were added followed by further distillation. When $AgNO_3$ testing indicated the absence of bromide, the distillation process was stopped. The resulting ionic liquid 1-butyl-3-ethylimidazolium methanesulfonate ([bei][OMs]) was a clear, light yellowish liquid with a quantitative yield. The determination of its melting point and viscosity is in progress. It should be noted that both [pmi][OMs] and [bei][OMs] have been successfully used as reaction media in the pinacol rearrangement of 1,2-diols in the presence of catalytic amounts of methanesulfonic acid (*19*).

Using the same protocol, we have investigated a variety of halide-based ionic liquids and acids that are readily available. Ethanol has been the favorable choice of alcohol for this process because it is cheap and is considered to be a green commodity chemical. The results have been compiled in Table 1. It is important to note that in order to obtain the highest purity for the final ionic liquid products, all volatile starting materials were purified prior to use. For all of the halide-free ionic liquids that were synthesized via the alcohol conversion route, no further purification was necessary upon removal of volatile starting materials and products. Although the 1-propy-3-methylimidazolium cation ([pmi]) forms solids with methanesulfonate and *p*-toluenesulfonate anions at room temperature, it nonetheless gives a room temperature ionic liquids with both camphorsulfonate anions (racemic) and (+)-camphosulfonate anions (enantiopure). The 1-butyl-3-methylimidazolium cation ([bmi]) also forms a solid methanesulfonate and a solid *p*-toluenesulfonate at room temperature, but $[bmi]HSO_4$ is a room temperature ionic liquid in our hand. Interestingly, the 1-butyl-3-ethylimidazolium cation ([bei]) gives a room temperature ionic liquid with methanesulfonate ([bei][OMs]). The 1-propyl-2,3-dimethylimidazolium cation ([pmmi]) and the 1-butyl-2,3-dimethylimidazolium cation ([bmmi]) give solids at room temperature with methanesulfonate and *p*-toluenesulfonate anions respectively, although it is not clear at this point if these cations will provide room temperature ionic liquids with other anions such as camphorsulfonate. The only pyridinium cation we tested so far is *N*-hexyl-3-picolinium, which forms a solid with the methanesulfonate. Both tetrabutylammonium and tetrabutylphosphonium cations give solids when combined with methanesulfonate anions.

Table 1. Green Synthesis of Ionic Liquids via Alcohol Conversion.

Ionic Liquids "ILX"	*Alcohols*	*Acids*	*Ionic Liquid Products*	*Appear-ance (at rt)*
[pmi]Br	1-propanol	methanesulfonic Acid	[pmi][OMs]	solid
[pmi]Br	1-propanol	*p*-toluenesulfonic Acid	[pmi][OTs]	solid
[pmi]Br	1-propanol	camphorsulfonic Acid	[pmi][CSA]	liquid
[pmi]Br	1-propanol	(+)-camphorsulfonic Acid	[pmi][(+)CSA]	liquid
[bmi]Br	1-propanol	methanesulfonic Acid	[bmi][OMs]	solid
[bmi]Br	1-propanol	*p*-toluenesulfonic Acid	[bmi][OTs]	solid
[bmi]Br	1-propanol	H_2SO_4	[bmi]HSO_4	liquid
[bei]Br	ethanol	methanesulfonic Acid	[bei][OMs]	liquid
[pmmi]Br	ethanol	methanesulfonic Acid	[pmmi][OMs]	solid
[bmmi]Br	ethanol	*p*-toluenesulfonic Acid	[bmmi][OTs]	solid
[hp]Br	ethanol	methanesulfonic Acid	[hp][OMs]	solid
Bu_4NBr	ethanol	methanesulfonic Acid	Bu_4N[OMs]	solid
Bu_4PBr	ethanol	methanesulfonic Acid	Bu_4P[OMs]	solid

NOTE: The yields are all quantitative.

In note below, "acronyms" for cations: "[pmi]" is 1-propyl-3-methylimidazolium, "[bmi]" is 1-butyl-3-methylimidazolium, "[bei]" is 1-butyl-3-ethylimidazolium, "[pmmi]" is 1-propyl-2,3-dimethylimidazolium, "[bmmi]" is 1-butyl-2,3-dimethylimidazolium, "[hp]" is *N*-hexyl-3-picolinium.

In note below, "acronyms" for anions: "[OMs]" is methanesulfonate anion, "[OTs]" is *p*-toluenesulfonate anion, "[CSA]" is camphorsulfonate anion (racemic), "[(+)CSA]" is (+)-camphorsulfonate anion (enantiopure), "HSO_4" is hydrogensulfate anion.

Thus from our preliminary results, we found that 1-butyl-3-ethylimidazolium methanesulfonate ([bei][OMs]), which is not very viscous, may be a good candidate as a room temperature ionic liquid to be further studied. We are in the process of determining the halide and water contents of all of the ionic liquids produced as well as their physical properties such as melting points, viscosity.

Future work

We plan to investigate the synthesis of ionic liquids with a variety of cation and anion combinations. Based on the reaction scheme illustrated in Figure 6, the use of readily available materials such as 1-alkylimidazole, 1,2-dialkylimidazole, pyridine derivatives, trialkylamines, trialkylphosphines and alkyl halides should provide a plethora of halide-based ionic liquids, some of which are commercially available chemicals. Utilization of various primary alcohols (particularly ethanol) and acids should contribute to the production of a variety of ionic liquids at reasonable costs. The fact that the only by-product is water that can be removed via distillation renders this approach a truly green methodology for production of ionic liquids.

Further investigation on the use of these ionic liquids in chemical transformations as well as in other technical areas should lead to the discovery of several classes of stable, economical and useful ionic liquids. Use of these ionic liquids for existing organic transformations will be expanded to seek out the most suitable ionic liquids for organic processes. The goal is to search for better reactivity and selectivity, facile product separation as well as reducing or eliminating the use of volatile organic solvents and production of toxic by-products. In the meantime, comparative studies in organic solvents will be carried out when necessary. It is hoped that more environmentally friendly reactions and processes can be discovered and be put to practice with the possibility of recycling and reuse of ionic liquids. Using the green synthetic method described above, the possibility of manufacturing several particularly versatile ionic liquids at tonnage scale is foreseen.

Conclusion

A safer, more efficient and more environmentally friendly chemical process for preparation of non-halide ionic liquids has been discovered and developed. The most efficient way to synthesize halide-free ionic liquids is to use an acid and a precursor halide-based ionic liquid, using an appropriate alcohol (ethanol).

Acknowledgement

The author wishes to thank the Department of Chemistry, Wesleyan University for support. An Undergraduate Summer Research Award (2002) from Bristol-Myers Squibb Co. is also gratefully acknowledged. I also thank the members in my group, Nicolas Blondin, Larisa D. Zueva, Yunting Luo, Jeff Xin Wu and Wei Ou for their efforts in this research.

References

1. Anastas, P. T.; Warner, J. C. *Green Chemistry*; Oxford: New York, 1998.
2. *The Wiley Encyclopedia of Environmental Pollution and Clean Up*; Meyers, R. A., Editor-in-Chief; Wiley: New York, 1999; Vol. 1-2.
3. (a) Lide, D. R. *Handbook of Organic Solvents*; CRC: Baton Rouge, 1995. (b) *Handbook of Solvents*; Wypych, G.; Ed.; ChemTec Publishing: New York, 2000.
4. Anderson, N. G. *Practical Process Research and Development*; Academic Press: San Diego, 2000.
5. (a) Loupy, A. *Topics in Current Chemistry* **1999**, *206*, 153. (b) Tanaka, K.; Toda, F. *Chem. Rev.* **2000**, *100*, 1025.
6. Brown, R. A.; Pollet, P.; McKoon, E.; Eckert, C. A.; Liotta, C. L.; Jessop, P. G. *J. Am. Chem. Soc.* **2001**, *123*, 1254.
7. Li, C.-J.; Chan. T.-H. *Organic Reactions in Aqueous Media*; Wiley: New York, 1997.
8. (a) Studer, A.; Hadida, S.; Ferritto, R.; Kim, S.-Y.; Jeger, P.; Wipf, P.; Curran, D. P. *Science* **1997**, *275*, 823. (b) Luo, Z.; Zhang, Q.; Oderaotoshi, Y.; Curran, D. P. *Science* **2001**, *291*, 1766.
9. (a) Dupont, J.; de Souza, R. F.; Suarez, P. A. Z. *Chem Rev.* **2002**, *published ASAP.* (b) Zhao, D.; Wu, M.; Kou, Y.; Min, E. *Catalysis Today* **2002**, *74*, 157. (c) Oliver-Bourbigou, H.; Magna, L. *J. Mol. Cat. A: Chem* **2002**, *182-183,* 419. (d) Brennecke, J. F. *AIChE Journal* **2002**, *47*, 2384. (e) Seddon, K. R. *J. Chem. Tech. Biotech*. **1997**, *68*, 351. (f) Olivier, H. *J. Mol. Cat., A: Chem*. **1999**, *146*, 285. (g) Welton, T. *Chem. Rev.* **1999**, *99*, 2071. (h) Wasserscheid, P.; Keim, W. *Angew. Chem., Int. Ed. Engl.* **2000**, *39*, 3772. (i) Holbrey, J. D.; Seddon, K. R. *Clean Products and Processes* **1999**, *1*, 223.
10. Blanchard, L. A.; Hancu, D.; Beckman, E. J.; Brennecke, J. F. *Nature* **1999**, *399*, 28.
11. Visser, A. E.; Swatloski, R. P.; Reichert, W. M.; Davis, J. H., Jr.; Rogers, R. D.; Mayton, R.; Sheff, S.; Wierzbicki, A. *Chem. Comm*. **2001**, 135.
12. *Ionic Liquids: Industrial Applications for Green Chemistry*; Rogers, R. D.; Seddon, K. R., Eds.; American Chemical Society: Washington, DC, 2002.

13. (a) Hurley, F. H. *U.S. Patent*. 2446331, **1948.** (b) Wier, T. P., Jr.; Hurley, F. H. *U.S. Patent* 2,446,349, **1948.**
14. Wilkes, J. S.; Levisky, J. A.; Wilson, R. A.; Hussey, C. L. *Inorg. Chem.* **1982**, *21*, 1263.
15. Wilkes, J. S.; Zaworotko, M. J. *J. Chem. Soc., Chem. Comm.* **1992**, 965.
16. Bonhote, P.; Dias, A.-P.; Papageorgiou, N.; Kalyanasundaram, K.; Gratzel, M. *Inorg. Chem*. **1996**, *35*, 1168.
17. Ren, R. X.; Wu, J. X. *Org. Lett.* **2001**, *3,* 3727.
18. Ren, R. X. "Process for Preparation of Ionic Liquids," *patent pending,* **2002**.
19. Blondin, N.; Ren, R. X. (Weselyan University, Middletown, CT 06459) *unpublished results*, **2002**.

Chapter 7

Expeditious Synthesis of Ionic Liquids Using Ultrasound and Microwave Irradiation

Rajender S. Varma

Clean Processes Branch, National Risk Management Research Laboratory, U.S. Environmental Protection Agency, MS 443, 26 West Martin Luther King Drive, Cincinnati, OH 45268

Environmentally friendlier preparations of ionic liquids have been developed that proceed expeditiously under the influence of microwave (MW) or ultrasound irradiation conditions using neat reactants, alkylimidazoles and alkyl halides. A number of useful ionic liquids have been prepared and the efficiency of this solvent-free approach has been extended to the preparation of other ionic salts bearing tetrafluoroborate anions that involves exposing *N, N'*-dialkylimidazolium chloride and ammonium tetrafluoroborate salt to MW irradiation. These novel solventless alternatives circumvent the common drawbacks in widely used preparative methods for ionic liquids that employ volatile organic solvents and relatively much longer reaction time. The utilization of these novel synthetic methods for ionic liquids generation may facilitate truly 'greener' applications of ionic liquids.

Introduction

Industrial chemistry in the new millennium is adopting the concept of green chemistry to meet the fundamental scientific challenges of protecting the human health and environment while maintaining commercial viability. The emerging area of green chemistry envisages minimum hazard as the performance criteria while designing new chemical processes. The reduction or replacement of volatile organic solvents from the reaction medium is of utmost importance with possible substitutions by nonvolatile or recyclable alternatives. One of the thrust areas for achieving this target is to explore alternative reaction media and optimize reaction conditions to accomplish the desired chemical transformations with minimized by-products or waste generation as well as eliminating the use of conventional organic solvents wherever possible. Among the emerging important tools the use of microwave (MW) as alternative energy source is becoming an attractive alternative especially under the solvent-free conditions (*1-4*).

To address the challenges posed by the waste minimization and pollution prevention efforts, several strategies have been developed. A significant development in the last few years has been the discovery and the use of ionic liquids in synthetic organic chemistry (*5*). Room temperature ionic liquids (RTIL's) comprise a wide range of anions that form low melting salts with several organic cations such as imidazolium, pyridinium, phosphonium and ammonium etc. although the most widely studied examples in the literature encompass 1,3-dialkylimidazolium cations. The second generation of ionic liquids predominantly comprise tetrafluoroborate and hexafluorophosphate as anions although several others have appeared recently that contain CF_3- or similar fluorinated alkyl groups (*6*) and some others that are completely halogen free. Their potential for recyclability, ability to dissolve a variety of materials, and importantly their non-volatile nature with barely measurable vapor pressure are some of their unique attributes responsible for newly found popularity.

Although originally used in electrochemical applications (*7*), these room temperature ionic liquids are currently being explored as environmentally benign solvent substitutes for conventional volatile solvents in a variety of applications such as chemical synthesis (*5*), liquid/liquid separations (*8*), extractions (*9*), dissolution (*10*), catalysis (*11*), biocatalysis (*12*) and polymerization (*13*).

Synthesis of Ionic Liquids Using Microwave Irradiation

The preparation of the 1,3-dialkylimidazolium halides via conventional heating method in refluxing solvents requires several hours to afford reasonable yields and also uses a large excess of alkyl halides/organic solvents as the reaction medium (*14*). In view of the emerging importance of the ionic liquids as reaction media (*5*), the high propensity of the polar reactants to couple to microwaves and our general interest in microwave-assisted chemical processes (*1-4*), we decided to explore the synthesis of ionic liquids using microwave (MW) irradiation under solvent-free conditions. To further simplify the preparation, we experimented by exposing neat reactants in open glass containers to microwaves in a household MW oven. In an unmodified household MW oven, however, it is not possible to effectively adjust the MW power. The reduction in power level simply means that the oven operates at its full power but for a reduced period of time. A recently introduced household MW oven (Panasonic) equipped with inverter technology provides a realistic control of the microwave power to a desirable level and was used throughout in our preparations.

Initially, the effect of microwave power on a set of reactions using alkyl halides and 1-methyl imidazole (MIM) as reactants was examined and it was found that operating the MW oven at reduced power level (240 Watts) afforded the highest yields. Upon microwave irradiation, the ionic liquid starts forming which increases the polarity of the reaction medium thereby increasing the rate of microwave absorption. It was observed that, at elevated power levels, the evaporation of alkyl halide and partial decomposition/charring of the ionic liquid occurs possibly due to the localized overheating of ionic liquid, which eventually results in lower yields. To circumvent this problem, the reactions were conducted with intermittent heating and mixing at a moderate power level to obtain better yields and cleaner ionic liquid formation. After the first irradiation for 30 sec at 240 Watts (~bulk temperature 70-100 ^{0}C), the homogeneity of the reaction mixture changed due to formation of a small amount of ionic liquid. The reaction mixture was then taken out, mixed for 10 sec on a vortex mixer, and then heated at same power level for additional 15 sec. This process was repeated until the formation of a clear single phase was obtained. Thus, the formation of ionic liquid could be monitored visibly in the reaction as it turned from clear solution to opaque and finally clear. Any unreacted starting materials were removed by washing with ether and the final product was dried under vacuum at 80 ^{0}C.

This solventless approach (*15*) requires only a few minutes of reaction time in contrast to several hours required by procedures using conventional heating, which uses an excess of reactants. A general schematic representation for the preparation of mono (**I**) and dicationic (**II**) 1,3-dialkylimidazolium halides is depicted in Figure 1.

This method is ideally suited for the preparation of ionic liquids with longer alkyl chains or with higher boiling points. Comparison of a series of ionic liquids prepared by microwave heating and similar preparation using conventional heating (oil bath at 80 ^{0}C) is summarized in Table 1. Most of the halides used in this study have higher boiling points and are converted efficiently to ionic liquids under microwave irradiation conditions. The relatively less reactive and low boiling reactants such as 1-bromobutane incurred losses due to evaporation and therefore required the use of more than stoichiometric amounts for best results. The reactivity trend of halides was found to be in the order $I^- > Br^- > Cl^-$. Highly reactive iodides afforded excellent yield of ionic liquids in all cases with minimum exposure time (*15*). In contrast, the conventional methods reported in the literature generally use a large excess of alkyl halide or tetrahydrofuran as solvents.

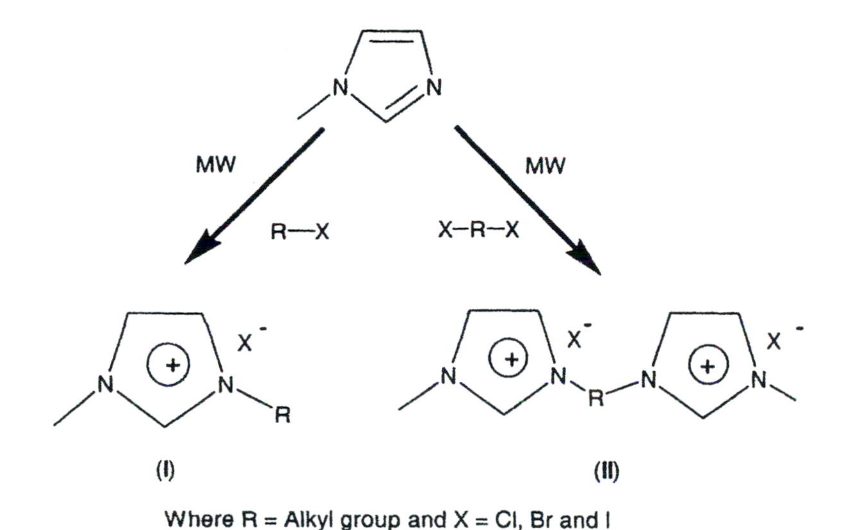

Figure 1. Microwave-assisted Preparation of Ionic Liquids

The butyl and hexyl dicationic salts are solids at room temperature. The corresponding octyl analogues bearing bromide/chloride as the anions are viscous liquids whereas the iodo compound is a solid. The NMR data analysis indicated that the dicationic salts generated from chloro and bromoalkanes were slightly contaminated with the corresponding monocationic intermediate (<5%). However, the diiodoalkanes, being reactive, expeditiously afforded pure products. The purity of ionic salts prepared via microwave heating were found to be superior to those prepared via conventional heating methods, presumably due to inefficient mixing in the latter, once the solid product (**II**) begins to form.

Table 1. Synthesis of Alkyl Imidazolium Halides (I, II) in Microwave Oven (240W).

Alkylhalide (RX)	RX mmol	MIM mmol	MW (min.)[a]	Yield (%)	Yield % (hr)[b]
1-bromobutane	2.2	2	1.25	86	76 (5)
1-chlorohexane	2.2	2	1.5	81	53 (5)
1-bromohexane	2.2	2	1.25	89	78 (5)
1-iodohexane	2.2	2	1.0	93	89 (3)
1-iodoheptane	2.2	2	1.0	94	95 (3)
1-bromooctane	2.2	2	1.25	91	73 (5)
1,4-dibromobutane	1	2.2	1.25	81	76 (5)
1,4-diiodobutane	1	2.2	0.83	91	89 (3)
1,6-dichlorohexane	1	2.2	1.5	82	56 (5)
1,6-dibromohexane	1	2.2	1.1	92	72 (5)
1,6-diiodohexane	1	2.2	0.83	85	97 (3)
1,8-dichlorooctane	1	2.2	1.5	78	72 (5)
1,8-dibromooctane	1	2.2	1.25	92	76 (5)
1,8-diiodooctane	1	2.2	0.83	94	93 (3)

[a]30 sec irradiation with 10 sec mixing time. [b]Alternative heating method (oil bath, 80 ^{0}C).

Water- and air-stable ionic liquids comprising 1,3-dialkylimidazolium cation and tetrafluoroborate anion are usually prepared via the anion metathesis method from 1,3-dialkylimidazolium halides and sodium tetrafluoroborate in organic solvents (*16*). Similarly, the preparation of hydrophobic (water-immiscible) ionic liquids containing hexafluorophosphate anions has also been demonstrated by anion metathesis in organic solvents or water (*6*).

The synthesis of ionic liquids bearing tetrafluoroborate anions has also been a beneficiary of the rapid MW protocol under solvent-free conditions wherein the preparation of RTILs involves exposing a mixture of imidazolium halides and ammonium tetrafluoroborate to microwaves in an oven. This solvent-free approach requires only a few minutes of reaction time, in contrast to several hours needed using conventional methods that require a large excess of organic solvent or expensive silver salts or aqueous tetrafluoroboric acid, which is difficult to handle (*17*).

Initially, the effect of MW power level on the formation of 1-butyl-3-methylimidazolium tetrafluoroborate ($[C_4MIM][BF_4]$) from 1-butyl-3-methylimidazolium chloride ($[C_4MIM]Cl$) and ammonium tetrafluoroborate as reactants was examined (Figure 2). The formation of ionic liquid could be monitored visibly in the reaction vessel when the ammonium tetrafluoroborate crystals reacted and formed fine precipitates of ammonium halide. At elevated power levels, a partial decomposition or charring of the ionic liquid occurred possibly due to the localized overheating of ionic liquid, resulting in lower yields. To circumvent this problem, the reactions were conducted with intermittent heating at a moderate power level with mixing to obtain better yields and cleaner product formation. After the initial exposure for 30 sec to microwaves at power level P3 (360 Watts, bulk temperature ~80-110 °C), the reaction mixture was taken out, mixed for 10 sec and then heated at same power level for additional 30 sec. This sequence was repeated until the formation of fine precipitate of ammonium halide ensued. At this stage, dry acetone was added, the ammonium salts were simply removed by filtration and the product was isolated by removal of acetone from the filtrate and then drying under vacuum at 60 °C. The ionic liquids with alkyl chain length hexyl or higher are insoluble in water and can be purified simply by washing with water followed by vacuum drying.

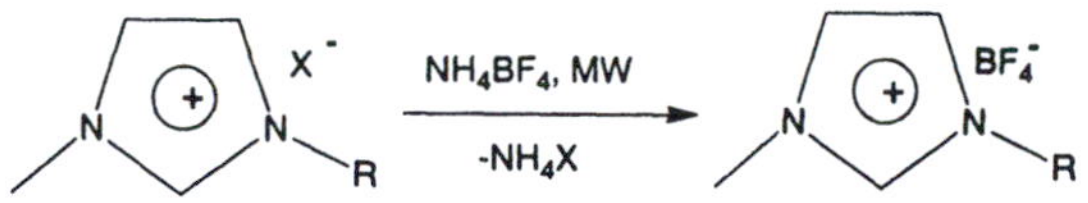

Figure 2. MW-assisted Preparation of Dialkyl Imidazolium Tetrafluoroborates

The MW-protocol is general in nature and is suitable for the preparation of ionic liquids bearing alkyl chains of varying lengths (C_n) or different halides (*18*). The MW-assisted syntheses of a series of ionic liquids and their comparison with the corresponding preparations under conventional heating (oil bath at 90 °C) are summarized in Table 2. The ionic liquids comprising relatively longer carbon chains such as the octyl group required extended reaction time. The tetrafluoroborates obtained from the corresponding iodides were slightly colored and resisted complete decoloration even after repeated washings. The main advantages of this high yield method are use of an inexpensive MW oven, faster generation of products, higher conversion and easier work up procedure compared to methods employing aqueous tetrafluoroboric acid (*16*).

Table 2. MW-assisted Preparation of 1,3-Dialkyl Imidazolium Tetrafluoroborates[a].

Substrate (1 mmol)	MW-power (Watts)	MW-time (min.)	Yield (%) MW (Oil bath)
$[C_4MIM]Cl$	360	2.5	90 (28)
$[C_4MIM]Br$	360	2.5	92 (36)
$[C_4MIM]I$	240	3.0	92[b]
$[i\text{-}C_4MIM]Br$	360	2.5	91
$[C_6MIM]Cl$	360	2.5	89
$[C_6MIM]Br$	360	2.5	88
$[C_7MIM]I$	240	3.0	88[b]
$[C_8MIM]Cl$	360	3.5	89 (35)
$[C_8MIM]Br$	360	3.5	88

[a]NH_4 BF_4 (1.05 mmol); [b]Colored even after several washings.

Sonochemical Synthesis of Ionic Liquids

Although the microwave-assisted preparation of 1,3-dialkylimidazolium halides and tetrafluoroborates has reduced the reaction time from several hours to a few minutes in protocols that avoid the use of a large excess of alkyl halides/organic solvents as the reaction medium (*15,18*), the continuous microwave heating may result in over heating, due to the exothermic nature of the reaction, that leads to the formation of colored products. Further, the subsequent reactions in the viscous ionic liquid medium pose adequate mixing challenges which require often diluting them with either aqueous or an organic solvent, which defeats the purpose of using these non-volatile solvents.

Ultrasound-promoted chemical reactions are well known and proceed *via* the formation and adiabatic collapse of the transient cavitation bubbles (*19*). Consequently, we decided to explore the preparation of ionic liquids using ultrasound as the energy source by simple exposure of neat reactants in closed container to irradiation using a sonication bath. Initially, the effect of ultrasound on a series of reactions comprising alkyl halides and 1-methylimidazole (MIM) was examined using a laboratory ultrasonic cleaning bath with the aim of preparing the commonly used 1,3-dialkylimidazolium halides, (**III, IV**) (Figure 3, Table 3).

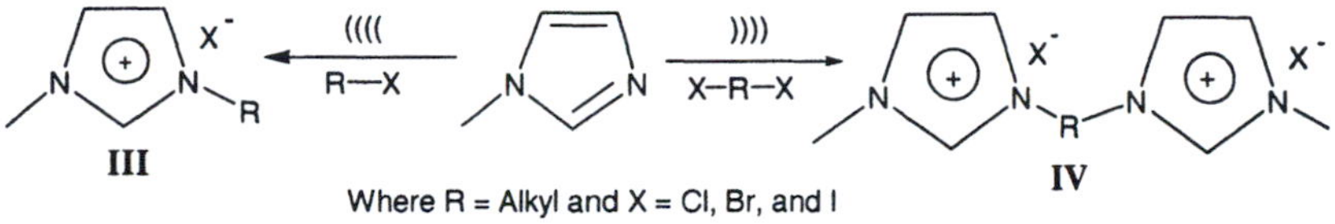

Where R = Alkyl and X = Cl, Br, and I

Figure 3. Ultrasound-accelerated Synthesis of Ionic Liquids

Table 3. Ultrasound-assisted Preparation of 1-Alkyl 3-Methylimidazolium Halides (III, IV).

No	Alkyl halide	RX, mmol	MIM, mmol	Time, (h)	Yield (%)[a]	Yield (%)[c] (h)
1	1-bromopropane	11	10	2	95	0(2)
2	1-chlorobutane	11	10	6	24	0(6)
		220	200	2	86[b]	0(2)
3	1-bromobutane	11	10	2	94	0(6)
4	1-iodobutane	11	10	0.5	94	30(0.5)
5	2-bromobutane	11	10	4	93	0(6)
6	1-chlorohexane	11	10	6	31	0(6)
		220	200	0.5	89[b]	0(2)
7	1-bromohexane	11	10	2	91	0(6)
8	1-iodohexane	11	10	0.5	93	32(0.5)
9	1-iodoheptane	11	10	0.5	91	28(0.5
10	1-chlorooctane	11	10	6	42	0(6)
		220	200	0.25	93[b]	0(2)
11	1-bromooctane	11	10	2	93	0(6)
12	1-iodooctane	11	10	0.5	92	26(0.5)
13	1,4-dibromobutane	5	11	3	93	0(6)
14	1,4-diiodobutane	5	11	2	94	28(3)
15	1,6-dibromohexane	5	11	3	94	0(3)
16	1,6-diiodohexane	5	11	2	93	33(2)
17	1,8-dichlorooctane	100	220	0.25[b]	92	0(2)
18	1,8-dibromooctane	5	11	3	92	0(3)
19	1,8-diiodooctane	5	11	2	94	25(2)
20	ethylchloroacetate	11	10	0.5	98	34(3)

[a]Ultrasonic bath, Fisher Scientific (FS 220); [b]Probe system, Sonic & Materials Inc., Model VCX 750, Power 750W, Frequency 20kHz, tip diameter 19mm, amplitude 97%). [c]reactions in oil bath at temperature reached in sonication bath.

During sonication, the formation of ionic liquid can be visibly monitored when the reaction contents turn from clear solution to opaque (emulsification) and finally a clear viscous phase or separation of solids occurs (*20*). Upon continuous irradiation for 2h, temperature of the bath rises from 22 °C to 40 °C. At this stage, the work-up only involves the removal of residual halide under vacuum or by washing with ethyl acetate/ether and drying (*21*).

To establish the generality of the reaction a series of ionic liquids were prepared *via* this sonication protocol and were then compared with the similar preparation using conventional heating in an oil bath (Table 3). Most of the halides used in this study were efficiently converted to the corresponding ionic liquids with the reactivity trend being in the order I^-> Br^-> Cl^-. The alkyl chlorides are relatively less reactive and required longer irradiation time and some heating except for reactive entities such as ethyl chloroacetate (entry 20). The higher reactivity of bromides and iodides provided excellent yields with minimum sonication time and the reactions were essentially completed at room temperature. For the preparation of chloride salts, the use of a probe system (Sonic & Materials Inc.) was explored wherein an ultrasonic field was produced directly in the reaction vessel. A number of chloride salts could be easily prepared via this modified approach (entries 2, 6, 10 and 17, Table 3). During sonication of 1-chlorobutane with MIM, the temperature of the reaction mixture rose from 22 °C to 80 °C and became constant at 80 °C. Upon formation of the ionic liquid, due to exothermicity of the reaction, the temperature increased to 115 °C. Similarly, in the case of reaction between 1-chlorooctane and MIM the temperature reached 100 °C and upon formation of the ionic liquid it further increased to 125 °C. Butyl, hexyl and octyl dicationic salts (entries 13-19, Table 3) were also prepared efficiently by this methodology. The diiodo and dibromoalkanes, being more reactive, rapidly afforded pure products.

The *in-situ* generation of ionic liquids and their subsequent utilization as reaction media by sequential addition of reactants in the same pot renders this an ideal approach for chemical transformation where higher viscosity of the ionic liquids impedes the efficient mixing of the reactants and catalysts. This ultrasound-accelerated method is adaptable for efficiently accomplishing a one-pot palladium acetate catalyzed Suzuki reaction of iodobenzene with 4-fluorophenylboronic acid at room temperature by subsequent addition of reactants to *in situ* generated ionic liquid. The ionic liquid and catalyst can be recycled 4 times without the loss of any reactivity (91 %). In brief, the developed sonochemical protocol provides the clean synthesis of ambient temperature ionic liquids that occurs nearly at room temperature.

Conclusion

Efficient, expeditious and environmentally friendlier chemical processes have been detailed for the preparation of ionic liquids that precludes the excessive use of volatile organic solvents and utilizes common microwave oven or the ultrasonic bath. Ionic liquids, being polar and ionic in character, couple to the microwave irradiation very efficiently and therefore are ideal microwave absorbing candidates (*22*). The surge of interest continues with this class of solvents especially under MW irradiation conditions where potential recycling of the catalyst is enhanced as demonstrated for a high-speed Heck reaction in ionic liquid media (*23*) and for expediting hetero Diels-Alder reactions of 2(1*H*)-pyrazinones in ionic liquid doped solvents (*24*). Further investigations on the use of such in situ generated ionic liquids via microwave or ultrasound (*19c,d*) activation should promote the utility and applications of ionic liquids in chemical transformations and syntheses.

Acknowledgement

The author thanks Dr. V. V. Namboodiri for his efforts in this research.

References

1. (a) Varma, R. S. in *Microwaves in Organic Synthesis*, Loupy, A. (Ed.) Chapter 6, pp 181-218, Wiley-VCH, Weinheim, (2002). (b) Varma, R. S. in *ACS Symposium Series No. 767/ Green Chemical Syntheses and Processes,* Anastas, P. T.; Heine, L.; Williamson, T. (Eds.), Chapter 23, pp 292-313, American Chemical Society, Washington DC (2000). (c) Varma, R. S. in *Green Chemistry: Challenging Perspectives,* Tundo, P. Anastas, P. T. (Eds.), Oxford University Press, Oxford, 2000, pp 221-244.
2. Varma, R. S. *Pure Appl. Chem.* **2001**, *73*, 193.
3. Varma, R. S. *Green Chem.* **1999**, *1*, 43.
4. (a) Varma, R. S. *Tetrahedron* **2002**, *58*, 1235. (b) Pillai, U.; Sahle-Demmessie, E.; Varma, R. S. *J. Mat. Chem.*, **2002**, *12*, 3199.
5. (a) Dupont, J.; de Souza, R. F.; Suarez, P. A. Z. *Chem Rev.* **2002**, *102*, 3667. (b) Zhao, D.; Wu, M.; Kou, Y.; Min, E. *Catalysis Today* **2002**, *74*, 157. (c). Brennecke, J. F. *AIChE Journal* **2002**, *47*, 2384. (d) Wasserscheid, P.; Keim, W. *Angew. Chem., Int. Ed. Engl.* **2000**, *39*, 3772. (e) Welton, T. *Chem. Rev.* **1999**, *99*, 2071. (f) Olivier, H. *J. Mol. Cat. A: Chem.* **1999**, *146*, 285. (g) Holbrey, J. D.; Seddon, K. R. *Clean Prod. Proc.* **1999**, *1*, 223.

6. Bonhôte, P.; Dias, A. -P.; Papageorgiou, N.; Kalyanasundaram, K.; Grätzel, M. *Inorg. Chem.* **1996**, *35*, 1168.
7. Hussey, C. L. *Molten Salt Chem.* **1983**, *5*, 185. (b) Wilkes, J. S.; Levisky, J. A.; Wilson, R. A.; Hussey, C. L. *Inorg. Chem* .**1982**, *21*, 1263.
8. Swatloski, R. P.; Visser, A. E.; Reichert, W. M.; Broker, G. A.; Farina, L. M. Holbrey, J. D.; Rogers, R. D. *Green Chem.* **2002**, *4*, 81.
9. Blanchard, L. A.; Hancu, D.; Beckmann, E. J.; Brennecke, J. F. *Nature*, **1999**, *399*, 28. (b) Huddleston, J. G.; Willauer, H. D.; Swatloski, R. P.; Visser A. E.; Rogers, R. D. *Chem. Commun.* **1998**, 1765.
10. Swatloski, R. P.; Spear, S. K.; Holbrey, J. D.; Rogers, R. D. *J. Am. Chem. Soc.* **2002**, 124, 4974.
11. (a) Cole, A. C.; Jensen, J. L.; Ntai, I.; Loan, K.; Tran, K. L.; Weaver, T. K. J.; Forbes, D. C.; Davis Jr., J. H. *J. Am. Chem. Soc.* **2002**, 5962. (b) Zhao, D.; Wu, M.; Kon, Y.; Min, E. *Cat. Today* **2002**, *74*, 157.
12. (a) Cull, S. G.; Holbrey, J. D.; Vargas-Mora, V.; Seddon, K. R.; Lye, G. J. *Biotechnol. Bioeng.* **2000**, *69*, 227. (b) Sheldon, R. A.; Maderia Lau, R.; Sorgedrager, M. J.; van Rantwijk, F.; Seddon, K. R. *Green Chem.* **2002**, *4*, 147.
13. (a) Hardacre, C.; Holbrey, J. D.; Katdare S. P.; Seddon, K. R. *Green Chem.* **2002**, *4*, 143. (b) Carmichael, A. J.; Haddleton, D. M.; Bon S. A. F.; Seddon, K. R. *Chem. Commun.* **2000**, 1237.
14. Volker, P.; Bohm, W.; Herrmann, W. A. *Chem. Eur. J.*, 2000, *6*, 1017.
15. (a) Varma, R. S.; Namboodiri, V. V. *Chem. Commun.* **2001**, 643. (b) Varma, R. S.; Namboodiri, V. V. *Pure Appl. Chem.* **2001**, *73*, 1309.
16. Wilkes, J. S.; Zaworotko, M. J. *J. Chem. Soc., Chem. Comm.* **1992**, 965.
17. Holbrey, J. D., Seddon, K. R., *J. Chem. Soc. Dalton. Trans.* **1999**, 2133.
18. Namboodiri, V. V.; Varma, R. S. *Tetrahedron Lett.* **2002**, *43*, 5381.
19. (a) *Synthetic Organic Chemistry*, Luche, J. L. (Ed.), Plenum Press, New York, 1998. (b) Gaplovsky, A.; Gaplovsky, M.; Toma S.; Luche, J. L. *J. Org. Chem.***2000**, 65, 8444. (c) Deshmukh, R. R.; Rajagopal, R.; Srinivasan, K. V. *Chem. Commun.* **2001**, 1544. (d) Leveque, J. M.; Luche, J. L.; Petrier, C.; Roux, R.; Bonrath, W. *Green Chem.* **2002**, *4*, 357.
20. Namboodiri, V. V.; Varma, R. S. *Org. Lett.* **2002**, *4*, 3161.
21. Huddleston, J. G.; Visser, A. E.; Reichert, W. M.; Willauer, H. D.; Broker, G. A.; Rogers, R. D. *Green Chem.* **2001**, *3*, 156.
22. Leadbeater, N. E.; Trenius, H. M. *J. Org. Chem.* **2002**, *67*, 3145.
23. Vallin, K. S. A.; Emilsson, P.; Larhed, M.; Hallberg, A. *J. Org. Chem.* **2002**, *67*, 6243.
24. Van der Eycken, E.; Appukkuttan, P.; De Borggraeve, W.; Dehaen, W.; Dallinger, D.; Kappe, C. O. *J. Org. Chem.* **2002**, *67*, 7904.

Chapter 8

Unprecedented Synthesis of 1,3-dialkylimidazolium-2-carboxylate

Applications in the Synthesis of Halogen-Free Ionic Liquids and Reactivity as Carbon Dioxide Transfer Agent to Active C–H Bonds

M. Aresta[1,2], I. Tkatchenko[3], and I. Tommasi[1,2,*]

[1]Department of Chemistry, University of Bari, Campus Universitario, via Orabona 4, 70126 Bari, Italy
[2]METEA Research Center, University of Bari, 70126 Bari, Italy
[3]Laboratoire de Synthèse et Electrosynthèse Organométallique, UMR 5632 CNRS, University of Burgundy, 9 ave A. Savary, BP 47870, 21078 Dijon, France

The paper reports the innovative synthesis of 1,3-dialkylimidazolium-2-carboxylates from 1-alkyl imidazoles and dimethylcarbonate. The use of carboxylates in the synthesis of "halogen free" ionic liquids and as CO_2-transfer agents to organic substrates is also described.

Introduction

Ionic liquids (ILs) have received a great deal of attention in the last few years because of their large utilization in catalysis (*1-4*). The availability of "high purity" ILs (i.e. ILs free from water or Cl^-, Br^-, I^- anions, essentially) is an important issue as impurities may cause significant inactivation of the catalytic species.

The use of organic carbonates as alkylating agents for organic synthesis has attracted the attention of the scientific community (*5, 6*). Moreover several patents have claimed their use for the preparation of N-onium compounds starting from amines and heterocycles (*7-9*).
We have investigated the alkylation reaction of 1-methyl imidazole with dimethyl carbonate and have isolated the unprecedented 1,3-dimethyl-imidazolium-2-carboxylate (*10, 11*).
In addition to their preparation, we report here the synthesis of high purity ILs which can be obtained in high yield by decarboxylation of 1,3-dialkylimidazolium-2-carboxylates in the presence of strong acids as proton sources.
We also report that the decarboxylation reaction carried out in the presence of a catalyst can be conveniently used for transferring the CO_2 moiety to an organic substrate containing an activated C-H bond, affording the relevant carboxylated product.

Synthesis of 1,3-dialkylimidazolium-2-carboxylates

The alkylation of 1-methylimidazole with dimethyl carbonate as alkylating agent afforded, with 89 % selectivity, the 1,3-dimethylimidazolium-2-carboxylate species.
The other product formed is 1,3-dimethylimidazolium-4-carboxylate with 11 % selectivity. Best yield (76 %) was obtained carrying out the reaction in an autoclave at 135 °C for 7 h in methanol as solvent.
As already mentioned, the use of dimethyl carbonate as alkylating agent of imidazole and imidazoline derivatives has been reported in two recent patents by Mitsubishi (*7*) and BASF (*9*). The Mitsubishi patent (*7*) describes the methylation of 1,2-dialkylimidazoline affording the methylmonocarbonate of 1-ethyl, 2,3-dimethylimidazolinium a ionic liquid (Eq.1).

$$\text{(1-ethyl-2-methylimidazoline: } N{-}CH_2{-}CH_3\text{, } CH_3\text{)} + CH_3O{-}C(=O){-}OCH_3 \longrightarrow CH_3{-}\overset{+}{N}\text{(ring)}N{-}CH_2{-}CH_3\text{ (}CH_3\text{)} \;\; {}^{-}O{-}C(=O){-}OCH_3 \qquad (1)$$

By reaction of this salt with an organic acid, the corresponding 1-ethyl, 2,3 dimethylimidazolinium salt of the organic acid is obtained (Eq. 2).

CH_3-N^+ (imidazolium ring, 2-CH_3) N-CH_2-CH_3 ^-O-C(=O)-OCH_3 + HY →

CH_3-N^+ (imidazolium ring, 2-CH_3) N-CH_2-CH_3 Y^- + CH_3OH + CO_2

(2)

Conversely the BASF patent (*9*) claims the formation of an acid stable 1,3-dimethylimidazolium-4-carboxylate product by alkylation of 1-methylimidazole with dimethyl carbonate at 140 °C (Eq. 3).

N-CH_3 (1-methylimidazole) + CH_3O-C(=O)-OCH_3 —(- CH_3OH)→ CH_3-N(+)N-CH_3 (COO^-)

(3)

We have found that 1-methylimidazole reacts with dimethylcarbonate as alkylating agent to afford the 1,3-dimethylimidazolium-2-carboxylate species, with 89 % selectivity in a relatively short reaction time (7 hours) when a solvent is used (Eq. 4). The zwitterionic salt was fully characterized by elemental analysis and IR and NMR spectroscopy.
It is worth to note that the synthesis of the analogous compound 1,3-diisopropyl-4,5-dimethylimidazolium-2-carboxylate through a more complex synthetic procedure has been very recently reported (*12, 13*).

N-CH_3 (1-methylimidazole) + CH_3O-C(=O)-OCH_3 —(- CH_3OH)→

CH_3-N(+)N-CH_3 (2-C(=O)O^-)

89 % selectivity

CH_3-N(+)N-CH_3 (COO^-)

11 % selectivity

(4)

The isolated 1,3-dimethylimidazolium-2-carboxylate can be isomerised into the 1,3-dimethylimidazolium-4-carboxylate by heating at temperatures higher than 140 °C (Eq. 5)

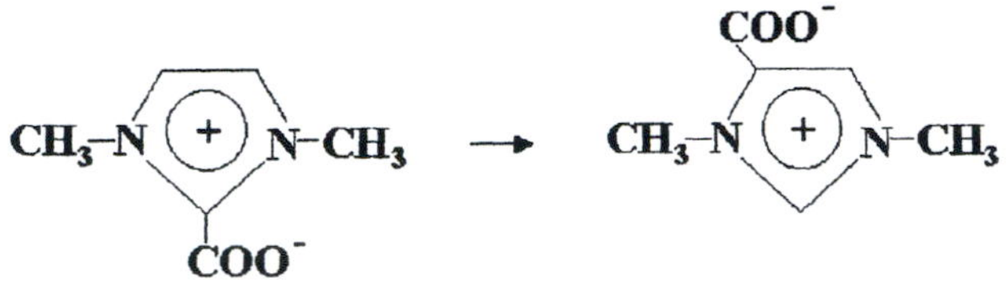

$$CH_3{-}N\,(+)\,N{-}CH_3 \;(COO^-) \longrightarrow CH_3{-}N\,(+)\,N{-}CH_3 \;(COO^-) \qquad (5)$$

Our synthetic approach can be extended to other 1-alkylimidazoles and alkylcarbonates. For example, 1-butyl, 3-methylimidazolium-2-carboxylate was prepared in 71 % yield from 1-butylimidazole.

Conversely, diethyl carbonate showed to be a poorer alkylating agent with respect to dimethylcarbonate as after 7 hours of reaction at 150 °C, a lower yield in alkylation products (14 %) was obtained, together with the formation of products resulting from alkyl-exchange.

Reactivity of 1,3-dimethylimidazolium-2-carboxylate with $[HOEt_2]BF_4$

The reactivity of 1,3-dimethylimidazolium-2-carboxylate towards $[HOEt_2]BF_4$ was investigated in detail. When 1,3-dimethylimidazolium-2-carboxylate was reacted with $[HOEt_2]BF_4$ in stoichiometric amount (Eq. 6) and the acid was added instantaneously, 2-hydroxycarbonyl-1,3-dimethylimidazolium tetrafluoroborate acid was quantitatively formed.

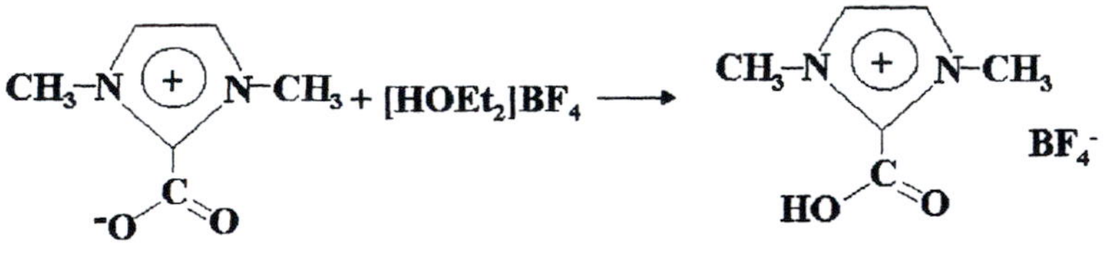

$$CH_3{-}N\,(+)\,N{-}CH_3\;({}^-O{-}C{=}O) + [HOEt_2]BF_4 \longrightarrow CH_3{-}N\,(+)\,N{-}CH_3\;(HO{-}C{=}O)\;\; BF_4^- \qquad (6)$$

Conversely, when 1,3-dimethylimidazolium-2-carboxylate was reacted with $[HOEt_2]BF_4$ in sub-stoichiometric amount (i.e. 1 to 0.8 ratio) almost 80 % of the product did undergo a decarboxylation reaction, affording 1,3-dimethylimidazolium tetrafluoroborate with 20 % remaining unreacted (Eq. 7).

$$CH_3\text{-}N(+)N\text{-}CH_3(CO_2^-) + 0.8\ [HOEt_2]BF_4 \xrightarrow{-CO_2} CH_3\text{-}N(+)N\text{-}CH_3\ BF_4^-\ (\sim 80\%) + CH_3\text{-}N(+)N\text{-}CH_3(CO_2^-)\ (\sim 20\ \%)$$

(7)

This observation indicated that the presence of a base was necessary in order to induce the decarboxylation reaction. The base catalysis was confirmed by reacting one mol of 2-hydroxycarbonyl-1,3-dimethylimidazolium tetrafluoroborate with 0.05 equivalents of NEt_3 (Eq. 8). Under these conditions 2-hydroxycarbonyl-1,3-dimethylimidazolium tetrafluoroborate was catalytically converted into 1,3-dimethylimidazolium tetrafluoroborate.

$$CH_3\text{-}N(+)N\text{-}CH_3(COOH)\ BF_4^- \xrightarrow[-\ CO_2]{NEt_3} CH_3\text{-}N(+)N\text{-}CH_3\ BF_4^-$$

(8)

1,3-Dimethylimidazolium tetrafluoroborate was obtained in a "one step" reaction by reacting 1,3-dimethylimidazolium-2-carboxylate with $[HOEt_2]BF_4$ in stoichiometric amount by slow addition of the acid in order to assure, during the reaction, the presence of the 1,3-dimethylimidazolium-2-carboxylate as a base (Eq. 9).

$$CH_3\text{-}N(+)N\text{-}CH_3(CO_2^-) \xrightarrow[-CO_2]{[HOEt_2]BF_4\ \text{added slowly}} CH_3\text{-}N(+)N\text{-}CH_3\ BF_4^-$$

(9)

Either the one step or two step reaction with $[HOEt_2]BF_4$ allows to obtain the absolutely "halogen free" 1,3-dimethylimidazolium tetrafluoroborate ionic liquid in quantitative yield.
Interestingly 1,3-dimethylimidazolium-4-carboxylate did not undergo the decarboxylation under the experimental conditions used for the 2-carboxylate isomer.

Transfer of the CO_2-Moiety to an Organic Substrate Containing an Activated C-H Bond.

By reacting 1-butyl,3-methyl imidazolium-2-carboxylate with acetophenone in the presence of a catalyst (without solvent addition) a quantitative trans-carboxylation reaction took place (Eq.10).

n-But–N(+)N–CH_3 (COO⁻) + C_6H_5–C(=O)–CH_3 —catalyst→ n-But–N(+)N–CH_3 C_6H_5–C(=O)–CH_2–C(=O)–O⁻

(10)

Benzoylacetic acid could be recovered in almost quantitative yield and fully characterized after acidification of the reaction mixture with H_2SO_4.
1,3-Dimethylimidazolium-2-carboxylate reacted, although to a lesser extent, with acetophenone in absence of the catalyst affording in very low yield the relevant carboxylation product. The scarce miscibility of the reactants may explain the low yield.
Extension of this carboxylic acids synthetic methodology to other organic compounds containing activated C-H bonds is under study.

Conclusions

We describe an unprecedented synthesis of 1,3-dialkyl-imidazolium-2-carboxylates by alkylation of 1-alkylimidazoles with dimethylcarbonate. The 2-

carboxylates undergo a decarboxylation reaction with a strong acid affording the relevant ionic liquid.
Also, they react with organic compounds containing activated C-H bonds to afford carboxylated derivatives. 1,3-dialkylimidazolium-2-carboxylates are shown to be versatile intermediates for several synthetic applications.

Aknowledgements

Authors aknowledge Region Bourgogne and Italian Ministry of Universities and Research, Project of National Interest n° MM03027791, for financial support.

References

1. Holbrey, J.D.; Seddon K. R., *Clean. Prod. Process.*, **1999**, *1*, 223-236.
2. Welton T., *Chem. Rev.*, **1999**, *99*, 2071-2083.
3. Dupont J. ; Consorti C.S. ; Spencer J., *J. Braz. Chem.*, **2000**, *11*, 337-344.
4. Wasserscheid P.; Keim W., *Angew. Chem., Internat. Ed. Engl.*, **2000**, *39*, 3772-3789.
5. Shaikh A.G., Sivaram S., *Chem. Rev.*, **1996**, *96*, 951-976.
6. Tundo P., *Pure App. Chem.*, **2001**, *73*, 1117-1124.
7. Mori S.; Ida K.; Ue M., Mitsubishi Chemical Co, Ltd., US Patent 4892944, **1990**.
8. Ue M., Takeda M.; Takahashi T.; Takehara M., US Patent 5856513, **1997**.
9. Fisher J.; Siegel W.; Bomm V.; Fisher M., Mundinger K., US Patent 6175019, **1999**.
10. Poinsot D.; Tkatchenko I.; Tommasi I., University of Burgundy, Fr. Patent Application F 02 12692, **2002**.
11. Holbrey, J.D.; Reichert, W. M.; Tkatchenko, I.; Bouajila, E.; Walter, O.; Tommasi, I.; Rogers, R., *JCS Chem Commun*, **2003**, 28-29.
12. Kuhn N.; Maichle-Mobmer C.; Weyers G., *Z. Anorg. Allg. Chem.*, **1999**, *625*, 851-856.
13. Kunh N.; Steimann M.; Weyers G., *Z. Naturforsch.*, **1999**, *54 b*, 427-433.

Chapter 9

Commercially Available Salts as Building Blocks for New Ionic Liquids

James H. Davis, Jr.[1,2,*] and Phillip A. Fox[1]

[1]Department of Chemistry, University of South Alabama, Mobile, AL 36688
[2]The Center for Green Manufacturing, The University of Alabama, Tuscaloosa, AL 35487

Imidazolium salts are by far the most widely used ionic liquids, and several of them are now commercially available. Nevertheless, there are a large number of other commercially available salts that have low melting points and that are of potential use as ionic liquids. In this article, we offer some suggestions as to possible avenues for their exploitation, and provide a tabulation of over one hundred of these salts.

Ask a dozen researchers working in the area to provide a definition for the term "ionic liquid" and you may get a dozen different answers. While virtually all will concur that the liquid must be composed *wholly* of ions, the temperature at which the material must become a liquid is less universally agreed upon. There is widespread agreement that any salt that melts at "room-temperature" or below meets the definition. There remains, however, some debate as to what temperature should be considered as the upper melting point limit for a salt to be described as an ionic liquid. Upper melting point limits of 100°C or 150°C are commonly favored. Advocates of the former point to the boiling point of water as being a universal thermal benchmark, while those favoring the latter do so because completely inorganic salts with melting points below that value are virtually unknown. Discussions of the issue can quickly take on a "how many angels can dance on the head of a pin" flavor.

So why all the fuss over a seemingly esoteric class of salts? It's because these unorthodox liquids have captured the imaginations of a large and still growing number of researchers, a fact manifested by the rapidly increasing number of papers published on the topic. (*1*) The interest in these salts appears to derive mostly from three factors. First, because ionic liquids usually have negligible vapor pressure, and since evaporative loss is a major mode of discharge of molecular solvents into the environment, the use of select ionic liquids as solvents may offer a degree of environmental benignity to industrial processes in which they are utilized. Secondly, a growing number of studies report that when used in place of molecular solvents in certain processes, ionic liquids provide improvements in product yields, selectivities and ease of product/solvent separation. Finally, these materials are wholly unconventional. By virtue of our training, most of us were simply not exposed to the notion that salts could be liquids under anything but rarified conditions. Hence, the prospect of working with materials that are so radically different from those to which we are accustomed is intriguing.

Probably the most widely used ionic liquids are 1-butyl-3-methyl imidazolium hexafluorophosphate, [bmim]PF_6, and a handful of close analogs with different n-alkyl chain lengths. To prove the point, consider the following figures. Of papers accepted or published by Elsevier imprint journals between January and May of 2002 dealing topically with ionic liquids, fourteen of twenty-one studies used [bmim]PF_6. A survey of papers published in or accepted for publication by ACS journals for the same period showed that of nineteen studies topically concerning ionic liquids, six used [bmim]PF_6. Figures from RSC journals are comparable. Much of the focus on this salt is doubtless driven by its demonstrated utility in earlier studies – a case of success begetting success. Too, this ionic liquid is one of the most frequently mentioned in trade magazine (e.g., *Chemical & Engineering News*, etc.) articles on ionic liquids; Anecdotally, such articles are the first exposures to the subject that many of us have had. Finally, and perhaps most significantly, using this and related salts for studies of known reactions is *easy*. Members of the imidazolium family of ionic liquid salts are readily prepared, and several of them (including [bmim]PF_6) are now commercially available.

There is great value in this ongoing, wide-ranging survey of reactions in [bmim]PF_6 and its congeners. These studies provide important insights into the comparative behavior of a more or less standard ionic liquid versus conventional molecular solvents. Such a survey will doubtless lead to discoveries that completely change the standard way in which we do one reaction or another.

Still, two irksome questions persist: *is this small family of ionic liquids the only one of possible utility as solvents* and *are these the only ionic liquids that are readily available for experimentation with as solvents?* We maintain that the answer to both questions is simple: *No*.

Due to the roundabout way in which we came to be involved in ionic liquid research, our focus from the outset has been on developing new ionic liquids that are not simple N,N'-dialkyl imidazolium salts. While many of the ionic liquids that we have developed have imidazolium rings as their locus of positive charge, our salts invariably feature somewhere within their structures a functional group such as an amine or sulfonic acid group. In other words, our interest is in the development of functionalized ionic liquids, capable of acting as not only solvents but also as reagents or catalysts for particular processes. We refer to these species as "task-specific" ionic liquids and we have had some gratifying successes in their development. Among other things, we have developed amine appended ionic liquids that can scrub CO_2 from gas streams, sulfonic acid appended ionic liquids that can catalyze an array of organic reactions, thiazolium ion based ionic liquids that solvate and catalyze the benzoin condensation, thiourea appended ionic liquids that can pull dissolved metal ions from aqueous co-phases, and fluorous ILs that function as surfactants. *(2, 3, 4, 5, 6)* Still, all of these projects involve the ground-up design and synthesis of the ionic liquids for the tasks at hand.

About a year ago, we noted that a compound of interest to us as a starting material for introducing a functional group into a new ionic liquid - chloroacetamidine hydrochloride - was not only commercially available but might possibly be categorizable as an ionic liquid itself, having a reported mp = 95°-98°C. Intrigued by this discovery, we set out to survey retail chemical catalogs item by item for other salts with low melting points. Most of these compounds are species pairing organic cations and small, hard inorganic anions, commonly halides. Given that it is commonly observed in imidazolium salts that the exchange of halide for anions like PF_6^- or BF_4^- gives rise to salts with lower melting points, these commercial species may constitute valuable platforms for the creation of new ionic liquids. This general principal of ionic liquid synthesis through the pairing of known ions has been previously articulated by Seddon, and has already been exploited by his group. *(7)* Nevertheless, fewer groups than might be expected have yet to explore the possibilities, and it is the purpose of this paper to renew the call to exploration, and to point out some pathways leading into the shadowy wood.

In our survey, we looked at every listing in the most recently published catalogs from Aldrich, Lancaster, and Fluka, three large fine chemicals retail outlets. We also requested and received lists of low melting quaternary ammonium and phosphonium salts from two specialty manufacturers, Sachem and Cytec, which they generously supplied. The latter are highly useful ionic liquids that have been around for ages but that have been largely overlooked in the current frenzy of ionic liquid research. From these sources we compiled a list of commercially available salts (and some salt hydrates) that have a melting point of 150°C or less, where the listed melting point is not noted as being a decomposition temperature. Because our objective is to encourage the development of new ionic liquids, we have omitted most commercially available imidazolium based species from the list. The results of our survey – a list of over 100 compounds – are compiled in Table I.

Certain general comments regarding the salts on our list bear making. First, their commercial availability is due to their *existing* use in some other type of application. For example, 1,8-diazabicyclo[5.4.0]undec-7-ene hydrotribromide (DBU-HBr_3, mp 119 – 122°C) is a superb reagent for high-yield aromatic brominations. (*8*) Naturally, the pre-existing applications point to potential uses as IL for the compounds themselves or salts related to or derived from them. Second, salts on the list or their derivatives need not necessarily be used in a single cation type – single anion type form. Ionic liquids composed of a single cation but incorporating two different anions have recently been reported. (*9*) In addition, combinations of certain higher melting salts can give rise to interesting ionic liquid eutectics. Davies has recently demonstrated that combinations of acetylcholine chloride and select inorganic salts form eutectics with melting points as low as room temperature. (*10*) These eutectics constitute interesting new ionic liquids with built-in, water-stable Lewis acidic character.

Many of the available salts on the list are biomolecular in nature. Examples include L-alanine ethyl ester hydrochloride (mp 78°C) and L-serine methyl ester hydrochloride (106°C). Like many other lower-melting salts, these are probably not ionic liquids in a strict sense. In water, they manifest equilibrium concentrations of neutral species, suggesting that their melts may also not be composed only of ions. Still, as melts in contact with a secondary, low-polarity organic phase, it is doubtful that any neutral melt component would partition into the latter, allowing the melt to *function* in a fashion similar to a true ionic liquid. Regardless, many of the ions in these low-melting, commercially available salts are potentially versatile skeletons that suggest themselves as starting points for modification into IL with built-in functional groups. (*11*)

The intrinsic utility of any of the salts in Table 1 as ionic liquids or ionic liquid precursors is dictated by nature. However, the unmasking of that intrinsic utility is up to the research community. For those willing to take the risks that accompany the process of exploration, the potential rewards are high. Now, without further commentary, we urge the reader to peruse the list and to imagine the possibilities!

Table I. Commercially Available Low Melting Salts

Compound Name	melting point (Celsius)
acetylcholine chloride	147-149
acetyl choline bromide	144-146
acetyl-β-methylcholine bromide	147-149
[2-(acryloyloxy)ethyl](4-benzoylbenzyl) dimethylammonium	118-123
DL-alanine ethyl ester hydrochloride	87-88
L-alanine ethyl ester hydrochloride	78-80
β-alanine ethyl ester hydrochloride	70-72
D-alanine methyl ester hydrochloride	108-110
L-alanine methyl ester hydrochloride	109-111
trcaprylylmethylammonium chloride [Aliquat 336]	below RT
O-allyl-N-benzylcinchodinium bromide	140-144
aluminum potassium sulfate dodecahydrate	92
aminoacetonitrile bisulfate	123-125
aminoethanethiol hydrochloride [cysteamine hydrochloride]	66-68
2-aminoethyl methacrylate hydrochloride [90%]	102-110
aminoguanidine nitrate	145-147
1-aminopyrrolidine hydrochloride	117-119
5-aminovaleric acid hydrochloride	95-97
ammonium formate	119-121
ammonium hydrogensulfate	121-145
ammonium sulfamate	131-135
ammonium trifluoroacetate	123-125
DL-arginine hydrochloride monohydrate	128-130
L-aspartic acid dimethyl ester hydrochloride	115-117
benzamidine hydrochloride hydrate	86-88
N-α-benzoyl-L-arginine ethyl ester hydrochloride	127-131
N-benzoyl-L-threonine methyl ester	97-99
benzylcetyldimethylammonium chloride monohydrate	62-64
benzyldimethylstearylammonium chloride monohydrate	67-69
benzyldimethyltetradecylammonium chloride dihydrate	63-65
3-benzyl-5-(2-hydroxyethyl)-4-methylthiazolium chloride	144-146
N-benzylhydroxylamine hydrochloride	108-110
benzyltributylammonium iodide	143-145
benzyltrimethylammonium dichloroiodate	126-128
benzyltrimethylammonium tribromide	99-101
N,N,N',N'-bis(pentamethylene)chloroformamidinium PF6	120-122
2-bromopyridine N-oxide hydrobromide	145-147

Table I. *Continued.*

Compound Name	melting point (Celsius)
2-bromopyridine N-oxide hydrochloride	131-134
calcium hypochlorite	100
2-chloroethylamine hydrochloride	143-146
2-(2-chloroethyl)-1-methylpyrrolidine hydrochloride	101-104
cresyl violet acetate	140-143
1-cyclohexyl-3-(2-morpholinoethyl)carbodiimide pTSA	113-115
cyclopropyldiphenylsulfonium tetrafluoroborate	136-138
decyltriphenylphosphonium bromide	90
1,8-diazabicyclo[5.4.0]undec-7-ene hydrotribromide	119-122
3,5-dichloro-1-fluoropyridinium triflate (85%)	107-112
2-(2,4-dichlorophenoxy)aniline hydrochloride	96-98
didecyldimethylammonium bromide	149-151
1,3-didecyl-2-methylimidazolium chloride	82
[1,4-dihydro-1-(trifluoroacetyl)Py]dimethylammonium TFA	102-103
dimethylamine hydrobromide	126-128
(dimethylaminomethylene)dimethylammonium chloride	130-139
1-[3-(dimethylamino)propyl]-3-ethylcarbodiimide	111-113
4-(dimethylamino)pyridinium tribromide	131-133
1,3-dimethyl-2-fluoropyridinium 4-toluenesulfonate	111-114
3,4-dimethyl-5-(2-hydroxyethyl)thiazolium iodide	85-97
N,O-dimethylhydroxylamine hydrochloride	112-115
N,N-dimethylmethylene ammonium chloride (90%)	146-148
dimethylphenylsulfonium fluoroborate	78-80
dodecylisothiouronium chloride	127-130
1-dodecylpyridinium chloride hydrate	66-70
dodecyltriphenylphosphonium bromide	85-88
(dodecyldimethyl-2-phenoxyethyl)ammonium bromide	117-119
1-heptyl-4-(4-pyridyl)pyridinium bromide	125-128
hexyltriphenylphosphonium iodide	131-133
L-cysteine ethyl ester hydrochloride	123-125
[N-methylbis(2-chloroethyl)amine hydrochloride	108-110
4-methoxybenzenediazonium tetrafluoroborate	142-144
1-methoxy-4-phenylpyridinium tetrafluoroborate	90-95
N-methylhydroxylamine hydrochloride	86-88
2-methylthio-2-imidazoline hydroiodide	144-146
methyltriphenoxyphosphonium iodide	142-146

Continued on next page.

Table I. *Continued.*

Compound Name	melting point (Celsius)
R-(-)-3-hydroxy-a-(methylaminomethyl)benzyl alcohol HCl	143-145
phenyltrimethylammonium tribromide	114-116
2-(chloromethyl)pyridine hydrochloride	125-129
3-(chloromethyl)pyridine hydrochloride	137-143
2-propylisoquinolinium bromide	143-146
(R)-(-)-3-pyrrolidinol hydrochloride	104-104
sarcosine tert-butyl ester hydrochloride	137-141
sarcosine ethyl ester hydrochloride	127-128
L-serine ethyl ester hydrochloride	130-132
D-serine methyl ester hydrochloride	163-166
DL-serine methyl ester hydrochloride	134-136
L-serine methyl ester hydrochloride	163 (dec)
serinol hydrochloride	106-108
(-)-sparteine sulfate pentahydrate	133-140
tetrabutylammonium borohydride	124-128
tetrabutylammonium bromide	102-106
tetrabutylammonium chloride hydrate	41-44
tetrabutylammonium iodide	144-146
tetrabutylammonium tribromide	74-76
tetraheptylammonium bromide	89-91
tetraheptylammonium chloride	38-40
tetrahexadecylammonium bromide	99-101
tetrahexylammonium bromide	99-100
tetrahexylammonium chloride	111-113
tetrahexylammonium hydrogensulfate	100-102
tetrakis(acetonitrile)silver(I) tetrafluoroborate	72-75
tetraoctadecylammonium bromide	103-105
tetraoctylammonium bromide	95-98
tetraoctylphosphonium bromide	38-43
tetrapentylammonium bromide	100-101
trihexyl(tetradecyl)phosphonium chloride	-68
trihexyl(tetradecyl)phosphonium tetrafluoroborate	26-28
trihexyl(tetradecyl)phosphonium hexafluorophosphate	29-31
trihexyl(tetradecyl)phosphonium decanoate	-8
trihexyl(tetradecyl)phosphonium dicyanamide	-18
trihexyl(tetradecyl)phosphonium bromide	< RT
tri-iso-butyl(methyl)phosphonium tosylate	31-32

References

1. Rogers, R. D. and Seddon, K. R. in *Ionic Liquids: Industrial Applications to Green Chemistry;* R. D. Rogers and K. R. Seddon, Eds.; ACS Symposium Series 818; American Chemical Society: Washington, D.C. 2002; xiii.
2. Bates, E. D.; Mayton, R. D.; Ntai, I. and Davis, J. H. Jr. *J. Am. Chem Soc.* **2002**, *124*, 926.
3. Cole, A. C.; Jensen, J. L.; Ntai, I.; Tran, K. L. T.; Weaver, K. J.; Forbes, D. C. and Davis, J. H. Jr. *J. Am. Chem. Soc.* **2002**, *124,* 5962.
4. Forrester, K. J. and Davis, J. H. Jr. *Tetrahedron Letters* **1999**, *40*, 1621.
5. (a) Visser, A. E.; Swatloski, R. P.; Reichert, W. M.; Mayton, R. D.; Sheff, S.; Wierzbicki, A.; Davis, J. H. Jr. and Rogers, R. D. *Envir. Sci. Technol.* **2002**, *36*, 2523; (b) Visser, A. E.; Swatloski, R. P.; Reichert, W. M.; Mayton, R. D.; Sheff, S.; Wierzbicki, A.; Davis, J. H. Jr. and Rogers, R. D. *Chemical Commun.* **2001**.
6. Merrigan, T. L.; Bates, E. D.; Dorman, S. C. and Davis, J. H. Jr. *Chemical Commun.* **2000**, 2051.
7. Earle, M. J.; McCormac, P. B. and Seddon, K. R. *Green Chem.,* **1999**, *1*, 23.
8. Muathen, H. A. *J. Org. Chem.,* 1992, **57**, 2740; A DBU derivative has recently been developed and used as an ionic liquid. See: Kitazume, T.; Zulfiqar, F.; Tanaka, G. *Green Chem.* **2000**, 106, 211.
9. Egashira, M.; Okada, S. and Yamaki, J.-I. *Solid State Ionics* **2002**, *148*, 457.
10. Abbott, A. P.; Capper, G.; Davies, D. L.; Munro, H. L.; Rasheed, L. K. and Tambyrajah, V. *Chem. Commun.,* **2001**, 2010.
11. See for example: Wasserscheid, P.; Bosman, A. and Bolm, C. *Chem. Commun.,* **2001**, 2484.

Characterization and Engineering

Chapter 10

Phase Equilibria of Gases and Liquids with 1-*n*-butyl-3-Methylimidazolium Tetrafluoroborate

Jennifer L. Anthony, Jacob M. Crosthwaite, Daniel G. Hert, Sudhir N. V. K. Aki, Edward J. Maginn, and Joan F. Brennecke

Department of Chemical Engineering, University of Notre Dame, South Bend, IN 46556

The solubility of carbon dioxide, methane, nitrogen, carbon monoxide and benzene in 1-*n*-butyl-3-methylimidazolium tetrafluoroborate ([bmim][BF_4]) at 25 °C are reported. In addition, we present densities, which compare very favorably with one set of literature values, for [bmim][BF_4]. Accurate values of density are needed in the analysis of gas solubility data. Finally, we report liquid-liquid equilibrium data for [bmim][BF_4] with 1-propanol and benzene. Information on vapor-liquid and liquid-liquid equilibrium of gases and liquids with ionic liquids is important for evaluating the use of ionic liquids in a wide variety of processes, including reactions and separations.

Introduction

Here we present preliminary results for the solubility of various gases in 1-*n*-butyl-3-methylimidazolium tetrafluoroborate ([bmim][BF_4]). Previously, we have measured the solubilities of a wide variety of gases in 1-*n*-butyl-3-methylimidazolium hexafluorophosphate ([bmim][PF_6]) (1). The goal of this work is to determine the influence of the nature of the anion on gas solubilities.

We know of no other published values of the solubility of gases in [bmim][BF_4]. CO_2 is the most soluble gas that we have found in [bmim][PF_6] (almost 2 mole % at 1 bar and 25°C), with C_2H_4, C_2H_6 and CH_4 being progressively less soluble (1,2). O_2 and Ar were just sparingly soluble and the solubility of H_2, CO and N_2 were below the detection limit of our apparatus (1). Our motivation for measuring gas solubilities is three-fold. First, various researchers have shown that ionic liquids (ILs) are excellent solvents for a wide variety of reactions (3-5) involving permanent gases, such as hydrogenations, oxidations and hydroformylations. From our previous results (1), it appears that the solubility of H_2, O_2 and CO in ILs is likely to be a limiting factor in these reactions. Second, the solubility of carbon dioxide (CO_2) in ionic liquids is important for evaluating the possibility of using supercritical CO_2 to extract solutes from ionic liquids, as we have proposed previously (6,7). Finally, we are interested in exploring the possibility of using water stable ILs for performing gas separations.

We also present measurements of the density of [bmim][BF_4] at temperatures between 20 and 70 °C. Accurate densities are required in the analysis of the gas solubility data (1). As described below, we measure the gas solubilities with a gravimetric microbalance. The mass uptake must be corrected for buoyancy effects, which requires the liquid density. There were several papers available in the literature that give densities for [bmim][BF_4] but the values were not consistent, necessitating the performance of our own measurements. Suarez et al. (8) reported values between 6 and 81 °C of 1.14-1.18 g/mL. Seddon et al. (9) reported values between 1.16 and 1.21 g/mL for temperatures between 20 and 90 °C. Rogers and coworkers (10) report the density of [bmim][BF_4] to be just 1.12 g/mL at 25 °C. One factor that could account for some variability is the water content. [bmim][BF_4] is completely miscible with water at room temperature (11) and is very hygroscopic. Suarez et al. (8) do not report the water content of their sample. The sample used by Seddon et al. was reported to contain just 0.03 wt% water, while the sample tested by Rogers and coworkers (10) contained 0.45 wt% water. While water content can affect density (12), it is unlikely that water content can fully explain the large discrepancies in the reported values.

Finally, we present the liquid-liquid equilibrium (LLE) of benzene and 1-propanol with [bmim][BF_4]. LLE is important for determining the solubility of reactants and products in ILs, as well as the amount of IL that will contaminate organic or aqueous phases that are brought into contact with the IL. Marsh has measured the LLE of a series of dialkylimidazolium hexafluorophosphates with ethanol, 1-propanol and 1-butanol (13). Najdanovic-Visak et al. have studied the ethanol/water/[bmim][PF_6] system, as well as the ethanol/[bmim][PF_6] and water/[bmim][PF_6] binaries (14). We have provided a few numbers on water/IL LLE (17) and some preliminary estimates of the solubility of a wide variety of organic liquids in [bmim][PF_6] (7). Data from Rogers and coworkers (15,16) for some water/IL systems show lower mutual solubilities than other published results (14,17). All of these systems show upper critical solution temperature (UCST) behavior; i.e. the mutual solubilities increase with increasing temperature.

Experimental

The gas solubility measurements reported here were made using a gravimetric microbalance (IGA 003, Hiden Analytical) that is normally used to measure gas adsorption onto solids. However, since the ILs are nonvolatile they could be used with this apparatus. This microbalance and the technique used to measure gas solubilities in ILs has been described in detail elsewhere (1,17,18). Briefly, a small IL sample is placed on the balance and the system is thoroughly evacuated to remove any volatile impurities, including water. The gas is introduced and the gas solubility is determined from the mass uptake.

The [bmim][BF_4] density was determined using a 1 mL pycnometer. At temperatures other than ambient, the sample was equilibrated for at least 30 - 40 minutes in a thermostatted gas chromatography oven. Care was taken to avoid any exposure to air that could result in water uptake. The water content of the sample tested was 0.2 wt%, as determined by Karl-Fisher titration.

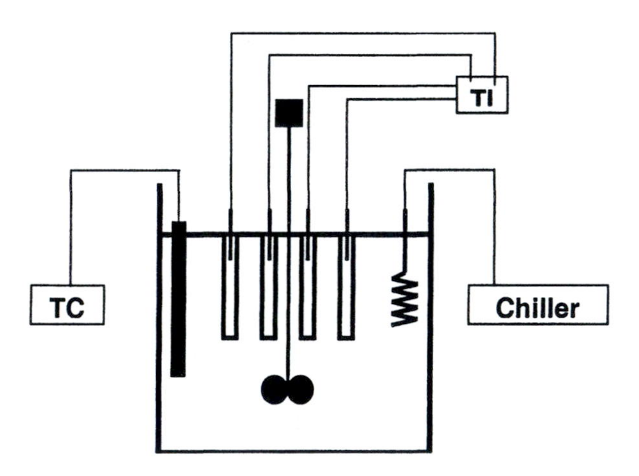

Figure 1. Experimental apparatus for determining liquid-liquid solubility of ionic liquid-organic systems; TC: Temperature controller for immersion heater, TI: Temperature indicator for samples.

A cloud point apparatus was used for determining the liquid-liquid equilibrium of IL/organic mixtures. This apparatus is shown schematically in Figure 1. In this technique, a sample of known composition was placed in a viewcell. The "cloud point temperature" is the temperature at which the sample initially changes from one phase to two phases; that phase transition was determined visually. In our apparatus, four solutions of ionic liquid and an organic liquid at different concentrations were prepared gravimetrically in 5 mL viewcells in a glovebox, sealed from the atmosphere, and placed in a water bath. The uncertainty in the compositions is estimated to be ±0.0001 mole fraction. The samples were heated in the water bath using a 1000 W immersion heater to about 5 °C above the highest expected cloud point temperature and maintained at that temperature for 10 minutes. The samples were cooled using a recirculating water chiller at a cooling rate of about 0.01 °C/s. The temperature of each sample was individually measured using a T-type thermocouple. The experiment was repeated four additional times to determine accurate and reproducible values (±0.5 °C).

The CH_4 was from Matheson Gas Products with a purity of 99.99+%. The CO_2 was from Scott Specialty Gases with a 99.99% purity. The CO and N_2 were both from Mittler Supply Co. and had purities of 99.97% and 99.99%, respectively. Benzene and 1-propanol were purchased from Aldrich with reported purities of 99.9+% and 99+%, respectively. Both benzene and 1-propanol were redistilled prior to use. [bmim][BF_4] was synthesized according to standard procedure (19) and was analyzed by NMR spectroscopy. The residual chloride content was < 10 ppm, as measured by a Cole-Parmer chloride ion specific electrode, calibrated with [bmim][Cl].

Results and Discussion

The solubility of CO_2, CH_4, N_2, and CO were measured in [bmim][BF_4] at 25 °C and pressures to 13 bar. The solubilities of N_2 and CO were below the detection limit of our instrument. The results for CO_2 and CH_4 are shown in Figures 2 and 3, below, where the solubilities are plotted as mole fraction of the gas in the IL at various pressures. The solubilities in [bmim][BF_4] are compared with those in [bmim][PF_6] (1). The solubility of CO_2 in the ILs is significantly greater than the solubility of CH_4. Moreover, it is obvious that the solubilities of each of these two gases in [bmim][BF_4] and [bmim][PF_6] are very similar. In other words, the nature of the anion (BF_4 versus PF_6) has very little effect on the solubility of these gases.

The simplest way to analyze gas solubility is in terms of a Henry's law constant. The Henry's law constant is defined as:

$$H(\mathrm{T}) = \lim_{x_i \to 0} \frac{f_i^L}{x_i}$$

Where H is the Henry's law constant, f_i^L is the fugacity of the solute in the liquid phase and x_i is the solute mole fraction in the liquid phase. For vapor/liquid equilibrium of IL/gas mixtures, the equifugacity criterion can be written as:

$$y_i \phi_i P = x_i H$$

Since the IL is nonvolatile, we assume that the vapor phase is pure gas; therefore, the mole fraction of gas, y_i, is unity. Since the Henry's law constant is determined at the lowest pressures, there is no need to correct for vapor phase nonidealities with a fugacity coefficient, ϕ_i. With these simplifications, the Henry's law constant is just the slope of the solubility versus pressure plots shown in Figures 2 and 3, evaluated in the limit of low solubility.

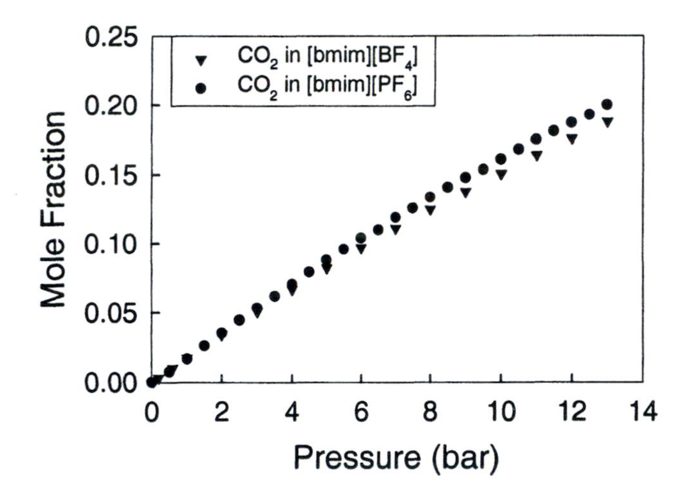

Figure 2. Solubility isotherms at 25 °C for CO_2 in [bmim][PF_6] and [bmim][BF_4].

At 25 °C, the Henry's law constant for CO_2 in [bmim][BF_4] is 56.5 ± 0.3 bar and for CH_4 in [bmim][BF_4] is 1560 ± 325 bar. By comparison, the values in [bmim][PF_6] are 53.4 ± 0.3 bar for CO_2, and 1690 ± 180 bar for CH_4. These Henry's law constants, as well as those for all the other gases studied in [bmim][PF_6] are shown in Figure 4. The striped bars in Figure 4 represent the estimated detection limits of the instrument for CO, N_2, and H_2.

As mentioned above, accurate density measurements are required for the buoyancy correction needed to determine the values in Figures 2 and 3 from the raw data of mass uptake. Our measurements for the density of [bmim][BF_4] at temperatures between 20 and 70 °C are shown in Figure 5, along with the data of Seddon et al. (9) and Suarez et al. (8).

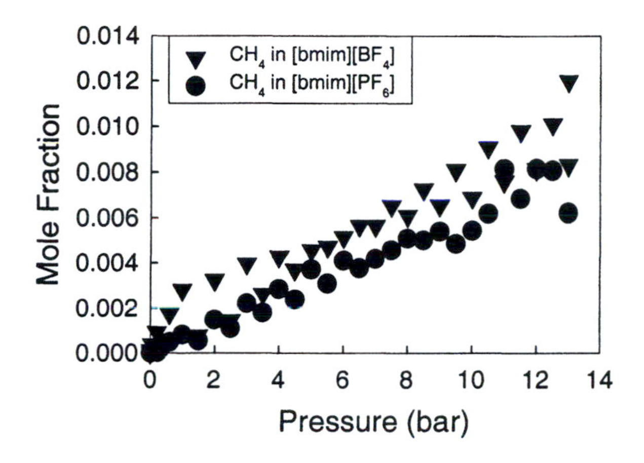

Figure 3. Solubility isotherms at 25 °C for CH_4 in [bmim][PF_6] and [bmim][BF_4].

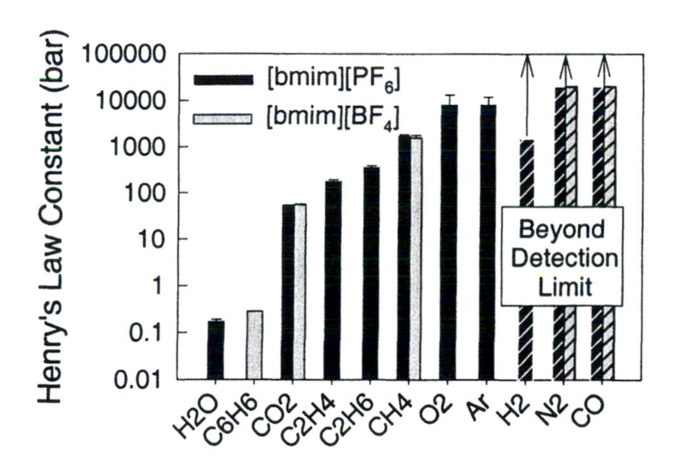

Figure 4. Henry's law constants for different gases in [bmim][PF_6] and [bmim][BF_4] at 25 °C.

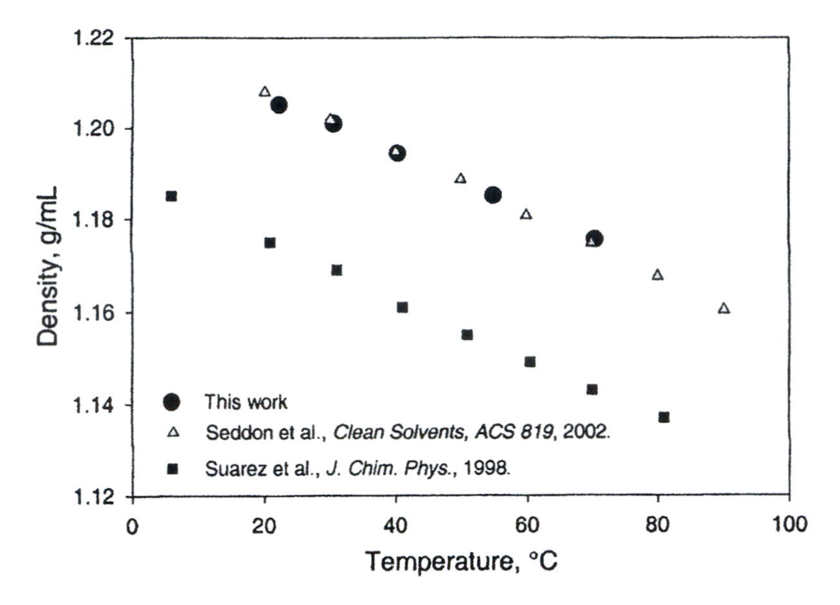

Figure 5 Temperature dependence of the density of [bmim][BF_4].

A linear regression through our data yields a temperature dependence given by the following equation: ρ (g/mL) = -6.43×10^{-4} T (°C) + 1.22. Our data is completely consistent with that of Seddon et al. (9). However, our values are significantly higher than those reported by Suarez et al. (8). Since those researchers did not report water content in that relatively early paper, it is likely that their densities are low due to large water content. As mentioned above, [bmim][BF_4] is totally miscible with water and is highly hygroscopic. However, the value reported by Rogers and coworkers (10) of 1.12 g/mL at room temperature cannot be explained by the water content, which was reported to be just 0.45 wt%. This value is inconsistent with all the other recent measurements. We used the linear regression through our data to determine the densities used in the buoyancy corrections to the raw gas solubility data.

Using the gravimetric microbalance, we have also measured the solubility of benzene vapor in [bmim][BF_4] at 25 °C. Measurements were made up to about 75% of the vapor pressure of benzene (P^{vap} = 0.126 bar at 25 °C). These results are shown in Figure 6, where the solubility of the benzene in the IL in mole fraction is plotted as a function of the pressure, which has been normalized by the vapor pressure. Benzene is very soluble in [bmim][BF_4], with a Henry's law constant of 0.25 bar. This value is shown in comparison to other gases above in the bar graph in Figure 4.

Also shown in Figure 6 is the solubility of liquid benzene in [bmim][BF_4]. The vapor-liquid equilibrium data match the liquid-liquid equilibrium data extremely well when extrapolated to the saturation pressure (i.e., $P/P^{vap} \rightarrow 1$), as expected.

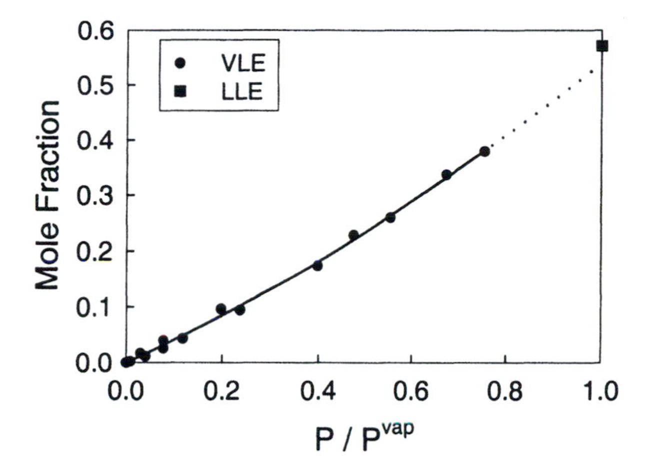

Figure 6. Solubility of benzene in [bmim][BF_4].

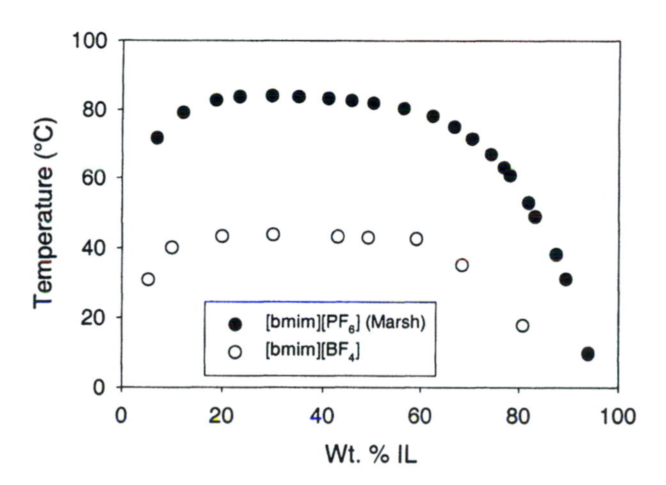

Figure 7 Liquid-liquid equilibrium of 1-propanol and [bmim][BF_4] as a function of wt% IL.

In addition, we have used the cloud point apparatus to measure the LLE of [bmim][BF_4] with 1-propanol. These results are shown in Figures 7 and 8, where the data are plotted both as a function of weight % and mole % IL.

For comparison, the data of Marsh (13) for 1-propanol with [bmim][PF_6] is also shown in the figures. Here the nature of the anion has a dramatic effect on the phase equilibria. The upper critical solution temperature is significantly lower for the [bmim][BF_4] system and at a given temperature the mutual solubilities are much greater.

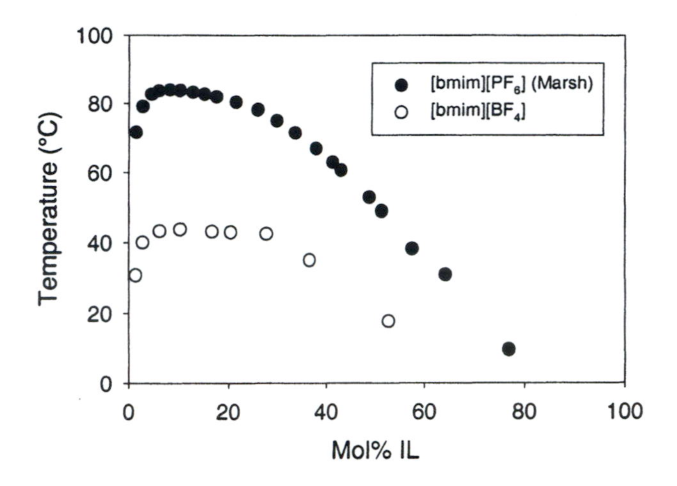

Figure 8 Liquid-liquid equilibrum of 1-propanol and [bmim][BF_4] as a function of mol % IL.

Summary

We have presented measurements for the solubility of gases in 1-*n*-butyl-3-methylimidazolium tetrafluoroborate ([bmim][BF_4]). The values are very similar to those for [bmim][PF_6], indicating that the nature of the anion has very little effect on the solubility of permanent gases in ILs. The Henry's law constants for CO_2 and CH_4 in [bmim][BF_4] at 25°C are 56.5 $\pm$ 0.3 bar and 1560 $\pm$ 325 bar, respectively. Our measurements of densities of [bmim][BF_4] as a function of temperature, which are needed to analyze the gas solubility data, closely match those of Seddon et al. (7) but are significantly greater than other published values (6, 8). Vapor-liquid and liquid-liquid equilibria measurements for benzene in [bmim][BF_4] are internally consistent and indicate that benzene is highly soluble in this IL. We also present liquid-liquid equilibria data for 1-propanol and [bmim][BF_4] that, when compared with LLE data for 1-propanol/[bmim][PF_6], suggest that the anion does plays a very important role in determining liquid-liquid equilibria.

Acknowledgments

Financial support from the National Science Foundation (grant CTS 99-87627), a Bayer Predoctoral Fellowship, and a National Science Foundation Graduate Research Traineeship Fellowship (grant 9452655) are gratefully acknowledged.

References

(1) Anthony, J. L.; Maginn, E. J.; Brennecke, J. F. Solubilities and Thermodynamic Properties of Gases in the Ionic Liquid 1-n-Butyl-3-methylimidazolium Hexafluorophosphate. *J. Phys. Chem. B* **2002**, *106*, 7315-7320.

(2) Blanchard, L. A.; Gu, Z. Y.; Brennecke, J. F. High-pressure phase behavior of ionic liquid/CO_2 systems. *J. Phys. Chem. B* **2001**, *105*, 2437-2444.

(3) Welton, T. Room-temperature ionic liquids. Solvents for synthesis and catalysis. *Chem. Rev.* **1999**, *99*, 2071-2083.

(4) Gordon, C. M. New developments in catalysis using ionic liquids. *Appl. Catal. A-Gen.* **2001**, *222*, 101-117.

(5) Wasserscheid, P.; Keim, W. Ionic liquids - New "solutions" for transition metal catalysis. *Angew. Chem.-Int. Edit.* **2000**, *39*, 3773-3789.

(6) Blanchard, L. A.; Hancu, D.; Beckman, E. J.; Brennecke, J. F. Green processing using ionic liquids and CO_2. *Nature* **1999**, *399*, 28-29.

(7) Blanchard, L. A.; Brennecke, J. F. Recovery of Organic Products from Ionic Liquids Using Supercritical Carbon Dioxide. *Ind. Eng. Chem. Res.* **2001**, *40*, 287-292.

(8) Suarez, P. A. Z.; Einloft, S.; Dullius, J. E. L.; De Souza, R. F.; Dupont, J. Synthesis and physical-chemical properties of ionic liquids based on 1-n-butyl-3-methylimidazolium cation. *J. Chim. Phys.* **1998**, *95*, 1626-1639.

(9) Seddon, K. R.; Stark, A.; Torres, M. J. Viscosity and density of 1-alkyl-3-methylimidazolium ionic liquids. In *Clean Solvents*; Abraham, M., Moens, L., Eds.; American Chemical Society: Washington D.C., 2002; ACS Symp. Ser. No. 819, pp 34-49.

(10) Huddleston, J. G.; Visser, A. E.; Reichert, W. M.; Willauer, H. D.; Broker, G. A.; Rogers, R. D. Characterization and comparison of hydrophilic and hydrophobic room temperature ionic liquids incorporating the imidazolium cation. *Green Chem.* **2001**, *3*, 156-164.

(11) Dullius, J. E. L.; Suarez, P. A. Z.; Einloft, S.; de Souza, R. F.; Dupont, J.; Fischer, J.; De Cian, A. Selective catalytic hydrodimerization of 1,3-butadiene by palladium compounds dissolved in ionic liquids. *Organometallics* **1998**, *17*, 815-819.

(12) Seddon, K. R.; Stark, A.; Torres, M. J. Influence of chloride, water, and organic solvents on the physical properties of ionic liquids. *Pure Appl. Chem.* **2000**, *72*, 2275-2287.

(13) Marsh, K. N., personal communication.

(14) Najdanovic-Visak, V.; Esperanca, J. M. S. S.; Rebelo, L. P. N.; Nunes da Ponte, M.; Guedes, H. J. R.; Seddon, K. R.; Szydlowski, J. Phase

behaviour of room temperature ionic liquid solutions: an unusually large co-solvent effect in (water + ethanol). *Phys. Chem. Chem. Phys.* **2002**, *4*, 1701-1703.

(15) Visser, A. E.; Holbrey, J. D.; Rogers, R. D. Hydrophobic ionic liquids incorporating N-alkylisoquinolinium cations and their utilization in liquid-liquid separations. *Chem. Commun.* **2001**, 2484-2485.

(16) Swatloski, R. P.; Visser, A. E.; Reichert, W. M.; Broker, G. A.; Farina, L. M.; Holbrey, J. D.; Rogers, R. D. On the solubilization of water with ethanol in hydrophobic hexafluorophosphate ionic liquids. *Green Chem.* **2002**, *4*, 81-87.

(17) Anthony, J. L.; Maginn, E. J.; Brennecke, J. F. Solution Thermodynamics of Imidazolium-Based Ionic Liquids and Water. *J. Phys. Chem. B* **2001**, *105*, 10942-10949.

(18) Anthony, J. L.; Maginn, E. J.; Brennecke, J. F. Gas Solubilities in 1-n-butyl-3-methylimidazolium hexafluorophosphate. In *Ionic Liquids*; Rogers, R. D., Seddon, K. R., Eds.; American Chemical Society: Washington D.C., 2002; ACS Symp. Ser. No. 818, pp 260-269.

(19) Cammarata, L.; Kazarian, S. G.; Salter, P. A.; Welton, T. Molecular states of water in room temperature ionic liquids. *Phys. Chem. Chem. Phys.* **2001**, *3*, 5192-5200.

Chapter 11

Heat Capacities of Ionic Liquids and Their Applications as Thermal Fluids

John D. Holbrey[1], W. Matthew Reichert[1,2], Ramana G. Reddy[1,3], and Robin D. Rogers[1,2]

[1]Center for Green Manufacturing and Departments of [2]Chemistry, and [3]Metallurgical and Materials Engineering, The University of Alabama, Tuscaloosa, AL 35487

The specific heat capacity of five common ILs containing 1-alkyl-3-methylimidazolium cations have been determined using modulated DSC. In each case, the specific heat capacities for the ILs were between 1.17-1.80 J g^{-1} K^{-1} at 100 °C, and increased linearly with temperature in the liquid region studied. The heat capacity was also determined for the related, higher melting organic salt, 1-butyl-3-methylimidazolium tetraphenylborate, in both the crystalline state and in the melt (above 140 °C), and it is significantly higher than for the ionic liquids. The results are compared with those of common organic thermal fluids and indicate that ILs could be considered as candidate thermal fluids for heat transfer applications.

Introduction

While the bulk of current research on room temperature ionic liquids (ILs) is focused on their use as solvents for chemical reactions (*1*), separations (*2*), and electrochemistry (*3*), ILs have properties which make them potentially excellent performance fluids for use in a wide range of engineering and materials applications. For example, the negligible vapor pressure exhibited by many ILs over a wide temperature range provides liquid stability under low pressure-high vacuum conditions, that may enable the use of IL lubricants (*4*) in space applications and as liquid substrates for mass spectroscopy (*5*) and electron microscopy. A number of engineering parameters need to be determined for the ILs in order to assess their applicability to materials applications, and for process design. In particular, the general absence of specific heat capacity (c_p) data is a significant hurdle for the design of chemical reactors and heat transfer systems, required if any IL processes are to be developed beyond the laboratory scale.

Heat capacities of pure substances are employed in many thermodynamic calculations, from thermochemistry to solution chemistry and are also required for the evaluation of other basic thermodynamic properties (*6*). The heat capacities of ILs and their mixtures are of importance in engineering work associated with the design and operation of reactors and heat pumps, required for scale-up, pilot-plant, and commercialization of IL technologies. Not only does the design of plant equipment require a knowledge of heat capacities over a wide range of operating temperatures, but this data is also helpful when storage or low temperature operation are considered.

The heat capacities (c_p) for a range of 1-alkyl-3-methylimidazolium ($[C_n\text{mim}]^+$) ILs containing either a common cation, or common anion, have been determined by modulated DSC. The magnitude of the heat capacities of ILs (with relevance to engineering and thermal fluids) and the effect of differing structural group-contributions on the results have been examined.

Experimental

1-Butyl-3-methylimidazolium (trifluoromethanesulfonyl)imide ($[C_4\text{mim}]$-$[NTf_2]$, Covalent Associates, MA, solvent grade 98%) was used as received. All other ILs (1-butyl-3-methylimidazolium chloride ($[C_4\text{mim}]Cl$), 1-butyl-3-methylimidazolium hexafluorophosphate ($[C_4\text{mim}][PF_6]$), 1-ethyl-3-methylimidazolium hexafluorophosphate ($[C_2\text{mim}][PF_6]$), 1-hexyl-3-methylimidazolium hexafluorophosphate ($[C_6\text{mim}][PF_6]$), and 1-butyl-3-methylimidazolium tetraphenylborate ($[C_4\text{mim}][BPh_4]$)) were prepared using literature procedures (*7, 8*) and were dried *in vacuo*. The water content (determined by

Karl-Fischer titration) of the ILs was < 0.2 wt %, except for the hexafluorophosphate, $[C_6mim][PF_6]$ which contained 0.9 wt % H_2O.

Modulated differential scanning calorimetry (MDSC) was performed using a TA Instruments model 2920 Modulated DSC (New Castle, DE) cooled with a liquid nitrogen cryostat. The calorimeter was calibrated for temperature and cell constants using indium (melting point 156.61 °C, ΔH 28.71 J g^{-1}), and for heat capacity using a standard sapphire sample. Data was collected at constant atmospheric pressure, using samples between 10-40 mg in aluminum sample pans sealed using pin-hole caps. Experiments were performed heating at 3 °C min^{-1} with a modulation amplitude of ±3 °C, the modulation period was fixed at 60 s. The DSC was adjusted so that zero heat flow was between 0 and -0.5 mW, and the baseline drift was less than 0.1 mW over the temperature range 0-180 °C. An empty sample pan was used as reference; matched sample and reference pans (within ± 0.20 mg) were used. The heat capacity (reversing flow component) was taken directly from the instrument.

Data was collected for three runs with different samples of each IL, then collated and averaged. Samples were loaded at ambient temperature and equilibrated at 25 °C for 5 min with selected modulation period and temperature amplitude. Data were collected on a heating ramp at 3 °C min^{-1} from 25 °C to 180 °C. For $[C_4mim][NTf_2]$, the sample was loaded at ambient temperature, then cooled to –80 °C in order to ensure initial crystallization of the sample, and data was collected on heating from –80 °C to 180 °C at 3 °C min^{-1}.

Results

Heat capacities of a series of $[C_4mim]X$ ILs with varying anion type and $[C_nmim][PF_6]$ ILs with varying cation substitution were determined between 0 and 180 °C by modulated DSC. The ILs studied, experimental temperature range, and measured c_p are shown in Table 1. The final results of the measurements in the liquid phase are plotted in Figures 1-3 and given for selected temperatures in Table 1.

The data followed a linear increase in heat capacity with temperature, and were fitted to a linear equation $c_p = k + xT$. Values for the parameters are shown in Table 1 along with the experimental temperature ranges. Solid-liquid transitions, in $[C_4mim][BPh_4]$, $[C_2mim][PF_6]$, $[C_4mim]Cl$, and $[C_4mim][NTf_2]$ are characterized by peaks in the heat capacity, and in increase in the underlying heat capacity on melting. However, the change in c_p on melting is pronounced only for $[C_4mim][BPh_4]$. Thermal degradation of the commercial sample of $[C_4mim][NTf_2]$ was indicated above 145 °C by deviations from linearity in the c_p profile. All the ILs were essentially anhydrous (< 0.2 wt % water), with the exception of $[C_6mim][PF_6]$ which contained 0.9 wt % water by Karl-Fisher titration, and was observed in the DSC heat flow trace as a peak corresponding to vaporization (onset temperature 137 °C, ΔH 1.376 J g^{-1}, equivalent to 1 wt % free water). It is notable that the c_p profile either side of this transition are comparable, indicating that the change in heat capacity response of the IL is essentially insensitive to low wt % impurities.

Table 1. Specific Heat Capacity (c_p, J g^{-1} K^{-1}) of the ILs, Determined at Selected Temperatures in the Liquid Region

T (°C)	$[C_4mim]Cl$	$[C_2mim][PF_6]$	$[C_4mim][PF_6]$	$[C_6mim][PF_6]$	$[C_4mim][NTf_2]$
20					1.04
30			1.15	1.36	1.06
40			1.17	1.38	1.08
50			1.21	1.41	1.11
60			1.23	1.44	1.13
70	1.71		1.25	1.47	1.15
80	1.74	1.13	1.28	1.50	1.18
90	1.77	1.15	1.31	1.52	1.20
100	1.80	1.17	1.33	1.55	1.22
110	1.83	1.20	1.35	1.58	1.25
120	1.86	1.22	1.38	1.60	1.27
130	1.89	1.24	1.40	1.63	1.29
140	1.92	1.27	1.43	1.66	1.32
150	1.95	1.29	1.45	1.69	1.32
160	1.98	1.31	1.48	1.72	1.34
170	2.01	1.33	1.49	1.74	1.35
180	2.03	1.35	1.52	1.77	1.37
k[a]	1.50	0.95	1.08	1.27	0.99
$x \times 10^3$	2.95	2.21	2.49	2.76	2.34
ΔT(°C)	60-180	70-180	25-180	25-180	20-145

[a] Data is fitted with a linear regression y = k + xT.

The salts that were solid at room temperature ($[C_2mim][PF_6]$, $[C_4mim]Cl$, and $[C_4mim][BPh_4]$) all melt within the sampling temperature range. A marked increase in slope occurs around the melting transition as an influence of the endothermic event on the heat capacity signal. For $[C_4mim][BPh_4]$, shown in Figure 1, an initial solid-solid transition at 95 °C was observed followed by melting at 130 °C, in agreement with the published melting point (*8*). The heat capacity in the liquid region was greater than in the solid. On melting, the specific heat capacity was 3.84 J g^{-1} K^{-1} and increased linearly to 4.0 J g^{-1} K^{-1} at 180 °C. On cooling rapidly from the molten state, the salt forms a glass which has a c_p profile identical to the initial liquid phase (dashed line in Figure 1). The heat capacity in the liquid phase for $[C_4mim][BPh_4]$ is remarkably high compared to most liquids, including the other ILs measured here (see later), with a value in the range 3.8-4.0 J g^{-1} K^{-1}, compared to 4.184 J g^{-1} K^{-1} for H_2O; however, the liquidus range is only *ca.* 70 °C (130-200 °C).

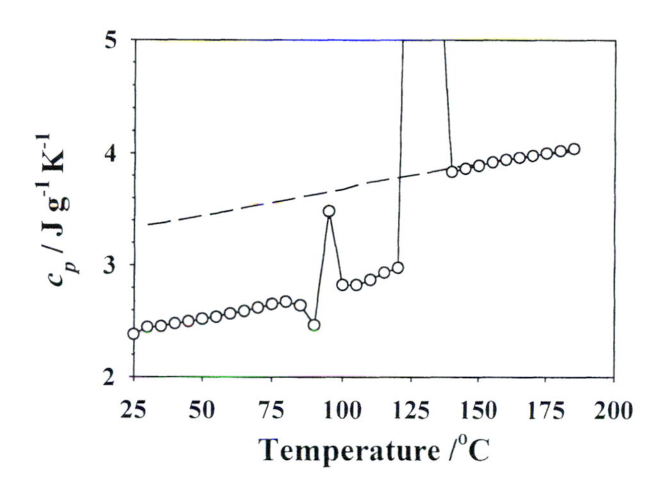

Figure 1. Specific heat capacity as a function of temperature for $[C_4mim][BPh_4]$ showing solid-solid and solid-liquid transformations.

Data for the ionic liquids $[C_2mim][PF_6]$, $[C_4mim][PF_6]$, $[C_6mim][PF_6]$, $[C_4mim]Cl$, and $[C_4mim][NTf_2]$ are shown in Figures 2 and 3, separated into two sets containing either a common cation (Figure 2), or common anion (Figure 3). Data is presented between 25-180 °C; a linear increase in c_p with T in the liquid region for each IL, and the melting transitions for $[C_4mim]Cl$ (57 °C) and $[C_2mim][PF_6]$ (59 °C) are shown, initial melting of solid $[C_4mim][NTf_2]$ at -4 °C is not shown. The variation in c_p with changes in IL become apparent comparing the two sets of data. The heat capacity increases linearly with an increase in temperature. Differences in heat capacity for the different ILs are due to changes in the anions and increasing alkyl-chain substitution in the cation.

All the ILs screened displayed a linear increase in heat capacity with temperature in the liquid region. For ILs with a common cation, the specific heat capacity varies with anion, and decreases following the order $[BPh_4]^- > Cl^- > [BF_4]^- > [PF_6]^- \approx [NTf_2]^-$. The heat capacities also increase with alkyl-chain substitution on the cation within the series of ILs containing a common anion. The two factors, chain length and anion type, appear to contribute independently to the c_p values of the ILs.

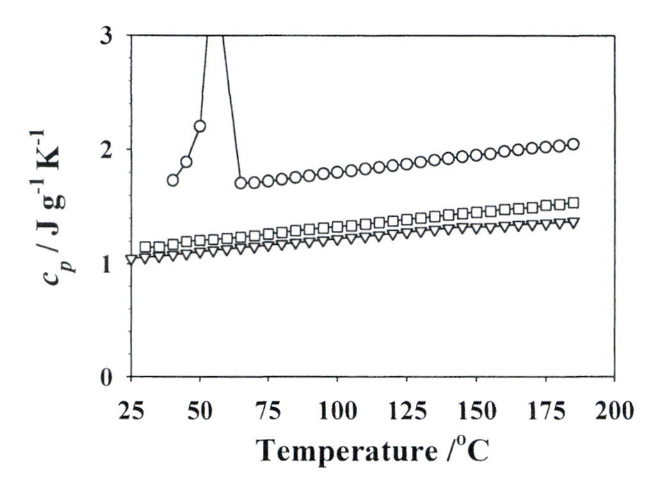

Figure 2. Specific heat capacity (c_p) as a function of temperature for [C_4mim]X ILs; [C_4mim]Cl (○), [C_4mim][PF_6](□), and [C_4mim][NTf_2] (▽).

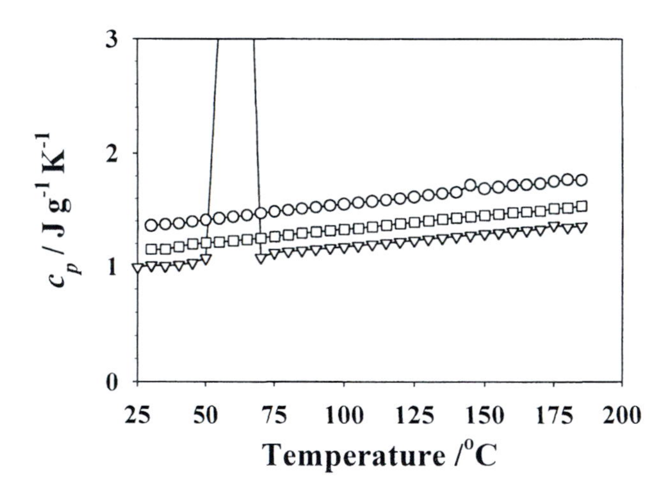

Figure 3. Specific heat capacity (c_p) as a function of temperature for the hexafluorophosphate-containing ILs; [C_2mim][PF_6] (▽), [C_4mim][PF_6](□), and [C_6mim][PF_6] (○).

For the three hexafluorophosphate ILs, the heat capacities increase with increasing alkyl-chain length, $[C_2mim][PF_6]$ < $[C_4mim][PF_6]$ < $[C_6mim][PF_6]$. After conversion to molar heat capacities (Figure 4), the increment in heat capacity for each methylene group was 41 J mol^{-1} per methylene group. This is greater than the increase anticipated based on changes in c_p with Δ_{CH2} for linear alkanes (*9*), possibly indicative of changes in liquid structuring interactions through the series of ILs that are more than purely van der Waals interactions observed for hydrocarbons.

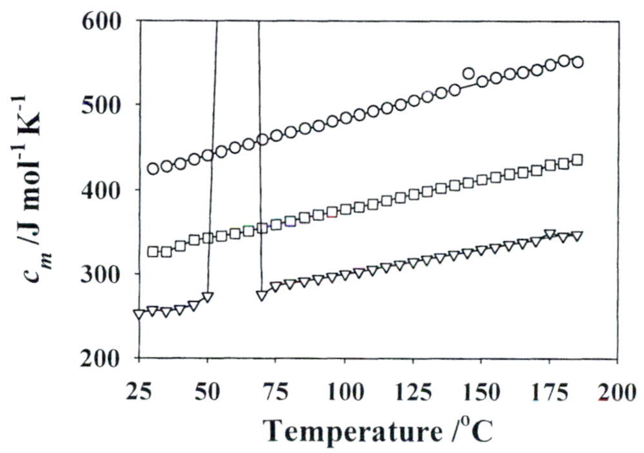

Figure 4. Molar heat capacity (c_m) as a function of temperature for the three hexafluorophosphate ILs; $[C_2mim][PF_6]$ (▽), $[C_4mim][PF_6]$(□), and $[C_6mim][PF_6]$ (○).

Anions with high hydrogen bond acceptor characteristics (i.e., the halide salts) allow strong and extensive hydrogen bonding networks to exist within the liquid structure (*10*). Increased alkyl-chain length provides additional rotational and vibrational modes within the cation that can also absorb energy. The results obtained here then, are related to the supramolecular liquid structure of these solvents *and* the molecular structure of the component ions. The data correlate in some aspects with viscosity, T_g, or melting point, and with crystal-structure/mesophase structure packing (*11*). It has been suggested that variations in layer spacing in the 2D-sheets structures of long-chain amphiphilic ILs correlates with the degree of hydrogen-bonding interactions between anions and cations (*11*), and so to the '*structuredness*' of the phase (both liquid and crystalline). Here, from the c_p data, we see another clear indication of the changes in structure-controlling interactions from the changes in c_p with anion type for the 1-butyl-3-methylimidazolium-containing ILs with the common cation.

Discussion

Liquids typically display specific heat capacities in the range $1.6 < c_p < 2.1$ J $g^{-1} K^{-1}$. Exceptions are strongly hydrogen-bonding liquids, such as water, liquid ammonia, liquid HF, and H_2SO_4. The results determined here indicate that ILs respond to temperature gradients much more like organic molecular solvents, than strongly hydrogen bonding liquids or high temperature molten salts and liquid metals. This is not an unreasonable observation, when the molecular composition of ILs is considered.

One of the most important and widely used applications of liquids as materials in industrial processes is as heat transfer fluids. Many industrial processes require a heat transfer fluid with an operational temperature range up to 300-370 °C. Materials for this range are either liquids (with low vapor pressure and heat transfer in the liquid phase) or liquid/vapor systems. The most widely used synthetic heat transfer fluids are: 73 % diphenyloxide-27 % biphenyl mixtures, hydrogenated terphenyl/quaterphenyl mixtures, and dibenzyltoluene. Choice depends on consideration of the thermal fluid characteristics and operating requirements. Common, representative thermal fluids and operating temperature ranges are shown in Table 2.

Table 2. Composition and Operating Temperature Ranges for Some Common, Commercial Thermal Fluids

Commercial Name	Composition	operating *T* range /°C
Dowfrost	propylene glycol/water	-45 to 120
Dowtherm MX	alkylaromatic	-25 to 330
Dowtherm G	bi/terphenyl	-6 to 360
Syltherm XLT	polysiloxane	-100 to 260
Fluorinert FC70	perfluorocarbon	-25 to 215
Therminol 55	hydrocarbon	-25 to 290
Therminol 59	alkylaromatic	-46 to 316
Therminol 66	terphenyl	0 to 345
Therminol 72	aromatic	-10 to 380
Therminol 75	ter/quatphenyl	80 to 385
Therminol D12	hydrocarbon	-85 to 230
Therminol VP1	biphenyl/phenyloxide	12 to 400
Therminol XP	mineral oil	-20 to 315

The properties, characteristics, and requirements of different thermal storage media, with reference to applications in solar energy power plants are described by Geyer (*12*). If the potential application of ILs is considered to be as an alternative to liquid thermal storage media, with applicability from ambient temperature and above, then an artificial, but desirable property set for an 'ideal' thermal fluid are: thermal storage density of 1.9 J cm^{-3} K^{-1}, upper temperature of 430 °C, no vapor pressure below maximum operating temperature, and materials compatibility with copper and stainless steel used in heat exchangers and piping. These criteria would define liquids with performance properties that surpass those of current mineral oil, aromatic, and siloxane thermal fluids. How do ILs compare?

The favorable materials properties of ILs that can be utilized to advantage include: lack of a measurable vapor pressure (advantage in high temperature applications, and in reducing hazards associated with flash points, flammability, etc.), wide thermal range (potentially from below ambient temperature to above 300 °C), linear thermal expansion characteristics, and reasonable thermal characteristics which may allow them to be substituted directly for hot-oil and synthetic aromatic thermal fluids. The long-term, high temperature thermal stability of some imidazolium-based ILs has been reported (*13*). Decomposition temperatures of ILs can vary, depending on the environment and the nature of the anion present, but in general, imidazolium salts decompose around 350-450 °C (*14*), via dealkylation of the imidazolium cation (*15*). Notably ILs with halide anions have lower decomposition temperatures.

A comparison of some potential thermal storage parameters and properties for ILs with common thermal heat transfer fluids is shown in Table 3. Calculated volumetric heat capacities (in J cm^{-3} K^{-1}) at 25 °C are also shown in Table 3. Potential thermal storage densities (E) for a standardized 100 °C temperature differential were calculated for the ILs studied here from the specific heat capacities calculated and published density data. In each case, the potential storage density of the liquids is in the range 150-200 MJ m^{-3}. This is comparable with oil/aromatic thermal fluids. Combining the heat capacity data collected for $[C_4mim][PF_6]$, $[C_6mim][PF_6]$, and $[C_4mim][NTf_2]$ with published density data (*16*) as a function of temperature from the literature, heat capacity characteristics of the ILs can be extrapolated and compared with available data for commercial heat transfer fluids. Figure 5 shows the volumetric heat capacities for $[C_4mim][PF_6]$, $[C_6mim][PF_6]$, and $[C_4mim][NTf_2]$ from ambient to 180 °C based on a linear extrapolation of the IL thermal expansion between 25-95 °C, compared with representative alkylaromatic, terphenyl, polysiloxane, and propylene glycol/water heat transfer fluids. The data indicate that the ILs have comparable heat adsorption characteristics to current thermal fluids.

It is worth noting that c_p and density of the ILs appear to vary in opposite senses between the ILs, so that the least dense IL ($[C_4mim]Cl$) has the highest value of c_p and the densest IL ($[C_4mim][NTf_2]$) exhibited the lowest specific heat capacity. A direct result of this, is that the volumetric heat capacity, $c_{p,v}$ (and subsequently, potential thermal storage density of the liquids) are approximately equal for all the 1-alkyl-3-methylimidazolium ILs investigated here. This may allow cost/stability issues to be treated without incurring other design and operating penalties.

Table 3. Calculated Volumetric Heat Capacity for ILs at 25 °C

Liquid	Mp /°C	Density /g cm^{-3}	Viscosity /cPs	c_p /J g^{-1} K^{-1}	$c_{p,v}$ /MJ cm^{-3} K^{-1}
Dowtherm HT		1.01	953	1.42	1.41
Thermal oil		0.89	1.9	1.69	1.90
CH_3NH_3Cl[a]	225-230			1.35	
$C_2H_5NH_3Br$[a]	107-108			0.85	
$(CH_3)_4NCl$[a]	>300			1.43	
$(CH_3)_4NBr$[a]	>300			1.04	
$(C_2H_5)_4NBr$[a]	285			1.18	
$(C_4H_9)_4NBr$[a]	102-106			1.39	
$[C_4mim]Cl$	57.1	~1.20[b]	--	1.58[c]	~1.90
$[C_2mim][BF_4]$	5.8	1.20	34	1.12[d]	1.34
$[C_4mim][BF_4]$	-71 (T_g)	1.20	115		
$[C_2mim][PF_6]$	60.5	1.10	--	1.00[c]	1.10
$[C_4mim][PF_6]$	6.5	1.37	389	1.14	1.56
$[C_6mim][PF_6]$	-80 (T_g)	1.30	688	1.34	1.75
$[C_4mim][NTf_2]$	-5.1	1.44	53	1.05	1.50

[a] Data from NIST for solid salts at 25 °C (*17*); [b] extrapolated for liquid from crystal structure density (1.39 g cm^{-3}); [c] c_p extrapolated to 25 °C from linear regression of higher temperature data; [d] Data from Ref. *18*.

Within limits, the properties of ILs can be varied by changes in the structure and nature of the components present. High temperature molten salt and formulation science, and liquid crystal technologies teach us that significant improvements to the properties of the IL systems can be made by the incorporation of additives/blends of ILs (*19*). From the data, the characteristics and thermal range of ILs are comparable with synthetic thermal fluids currently commercially available. Thus, the energy storage characteristics should also allow direct substitution in applications.

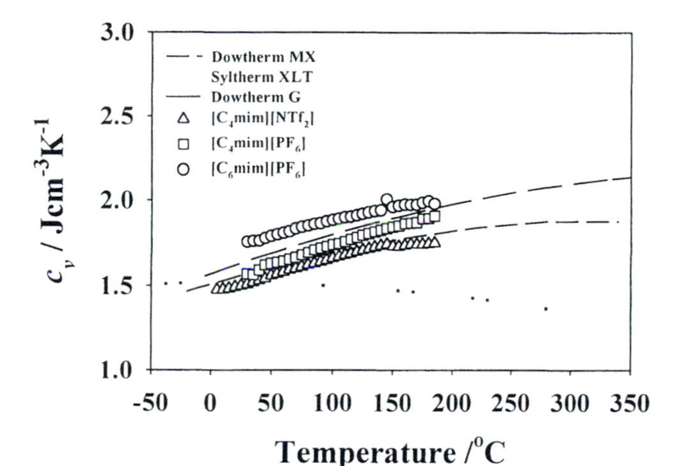

Figure 5. A comparison of volumetric heat capacity performance characteristics with temperature for the ILs [C4mim][PF6], [C6mim][PF6], and [C4mim][NTf2], and commercial heat transfer fluids: Dowtherm MX (alkylated aromatics), Dowtherm G (di- and tri-aryl compounds), and Syltherm XLT (polysiloxane).

Conclusions

Heat capacity data for the ILs, $[C_4mim]Cl$, $[C_4mim][PF_6]$, $[C_2mim][PF_6]$, $[C_6mim][PF_6]$, and $[C_4mim][NTf_2]$ have been determined; the values are comparable to those for other organic liquids, and are lower than those of strongly hydrogen-bonding liquids such as water, HF, or H_2SO_4. The c_p profile determined for the higher melting salt, $[C_4mim][BPh_4]$ was greater, however this salt has only a limited, high temperature liquidus range. The ILs investigated have volumetric heat capacity profiles that are comparable to those of conventional synthetic thermal fluids, used in heat transfer operations and it may be concluded that ILs have potential for use as heat transfer fluids (*20*).

Since many ILs show typical decomposition temperatures of 400+ °C (recorded dynamically by TGA), an operating range from ambient to 350 °C may be feasible. However, long term stability under high temperature operating conditions needs further evaluation. Lower temperature operation can be achieved with a suitable formulation with desirable temperature/viscosity characteristics. The operational temperature range for the ILs may depend on the limitations for turbulent pumping of these relatively viscous liquids; at higher temperatures, the viscosity decreases significantly, thus using

$[C_2mim][NTf_2]$ as an example, the viscosity varies from 30 cPs at 25 °C to 6 cPs at 100 °C. Similarly, data on the vapor pressure of ILs at high temperatures is needed. If ILs do, generically, prove to have exceptionally low to non-measurable vapor pressures then these liquids can offer some exceptional advantages over conventional thermal oils as heat transfer fluids, eliminating the hazards associated with vapor above the liquids.

Among the unresolved issues with an IL system is the assessment of economics. Many ILs currently known are intrinsically complex salts, containing expensive groups, for example, the bis(trifluoromethane sulfonyl)imide anion. Attempts to address this by identifying simpler, lower cost cations and anions, or manufacturing routes are required to make these materials economically realizable, particularly on the scales that could be envisaged for this sort of application.

Acknowledgments

This research has been supported by the U.S. Environmental Protection Agency STAR program through grant number R-82825701-0 (Although the research described in this article has been funded in part by EPA, it has not been subjected to the Agency's required peer and policy review and therefore does not necessarily reflect the views of the Agency and no official endorsement should be inferred). Additional support was provided to the Center for Green Manufacturing from the National Science Foundation Grant EPS-9977239 and the Environmental Management Science Program of the Office of Environmental Management, U.S. Department of Energy, grant DE-FG07-01ER63296.

References

1 Welton, T.; *Chem. Rev.* **1999**, *99*, 2071; Holbrey, J. D.; Seddon, K. R. *Clean Prod. Proc.* **1999**, *1*, 223; Wasserscheid, P.; Keim, W. *Angew. Chem. Int. Ed.* **2000**, *39*, 3772; Sheldon, R. *Chem. Commun.* **2001**, 2399; Gordon, C. M. *Appl. Catal. A* **2001**, *222*, 101.

2 Huddleston, J. G.; Willauer, H. D.; Swatloski, R. P.; Visser, A. E.; Rogers, R. D. *Chem. Commun.* **1998**, 1765; Visser, A. E.; Swatloski, R. P.; Reichert, W. M.; Griffin S. T.; Rogers, R. D. *Ind. Eng. Chem. Res.* **2000**, *39*, 3596: Visser, A. E.; Swatloski, R. P.; Griffin S. T.; Hartman D. H.; Rogers, R. D. *Sep. Sci. Technol.* **2001**, *36*, 785.

3 Endres, F. *Chem. Phys. Chem*. **2002**, *3*, 144.

4 Ye, C.; Lui, W.; Chen, Y.; Yu, L. *Chem. Commun.* **2001**, 2244.
5 Gannon, T. J.; Law, G.; Watson, P. R.; Carmichael, A. J.; Seddon, K. R. *Langmuir* **1999**, *15*, 8429.
6 Gokcen N. A.; Reddy, R. G. *Thermodynamics*, Plenum, New York, 1996; Honig, J. M. *Thermodynamics*, Academic Press, San Diego, 1999.
7 Huddleston, J. G.; Visser, A. E.; Reichert, W. M.; Willauer, H. D., Broker, G. A.; Rogers, R. D. *Green Chem.* **2001**, *3*, 156.
8 Dupont, J.; Suarez, P. A. Z.; De Souza, R. F.; Burrow, RF.; Kintzinger, J.-P. *Chem. Eur. J.* **2000**, *6*, 2377.
9 *CRC Handbook of Chemistry and Physics*, Lide, D. R., Ed; 71 edition, CRC, Boston, 1991.
10 Elaiwi, A.; Hitchcock, P. B.; Seddon, K. R.; Srinivasan, N.; Tan, Y. M.; Welton, T.; Zora, J. A. *J. Chem. Soc., Dalton Trans.* **1995**, 3467.
11 Bradley, A. E.; Hardacre, C.; Holbrey, J. D.; Johnston, S.; McMath, S. E. J.; Nieuwenhuyzen, M. *Chem. Mater.* **2002**, *14*, 629.
12 Geyer, M. A. in *Solar Power Plants*, Winter, C.-J.; Sizmann, R. L.; Vant-Hull, L. L., Eds.; Springer-Verlag Berlin, 1991; p 199.
13 Zhang, Z.; Reddy, R. G. in *EPD Congress 2002*, Taylor, P. R. ed.; TMS, Warrendale PA, 2002; p 199.
14 Ngo, H. L.; LeCompte, K.; Hargens, L.; McEwan, A. B. *Thermochim. Acta.* **2000**, *357-358*, 97: Bonhôte, P.; Dias, A. P.; Papageorgiou, N.; Kalyanasundaram, K.; Grätzel, M. *Inorg. Chem.* **1996**, *35*, 1168: Holbrey, J. D.; Seddon, K. R. *J. Chem. Soc., Dalton Trans.* **1999**, 2133: Takahashi, S.; Koura, N.; Kohara, S.; Saboungi, M.-L.; Curtiss, L. A. *Plasmas & Ions* **1999**, *2*, 91.
15 Chan, B. K. M.; Chang, N.-H.; Grimmett, M. R. *Aust. J. Chem.* **1977**, *30*, 2005.
16 Seddon, K. R.; Stark, A.; Torres, M.-J. in *Clean Solvents: Alternative Media for Chemical Reactions and Processing*, Abraham, M. A.; Moens, L., eds.; ACS Symposium Series 819, American Chemical Society, Washington DC, 2002; p 34.
17 Linstrom, P. J; Mallard, W. G. Eds., *NIST Chemistry WebBook, NIST Standard Reference Database Number 69*, July 2001, National Institute of Standards and Technology, Gaithersburg MD, 20899 (http://webbook.nist.gov).
18 Wilkes, J. S. in *Proceedings of the 13th International Molten Salt Symposium*, The Electrochemical Society, Philadelphia, PA, 2003, in press.
19 Sun, J.; Forsyth, M.; MacFarlane, D. R. *J. Phys. Chem. B* **1998**, *102*, 8858.
20 Wu, B.; Reddy, R. G.; Rogers, R. D. in *Proceedings of Solar Forum 2001, Solar Energy: The Power to Choose*, ASME, Washington, DC, April 21-25, 2001.

Chapter 12

Thermodynamic Properties of Liquid Mixtures Containing Ionic Liquids

Andreas Heintz, Jochen K. Lehmann, and Sergey P. Verevkin

Department of Physical Chemistry, University of Rostock, D–18055 Rostock, Germany

The systematic study of thermodynamic properties of ionic liquids and their mixtures with organic solvents is an important task for understanding the physical chemistry of systems containing ionic liquids. Results of activity coefficients of solutes in three ionic liquids mixed with a series of alkanes, alkylbenzenes, and alcohols are reported as well as liquid-liquid equilibrium curves and excess volumes. The experimental equipment is described and the results are discussed on a qualitative basis.

The study of ionic liquids that are air and moisture stable has become the subject of an increasing number of scientific investigations documented in the literature and the other chapters of this book (1 – 11). Most work has been invested in the elaboration of the synthetic methods, application in catalytic processes, electrochemistry and other more specialized fields. However, the physico-chemical properties of ionic liquids and their mixture with other fluids have not been studied systematically, this holds in particular for thermodynamic properties. For example, ionic liquids are claimed to be potential solvents for many organic, inorganic, and polymeric substances (2), but even the simple question as to what degree any compound of interest is soluble in any kind of ionic liquid has received almost no answer so far due to the lack of experimental data. Only a very restricted number of investigations of the liquid-liquid equilibrium (7, 12, 13), solubility of gases in ionic liquids (14 – 16) and the viscosity of mixtures (8) containing ionic liquids are available in the literature.

Our interest in mixtures containing ionic liquids is focussed on the following thermodynamic properties:

- Activity coefficients of solutes in ionic liquids
- Liquid-liquid equilibria in mixtures of ionic liquids and organic solvents
- Excess properties such as excess volumes and excess enthalpies (heats of mixing).

The systematic knowledge of these properties is important for designing chemical engineering processes such as homogeneous catalysis, biphase catalysis or biocatalysis where ionic liquids are involved. In this chapter experimental methods will briefly be described and the results will be discussed. The ionic liquids 4-methyl-N-butyl-pyridinum tetrafluoroborate ([4-M-nBP][BF_4]); 1-methyl-3-ethyl-imidazolium bis(trifluoromethyl-sulfonyl) amide ([emim][NTf_2]), and 1,2-dimethyl-3-ethyl-imidazolium bis(trifluoromethyl-sulfonyl) amide ([emmim][NTf_2]) discussed in this chapter are characterized in Fig. 1.

1. Activity coefficients and solubilities of solutes in ionic liquids at infinite dilution

Activity coefficients γ_i^∞ in ionic liquids at infinite dilution of the solute allow to determine the solubility of the solute *i* expressed by the mole fraction x_i at low pressures according to Henry's law:

$$P_i = H_i \cdot x_i \tag{1}$$

with

$$H_i = P_{io} \cdot \gamma_i^\infty \tag{2}$$

where H_i in eq. (1) is Henry's coefficient strictly defined as

$$H_i = \lim_{x_i \to 0} \frac{f_i}{x_i} \tag{3}$$

f_i is the fugacity in the vapor phase, P_{io} is the vapor pressure of the pure liquid solute i and P_i is the (partial) pressure of the solute dissolved in the ionic liquid. Eq. (1) is a good approximation as long as values of x_i are small enough. Due to the fact that ionic liquids have no detectable vapor pressure the special technique of gas-liquid chromatography is most suitable for determining values of γ_i^∞ in ionic liquids which are used as stationary phase in the chromatographic process. Chromosorb W/AW-DMCS mesh was used as solid support for the ionic liquid in the GC column. This support material is covered by the ionic liquid of accurately known mass and filled into the GC column. The retention time t_r of solute samples injected into the carrier gas nitrogen has been measured. t_r is related to the retention volume V_N:

$$V_N = J \cdot U_0 (t_r - t_G) \frac{T_{col}}{T_f} \left[1 - \frac{P_{ow}}{P_{out}} \right] \tag{4}$$

where t_G is the dead time of the column. U_0 is the volume flow rate, measured by a soap bubble flow meter, T_{col} is the temperature of the column, and T_f is the temperature of the flow meter. P_{ow} is the saturation pressure of water at T_f and P_{out} is the pressure at the column outlet. The quantity J in eq. (4) is a correction factor accounting for the influence of the drop of gas pressure $P_{in} - P_{out}$ inside the column where P_{in} is the pressure of the carrier gas nitrogen measured at the inlet of the column. J is given by (17):

$$J = \frac{3}{2} \frac{(P_{in} / P_{out})^2 - 1}{(P_{in} / P_{out})^3 - 1} \tag{5}$$

The desired value of γ_i^∞ is given by

$$\ln \gamma_i^\infty = \ln \left(\frac{n_{IL \cdot RT}}{V_N \cdot P_1^0} \right) - \frac{B_{11} - V_1^0}{RT} P_1^0 + \frac{2B_{12} - V_1^\infty}{RT} J \cdot P_0 \tag{6}$$

where n_{IL} is the number of moles of the ionic liquid in the column used as stationary phase and P_1^0 is the vapor pressure of the pure liquid solute. The second and third term in eq. (6) are correction terms containing the second virial coefficient of the solute B_{11} and the cross virial coefficient B_{12} of the solute and nitrogen. It turns out that these correction terms contribute less than 3 % to the value of γ_i^∞ obtained with eq. (6) (18). 38 solutes in ionic liquids have been studied at different temperatures so far (18 - 20). Selected results are shown graphically in Figs. 2 – 4. Values of γ_i^∞ increase with increasing size of the alkyl rests in all three classes of solutes indicating a decreasing solubility in the ionic liquids with the number of C-atoms. Generally values of γ_i^∞ for alkylbenzenes and alkanols are distinctly lower than for n-alkanes. This indicates a higher molecular affinity of aromatic rings and polar groups such as the OH-groups to ionic liquids than for the unpolar and less polarizable n-alkanes. Similar results are obtained with other solutes (18 – 20).

From the temperature dependence of γ_i^∞ the heat of solution of the liquid solute at infinite dilution in the ionic liquid $H_i^{E\infty}$ can be obtained:

$$\left(\frac{\partial \ln \gamma_i^\infty}{\partial \left(\frac{1}{T} \right)} \right)_P = \frac{H_i^{E\infty}}{R} \tag{7}$$

Selected results are shown in Table 1. Not unexpected the values of $H_i^{E\infty}$ of alkylbenzenes are lower than those of n-alkanes and in some cases even a negative sign is observed indicating an exothermic solution process.

Somewhat surprising is the fact the values of $H_i^{E\infty}$ of alkanols are relatively high. Most probably the reason is that hydrogen bonding of pure alkanols in the liquid state has to be broken when alkanols are dissolved in the ionic liquid. Breaking hydrogen bonds exhibits an energy consuming i.e. positive contribution to the solution process.

2. Activity coefficients of solutes in ionic liquids in the whole concentration range

Activity coefficients covering the whole range of concentration in a mixture of an ionic liquid with an organic solute can be measured using a recently developed method (21). This method is based on the so-called transpiration technique which is particularly suitable when the vapor pressure of the solute is low. The principle of this method is shown in Fig. 5. Small glass beads are covered with a liquid mixture consisting of the ionic liquid and an organic solute of exactly known composition and are filled into an U-shaped glass tubing which

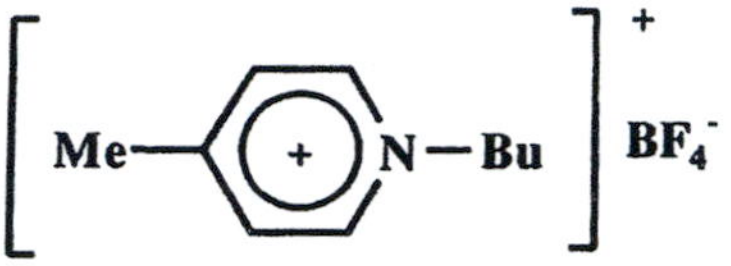

$[4\text{-M-nBP}][BF_4]$;

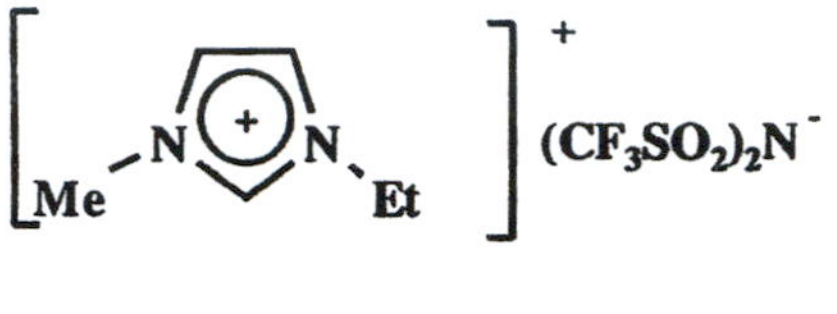

$[emim][NTf_2]$ 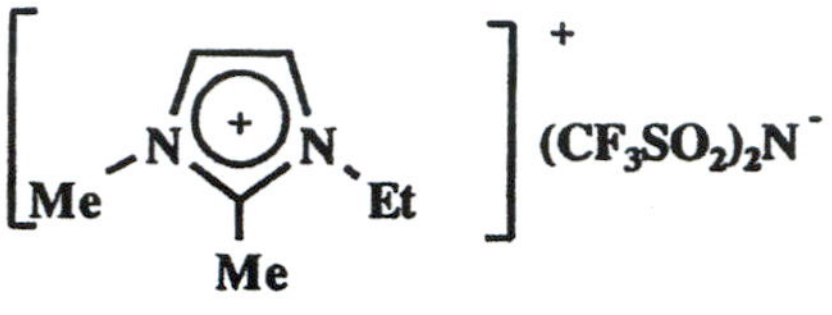 $[emmim][NTf_2]$

Figure 1. Ionic liquids discussed in this chapter

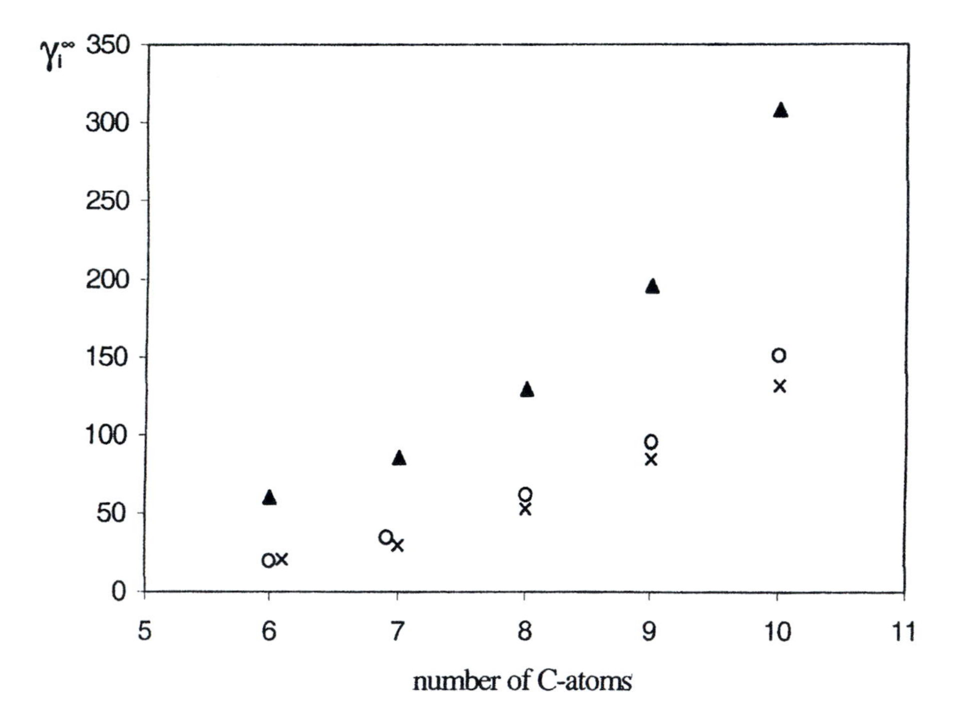

Figure 2. Limiting activity coefficients γ_i^∞ of n-alkanes in ionic liquids at 313K

▲= $[4\text{-M-nBP}][BF_4]$; ○= $[emim][NTf_2]$ x = $[emmim][NTf_2]$

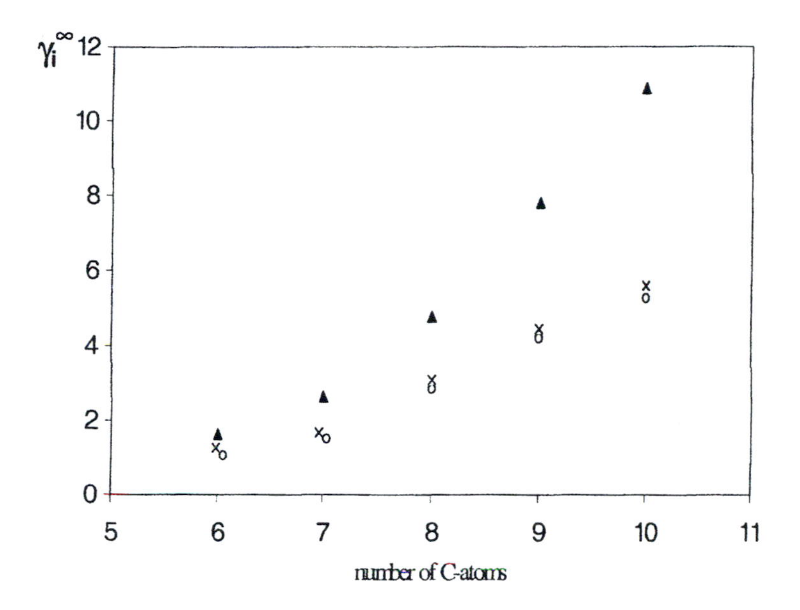

Figure 3. Limiting activity coefficients γ_i^∞ of alkylbenzenes in ionic liquids at 313 K
▲= [4-M-nBP][BF_4]; ○= [emim][NTf_2] x = [emmim][NTf_2]

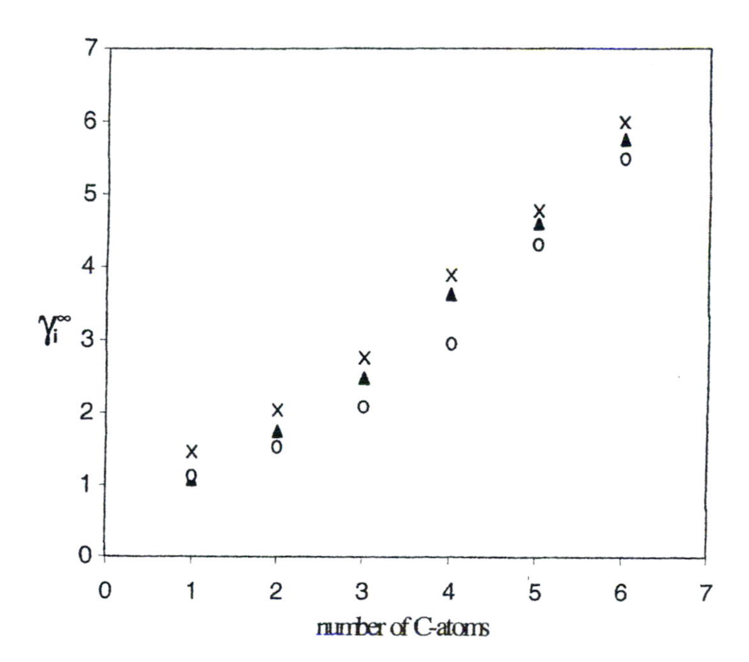

Figure 4. Limiting activity coefficients γ_i^∞ of alkanols in ionic liquids at 313 K
▲= [4-M-nBP][BF_4]; ○= [emim][NTf_2] x = [emmim][NTf_2]

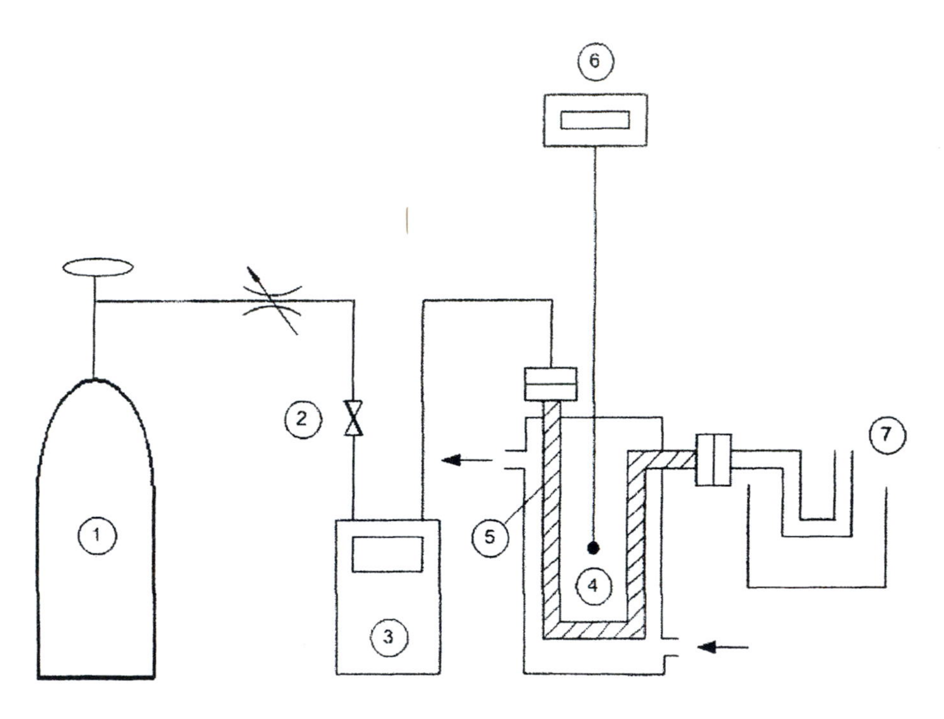

Figure 5. Schematic diagramm of the transpiration apparatus: (1) carrier gas cylinder; (2) flow valve; (3) flow meter; (4) equilibrium cell; (5) U-shaped tube filled with the sample; (6) thermometer; (7) cooling trap at −30°C

Table 1. Values of $H_i^{E,\infty}$ in J·mol^{-1}

Solute i	[4-M-nBP][BF_4];	[emim][NTf_2]	[emmim][NTf_2]
Alkanes			
Hexane	4037	4200	5325
Heptane	5160	6608	6895
Octane	6137	7725	8216
Nonane	8028	9777	9683
Decane	8136	11803	11224
Cyclohexane	6667	5579	7162
Alkylbenzenes			
Benzene	-229	-97	-215
Toluene	-631	-683	-431
Ethylbenzene	447	718	1069
Isopropylbenzene	1352	1289	2018
Tertbutylbenzene	2245	1525	2397
Alcohols			
Methanol	7545	7787	9045
Ethanol	8340	7453	9375
1-Propanol	9134	7946	9949
1-Butanol	10194	9307	11061
1-Pentanol	10630	11217	13148
1-Hexanol	9275	10812	11825
tert-Butanol	8477	7240	8993
Cyclohexanol	6291	6434	7685

is kept at a certain temperature using the circulating bath of a thermostat. A slow stream of N_2-gas flows through the tubing eluating continuously the vapor phase in the glass tubing. Due to the negligible vapor pressure of the ionic liquid the vapor phase consists exclusively of the solute and is condensed in a cooling trap. The mass of solute collected within a certain time interval is determined by dissolving it in a suitable solvent of known amount. This solution is analysed using a gas chromatograph. The peak area of the solute is a direct measure of the mass of the solute condensed into the cooling trap provided a calibration run has been made. From these information the partial pressure of the solute in the glass tubing can be determined. If the amount of solute condensed is small compared to its contents in the liquid phase inside the tubing the change of concentration in the liquid mixture is negligible during such an experiment and the partial pressure of the solute can be assigned to the known composition of the liquid mixture being in thermodynamic equilibrium with the vapor phase. Details of the technique can be found elsewhere (21). Test measurements including thermodynamic consistency tests with the mixture (n-pentanol + decane) have been made showing excellent agreement of the results of γ_i with literature data obtained by a different method (22). Measurements of γ_{solute} covering the whole range of concentration of solute + ionic liquid mixtures have been performed. A series of ketones, aldehydes and diols mixed with ionic liquids has been studied (21). As an example Fig. 6 shows the results of two diols where the pressure as well as the activity coefficients γ_{solute} of the solutes are presented as function of the mole fraction of the solute. Values of γ_{solute} have been obtained according to

$$\gamma_{solute} = \frac{P}{P_{io}} \cdot \frac{1}{x_{solute}} \tag{8}$$

where P is the measured pressure of the solute at the mole fraction x_{solute} and P_{io} is the vapor of the pure solute, this is the value of P at $x_{solute} = 1$. While vapor pressures p of the two diols are quite different, the results of the activity coefficients are similar showing positive values decreasing from 10 to 12 in the infinite diluted region to 1 in the pure liquid state of the diols.

3. Liquid-liquid equilibria containing ionic liquids

Since only small amounts of ionic liquids can be utilized for experimental work due to the relatively high costs of ionic liquids an apparatus designed for the determination of liquid-liquid equilibria of small samples has to be used (see Fig. 7). The onset of phase separation is detected by measuring the intensity of scattered laser light. The optical set-up consists of a laser light source and a photodiode used as light receiving detector. The light intensity is measured at an

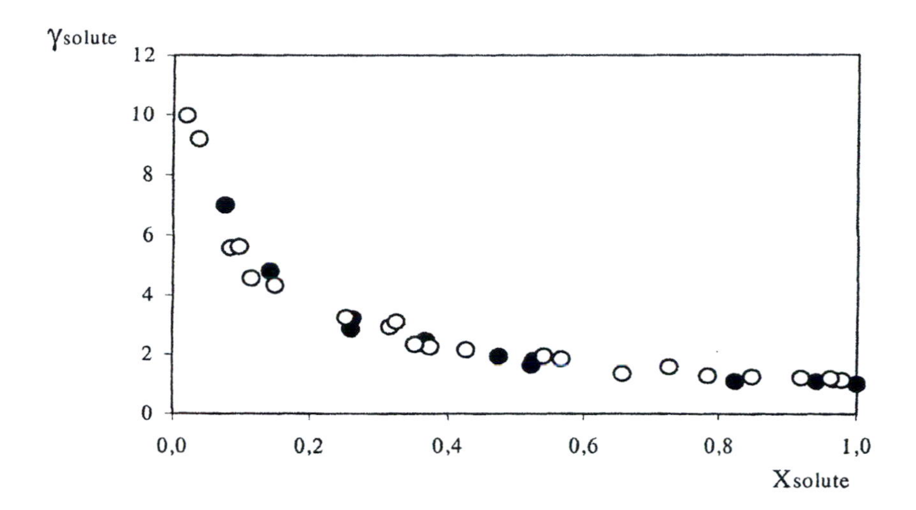

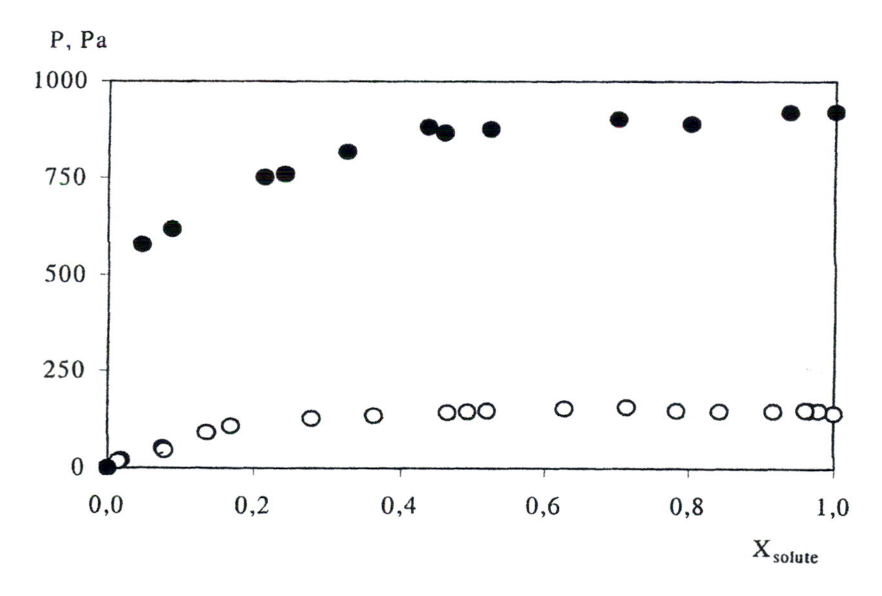

Figure 6. Activity coefficients γ_{solute} and vapor pressures P of the diols in (diol + [emim][NTf_2]) mixture at 353.7 K obtained by transpiration method

○ =1,4-butanediol; • =1,2-ethanediol

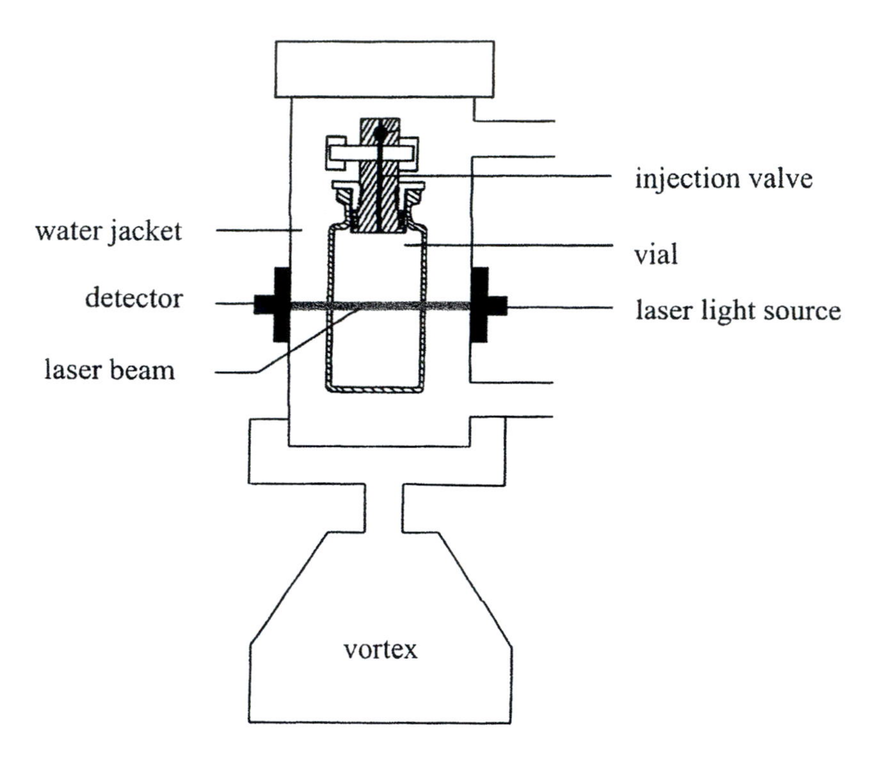

Figure 7. Experimental setup of the liquid-liquid equilibrium

angle of 135° with respect to the direction of the incident laser light beam. Laser and detector are mounted on a sleeve which is fixed tightly surrounding the jacket of the liquid-liquid equilibrium cell. The mixtures are prepared by injecting directly the two components into a probe vial of 4 ml. The composition is determined gravimetrically. The vial is mounted inside the temperature controlled jacket. The complete jacket with the vial and the optical setup is shaken in a vortex which leads to a much more effective generation of small droplets at the onset of phase splitting resulting in a sharper and better reproducible change of the scattered light intensity than in case of using conventional stirring devices.
Starting in the homogeneous region of the liquid mixture of given composition the temperature is lowered stepwise in small intervals until the cloud point is observed. Details of the experimental procedure are described elsewhere (23). Experimental results of so-called binodal curves describing the liquid-liquid miscibility gap of mixtures containing the ionic liquid [emim][NTf_2] with propan-1-ol, butan-1-ol, and pentan-1-ol are shown in Fig. 8. All systems show an upper critical solution temperature (UCTS). It is obvious that the tendency of demixing increases with decreasing polar character of the alcohol molecule in the order propanol, butanol, pentanol. This fact corresponds to the results of γ_i^∞ of alcohols in the same ionic liquid which increase with increasing chain length of the alcohol indicating smaller mutual solubility with decreasing polar character of the alcohols (see Fig. 3 and ref. 19).

4. Densities and excess volumes of mixtures containing ionic liquids

Density measurements of pure ionic liquids are available in the literature (24, 25). Precise measurements of densities ionic liquids mixed with organic compounds allow to determine the molar excess volume V^E which is defined in case of binary mixtures

$$V^E = x_1 M_1 \left(\frac{1}{\rho_{Mix}} - \frac{1}{\rho_1} \right) + x_2 M_2 \left(\frac{1}{\rho_{mix}} - \frac{1}{\rho_2} \right) \tag{9}$$

Eq. (9) describes the deviation of the additive behaviour of the volume of the mixture upon mixing two pure liquid components related to one mol of the mixture. Eq. (9) also reveals that the measurement of the density of the mixture ρ_{mix} and the density of the pure liquid component ρ_i (i = 1, 2) have to be determined with high precision to obtain accurate results since (ρ_{mix}^{-1} - ρ_i^{-1}) appear in eq. (9) as a small difference of relatively large numbers. x_i (i = 1, 2) and M_i (i = 1, 2) are the mole fractions and molar masses of the components

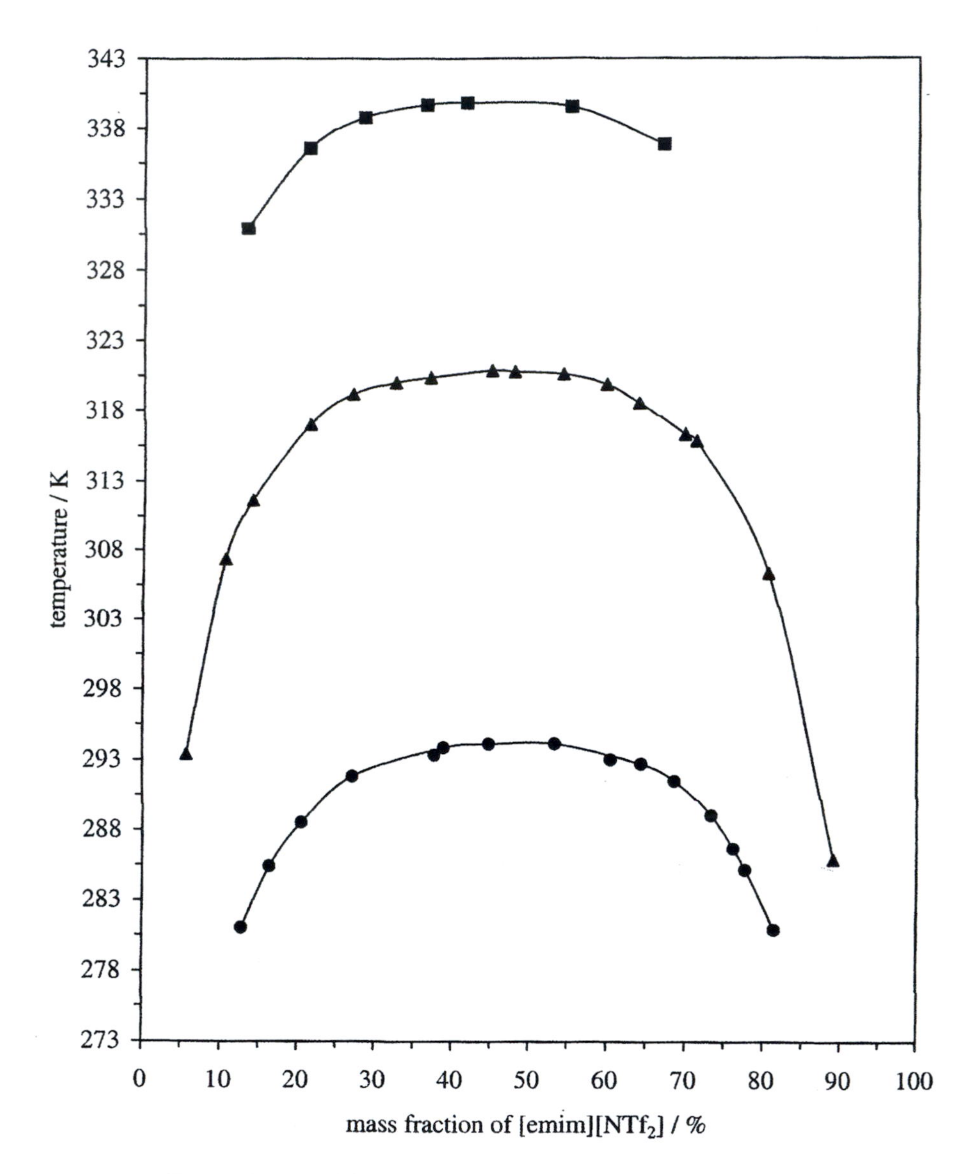

Figure 8. Liquid-liquid coexistence curves for [emim][NTf_2] + propan-1-ol (●), butan-1-ol (▲), pentan-1-ol (■).

respectively. To obtain densities as function of the mole fraction which are precise enough the method of the vibrating tube can be used. A vibrating tube densitometer consists of an U-shaped glass tubing excited at its resonance frequency by a coupled magnetic device. The resonance frequency depends very sensitively on the mass of liquid sample filled into the tubing. After suitable calibration densities of liquid systems can be determined with an accuracy of $\pm 10^{-5}$ g· cm^{-3}. Such an apparatus has been used to measure densities and molar excess volumes of ionic liquids mixed with organic solvents. Details of the procedure and of the results are reported in the literature (26). As an example the results of V^E of [4-M-nBP][BF_4] mixed with methanol are shown in Fig. 9. The molar excess volume is negative over the whole range of concentration. This means that a certain contraction of the liquid mixture is observed upon mixing. V^E becomes even more negative with increasing temperature which is an unusual behaviour rarely observed in liquid mixtures. The negative values of V^E indicate that the relatively small methanol molecules fit into the free volume between the relatively large ions of [4-M-nBP][BF_4] upon mixing. In addition the methanol molecules are most likely energetically stabilized by ion-dipol interactions in the mixture.

Conclusions

The study of thermodynamic properties of mixtures containing ionic liquids is an important task in the rapidly growing research field of ionic liquids. Systematic investigation of activity coefficients in this kind of mixtures allow to develop reliable methods of predicting solubilities of gases and vapors in ionic liquid where direct experimental data are not available. The knowledge of heats of mixing enable scientists working on the application of ionic liquids in chemical and separation processes to determine the temperature dependence of solubility data. Liquid-liquid equilibria are important in such technical and chemical processes where two liquid phases play a role. Examples are extraction processes, e. g., in the biphase catalysis. Experimental data such a molar excess volumes of ionic liquids mixed with organic compounds or water will provide a better insight into the molecular structure and the nature of intermolecular interactions between the ionic species and the neutral molecules of the solute. Such questions are of interest for theoreticians dealing with electrolyte systems and for comparison of computer based calculations of the molecular dynamics with experimental data. Future work should concentrate on filling the gap of experimental data of thermodynamic properties. Also studying the influence of changing e. g. systematically the chain length of alkyl rests in the ionic liquid on phase equilibria and solubilities will be the subject of future projects.

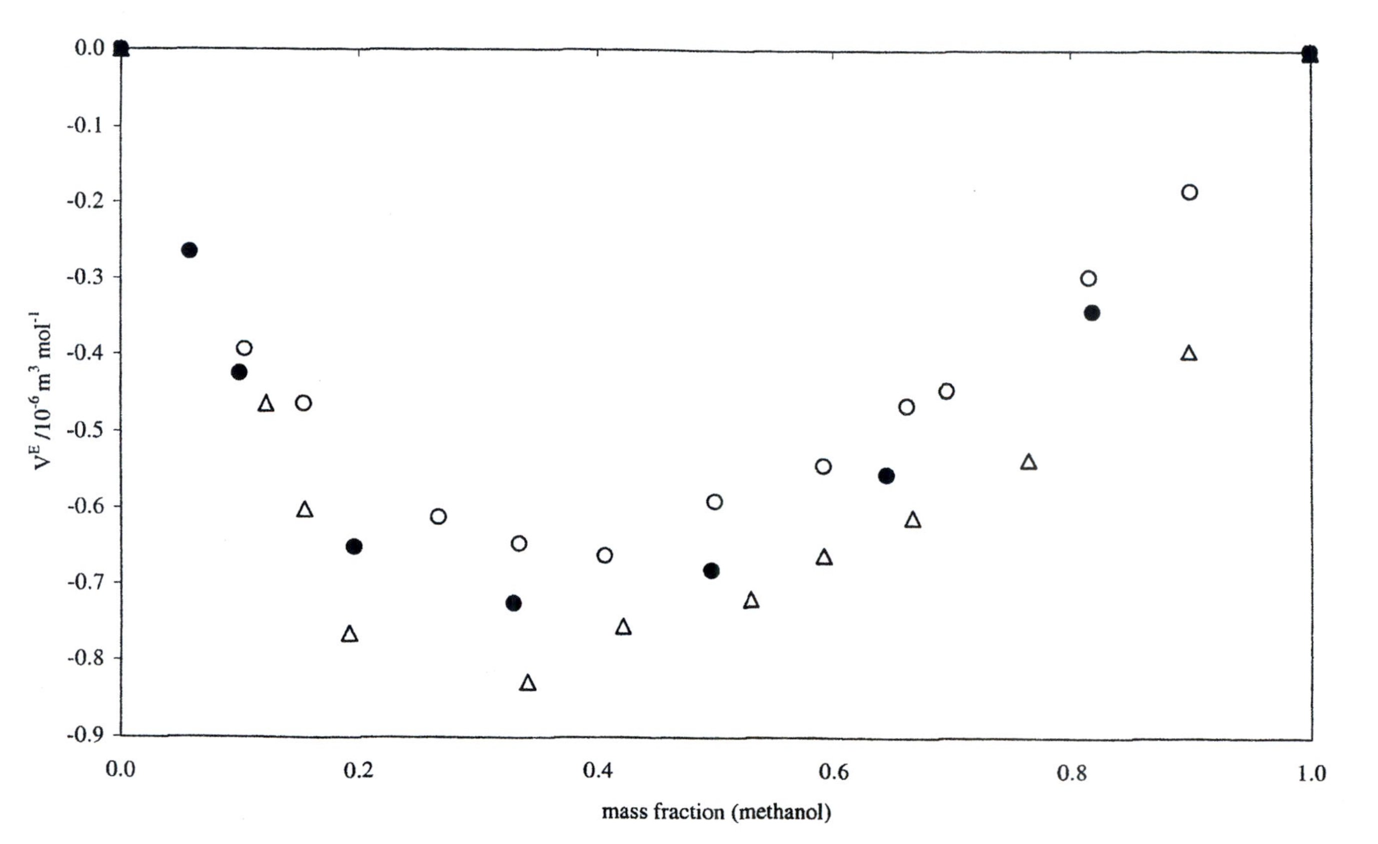

Figure 9. Molar excess volumes V^E of $[4\text{-M-}n\text{BP}][BF_4]$ + methanol at 298 K (○), 313 K (●), and 323 K (Δ) vs. the mass fraction of methanol

References

1. C. L. Hussey. Pure Appl. Chem. **1988,** 60, 1763 - 1772
2. P. Wasserscheid and W. Keim. Angew. Chem. Int. Ed. **2000,** 39, 3772 - 3789
3. C. I. Adams, M. I. Earle, G. Roberts, K. R. Seddon. Chem. Commun. **1998,** 2097 – 2098
4. K. K. Laali, V. J. Gettwert. J. Org. Chem. **2001,** 65, 35 - 40
5. A. S. Larsen, J. D. Holbrey, F. S. Tham, C. A. Reedm. J. Am. Chem. Soc. **2000,** 122, 7264 - 7272
6. H. L. Ngo, K. LeCompte, L. Hargens, A. B. McEwen. Thermochim. Acta **2000,** 357 – 358, 97 - 102
7. M. S. Selvan, M. McKinley, M. D. Dubois, J. L. Atwood. J. Chem. Eng. Data **2000,** 45, 841 - 845
8. Q. Liao, C. L. Hussey. J. Chem. Eng. Data **1996,** 41, 1126 – 1130
9. J. F. Brennecke and E. J. Maginn. AICh Journal **2001,** 47, 2384 - 2389
10. R. D. Rogers and K. R. Seddons (Editors), Ionic Liquids – Industrial Applications to Green Chemistry, ACS Symposium Series 818, American Chemical Society, Washington DC, **2002**
11. H. Olivier-Bourbigou and L. Magna. J. Mol. Catalysis A **2002,** 182 – 183, 419 - 437
12. D. S. H. Wang, J. P. Chen, J. M. Chang, and C. H. Chou. Fluid Phase Eq. **2002,** 194 – 197, 1089 - 1095
13. V. Najdanovic-Visak, J. M. S. S. Esperanza, L. P. N. Rebelo, M. Nunes da Ponte, H. J. R. Guedes, K. R. Seddon, and J. Szydlowski. Phys. Chem. Chem. Phys. **2002,** 4, 1701 - 1703
14. L. A. Blanchard, Z. Gu, J. F. Brennecke. J. Phys. Chem. B **2001,** 105, 2437 - 2444
15. J. L. Anthony, E. J. Maginn, J. F. Brennecke. J. Phys. Chem. B **2001,** 105, 10942 - 10949
16. J. L. Anthony, E. J. Maginn, J. F. Brennecke. J. Phys. Chem. B **2002,** 106, 7315 - 7320
17. D. W. Grant, Gas-Liquid Chromatography, van Nostrand-Reinhold, London, U.K., 1971
18. A. Heintz, D. V. Kulikov, S. P. Verevkin. J. Chem. Eng. Data **2001,** 46, 1526 - 1529
19. A. Heintz, D. V. Kulikov, S. P. Verevkin. J. Chem. Eng. Data **2002,** 47, 894 - 899
20. D. V. Kulikov, S. P. Verevkin, A. Heintz. J. Chem. Thermodyn. **2002,** 34, 1341-1347
21. T. Vasiltsova, S. P. Verevkin, A. Heintz. Fluid Phase Eq. **2003,** in press

22. T. Treszczanowicz and J. Treszczanowicz. Bull. Acad. Pol. Sci. **1979,** 27, 689
23. C. Wertz, J. K. Lehmann, A. Heintz. J. Chem. Eng. Data **2003,** in press
24. K. R. Seddon, A. Stark, M.-I. Torres, in: ACS Symposium Series 818, American Chemical Society, Washington DC, **2002**
25. Z. Gu, J. F. Brennecke. J. Chem. Eng. Data **2002,** 47, 339 - 345
26. A. Heintz, D. Klasen, J. K. Lehmann. J. Sol. Chem. **2002,** in press

Chapter 13

Liquid and Solid-State Structures of 1,3-Dimethylimidazolium Salts

D. T. Bowron[1], C. Hardacre[2,3], J. D. Holbrey[4], S. E. J. McMath[2], M. Nieuwenhuyzen[2], and A. K. Soper[1]

[1]Rutherford Appleton Laboratory, Chilton, Didcot, Oxon OX11 0QX, United Kingdom
[2]School of Chemistry and the [3]QUILL Center, The Queen's University of Belfast, Belfast BT9 5AG, Northern Ireland
[4]Current address: Center for Green Manufacturing, The University of Alabama, Tuscaloosa, AL 35487

Using neutron and single crystal X-ray diffraction the structures of 1,3-dimethylimidazolim chloride and hexafluorophosphate salts have been determined in the liquid and the solid-state. The relative hydrogen bonding characteristics and sizes of the two anions force the ions to pack differently. In each case, a strong correlation between the crystal structure and liquid structure is found.

Introduction

Although the liquid structure of molecular liquids and simple salts has been studied extensively, the structure of complex molten salts is relatively new. Those studies which have been performed have been concerned with binary mixtures associated with examining the first generation ionic liquids. For example, using neutron diffraction, Lee *et al.* (*1*) investigated a 1:1 LiSCN-$AlCl_3$ mixture using Li isotope substitution. Here the data showed that the aluminium in the liquid was tetrahedrally co-ordinated by three chlorines and the isocyanate group attached with nitrogen. Similarly, Takahashi *et al.* determined the liquid structure of $AlCl_3$: $[C_2mim]Cl$ mixtures ranging 46 to 67 mol% $AlCl_3$ (*2*). In agreement with many studies, in the basic composition $[AlCl_4]^-$ species were found, however, although, $[Al_2Cl_7]^-$ was present in the acidic melts, as expected, the structure of the complex anion was distorted indicating a strong interaction with the imidazolium cation. Other systems based on [emim]Cl-HCl, rather than $AlCl_3$, have also shown that the $[HCl_2]^-$ anion forms (*3,4*).

Compared with the liquid structure, extensive research has been undertaken on examining the crystal structures of many ionic liquid related species. These have tended to concentrate on the $[emim]^+$ cation with a variety of anions. A search of the Cambridge Database (*5*) revealed 29 structures containing $[emim]^+$ cations and 11 with other 1-alkyl-3-methylimidazolium cations. However, to date little focus has been placed on understanding the simplest of the dialkyl-imidazolium based salts, *i.e.* the 1,3-dimethylimidazolium cation ($[dmim]^+$). Arduengo *et al* (*6*) have reported the crystal structure of 1,3-dimethylimidazolium chloride, however, no other crystal structures using this cation have been investigated, to date.

In this paper we report on the crystal structure of 1,3-dimethylimidazolium hexafluorophosphate and compare it with the analogous chloride salt. In addition the close interactions found in the liquid state of both salts are compared with the crystal structures as determined by neutron diffraction.

Experimental

Single Crystal X-ray Diffraction

For the single crystal x-ray diffraction studies hexafluorophosphate salt was prepared using the standard literature procedure and recrystallised from methanol (*7*). The atomic connectivity and numbering scheme of the crystal structure of $[dmim][PF_6]$ is illustrated in Figure 1. Data were collected a Bruker-AXS SMART diffractometer using the SAINT-NT (*8a*) software with graphite monochromated Mo-K_α radiation. A crystal was mounted on to the diffractometer at low temperature under nitrogen at *ca.* 120K. The structure was solved using direct methods and refined with the SHELXTL version 5 (*8b*) and the non-hydrogen atoms were refined with anisotropic thermal parameters. Hydrogen-atom positions were added at idealised positions with a riding model and fixed thermal parameters ($U_{ij} = 1.2U_{eq}$ for the atom to which they are bonded(1.5 for Methyl)). The function minimised was $\Sigma[w(|F_o|^2 - |F_c|^2)]$ with reflection weights $w^{-1} = [\sigma^2 |F_o|^2 + (g_1P)^2 + (g_2P)]$ where $P = [\max |F_o|^2 + 2|F_c|^2]/3$. Additional material available from the Cambridge Crystallographic

Data Centre comprises relevant tables of atomic coordinates, bond lengths and angles, and thermal parameters. *Crystal Data for* $C_5H_9N_2PF_6$: - M = 242.11, orthorhombic, space group *P*bca, a = 11.254 (2) Å, b = 9.3606 (19) Å, c = 17.986 (4) Å, U = 1894.7 (7) $Å^{-3}$, Z = 8, μ = 0.347 mm^{-1}, R_{int} = 0.0689. A total of 10979 reflections were measured for the angle range $4.5 < 2\theta < 52.7$ and 1934 independent reflections were used in the refinement. The final parameters were wR2 = 1528 and R1 = 0.0529 [$I > 2\sigma I$].

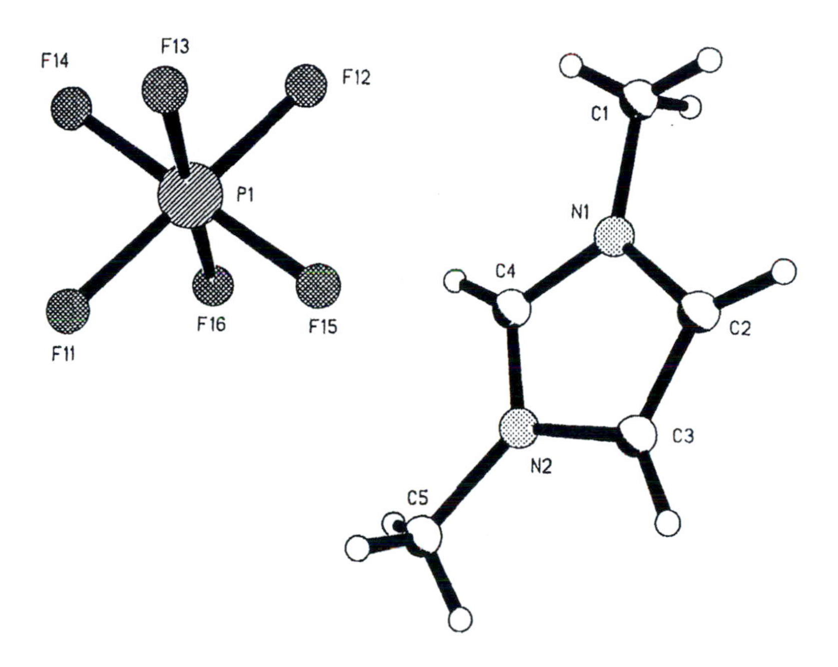

Figure 1. The atomic connectivity and numbering scheme for [dmim][PF_6].

Neutron Diffraction

For the neutron data, the chloride salts were synthesised using methods described previously (*9*) and the hexafluorophosphate salt synthesized from the chloride *via* metathesis with NH_4PF_6, or in the case of the fully deuteriated and ring deuteriated salts, ND_4PF_6 in D_2O was used to ensure that no proton exchange occurred and the imidazolium ring remained deuteriated. Each sample was a white crystalline solid and was analysed by ^1H and ^2H NMR, infrared and elemental analysis and showed > 97 % isotopic exchange. All samples were dried as liquids under high vacuum, cooled and stored in a N_2 filled dry box prior to being loaded into the sample cells.

The neutron data was take at 425 K and 400 K for the chloride and hexafluorophosphate salts, respectively. In the former, the samples were run in a Ti-Zr sample holder (40 mm deep, 35 mm wide, 1 mm thick) with a PTFE seal and the latter an aluminium sample holder with 0.2 mm thick vanadium windows sealed using a PTFE coated Viton 'o' ring was used. In each case, the cells were heated in the vacuum chamber and left for 10 minutes to equilibrate before measurements were taken. The temperature was maintained to $\pm$ 0.1° using a Eurotherm PID temperature controller. After normalisation with respect to the cell, spectrometer and the vanadium window, good reproducibility was found between the sample cells. The neutron diffraction data was taken using the SANDALS diffractometer at the ISIS pulsed neutron source, Rutherford Appleton Laboratory, U.K. This instrument has a wavelength range 0.05 to 4.5 Å and data is taken over the Q range 0.05 to 50 $Å^{-1}$.

Data analysis was performed using the ATLAS package (*10*). For each anion, differential scattering cross sections for four samples with different hydrogen isotope substitutions were extracted in order to highlight the ion-ion structure. The [dmim]Cl samples used were fully protiated, fully deuteriated, ring deuteriated and a 1:1 fully deuteriated:fully protiated mixture. In the case of [dmim][PF_6], fully protiated, fully deuteriated, ring deuteriated and side chain were used. The sample density of the liquid was taken as 0.090 and 0.085 atoms/$Å^3$ for the Cl^- and $[PF_6]^-$ salts, respectively (*11*).

Table 1 Lennard-Jones and Coulomb parameters used for the reference potential of the simulation of liquid 1,3 dimethylimidazolium chloride and hexafluorophosphate salts. The charges were derived from Hanke *et al* (*11*). The carbon and nitrogen atoms are defined as their position on the *molecular* skeleton with M referring to the methyl groups. The hydrogen atoms, H*x*, are defined as being bonded to the C*x* atom in Figure 1. The Lennard Jones parameters for interactions between unlike atoms were calculated by employing the usual Lorentz-Berthelot mixing rules.

Atom	$\sigma / Å$	$\varepsilon / kJ\ mol^{-1}$	q / e
H4	0.0	0.0	0.097
C4	3.9	0.79396	0.407
H2,H3	0.0	0.0	0.094
C2,C3	3.9	0.79396	0.105
N1,N2	3.4	0.57341	-0.267
H1,H5	0.0	0.0	0.064
C1,C5	3.9	0.79396	0.124
Cl	4.2	0.49931	-1
P	2.2	7.177	0.740
F	2.66	0.59560	-0.290

In order to model the diffraction data, the normalized data was subjected to an empirical potential structure refinement process (EPSR). By using Lennard-Jones potential and running reverse Monte-Carlo simulations, configurations of the ions are produced which are interated with the experimental data to obtain structures which are consistent with the diffraction data. The full details of this method have recently been reported (*12*). The full parameters of this reference potential used for both salts are given in Table 1. It should be noted that although there is always a question concerning the uniqueness of the EPSR fits, and therefore the minor details need to be carefully evaluated, the ion-ion correlations upon which the results of this paper are based are thought to be represented correctly.

Results and Discussion

Crystal Structures of [dmim]Cl and [dmim][PF_6]

The structures of these two materials are very similar. In fact the cations of both [dmim]Cl and [dmim][PF_6] are arranged in a dimeric motif where adjacent cations are hydrogen bonded to one another *via* $CH_3 \ldots \pi$ interactions. The differences in the two structures arise from how these dimers pack with respect to one another and the different hydrogen bonding between the cations and anions. In [dmim]Cl the dimers are arranged into 'layers' in which they have a herringbone type arrangement such that the closest contacts between dimers are *via* the methyl groups. The chloride ions lie in the regions between these 'layers'. In [dmim][PF_6] the dimers are also arranged in a herringbone pattern with the closest contacts also being between the methyl groups. However, in [dmim][PF_6] the dimeric motif forms a three dimensional array of 'holes' into which the $[PF_6]^-$ anions are incorporated. The chloride anions, because of their relatively localised charge, are much better hydrogen bond acceptors than the hexafluorophosphate anions, where the charge is distributed over the six fluorine atoms. Furthermore, the larger size of the hexafluorophosphate anions pushes the cation dimers further apart from one another and this requires a rearrangement of cations in [dmim][PF_6] as compared to [dmim]Cl. Thus the differences between cation anion packing in the two crystal structures, [dmim]Cl and [dmim][PF_6], can be rationalised as a result of both differeing anion size and differing hydrogen bonding ability.

Liquid Structures of [dmim]Cl and [dmim][PF_6]

The experimental and EPSR modeled differential cross sections as a function of Q, for both liquid [dmim]Cl and [dmim][PF_6] samples show good agreement between the model and experimental data examined, the data for the [dmim]Cl is shown as an example in Figure 2 below.

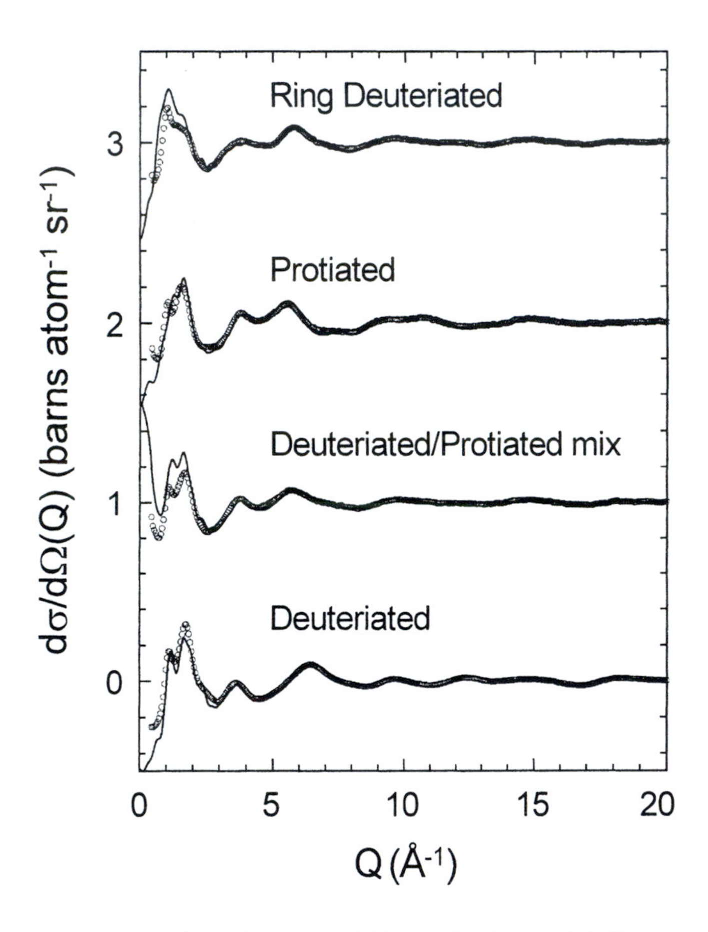

Figure 2. Experimental (circles) and EPSR fitted (solid line) differential cross sections as a function of Q for the fully protiated, fully deuteriated, a 1:1 fully deuteriated:fully protiated mixture and ring deuteriated 1,3 dimethylimidazolium chloride.

From the EPSR model, the partial radial distribution functions for the anions and cations surrounding a central imidazolium cation have been extracted in each case and are compared in Figure 3. It is clear from the partial radial distribution functions that the liquids are charge ordered out to at least two cation/anion shells. As expected the oscillations are weaker for the cation-cation distribution than for the anion-cation and in both cases, again as predicted, the order becomes less well defined as distance increases. Grossly, the radial distribution functions in each salt are similar, however, a small but distinct expansion from 4.3 Å to 5.0 Å in the first shell anion-cation distance is observable on comparing the Cl^- salt with the $[PF_6]^-$. With increasing distance/shell, the expansion in the liquid network becomes more pronounced and is simply the consequence of the size of the two anions. Since the chloride smaller than the hexafluorophosphate, the ions can pack more closely together.

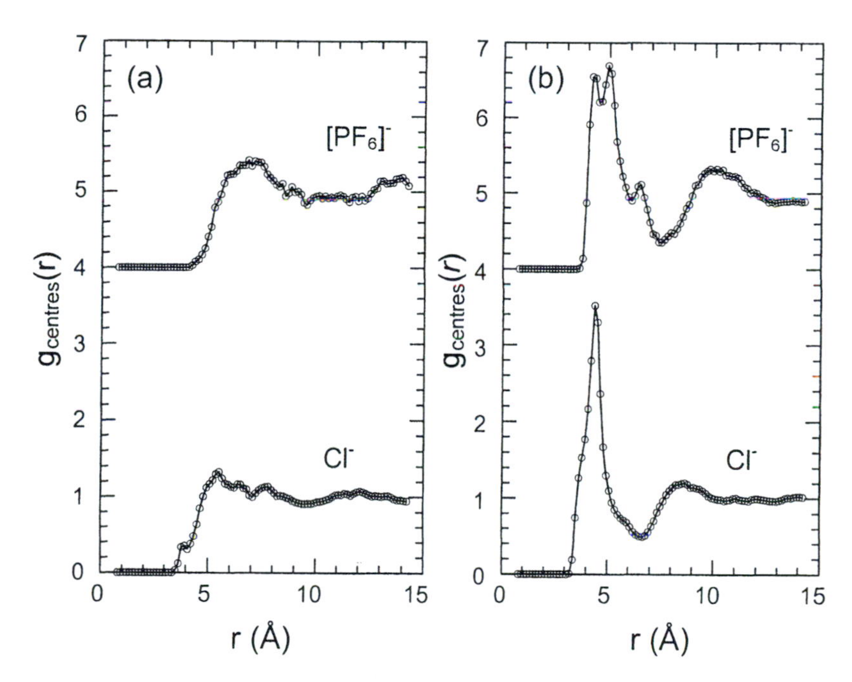

Figure 3. Comparison of the partial radial distribution functions for (a) the cation-cation distribution and (b) the cation-anion distribution for the 1,3 dimethylimidazolium hexafluorophosphate and chloride salts derived from the EPSR model. Each radial distribution function is calculated from centre of the imidazolium ring and, in the case of $[PF_6]$, from the phosphorus atom.

Three dimensional probability distributions can also be extracted from the model and the comparison of the analogous distributions to the radial distribution functions described above are shown in Figures 4 and 5. At high probabilities, the anion distributions are quite different. In the chloride the strong hydrogen bonding ability dominates over coulombic interactions and the main interaction is found with the most acidic hydrogen on the imidazolium cation, *i.e.* the H4 position, at an approximate distance of 2.65 Å. This compares with the much weaker hydrogen bonding ability of the $[PF_6]^-$ anion which interacts initially with the region of highest positive charge, *i.e.* the ring. These differences are subtle, however, and only exist at very high probability levels.

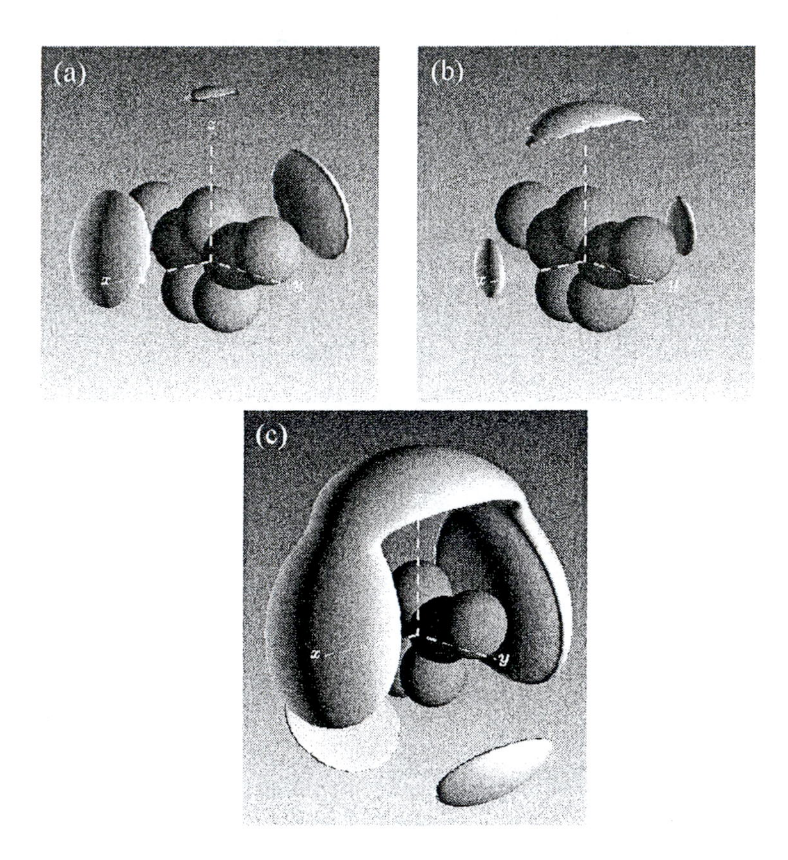

Figure 4. Probability distribution of (a) the hexafluorophosphate anion, (b) and (c) the chloride anion around a central imidazolium cation in [dmim][PF_6] and [dmim]Cl, respectively,. The distributions are the probability of finding 2.5 % ((a) and (b)) and 20% ((c)) of the ions out to a distance of 9 Å.

As the probability decreases, both anion distributions become similar and form a band around the imidazolium cation with little density associated with the methyl groups. The chloride structure at lower probability is shown in Figure 4c.

Unlike the anion-cation distributions, the cation-cation distributions, shown in Figure 5, are different for each salt. In the chloride, the cation distribution is essentially even around the central cation. In the hexafluorophosphate salt, however, strong structure is observed. If the anion and cation distributions in the hexafluorophosphate salt are compared, it is clear that the two ions effectively occupy mutually exclusive positions. This is again due to steric factors as described above. The hexafluorophosphate is too big and prevent the cations from pack tightly, in regions of high anion density whereas in the chloride salt any steric effects are significantly reduced.

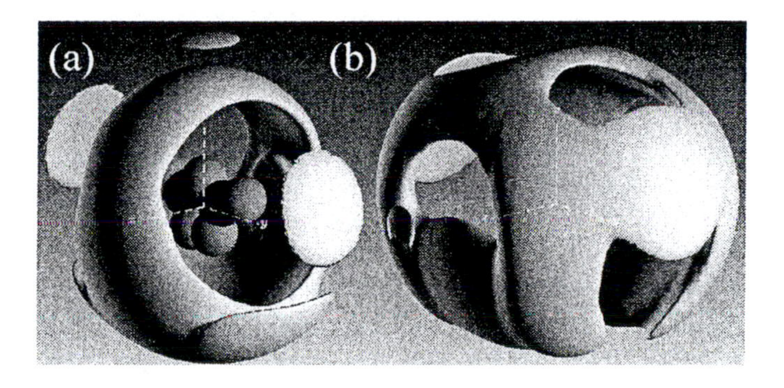

Figure 5. Probability distribution of (a)chloride and (b) hexafluorophosphate anion around an imidazolium cation derived from the EPSR model. In each case the distributions are the probability of finding 20 % of the ions out to a distance of 9 Å.

The large difference in the cation-cation distributions is also shown in the hydrogen-hydrogen intermolecular pair distributions functions. Few significant correlations are observed in either salt but in both salts the close contacts are associated with the methyl hydrogens. In the chloride salt, both methyl hydrogen-methyl hydrogen and methyl hydrogen-H4 contacts are observed whereas in the hexafluorophosphate salt only methyl hydrogen-methyl hydrogen are present. This again reflects the fact that the larger $[PF_6]^-$ anion is distributed around the ring hydrogens and prevents the ring hydrogens on cations from interacting, even with a methyl hydrogen.

Hanke *et al* have also examined the liquid structures of these salts using molecular dynamics simulations (*11*). In general, good agreement between the experimental study and the simulated data is found. The most striking difference is associated with the anion interaction with the ring. Even at very low probability, in the chloride salt for example, little anion density is found facing the ring with the density exclusively interacting with the ring hydrogens. Although at high probability levels, this is true, for the chloride, in both cases at lower probabilities significant ring-anion contact is observed. Although, strong anion-ring hydrogen interactions have been commonly reported for other imidazolium halide ionic liquids using NMR and IR (*13-15*), there is also some precedent for ring-anion interactions. Dieter *et al* (*14*) have also calculated that, in the basic melt of $[C_2mim]Cl-AlCl_3$, a structure consisting of anions sandwiched between "stacked" parallel cation rings was stable.

Liquid and Crystal Structures compared.

In both salts, the crystal structures and the structure found in the liquid are strongly correlated. For example, in the crystal structures and in the liquid, the cation-cation contacts in the chloride are dominated by the methyl hydrogen-methyl hydrogen and methyl hydrogen-H4 interactions whereas in the hexafluorophosphate only methyl hydrogen-methyl hydrogen short distances are found. Likewise, it is interesting that in both liquid and solid-state, gaps are found in the cation-cation distribution where the anions reside whereas a much more close packed arrangement is found for the chloride salt.

Conclusions

Although the general anion-cation interactions are similar in both salts, the cation-cation interactions differ strongly due to the size of the anion. This controls the structure in the crystal structure, in combination with the hydrogen bonding ability of the anion, and in the liquid structure For both chloride and hexafluorophosphate salts, the crystal and liquid structures correlate well.

Acknowledgements

The authors would like to thank DENI and QUILL (SEJMcM) for financial support and the EPSRC under grant GR/N02085 for beamtime allocation at ISIS.

References

1 Lee, Y.-C.; Price, D.L.; Curtiss, L.A.; Ratner, M.A.; Shriver, D.F. *J. Chem. Phys.*, **2001** *114*, 4591.

2 Takahashi, S.; Suzuya, K.; Kohara, S.; Koura, N.; Curtiss, L.A.; Saboungi, M.-L. *Z. Phys. Chem.* **1999**, *209*, 209.

3 Truelove, P.C.; Haworth, D.; Carlin, R.T.; Soper, A.K.; Ellison, A.J.G.; Price, D.L. *Proc. 9th Int. Symp. Molten Salts, San Fransisco* (Pennington, NJ: Electrochem. Soc.) **1994**, *3*, 50.

4 Trouw, F.R.; Price, D.L. *Ann. Rev. Phys. Chem.* **1999**, *50*, 571.

5 The Cambridge Structural Database, Version 5.23, June **2002**.

6 Arduengo, A.J.; Dias, H.V.R.; Harlow, R.L.; Kline, M. *J. Am. Chem. Soc.* **1992**, *114*, 5530.

7 Gordon, C.M.; Holbrey, J.D.; Kennedy A.R.; Seddon, K.R. *J. Mater. Chem.* **1998**, *8*, 2627.

8 a) SAINT-NT, program for data collection and data reduction, Bruker-AXS, Madison, WI, **1998**; b) SHELXTL Version 5.0, A System for Structure Solution and Refinement, Sheldrick, G. M. Bruker-AXS, Madison, WI, **1998**.

9 Hardacre, C.; Holbrey, J.D.; McMath, S.E.J. *J. Chem. Soc. Chem. Commun.* **2001**, 367.

10 *ATLAS - Analysis of Time-of-Flight Diffraction Data from Liquid and Amorphous Samples,* Soper, A.K.; Howells, W.S.; Hannon, A.C. **1989** *Rutherford Appleton Laboratory Report* RAL-89-046.

11 Hanke, C.G.; Price, S.L.; Lynden-Bell, R.M. *Mol. Phys.* **2001**, *99*, 801.

12 Soper, A.K. *Chem. Phys.* **1996**, *202*, 295; Soper, A.K. *Chem. Phys.* **2000**, *258*, 121; Soper, A.K. *Mol. Phys.* **2001**, *99*, 1503.

13 Fannin, A.A.; King, L.A.; Levisky, J.A.; Wilkes, J.S., *J. Phys. Chem.* **1984**, *88*, 2609.

14 Dieter, K.M.; Dymek, C.J.; Heimer, N.E.; Rovang, J.W.; Wilkes, J.S. *J. Am. Chem. Soc.* **1988**, *110*, 2722.

15 Dymek, C.J.; Stewart, J.J.P. *Inorg. Chem.* **1989**, *28*, 1472.

Chapter 14

Molecular Structure of Various Ionic Liquids from Gas Phase Ab Initio Calculations

Timothy I. Morrow and Edward J. Maginn*

Department of Chemical Engineering, University of Notre Dame, South Bend, IN 46554

Ab initio molecular structures are computed for the gas phase $bmim^+$ (1-butyl-3-methyl-imidazolium) ion paired with the PF_6^- (hexafluorophosphate), BF_4^- (tetrafluoroborate), NO_3^- (nitrate), $CF_3SO_3^-$ (trifluoromethanesulfate), $CF_3CO_2^-$ (trifluoromethaneacetate), $(CH_3)_3CCOO^-$ (pivalate), $CH_3C_6H_4SO_3^-$ (*p*-toluenesulfonate), $CH_3(CH_2)_2COO^-$ (butanoate), and $(CF_3SO_2)_2N^-$ (bis(trifyl)amide) anions. The ab initio calculations include optimized structures at the RHF/6-31G*, B3LYP/6-31G*, and B3LYP/6-311+G* levels as well as partial atomic charges and vibrational analyses at the B3LYP/6-311+G* level. Hydrogen bond tendencies are identified from bond frequencies.

Ionic liquids have attracted a great deal of interest as possible environmentally benign solvents for industrial processes. Studies have shown that ionic liquids do not evaporate, and that they can dissolve both polar and non-polar compounds. Many of the ionic liquids that have been studied to date have been based upon cations that contain imidazolium or pyridinium rings, and are commonly paired with the PF_6^-, BF_4^-, $CF_3SO_3^-$, and $(CF_3SO_2)_2N^-$ anions. Unfortunately for these ionic liquids, their possible ecotoxicity is an important concern which may render them undesirable for large-scale industrial use. With this concern in mind it is desirable to design new ionic liquids which will have less negative impact on the environment and will still maintain attractive physical properties, including negligible vapor pressure and low melting point.

The physical properties of a new ionic liquid can be crafted for use in a specific industrial process by careful selection of the cation, anion, and any substituents. However, without a solid understanding of the relationship between the molecular structure of an IL and its physical properties, one could spend a lifetime synthesizing different ILs and measuring their properties, and still have not reached the goal of finding an ionic liquid that can be used for a specific industrial task, will not pose a serious health hazard if released in a chemical spill, and can be produced in large quantities at a low cost. By using computer simulation techniques we can look at and analyze the microscopic structure and dynamics of ILs and begin to understand why they behave the way they do. This knowledge will provide us with a road map to the design of new, environmentally benign, task-specific ionic liquids.

The present work reports the results of single molecule gas phase quantum chemical calculations in which the 1-*n*-butyl-3-methylimidazolium cation, shown below in Figure 1, was paired with nine different anions. Six of the nine ionic liquids studied in this work have already been synthesized, while the other three are new and their physical properties are thusfar unknown.

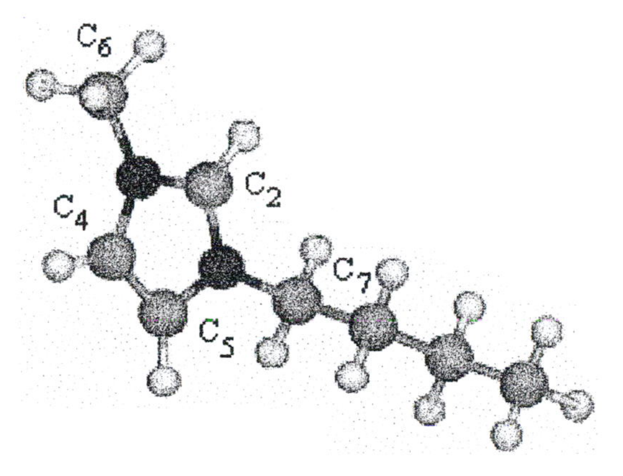

Figure 1. Atom numbering convention used for the [bmim] cation.

Methodology

The minimum energy geometry of the cation and anions studied in this work were determined by performing calculations at the B3LYP/6-311+G* level of theory. Full geometry optimization of these systems at the B3LYP/6-311+G* level is computationally difficult. To make the process tractable, geometry optimizations of the cation-anion pairs were carried out in a sequential process at the RHF/6-31G*, B3LYP/6-31G*, and B3LYP/6-311+G* levels of theory. To begin a calculation, the anion was placed well away (over 1 nm) from the cation facing the imidazolium C_2 carbon. (Figure 1 shows a definition of the cation atom labels used throughout this paper). The geometry of this cation-anion pair was then optimized at the RHF/6-31G* level. The resulting RHF/6-31G* optimized structures were used as the initial structures for subsequent B3LYP/6-31G* geometry optimizations, and these optimized structures were in turn used as initial structures for the B3LYP/6-311+G* optimizations. Partial atomic charges were derived from the ion pair geometries using the CHELPG (*1*) method. Vibrational analyses were performed only on the ion pair structures that had been optimized at the B3LYP/6-311+G* level of theory. The ab initio calculations were carried out using GAUSSIAN98 (*2*) on SGI Origin 3000 and Dell Precision 530 computers.

Results and Discussion

In this section the optimized cation-anion pair geometries are presented and the sites of relevant non-bonded interactions between the cation and anion are noted. The computed partial atomic charges of the cation, ion pair dipole moments, and computed vibrational frequencies of the cation-anion pairs are presented and discussed.

Cation-Anion Pair Geometries and Partial Atomic Charges

Previously Synthesized Ionic Liquids

The ab initio structures of the respective cation-anion pairs are virtually identical at all levels of theory. This is not surprising since the higher level calculations were started from lower level optimized structures as discussed above. The [bmim][PF_6] structures are very similar to those calculated by Meng, et. al.[3], and display non-bonded interactions between the [PF_6] and the hydrogens at the C_2, C_6 (methyl), and C_7 (butyl) positions as shown in Figure 2. The $[PF_6]$ anion is positioned slightly below the imidazolium ring. This is consistent with the molecular simulation results of Hanke, et. al.[4], and Shah, et. al.[5], whose calculations predicted anion interactions with the imidazolium cation above and below the plane of the imidazolium ring. The C--H..F interatomic distances, in Angstroms, for this ionic liquid are also shown in Figure 2. Note that the terminal methyl group of the butyl side chain is facing the $[PF_6]$ anion. The butyl side chain was arbitrarily positioned in this manner when the initial $[bmim][PF_6]$ structure was created (i.e. the positions of the atoms of the butyl chain did not greatly change during geometry optimization). This initial cation geometry, in which the butyl chain appears rotated, was used in all the calculations except for $[bmim][CF_3SO_3]$. To determine the importance of initial configuration on the final energy and butyl chain conformation, the $[bmim][PF_6]$ optimization was redone using an initial cation structure having an extended butyl chain, such as that shown in Figure 6. The optimized geometry retained its extended butyl chain. This indicates that the gas phase energies are insensitive to the conformation of the alkyl chain, although it is known that substituent groups play a major role in determining liquid phase properties. These results suggest that it is not through specific energetic interactions, but rather through excluded volume and packing effects, that substituent groups impact the properties of ionic liquids.

The magnitude of the computed dipole moment for $[bmim][PF_6]$ is 14.82 D. This large dipole moment is not surprising considering that the charges in the ion pairs are strong and are distributed over such a large area. The total charges on [bmim] and $[PF_6]$ are +.904 and -.904, respectively. This indicates that there is some charge transfer between the two ions in the liquid. The partial charges at the C_2, C_4, and C_5 sites of the imidazolium ring are 0.234, -0.010, and 0.040, repectively. These numbers are the sum of the charges on the carbon atom and its associated hydrogen atom.

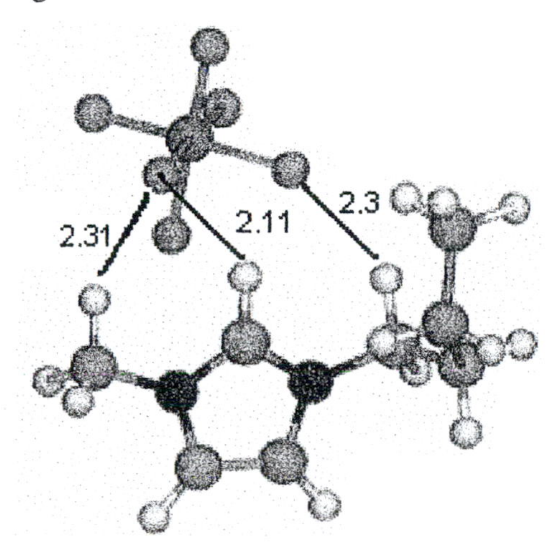

Figure 2. $[bmim][PF_6]$ structure

The ab initio structure of [bmim][BF_4] is shown in Figure 3. In this structure the [BF_4] anion is positioned slightly above the plane of the imidazolium ring, and displays non-bonded interactions between the anion and the C_2, C_6, and C_7 positions of the cation. Notice that the C--H..F interatomic distances in Figure 3 are slightly lower than those of the [bmim][PF_6] structure. This could be because the negative charge on the [BF_4] anion is delocalized over fewer F atoms than in the [PF_6] anion. This would allow for slightly stronger Coulombic interactions between the [BF_4] fluorines and the C_2 carbon site of the cation.

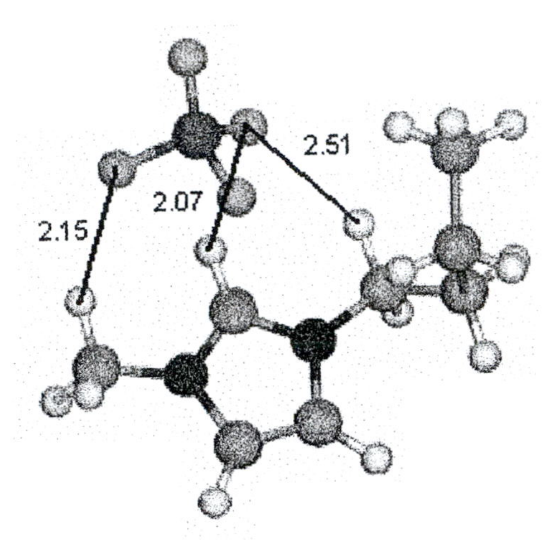

Figure 3. [bmim][BF_4] structure

The magnitude of the computed dipole moment for [bmim][BF_4] is 12.92 D. The total charges on [bmim] and [BF_4] are +.856 and -.856, respectively. The partial charges at the C_2, C_4, and C_5 sites of the imidazolium ring are 0.136, -0.004, and 0.016, repectively.

The ab initio structure of [bmim][NO_3] is shown in Figure 4. In this structure, the [NO_3] anion is aligned with the plane of the imidazolium ring, which indicates that there is little or no interaction of the π electrons with the anion. Non-bonded interactions are displayed between the anion and the C_2, C_6, and C_7 positions of the cation. The C--H..F interatomic distances for this ion pair are slightly lower than those of the [bmim][BF_4] structure, which are the result of stronger negative charges on the oxygen atoms as compared to the fluorine atoms of [BF_4]. In this particular case, the average charge on a NO_3 oxygen is –0.72, while that on the BF_4 fluorine is –0.55.

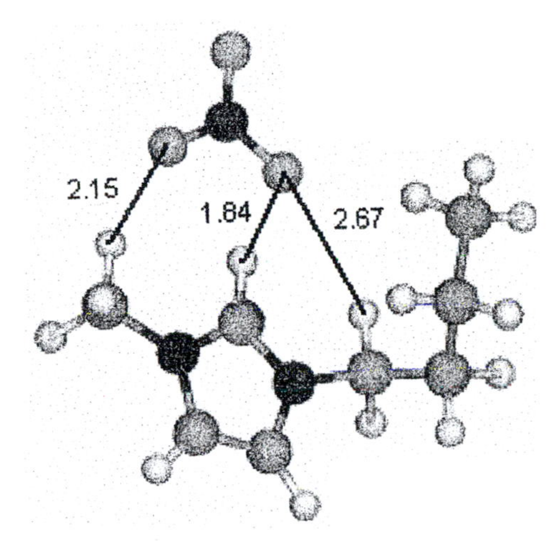

Figure 4. [bmim][NO_3] structure

The magnitude of the computed dipole moment for [bmim][NO_3] is 14.13 D. The total charges on [bmim] and [NO_3] are +.886 and -.886, respectively. The partial charges at the C_2, C_4, and C_5 sites of the imidazolium ring are 0.195, -0.008, and 0.020, repectively.

The ab initio structure of [bmim][CF_3CO_2] is shown in Figure 5. In this structure the [CF_3CO_2] anion is aligned with the plane of the imidazolium ring, and displays non-bonded interactions between the anion and the C_2, and C_6 positions of the cation. The C--H..O interatomic distances in Figure 5 are similar to those of the [bmim][NO_3] structure.

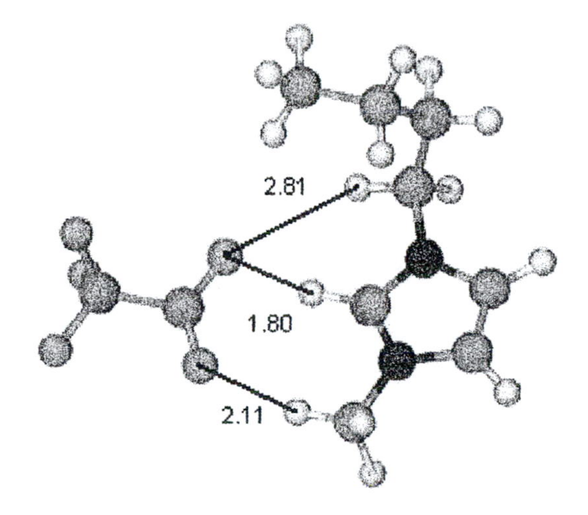

Figure 5. [bmim][CF_3CO_2] structure

The magnitude of the computed dipole moment for [bmim][CF_3CO_2] is 14.14 D. The total charges on [bmim] and [CF_3CO_2] are +.902 and -.902, respectively. The partial charges at the C_2, C_4, and C_5 sites of the imidazolium ring are 0.196, -0.007, and 0.010, repectively.

The ab initio structure of [bmim][CF_3SO_3] is shown in Figure 6. In this structure the -SO_3 group is aligned with the plane of the imidazolium ring and the –CF_3 group extends above the ring plane. Non-bonded interactions are displayed between the anion and the C_2, and C_6 positions of the cation. The C--H..O interatomic distances in Figure 6 are larger than those of the [bmim][CF_3CO_2] structure. The CF_3 group is above the imidazolium plane, while the SO_3 group aligns itself with the C_2 carbon group in a fashion almost identical to that of NO_3^-.

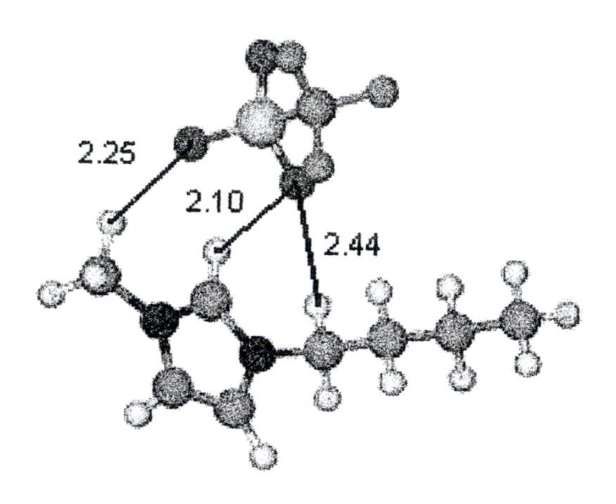

Figure 6. [bmim][CF_3SO_3] structure

The magnitude of the computed dipole moment for [bmim][CF_3SO_3] is 14.49 D. The total charges on [bmim] and [CF_3SO_3] are +.868 and -.868, respectively. The partial charges at the C_2, C_4, and C_5 sites of the imidazolium ring are 0.163, 0.027, and -0.009, repectively.

The ab initio structure of [bmim][$(CF_3SO_2)_2N$] is shown in Figure 7. In this structure the nitrogen atom of the anion is aligned with the plane of the imidazolium ring, the –SO_2 group on the lower part of the figure is slightly above the ring plane, and the –SO_2 group above the anion N atom is slightly below the ring plane. Non-bonded interactions between the anion and the C_2, C_6, and C_7 positions of the cation are displayed. The C--H..O interatomic distances in Figure 7 are slightly smaller than those of the [bmim][PF_6], [bmim][BF_4], and [bmim][CF_3SO_3] structures, and slightly larger than the [bmim][NO_3] and [bmim][CF_3CO_2] structures.

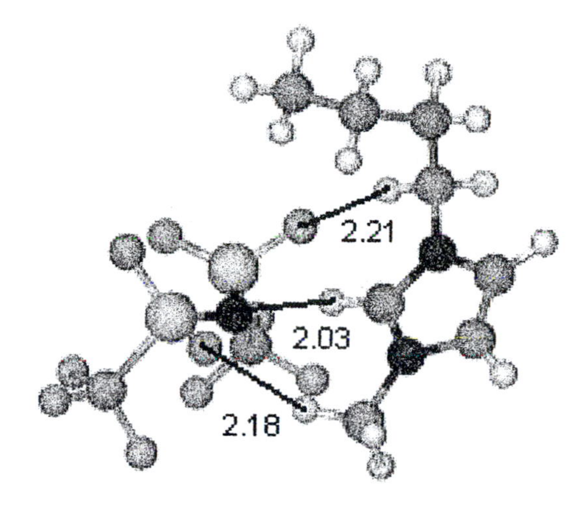

Figure 7. [bmim][$(CF_3SO_2)_2N$] structure

The magnitude of the computed dipole moment for [bmim][$(CF_3SO_2)_2N$] is 14.71 D. The total charges on [bmim] and [$(CF_3SO_2)_2N$] are +.945 and -.945, respectively. The partial charges at the C_2, C_4, and C_5 sites of the imidazolium ring are 0.245, 0.048, and 0.004, repectively.

New ionic liquid compounds

All of the structures described so far are for ionic liquid compounds that have already been synthesized and whose properties have been described in the open literature. The remaining three structures for which calculations were performed contain anions that, to our knowledge, had not yet been synthesized with the [bmim] cation prior to our peforming these calculations. Two of the compounds discussed in this section have since been made in our laboratories, as noted below. The anions for these three compounds consist of relatively bulky species derived from organic acids. By comparing the geometries and charge distributions with currently available ionic liquids, some insight into their possible performance characteristics may be determined.

The ab initio structure of the *p*-toluenesulfonate anion system, [bmim][*p*-TSA], is shown in Figure 8. In this structure the sulfur atom of the anion is aligned with the plane of the imidazolium ring, one of the oxygen atoms is also aligned with the ring plane, and the other two oxygens lie above and below the ring plane. Non-bonded interactions between the anion and the C_2, C_6, and C_7 positions of the cation are displayed. The C--H..O interatomic distances in Figure 8 are slightly higher than those of the [bmim][$(CF_3SO_2)_2N$] structure.

The magnitude of the computed dipole moment for [bmim][*p*-TSA] is 8.39 D, which is less than any of the compounds studied here. The total charges on [bmim] and [*p*-TSA] are +.904 and -.904, respectively. The partial charges at the C_2, C_4, and C_5 sites of the imidazlium ring are 0.247, 0.003, and -0.001, repectively.

The ab initio structure of [bmim] pivalate is shown in Figure 9. The $-CO_2$ group of the anion is aligned with the plane of the imidazolium ring much like the other anions that contain the $-CO_2$ and $-SO_3$ groups. Non-bonded interactions between the anion and the C_2 and C_6 positions of the cation are displayed. The C--H..O interatomic distances in Figure 9 are the shortest of all the anions studied in this work.

The magnitude of the computed dipole moment for [bmim] pivalate is 10.76 D. The total charges on [bmim] and pivalate are +.832 and -.832, respectively. The partial charges at the C_2, C_4, and C_5 sites of the imidazolium ring are 0.209, -0.015, and 0.009, repectively.

Finally, the ab initio structure of [bmim] butanoate is shown in Figure 10. As expected, this structure is very similar to that of [bmim] pivalate; the $-CO_2$ group of the anion is aligned with the plane of the imidazolium ring, and the C--H..O interatomic distances in Figure 10 are essentially identical to those of the [bmim] pivalate structure.

The magnitude of the computed dipole moment for [bmim] butanoate is 10.63 D. The total charges on [bmim] and butanoate are +.822 and -.822, respectively. The partial charges at the C_2, C_4, and C_5 sites of the imidazolium ring are 0.209, -0.015, and 0.010, repectively.

Vibrational frequencies

Figure 11 shows the computed frequency of the C_2-H bond stretch in the imidazolium ring versus the type of anion. The location of this peak in the vibrational spectrum gives insight into the relative strength of the hydrogen bonding between the cation and anion at the acidic C_2 carbon site. A peak at a lower frequency indicates that the strength of the carbon-hydrogen bond has been weakened, which could be caused by the presence of a strong hydrogen bond to another atom. Figure 11 shows that the anions containing the $-CO_2$ group appear to have the strongest hydrogen bonding ability with the cation, followed by [NO_3], [CF_3SO_3], [*p*-TSA], and lastly the two fluorinated anions [BF_4] and [PF_6]. This does not suggest that [BF_4] and [PF_6] do not hydrogen bond with [bmim], only that the other anions may bind even stronger.

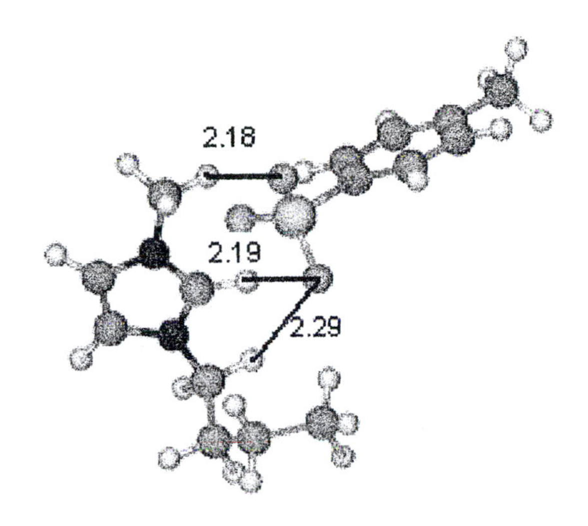

Figure 8. [bmim][p-TSA] structure

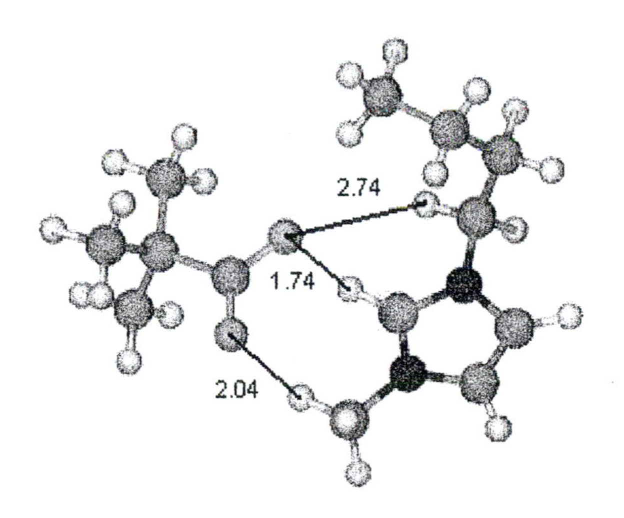

Figure 9. [bmim] pivalate

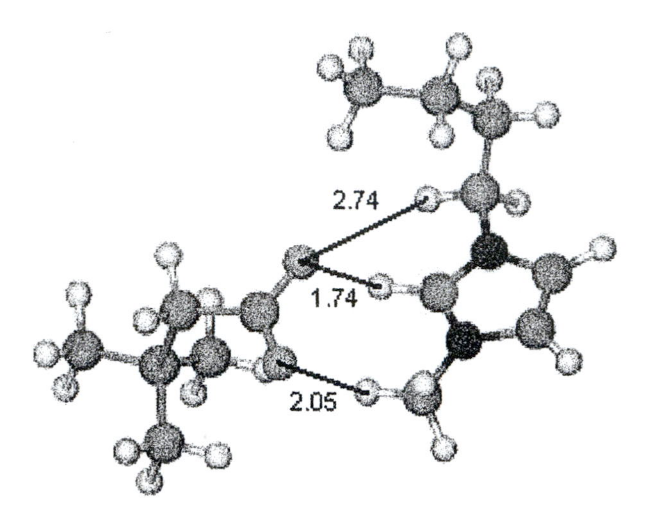

Figure 10. [bmim] butanoate

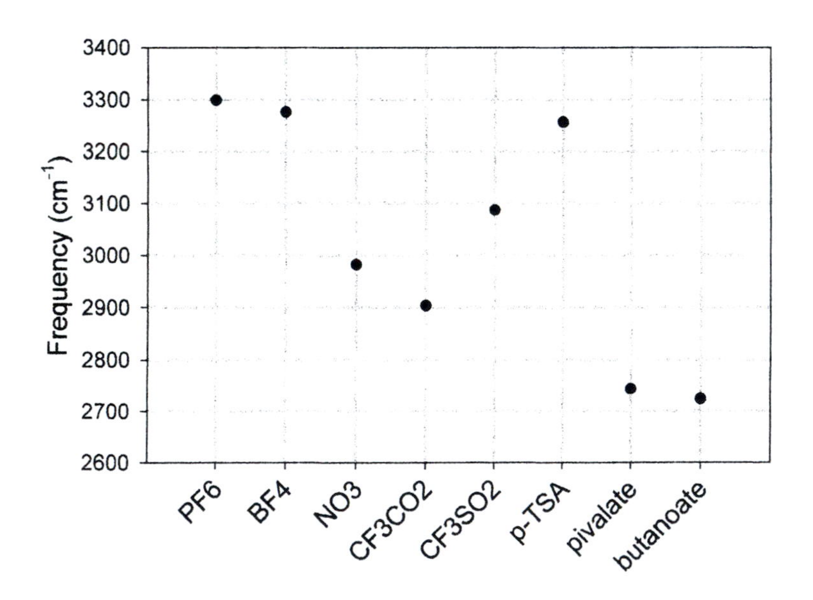

Figure 11. Computed C_2-H stretching frequencies

Conclusions

Results of gas phase density functional theory calculations on ionic liquids containing the [bmim] cation and nine different anions have been presented. Minimized structures and charge distributions were computed and compared. Bulky organic anions such as [*p*-TSA], pivalate, and butanoate tend to result in dipole moments that are lower than those found with small inorganic anions. Vibrational frequencies for the C_2-H bond indicate that hydrogen bonding is strongest with anions containing a $-CO_2^-$ group, followed by NO_3^-. The anions that showed the lowest hydrogen bonding tendency of the group are PF_6^- and BF_4^-.

Acknowledgements

Support for this work was provided by the National Science Foundation under grants CTS-9987627 and DMR-0079647. T.I.M. wishes to thank Dr. Jonas Oxgaard for many helpful discussions about density functional theory.

References

1. Brenneman, C.M.; Wiberg, K.B.; *J. Comp. Chem.*, **1990**, *11*, 361.
2. Gaussian 98 (Revision A.9), M. J. Frisch, G. W. Trucks, H. B. Schlegel, G. E. Scuseria, M. A. Robb, J. R. Cheeseman, V. G. Zakrzewski, J. A. Montgomery, Jr., R. E. Stratmann, J. C. Burant, S. Dapprich, J. M. Millam, A. D. Daniels, K. N. Kudin, M. C. Strain, O. Farkas, J. Tomasi, V. Barone, M. Cossi, R. Cammi, B. Mennucci, C. Pomelli, C. Adamo, S. Clifford, J. Ochterski, G. A. Petersson, P. Y. Ayala, Q. Cui, K. Morokuma, D. K. Malick, A. D. Rabuck, K. Raghavachari, J. B. Foresman, J. Cioslowski, J. V. Ortiz, A. G. Baboul, B. B. Stefanov, G. Liu, A. Liashenko, P. Piskorz, I. Komaromi, R. Gomperts, R. L. Martin, D. J. Fox, T. Keith, M. A. Al-Laham, C. Y. Peng, A. Nanayakkara, C. Gonzalez, M. Challacombe, P. M. W. Gill, B. G. Johnson, W. Chen, M.W. Wong, J. L. Andres, M. Head-Gordon, E. S. Replogle and J. A. Pople, Gaussian, Inc., Pittsburgh PA, 1998.
3. Meng, Z.; Dölle, A.; Carper, W. R. *J. Molec. Struct. (Theochem),* **2002**, 585, 119.
4. Hanke, C. G.; Price, S. L.; Lynden-Bell, R. M. *Mol. Phys.*, **2001**, 99, 801.
5. Shah, J. K.; Brennecke, J. F.; Maginn, E. J. *Green Chem.*, **2002**, 4, 112.

Chapter 15

Transition Structure Models of Organic Reactions in Chloroaluminate Ionic Liquids

Cyclopentadiene and Methyl Acrylate Diels–Alder Reaction in Acidic and Basic Melts of 1-Ethyl-3-methylimidazolium Chloride with Aluminum(III) Chloride

Orlando Acevedo and Jeffrey D. Evanseck*

Center for Computational Sciences and Department of Chemistry and Biochemistry, Duquesne University, 600 Forbes Avenue, Pittsburgh, PA 15282

The four-stereospecific Diels-Alder transition structures for the cyclopentadiene and methyl acrylate cycloaddition have been computed using the Becke three-parameter density functional theory with the 6-31G(d) basis set within basic and acidic ionic melt approximations. The computational model includes 1-ethyl-3-methylimidazolium cation (EMI^+) and chloroaluminates ($AlCl_4^-$ and $Al_2Cl_7^-$) in a stacked configuration which exploits lessons learned from other Lewis acid catalyzed Diels-Alder reactions. Several orientations and possible interactions have been attempted in stationary point determinations. The computed activation energies, structures and vibrations indicate that stacked models may explain experimental data. Large scale density functional computations are necessary to probe larger models including several anion and cation layers, and the inclusion of bulk phase effects.

Introduction

The observed effects of ionic liquids on chemical reactions range from weak to powerful.[1] Unfortunately, only a few systematic studies addressing the microscopic details on how ionic liquids influence chemical reactivity and selectivity have been reported.[2,3] Consequently, the molecular factors that endow room temperature ionic liquids with the ability to rate enhance and control a vast range of organic, inorganic, and enzymatic reactions in an environmentally safe manner are largely unknown.

Osteryoung, Wilkes and Hussey have developed ionic liquids that are fluid at room temperature.[4-6] The most commonly used salts are those with alkylammonium, alkylphosphonium, N-alkylpyridinium, and N,N'-dialkylimidazolium cations. Our interest is in 1-ethyl-3-methylimidazolium chloride (EMIC), and N-1-butylpyridinium chloride (BPC), shown in Scheme 1.

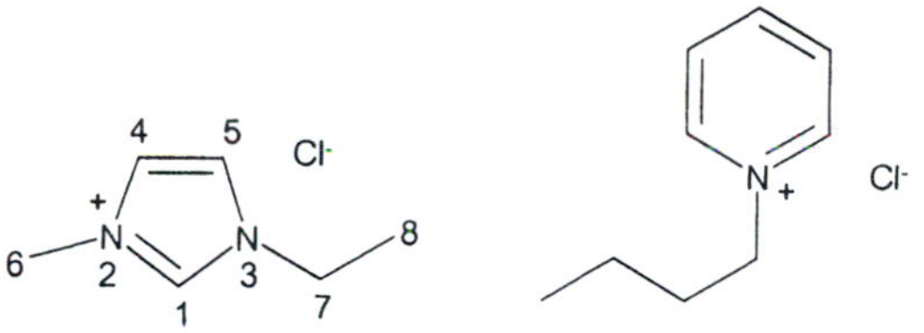

Scheme 1*. Common ionic liquids 1-ethyl-3-methylimidazolium chloride (EMIC),* ***1****, and N-1-butylpyridinium chloride (BPC),* ***2****.*

Several anions are typically used when forming an ionic liquid, such as BF_4^-, PF_6^-, ClO_4^- and of particular interest are the chloroaluminates from $AlCl_3$. An attractive property of using $AlCl_3$ is that the Lewis acidity of the melt can be varied with the composition of the liquid.[5] Raman,[6] ^{27}Al NMR[7] and mass spectra[8] all indicate that when $AlCl_3$ comprises <50% by mol of the melt in a pure ionic liquid, $AlCl_4^-$ is the only chloroaluminate(III) species present. These are known as "basic melts" because they contain $AlCl_4^-$ along with chloride ions that are not bound to aluminum. When an excess of >50% ratio of $AlCl_3$ to cation exists, this is referred to as an "acidic melt". ^{27}Al NMR[9] and negative-ion FAB mass spectra[10] have shown that $AlCl_4^-$ and $Al_2Cl_7^-$ are the principal constituents of the system.

Our approach in understanding how ionic liquids influence chemical reactions is to build upon the extensive body of recent experimental[11-16] and computational[17-22] data that yields insight into the microscopic factors at the origin of increased rate and *endo/exo* selectivities of aqueous phase Diels-Alder reactions. "Enhanced hydrogen bonding" between solvent and the transition structure relative to the initial state,[13,14] and "enforced hydrophobic interactions" as the reactant hydrophobic surface area decreases during the activation process are two major effects.[12,15,16] Several efforts have been aimed at separating and quantifying the relative contribution of each interaction type to aqueous acceleration.[11-19,22] Jorgensen and coworkers have probed enhanced hydrogen bonding effects by employing an explicit treatment of solvent using the OPLS force-field and Monte Carlo simulations.[17,22] The Gibbs energy of solvation (ΔG_{sol}) for the cyclopentadiene and methyl vinyl ketone Diels-Alder reaction has been computed.[17,20,22] Analysis revealed *ca.* 2-2.5 hydrogen bonds with the dienophile along the reaction coordinate. The polarized carbonyl group produced stronger hydrogen bonds (1-2 kcal/mol per hydrogen bond) at the transition structure. *Ab initio* calculations support the idea that the observed rate enhancements for Diels-Alder reactions arise from the hydrogen bonding effect in addition to a relatively constant hydrophobic contribution (factor of *ca.* 10 on the rate).[17,20,22] However, the special effect of water on Diels-Alder reactions beyond hydrogen bonding has been shown by the reduced rate constants in fluorinated alcohol solvents, which have stronger hydrogen bond donor capacity than water.[12,13] Therefore, factors other than hydrogen bonding are in operation. We have recently shown that local hydrogen bonding effects account for roughly one-half of the observed rate and *endo/exo* selectivity enhancements, while the remaining contributions are derived from the bulk phase.[18] In addition, we have found that novel interactions at the transition structure impact the rate and selectivity of some Diels-Alder reactions.[23]

Breslow first suggested that hydrophobic packing of the diene and dienophile is responsible for observed rate accelerations.[15,24] Subsequently, Engberts has articulated the difference between hydrophobic packing and enforced hydrophobic interactions. The term "enforced" is used to stress that hydrophobic interactions are an integral part of the activation process.[12,15,16] As with enhanced hydrogen bonding, the focus is on the relative stabilization between the initial state and the transition structure, which manifests in the available hydrophobic surface area along the reaction coordinate. In mixed solvent systems, Breslow has recently reported antihydrophobic cosolvent effects, which is a relationship between reaction rate and inaccessible hydrophobic surface area of the transition structure.[25,26] Briefly, an alcohol cosolvent does not interfere with water's ability to stabilize polarizable transition structures, rather it enhances transition structure solvation by stabilizing hydrophobic regions.[25]

Since the inception of aqueous rate accelerations and enhanced stereoselectivities of organic reactions,[27] many interpretations beyond enhanced hydrogen bonding and enforced hydrophobic effects have been advanced and recently discussed.[12] Most notably are internal pressure,[28] cohesive energy density of water,[29] micellar,[30] and medium polarity and solvophobic effects.[16,31] Even though electrostatic effects through enhanced hydrogen bonding between water and the transition structure account for an important part of the Diels-Alder rate acceleration, the remainder of enhancements by hydrophobic and other effects continue to be controversial and an active area of research.

In the same spirit as considering water as a weak Lewis acid, the plan is to examine the potential of the 1-ethyl-3-methylimidazolium cation (EMI^+) as a Lewis acid as it is modulated by the surrounding chloroaluminate counterions. This chapter focuses upon efforts to understand the local microscopic interactions of ion pairs at the transition structure that give ionic liquids the ability to impact the rates and *endo/exo* selectivities of organic reactions.

Methods

All *ab initio* and density functional theory (DFT) calculations have been performed with the GAUSSIAN 98 software package,[32] using a 16-node IBM RS/6000 super computer located at the Center for Computational Sciences (CCS) at Duquesne University. DFT includes electron correlation effects while remaining computationally feasible.[33] Compared to more sophisticated and resource demanding computational methods, DFT methods have shown to give comparable energetic and structural accuracy at a fraction of the cost. Specifically, the Becke three parameter exchange functional and nonlocal correlation functional of Lee, Yang, and Parr[33] with the 6-31G(d) basis set[34] has been shown to produce realistic structures and energetics for pericyclic reactions.[35] DFT has been used to include the effects of electron correlation in describing molten salts.[2] Recently we have shown that larger basis sets, such as 6-311G(3df), using B3LYP do not change the energetics or structures of chloroaluminate containing systems as compared to the smaller 6-31G(d) basis set.[23] Specifically, the activation enthalpies and enthalpies of reaction are in excellent agreement between the two basis sets for the isoprene and acrolein Diels-Alder reaction when catalyzed by Et_2AlCl.[23] Consequently, all energy optimizations, frequency analyses and solvation computations are carried out at the B3LYP/6-31G(d) level of theory. Vibrational frequency calculations have been used to confirm all stationary points as either minima or transition structures, provide thermodynamic and zero-point energy corrections, and reproduce the vibrational spectrum.

Results and Discussion

The B3LYP/6-31G(d) level of theory has been employed to compute the isolated ions which form in the melts ($AlCl_4^-$, $Al_2Cl_7^-$ and EMI^+) along with their 1:1 complexes ($AlCl_4^- \cdots EMI^+$ and $Al_2Cl_7^- \cdots EMI^+$). It has been determined that the binding energy for the $AlCl_4^{-}\cdots EMI^{+}$ complex is 69.9 kcal/mol, while the binding energy of the acidic melt, $Al_2Cl_7^{-}\cdots EMI^{+}$ is 63.9 kcal/mol, as shown in Table I. The computed B3LYP/6-31G(d) energetics compare well to results previously reported using the HF method and the 6-31G(d) basis set.[3] We have also determined the $Al_3Cl_{10}^{-}\cdots EMI^{+}$complex, found in highly acidic melts.

Table I. Binding enthalpies of ionic liquid complexes using the HF/ 6-31G(d) and B3LYP/6-31G(d) levels of theory (binding enthalpies in kcal/mol).

HF	*Binding Enthalpy*
$AlCl_4^{-}\ldots EMI^{+}$	68.0[a]
$Al_2Cl_7^{-}\ldots EMI^{+}$	63.2[a]
B3LYP	
$AlCl_4^{-}\ldots EMI^{+}$	69.9
$Al_2Cl_7^{-}\ldots EMI^{+}$	63.9
$Al_3Cl_{10}^{-}\ldots EMI^{+}$	60.5

[a] Taken from ref. 3.

The B3LYP/6-31G(d) level of theory predicts a slightly greater binding energy for both the $AlCl_4^{-}\cdots EMI^{+}$ and $Al_2Cl_7^{-}\cdots EMI^{+}$ complexes as compared to HF/6-31G(d). The trend exists for both levels of theory, where the binding energy decreases significantly as the chloroaluminate ion becomes larger. The results are important, since as the binding energy decreases between EMI^{+} and the chloroaluminate ion, the more likely a Lewis acid adduct with the dienophile will form. This has been a serious topic of debate, especially for the more acidic forms of the chloroaluminates.[36] In this study, the assumption is that the ion pairs between $AlCl_4^{-}\cdots EMI^{+}$ (basic melt model) and $Al_2Cl_7^{-}\cdots EMI^{+}$ (acidic melt model) stay in tact, where adduct formation does not take place. In collaboration with Dan Singleton, on-going research is aimed at addressing the possibility of adduct formation through a combined experimental and computational effort in the determination of secondary kinetic isotope effects.[23,37]

B3LYP/6-31G(d) calculations have been performed on the anions in the molten chloroaluminate salts to determine their geometry. The geometry of the anion, $AlCl_4^{-}$, is confirmed to be tetrahedral in accordance with x-ray crystallographic results.[38] The geometry of $Al_2Cl_7^{-}$ is found to be C_2 symmetric confirmed by the observed IR spectrum.[39] The computed structures are consistent with results presented earlier from MNDO calculations.[39]

An x-ray crystal structure of the basic melt involving EMIC and $AlCl_3$, where the mole fraction of $AlCl_3$ compared to EMIC was less that 0.5, has been previously reported.[40] The geometry has been compared to HF/6-31G(d) calculations by Takahashi and coworkers.[41] The bond lengths and angles for EMI^+ are presented in Table II, along with our calculations using the B3LYP/6-31G(d) and second-order MP2/6-31G(d) levels of theory. In general the comparison between experimental and both levels of theory is good. However, there seems to be an unusual distortion at the N_3 ethyl group in the crystal structure. The crystallographic C_7C_8 distance is shorter than that expected for a single bond, the N_3C_7 length is greater than the typical nitrogen-carbon single bond, and the $N_3C_7C_8$ angle is compressed from the usual sp^3 tetrahedral value, as shown in Table II. Crystal packing forces could be in operation.

Table II. Comparisons of x-ray and theoretical data for EMI^{+a}.

Geometry	*x-ray^b*	*HF*	*B3LYP*	*MP2*
C_1N_2	1.30	1.32	1.34	1.34
C_1N_3	1.27	1.31	1.34	1.34
N_2C_4	1.39	1.38	1.38	1.37
N_3C_5	1.41	1.38	1.38	1.37
C_4C_5	1.38	1.34	1.36	1.37
N_2C_6	1.48	1.47	1.47	1.46
N_3C_7	1.55	1.48	1.48	1.47
C_7C_8	1.43	1.52	1.52	1.52
$\angle N_2C_1N_3$	110	110	109	108
$\angle C_1N_2C_4$	111	108	108	109
$\angle C_1N_3C_5$	109	108	108	109
$\angle N_2C_4C_5$	104	107	107	107
$\angle N_3C_5C_4$	106	107	107	107
$\angle C_1N_2C_6$	126	126	126	126
$\angle C_1N_3C_7$	130	126	126	126
$\angle N_3C_7C_8$	104	112	112	111

[a] Bond distances (Å) and bond angles (°) defined in Scheme 1. HF/6-31G(d) total energy of EMI^+ is -342.31340 hartrees, B3LYP/6-31G(d) total energy is -344.5497092 hartrees, MP2/6-31G(d) total energy is -343.4048579. [b] Taken from ref. 6.

An infrared study of two separate ionic liquids (EMIC and BPC) was carried out by Osteryoung, which can be used to compare specific vibrational data with computed values.[42] A frequency analysis has been carried out and reported using the semi-empirical method AM1.[43] The vibrations were later investigated using MNDO and HF/6-31G(d) on the individual ions.[39,44] We carried out the frequency analysis on the EMI^+ ion and the acidic and basic melt models at B3LYP/6-31G(d) and compared specific stretches to experimental data in Table III. The scaled vibrational frequencies are within 2% of experimental frequencies of vibration. The B3LYP/6-31G(d) level of theory, in agreement with previous structural, vibrational and energetic ionic melt studies, is an appropriate level of theory to describe chloroaluminate ionic liquids.

Table III. Comparison of peak frequency (cm^{-1}) changes occurring with melt acidity, calculated at the B3LYP/6-31G(d) level of theory.

melt	*aromatic C-H str*	*aliphatic C-H str*	*-C=N- str*	*ring str*	*ring i/p b*	*ring o/p b*
EMI^{+} [a]	3183	3039	1550	1139	794	717
acidic						
$EMI\text{-}Al_2Cl_7$ [a]	3176	2970	1547	1140	837	710
1.50:1 melt[b]	3161	2992	1595	1169	836	747
basic						
$EMI\text{-}AlCl_4$ [a]	3152	2971	1549	1140	867	712
0.55:1 melt[b]	3154	2983	1590	1175	838	758

[a] Scaled by 0.9613.[45] [b]Experimental data taken from ref. 42.

The influence of solvent on the *endo/exo* selectivity and reaction rate of the reaction of cyclopentadiene (**3**) and methyl acrylate (**4**) to give the *endo* (**5**) and *exo* (**6**) bicyclic products (Scheme 2) has been recently reported in a range of molecular solvents, including different ionic liquids.

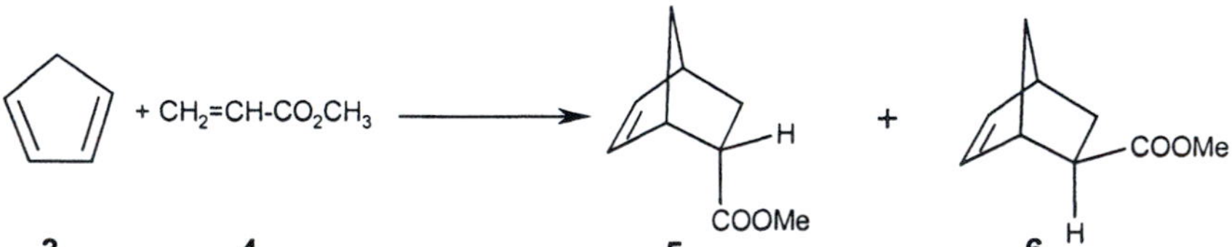

3 **4** **5** **6**

Scheme 2*. Diels-Alder reaction between cyclopentadiene (3) and methyl acrylate (4) to give the endo (5) and exo (6) bicyclic products.*

As shown by Carlos W. Lee, when an acidic melt (51% $AlCl_3$) of EMIC was used as the solvent for the equimolar reaction of **3** and **4**, the rate of reaction is 10 times faster than in water and 175 faster than in ethyl ammonium nitrate.[46] The results are significant, since the reaction rate is already 20 times faster in water as compared to nonpolar solvents.[47] However, when the basic melt (48% $AlCl_3$) of EMIC was used as the solvent, Lee found that the rate was significantly slower. Water has a reaction rate 2.4 times faster that the basic melt. *Endo* selectivity is enhanced with good yield when using the EMIC ionic liquid. A 6.7:1 *endo* to *exo* ratio is observed for the solvent ethyl ammonium nitrate

(EAN), the basic ionic liquid melt has a ratio of 5.25:1, while the acidic melt ionic liquid EMIC has a ratio of 19:1. In nonpolar solvents, *endo/exo* selectivity is 2:1.[46,48]

Four possible reaction pathways are possible for the Diels-Alder reaction between cyclopentadiene and methyl acrylate. Consistent with previous conventions, the computed four-stereospecific transition structures are denoted as NC (*endo*, *s-cis* methyl acrylate), XC (*exo*, *s-cis* methyl acrylate), NT (*endo*, *s-trans* methyl acrylate) and XT (*exo*, *s-trans* methyl acrylate), as shown in Figure 1.

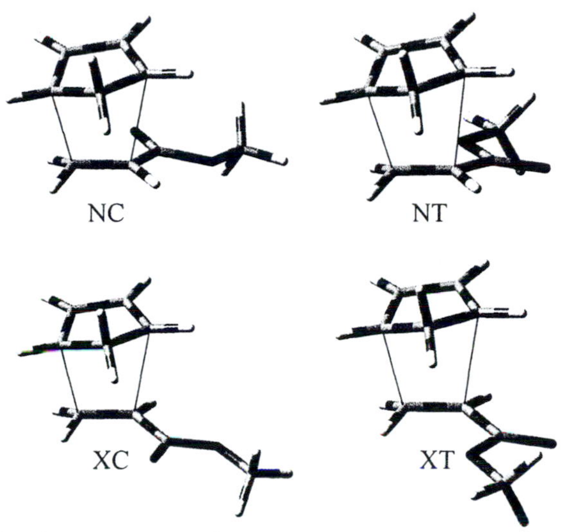

***Figure 1**. The four stereospecific transition structures optimized using the B3LYP/6-31G(d) level of theory*

The computed activation energies, enthalpies and Gibbs energies in vacuum are presented in Table IV. The computed B3LYP/6-31G(d) activation enthalpy is 16.9 kcal/mol in vacuum at 298 K for the parent reaction. The experimental activation enthalpy and entropy is 15.1 kcal/mol and –29.7 cal/mol-K when determined in toluene.[47] To provide a rough comparison, we have previously computed that the activation enthalpy drops by 1.5 kcal/mol when comparing the barriers from vacuum and in benzene (similar to toluene) for the similar Diels-Alder reaction of acrolein and butadiene.[18] Consequently, our computed activation enthalpy is in reasonable agreement (16.9 – 1.5 = 15.4 kcal/mol) with the experimental value of 15.1 kcal/mol. Furthermore, we expect an additional drop of 4.9 kcal/mol in the activation enthalpy when the reaction is carried out in a methanol/water mixture from benzene.[18] The observed activation enthalpy and entropy for the cyclopentadiene and methyl acrylate reaction is 10.2 kcal/mol

Table IV. Computed thermodynamic activation parameters for the parent Diels-Alder reaction from the B3LYP/6-31G(d) level of theory.

TS	$\Delta E^{\ddagger}_{0}$	$\Delta E^{\ddagger}_{298}$	$\Delta H^{\ddagger}_{298}$	$\Delta G^{\ddagger}_{298}$
NC	17.7	17.5	16.9	30.6
NT	19.2	19.0	18.4	32.2
XC	17.3	17.2	16.6	30.3
XT	19.3	19.1	18.5	32.2

$\Delta E^{\ddagger}_{o}$ is the activation energy given by the electronic energy corrected by the zero-point energy at zero K. $\Delta E^{\ddagger}_{298}$, $\Delta H^{\ddagger}_{298}$ and $\Delta G^{\ddagger}_{298}$ are the activation energy, activation enthalpy and activation free energy at 298 K, respectively.

and –40.9 cal/mol-K, respectively, in a methanol/water mixture, which is in excellent agreement with the B3LYP/6-31G(d) value of (16.9 – 1.5 – 4.9 = 10. 5 kcal/mol).[47]

The most recent experimental work by Lee suggests that basic melts suppress the rate and *endo* selectivity of the cyclopentadiene and methyl acrylate Diels-Alder reaction, while the acidic melt enhances both.[46] Consequently, it is necessary to worry about the principal constituents of the basic system, $[EMI]^{+}[AlCl_4]^{-}$ and $[EMI]^{+}Cl^{-}$, and of the acidic system, $[EMI]^{+}[AlCl_4]^{-}$ and $[EMI]^{+}[Al_2Cl_7]^{-}$. A full density functional theory treatment on a box of ionic species is currently beyond the limit of computer resources available. Consequently for this study, three levels of approximation are made to understand how ionic liquids impact chemical reactivity and stereoselectivity. First, only the EMI^{+} cation is used to activate the dienophile as a Lewis acid. This represents an unrealistic situation, since the positive charge is not diminished by surrounding counter anions and it does not serve to differentiate between the enhanced and suppressed chemical reactivity observed by the acidic and basic experiments. For both the second and third approximations, the assumption is that the ion pairs between $AlCl_4^{-}\cdots EMI^{+}$ (basic melt model) and $Al_2Cl_7^{-}\cdots EMI^{+}$ (acidic melt model) stay in tact, where adduct formation does not take place. These models are still rudimentary, yet form the basis of this current work and future simulations which require extensively more resources.

NC transition structures have been located for the cyclopentadiene and methyl acrylate Diels-Alder reaction in four different configurations with the EMI^{+} cation, as shown in Figure 2. The configurations were selected to orient the acidic imidazolium proton towards the carbonyl oxygen of methyl acrylate (**8** and **9**), or away from methyl acrylate (**7** and **10**). Additionally, the computations resulted with the carbonyl oxygen forming a weaker second interaction with either the methyl proton (**7** and **8**) or the proton from the ethyl group (**9** and **10**).

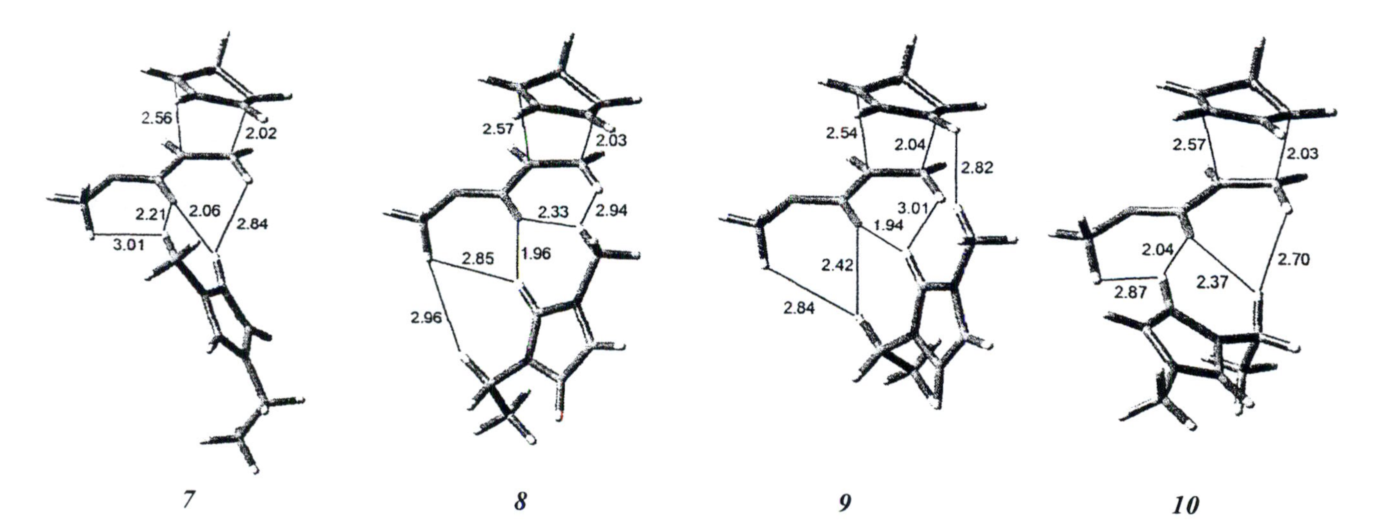

Figure 2. *NC transition structures with four different configurations of the EMI^+ cation optimized using the B3LYP/6-31G(d) level of theory.*

The four transition structures are concerted and asynchronous, as indicated by the lengths of the breaking and forming bonds (~2.03 and ~2.56 Å). The four transition structures are similar, where the breaking and forming bonds are within 0.03 A. However, the interaction between EMI^+ and the transition structure varies. As expected, the orientations with the acidic imidazolium proton towards the carbonyl oxygen of methyl acrylate (**8** and **9**) give a stronger energy of interaction by roughly 2 kcal/mol, and result with the key carbonyl oxygen and imidazolium proton interaction distance being about 0.1 Å closer. The favored configuration with this model is to have the acidic imidazolium proton point towards the carbonyl oxygen of methyl acrylate and have a second interaction form with a methyl hydrogen. The results match that expected from chemical intuition, however this analysis is an oversimplification of the environment found in an ionic liquid. In this explicit ionic liquid cation model, the activation energies in Table V are computed by subtracting the isolated cyclopentadiene and the hydrogen bonded methyl acrylate and ion complex energies from the total transition structure energy.

In calculating activation energies for the transition structure of cyclopentadiene and methyl acrylate in a basic and acidic environment, as shown in Figure 3, two different approaches were used. The first method is identical to reports, where the activation energies are computed by subtracting the isolated cyclopentadiene and the hydrogen bonded methyl acrylate and ion complex energies from the total transition structure energy. The second method is different, where the isolated cyclopentadiene, isolated methyl acrylate, and the ion complex energies are subtracted from the total transition structure energy.

For a weak Lewis acid, such as water, the first method delivers a realistic picture of the energetic behavior.[18] However, using this method in a truncated ionic environment leads to binding energies that are superficially large because the environment does not partially quench the full ionic charges on EMI^+ or the chloroaluminate. This enhanced coordination results in larger activation enthalpies for both the acidic (13.9 kcal/mol) and basic melts (14.7 kcal/mol) than expected. Using the second method, the computed lowering of the activation enthalpies of 9.1 and 8.1 kcal/mol for the basic and acidic melts, respectively, leads to gross overestimations of the impact on the activation barrier lowering.

The experimental activation barrier lowering is between the two extremes of 9.1 to 14.7 kcal/mol for the basic melt and 8.1 to 13.9 kcal/mol for the acidic melt activation enthalpies as seen in Table VI. The relative difference between basic and acidic melts follows the same general trend as experiment. Where the acidic melt is lower in energy, in this case by 1 kcal/mol, than the basic melt.

A single kcal/mol is not the energy difference measured experimentally and consequently exhibits a weakness in our model. This suggests that further bulk phase or additional ion participation is necessary to explain the experimental phenomenon. Since we compute the correct experimental trend, we conclude that local effects impact the rate enhancement of Diels-Alder reactions, but that other

Table V. Computed activation energies for EMI^+ and the four stereospecific transition structures from the B3LYP/6-31G(d) level of theory.

Structure	$\Delta E^{\ddagger}_{0}$	$\Delta E^{\ddagger}_{298}$	$\Delta H^{\ddagger}_{298}$	$\Delta G^{\ddagger}_{298}$
7				
NC	17.72	17.59	16.96	31.59
NT[a]	17.20	16.42	15.79	33.33
XC	17.89	17.74	17.10	32.25
XT[a]	17.62	16.87	16.23	33.47
8				
NC[a]	15.52	14.72	14.09	31.99
NT	14.61	14.39	13.75	29.25
XC	15.40	15.24	14.61	29.69
XT	15.11	14.90	14.26	29.37
9				
NC	15.98	15.92	15.28	30.21
NT	14.79	14.70	14.07	28.19
XC	15.61	15.54	14.90	29.63
XT	15.39	15.26	14.63	29.68
10				
NC[a]	17.65	16.99	16.35	33.54
NT	17.42	17.38	16.74	31.13
XC	18.08	18.05	17.41	32.10
XT[a]	17.73	17.05	16.41	33.64

[a] 2-imaginary frequencies. $\Delta E^{\ddagger}_{o}$ is the activation energy given by the electronic energy corrected by the zero-point energy at zero K. $\Delta E^{\ddagger}_{298}$, $\Delta H^{\ddagger}_{298}$ and $\Delta G^{\ddagger}_{298}$ are the activation energy, activation enthalpy and activation free energy at 298 K, respectively.

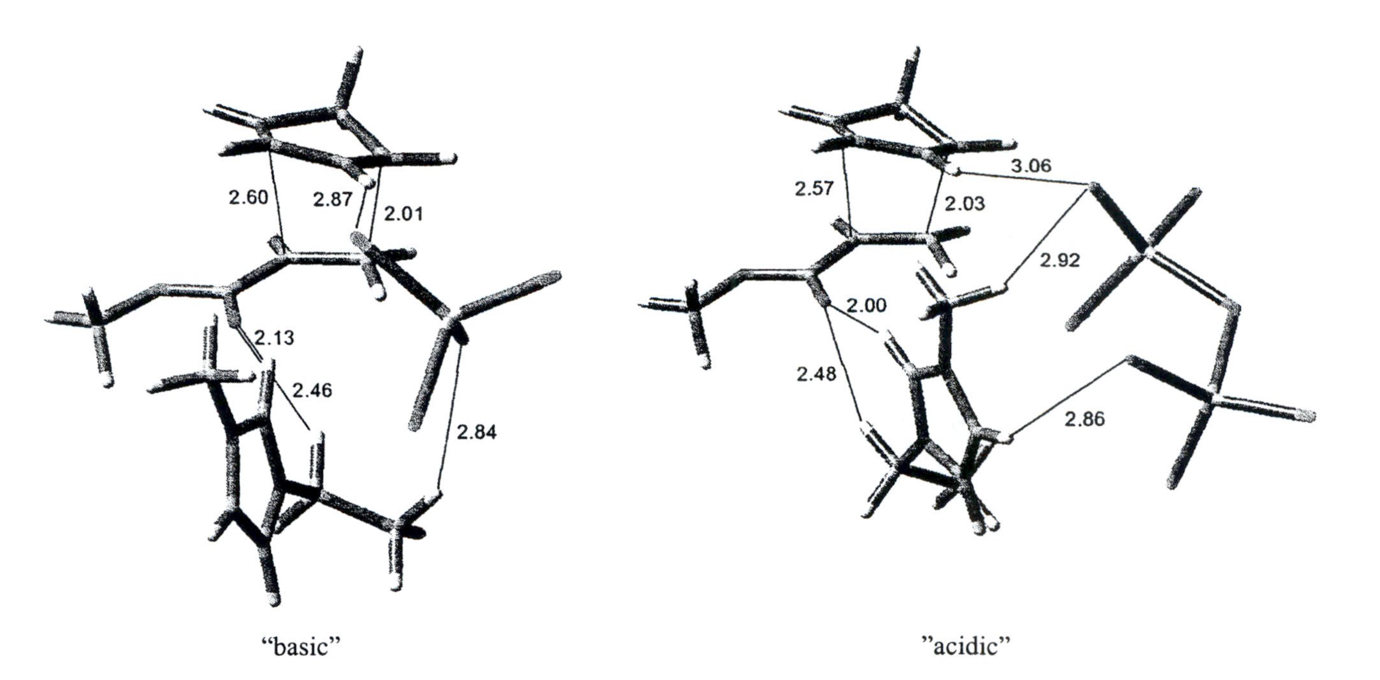

Figure 3. *The NC transition structures for the cyclopentadiene and methyl acrylate with a basic and acidic approximated environment computed using the B3LYP/6-31G(d) level of theory.*

Table VI. Comparison of computed thermodynamic activation parameters from the B3LYP/6-31G(d) level of theory.

TS	$\Delta E^{\ddagger}_{o}$	$\Delta E^{\ddagger}_{298}$	$\Delta H^{\ddagger}_{298}$	$\Delta G^{\ddagger}_{298}$
NC				
TS+EMI^{+}+AlCl$_4^-$	15.4/9.0	15.3/10.3	14.7/9.1	29.1/32.4
TS+EMI^{+}+Al$_2$Cl$_7^-$	14.6/8.1	14.5/9.3	13.9/8.1	28.9/31.9

$\Delta E^{\ddagger}_{o}$ is the activation energy given by the electronic energy corrected by the zero-point energy at zero K. $\Delta E^{\ddagger}_{298}$, $\Delta H^{\ddagger}_{298}$ and $\Delta G^{\ddagger}_{298}$ are the activation energy, activation enthalpy and activation free energy at 298 K, respectively.

unexplored effects make significant contributions. These other microscopic effects are the topics of our ongoing research efforts.

References

(1) (a) Welton, T. *Chem. Rev.* **1999**, *99*, 2071; (b) Seddon, K. R. *J. Chem. Tech. Biotechnol.* **1997**, *68*, 351; (c) Freemantle, M. In *C&EN*, 2001, pp 21.

(2) (a) Hébant, P.; Picard, G. *J. Mol. Struc. (Theochem)* **1995**, *358*, 39; (b) Picard, G.; Bouyer, F. C.; Leroy, M.; Bertaud, Y.; Bouvet, S. *J. Mol. Struc. (Theochem)* **1996**, *368*, 67; (c) Ribeiro, M. C. C.; Almeida, L. C. J. *J. Chem. Phys.* **2000**, *113*, 4722; (d) Bock, C. W.; Trachtman, M.; Mains, G. J. *J. Phys. Chem.* **1994**, *98*, 478.

(3) Takahashi, S.; Suzuya, K.; Kohara, S.; Koura, N.; Curtiss, L. A.; Saboungi, M. L. *Z. Phys. Chem.* **1999**, 209.

(4) Wilkes, J. S.; Leveisky, J. A.; Wilson, R. A.; Hussey, C. L. *Inorg. Chem.* **1982**, *21*, 1263.

(5) Hussey, C. L. *Pure & Appl. Chem.* **1988**, *60*, 1763.

(6) Gale, R. J.; Gilbert, B. P.; Osteryoung, R. A. *Inorg. Chem.* **1978**, *17*, 2728.

(7) Wilkes, J. S.; Frye, J. S.; Reynolds, G. F. *Inorg. Chem.* **1983**, *22*, 3870.

(8) (a) Ackermann, B. L.; Tsarbopoulos, A.; Allison, J. *Anal. Chem.* **1985**, *57*, 1766; (b) Wicelinski, S. P.; Gale, R. J.; Pamidimukkala, K. M.; Laine, R. A. *Anal. Chem.* **1988**, *60*, 2228.

(9) Gray, J. L.; Maciel, G. E. *J. Am. Chem. Soc.* **1981**, *103*, 7147.

(10) Franzen, G.; Gilbert, B. P.; Pelzer, G.; Depauw, E. *Org. Mass Spectrom.* **1986**, *21*, 443.

(11) Garner, P. P. In *Organic Synthesis in Water*; Grieco, P. A., Ed.; Blackie Academic and Professional: London, 1998.

(12) Meijer, A.; Otto, S.; Engberts, J. B. F. N. *J. Org. Chem.* **1998**, *63*, 8989.

(13) Wijnen, J. W.; Engberts, J. *Journal of Organic Chemistry* **1997**, *62*, 2039.
(14) (a) van der Wel, G. K.; Wijnen, J. W.; Engberts, J. B. F. N. *J. Org. Chem.* **1996**, *61*, 9001; (b) Engberts, J. B. F. N. *Pure & Appl. Chem.* **1995**, *67*, 823; (c) Otto, S.; Blokzijl, W.; Engberts, J. B. F. N. *J. Org. Chem.* **1994**, *59*, 5372; (d) Blokzijl, W.; Engberts, J. B. F. N. In *Structure and Reactivity in Aqueous Solution*; Cramer, C. J., Truhlar, D. G., Eds.; American Chemical Society: Washington, DC, 1994; Vol. 568, p 303; (e) Blokzijl, W.; Engberts, J. B. F. N. *J. Am. Chem. Soc.* **1992**, *114*, 5440.
(15) Blokzijl, W; Engberts, J. B. F. N. *Angew. Chem. Int. Ed. Engl.* **1993**, *32*, 1545.
(16) Blokzijl, W.; Blandamer, M. J.; Engberts, J. B. F. N. *J. Am. Chem. Soc.* **1991**, *113*, 4241.
(17) Chandrasekhar, J.; Shariffskul, S.; Jorgensen, W. L. *J. Phys. Chem. B* **2002**, *106*, 8078.
(18) Kong, S.; Evanseck, J. D. *J. Am. Chem. Soc.* **2000**, *122*, 10418.
(19) (a) Furlani, T. R.; Gao, J. *J. Org. Chem.* **1996**, *61*, 5492; (b) Jorgensen, W. L.; Blake, J. F.; Lim, D.; Severance, D. L. *J. Chem. Soc., Farady Trans.* **1994**, *90*, 1727.
(20) Blake, J. F.; Lim, D.; Jorgensen, W. L. *J. Org. Chem.* **1994**, *59*, 803.
(21) Jorgensen, W. L.; Lim, D.; Blake, J. F. *J. Am. Chem. Soc.* **1993**, *115*, 2936.
(22) Blake, J. F.; Jorgensen, W. L. *J. Am. Chem. Soc.* **1991**, *113*, 7430.
(23) Acevedo, O.; Evanseck, J. D. *Org. Lett.* **2003**, *(in press)*.
(24) (a) Breslow, R. *Acc. Chem. Res.* **1991**, *24*, 159; (b) Breslow, R.; Rizzo, C. J. *J. Am. Chem. Soc.* **1991**, *113*, 4340; (c) Breslow, R.; Guo, T. *J. Am. Chem. Soc.* **1988**, *110*, 5613; (d) Breslow, R.; Rideout, D. C. *J. Am. Chem. Soc.* **1980**, *102*, 7816; (e) Breslow, R.; Maitra, U. *Tetrahedron Lett.* **1984**, *25*, 1239; (f) Breslow, R.; Maitra, U.; Rideout, D. *Tetrahedron Lett.* **1983**, *24*, 1901; (g) Hunt, I.; Johnson, C. D. *J. Chem. Soc., Perkin Trans. 2* **1991**, 1051.
(25) (a) Breslow, R.; Groves, K.; Mayer, M. U. *Org. Lett.* **1999**, *1*, 117; (b) Breslow, R.; Connors, R.; Zhu, Z. *Pure Appl. Chem.* **1996**, *68*, 1527; (c) Breslow, R.; Zhu, Z. *J. Am. Chem. Soc.* **1995**, *117*, 9923.
(26) Breslow, R. *Struct. React. Aq. Sol.* **1994**, *568*, 291.
(27) Rideout, D. C.; Breslow, R. *J. Am. Chem. Soc.* **1980**, *102*, 7816.
(28) (a) Grieco, P. A.; Nunes, J. J.; Gaul, M. D. *J. Am. Chem. Soc.* **1990**, *112*, 4595; (b) Lubineau, A. *J. Org. Chem.* **1986**, *51*, 2142; (c) Lubineau, A.; Quenau, Y. *J. Org. Chem.* **1987**, *52*, 1001; (d) Grieco, P. A.; Brandes, E. B.; McCann, S.; Clark, J. D. *J. Org. Chem.* **1989**, *54*, 5849; (e) Dack, M. R. J. *Chem. Soc. Rev.* **1975**, *4*, 211.
(29) (a) Gajewski, J. J.; Peterson, K. B.; Kagel, J. R. *J. Am. Chem. Soc.* **1992**, *109*, 5545; (b) Grieco, P. A. *Tetrahedron Lett.* **1986**, *27*, 1975.
(30) (a) Otto, S.; Engberts, J.; Kwak, J. C. T. *Journal of the American Chemical Society* **1998**, *120*, 9517; (b) Yoshida, K.; Garner, P. *J. Org. Chem.* **1983**, *48*, 3137; (c) Grieco, P. A.; Garner, P.; Zhen-min, H. *Tetrahedron Lett.* **1983**, *24*, 1897; (d) Grieco, P. A. *J. Org. Chem.* **1983**, *48*, 3137.

(31) (a) Cativiela, C.; García, J. I.; Gil, J.; Martínez, R. M.; Mayoral, J. A.; Salvatella, L.; Urieta, J. S.; Mainar, A. M.; Abraham, M. H. *J. Chem. Soc., Perkin Trans. 2* **1997**, 653; (b) Cativiela, C.; García, J. I.; Mayoral, J. A.; Royo, A. J.; Salvatella, L.; Assfeld, X.; Ruiz-lopez, M. F. *J. Phys. Org. Chem.* **1992**, *5*, 230; (c) Cativiela, C.; García, J. I.; Mayoral, J. A.; Avenoza, A.; Peregrina, J. M.; Roy, M. A. *J. Phys. Org. Chem.* **1991**, *4*, 48; (d) Sangwan, N. K.; Schneider, H.-J. *J. Chem. Soc., Perkin Trans. 2* **1989**, 1223; (e) Schneider, H.-J.; Sangwan, N. K. *Angew. Chem. Int. Ed. Engl.* **1987**, *26*, 896.

(32) *Gaussian 98* (Revision A.3), Frisch, M. J.; Trucks, G. W.; Schlegel, H. B.; Scuseria, G. E.; Robb, M. A.; Cheeseman, J. R.; Zakrzewski, V. G.; Montgomery Jr., J. A.; Stratmann, R. E.; Burant, J. C.; Dapprich, S.; Millam, J. M.; Daniels, A. D.; Kudin, K. N.; Strain, M. C.; Farkas, O.; Tomasi, J.; Barone, V.; Cossi, M.; Cammi, R.; Mennucci, B.; Pomelli, C.; Adamo, C.; Clifford, S.; Ochterski, J.; Petersson, G. A.; Ayala, P. Y.; Cui, Q.; Morokuma, K.; Malick, D. K.; Rabuck, A. D.; Raghavachari, K.; Foresman, J. B.; Cioslowski, J.; Ortiz, J. V.; Stefanov, B. B.; Liu, G.; Liashenko, A.; Piskorz, P.; Komaromi, I.; Gomperts, R.; Martin, R. L.; Fox, D. J.; Keith, T.; Al-Laham, M. A.; Peng, C. Y.; Nanayakkara, A.; Gonzalez, C.; Challacombe, M.; Gill, P. M. W.; Johnson, B.; Chen, W.; Wong, M. W.; Andres, J. L.; Gonzalez, C.; Head-Gordon, M.; Replogle, E. S.; Pople, J. A., Gaussian, Inc., Pittsburgh PA, 1998.

(33) Lee, C.; Yang, W.; Parr, R. G. *Phys. Rev.* **1988**, *37*, 785.

(34) Petersson, G. A.; Al-Laham, M. A. *J. Chem. Phys.* **1991**, *94*, 6081.

(35) (a) Wiest, O.; Montiel, D. C.; Houk, K. N. *J. Phys. Chem. A* **1997**, *101*, 8378; (b) Froese, R. D. J.; Humbel, S.; Svensson, M.; Morokuma, K. *J. Phys. Chem. A* **1997**, *101*, 227.

(36) Personal *communication with Tom Welton*.

(37) (a) Collaboration *with Dan Singleton*; (b) Singleton, D. A.; Merrigan, S. R.; Beno, B. R.; Houk, K. N. *Tetrahedron Lett.* **1999**, *40*, 5817.

(38) Couch, T. W.; Lokken, D. A.; Corbett, J. D. *Inorg. Chem.* **1972**, *11*, 357.

(39) Davis, L. P.; Dymek Jr., C. J.; Stewart, J. J. P.; Clark, H. P.; Lauderdale, W. J. *J. Am. Chem. Soc.* **1985**, *107*, 5041.

(40) Dymek Jr., C. J.; Grossie, D. A.; Fratini, A. V.; Adams, W. W. *J. Mol. Struc.* **1989**, *213*, 25.

(41) Takahashi, S.; Curtiss, L. A.; Gosztola, D.; Koura, N.; Saboungi, M. L. *Inorg. Chem.* **1995**, *34*, 2990.

(42) Tait, S.; Osteryoung, R. A. *Inorg. Chem.* **1984**, *23*, 4352.

(43) Dieter, K. M.; Dymek Jr., C. J.; Heimer, N. E.; Rovang, J. W.; Wilkes, J. S. *J. Am. Chem. Soc.* **1998**, *110*, 2722.

(44) Blander, M.; Bierwagen, E.; Calkins, K. G.; Curtiss, L. A.; Price, D. L.; Saboungi, M. L. *J. Chem. Phys.* **1992**, *97*, 2733.
(45) Foresman, J. B.; Frisch, A. *Exploring Chemistry with Electronic Structure Methods*; Second Edition ed.; Gaussian, Inc.: Pittsburgh, PA, 1996.
(46) Lee, C. W. *Tetrahedron Lett.* **1999**, *40*, 2461.
(47) Ruiz-López, M. F.; Assfeld, X.; García, J. I.; Mayoral, J. A.; Salvatella, L. *J. Am. Chem. Soc.* **1993**, *115*, 8780.
(48) Jaeger, D. A.; Su, D. *Tetrahedron Lett.* **1999**, *40*, 257.

Biotechnology in Ionic Liquids

Chapter 16

Biotransformations in Ionic Liquids: An Overview

Roger A. Sheldon, F. van Rantwijk, and R. Madeira Lau

**Department of Biocatalysis and Organic Chemistry,
Delft University of Technology, Delft, The Netherlands**

The current state of the art with regard to biotransformations in ionic liquids is reviewed. Research on this subject has rapidly expanded following the first publication in 2000. Performing biotransformations in ionic liquid media has been shown to have potential benefits with regard to activity, stability and (enantio)selectivity. The scope of the methodology with regard to type of ionic liquid, substrate, enzyme and reaction is discussed.

Introduction

During the last decade increasing attention has been focused on the use of room temperature ionic liquids as novel reaction media for organic synthesis, in particular for catalytic processes *(1)*. Their non-volatile character coupled with thermal stability makes them potentially 'green' alternatives for environmentally unattractive, volatile organic solvents, such as chlorinated hydrocarbons. Moreover, their hydrophobic/hydrophilic character and solubility parameters can be tuned by appropriate modifications of the cation or anion and they have been called 'designer solvents' *(2)*.

Depending on their structure, they are immiscible with water, or, *e.g.* alkanes, which provides the possibility of performing reactions in biphasic media, thus facilitating product recovery and catalyst recycling. Until recently, most studies have involved the use of 1,3-dialkylimidazolium cations, *e.g.* 1-butyl-3-methylimidazolium [bmim] in combination with tetrafluoroborate [BF_4] or hexafluorophosphate [PF_6] as the anion (see Fig. 1). The tetrafluoroborate is water miscible while the hexafluorophosphate is immiscible with water. The choice of BF_4^- or PF_6^- was generally dictated by the need for weakly coordinating anions since the presence of strongly coordinating anions, *e.g.* halides, would inhibit many metal-catalyzed processes. However, one can discern a distinct trend towards the use of alternative cations and anions, generally motivated by cost considerations and/or the limited hydrolytic stability of BF_4 and PF_6 salts (which can result in the generation of HF). Moreover, functionalization of the cation, *e.g.* with an oxygen functionality can make ionic liquids suitable solvents for highly polar organic molecules, such as carbohydrates (see later), thus extending their scope.

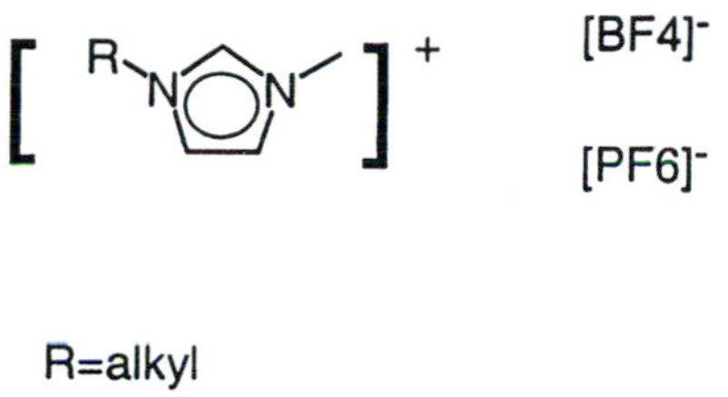

[BF4]$^-$

[PF6]$^-$

R=alkyl

Figure 1

Biotransformations in Ionic Liquids

The first example of a biotransformation in an ionic liquid was reported by Lye and coworkers in 2000 *(3)*. It involved a whole cell biotransformation of 1,3-dicyanobenzene to 3-cyanobenzamide, with a *Rhodococcus sp.* in a biphasic [bmim][PF_6]/H_2O medium. The ionic liquid essentially acts as a reservoir for the substrate and product, thereby decreasing the substrate and product inhibition observed in water. In principle, an organic solvent could be used for the same purpose but it was found that the ionic liquid caused less damage to the microbial cells than, for example, toluene. More recently, 1-octyl-3-methylimidazoliumhexafluorophosphate, [omim][PF_6] was used to enhance the recovery of n-butanol from a fermentation broth *(4)*. Subsequent recovery of the n-butanol from the ionic liquid by pervaporation provided an attractive alternative to the conventional, energy-intensive separation from water by distillation.

The use of an isolated enzyme in an ionic liquid was reported in 2001 by Erbeldinger and coworkers *(5)*. They showed that the thermolysin-catalyzed synthesis of *Z*-aspartame in [bmim][PF_6]/H_2O (95/5, v/v) afforded comparable reaction rates to those observed in ethylacetate/H_2O. Furthermore, the enzyme exhibited a higher stability in the ionic liquid/water medium although the small amount (3.2 $mg.ml^{-1}$) of enzyme that dissolved in the ionic liquid was catalytically inactive. At the same time, we showed *(6)* that *Candida antarctica* lipase B (CaL B), either as the free enzyme (SP525) or in an immobilized form (Novozym 435) is able to catalyze a variety of biotransformations in [bmim][BF_4] or [bmim][PF_6] , in the total absence of added water. The ionic liquids were stored over P_2O_5 and the enzyme was essentially anhydrous, *e.g.* lyophilized in the case of the free enzyme. Transesterifications (Reaction 1), for example, proceeded with rates comparable to those observed in *tert*-butyl alcohol, a commonly employed solvent for lipase-catalyzed processes. The immobilized enzyme (Novozym 435) gave higher rates than the free enzyme (SP525) suspended in the ionic liquid.

Similarly, CaL B catalyzed the ammoniolysis of octanoic acid (Reaction 2) in [bmim][BF_4] *(6)*. Complete conversion occurred in 4 days with Novozym 435 at 40°C, compared to the 90-100% conversion in 17 days observed with ammonium carbamate in methylisobutyl ketone *(7)*.

$$R^1CO_2Et + R^2OH \underset{\text{[bmim][PF}_6\text{] or [bmim][BF}_4\text{]}}{\overset{\text{CaL B, 40}^\circ\text{C}}{\rightleftharpoons}} R^1CO_2R^2 + EtOH \qquad (1)$$

$$RCO_2H + NH_3 \underset{\text{[bmim][BF}_4\text{]}}{\overset{\text{CaL B, 40}^\circ\text{C}}{\rightleftharpoons}} RCONH_2 + H_2O \qquad (2)$$

$R = C_7H_{15}$

A third reaction, which was shown to be feasible in the ionic liquid *(6)*, is the *in situ* generation of a peroxycarboxylic acid *via* CaL B-catalyzed perhydrolysis of the corresponding carboxylic acid. Thus, the epoxidation of cyclohexene by peroxyoctanoic acid, generated *in situ* by Novozym 435-catalyzed reaction of octanoic acid with commercially available 60% aqueous hydrogen peroxide in [bmim][BF_4], afforded cyclohexene oxide in 83% yield in 24h (Reaction 3). For comparison, a yield of 93% was observed in 24h in acetonitrile, which we previously showed *(8)* to be the optimum organic solvent for this reaction.

83% yield

RCOOOH RCOOH

CaL B

H_2O H_2O_2

[bmim][BF_4]

(3)

Scope and Added Value of Biotransformations in Ionic Liquids

The seminal publications referred to above demonstrate the feasibility of performing (some) biotransformations in ionic liquids, a result which *a priori* may not have been predicted. However, this generally provokes the remark: so what. The questions on most people's lips is: What are the benefits of performing biotransformations in ionic liquids? Having shown that enzymes can function in an ionic liquid medium the challenge for further research is clearly to demonstrate that there is an added value in doing so. One could say that enzymes in ionic liquids are at the same stage that enzymes in organic solvents were twenty years ago when Klibanov and coworkers published their seminal contributions *(9)*. There is one important difference, however. Biotransformations in organic media had to compete with the corresponding reactions in water while biotransformations in ionic liquids have to compete with the corresponding processes in water and in organic solvents. Another important issue is the scope of biotransformations in ionic liquids. Initial studies have focused on a few hydrolases and almost nothing is known with regard to the use of *e.g.* oxidoreductases and lyases in ionic liquids.

What could be the added value of biotransformations in ionic liquids? Potential benefits that readily come to mind are: enhanced activities, (enantio)selectivities and stabilities. It is well-known that the activities of enzymes are generally much lower in organic media compared to water and enzyme stability and/or selectivity is an issue in many biotransformations. Ionic liquids could also have added value for performing biotransformations with highly polar substrates which are sparingly soluble in most organic solvents (see later).

Enantioselectivity

Several groups have investigated the effect of performing lipase-catalyzed reactions in ionic liquids on the enantioselectivities of these transformations. For example, Kragl and coworkers *(10)* investigated the kinetic resolution of 1-phenylethanol (Reaction 4) with nine different lipases in ten different ionic liquids.

OH Lipase/24°C/3 days OAc CH_3CHO OH + OAc (4)

Good activities and, in many cases, improved enantioselectivities were observed compared to the corresponding reaction in methyl *tert*-butyl ether (MTBE). The rates and enantioselectivities were dependent on the particular lipase used and the nature of both the cation and anion of the ionic liquid. The best results were generally observed with CaL B while some lipases, *e.g. Candida rugosa* (CRL) and *Thermomyces lanuginosus* lipases showed almost no activity. The highest conversions and enantioselectivities were observed in [bmim][CF_3SO_3], [bmim][$(CF_3SO_2)_2N$] and [omim][PF_6]. Surprisingly, virtually no reaction was observed in [bmim][BF_4] and [bmim][PF_6], which contrasts with what we and others have observed (see later).

More recently, the same group *(11)* compared the enantioselectivities of Reaction 4 with a *Pseudomonas sp.* lipase in [bmim][$(CF_3SO_2)_2N$], MTBE and n-hexane at fixed water activities (a_w). At low water activities ($a_w < 0.53$) the enantioselectivity was higher in the ionic liquid. When the reaction temperature was increased from 25°C up to 90°C the enantioselectivity in MTBE decreased dramatically (from E = 200 to E = 4) while in the ionic liquid it remained high, E decreasing from 200 to 150. In both solvents the decrease in enantioselectivity was observed at the boiling point of either the solvent (MTBE) or the vinyl acetate.

Park and Kazlauskas *(12)* investigated Reaction 4 with *Pseudomonas cepacia* (PCL) lipase in a series of ionic liquids containing different cations in conjunction with BF_4 as the anion. They observed that treatment of the ionic liquid with sodium carbonate resulted in a dramatic increase in rate. This was attributed to the removal of residual silver ions and acidic impurities in the ionic liquid, remnants of the synthesis procedure.

Kim and coworkers *(13)* studied the CaL B and PCL catalyzed transesterification of four different chiral secondary alcohols with vinyl acetate in [emim][BF_4] and [bmim][PF_6]. They observed enhanced enantioselectivities, compared with the same reactions in toluene or THF, in all cases.

Itoh and coworkers *(14)* observed only a minor effect on the enantioselectivity of the Novozym 435 catalyzed transesterification of a chiral allylic alcohol (Reaction 5) in ionic liquids, compared to the same reaction in di-isopropylether. We note, however, that the enantioselectivity was very high (E > 200) in all cases which makes it difficult to observe differences.

OH Ph Novozym 435/IL OAc CH_3CHO OH Ph + OAc Ph (5)

E > 200

The rate of Reaction 5 was strongly dependent on the nature of the anion in [bmim][X]. The best results were obtained with X = BF_4 or PF_6. Much slower reactions were observed with X = CF_3CO_2, CF_3SO_3 and SbF_6. These results contrast with those reported by Kragl and coworkers (see earlier). We note, however, that the purity of ionic liquids used in these studies may be an important source of differences in the observed results. A comparison of different lipases in [bmim][PF_6] revealed that Novozym 435 gave the highest rate, followed by an *Alcaligenes sp.* lipase and PCL while CRL and porcine pancreas lipase (PPL) gave no reaction. It was further shown that the product could be extracted with ether and the ionic liquid, containing the suspended enzyme, could be recycled albeit with a dramatic decrease in activity after the second recycle.

More recently, lipase-catalyzed transesterifications in ionic liquids have been extended to chiral primary alcohols as shown in Reactions 6 *(15)* and 7 *(16)*.

Reaction 6 proceeded up to six times more enantioselectively in [bmim][PF_6] compared to common organic solvents *(15)*. In stark contrast, when the reaction was performed in [bmim][BF_4] almost no enantioselectivity was observed.

lipase/IL (6)

R1 = Ph or Et
R2 = t-Bu, MeO, EtO, i-PrO

lipase/IL (7)

Astonishingly, the study of Reaction 7 *(16)* does not contain any data on the enantioselectivity of the transformation.

Activity and Stability of Enzymes In Ionic Liquids

Activities and stabilities of enzymes are obviously interrelated parameters. It is often difficult to assess whether higher turnover numbers (TON) or turnover frequencies (TOF) are a result of a higher intrinsic activity or a higher stability of the enzyme (or both). For this reason it is convenient to treat these two properties together.

From both a practical and a theoretical viewpoint it is important to establish the stability of enzymes, in various formulations, in ionic liquids compared to common organic solvents and water. Another important question is whether or not enzymes can retain their activity when dissolved in a water-free ionic liquid (all of the above described examples pertain to enzymes dissolved in an ionic liquid/water mixture or suspended in an ionic liquid). To this end we investigated CaL B-catalyzed transesterifications in a range of ionic liquids *(17)*. Four different enzyme formulations were studied: free CaL B (SP525), immobilized on a support (Novozym 435), cross-linked enzyme crystals (CLEC) *(18)* and cross-linked enzyme aggregates (CLEA) *(19)*. Reaction rates in [bmim][PF_6], [bmim][BF_4] and [bmim][CF_3SO_3] were comparable to those observed in *tert*-butanol. In contrast, no reaction (< 5% conversion) was observed in ionic liquids – [bmim][NO_3], [bmim][lactate], [emim][$EtSO_4$] and [$EtNH_3$][NO_3] – in which the free enzyme (SP525) was shown to dissolve. However, dissolution as such is not a prerequisite for loss of activity since the immobilized preparations were also inactive in these ionic liquids. In the case of Novozym 435, in which the enzyme is ionically bonded to the surface of an ion exchange resin, the enzyme is probably leached from the support, as is the case in aqueous media. With the CLEC and the CLEA, however, dissolution of the enzyme is unlikely. We tentatively conclude that coordination of the anions – $CH_3SO_4^-$, NO_3^- and lactate – to the enzyme surface is responsible for the loss of activity (the weakly coordinating BF_4^-, PF_6^- and $CF_3SO_3^-$ do not cause any deactivation).

Interestingly, when the solutions of CaL B in the ionic liquids were diluted with aqueous buffer the activity of the enzyme was (partially) restored, *e.g.* 73% activity was recovered in the case of [bmim][NO_3]. In this context, relevant observations have recently been reported by Summers and Flowers *(20)*. They showed that thermally denatured lysozyme did not precipitate from a solution in [$EtNH_3$][NO_3] after heating to 100°C, while the denatured enzyme precipitated from aqueous buffer under the same conditions. Furthermore, the addition of 5% [$EtNH_3$][NO_3] to the solution of lysozyme in aqueous buffer was shown to stabilize the enzyme against irreversible thermal denaturation. The authors concluded that, although the ionic liquid acts as a denaturant, the denatured protein does not undergo irreversible aggregation. Consequently, on the addition of water, to the cooled solution of the enzyme in the ionic liquid, refolding occurs to afford the active enzyme (87% refolding was determined using differential scanning calorimetry). Based on these results the authors proposed the use of [$EtNH_3$][NO_3] as a refolding additive in protein renaturation. Hence, we tentatively conclude that CaL B undergoes reversible denaturation on dissolution in ionic liquids and subsequent dilution with aqueous buffer results in refolding with restoration of catalytic activity.

We also studied *(17)* the thermal stability of different CaL B preparations by suspending them in an organic solvent or [bmim][PF_6], allowing them to stand at 80°C and taking aliquots at various time intervals and measuring the residual activity, in triacetin hydrolysis, after dilution with water. Incubation of free enzyme (SP525) in [bmim][PF_6] at 80°C actually resulted in an increase in activity to 120% of that of the untreated enzyme. This activity did not decrease up to an incubation time of 100h whereas in *tert*-butanol a roughly linear deactivation in time was observed. With Novozym 435 an even greater increase in activity was observed, up to a maximum of 350%, and 210% activity was still observed after 5 days incubation at 80°C.

In contrast, the CLEC and CLEA from CaL B exhibited a decrease in activity with time, analogous to that observed with SP525 and Novozym 435 in *tert*-butanol. We attribute the stabilizing/activating effect of the ionic liquid on the free and supported lipase to a protecting action of a coating of ionic liquid on the microenvironment (hydration layer) of the enzyme. The reason for the lack of a stabilizing effect on the CLEC and CLEA preparations is unclear and is being further investigated.

More recently, other authors have reported similar stabilizing effects of ionic liquids on lipases. Kirk and coworkers *(21)* reported excellent stabilities for a variety of lipases in [bmim][PF_6], although the highest activity was observed in n-hexane. Iborra and coworkers *(22)* found that the half-life of CaL B increased more than 2300 times when it was incubated at 50°C in the presence of the substrates (vinyl butyrate and 1-butanol) in [bmim][PF_6] at 2% (v/v) water content. In the absence of the substrates, in contrast, no stabilizing effect was observed compared to n-hexane or 1-butanol as the incubation medium. The authors concluded that ionic liquids provide an adequate microenvironment for the catalytic action of the enzyme and that they should be viewed as both an immobilization support and a reaction medium.

Proteases in Ionic Liquids

The first isolated enzyme to be studied in an ionic liquid (containing 5% v/v water) was a protease: thermolysin. More recently, two other groups have studied the performance of another protease, α-chymotrypsin, in ionic liquids. Laszlo and Compton *(23)* showed that the α-chymotrypsin catalyzed transesterification of *N*-acetyl-L-phenylalanine ester with 1-propanol (Reaction 8) proceeded with moderate rates in [bmim][PF_6] and [omim][PF_6] in the presence of 0.25-0.75% (v/v) water. Activities were substantially higher in [omim][PF_6] compared to [bmim][PF_6]. In this study the α-chymotrypsin was freeze-dried together with K_2HPO_4, KCl or poly(ethylene glycol) prior to use, a procedure known to stimulate activity in, *e.g.* hexane. No reaction was observed in the total absence of added water (< 0.03% w/v water). Interestingly, added water was not necessary for enzyme activity when the ionic liquid was combined with supercritical carbon dioxide. The latter is an attractive cosolvent for biotransformations in ionic liquids as it both modifies the solvent properties and can be used to extract reaction products (see later).

OEt + ROH —α-chymotrypsin, IL/H_2O→ OR + EtOH (8)
NHAc NHAc

Adlercreutz and coworkers *(24)* studied the kinetics of Reaction 8 at fixed water activity (a_w) in [bmim][$(CF_3SO_2)_2N$], [emim][$(CF_3SO_2)_2N$], MTBE and ethyl acetate, noting that if solvents are not compared at the same water activity observed effects may be due to solvent effects and/or differences in enzyme hydration. The four solvents behaved quite similarly; no dramatic differences in activity were observed. These two studies show that although α-chymotrypsin can function in ionic liquid media there appears to be no increase in activity/stability compared to commonly used organic solvents. It should be noted, however, that the much higher solubility of the substrate in the ionic liquid can provide for higher reaction rates than in organic solvents.

Glycosidases in Ionic Liquids

Kragl and coworkers *(25,26)* have reported the successful use of ionic liquid/water mixtures as reaction media for the β-galactosidase catalyzed synthesis of *N*-acetyllactosamine (Reaction 9). The addition of 25% (v/v) [mmim][$MeSO_4$] as a water-miscible ionic liquid suppressed secondary hydrolysis of the *N*-acetyllactosamine product, resulting in an increase in yield from 30 to 58%. They further showed that the enzyme could be recycled several times, after ultrafiltration of the reaction mixture, without loss of activity.

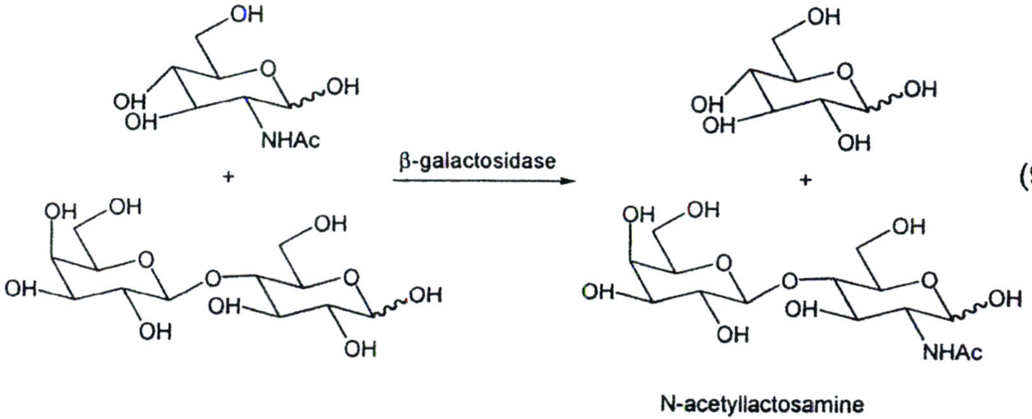

β-galactosidase

(9)

N-acetyllactosamine

Solvent	Yield (%)
Aqueous buffer	30
[mmim][$MeSO_4$]/H_2O (75/25 v/v (a_w=0.83)	58

Acylations of Carbohydrates in Ionic Liquids

As noted earlier ionic liquids could have added value in performing biotransformations of highly polar substrates which are sparingly soluble in most organic solvents. A typical example is provided by carbohydrates, which are highly soluble in water but sparingly soluble in common organic solvents. Consequently, enzymatic acylations of carbohydrates have generally been performed in DMF or pyridine *(27)* which have obvious disadvantages.

Park and Kazlauskas *(12)* recently showed that the CaL B catalyzed regioselective acetylation of glucose (Reaction 10) proceeded smoothly in ionic liquids. The best ionic liquid for this purpose was 3-(2-methoxyethyl)-1-methylimidazolium fluoroborate which dissolves ca. 100 times more glucose (5g/L at 55°C) than acetone or THF.

OAc
Nov 435
55 °C /36 h
99% conv. (10)

Solvent	Selectivity (%)	
N N OMe BF_4	93	7
THF	53	47

An additional benefit was the much higher selectivity to the mono-acylated product compared to that in the organic solvents. In the latter solvents, owing to the poor solubility of glucose (0.02-0.04g/L), the more soluble mono-acetylated glucose undergoes further acetylation to the 3,6-*O*-diacetylglucose. The authors noted that the disaccharide, maltose, also undergoes regioselective acylation in ionic liquids.

It will be interesting to see if even less soluble disaccharides, such as sucrose and lactose or even polysaccharides, can be readily acylated in ionic liquids. We note, in this context, the recently reported solubility of cellulose in ionic liquids *(28)*.

Oxidoreductases in Ionic Liquids

Much less attention has been paid, as yet, to oxidoreductases in ionic liquids. Recently, the immobilized baker's yeast reduction of ketones in a 10:1 [bmim][PF_6]/water was reported *(29)*. As in the previously reported whole cell biotransformations (see earlier) the ionic liquid acts as a reservoir for the substrate and product and the reaction occurs in the aqueous phase.

Kragl and coworkers *(25)* showed that formate dehydrogenase (FDH) from *Candida boidinii* exhibited 98% residual activity (compared to aqueous buffer) in a 3:1 (v/v) [mmim][$MeSO_4$]/H_2O mixture. Surprisingly, the carbonyl reductase from *Candida parapsilosis* (CPCR) showed no activity in 1:3 [mmim][$MeSO_4$]/H_2O in the reduction of acetophenone (for which FDH can be used for cofactor regeneration). Similarly, yeast alcohol dehydrogenase (YADH) displayed only a very low activity in ethanol oxidation in ionic liquid/water mixture (*cf.* the yeast mediated reductions in ionic liquid/water mixtures described above).

Based on the practical importance of oxidoreductases we can expect many more studies of these enzymes in ionic liquids. The studies referred to above all involve dehydrogenases and, in the future, they will presumably be extended to include oxidases, peroxidases and oxygenases.

Product Recovery and Catalyst Recycling

Another important issue which has to be addressed when performing reactions in ionic liquids is product recovery and (bio)catalyst recycling. One method which is often used is extraction with an organic solvent. At first sight this would appear to be self-defeating: one reason for using an ionic liquid was to avoid the use of volatile organic solvents. However, an ionic liquid could be used to avoid using an environmentally unattractive solvent as the reaction medium, followed by product extraction into a more environmentally attractive solvent.

An interesting solvent, in this respect, is supercritical carbon dioxide ($scCO_2$) which forms a biphasic mixture with ionic liquids *(30)*. Quite remarkably, $scCO_2$ is highly soluble (up to 60mol%) in [bmim][PF_6] while the latter is insoluble in $scCO_2$. The successful use of biphasic ionic liquid/$scCO_2$ mixtures for performing lipase catalyzed transesterifications, in batchwise or continuous flow operation (using $scCO_2$ as the mobile phase) has been independently reported by two groups *(31,32)*. For example, the CaL B catalyzed acylation of racemic 1-phenylethanol (Reaction 4) was shown to proceed with good activity, enantioselectivity and operational stability in [bmim][$(CF_3SO_2)_2N$]/$scCO_2$ mixtures *(31,32)*. This would appear to constitute an attractive methodology for performing biotransformations with *in situ* product recovery.

Conclusion and prospects

Biotransformations in ionic liquids constitutes a new, exciting and rapidly expanding area of research that offers potential benefits for performing such processes on an industrial scale. Up till now, emphasis has been mainly on hydrolases and in the future we can expect that the scope will be further broadened to include various types of oxidoreductases and lyases. The ionic liquids used have, up till now, been largely limited to a few dialkylimidazolium salts and anions, *e.g.* PF_6, BF_4. In the future we can expect this will be extended to other ionic liquids containing even less expensive cations and anions *(33)*. Similarly, the type of substrate and reaction will be further extended. For example, the first enzymatic polymerization (lipase catalyzed polyester synthesis) in ionic liquids has recently been reported *(34)*. In the final analysis, however, in order to be viable biotransformations in ionic liquids must deliver economic and environmental benefits.

References

1. For recent reviews see: Sheldon, R. A. *Chem. Commun.* **2001**, 2399; Gordon, C. M. *Appl. Catal. A: General* **2001**, *222*, 101; Zhao, D.; Wu, M.; Kou, Y.; Min, E. *Catalysis Today* **2002**, *74*, 157; Wasserscheid, P.; Keim, W. *Angew. Chem. Int. Ed.* **2000**, *39*, 3772.
2. Freemantle, M. *Chem. Eng. News*, **March 30, 1998**, p. 32; **May 15, 2000**, p. 37; **January 1, 2001**, p. 21.
3. Cull, S. G.; Holbrey, J. D.; Vargas-Mora, V.; Seddon, K.R.; Lye, G.J. *Biotechnol. Bioeng.* **2000**, *69*, 227.
4. Fadeev, A. G.; Meagher, M. M. *Chem. Commun.* **2001**, 295.
5. Erbeldinger, M.; Mesiano, M.; Russell, A.J. *Biotechnol. Prog.* **2000**, *16*, 1129.
6. Madeira Lau, R. ; Rantwijk, F. van ; Seddon, K. R.; Sheldon, R. A. *Org. Lett.* **2000**, *2*, 4189.
7. Litjens, M. J.; Straathof, A. J. J.; Jongejan, J. A.; Heijnen, J. J. *Chem. Commun.* **1999**, 1255.
8. Zoete, M. C. de Ph.D. thesis, Delft University of Technology, The Netherlands, **1995**.
9. Zaks, A.; Klibanov, A. M. *Proc. Natl. Acad. Sci. USA* **1983**, *82*, 3392; Klibanov, A. M. *CHEMTECH* **1986**, 354 and references cited therein.

10. Schöfer, S. H.; Kaftzik, N.; Wasserscheid, P.; Kragl, U. *Chem. Commun.* **2001**, 425.
11. Eckstein, M.; Wasserscheid, P.; Kragl, U. *Biotechnol. Lett.* **2002**, *24*, 763.
12. Park, S.; Kazlauskas, R. J. *J. Org. Chem.* **2001**, *66*, 8395.
13. Kim, K. W.; Song, B.; Choi, M.-Y.; Kim, M.-J. *Org. Lett.* **2001**, *3*, 1507.
14. Itoh, T.; Akasaki, E.; Kudo, K.; Shirakami, S. *Chem. Lett.* **2001**, 162.
15. Pielbasinki, P.; Albrycht, M.; Luczak, J.; Mikolaczyk, M. *Tetrahedron Asymm.* **2002**, *13*, 735.
16. Nara, S. J.; Harjani, J. R.; Salunkhe, M. M. *Tetrahedron Lett.* **2002**, *43*, 2979.
17. Sheldon, R. A.; Madeira Lau, R.; Sorgedrager, M. J.; Rantwijk, F. van; Seddon, K. R. *Green Chem.* **2002**, *4*, 147.
18. Margolin, A. L;. Navia, M. *Angew. Chem. Int. Ed.* **2001**, *40*, 2204.
19. Cao, L.; Rantwijk, F. van; Sheldon, R. A. *Org. Lett.* **2000**, *2*, 1361; Cao, L.; Langen, L. M. van; Rantwijk, F. van; Sheldon, R. A. *J. Mol. Catal. B: Enzym.* **2001**, *11*, 665; Lopez-Serrano, P.; Cao, L.; Rantwijk, F. van; Sheldon, R. A. *Biotechnol. Lett.* in press.
20. Summers, C. A.; Flowers, R. A. *Protein Sci.* **2000**, *9*, 2001.
21. Husum, T. L.; Jorgensen, C. T.; Christensen, M. W.; Kirk, O. *Biocatal. Biotrans.* **2001**, *19*, 331.
22. Lozano, P.; Diego, T. De ; Carrie, D.; Vaultier, M.; Iborra, J. L. *Biotechnol. Lett.* **2001**, *23*, 1529.
23. Laszlo, J. A.; Compton, D. L. *Biotechnol. Bioeng.* **2001**, *75*, 181.
24. Eckstein, M.; Sesing, M.; Kragl, U.; Adlercreutz, P. *Biotechnol. Lett.* **2002**, *24*, 867.
25. Kaftzik, N.; Wasserscheid, P.; Kragl, U. *Org. Proc. Res. Dev.* **2002**, *6*, 553.
26. Kragl, U.; Kaftzik, N.; Schöfer, S. H.; Eckstein, M.; Wasserscheid, P.; Hilgers, C. *Chemistry Today*, **July/August 2001**, p. 22.
27. See for example, Riva, S.; Chopineau, J.; Kieboom, A. P. G.; Klibanov, A. M. *J. Am. Chem. Soc.* **1988**, *61*, 1761.
28. Swatloski, R. P.; Spear, S. K.; Holbrey, J. D.; Rogers, R. D. *J. Am. Chem. Soc.* **2002**, *124*, 4974.
29. Howarth, J.; James, P.; Dai, J. *Tetrahedron Lett.* **2001**, *42*, 7517.
30. Blanchard, L. A.; Hancu, D.; Beckman, E. J.; Brennecke, J. F. *Nature* **1999**, *399*, 28; Blanchard, L. A.; Brennecke, J. F. *Ind. Eng. Chem. Res.*, **2001**, *40*, 287.
31. Lozano, P.; Diego, T. De ; Carrie, D.; Vaultier, M.; Iborra, J. L. *Chem. Commun.* **2002**, 692.
32. Reetz, M. T.; Wiesenhöfer, W.; Francio, G.; Leitner, W. *Chem. Commun.* **2002**, 992.
33. For example see: Golding, J.; Forsyth, S.; MacFarlane, D. R.; Forsyth, M.; Deacon, G. B. *Green Chem.* **2002**, *4*, 223.
34. Uyama, H.; Takamoto, T.; Kobayashi, S. *Polym. J.* **2002**, *34*, 94.

Chapter 17

Enzymatic Condensation Reactions in Ionic Liquids

Nicole Kaftzik[1], Sebastian Neumann[2], Maria-Regina Kula[2], and Udo Kragl[1,*]

[1]Department of Chemistry, Rostock University, 18051 Rostock, Germany
[2]Heinrich Heine University at Düsseldorf, Institute of Enzyme Technology, 52426 Jülich, Germany

In an aqueous environment glycosidases and peptide amidases usually hydrolyse glycosidic bonds or amides, respectively. The reaction can be reversed by incubating the enzyme at lower water activity in the presence of ionic liquids, resulting in a higher yield of disaccharide or peptide amide. β-Galactosidase from *Bacillus circulans* can be applied in nearly anhydrous ionic liquids for reverse hydrolysis with yields of lactose of up to 17%. Peptide amidase from *Stenotrophomonas maltophilia* is used for the direct C-terminal peptide amidation of H-Ala-Phe-OH.

There is no question that biotransformations are an established method either in the lab or for industrial production of bulk and fine chemicals (1-3). Nevertheless, there are still problems with substrate solubility, yield or selectivity.

Some progress has been made by the use of microemulsions (4), supercritical fluids (5) or use of organic solvents (6, 7). The use of hydrolytic enzymes in anhydrous organic media has become a valuable addition to the synthetic repertoire. Reactions may be performed that are impossible in water and the enzymes may even show enhanced thermostability (8, 9). During the past decade, ionic liquids have gained increasing attention for performing all types of reactions with sometimes remarkable results (10). Ionic liquids are salts with melting points below 100 °C; they have no measurable vapor pressure making them ideal tools for clean and sustainable processes (11). Most of the studies in this area have focused on transformations using transition-metal catalysts (10, 11). Recently the application of lipases (12-16), thermolysin or α-chymotrypsin (17, 18) in ionic liquids has been reported. We herein wish to report the results from our studies on the synthesis of disaccharides using the reverse hydrolysis activity of β-galactosidase from *Bacillus circulans* (19) with enhanced yield in the presence of ionic liquids. The second example is the use of a peptide amidase from *Stenotrophomonas maltophilia* (20) for the direct C-terminal peptide amidation of peptides using different ionic liquids instead of organic solvents.

β-Galactosidase Catalyzed Synthesis Of Disaccharides

D-Galacto-oligosaccharides have been synthesized mainly by utilizing the transglycosylation activity of β-galactosidase of various origins (19, 21). However, it is difficult to separate the products from the starting disaccharides and to determine the optimum time to terminate the reaction in order to obtain high yields of the desired compounds. Nevertheless we have studied the behaviour of the β-galactosidase from *Bacillus circulans* in this type of reaction, in the presence of ionic liquids. These results are presented elsewhere (22, 23).

Here we studied the reverse hydrolytic activity of β-galactosidase from *Bacillus circulans* (Figure 1). The reaction was performed in both aqueous media containing different amounts of the ionic liquid and the pure ionic liquid 1,3-dimethylimidazol methylsulfate ([MMIM] $MeSO_4$) **6**. To favour the formation of disaccharides **4** and **5**, high concentrations of the monosaccharides **1** and **2** are necessary. Ionic liquids offer the advantage of dissolving carbohydrates very well allowing higher concentrations than in water and especially mixtures of water and organic solvents as cosolvents. Furthermore, the ratio of substrates **1** and **3** must be as high as possible in order to minimize the formation of D-galactosyl-D-galactoses. Therefore a solution containing **3** and **1** in a ratio of 1:5 was used together with β-galactosidase from *Bacillus circulans* (24). The reaction mixture was incubated for 24 h at 20, 35 and 50 °C. After removal of the heat-denaturated enzyme the filtrate was analysed by HPLC (conditions: BioRad Aminex HPX-87H, UV (208 nm) and RI detection, 0.006 M H_2SO_4 with 0.8 ml/min at 65 °C). The yield of disaccharide obtained is shown in Figure 2.

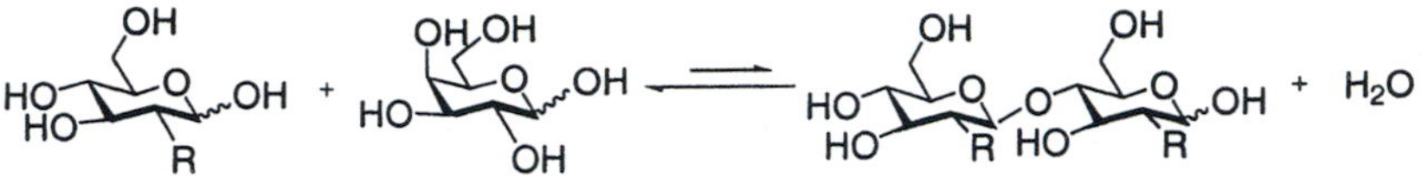

1 R: OH
2 R: NHAc

3

4 R: OH
5 R: NHAc

6 R: Me A⁻: $MeSO_4$ [MMIM]$MeSO_4$
7 R: Bu A⁻: $MeSO_4$ [BMIM]$MeSO_4$
8 R: Et A⁻: BF_4 [EMIM]BF_4
9 R: Bu A⁻: $[CF_3SO_2)_2N$ [BMIM]BTA
10 [Et_3NH]$MeSO_4$
11 [Et_3NMe]$MeSO_4$

Figure 1. β-Galactosidase catalyzed synthesis of disaccharides and ionic liquids used.

The highest yield of **4** was obtained at 35 °C and in almost pure ionic liquid containing only 0.6% v/v water. There was no further increase in the product concentration after 24 h. Compared to data published previously, where the maximum yield was reached after 5 d, the reaction velocity is much faster in the presence of ionic liquids (19). Kren and coworkers reported similary results using high salt concentrations in a thermodynamically controlled reaction, but only after a reaction time of 6 days (25).

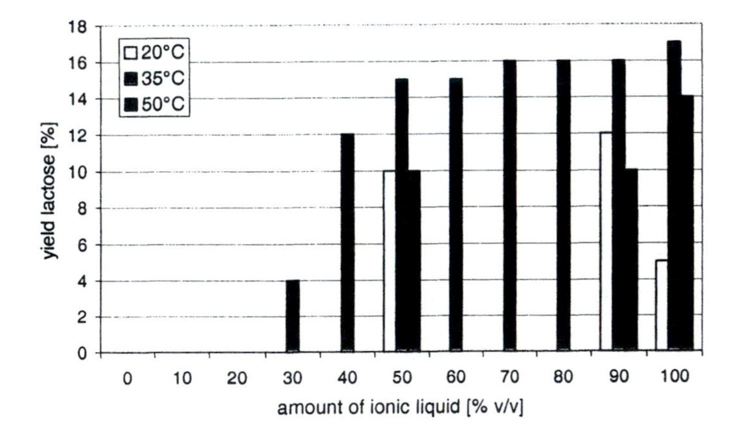

*Figure 2. Yield of lactose **4** as function of amount of the ionic liquid [MMIM] $MeSO_4$ **6**; conditions: 100 mmol/l **1**, 20 mmol/l **3**, 2 mg/ml galactosidase.*

Peptide Amidase Catalyzed Amidation

Peptides and their synthetic analogues represent an interesting class of biologically active compounds, which find application in all areas of immunology, pharmaceutical research and therapy. Many bioactive peptides and more than half of all peptides hormones are amidated at their C-terminal carboxyl group to exhibite full biological activity (26, 27). A peptide amidase that hydrolyses exclusively C-terminal amide groups in peptides, without affecting the internal peptide bonds, was first described by Steinke and Kula (28). This enzyme is also able to catalyze the reverse reaction, the direct C-terminal peptide amidation (29) (Figure 3). The best results of amide syntheses using model substrates of a large series of N^{α}-protected di-, tri-, tetra- and penta-peptide were achieved in a medium consisting of acetonitrile with 25 % v/v of dimethylformamide and 3 % v/v of water (30). The aim of this work was to investigate the synthetic possibilities of peptide amidase-catalyzed peptide amidation with a microbial peptide amidase from *Stenotrophomonas maltophilia* (20) using H-Ala-Phe-OH **12** as model substrate in the presence of NH_4HCO_3 **13** as the ammonium source in ionic liquids as reaction media.

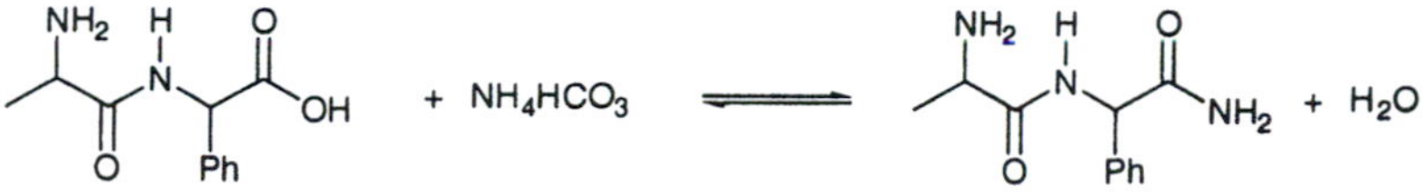

12 + NH_4HCO_3 (13) ⇌ 14 + H_2O

Figure 3. Peptide amidation using peptide amidase from Stenotrophomonas maltophilia.

A series of different ionic liquids **6 - 11** (Scheme 1) from total water miscible to water insoluble ionic liquids was investigated. Solubility properties and amidation rates were compared with the results obtained in oganic solvents like acetonitrile, glycerine or ethylenglycol. Amongst the ionic liquids tested best results were obtained with **7**. Reactions were carried out in sealed 2 ml screw-cap plastic tubes incubated in a rotatory shaker. The peptide (0.025 mmol) was dissolved in a mixture of ionic liquid (125 μl, 250 μl, 375 μl, 450 μl) and aqueous buffer-solution (325 μl, 200 μl, 75 μl, 50 μl) and in pure ionic liquid. After the addition of 1 mg **13** (0.035 mmol of NH_4^+) and lyophilized amidase (6 mg) respectively 50 μl enzyme solution (10 mM in buffer-solution) the reaction-mixture was shaken at 37 °C for 12 h. Both substrates and the enzyme were apparently soluble in **6**, **7**, **10** and **11** as well as in glycerine and in ethylenglycol. The influence of ionic liquid and water concentration on the peptide amidase-catalyzed amidation of **12** was analyzed by HPLC (conditions: RP 18 Hypersil OPS 5, UV detection 220 nm, 90% water, 10% acetonitrile, 0.1% trifluoracetic acid with 1 ml/min at room temperature).

Lowering the total amount of water to almost 0 % v/v results in a maximum yield of 15% in **7.** This result is in the same order of magnitude obtained in acetonitrile containing 25 % v/v DMF and 4 % v/v of water (31). Further studies have to show whether the ionic liquids will improve the enzyme stability as well.

Conclusions

We have shown that ionic liquids can serve as sole solvents for enzymes others than lipases and proteases reported so far. They may be especially useful in thermodynamically controlled reactions where proper control of the water activity in the medium is of great importance. For the β-galactosidase-catalyzed formation of disaccharides we have found an improved enzyme stability also at higher temperatures and an increased formation of disaccharide by lowering the total water content in the reaction media. Reaction rates of peptide amidase-catalyzed amidation were comparable with those observed in conventional organic reaction media. In both cases ionic liquids show good or even better solubility properties for both substrates and enzymes. To represent a real alternative to synthesis in organic solvent for example, it is important to demonstrate that the ionic liquid can be recycled after the reaction. This aspect is subject to further studies. Additionally, other influencing factors such as the nature of the medium, diffusion rates, the water content and the question which of the ions are responsible for the effects have to be investigated as well.

Acknowledgments

We thank P. Wasserscheid and V. Cerovsky for fruitful discussions, Solvent Innovation, Cologne, for the gift of ionic liquids and the "Fonds der Chemischen Industrie" for partial financial support.

References

1. Faber, K. Biotransformation in Organic Chemistry, Springer, Berlin **2000.**
2. Bornscheuer, U. T.; Kazlauskas, R. J. Hydrolases in Organic Synthesis, Wiley-VCH, Weinheim **1999.**
3. Liese, A.; Seelbach, K.; Wandrey, C. Industrial Biotransformations, Wiley-VCH, Weinheim **2000.**
4. Orlich, B.; Schomäcker, R. *Biotech. Bioeng.* **1999**, 65, 357-362.
5. Hartmann, T.; Schwabe, E.; Scheper, T. "Enzyme catalysis in supercritical fluids" in R. Pathel, Stereoselective Biocatalysis, Marcel Decker **2000**, pp. 799-838.
6. Carrea, G.; Riva, S. *Angew. Chem. Int. Ed.* **2000**, 39, 2226-2254.
7. Cabral, J. M. S.; Aires-Barros, M. R.; Pinheiro, H.; Prazeres, D. M. F. *J. Biotechnol.* **1997**, 59, 133.
8. Dai, L.; Klibanov, A. M. *Proc. Natl. Acad. Sci.* USA **1999**, 96, 9475-9478.

9. Sellek, G. A.; Chaudhuri, J. B. *Enzyme Microb. Technol.* **1999**, 25, 471-482.
10. Welton, T. *Chem. Rev.* **1999**, 99, 2071-.
11. Wasserscheid, P.; Keim, W. *Angew. Chem. Int. Ed.* **2000**, 39, 3773-3789.
12. Lau, R. M.; Van Rantwijk, F.; Seddon, K. R.; Sheldon, R. A. *Org. Lett.* **2000**, 2, 4189-4191.
13. Schöfer, S.; Kaftzik, N.; Wasserscheid, P.; Kragl, U. *Chem. Commun.* **2001**, 425-426.
14. Kwang-Wook Kim, Boyoung Song, Min-Young Choi, Mahn-Joo Kim *Org. Lett.* **2001**, 3, 1507-1509.
15. Itoh, T.; Akasaki, E.; Kudo, K.; Shirakami, S. *Chem. Lett.* **2001**, 3, 262-263.
16. Lozano, P., DeDiego, T.; Carrie, D.; Vaultier, M., Iborra, J. L. *Biotechnol. Lett.* **2001**, 23, 1529-1533.
17. Erbeldinger, M.; Mesiano, A. J.; Russel, A. *Biotechnol. Prog.* **2000**, 16, 1131-.
18. Laszlo, J. A.; Compton, D. L.; *Biotechnol. Bioeng.* **2001**, 75, 181-186.
19. Ajisaka, K.; Fujimoto, H.; Nishida, H. *Carbohydr. Res.* **1988**, 180, 35-42.
20. Cerovsky, V.; Kula, M.-R. *Biotechnol. Appl. Biochem.* **2001**, 33, 183-187.
21. Crout, D. H. G.; Vic, G.; *Curr. Opin. Chem. Biol.* **1998**, 2, 98-111.
22. Kragl, U.; Kaftzik, N.; Schöfer, S. H.; Eckstein, M.; Wasserscheid, P.; Hilgers, C.; *Chim. Oggi* **2001**, 19, 22-24.
23. Kaftzik, N.; Wasserscheid, P.; Kragl, U., *Org. Process Res. Dev.* **2002**, *6*, 553-557.
24. β-Galactosidase from *Bacillus circulans* is commercial available from Daiwa Kasei, Osaka, Japan.
25. Rajnochova, E.; Dvorakova, J.; Hunkova, Z.; Kren, V.; *Biotechnol. Lett.* **1997**, 19, 869-872.
26. Merkler, D.J. *J. Enzyme Microb. Technol.* **1999**, 16, 450-456.
27. Pigge, S. T.; Mains, R. E.; Eipper, B. A.; Amzel, L. M. *Cell. Mol. Life. Sci.* **2000**, 57, 1236-1259.
28. Steinke, D.; Kula, M.-R. *Angew. Chem. Int. Ed. Engl.* **1990**, 29, 1139-1140.
29. Stelkes-Ritter, U.; Wyzgol, K.; Kula, M.-R. *Appl. Microb. Biotechnol.* **1995**, 44, 393-398.
30. Cerovsky, V.; Kula, M.-R. *Angew. Chem. Int. Ed. Engl.* **1998**, 37, 1885-1887.
31. Neumann, S., Kula, M. R. *Appl Microbiol Biotechnol.* **2002**, *58*, 772-780.

Chapter 18

Aspects of Chemical Recognition and Biosolvation within Room Temperature Ionic Liquids

Gary A. Baker[1,*], Sheila N. Baker[2], T. Mark McCleskey[2], and James H. Werner[1]

[1]Michelson Resource, Bioscience Division and [2]Chemistry Division, Los Alamos National Laboratory, Los Alamos, NM 87545

With near-exponential growth in the number of reports in the area, room temperature ionic liquids have become the poster child for Green Chemistry. From an industrial standpoint, however, the replacement of current organic solvents is a protracted and uncertain one. Current impediments to their broader use include the lack of toxicological knowledge as well as a paucity in knowledge about solvent dynamics, diffusive processes, and biomolecular structure within ionic liquids. Here, we introduce experimental approaches to the study of 1) probe translational mobility 2) solute self-assembly and 3) protein thermal stability, all within ionic liquids.

Introduction

Current applications for room temperature ionic liquids are on the rise due to the prospects for 'green' industrial catalysis and the relative ease of generating tailored families of these novel fluids for disparate aims (*1-7*). Some fertile areas that are targeted for further development include, for example, the concept of task-specific ionic liquids (*8,9*) as designer solvents as well as the implementation of compressed or supercritical carbon dioxide with ionic liquids to perform phase-separable catalysis (*10-13*), to reduce the melting temperature of ionic liquids (*14*) and even as a switch for the phase separation of ionic liquids from molecular organic phases (*15*).

As early as 1993, Pagni and Mamantov discovered unexpected photochemical behavior for the photolysis of anthracene in a 1-ethyl-3-methylimidazolium chloride/$AlCl_3$ melt. More recent work on non chloroaluminate ionic liquids by other researchers reveal them to be unique media for the generation of labile photoproducts (*16-19*) and as hosts for novel photophysics (e.g., selective alternate polycyclic aromatic hydrocarbon quenching by nitromethane (*20*)).

In 2000, Erbeldinger et al. reported successful thermolysin-catalyzed *Z*-aspartame formation within the ionic liquid 1-butyl-3-methylimidazolium hexafluorophosphate, heralding a new era in non-aqueous enzymology (*21*). The same year, Cull et al. showed that the same ionic liquid could be applied to liquid liquid extraction of the antibiotic erythromycin-A as well as for the *Rhodococcus* R312 catalyzed two phase biotransformation of 1,3-dicyanobenzene (*22*). Since these pioneering studies, several prominent research groups have made significant advances in our understanding of ionic liquids as hosts for biocatalysis (*12,23-29*) and even as suitable media for bioelectrocatalysis (*30*).

In all of the applications discussed above, the solvent properties of the ionic liquid in question play a deterministic role. Thus, as pointed out by the leaders in the field, in order to implement room temperature ionic liquids into existing chemical processes, data on their solvent properties are of the utmost importance. The key to understanding this solvency and the manifold interactions with solute species of interest lies in revealing the subtleties of their physicochemical properties, including dynamical aspects. While studies on the static solvent properties of the more common ionic liquids are cumulating (*31-37*), with few exceptions (*38,39*), there is virtually no reliable, quantitative information on solvation dynamics, reaction rates, or mass transport within ionic liquids. Curent research in our group seeks to redress this deficiency. In this chapter, we briefly discuss recent results for the following within ionic liquids: 1) translational mobility of fluorescent reporters 2) self-assembly of porphyrin aggregates and 3) the thermal stabilization/denaturation of a model protein.

Experimental

Materials

The ionic liquids 1-ethyl-3-methylimidazolium bis(trifluoromethane sulfonyl)imide, [emim][Tf_2N], 1-butyl-3-methylimidazolium hexafluorophosphate, [bmim][PF_6], and *N,N*-butylmethylpyrrolidinium bis(trifluoromethane sulfonyl)imide, [bmp][Tf_2N], were synthesized following the methods of Bonhote et al. (*40*), the Rogers group (*41*), and MacFarlane and co-workers (*42*), respectively. The ionic liquids were dried under constant stirring with heating at *ca.* 70 °C *in vacuo* for no less than 24 h until no water overtones were detectable by near-infrared spectroscopy for a 4 mm pathlength. For [bmim][PF_6], this step was preceeded by drying under vacuum without application of heat for 24 h in order to avoid hydrolysis of the anion. To avoid the sorption of environmental moisture, all ionic liquids were stored within evacuated glass EPA vials equipped with PTFE/silicone septa in an argon filled dry box. These methods have previously been shown to provide ionic liquids that are 'operationally' dry (*34,35*).

Fluorescein, rhodamine 110 (R110), and rhodamine 6G (R6G) were obtained from Aldrich, 4-dicyanomethylene-2-methyl-6-*p*-dimethylaminostyryl-4-H-pyran (DCM) was purchased from Exciton and *meso*-tetrakis(4-sulfonatophenyl)porphine (TSPP) was provided by Porphyrin Products, Inc. (Logan, Utah). Monellin (*Dioscoreophyllum cumminsii*) was purchased from Sigma and used as received. All water used was polished to a specific resistivity of at least 18.2 M cm with a Barnstead NANOpure system (model 04741).

Instrumentation

Fluorescence correlation spectroscopic (FCS) measurements were carried out on a home built inverted confocal microscope (see Figure 1). Briefly, a small fraction of the 488 nm line from an argon ion laser was coupled into a single mode fiber (SMF) by a microscope objective (MO1) in order to improve the mode quality of the laser and to eliminated spatial drift of the excitation laser beam upon entering the back of an objective. The beam exiting the fiber was collimated by a 10x microscope objective (MO2) and focused by a 125 mm lens (L1) to a spot approximately 160 mm away in order to slightly overfill the back of a 100x 1.1 NA oil immersion microscope objective (MO3) (Zeiss, Thornwood, NY). The 488 nm excitation light was then focused by this objective to a diffraction limited spot within a 4 μL sample droplet placed on a

#1 coverslip above the objective (CS). All measurements were taken 10 μm above the droplet coverslip interface. The fluorescence was collected by the same 100x objective and was reflected by a dichroic (DC) onto a 75 μm pinhole (PH) placed in the image plane of the 100x objective. This spatially filtered light then passed through an appropriate band pass filter (F) (e.g., 530 30 nm bandpass for fluorescein) before being focused by another lens (L2) onto the active area of a single photon counting avalanche photodiode (APD) (SPCM 200 PQ, Perkin Elmer Optoelectronics, Quebec, Canada).

The output pulses from the detector were suitably conditioned prior to being fed into the input of a digital correlator card (ALV-5000E, ALV-Laser Vertriebsgesellschaft, m-b.H., Langen FRG). Normalized single-channel autocorrelations were performed using ALV-5000 software. All measured autocorrelation functions were collected for 5 minutes with approximately 150 μW of excitation laser power on a 4 μL sample.

All UV-vis measurements were made using a Hewlett Packard (Model 8453) spectrometer.

Steady state fluorescence measurements were carried out on a custom order PTI spectrofluorometer. Monellin samples were excited at 295 nm, and emission was collected from 300 nm to 450 nm in 1 nm increments with an integration time of 0.50 s, using 4 nm spectral bandpasses in both the excitation and emission paths.

Temperature dependent data from 20 to 140 C were acquired at equilibrium following slow (1-2 C/min) sample heating by using a custom thermostated circulation bath/heater cartridge system. The actual sample temperature was measured directly prior to each measurement by using a solid state Omega 450 type E thermocouple to account for thermal losses.

Measuring Translational Diffusion within Ionic Liquids

We have recently demonstrated fluorescence correlation spectroscopy (FCS) as a powerful method to rapidly and accurately determine diffusion coefficients for fluorescent molecular probes within minute ionic liquid samples. Since FCS measures spontaneous intensity fluctuations within the focal volume of a tightly focused laser beam, it is exquisitely sensitive to subtle differences in hydrodynamic volume and can provide information on translational diffusion, the number of molecules under observation (down to the single molecule level), and the relative fluorescence quantum yield of species in inhomogeneous solution. Several excellent references describing FCS methods and applications exist (*43-45*).

Experimentally, the time dependent fluorescence intensity $I(t)$ is analyzed in terms of its temporal autocorrelation function $G(\ )$ which compares the fluorescence intensity at time t with the intensity after a time delay :

$$G(\tau) = \frac{\langle \delta I(t) \bullet \delta I(t+\tau) \rangle}{\langle I(t) \rangle^2} \quad (1)$$

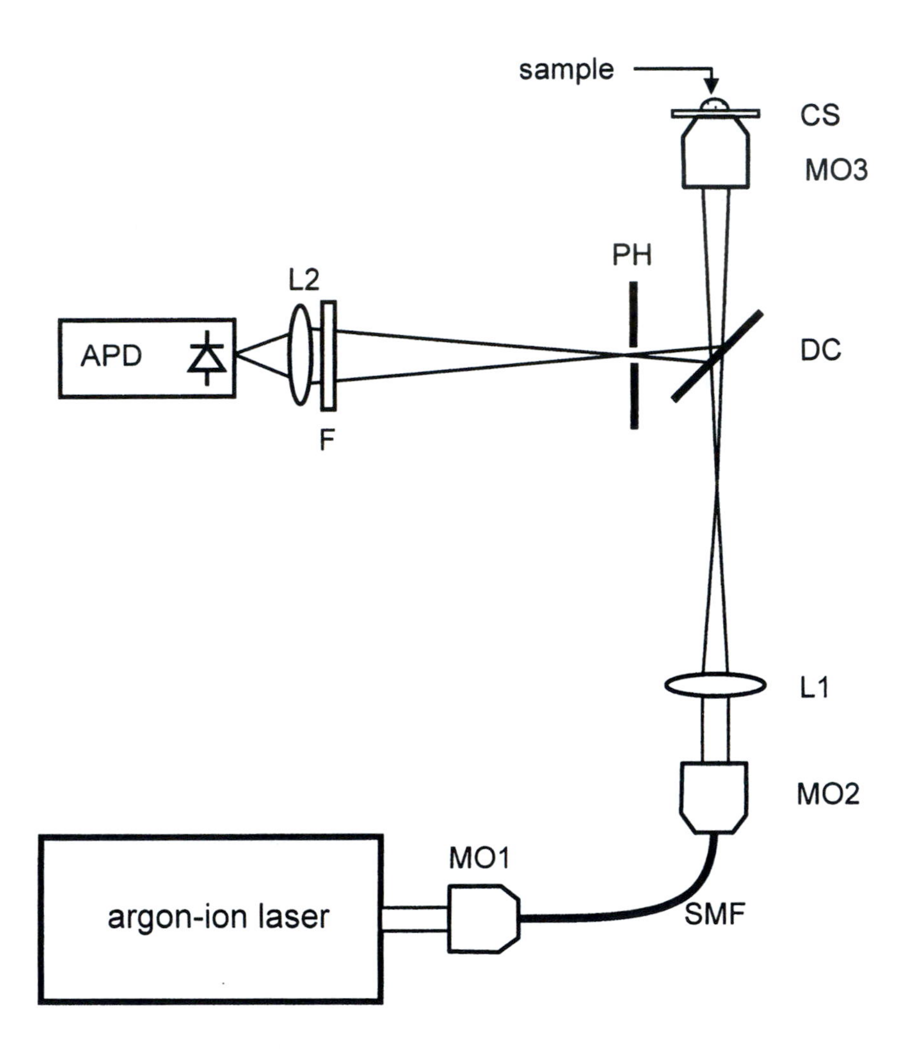

Figure 1. Simplified schematic of our confocal FCS system. See text for further details and abbreviations.

Here, angle bracketed quantities denote time averages where deviations in the fluorescence light intensity from the average ($\delta I(t) = I(t) \quad I(t)$) can result from changes in either the number or the fluorescence yield of molecules resident in the detection volume. The expected correlation function due to the diffusion of molecules into and out of a three-dimensional ellipsoidal Gaussian probe volume was treated by Aragon and Pecora (*46*). We have fit our measured correlation functions to this model and have accounted for triplet or dark state transitions by the method of Widengren et al. (*47*). The following functional form was used to fit our measured data:

$$G(\tau) = \frac{1}{N} \cdot \frac{1}{1 + \frac{\tau}{\tau_c}} \cdot \frac{1}{\sqrt{1 + \left(\frac{\omega_0}{\omega_z}\right)^2 \frac{\tau}{\tau_c}}} \cdot \left[1 - f + f \cdot \exp\left(\frac{-\tau}{\tau_{tr}}\right)\right] + dc \qquad (2)$$

where the five parameters of the fit are: N, the mean number of molecules in the detection volume; τ_c, the correlation time; (ω_o/ω_z), the ratio of the $1/e^2$ radii for the lateral and axial dimensions; f, the fraction of irradiated molecules in the triplet state; τ_{tr}, the triplet lifetime; and dc, a baseline offset. Based on the autocorrelation functions generated for 10 nM solutions of Rhodamine 110 in water, the ratio of the radial to axial dimensions of the probe volume, (ω_o/ω_z) was held fixed at 0.143 in all subsequent fitting procedures.

From this work, the main parameter of interest recovered from the fit to the experimental autocorrelation curves is the correlation time, τ_c, which relates inversely to the diffusion constant, D, as $\tau_c = \omega_o^2/4D$. From the measured correlation time of R110 in water (35 μs) and its known diffusion constant (280 μm^2/sec (*48*)), ω_o is directly determined to be 0.20 μm for our FCS apparatus.

As illustration of the utility of FCS toward assessing room temperature ionic liquid diffusional properties, as well as cosolvent and solute charge effects, we studied the effects of controlled 'wetting' (3.0 % v/v water) on the diffusion coefficients measured for model cationic (R6G), neutral (DCM), and anionic (fluorescein) fluorescent reporters (all at 10 nM) within [bmim][PF_6]. The resultant autocorrelation curves for 10 nM R6G in dry and wet [bmim][PF_6], as shown in Figure 2, clearly show the water induced diffusional acceleration leading to shorter correlation times. From the recovered correlation times, the diffusion constant D for R6G in [bmim][PF_6] was found to increase from *ca.* 1.54 μm^2/sec to 2.50 μm^2/sec upon 'wetting'. These diffusion coefficients are about two orders of magnitude lower than that of R6G in water, and are largely explained by viscosity considerations (*34*). Similar behavior is observed for DCM and fluorescein in [bmim][PF_6] where D_{dry}/D_{wet} values are 1.85/3.23 μm^2/sec and 2.27/4.55 μm^2/sec, respectively. Unexpectedly, the anionic probe not only exhibits the fastest diffusion in [bmim][PF_6] but also experiences the greatest acceleration, doubling upon wetting. This result is surprising since we expect the anionic probe to associate with the bulky

$[bmim]^+$. One possible explanation is that the symmetrical distribution of negative charge on $[PF_6]$ permits interaction with several of the surrounding cations (*38*), leading to ionic 'cross linking'. At this point, however, we caution against generalizing this trend until this aspect is explored in further detail.

Using cyclic voltammetry, Marken and coworkers measured the diffusion coefficients of the neutral redox probe *N,N,N ,N* -tetramethyl-*p*-phenylenediamine (TMPD) in $[bmim][PF_6]$ and found that *D* doubled from 2.6 to 5.2 μm²/sec with the addition of 5.0 % (w/w) water (*49*). They also claim that *D* increases from 1.1 to 10 μm²/sec for methyl viologen (MV^{2+}) in $[bmim][PF_6]$ upon the addition of 6.0 % (w/w) water. These authors offer no data for the diffusion of an anionic species in $[bmim][PF_6]$. While their results may suffer from the large probe concentrations necessary (10 and 20 mM for TMPD and MV^{2+}, respectively) and the fact that they far exceed the solubility limit for water in $[bmim][PF_6]$ (*35,50*), taken in conjuction with our results, it is clear that while diffusional rates are intimately entwined with probe structure, trace water levels result in universal solute acceleration. The discovery by the Seddon group that the viscosity of ionic liquid organic solvent mixtures is dependent mainly on the mole fraction of added molecular solvent, and only to a lesser extent upon their identity (*50*), may ultimately allow diffusional changes to be entirely predictable.

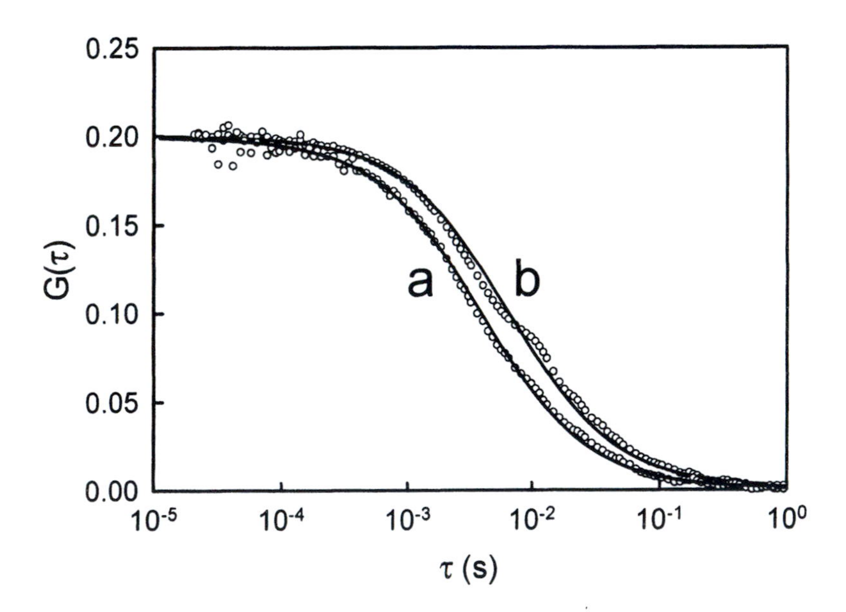

Figure 2. Fluorescence autocorrelation curves and the corresponding fits to equation (2) for 10 nM R6G in (a) wet and (b) dry $[bmim][PF_6]$.

Porphyrin Self-Assembly within an Ionic Liquid

Although ionic liquids offer a unique nanoenvironment for performing novel chemistries, we were surprised to identify only a single report on exogenous self assembly within an ionic liquid (*51*). By dispersing amide group enriched glycolipids within ether containing 'sugar-philic' ionic liquids, Kimizuka and Nakashima were able to form stable, thermally reversible bilayer membranes, 'self assembling ionogels'. Even in this case, however, the ionic liquid is a participant in the aggregate formed. In order to show that ionic liquids are viable 'spectator' media for fabricating novel assemblies, we turned our attention to *meso*-tetrakis(4-sulfonatophenyl)porphine (TSPP). Under favorable conditions of concentration, pH, and ionic strength, the zwitterionic TSPP can form periodic, unidimensional side by side structures, so-called *J-aggregates*, mediated by electrostatic interaction between the negatively-charged sulfonates and the positive porphine macrocycle (*52-56*). Such assemblies are of great interest as opto electronic materials and as models of photosynthetic antenna complexes. Moreover, the possibility of charge screening by the ionic liquid proves to be an interesting prospect. In order to monitor the kinetics of J-aggregation, we followed the disappearance of the diacid monomeric TSPP Soret-band (*ca.* 434 in water) by UV-vis absorption. Results for TSPP J-aggregation in [emim][Tf_2N] are shown in Figure 3. In this experiment, time zero indicates the addition of 2 μL hydrochloric acid to an equilibrated mixture of 5 μM TSPP in [emim][Tf_2N] containing 2.0 % water, to result in a *bulk* pH of 2.0. The observed kinetics are biphasic with a fast rate of 0.0092 s^{-1} and a slower rate near 0.0003 s^{-1}. As shown in the inset of Figure 3, the evolution in the absorption maximum for the aggregate species (ca. 491 nm in water) confirms the fast rate observed for monomeric loss. It is interesting to note that in water no J-aggregation occurs at a pH of 2.0. Even at pH 0.5, where J-aggregation of TSPP does proceed, pseudo first-order kinetics are still 3-4 fold slower than the fast rate in [emim][Tf_2N]. These results make conspicuous the nanoheterogeneous nature of ionic liquids and may open new dimensions for materials engineering within room temperature ionic liquids.

Native Fluorescence as a Probe of the Thermodynamic Stability of a Single Trp Protein within an Ionic Liquid

There is clear indication that novel chemistries can be pursued within ionic liquids which have no peer among conventional solvents. One burgeoning area where this is evident is in the area of biocatalysis where performing enzymatic catalysis within ionic liquids has, in some cases, been shown to provide higher rates and/or improved regio- and enantioselectivity relative to aqueous media *or* conventional organic solvents (*29*). While there is clearly the potential for another evolutionary step in non-aqueous enzymology, set in motion two

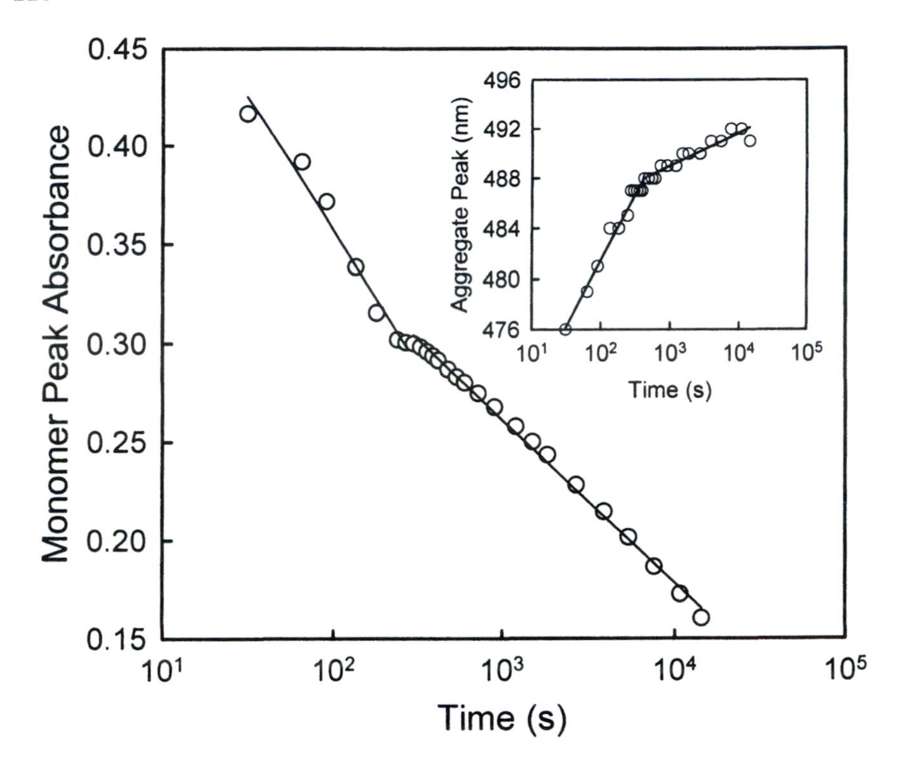

Figure 3. Kinetics of acid-triggered TSPP J-aggregation in [emim][Tf_2N] showing the loss of monomeric species. The inset shows time dependent changes in the emission maximum of the J-aggregate.

decades ago by the pioneering work of Klibanov (*57-59*), there is a scarcity of data on biomolecular structure within *any* ionic liquid. Of course, the maintenance of a nativelike enzyme structure is considered essential to attaining optimum activity in a given solvent. In fact, it has even been suggested that evolutionary adaptation has adjusted enzyme conformational mobility as a key parameter allowing such optimization at a given temperature (*60,61*). Thus, it is clear that rational design of an effective biocatalytic process within an ionic liquid requires that one understand the influence of the solvent system, as a whole, on enzyme structure.

As a first approach, we elected to monitor the stability of native protein in ionic liquid simply by following its intrinsic tryptophan emission, a practice common in studies of protein structure, dynamics, stability, and folding. We selected monellin as our archetypical protein for study since it represents a well-studied + protein, it contains a single tryptophan (Trp) residue, it exists as a monomer in solution, and it has been investigated in a number of exotic environs.

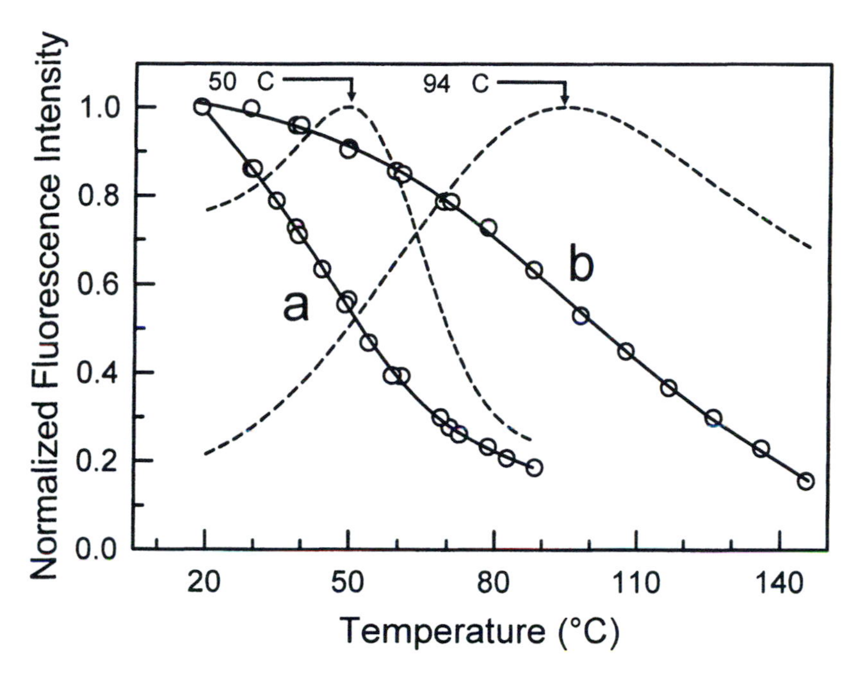

Figure 4. Thermal denaturation curves for monellin in (a) water and (b) [bmp][Tf_2N] with 2.0 % water (v/v). The dashed curves are the corresponding inverted, normalized first derivative spectra, where T_u can be estimated from the maxima.

Our results for monellin in [bmp][Tf_2N] proved to be very interesting (Figure 4). While enzyme immobilization has been shown to increase thermal stability (*62*), we observed an increase in the unfolding temperature (T_u) from 50 C for monellin in water to ca. 94 C for monellin freely *dissolved* in [bmp][Tf_2N] containing 2.0 % water (v/v). These results show how remarkably the ionic liquid is able to protect the monellin from thermal degradation. This remarkable stabilization against thermal inactivation has important implications in the use of 'ordinary' enzymes in high temperature biocatalytic processes. In particular, such an approach may provide alternatives to enzymes obtained from thermophilic bacteria or engineered proteins. The high stabilities may also prove useful for biosensory applications within hostile environments. Further research should reveal which interactions and factors are important for preserving native elements within ionic liquids and, ultimately, the means by which to control enzyme stability and extend the possibilities for biocatalysis itself within this unique class of solvent.

Conclusions

In this chapter, we offer several approaches to the study of ionic liquids, with an eye toward features of molecular recognition, mobility, and biological structure. We show that: (a) analysis of the spontaneous fluctuations of small molecular ensembles by way of FCS provides a rapid, accurate means for diffusion coefficient determination of microliter volumes of ionic liquid (b) ionic liquids provide a unique environment for nanochemistry and material design (c) ionic liquids may offer a general means for the thermostabilization of proteins against thermal inactivation, opening new potential in high temperature biocatalysis and use in nonnatural environments.

Acknowledgements

The authors are indebted to Professor Siddharth Pandey (New Mexico Institute of Mining and Technology) for many fruitful discussions. GAB would also like to thank Los Alamos National Laboratory for the award of a Director's Fellowship. LAUR # 02-4566.

References

1. Seddon, K. R. *Kinet. Catal.* **1996**, *37*, 693-697.
2. Seddon, K. R. *J. Chem. Technol. Biotechnol.* **1997**, *68*, 351-356.
3. Welton, T. *Chem. Rev.* **1999**, *99*, 2071-2083.
4. Gordon, C. M. *Appl. Catal., A* **2001**, *222*, 101-117.
5. Olivier-Bourbigou, H.; Magna, L. *J. Mol. Catal. A:Chem.* **2002**, *182*, 419-437.
6. Huddleston, J. G.; Visser, A. E.; Reichert, W. M.; Willauer, H. D.; Broker, G. A.; Rogers, R. D. *Green Chem.* **2001**, *3*, 156-164.
7. Zhao, D. B.; Wu, M.; Kou, Y.; Min, E. *Catal. Today* **2002**, *74*, 157-189.
8. Bates, E. D.; Mayton, R. D.; Ntai, I.; Davis, J. H. *J. Am. Chem. Soc.* **2002**, *124*, 926-927.
9. Visser, A. E.; Swatloski, R. P.; Reichert, W. M.; Mayton, R.; Sheff, S.; Wierzbicki, A.; Davis, J. H.; Rogers, R. D. *Environ. Sci. Technol.* **2002**, *36*, 2523-2529.
10. Liu, F. C.; Abrams, M. B.; Baker, R. T.; Tumas, W. *Chem. Commun.* **2001**, 433-434.
11. Brown, R. A.; Pollet, P.; McKoon, E.; Eckert, C. A.; Liotta, C. L.; Jessop, P. G. *J. Am. Chem. Soc.* **2001**, *123*, 1254-1255.
12. Reetz, M. T.; Wiesenhofer, W.; Francio, G.; Leitner, W. *Chem. Commun.* **2002**, 992-993.
13. Lozano, P.; de Diego, T.; Carrie, D.; Vaultier, M.; Iborra, J. L. *Chem. Commun.* **2002**, 692-693.

14. Kazarian, S. G.; Sakellarios, N.; Gordon, C. M. *Chem. Commun.* **2002**, 1314-1315.
15. Scurto, A. M.; Aki, S. N. V. K.; Brennecke, J. F. *J. Am. Chem. Soc.* **2002**, *124*, 10276-10277.
16. Gordon, C. M.; McLean, A. J. *Chem. Commun.* **2000**, 1395-1396.
17. Muldoon, M. J.; McLean, A. J.; Gordon, C. M.; Dunkin, I. R. *Chem. Commun.* **2001**, 2364-2365.
18. Marcinek, A.; Zielonka, J.; Gebicki, J.; Gordon, C. M.; Dunkin, I. R. *J. Phys. Chem. A* **2001**, *105*, 9305-9309.
19. Alvaro, M.; Ferrer, B.; Garcia, H.; Narayana, M. *Chem. Phys. Lett.* **2002**, *362*, 435-440.
20. Fletcher, K. A.; Pandey, S.; Storey, I. K.; Hendricks, A. E. *Anal. Chim. Acta* **2002**, *453*, 89-96.
21. Erbeldinger, M.; Mesiano, A. J.; Russell, A. J. *Biotechnol. Progr.* **2000**, *16*, 1129-1131.
22. Cull, S. G.; Holbrey, J. D.; VargasMora, V.; Seddon, K. R.; Lye, G. J. *Biotechnol. Bioeng.* **2000**, *69*, 227-233.
23. Lau, R. M.; van Rantwijk, F.; Seddon, K. R.; Sheldon, R. A. *Org. Lett.* **2000**, *2*, 4189-4191.
24. Lozano, P.; de Diego, T.; Guegan, J. P.; Vaultier, M.; Iborra, J. L. *Biotechnol. Bioeng.* **2001**, *75*, 563-569.
25. Schofer, S. H.; Kaftzik, N.; Wasserscheid, P.; Kragl, U. *Chem. Commun.* **2001**, 425-426.
26. Laszlo, J. A.; Compton, D. L. *Biotechnol. Bioeng.* **2001**, *75*, 181-186.
27. Itoh, T.; Akasaki, E.; Kudo, K.; Shirakami, S. *Chem. Lett.* **2001**, 262-263.
28. Lozano, P.; De Diego, T.; Carrie, D.; Vaultier, M.; Iborra, J. L. *Biotechnol. Lett.* **2001**, *23*, 1529-1533.
29. Sheldon, R. A.; Lau, R. M.; Sorgedrager, M. J.; van Rantwijk, F.; Seddon, K. R. *Green Chem.* **2002**, *4*, 147-151.
30. Laszlo, J. A.; Compton, D. L. *J. Mol. Catal. B:Enzym.* **2002**, *18*, 109-120.
31. Aki, S. N. V. K.; Brennecke, J. F.; Samanta, A. *Chem. Commun.* **2001**, 413-414.
32. Muldoon, M. J.; Gordon, C. M.; Dunkin, I. R. *J. Chem. Soc., Perkin Trans. 2* **2001**, 433-435.
33. Fletcher, K. A.; Storey, I. A.; Hendricks, A. E.; Pandey, S. *Green Chem.* **2001**, *3*, 210-215.
34. Baker, S. N.; Baker, G. A.; Kane, M. A.; Bright, F. V. *J. Phys. Chem. B* **2001**, *105*, 9663-9668.
35. Baker, S. N.; Baker, G. A.; Bright, F. V. *Green Chem.* **2002**, *4*, 165-169.
36. Fletcher, K. A.; Pandey, S. *Appl. Spectrosc.* **2002**, *56*, 266-271.
37. Dzyuba, S. V.; Bartsch, R. A. *Tetrahedron Lett.* **2002**, *43*, 4657-4659.
38. McLean, A. J.; Muldoon, M. J.; Gordon, C. M.; Dunkin, I. R. *Chem. Commun.* **2002**, 1880-1881.

39. Karmakar, R.; Samanta, A. *J. Phys. Chem. A* **2002**, *106*, 4447-4452.
40. Bonhote, P.; Dias, A. P.; Papageorgiou, N.; Kalyanasundaram, K.; Gratzel, M. *Inorg. Chem.* **1996**, *35*, 1168-1178.
41. Huddleston, J. G.; Willauer, H. D.; Swatloski, R. P.; Visser, A. E.; Rogers, R. D. *Chem. Commun.* **1998**, 1765-1766.
42. MacFarlane, D. R.; Meakin, P.; Sun, J.; Amini, N.; Forsyth, M. *J. Phys. Chem. B* **1999**, *103*, 4164-4170.
43. Thompson, N. L. In *Topics in Fluorescence Spectroscopy*; Lakowicz, J. R., Ed.; Plenum Press: New York, 1991; Vol. 1; pp 337-378.
44. *Fluorescence Correlation Spectroscopy, Theory and Applications*; Rigler, R.; Elson, E. S., Eds.; Springer-Verlag: Heidelberg, 2001.
45. *Single-Molecule Detection in Solution, Methods and Applications*; Zander, C.; Enderlein, J.; Keller, R. A., Eds.; Wiley-VCH: Berlin, 2002.
46. Aragon, S. R.; Pecora, R. *J. Chem. Phys.* **1976**, *64*, 1791-1803.
47. Widengren, J.; Mets, U.; Rigler, R. *J. Phys. Chem.* **1995**, *99*, 13368-13379.
48. Rigler, R.; Mets, U.; Widengren, J.; Kask, P. *Eur. Biophys. J.* **1993**, *22*, 169-175.
49. Schroder, U.; Wadhawan, J. D.; Compton, R. G.; Marken, F.; Suarez, P. A. Z.; Consorti, C. S.; de Souza, R. F.; Dupont, J. *New J. Chem.* **2000**, *24*, 1009-1015.
50. Seddon, K. R.; Stark, A.; Torres, M. J. *Pure Appl. Chem.* **2000**, *72*, 2275-2287.
51. Kimizuka, N.; Nakashima, T. *Langmuir* **2001**, *17*, 6759-6761.
52. Ohno, O.; Kaizu, Y.; Kobayashi, H. *J. Chem. Phys.* **1993**, *99*, 4128-4139.
53. Maiti, N. C.; Mazumdar, S.; Periasamy, N. *J. Phys. Chem. B* **1998**, *102*, 1528-1538.
54. Pasternack, R. F.; Fleming, C.; Herring, S.; Collings, P. J.; dePaula, J.; DeCastro, G.; Gibbs, E. J. *Biophys. J.* **2000**, *79*, 550-560.
55. VanPatten, P. G.; Shreve, A. P.; Donohoe, R. J. *J. Phys. Chem. B* **2000**, *104*, 5986-5992.
56. Andrade, S. M.; Costa, S. M. B. *Biophys. J.* **2002**, *82*, 1607-1619.
57. Zaks, A.; Klibanov, A. M. *Science* **1984**, *224*, 1249-1251.
58. Zaks, A.; Klibanov, A. M. *Proc. Natl. Acad. Sci. U. S. A.* **1985**, *82*, 3192-3196.
59. Klibanov, A. M. *Trends Biochem. Sci.* **1989**, *14*, 141-144.
60. Jaenicke, R.; Zavodszky, P. *FEBS Lett.* **1990**, *268*, 344-349.
61. Zavodszky, P.; Kardos, J.; Svingor, A.; Petsko, G. A. *Proc. Natl. Acad. Sci. U. S. A.* **1998**, *95*, 7406-7411.
62. Klibanov, A. M. *Anal. Biochem.* **1979**, *93*, 1-25.

Chapter 19

Ionic Liquids Create New Opportunities for Nonaqueous Biocatalysis with Polar Substrates: Acylation of Glucose and Ascorbic Acid

Seongsoon Park[1], Fredrik Viklund[2], Karl Hult[2], and Romas J. Kazlauskas[1,2,*]

[1]Department of Chemistry, McGill University, 801 Sherbrooke Street West, Montreal, Quebec H3A 2K6, Canada
[2]Royal Institute of Technology (KTH), Department of Biotechnology, AlbaNova University Centre, SE–106 91 Stockholm, Sweden

Lipase-catalyzed reactions of polar substrates are inefficient in organic solvents. Nonpolar organic solvents do not dissolve polar substrates, while polar organic solvents inactivate lipases. Ionic liquids such as 1-alkyl-3-methyl imidazolium tetrafluoroborate are as polar as *N*-methyl formamide or methanol, but, unlike these solvents, ionic liquids do not inactivate lipases. This unusual feature creates opportunities for nonaqueous biocatalysis with polar substrates. First, we describe a simple purification involving filtration through silica gel, which yields ionic liquids that work reliably as solvents in lipase-catalyzed reactions. Next, we report two examples that exploit these unique advantages of ionic liquids. First, lipase-catalyzed acetylation of glucose was up to twelve times more regioselective in ionic liquids than in acetone. Second, lipase catalyzed the acylation of ascorbic acid to make fat-soluble antioxidants. In some cases, reactions in ionic liquids were comparable or slower than in *tert*-amyl alcohol, but in typical cases, the reactions in ionic liquids were twice as fast and proceeded to higher conversion. Ionic liquids also offer the possibility to use vacuum to remove water formed by the esterification and drive the equilibrium even further toward product.

Introduction

A long-standing problem in biocatalysis is reactions of polar substrates under nonaqueous conditions. Reactions such as acylation of an alcohol require nonaqueous conditions to avoid competing hydrolysis. However, polar substrates such as sugars dissolve only in the most polar organic solvents such as dimethylsulfoxide. Unfortunately enzymes such as lipases inactivate in such polar organic solvents. Current solutions involve compromises. In some cases, researchers use moderately polar organic solvents, where the substrate dissolves slightly and the enzymes retain some activity. Such reactions are usually too slow for preparative use. Another alternative is to modify the substrates (e.g., use an alkyl glycoside instead of a glycoside) and use a less polar organic solvent where the enzyme remains active. However, this approach yields an analog of the desired product.

In spite of these difficulties, the ability to catalyze reactions on polar substrates in nonaqueous media is becoming increasingly important. Natural building blocks - peptides, sugars, nucleotides, biochemical intermediates – are important starting materials for pharmaceuticals, fine chemicals and materials. These building blocks are becoming increasingly important in a bio-based economy, where chemicals and materials come from plants and microorganisms.

This paper focuses on room temperature ionic liquids (*1*) as a solution to biocatalysis reactions with polar substrates under nonaqueous conditions. Ionic liquids are polar solvents (comparable to methanol) and readily dissolve polar substrates. However, for reasons that are still not clear, ionic liquids do not denature lipases, as would an organic solvent of comparable polarity. For this reason lipase-catalyzed reactions of polar substrates proceed more efficiently or more selectively in ionic liquids. Several groups have reported enzyme-catalyzed reactions in ionic liquids (*2-5*). The advantages of using ionic liquids over an organic solvent varied for each case and included increased enantioselectivity (*3*), increased stability of the enzyme (*4*) or increased molecular weight of the product polymer (*5*). Here we focus on advantages related to reactions of polar substrates.

The first example is the acetylation of glucose with vinyl acetate catalyzed by lipase B from *Candida antarctica* (CAL-B) (*6*). This acetylation is more regioselective in ionic liquids than in moderately polar organic solvents such as tetrahydrofuran. This increased regioselectivity yields only 6-*O*-acetyl D-glucose instead of a mixture of 6-*O*-acetyl- and 3,6-*O*-diacetyl D-glucose. The increased solubility of glucose relative to 6-*O*-acetyl D-glucose in ionic liquids accounts for the increased regioselectivity.

The second example is the CAL-B-catalyzed acylation of L-ascorbic acid (vitamin C) with unactivated fatty acids to make fat-soluble antioxidants, such as 6-*O*-ascorbyl palmitate or 6-*O*-ascorbyl oleate. Reaction of the polar L-ascorbic acid with nonpolar fatty acids proceeds to higher conversion in ionic liquids than in an organic solvent (*tert*-amyl alcohol). In addition, since ionic liquids are nonvolatile, they offer the possibility of using vacuum to remove water and shift the equilibrium of the reaction more toward product formation.

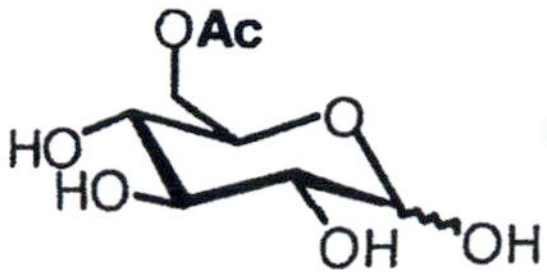

6-*O*-acetyl D-glucose

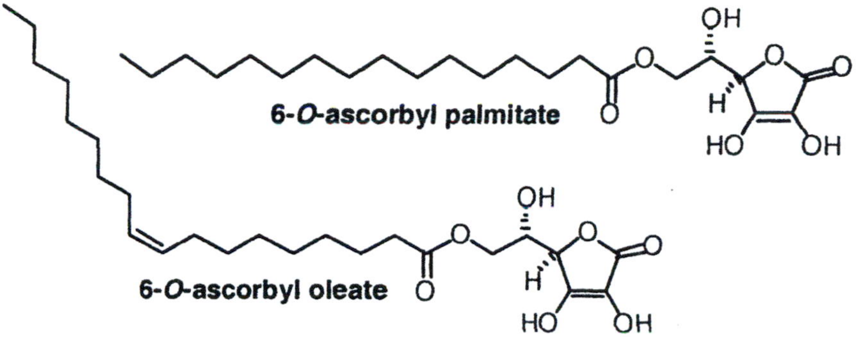

6-*O*-ascorbyl palmitate

6-*O*-ascorbyl oleate

Results

Ionic liquids, prepared either by literature procedures (*7*) or straightforward modifications, did not work reliably as solvents for lipase-catalyzed reactions. In some cases, reactions proceeded well; in other cases reactions proceeded slowly or not at all. Since the structures of the ionic liquids were similar, we suspected that impurities might cause the unpredictable behavior.

The synthesis of ionic liquids involved initial preparation of the halide salt followed by exchange of the halide with tetrafluoroborate, Scheme 1. A likely impurity in ionic liquids is the halide salt due to incomplete exchange. Indeed, ionic liquids gave a precipitate with silver nitrate solution, thereby confirming the presence of halide. For this reason, we purified all ionic liquids to remove halide salts.

Purification of Ionic Liquids

Purification involved filtration of the diluted ionic liquid through silica gel to remove traces of 3-alkyl-1-methylimidazolium halide and then washing with saturated aqueous sodium carbonate to remove cloudiness due to fine particles of silica gel. Finally, drying the solution over magnesium sulfate and removing the diluent (methylene chloride) by vacuum yielded the purified ionic liquid. An alternative purification replaced the sodium carbonate wash and drying with magnesium sulfate with a filtration through neutral alumina. This second method avoided traces of the basic carbonate anion in the ionic liquid. Both methods yielded ionic liquids that work reliably in all lipase-catalyzed reactions that we tested. These procedures yielded six 3-alkyl-1-methylimidazolium tetrafluoroborate ionic liquids, Scheme 1.

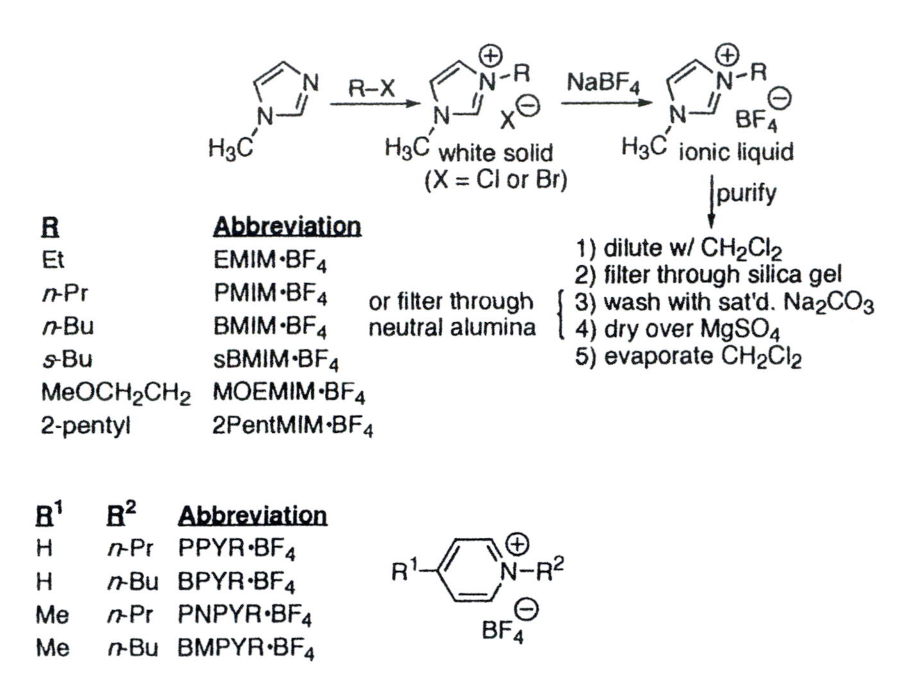

Scheme 1. *Preparation and purification of 3-alkyl-1-methylimidazolium tetrafluoroborate ionic liquids for biocatalysis. Similar methods yielded the related* N*-alkylpyridinium tetrafluoroborate ionic liquids.*

Similar reactions and purifications gave five other ionic liquids: one hexafluorophosphate salt, 3-*n*-butyl-1-methylimidazolium hexafluorophosphate, and the four based on pyridinium and 4-methylpyridinium cations shown below.

Polarity of Ionic Liquids is Similar to that for Polar Organic Solvents

To compare the polarity of ionic liquids and organic solvents, we measured their polarities with Reichardt's dye (*8*). Polar solvents stabilize the polar ground state of Reichardt's dye thereby shifting its color to shorter wavelengths. We

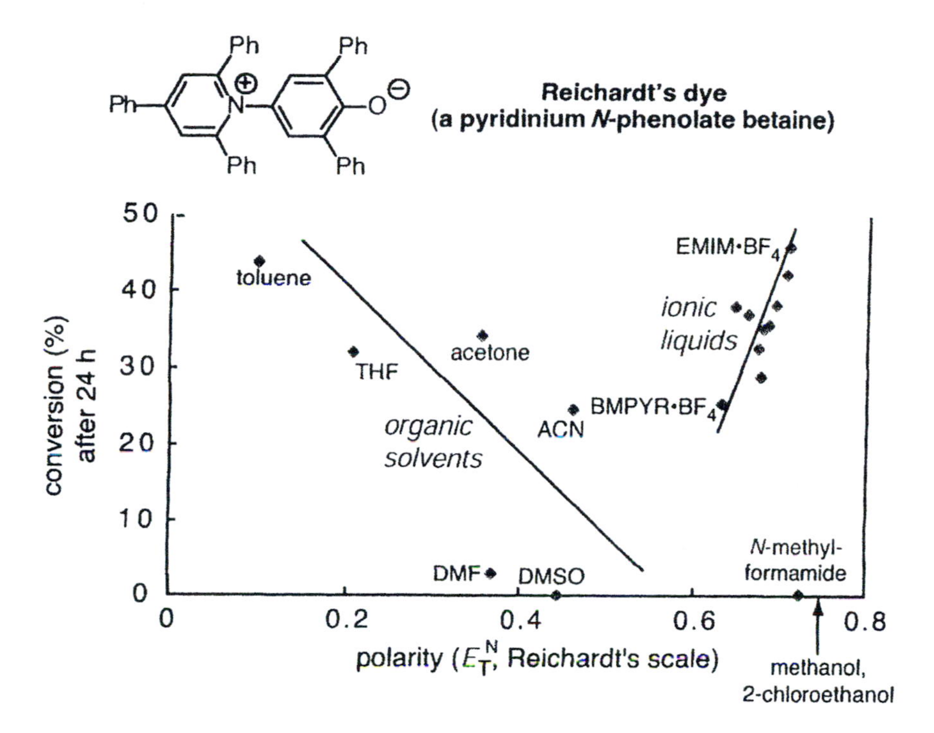

Figure 1. *The conversion for* Pseudomonas-cepacia-*lipase catalyzed acetylation of racemic 1-phenylethanol with vinyl acetate decreased in polar organic solvents, but remained high for ionic liquids in spite of their highly polar nature. This reaction is highly enantioselective, so the maximum conversion is 50%. The trend lines are not a fit to theory, but only to guide the eye.*

compared the polarities of the different solvents using Reichardt's normalized scale where tetramethylsilane has a value of zero and water has a value of one. The polarity values for the ionic liquids we used ranged from 0.63 to 0.71 with the most polar being EMIM•BF_4 and the least polar being BMPYR•BF_4, see x-axis of Figure 1. BMIM•BF_4 is more polar that BMIM•PF_6.

Organic solvents with polarities similar to that of the ionic liquids include: methanol, 2-chloroethanol, *N*-methyl formamide, diethylene glycol and 1,2-propanediol. Most of these are hydroxylic solvents, which are not suitable for acylation reactions since the solvent would compete with the substrate for the acyl group.

Others also measured the polarity of ionic liquids using another solvatochromic dye, Nile Red (*9*) or using fluorescent probes (*10*). Although only a few ionic liquids are the same as the ones we measured, the relative ranking of the polarities is the same.

High Activity of Lipases in Ionic Liquids in Spite of their High Polarity

Lipases showed good activity in ionic liquids even though they showed no, or little, activity in normal organic solvents with similar polarities. As a model reaction, we used the acetylation of racemic 1-phenylethanol with vinyl acetate catalyzed by lipase from *Pseudomonas cepacia*, PCL, equation 1. This reaction is highly enantioselective so the maximum conversion was 50%. We compared the rates of reaction and enantioselectivities in ionic solvents to those in normal

OH + OAc —PCL / solvent→ OAc + OH + O (1)

organic solvents. In all cases, the enantioselectivity of the reaction remained high: E >200. Surprisingly, the conversion after 24 h also remained high in ionic liquids despite their high polarity, Figure 1 above.

For normal organic solvents, the acetylation reaction proceeds well in nonpolar solvent, but not in polar solvents. The reaction is nearly complete in toluene, partially complete in tetrahydrofuran (THF), acetone or acetonitrile (ACN), but proceeds very slowly or not at all in the more polar *N,N*-dimethylformamide (DMF), dimethylsulfoxide (DMSO), or *N*-methylformamide. Although the ionic liquids are highly polar (similar to *N*-methyl-formamide), the acetylation reaction proceeds well in all ionic liquids tested. The reaction is nearly complete in EMIM•BF_4 and MOEMIM•BF_4 and partially complete in all other ionic liquids. In addition, the trend for ionic liquids is for higher degrees of conversion as the polarity of the ionic liquid increases, while the trend for organic solvents is the opposite – lower degrees of conversion as the polarity increases.

Since the substrates for this model reaction dissolve in both nonpolar organic solvent and in ionic liquids, there is no obvious advantage to using ionic liquids in this case.

More Regioselective Acylation of Glucose in Ionic Liquids

Since ionic liquids are polar solvents that do not denature lipases, they may be ideal for lipase-catalyzed transformations of polar substrates. As a first example, we examined the lipase-catalyzed 6-*O*-acetylation of glucose catalyzed by lipase B from *Candida antarctica* (CAL-B), equation 2, Table I.

β-D-glucose → (OAc vinyl acetate, CAL-B) 6-O-acetyl + 3,6-O-diacetyl glucose

organic solvents: **2-3: 1**
ionic liquids: **>13: 1**

In organic solvents such as acetone and tetrahydrofuran (THF), the 6-*O*-acetylation reaction proceeded along with further acetylation of the 3-position. In acetone, acetylated products formed in 72% yield, of which 76% was the desired 6-*O*-acetyl compound (~3:1 selectivity). In THF, glucose reacted completely, but only 53% was the desired 6-*O*-acetyl compound (~2:1 selectivity). Even at a lower extent of conversion, the regioselectivity remained low. In acetone at 42% conversion, 82% was the 6-*O*-acetyl compound (~5:1 selectivity), while in THF at 50% conversion, 85% was the 6-*O*-acetyl compound (~6:1 selectivity). The low selectivity is likely related to the poor solubility of glucose in these organic solvents (0.02-0.04 mg/mL at 60 °C (*11*)). Glucose remains a suspended solid and the initial 6-*O*-acetylation yields a more soluble compound, which then undergoes further acetylation to the 3,6-*O*-diacetyl derivative.

Table I. Regioselective CAL-B-Catalyzed Acetylation of Glucose[a]

Solvent	*Final composition of reaction mixture*			*Conversion %*	*Mono-acylation, %*
	D-Glucose, %	*6-O-Acetyl-D-Glucose, %*	*3,6-O-Diacetyl-D-Glucose, %*		
EMIM•BF_4	49.6	50.3	0.0	50.4	99.9
MOEMIM•BF_4	0.0	93.0	6.9	99.9	93.1
PMIM•BF_4	72.2	27.8	0.0	27.8	99.8
BMIM•BF_4	22.4	68.9	8.7	77.6	88.8
sBMIM•BF_4	9.8	79.2	10.8	90.1	87.9
BMIM•PF_6	70.5	11.3	18.1	29.5	38.5
BPYR•BF_4	58.0	37.3	4.7	42.0	88.7
PPYR•BF_4	56.4	38.6	5.0	43.6	88.4
Acetone	27.7	55.0	17.4	72.3	76.1
THF	0.0	52.6	47.4	99.9	52.6

[a]Conditions: 0.5 mmol β-D-glucose, 1 mmol vinyl acetate, 1 mL solvent, 30 mg Novozyme SP435, 36 h, 55 °C, Data from reference 6. After the reaction, both remaining glucose and the acetylated products were a mixture of anomers. The conversion was measured by gas chromatography after derivatization with chlorotrimethylsilane and 1,1,1,3,3,3-hexamethyldisilazane (*12*). The acylation positions were determined by COSY experiments.

On the other hand, acetylation of glucose proceeded with much higher selectivity for monoacetylation in ionic liquids than in organic solvents. In the seven ionic liquids containing a tetrafluoroborate anion, the 6-*O*-acetylation proceeded with 42-99% conversion, of which 88-99% was the desired 6-*O*-acetyl glucose (~7:1 to ~100:1 selectivity). The best ionic liquid was MOEMIM•BF_4, where all the glucose was acetylated and 93% was the desired 6-*O*-acetyl compound (~13:1 selectivity). The one ionic liquid with a hexafluorophosphate anion, BMIM•PF_6, showed both slow reaction (29% conversion) and low selectivity (39% monoacetyl, ~0.6:1 selectivity).

The higher solubility of glucose in ionic liquids correlates with the higher regioselectivity in these solvents. Approximately 100 times more glucose dissolves in the best ionic liquid, MOEMIM•BF_4 ~5 mg/mL at 55 °C, than in acetone or THF. On the other hand, glucose is not very soluble in the worst ionic liquid, BMIM•PF_6, <1 mg/mL at 55 °C. The increased solubility of glucose increases the relative concentration of the desired reactant, glucose, relative to the undesired reactant, 6-*O*-acetyl glucose.

Initial experiments also showed that CAL-B catalyzes the regioselective acylation of maltose monohydrate, a disaccharide that is even more polar than glucose. Using the reaction conditions in Table I, but only half the amount of maltose (0.25 mmol instead of 0.5 mmol) and MOEMIM•BF_4 as the solvent, yielded 50% of acetylated products.

Regioselective Acylation of Ascorbic Acid in Ionic Liquids

Another example of a lipase-catalyzed acylation of a polar substrate is the acylation of ascorbic acid (vitamin C) with a fatty acid to make a fat-soluble antioxidant, equation 2. The choice of solvent for this reaction is more difficult because one reactant is polar (ascorbic acid), the other is nonpolar (fatty acid) and the product is amphiphilic.

palmitic acid or oleic acid + ascorbic acid —(CAL-B, molecular sieves, solvent, 60 °C)→ **6-*O*-ascorbyl palmitate or 6-*O*-ascorbyl oleate** (2)

Although ascorbic acid dissolved readily in all ionic liquids (e.g., >130 mg/mL of sBMIM·BF_4 at 60 °C), the other reactant, palmitic or oleic acid, dissolved only in the most hydrophobic ionic liquids, sBMIM•BF_4 or 2PentMIM•BF_4. Not surprisingly therefore, the initial reaction rate was 1.8-8 times faster in sBMIM•BF_4 or 2PentMIM•BF_4 than in other ionic liquids. (Data not shown.) Reactions in the more hydrophobic 2PentMIM•BF_4, which we prepared specifically for this reaction, showed higher conversions than reactions in sBMIM•BF_4. (Compare entries 1 and 3 of Table II).

In the best cases, the rates of reaction and conversion were slightly better in an organic solvent, *tert*-amyl alcohol, than in ionic liquids (entry 10, Table II). However, these best cases were difficult to reproduce and more typical reactions in *tert*-amyl alcohol were up to two times slower than in ionic liquids and reached only 25-40% conversion (entry 9, Table II). We suspect that incomplete drying of the *tert*-amyl alcohol causes the lower conversions. Acylation occurred only at the primary alcohol position of ascorbic acid in all cases.

Consistent with their role as antioxidants, the product ascorbyl fatty acid esters were very sensitive to oxidation. For this reason, the conversion (amount of ascorbic acid consumed) was always higher than the yield (amount of product formed). For example, entry 2 in Table II shows 42% conversion, but only 16% yield. To minimize this oxidation, we used an alternate purification method for the ionic liquid, filtration through neutral alumina, which avoids the wash with the strongly basic sodium carbonate solution. This modified purification gave a higher yield: 43% conversion, 40% yield (entry 1, Table II).

Table II. CAL-B-Catalyzed Acylation of Ascorbic Acid[a]

Entry	*Solvent*	*Additive*	*Acylating acid*	*Conv., %*	*Approx. Rate[b]*	*Yield, %*
1	sBMIM•BF_4	*c*	Palmitic	43	0.090	40
2	sBMIM•BF_4[c]	*c, d*	Palmitic	42	0.088	16
3	2PentMIM•BF_4	*c*	Palmitic	74	0.15	53
4	sBMIM•BF_4	polypropylene[c]	Palmitic	63	0.13	43
5	sBMIM•BF_4	10 vol% hexane	Palmitic	66	0.15	54
6	sBMIM•BF_4	10 vol% hexane	Oleic	72	0.16	44
7	2PentMIM•BF_4	10 vol% hexane	Palmitic	73	0.16	62
8	2PentMIM•BF_4	10 vol% hexane	Oleic	78	0.17	65
9	*t*-Amyl alcohol	*e*	Palmitic	25-40	0.079 – 0.13	nd
10	*t*-Amyl alcohol	*f*	Palmitic	71-86	0.22 – 0.27	nd

[a]Conditions: 200 mM (100 μmol) of ascorbic acid, 240 mM of palmitic acid, 0.5 mL of solvent, 45 mg of CAL-B, 50 mg of molecular sieve 4A, 2 mg of internal standard (9-fluorenone), 60 °C, 10 h, under nitrogen, stirred with magnetic stirring bar. The conversion (amount of starting material consumed) and yield (amount of product formed) were

determined using HPLC by comparison with internal standard. Unless otherwise noted, ionic liquids were purified by the filtration-through-neutral-alumina method. [b] μmol/h/mg CAL-B for the consumption of starting material. [c]Only 20 mg of CAL-B were used, but the time was extended to 24 h. [d]Ionic liquid was purified by filtration through silica gel and washing with saturated sodium carbonate solution. [e]Typical yields using the following conditions: 57 mM (570 μmol) of ascorbic acid, 57 mM of palmitic acid, 10 ml solvent, 50 mg of CAL-B, 36 h, 100 mg of molecular sieve 3A. [f]Best yields under the conditions in note e.

Another factor that limited conversion was the poor solubility of ascorbyl palmitate in ionic liquids (<10 mg/mL of 2PentMIM·BF_4). In some cases, this precipitate occluded the catalyst and stopped the reaction. To minimize the interaction between the product ester and the hydrophobic support of the lipase, we added either hexane or polypropylene beads. The hexane did not dissolve the ascorbyl palmitate, but prevented the oiling out on the immobilized lipase. For example, under similar conditions condensation of ascorbic acid with palmitic acid in sBMIM•BF_4 with no additive gave 43% conversion (entry 1), while addition of 10 vol% hexane increased the conversion to 73% (entry 5). As an alternative, addition of 50 wt% polypropylene beads increased the conversion to 63% (entry 4).

Discussion

Solvent purification is a key first step in most organic synthesis laboratories. Not surprisingly, solvent purification is also critical for reproducible results when working with ionic liquids. The purification methods outlined here involve filtration through silica gel followed by either a wash with aqueous sodium carbonate or a filtration through neutral alumina. These methods remove traces of chloride salts from the ionic liquids and possibly other unidentified impurities. Ionic liquids prepared in this manner worked reliably and consistently as solvents for lipase-catalyzed reactions.

Besides potential environmental benefits, ionic liquids also expand the accessible solvent polarity range for lipase-catalyzed reactions. Lipase-catalyzed acylations did not proceed in a polar organic solvent like *N*-methylformamide, but did proceed in ionic liquids with similar polarities. As researchers have previously noted (*13*), ionic liquids can dissolve polar molecules such as carbohydrates. Lipase-catalyzed acylations of these polar substrates work better in ionic liquids than in organic solvents, but the precise advantage differs for each case.

Acetylation of 1-phenylethanol (a non polar substrate) was as fast in ionic liquids as in organic solvents like toluene. There was no obvious advantage to carrying out this reaction in an ionic liquid.

Acetylation of glucose was more regioselective for the 6-hydroxyl group in ionic liquids because of the higher solubility of glucose. Although the catalyst usually controls the regioselectivity of a reaction, with poorly soluble substrates and products such as glucose and its derivatives, the relative solubility also contributes. The product 6-*O*-acetyl glucose is much more soluble in organic solvents than glucose and therefore underwent further acetylation. In ionic liquids, glucose is more soluble so acetylation of glucose is fast enough to compete with the further acetylation of the product. A less attractive alternative is a dilute reaction mixture where the glucose dissolves. For example, *tert*-butyl alcohol dissolves glucose to 2.4 mg/mL at 45 °C and CAL-B was highly regioselective for the primary alcohol position in this solvent (*14*), but the conditions are about forty times more dilute than our conditions in ionic liquid.

The higher solubility of maltose, a disaccharide, in ionic liquids also facilitates acetylation. Previous lipase-catalyzed acylations of maltose required refluxing *tert*-butyl alcohol as the solvent (*15*).

Acylation of ascorbic acid with fatty acids proceeded to higher conversion in ionic liquids that dissolved both substrates than a typical reaction in *tert*-amyl alcohol. The advantage of this reaction in ionic liquids as compared to previous reports in organic solvents is that fatty acids can be used directly as acyl donors (*16*). Further, since ionic liquids are not volatile, one could shift equilibrium toward synthesis by vacuum removal of water. Researchers previously used the vacuum removal of water to increase the molecular weight of condensation polymers (*17*).

The best ionic liquid differed for each reaction, but was usually the one that best dissolved the substrates. For glucose, acetylation was fastest in MOEMIM,BF_4 and slowest, by about a factor of 3, in either PMIM•BF_4 or BMIM•PF_6. The regioselectivity was high in all ionic liquids except for one, BMIM•PF_6. For acylation of ascorbic acid with fatty acids, the best ionic liquid was 2PentMIM•BF_4. This liquid dissolved both the ascorbic acid and the fatty acid and reactions were up to eight times faster than in other ionic liquids, which dissolved the ascorbic acid, but not the fatty acid. In both cases, the ability to dissolve the substrates was a key parameter, but perhaps not the only one.

This research focused on efficient reactions and did not address how best to isolate products from ionic liquids. Possibilities include crystallization, extraction with a polar organic solvent or even a supercritical fluid (*18*).

Experimental Section

General. ^1H NMR spectra were recorded in acetone-d_6 or $CDCl_3$ at 400 MHz (M400, Varian) and 500 MHz (Bruker). An immobilized form of lipase B from *Candida antarctica* (Novozym SP435) was donated from Novo Nordisk (Denmark). Other chemicals were purchased from Sigma-Aldrich.

Synthesis of ionic liquids. 1-Alkyl-3-methyl-imidazolium bromide was prepared from *N*-methylimidazole and alkyl bromide by literature methods (*7*). The tetrafluoroborate salts were prepared by a slight modification of literature procedures. 1-Alkyl-3-methylimidazolium-bromide (0.40 mol) was added to a suspension of $NaBF_4$ (1.2 equiv, 52.7 g, 0.48 mol) in acetone (150 mL). After the mixture was stirred for 48 h at room temperature, the sodium bromide precipitate was removed by filtration and the filtrate concentrated by rotary evaporation to an oil (~100 mL). This oil gave a precipitate when mixed with aqueous silver nitrate indicating that it still contained some some 1-alkyl-3-methyl imidazolium halide.

Purification of MOEMIM·BF_4. The crude ionic liquid was diluted with methylene chloride (200 mL) and filtered through silica gel (~100 g). This step removed the 1-alkyl-3-methyl imidazolium halide since the filtrate no longer gave a precipitate mixed with aqueous silver nitrate. The solution was washed twice with sat'd sodium carbonate aqueous solution (40 mL) and dried over anhydrous magnesium sulfate. Removal of solvent under vacuum yielded a pale yellow oil, 50~70% yield. ^{1}H-NMR: δ 8.95 (1H, s); 7.71 (1H, dd); 7.68 (1H, dd); 4.51 (2H, t); 4.05 (3H, s); 3.80 (2H, t); 3.34 (3H, s).

Purification of (±)-2PentMIM•BF_4. The crude ionic liquid was diluted with methylene chloride (200 mL), filtered through silica gel (~100 g) and then filtered through neutral aluminum oxide (~50 g) to remove traces of silica gel. Removal of solvent under vacuum yielded a pale yellow oil, 60% yield. ^{1}H NMR (400 MHz, acetone-d_6): δ 9.04 (s, 1H); 7.82 (dd, 1H); 7.72 (dd, 1H); 4.66 (m, 1H); 4.04 (s, 3H); 1.92 (m, 2H); 1.60 (d, 3H); 1.30 (m, 2H); 0.92 (t, 3H). ^{13}C NMR: 136.89, 124.22, 120.64, 57.50, 38.70, 36.01, 20.87, 19.15, 13.28.

Transesterification of *sec*-phenethyl alcohol. Vinyl acetate (92 μL, 1.0 mmol) and *sec*-phenethyl alcohol (13 μL, 1.0 mmol) were added to a suspension of lipase from *Pseudomonas cepacia* (PCL, Amano Pharmaceutical Co. Nagoya, Japan, 20.0 mg) in solvent (1.0 mL of either organic solvents or ionic liquids) and stirred at 25 °C. The reactions were monitored by TLC (ethyl acetate:hexane, 1:3). After 24 h, the reaction mixture was extracted with hexane (3 mL) and the hexane extract was analyzed by gas chromatography on a Chiralsil-Dex CB column (Chromopak). The conversion, c, was calculated from the enantiomeric excess of the product, ee_p, and of the starting material, ee_s, using the equation below (*19*).

$$c = \frac{ee_S}{ee_S + ee_P}$$

Acetylation of glucose. Vinyl acetate (92 μL, 1.0 mmol), β-D-glucose (90 mg, 0.5 mmol), and Novozyme SP435 (30 mg) were mixed with solvent (1.0 mL of either organic solvents or ionic liquids) and stirred at 55 °C. After 36 h, pyridine (2 mL), 1,1,1,3,3,3-hexamethyldisilazane (1 mL) and chloromethylsilane (1 mL) were added to the reaction mixture. The mixture was extracted with hexane (5 mL) and analyzed by gas chromatography on the column above. Temperature program: initial temperature 180 °C for 2 min, increase to 190 °C over 10 min, and hold for 28 min.

Acylation of ascorbic acid. Oleic acid (38 μL, 0.12 mmol) or palmitic acid (31 mg, 0.12 mmol), ascorbic acid (18 mg, 0.1 mmol), 9-fluorenone (2 mg, internal standard), molecular sieve (50 mg), and Novozym SP435 (20 mg) were mixed with ionic liquid (0.5 mL) and stirred at 60 °C under nitrogen. After 24 h, methanol (10 mL) was added to the reaction mixture. Analysis of the mixture was performed by high performance liquid chromatography on a C-18 column (4.6 mm id x 25 cm) eluted with 95% methanol/5%water containing 0.5% acetic acid at 1 mL/min. Peaks were detected by UV absorbance at 254 nm. Retention time: ascorbic acid, 1.92 min; internal standard (9-fluorenone), 2.41 min; ascorbyl oleic acid ester, 3.66 min; ascorbyl palmitic acid ester, 3.60 min.

6-*O*-L-Ascorbyl oleate. ^{1}H NMR (500 MHz, $CDCl_3$): δ 5.40 (m, 2H); 4.86 (d, 1H); 4.45 (m, 1H); 4.26 (d, 2H); 2.81 (m, 3H); 2.40 (t, 2H); 2.15 (m, 4H); 1.65 (m, 2H); 1.30 (br s, 20H); 0.90 (t, 3H).

References

1. Review: Welton, T. *Chem. Rev.* **1999**, *99*, 2071-2083.
2. Lau, R. M.; van Rantwijk, F.; Seddon, K. R.; Sheldon, R. A. *Organic Lett.* **2000**, *2*, 4189-4191; Itoh, T.; Akasaki, E.; Kudo, K.; Shirakami, S. *Chem. Lett.* **2001**, 262-263; Husum, T. L.; Jorgensen, C. T.; Christensen, M. W.; Kirk, O. *Biocatal. Biotransform.* **2001**, *19*, 331-338.
3. Kim, K.-W.; Song, B.; Choi, M.-Y.; Kim, M.-J. *Org. Lett.* **2001**, *3*, 1507-1509; Schofer, S. H.; Kaftzik, N.; Wasserscheid, P.: Kragl, U. *Chem. Commun.* **2001**, 425-426; Kielbasinski, P.; Albrycht, M.; Luczak, J.; Mikolajczyk, M. *Tetrahedron Asymm.* **2002**, *13*, 735-738.
4. Erbeldinger, M.; Mesiano, A. J.; Russell, A. J. *Biotechnol. Prog.* **2000**, *16*, 1131-1133; Lozano, P.; de Diego, T.; Guegan, J.-P.; Vaultier, M.; Iborra, J. L. *Biotechnol. Bioeng.* **2001**, *75*, 563-569; Kaftzik, Nicole; Wasserscheid, P.; Kragl, U. *Org. Process Res. Dev.* **2002**, *6*, 553-557.
5. Uyama, H.; Takamoto, T.; Kobayashi, S. *Polymer J. (Tokyo)* **2002**, *34*, 94-96.
6. Park, S.; Kazlauskas, R. J. *J. Org. Chem.* **2001**, *66*, 8395-8401.
7. Huddleston, J. G.; Willauer, H. D.; Swatloski, R. P.; Visser, A. E.; Rogers, R. D. *J. Chem. Soc., Chem. Commun.* **1998**, 1765-1766; Suarez, P. A. Z.; Dullius, J. E. L.; Einloft, S.; De Souza, R. F.; Dupont, J. *Polyhedron*, **1996**, *15*, 1217-1219. Also see: Dupont, J.; Consorti, C. S.; Suarez, P. A. Z.; de Sousa, R. F. *Org. Synth.* **2002**, *79*, 236-243; Law, M. C.; Wong, K. Y.; Chan, T. H. *Green Chem.* **2002**, *4*, 328-330.
8. Reichardt, C. *Chem. Rev.* **1994**, *94*, 2319-2358.
9. Carmichael, A. J.; Seddon, K. R. *J. Phys. Org. Chem.* **2000**, *13*, 591-595.
10. Aki, S. N. V. K.; Brennecke, J. F.; Samanta, A. *Chem. Commun.* **2001**, 413-414.
11. Cao, L., Fischer, A., Bornscheuer, U. T., Schmid, R. D. *Biocatal. Biotransform.* **1997**, *14*, 269-283.
12. Sweeley, C. C.; Bentley, R.; Makita, M.; Wells, W. W. *J. Am. Chem. Soc.* **1963**, *85*, 2497-2507.
13. Kimizuka, N.; Nakashima, T. *Langmuir* **2001**, *17*, 6759-6761; Swatloski, R. P.; Spear, S. K.; Holbrey, J. D.; Rogers, R. D. *J. Am. Chem. Soc* **2002**, *124*, 4974-4975; Khan, N.; Moens, L. In *Ionic Liquids*; Rogers, R. D.; Seddon,

K. R., Eds. ACS Symposium Series 818; American Chemical Society: Washington, DC, 2002, pp 360-372.

14. Degn, P.; Pedersen, L. H.; Duus, J. Ø.; Zimmermann, W. *Biotechnol. Lett.* **1999**, *21*, 275-280.

15. Woudenberg-van Oosterrom, M.; van Rantwijk, F.; Sheldon, R. A. *Biotechnol. Bioeng.* **1996**, *49*, 328-333; Revew: Plou, F. J.; Crucesa, M. A.; Ferrera, M.; Fuentesa, G.; Pastora, E.; Bernabé, M.; Christensen, M.; Comelles, F.; Parrad, J. L.; Ballesteros, A. *J. Biotechnol.* **2002**, *96*, 55-66.

16. Humeau, C.; Girardin, M.; Rovel, B.; Miclo, A. *J. Biotechnol.* **1998**, *63*, 1-8; *idem, J. Mol. Catal. B: Enzymatic* **1998**, *5*, 19-23; Yan, Y.; Bornscheuer. U. T.; Schmid, R. D. *Biotechnol. Lett.* **1999**, *21*, 1051-1054; Watanabe, Y.; Minemoto, Y.; Adachi, S.; Nakanishi, K.; Shimada, Y.; Matsuno, R. *Biotechnol. Lett.* **2000**, *22*, 637-640; Luhong, T.; Hao, Z.; Shehate, M. M.; Yunfei, S. *Biotech. Appl. Biochem.* **2000**, *32*, 35-39.

17. Brazwell, E. M.; Filos, D.; Morrow, C. J. *J. Polym. Sci., A,* **1995**, *33*, 89-95.

18. Blanchard, L. A.; Hancu, D.; Beckman, E. J.; Brennecke, J. F. *Nature* **1999**, *399*, 28-29.

19. Chen, C. S.; Fujimoto, Y.; Girdaukas, G.; Sih, C. J. *J. Am. Chem. Soc.* **1982**, *104*, 7294-7299.

Chapter 20

Enzymatic Catalysis in Ionic Liquids and Supercritical Carbon Dioxide

Pedro Lozano[1], Teresa De Diego[1], Daniel Carrié[2], Michel Vaultier[2], and José L. Iborra[1]

[1]Departamento de Bioquímica y Biología Molecular B e Inmunología, Facultad de Química, Universidad de Murcia, P.O. Box 4021, E-30100 Murcia, Spain
[2]Université de Rennes, Institut de Chimie, UMR CNRS 6510, Campus de Beaulieu. Av. Général Leclerc, 35042, Rennes, France

The suitability of ionic liquids (ILs) and supercritical carbon dioxide ($scCO_2$) as reaction media for enzyme catalyzed synthesis was studied. Ionic liquids exhibited an over-stabilizing effect (up 2,500-times) on α-chymotrypsin and *Candida antarctica* lipase B during thermal deactivation. All the assayed ILs were adequate media for lipase-catalyzed transesterification. A continuous biphasic reactor combining free lipase-ILs (catalytic phase) and $scCO_2$ (extractive phase) was used for butyl butyrate synthesis and the kinetic resolution of *rac*-1-phenylethanol. The synthetic activity and operational stability of this green bioreactor was dependent on the supercritical conditions, revealing high catalytic efficiency even at extreme temperatures (120 and 150 °C).

Introduction

One active area of current research in biotechnology is biocatalysis in non-conventional media, involving all reaction systems with a reduced water content, *e.g.* organic solvents *(1)*. However, in many cases, the use of organic solvents in biocatalytic reactions is seriously limited by their denaturative action on proteins, and it is necessary to use enzyme stabilization strategies at the same time (*e.g.* immobilization, chemical modification, presence of polyols, etc) *(2,3)*. Furthermore, organic solvents are usually volatile liquids that evaporate into the atmosphere with detrimental effects to the environment and human health.

Enzymatic reactions based on neoteric solvent, *e.g.* ionic liquids (ILs) and supercritical fluids, appear to be promising alternative media for developing integral green chemical processes, because of the physical and chemical characteristics that these solvents possess. ILs are organic salts composed of cations (*e.g.* alkylimidazolium, alkylammonium, etc.) and anions (*e.g.* BF_4^-, PF_6^-, Tf_2N^-, etc.) that are liquids at room temperature, with negligible vapour pressure and excellent chemical and thermal stability. Additionally, their solvent properties can be finely tuned by changing either the anion or the alkyl substituents in the cation *(4)*. Recently, the use of ionic liquids as reaction media has been extended to biocatalytic processes, and excellent results have been obtained for many different synthetic reactions catalyzed by lipases and proteases *(5-9)* Supercritical fluids, *i.e.* fluids at temperatures and pressures slightly above the critical points (*e.g.* 31 °C and 7.38 MPa for CO_2), exhibit unique combined properties, such as liquid-like density, gas-like diffusivity and viscosities. Their solvent characteristics can be reliably controlled by changing the pressure or the temperature, and they have been successfully used for several large-scale extraction industrial processes (*e.g.* caffeine-free coffee) *(10)*. Supercritical CO_2 has been widely tested as a reaction medium for biocatalysis, but it has been shown to have a direct adverse effect on enzyme activity, including decrease in pH of the enzyme microenvironment *(11)*, covalent modification of free amino groups on enzymes to form carbamates *(12)*, and deactivation by pressurization/depressurization cycles *(13)*.

In this work, the effect of different ILs on the stability of two enzymes (*Candida antarctica* lipase B, CALB, and α-chymotrypsin) is shown, and a continuous green bioreactor, based on the use of CALB-ILs and $scCO_2$, is studied. Two different reactions (butyl butyrate synthesis and the kinetic resolution of rac-1-phenylethanol) are used as reaction models to analyze the catalytic efficiency of this system, including under extreme conditions (*e.g.* at 120 and 150 °C).

Materials and Methods

Materials

Soluble CALB (Novozym 525, EC 3.1.1.3, from Novo Nordik S.A.) was washed by ultrafiltration to eliminate all the low molecular weight additives, obtaining an enzyme solution of 5.75 mg/mL, as determined by Lowry's method. α-Chymotrypsin (EC 3.4.21.1) type II from bovine pancreas, substrates, solvents and other chemicals were purchased from Sigma-Aldrich-Fluka Chemical Co, and were of the highest purity available.

Synthesis of ionic liquids

1-Ethyl-3-methylimidazolium tetrafluoroborate [Emim][BF_4] and 1-butyl-3-methylimidazolium tetrafluoroborate, [Bmim][BF_4] were prepared according to Suarez *et al (14)*. 1-Ethyl-3-methylimidazolium triflimide [Emim][Tf_2N] and 1-buthyl-3-methylimidazolium triflimide, [Emim][Tf_2N], were synthesized following the procedure described by Bonhôte *et al (15)*. 1-Butyl-3-methylimidazolium hexafluorphosphate, [Bmim][PF_6] was prepared according to Huddleston *et al (16)*.

Stability of α-chymotrypsin in liquid media

Twenty microliters of 0.5% w/v α-chymotrypsin were added to different screw-capped vials of 1 mL total volume containing 980 μL of water, 3.2 M sorbitol in water, [Bmim][PF_6], or 0.1M N-acetyl-L-tyrosine ethyl ester (ATEE) in [Bmim][PF_6]. The resulting solutions were homogenized by shaking, and then incubated at 50 °C. At regular intervals, aliquots of 50 μL were extracted and the residual esterase activity was measured, as described previously *(3)*.

Stability of CALB in liquid media

Ten microliters of 0.3% w/v CALB solution in water were mixed with 400 μl of 1-butanol, or ILs ([Emim][BF_4], [Emim][Tf_2N], or [Bmim][PF_6]) containing (or not) 30 μL vinyl butyrate in different screw-capped vials, and the resulting solutions were incubated at 50°C. Then, at selected incubation time, 90 μl of substrate solution (1.68 M vinyl butyrate in 1-butanol) were added to each vial, and the synthetic activity was followed and analyzed by GC (see next sections).

Butyl butyrate synthesis in ILs

Thirty microliters (236 μmol) of vinyl butyrate and 110 μL (1.21 mmol) of 1-butanol were added to different screw-capped vials of 1 mL total capacity containing 300 μl of ILs ([Emim][BF_4], [Emim][Tf_2N], [Bmim][PF_6], or [Bmim][Tf_2N]), or hexane, or 1-butanol. The reaction was started by adding 10 μL of 0.3% w/v CALB solution in water, and run at 50 °C in an oil bath, shaking for 1 h. At regular time intervals, 20 μl aliquots were taken and suspended in 1 mL of hexane. The biphasic mixture was strongly shaken for 3 min to extract all substrates and product into the hexane phase. For hexane and 1-butanol reaction media, aliquots were dissolved in acetone:HCl (99:1 v/v) to stop the reaction. Samples were analyzed by GC.

Enzymatic reactions in ILs-scCO$_2$

Sixty-five microliters of 0.9% w/v CALB solution in water was mixed with 2 mL of ILs ([Emim][Tf$_2$N] or [Bmim][Tf$_2$N]) in a test-tube, and then 3 g of dry Celite were added to absorb the enzyme-IL solution. The final mixture was placed in the cartridge of an ISCO 220SX high pressure extraction apparatus of 10 mL total capacity. Reactions (butyl butyrate synthesis or the kinetic resolution of *rac*-1-phenylethanol) were carried out by continuous pumping of a substrate solution (0.38 M vinyl butyrate and 0.76 M 1-butanol in hexane for the former, or 50 mM vinyl propionate and 100mM *rac*-1-phenylethanol in hexane for the latter) at 0.1 mL/min, and mixed with the scCO$_2$ flow of the system at different pressures and temperatures (Figure 1). The reactor was continuously operated for 4 h.

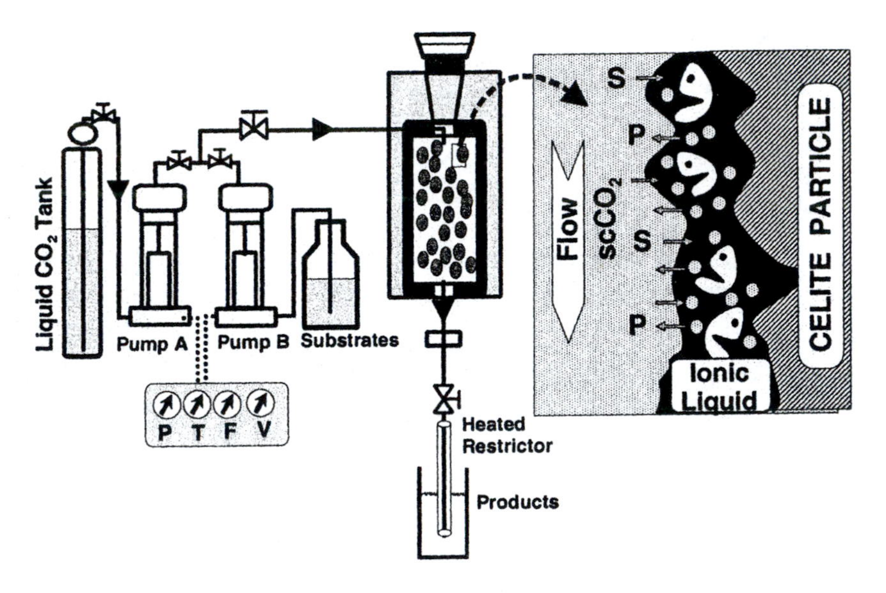

Figure 1. Experimental set-up of the bioreactor with ILs-scCO$_2$.

Substrates and products were fully soluble in scCO$_2$, and the reaction mixtures were recovered by continuous depressurising through a calibrated heated restrictor (1 mL/min, 70 °C) for 30 minute steps. Samples were analysed by GC. In all cases, the synthetic activity of each enzyme-IL system was continuously tested by operation/storage cycles. During the storage steps (20 h/d), the cartridge was maintained under dry conditions at room temperature.

GC analysis

Analyses were performed with a Shimadzu GC-17A instrument equipped with a FID detector. Samples from the butyl butyrate synthetic reactions were analyzed by a Nukol™ column (15 m x 0.53 mm, Supelco), using propyl acetate as internal standard and the following conditions: carrier gas (nitrogen) at 8 kPa (20 mL/min total flow); temperature program: 45 °C, 4 min, 8 °C/min, 133 °C, split ratio, 5:1; detector, 220 °C. One unit of activity was defined as the amount of enzyme that produces 1 μmol of butyl butyrate per min. Samples from the kinetic resolution of *rac*-1-phenylethanol were analyzed by a Beta DEX-120 column (30 m x 0.25 mm x 0.25 μm, Supelco), using butyl butyrate as internal standard and the following conditions: carrier gas (He) at 1 MPa (205 mL/min total flow); temperature programme: 60°C, 10°C/min, 130°C; split ratio, 100:1; detector, 300°C. One unit of activity was defined as the amount of enzyme that produces 1 μmol of *(R)*-1-phenylethyl propionate per min.

Results and Discussion

Stability and activity of α-chymotrypsin and CALB in ILs

In all cases, both α-chymotrypsin and CALB thermal stabilities were followed by a first-order deactivation kinetic, allowing determination of half-life of the enzymes at 50 °C. The influence of the different media on enzyme stability could be comparatively analyzed by a parameter called "protective effect", defined as the ratio of enzyme half-life time in the presence of ILs (or solvent) to the half-life time without any ILs (using water for α-chymotrypsin and 1-butanol for CALB) (see Figure 2).

In the case of α-chymotrypsin, [Bmim][PF_6] has a clearly stabilizing effect (140-times), although lower than that of classical polyhydric additives used for enzyme stabilization, such as sorbitol (700-times) (see Figure 2 I). The protective effect might be explained by changes in the microenvironment of the enzyme produced by the presence of either IL or polyol. Under thermal stress, a greater degree of protein unfolding takes place, resulting in a disruption of all interactions responsible for the maintenance of the tertiary structure (*e.g.* hydrogen, van der Walls, etc). Sorbitol increased the organization of water molecules around the enzyme by means of hydrogen bonds, maintaining the solvophobic interactions inside the protein core, which resulted in the clear enhancement of enzyme stability *(3)*. For small molecules (*e.g.* water, methanol, etc) dissolved in ILs, the local solute structure around ILs is determined by competition between the solute-IL interactions and the interaction between the solvent ions *(17)*. However, in the case of proteins, ILs should be considered as a

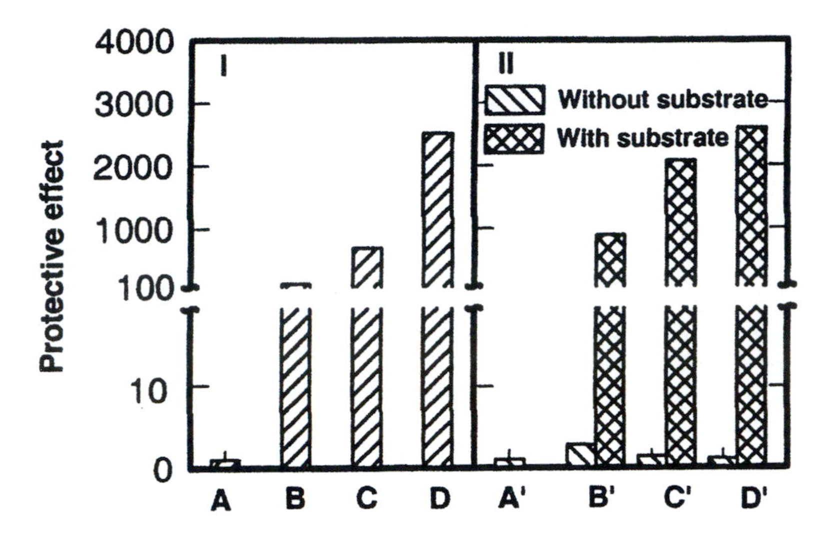

Figure 2. Protective effect of different media on the α-chymotrypsin (I) and CALB (II) stability at 50°C. A, water; B, [Bmim][PF$_6$]; C, 3.2M Sorbitol; D, 100 mM ATEE in [Bmim][PF$_6$]; A', 1-butanol; B', [Emim][BF$_4$]; C', [Emim][NTf$_2$]; D', [Bmim][PF$_6$].

liquid immobilization support, rather than as a solvent, because multipoint enzyme-IL (*e.g.* ionic, hydrogen, van der Walls, etc) interactions may occur. In this way, ILs resulted as an external and flexible supramolecular net able to maintain the active protein conformation under denaturing conditions *(7)*.

In presence of the substrate, the symmetrical charge distribution produced in the active centre of α-chymotrypsin produced ensures internal rigidification of the protein structure, which results in over-stabilization (up to 2500-times). These phenomena were also observed for CALB (Figure 2 II), where the three ILs assayed exhibited a poor protective effect against enzyme deactivation, whereas in the presence of substrate, all the ILs demonstrated an extremely good protective effect, which correlated with their polarity as follows: [Emim][BF$_4$] < [Emim] [Tf$_2$N] < [Bmim][PF$_6$] *(8)*.

CALB-catalyzed butyrate synthesis in ILs

The ability of CALB to catalyze butyl butyrate synthesis from butyl vinyl ester and 1-butanol in these ILs was also studied at 50 °C and 2% v/v water content. Since synthetic activity is a kinetically controlled process, the efficiency of the enzyme activity can be expressed by the synthetic activity and the selectivity parameter (ratio between the synthetic and hydrolytic activities). Figure 3 shows the synthetic activity exhibited by the enzyme in different ILs

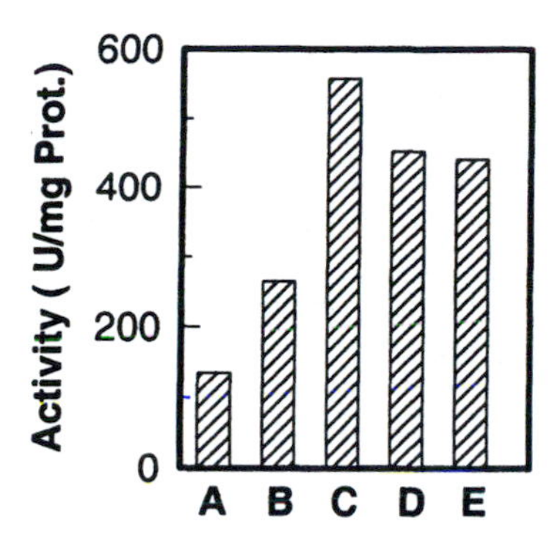

Figure 3. Activity of CALB for butyl butyrate synthesis in different reaction media. A, hexane; B, 1-butanol; C, [Emim][BF_4]; D, [Emim][NTf_2]; E, [Bmim][PF_6].

([Emim][BF_4], [Emim][Tf_2N] and [Bmim][PF_6]) and organic solvents (hexane and 1-butanol). As can be seen, all the ILs acted as suitable reaction media for the enzymatic synthesis, which involves an appropriate solubilization of substrates and products, and an active enzyme conformation.

Thus, for all ILs, the synthetic activity was (up to 2-times) higher than that observed in the organic solvents assayed. Also, for all cases, the selectivity parameter was higher than 94% due to the low water content of these reaction media.

CALB-catalyzed butyl butyrate synthesis in ILs-$scCO_2$

A continuous biphasic enzymatic reactor was designed, taking into account the excellent properties of ILs to maintain enzymes in an active conformation, as well as the ability of $scCO_2$ to transport hydrophobic substrates and products (see Figure 1) *(18)*. As has been described, $scCO_2$ is highly soluble in certain ILs (up to 0.6 mol fraction), while ILs are insoluble in $scCO_2$ *(19)*. Thus, a new concept for a continuous phase-separable biocatalytic reactor is proposed, whereby a homogeneous free enzyme solution is "dissolved" into the IL phase (catalytic phase), and substrates and/or products reside largely in the supercritical phase (extractive phase). Two kinetically-controlled synthetic processes catalyzed by CALB were considered as reaction models to analyze the efficiency of this proposed enzymatic reactor, as follows: a simple transesterification reaction towards a primary alcohol (1-butanol), and the kinetic resolution of a *sec*-alcohol (*rac*-1-phenylethanol).

Figure 4 shows the evolution of the activity and half-life time of CALB-[Bmim][Tf_2N] system for the butyl butyrate synthesis as a function of the $scCO_2$ density, determined by the experimental conditions (40°C, 15.0 MPa; 50 °C 12.5 MPa; and 100 °C, 15.0 MPa, respectively). As can be seen, lipase was able to catalyze the transesterification reaction in all the assayed supercritical conditions, being the activity increasing as $scCO_2$ density decreased. This may be related with an improvement in the mass-transport efficiency of the supercritical fluid for exposing substrates to the enzyme microenvironment, thus enhancing the synthetic activity. The negative effect of CO_2 on the enzyme action was also observed because the synthetic activity was 10-times lower than that observed in ILs without $scCO_2$ (see Fig. 3). On the other hand, the enzyme stability was reduced by a decrease in $scCO_2$ density, probably due to the associate increase in temperature (from 40 to 100 °C).

Two additional observations were made, firstly that butyl butyrate was not synthesized in the absence of enzyme, and secondly, no enzyme activity was detected at the exit of the reactor, which means that the enzyme molecules "dissolved" into the IL must be considered as immobilized. Once again, the selectivity parameter in all cases was higher than 96%, which remained constant throughout all the operation periods due to the low water content (2% v/v in the IL phase).

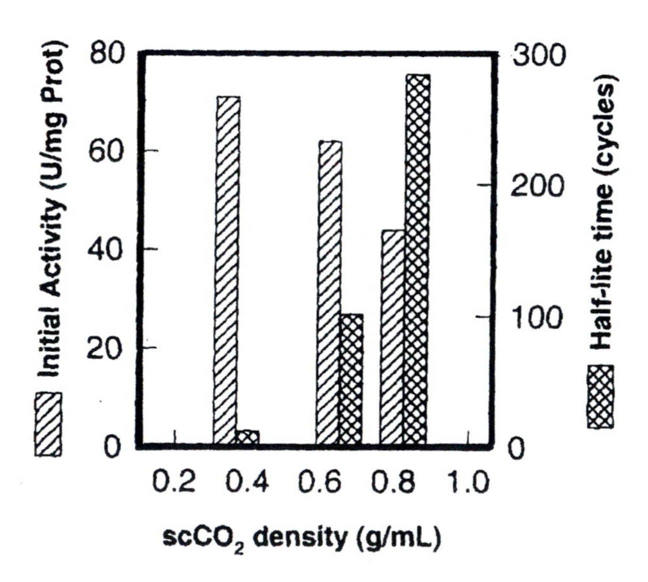

Figure 4. Effect of $scCO_2$ density on the activity and half-life time of CALB-[Bmim][Tf_2N] for butyl butyrate synthesis.

CALB-catalyzed kinetic resolution of *rac*-1-phenylethanol in ILs-scCO$_2$

Another enzymatic property, enantioselectivity, was also tested for this continuous lipase-IL-scCO$_2$ reactor using two different ILs, [Emim][Tf$_2$N] and [Bmim][Tf$_2$N], in four supercritical conditions (50 °C, 15 MPa; 100°C 15 MPa; 120°C, 10 MPa; and 150 °C, 10 MPa). In all cases, similar catalytic behaviour was observed. As an example, Figure 5 depicts the activity, selectivity and enantioselectivity profiles during operation cycles, shown by the CALB-[Emim][Tf$_2$N]-catalyzed kinetic resolution of *rac*-1-phenylethanol in scCO$_2$ at 120 °C and 10 MPa.

As can be seen, for all the assayed cycles the enantioselectivity of the enzyme was maintained at the maximum level (ee> 99.9%), because the (S)-1-phenylethyl propionate isomer was not detected at all, and the (R)-1-phenylethyl propionate isomer was always obtained at a rate of conversion higher than 35 %. However, the initial synthetic activity and selectivity of the enzyme for this synthetic reaction were lower than in the case of butyl butyrate synthesis because of the drop in nucleophilic power of the hydroxyl group from 1-butanol (primary position) to *rac*-1-phenylethanol (secondary position). Furthermore, two interesting phenomena were observed in Figure 5. Firstly, the enzyme was active and stable ($t_{1/2}$= 13 cycles) at this extreme temperature (120 °C), demonstrating the important protective effect of this IL against enzyme deactivation by heat and/or CO$_2$. Secondly, the selectivity profile showed a hyperbolic increase (from 35% to 98%) with an increasing number of operation cycles, which could be related with the non-optimized initial water content in the IL phase (2% v/v).

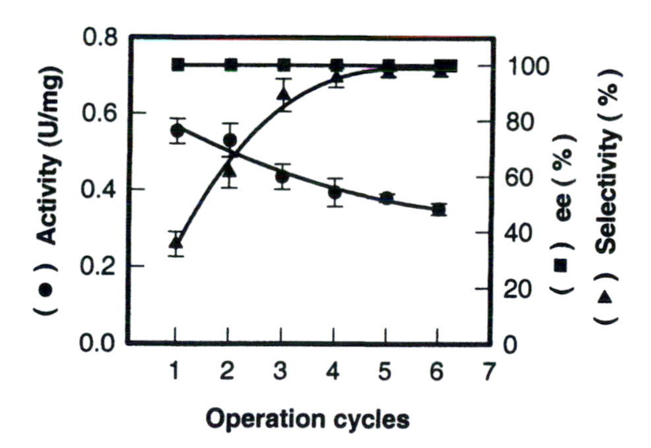

Figure 5. Activity, selectivity and enantioselectivity profiles of CALB-[Emim][Tf$_2$N] system for continuous kinetic resolution of rac-1-phenylethanol at 120 °C and 10 MPa.

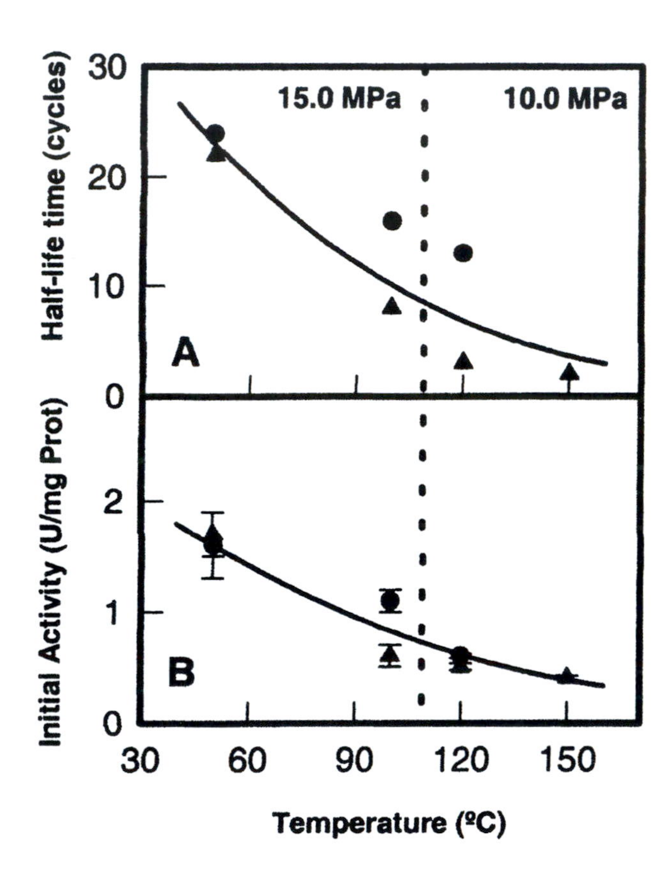

Figure 6. Effect of temperature on the synthetic activity and half-life time of different CALB-IL (●, [Emim][Tf$_2$N]; ▲, [Bmim][Tf$_2$N]) systems for continuous kinetic resolution of rac-1-phenylethanol in scCO$_2$ at 15.0 and 10.0 MPa, respectively.

The decrease in nucleophilicity of the *sec*-alcohol, together with an excess in water content, may have involved a decrease in the selectivity of the catalytic action of the enzyme compared with the butyl butyrate synthesis. In this case, the continuous "drying" of the IL-phase by the flow of substrate-$scCO_2$, as well as the consumption of free water molecules by the hydrolytic activity during operation cycles, would produce an enhancement in selectivity. These results clearly pointed to the need to optimize the initial water content in the IL-enzyme system, as well as its control during continuous operation processes.

Figure 6 shows the evolution of synthetic activity and half-life time of both CALB-IL ([Emim][Tf_2N] and [Bmim][Tf_2N]) systems, for the kinetic resolution of *rac*-1-phenylethanol, as a function of the temperature of $scCO_2$ at two pressures (10.0 and 15.0 MPa). As can be seen, the biocatalytic system was active in all the assayed conditions, exhibiting a decrease in both catalytic parameters as temperature increased. The initial activity profile was practically independent of the nature of the assayed IL. However, it is necessary to point to the significant protective effect these media had on the enzyme activity even in extreme thermal conditions, [Emim][Tf_2N] providing the best stability. In this way, a mesophilic biological macromolecule designed to work at room temperature can maintain its catalytic activity for several daily cycles ($t_{1/2}$ = 2 cycles) of continuous operation at 150 °C.

Conclusions

ILs have been shown to act as excellent agents for stabilizing α-chymotrypsin and CALB. For both enzymes, the presence of substrate greatly improved their half-life time, resulting in an over-stabilizing effect (up to 2500-times). Also, ILs appeared to be adequate media for carrying out synthetic enzymatic processes under in low water conditions. A continuous biphasic reactor based on CALB-IL systems and $scCO_2$ have been tested for two different synthetic reactions with excellent results. The system exhibited high activity, enantioselectivity and stability in continuous operation, even in very extreme conditions (*e.g.* 120 and 150°C). These results clearly show how clean synthetic chemical processes, running as a "black box" directly providing pure products, can be designed easily. In this context, the door for the green chemical industry of the near future is open.

Acknowledgements

This work was partially supported by the CICYT grants BIO99-0492-C02-01 and PPQ2002-03549. We like to thanks Ms. C. Saez for technical assistance, and Mr. R. Martínez from Novo España, S.A. the gift of Novozym 525.

References

(1) Halling, P. *Enzyme Microb. Technol.* **1994**, *16*, 178-206.
(2) Dordick, J. S. *Biotechnol Prog.*, **1992**, *8*, 259-267.
(3) Lozano, P.; Combes, D.; Iborra, J.L. *J. Biotechnol.*, **1994**, *35*, 9-18.
(4) Carmichael, A.J.; Seddon, K.R. *J. Phys. Org. Chem.*, **2000**, *13*, 591-595.
(5) Erbeldinger, M.; Mesiano, A.J.; Russel, A.J. *Biotechnol. Prog.* **2000**, *16*, 1129-1131.
(6) Madeira Lau, R; van Rantwijk, F.; Seddon, K.R.; Sheldon, R.A. *Org. Lett.*, **2000**, *2*, 4189-4191.
(7) Lozano, P.; De Diego, T.; Guegan, J.P.; Vaultier, M.; Iborra, J.L. *Biotechnol. Bioeng.*, **2001**, *75*, 563-569.
(8) Lozano, P.; De Diego, T.; Carrié, D.; Vaultier, M.; Iborra, J.L. *Biotechnol. Lett.*, **2001**, *23*, 1529-1533.
(9) Itoh, T.; Akasaki, E.; Kudo, K.; Shirakami, S. *Chem. Lett.*, **2001**, 262-263.
(10) Jarzebski, A.B.; Malinowski, J.J. *Process Biochem.*, **1995**, *30*, 343-352.
(11) Nakamura, K. *Trends Biotechnol.*, **1990**, *8*, 288-292.
(12) Kamat, S.; Critchley, G.; Beckman, E.J.; Russel, A.J. *Biotechnol. Bioeng.*, **1995**, *46*, 610-620.
(13) Lozano, P.; Avellaneda, A.; Pascual, R.; Iborra, J.L. *Biotechnol. Lett.*, **1996**, *18*, 1345-1350.
(14) Suarez, P.A.Z.; Dullius, J.E.L.; Einloft, S.; De Souza, R.F.; Dupont, J. *Polyhedron*, **1996**, *15*, 1217,-1219.
(15) Bonhôte, P.; Dias, A.P.; Papageorgiu, N.; Kalyanasundaram, K.; Grätzel, M. *Inorg. Chem.*, **1996**, *35*, 1168-1178.
(16) Huddleston, J.G.; Willauer, H.D.; Swatloski, R.P.; Visser, A.E.; Rogers, R.D. *Chem. Commun.* **1998**, 1765-1766.
(17) Hanke, C.G.; Atamas, N.A.; Lynden-Bell, R.M. *Green Chem.*, **2002**, *4*, 107-111.
(18) Lozano, P.; De Diego, T.; Carrié, D.; Vaultier, M.; Iborra, J.L. *Chem. Commun.*, **2002**, 692-693.
(19) Blanchard, L.A.; Hancu, D.; Beckman, E.J.; Brennecke, J.F. *Nature*, **1999**, *399*, 28-29.

Chapter 21

Efficient Lipase-Catalyzed Enantioselective Acylation in an Ionic Liquid Solvent System

Toshiyuki Itoh[1,*], Yoshihito Nishimura[1], Masaya Kashiwagi[2], and Makoto Onaka[2]

[1]Department of Materials Science, Faculty of Engineering, Tottori University, 4–101 Koyama Minami, Tottori 680–8552, Japan
[2]Graduate School of Arts and Sciences, The University of Tokyo, Komaba, Meguro-ku, Tokyo 153–8902, Japan

The lipase-catalyzed enantioselective acylation of allylic alcohols in an ionic liquid solvent was demonstrated; the reaction was significantly dependent on the counter anion of the imidazolium salt and good results were obtained when the reaction was carried out in [bmim]PF_6 or [bmim]BF_4 as the solvent. The lipase-catalyzed transesterification was then investigated using methyl esters as acyl donors, especially under reduced pressure in an ionic liquid ([bmim]PF_6) solvent system. The transesterification of 5-phenyl-1-penten-3-ol took place smoothly under reduced pressure at 20 Torr and 40 °C when methyl phenylthioacetate was used as the acyl donor in [bmim]PF_6, and we succeeded in obtaining the corresponding acylated compound in optically pure form; this makes it possible to repeatedly use the lipase because there was no drop in the reaction rate despite five repetitions of the process.

Introduction

To meet the challenge in chemistry of developing practical processes, the proper choice of a reaction medium is very important. A breakthrough has sometimes occurred with the invention of the reaction medium in chemical reactions and this is true even in enzymatic reactions; lipase-catalyzed transesterification in an organic solvent system is now well recognized as a very useful means of synthesizing optically active compounds,[1] it had been long believed, however, that an enzymatic reaction could proceed only in aqueous medium before Klibanov and his co-workers first demonstrated lipase-catalyzed trans-esterification of alcohols in an organic solvent system.[2] Ionic liquids are a new class of solvents which have attracted growing interest over the past few years because of their unique physical and chemical properties. [3] Because lipase tolerates non-natural reaction conditions, it was believed that the lipase-catalyzed reaction might occur in the ionic liquid solvent.[4,5,6] We describe the results of the lipase-catalyzed reaction in this unique novel reaction medium system.

Experimental

Normal pressure conditions. Typically, the reaction was carried out as follows: To a mixture of lipase in the ionic liquid were added (±)-5-phenyl-1-penten-3-ol (**1**) as a model substrate and vinyl acetate (1.5 eq) as the acyl donor. The resulting mixture was stirred at room temperature (ca. 25 °C) and the reaction course was monitored by GC analysis. The reaction was stopped by the addition of 3 ml of ether when the molar ratio of 5-phenyl-1-penten-3-yl acetate (**2a**) and alcohol **1** became equal. The reaction mixture was filtered through a glass-sintered filter with a celite pad to remove the enzyme and product, and unreacted alcohol was isolated from the filtrate. It is noteworthy that the ionic solvent was recovered without any loss in the amount after the work-up process and it was possible to reuse it after washing with water and dried under vacuum for several hours at 50 °C. The optical purities of the acetate (*S*)-**2a**[7] produced and the remaining alcohol (*R*)-**1**[7] were determined by capillary GC analysis or HPLC using a chiral column. Chiraldex G-TA was used for GC analysis : ϕ0.25 mm x 20 m, Carrier gas: He 40 ml/min. Temp (°C): 100, Inlet Pressure:1.35kg/cm^2, Amount 400 ng, Detection; FID. HPLC analysis: Chiralcel OD (ϕ4.6 mm x 250 mm), Hexane: 2-propanol (10:1~8:1), 35°C, 1.0 ml/min, 254 nm.

Under reduced pressure conditions. To a mixture of lipase (25 mg) in the ionic liquid (1.5 ml) were added racemic (±)-**1** (50 mg, 0.30 mmol) as a model substrate and methyl phenylthioacetate (27 mg, 0.15 mmol, 0.5 equiv.) as the acyl donor. The resulting mixture was stirred at 40 °C at 20

Torr for 13 h. The reduced pressure was broken and the reaction was stopped by the addition of 3 ml of ether to the reaction mixture to form the biphasic state. The desired products and unreacted alcohol were quantitatively extracted from the ether. To the remaining ionic liquid phase, which was placed under reduced pressure for 15 minutes to remove the ether, a mixture of the substrate and methyl phenylthioacetate was again added. This mixture was stirred at 40 °C and 20 Torr. The optical purities of the produced ester **2** and the remaining alcohol (*R*)-**1**[7] were determined by capillary GC analysis using a chiral column (Chiraldex G-TA).

In addition, the ionic solvent was sometimes significantly acidified and lowered to less than pH 2 due to partial hydrolysis of the salt by the moisture. Therefore, the pH values of the solvent should be checked prior to use in the reaction. We developed two good methods of restoring damaged ionic liquid: the solvent is washed with a mixture of hexane and ethyl acetate (1:1) and treated with the ionic exchange resin IRA 400, or is washed with the same mixed solvent followed by treatment with neutral activated alumina type I.[5]

Results and Discussion

1. Lipase-catalyzed reaction system anchored in an ionic liquid solvent.

We chose the imidazolium salts as the solvent for our enzymatic reaction among the various types of ionic liquids based on two criteria. The first is that the imidazolium salts are stable under atmospheric conditions and especially tolerant to water.[3] The second is that we are systematically able to investigate the suitable combination of the imidazolium cation and counter anion of the salt for the enzymatic reaction. First, we investigated the *Candida antactica* lipase (Novozym435)-catalyzed enantioselective transesterification of 5-phenyl-1-penten-3-ol (**1**) [7] in five types of butylmethyl imidazolium salts (Eq. 1). It was found that the acylation rate was strongly dependent on the anionic part of the solvent, while the CAL catalyzed acylation proceeded with high enantioselectivity in all tested solvents (Table 1). [4] The best result was recorded when [bmim]BF_4[8a] was employed as the solvent (Entry 1) and the reaction rate was nearly equal to that of the reference reaction in *i*-Pr_2O (Entry 8). The second choice of solvent was [bmim]PF_6[8b] and the acetate **2a** was obtained in excellent enantioselectivity, though the reaction rate was slightly inferior to the reaction in [bmim]BF_4 (Entry 2). On the contrary, a significant drop in the reaction rate was obtained when the reaction was carried out in [bmim]TFA[8c] (Entry 3), [bmim]OTf [8d] (Entry 4) or [bmim]SbF_6[8e] (Entry 5). From these obtained results, it was concluded that [bmim]BF_4 and [bmim]PF_6 are suitable solvents for the reaction. Although the acylation rate for the reaction in [bmim]PF_6 was slightly inferior to that in [bmim]BF_4, we chose [bmim]PF_6 as the best solvent for the lipase-catalyzed reaction system, because a very easy work-up process was realized in the [bmim]PF_6 solvent system due to the insolubility of this salt in both water and ether. On the contrary, [bmim]BF_4 was quite soluble in water and therefore it was difficult to remove the by-product such as acetic acid by simple work-up processes. We next investigated the enantioselective acylation of (±)-**1** using five types of lipases in the [bmim]PF_6 solvent system. A poor reactivity was observed for lipase from *Alcaligenes* sp. (QL) and *Pseudomonas cepacia* lipase (PS), though the desired acetate **2a** obtained with an extremely high enantioselectivity for all these enzymes (Entries 6 and 7). On the other hand, no reaction took place when *Candida rugosa* lipase (CRL) or porcine liver lipase (PPL) was used as the catalyst in the [bmim]PF_6 solvent system.

(±)-**1** → vinyl acetate; Novozym435 (50 wt%); Ionic Liquid: [bmim]X^- → **2a** + **1** (1)

Table 1. Lipase-catalyzed transesterification in ILs

Entry	Lipase[a]	Solvent	Time /h	%ee of **2a** (Yield/%)[b]	Conv. /c	Relative Rate[c]	E value[d]
1	Novozym435	[bmim]BF_4	3.5	>99 (44)	0.48	14	>640
2	Novozym435	[bmim]PF_6	5	>99 (45)	0.47	9.4	>580
3	Novozym435	[bmim]TFA	48	91 (19)	0.12	0.25	230
4	Novozym435	[bmim]OTf	24	>99 (34)	0.43	1.8	>450
5	Novozym435	[bmim]SbF_6	48	>99 (31)	0.37	0.77	>360
6	QL	[bmim]PF_6	25	94 (49)	0.41	1.6	65
7	PS	[bmim]PF_6	168	>99 (17)	0.19	0.11	>250
8	CAL	*i*-Pr_2O	3	>99 (47)	0.50	17	>1000

a) Novozym435: *Candida antarctica*; QL: *Alcaligenes* sp.; PS: *Pseudomonas cepacia* (Amano) b) Isolated yield. c) Relative Rate: %conv./reaction time (h). d) See Ref. 13.

Lipase PS is well respected and one of the most widely used enzymes applicable for various substrates, however, poor reactivity was obtained when commercial lipase PS was used for the acylation of the mandelic acid methyl ester (±)-**3**, though desired acetate **4** was obtained in optically pure form as shown in Table 2 (Entry 1). We found that commercial lipase PS gradually lost its activity in the [bmim]PF_6 solvent system, so, because this lipase is immobilized by Celite, we next attempted to improve the reactivity by changing the supporting materials in this solvent system (Eq. 2). The results were strongly dependent on the supporting materials and the reaction rate was drastically improved when Toyonite 200M [9] immobilized lipase PS was used for the reaction (Entry 2). Enantioselectivity was also modified by the supporting materials. The best enantioselectivity was recorded for methacryoxypropyl SBA-15 [10] (Entry 6). On the contrary, the reaction rate was significantly reduced for Toyonite 200A[9] and aminopropyl SBA-15 [10] (Entries 4 and 7). A reduced enantioselectivity was obtained when lipase PS was immobilized by Toyonite 200 [9] (Entry 5).

(±)-**3** (mandelic acid methyl ester, OH, COOMe) —[Vinyl acetate / Immobilized Lipase PS / [bmim]PF_6]→ **4** (OAc, COOMe) + **3** (OH, COOMe) (2)

Table 2. Evaluation of supporting materials for lipase PS-catalyzed transesterification in [bmim]PF_6 solvent system

Entry	Supporting material	Time (h)	%ee of **4** (%Yield)[a]	%conv.	Relative Rate [b]	E value [c]
1	Celite	72	>99 (3)	9.0	0.13	>220
2	Toyonite 200M	5	97 (20)	13	2.60	80
3	Toyonite 200P	48	>99 (10)	20	0.41	>250
4	Toyonite 200A	168	>99 (9)	0.4	0.002	>200
5	Toyonite 200	24	80 (12)	7	0.29	10
6	Methacryoxypropyl SBA-15	48	>99 (14)	22	0.46	>260
7	Aminopropyl SBA-15	168	>99 (10)	20	0.12	>250

[a] Isolated yield. [b] Relative Rate:%conv./time (h). [c] Ref 13.

It was established that proper immobilization of the enzyme makes it possible to extend its applicability, though it has not yet clear how the origin of these supporting materials affects enzyme reactivity.

Since it was anticipated that lipase might be anchored by the ionic liquid solvent and might remain in it after the extraction work-up of the products, we next evaluated the repeated use of Novozym435 in the [bmim]PF_6 solvent system (Figure 1). A mixture of the substrate, lipase, and vinyl acetate in the [bmim]PF_6 solvent was stirred for 3 h at rt, and then ether was added to the reaction mixture to form the biphasic state. The desired products and unreacted alcohol were quantitatively extracted from the ether (upper layer). To the remaining ionic liquid phase, which was placed under reduced pressure for 15 minutes to remove the ether, a mixture of the substrate and vinyl acetate was again added. This mixture was stirred at rt. As expected, the acylation reaction took place smoothly and the product was obtained without any loss in enantioselectivity. It was thus confirmed that the enzyme was, in fact, anchored in the ionic liquid solvent after the work-up process. By repeating the same process, we showed that recycling of the enzyme was indeed possible in our ionic liquid solvent system, though the reaction rate gradually dropped with repetition of the reaction process (Figure 2). [4]

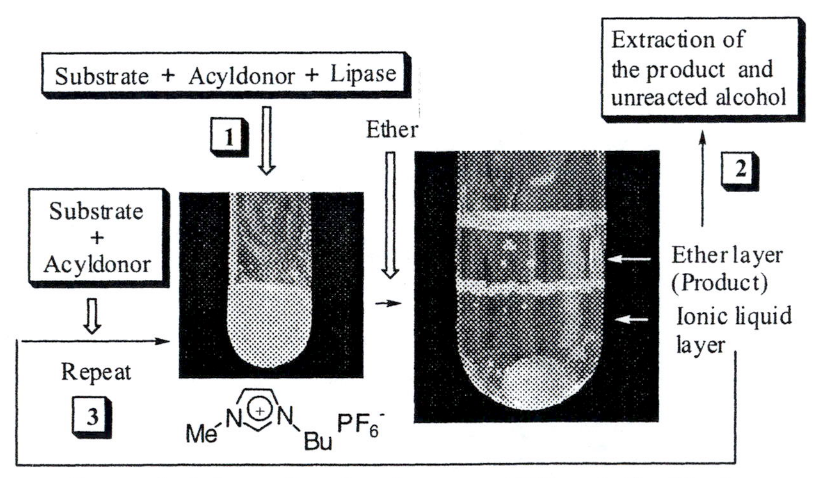

Figure 1. Lipase-catalyzed reaction system anchored to the solvent.

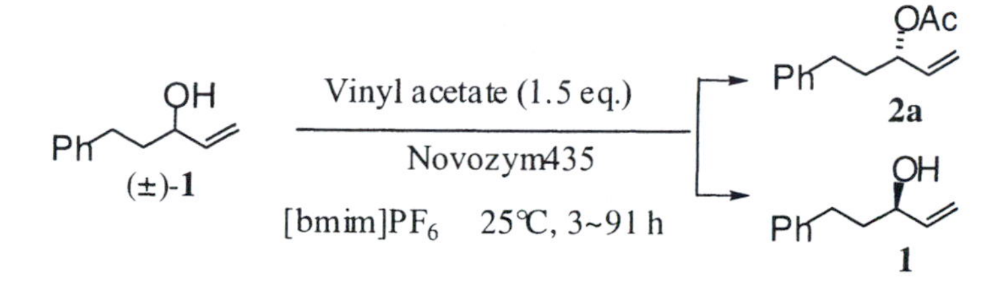

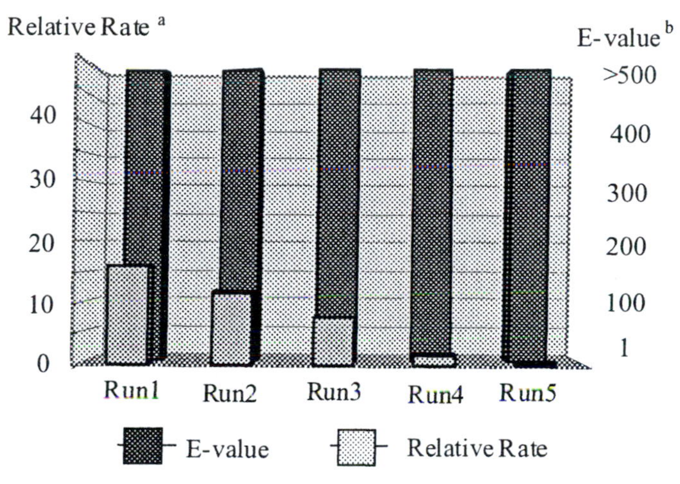

[a] Relative Rate: %conv./reaction time (h). [b] Ref.13

Figure 2. Results of the repeated use of lipase in the [bmim]PF_6 solvent system

2. Lipase-catalyzed reaction under reduced pressure in an ionic liquid solvent system.

The repeated use of lipase in [bmim]PF_6 solvent system has been obtained. However, as stated, the reaction rate gradually dropped with repetition of the reaction process.[4] This drop in reactivity was assumed to be caused by the inhibitory action of the acetaldehyde oligomer that accumulated in the solvent system based on a ^{1}H NMR analysis. One of the most important characteristics of ionic liquids is their wide temperature range for liquid phase, and ionic liquids have no vapor pressure; we therefore decided to look at the lipase-catalyzed reaction under reduced pressure conditions in the [bmim]PF_6 solvent system. We found that the transesterification proceeds efficiently under reduced pressure [11] in an ionic liquid solvent system using methyl esters as acyl donors.

To realize the transesterification under reduced pressure conditions, the proper choice of the acyl donor ester is very important; it is essential to use an acyl donor ester which has a sufficiently higher boiling temperature than the corresponding alcohol which is produced by transesterification with the substrate ester. Thus, methyl esters seem appropriate as the acyl donors for the lipase-catalyzed transesterification under reduced pressure. This is true even though ordinary methyl esters are recognized as not being suitable for the lipase-catalyzed transesterification as acyl donors because reverse reaction with produced methanol easily takes place.[5] However, we would be able to avoid such a difficulty when the reaction was carried out under reduced pressure even if the methyl esters were used as an acyl donor, because the produced methanol would be immediately removed from the reaction mixture and thus the reaction equilibrium would shift to produce the desired product (Eq 3). Thus several types of methyl esters were evaluated as acyl donors for the lipase-catalyzed reaction (Table 3).

We tested the transesterification of (±)-**1** as a model substrate using methyl pentanoate as the acyl donor at 100 Torr and 27 °C. However, the reaction rate was very slow and only 6 % of the product **2b** (R=n-C_4H_9) was obtained with poor enantioselectivity after 48 h reaction using 1.5 equiv. of methyl pentanoate as the acyl donor (Entry 1, Table 3). Fortunately, the desired reaction was efficiently accomplished when the reaction was carried out using methyl nonanoate as the acyl donor at 100 Torr and 32 °C (Entry 2); transesterification proceeded very smoothly and the desired ester **2c** (R=n-C_8H_{17}) was obtained with >99% ee. Other methyl esters can also be used as acyl donors for these reactions. Methyl phenoxyacetate,[12] methyl methylthioacetate,[12] and methyl phenylthioacetate[12] also worked very well, and the esters **2d** (R= CH_2OPh), **2e** (R= CH_2SMe), and **2f** (R= CH_2SPh) were obtained with perfect enantiomeric excess, respectively (Entries 3 to 5). It was very easy to monitor the reaction course by silica gel thin layer chromatography (TLC) when phenoxyacetate or phenylthioacetate was used as the acyl donor. It was also possible to reduce the amount of the acyl donor to 0.5 equiv. versus the substrate alcohol when these esters were used in the reaction; this is the least recorded amount of an acyl donor used in this type of lipase-catalyzed transesterification (Entry 6).

Due to the large difference in boiling points between methyl phenylthioacetate and methanol, this ester was indeed useful for the lipase recycling system. The transesterification smoothly took place under reduced pressure at 20 Torr and 40 °C when 0.5 equivalent of methyl phenylthioacetate was used as the acyl donor, and we were able to obtain an ester **2f** in optically pure form. As shown in Figure 3, five repetitions using this process showed no drop in the reaction rate.

(±)-**1** → Novozym435 (50wt%), [bmim]PF_6 40°C, 20-100 Torr; RCO_2Me, −MeOH → **2** + **1** (3)

Table 3. Lipase-catalyzed enantioselective transesterification under reduced pressure conditions in [bmim]PF_6 solvent system

Entry	Acyl donor[a] R	Time /h	%ee of **2** (Yield/%)[b]	Conv. /c	Relative Rate[c]	E value[d]
1	C_4H_9	48	37 (6)	0.15	0.3	2
2	n-C_8H_{17}	5	>99 (43)	0.42	8.4	>420
3	$PhOCH_2$	9	>99 (35)	0.42	4.7	>470
4	$MeOCH_2$	29	>99 (30)	0.33	1.1	>580
5	$PhSCH_2$	13	>99 (30)	0.46	3.5	>530
6	$PhSCH_2$[e]	12	>99 (30)	0.45	3.8	>500

a) 1.5 equiv. to the substrate. b) Isolated yield. c) Relative Rate: %conv./ reaction time (h). d) Ref. 13. e) 0.5 equiv. to the substrate.

Conclusions

We demonstrated the lipase-catalyzed enantioselective transesterification of an allylic alcohol in the [bmim]PF_6 solvent system under reduced pressure conditions and showed that it was possible to repeatedly use the enzyme in this system. It is assumed that a good acyl donor must be selected depending on the substrate. We do believe, however, that this might be a very important means of lipase-catalyzed enantioselective acylation in the ionic liquid solvent system. Further investigation of the scope and limitations of this reaction, especially optimization of the reaction conditions for the lipase recycling system in the ionic solvent system, will make it even more beneficial.

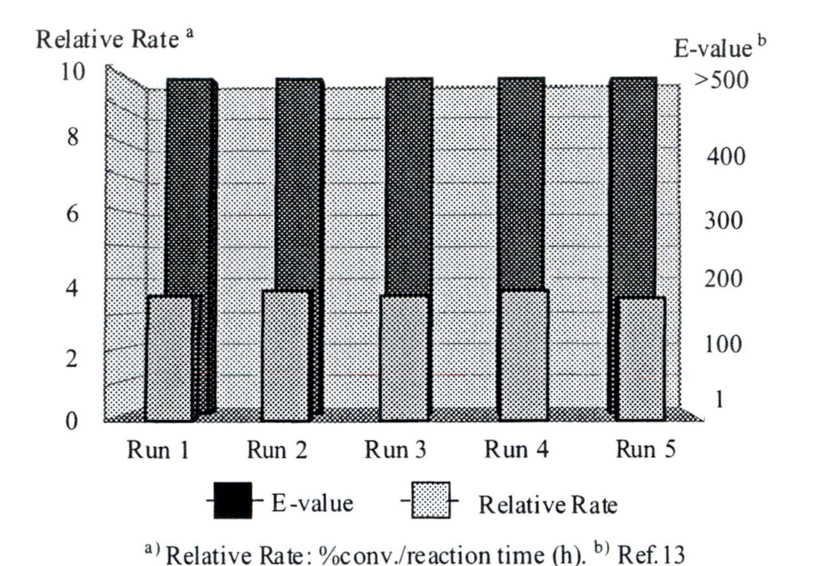

Figure 3. Results of the repeated use of lipase in the [bmim]PF_6 solvent system under reduced pressure conditions

Acknowledgments

The authors are grateful to Professor Tomoya Kitazume at the Tokyo Institute of Technology for his helpful discussions throughout this study. They also thank Novo Nordisk Bioindustry Co., Ltd., Meito Sangyo Co., Ltd., and Amano Pharmaceutical Co., Ltd. for providing the lipases. The authors are grateful to Mr. Masanobu Kamori of Toyodenka Co., Ltd. for providing Toyonite.

References

1. Reviews see: a) Wong C. H.; Whitesides, G. M. Enzymes in Synthetic Organic Chemistry, Tetrahedron Organic Chemistry Series, Vol. 12, ed. by J. E. Baldwin and P. D. Magnus, Pergamon (1994). b) Theil, F. *Chem. Rev.* **1995**, *95,* 2203. c) Itoh, T.; Takagi, Y. Tsukube, H. *Trends in Organic Chemistry*, **1997,** *6*, 1. d) Theil, F. *Tetrahedron*, **2000**, *56*, 2905.
2. Klibanov, A. M. *Acc. Chem. Res.* **1990**, *23*, 114.
3. A review, see : Welton, T. *Chem. Rev.* **1999**, *99*, 2071.
4. Itoh, T.; Akasaki, E.; Kudo, K.; Shirakami, S. *Chem. Lett.*, **2001**, 262.
5. Itoh, T.; Akasaki, E.; Nishimura, Y.*Chem. Lett.* **2002**, 154.
6. For examples of the enzymatic reactions in an ionic liquid solvent system, see. a) Cull, S. G.; Holbrey, J. D.; Vargas-More, V.; Seddon, K. R.; Lye, G. J. *Biotechnol. Bioeng.* **2000**, *69*, 227. b) Erbeldinger, M.; Mesiano, A. J.; Russell, A. J. *Biotechnol. Prog.* **2000**, *16*, 1131. c) Lau, M. R.; Rantwijk, F. v.; Seddon, K. R.; Sheldon, R. A. *Org. Lett.* **2000**, *2*, 4189. d) Schöfer, S. H.; Kaftzik, N.; Wasserscheid, P.; Kragl, U. *Chem. Commun.* **2001**, 425. e) Kim, K-W.; Song, B.; Choi, M-Y.; Kim, M-J. *Org. Lett.* **2001**, *3*, 1509. f) Howarth, J. James, P.; Dai, J. *Tetrahedron Lett.* **2001**, *42,* 7517. g) Park, S.; Kazlauskas, R. J. *J. Org. Chem.* **2001**, *66*, 8395. h) Lozano, P.; Diego, T. de.; Carri , D.; Vaultier, M.; Iborra, J. L. *Chem. Commun.* **2002**, 692. h) Nara, S. J.; Harjani, J. R.; Salunkhe, M. M. *Tetarhedron Lett.* **2002**, *43*, 2979.
7. Takagi, Y.; Nakatani, T.; Itoh, T.; Oshiki, T. *Tetrahedron Lett.* **2000**, *41*, 7889.
8. a) [bmim]BF_4: Suarez, P. A. Z.; Dullius, J. E. L.; Einloft, S.; Souza, R. F. de.; Dupont, J. *Polyhedron*, **1996**, *15*, 1217. b)[bmim]PF_6: Huddleston, J. G.; Willauer, H. D.; Swatloski, R. P.; Visser, A. E.; Rogers, R. D. *Chem. Commun.* **1998**, 1765. c) [bmim]TFA: The synthesis of this salt was similar to that of [bmim]PF_6 with the exception that CF_3COONa was used in place of $NaPF_6$. d) [bmim]OTf: Bonhote, P.; Dias, A.-P.; Papageorgiou, N.; Kalyanasundaram, K.; Gr tzel, M. *Inorg. Chem.* **1996**, *35*, 1168. e) [bmim]SbF_6: Song, C. E.; Oh, C. R.; Roh, E. J.; Choo, D. J. *Chem. Commun.* **2000**, 1743.
9. Toyonite is a porous ceramics prepared from a kaolinite: Toyodenka Co., Ltd. Phone: +81-888-31-1241. E-mail: m-kamori@toyodenka.com
10. Zhao, D.; Feng, J.; Stucky, G. D. *Science*, **1998**, *279*, 548.
11. Examples of lipase-catalyzed reaction under reduced pressure conditions, see. a) Haraldsson, G. G.; Gudmundsson, B. Ö.; Almarsson, Ö. *Tetrahedron Lett.* **1993**, *34*, 5791. b) Haraldsson, G. G.; Thorarensen, A. *Tetrahedron Lett.* **1994**, *35*, 7681. c) Sugai, T.; Takizawa, M.; Bakke, M.; Ohtsuka, Y.; Ohta, H. *Biosci. Biotech. Biochem.* **1996**, *60*, 2059. d) Cordova A.; Janda, K. D. *J. Org. Chem.* **2001**, *66*, 1906.
12. Itoh, T.; Takagi, Y.; Nishiyama, S. *J. Org. Chem.* **1991**, *56*, 1521.
13. Chen, C. -S.; Fujimoto, Y.; Girdauskas, G.; Sih, C. J. *J. Am. Chem. Soc.***1982**, *102*, 7294.

Non-Catalytic and Calalytic Chemistry

Chapter 22

Acids and Bases in Ionic Liquids

Douglas R. MacFarlane and Stewart A. Forsyth

School of Chemistry, Monash University, Clayton, Victoria 3800, Australia

Salts that are liquid at ambient temperatures are of interest in an enormous range of applications from 'green' synthesis of chemicals, to electrolytes in devices such as artificial muscles and electrochromic windows, to media for biochemical and biological processes. All of these applications require the ionic liquid (IL) to take on the role as solvent for a variety of dissolved species, yet an understanding of the range of behaviour of these new liquids as solvents is only beginning to emerge. The commonplace assumption that they are inert may not be correct in all cases. Drawing on a range of ionic liquids we present an investigation of their solvent properties as Lewis acids and bases using a number of probe acids to determine the relative state of dissociation of the acids in the IL. This allows us to classify ionic liquids into a number of distinct classes, each expressing distinctly and importantly different behaviours. We demonstrate that the Lewis base class may exhibit a general base catalysis phenomenon which may be of importance in synthetic use of ionic liquids. The role of low levels of water contamination is also discussed. It is clear that there is no single "ionic liquid" behaviour; rather there are a range of properties possible across the ever growing range of ionic liquids discovered and yet to be discovered.

Ionic liquids have been known for quite some time, some of the first reports dating back to 1914.[1] They became of interest to the electrochemist as a result of their high ionic conductivity which made them interesting as solvents for electrochemically active species.[2] A range of familiar and important compounds including various acids and bases are quite soluble in ionic liquids and this has stimulated interest in their use as reaction media for a range of important chemical and biochemical processes.[3] More recently they have been recognized as superb media for "green" chemical synthesis by virtue of their virtually non-existent vapour pressure.[4] This lack of volatility is a general feature of ionic liquids that do not contain an active proton. Other investigations have recognized their special properties as electrolytes for electrochemical devices,[5] and as solvents for in-vitro enzymatic and biochemical processes.[6]

The majority of ionic liquids that are commonly investigated consist of an organic cation based on a quaternized nitrogen such as those shown in Scheme 1. Other cations of similar structures to those in Scheme 1, but having a N-H type group (R=H) are also known, but the components of these tend to be volatile to some extent.

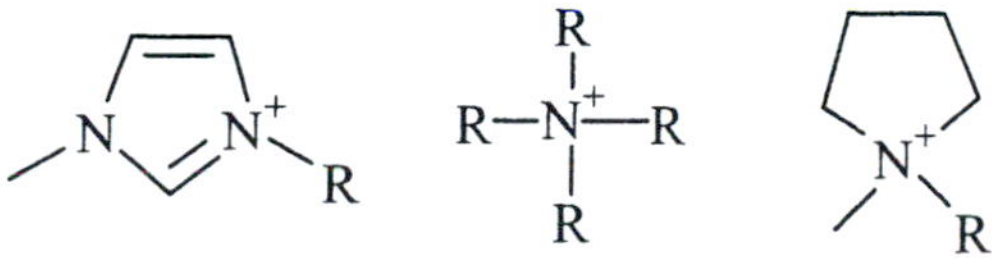

[1] [2] [3]

Scheme 1: Quaternized nitrogen cations; imidazolium [**1**], quaternary ammonium [**2**] and pyrrolidinium [**3**].

The anions are typically drawn from a set of well-known "very weakly basic" anions (Scheme 2). The weak electrostatic interactions between these anions and the cation tend to promote low melting points. Such ionic liquids include well known examples such as 1-butyl-3-methylimidazolium hexafluorophosphate [bmIm][PF_6] and 1-ethyl-3-methylimidazolium bis(trifluoromethane)sulfonimide [emIm][TFSI]. Other recent examples of such anions include nitrate [6],[7] mesylate [7],[8] thiocyanate [8],[9] tricyanomethide [10] and tosylate [11][10] ionic liquids. Later in this paper we will describe these salts as Class I ionic liquids. These salts are typically extremely stable in all respects: their decomposition temperatures as measured by thermogravimetric analysis are typically > 300°C[11] and their electrochemical stability is usually in excess of +2 V (vs Ag/Ag^+) with respect to oxidation and –2V with respect to reduction.[12-14] For these reasons the ionic liquids are often described

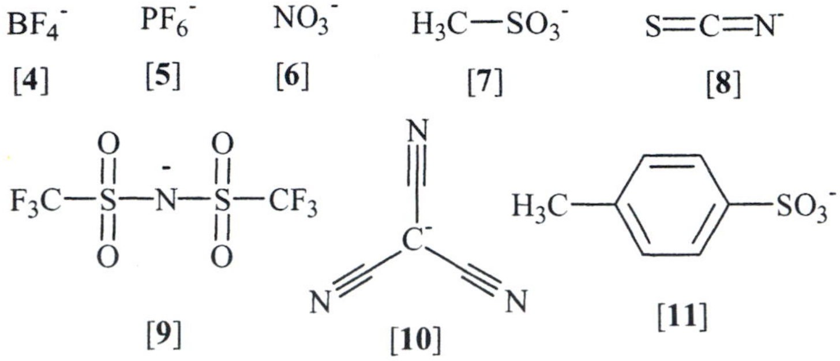

BF_4^- [4] PF_6^- [5] NO_3^- [6] $H_3C—SO_3^-$ [7] $S{=}C{=}N^-$ [8]

[9] [10] [11]

Scheme 2: Weakly basic anions; tetrafluoroborate [**4**], hexafluorophosphate [**5**], nitrate [**6**], mesylate [7], thiocyanate [**8**], bis(trifluoromethane)sulfonimide [**9**], tricyanomethide [**10**] and tosylate [**11**].

as being inert and hence of exceptional interest as potential solvents. Nonetheless many aspects of their ability to dissolve and interact, or not, with solutes is, as yet, far from understood.

More recently ionic liquids have been discovered involving cations bearing other functional groups, eg ether functionality [**13**],[15] and anions of varying properties, for example the dicyanamide anion [**14**] (scheme 3).[16] These ionic liquids express a variety of properties that are not easily accommodated in the description of the Class I ionic liquids given above and serve to illustrate the fact that there is a rich diversity of solvent properties available within the broad family of ionic liquids. A number of recent reports have described the preparation and properties of families of ionic liquids involving anions that would not normally be described as "very weak bases". These include the acetate[**15**],[7] and dicyanamide[**14**][16] anions (Scheme 3). While expected to be stronger bases than the anions in Scheme 2, these salts would nonetheless only produce slightly basic solutions in water. This difference in basicity can be expected to produce different solvent properties in the ionic liquids of these anions.

[13] [14] [15]

Scheme 3: Ether functionality in cation, 1-methyl-3-ethylmethoxyimidazolium [**13**] and basic anions dicyanamide [**14**] and acetate [**15**].

One of the important and very revealing properties of any solvent, and one that is very familiar to all of the fields in which water is a key solvent, is the way that the solvent responds to the presence of an acidic or basic solute. The state of a traditional acid or base compound, such as acetic acid or ammonia, in an ionic liquid solvent is not immediately clear and the question challenges traditional understanding of acid/base behaviour. In this paper we describe a number of experiments designed to elucidate this important behaviour in a broad range of ionic liquids. It appears that some ionic liquids can turn a strong acid into a weaker acid, while other ionic liquids turn weak into strong. These results suggest a classification of ionic liquids with respect to their acid/base properties as discussed further below.

Acids in Ionic Liquids

In traditional solvents, acids and bases are categorized in terms of *Brønsted-Löwry* acidity which relates to the ability of the acid HA to donate a proton to a molecule of the solvent, S.

$$HA + S \rightleftharpoons A^- + HS^+$$

The equilibrium constant for this reaction, $pKa(HA)_s$, is a measure of the acidity of the acid, but it clearly has a dependence on the nature of the solvent S. The more general *Lewis* definition of acidity and basicity involves simply the ability of the molecule to accept or donate electrons. In this case the proton, notionally free of A^-, is the Lewis acid (electron acceptor) and A^- is the Lewis Base (electron donor). The role of the solvent in the Lewis perspective can be to act as the predominant base and thereby to form HS^+ (eg H_3O^+ in water). Water is amphoteric; it is a very weak acid (forming OH^-) and is also a weak base (forming H_3O^+, which is a relatively strong acid). The experiments described below show that some ionic liquids can act as both *Lewis* and *Brønsted-Löwry* bases.

The exact values of acid dissociation constants, pKa, in ionic liquids are unknown at present and are not easily measurable by direct means. In the case of an ionic liquid involving the extreme of a very weakly basic anion such as $CH_3SO_3^-$, some progress towards understanding the relative magnitude of pKa values can be achieved by asking the question: Is the ionic liquid anion less basic than water? Given that the anion is intentionally chosen at the extreme of the very weakly basic anions, the answer to this question is certainly yes. The corresponding conjugate acid is therefore much more acidic than the hydronium ion. Since the anion is a weaker Lewis base than water, one might immediately expect that an acid dissolved in such an ionic liquid would exhibit a lower degree of dissociation than in water. Thus the Brønsted acidity of a solute acid is expected to be lower in such ionic liquids than in water. The anions in Scheme 2 typically are conjugate bases of acids having pKa's ranging from similar to H_3O^+ to very much larger than H_3O^+.

On the other hand, the anions in Scheme 3 are typically more basic than water and therefore can be expected to act as stronger proton acceptors than water. Acid dissolved in such an ionic liquid may then exhibit a higher degree of dissociation than in water. Thus the Brønsted acidity of a solute acid may be higher in such ionic liquids than in water. The extent to which this is the case is a matter of the relative acceptor strength of the ionic liquid anion and the conjugate base anion of the solute acid.

This understanding is confirmed by the data in Figure 1, which shows the visible spectrum of a weak organic acid, *m*-cresol purple [**16**] (Scheme 4), in water and in a number of ionic liquids in which it is soluble.
The ionic liquids studied in this work include emlm acetate, emlm nitrate, emlm tricyanomethide, emlm dicyanamide, emlm TFSI, bmlm hexafluorophosphate and bmlm tetrafluoroborate. These were prepared, purified and characterized as described previously[5,7,14,16]. Samples were dried under vacuum before use. Metacresol purple is chosen as a typical probe indicator acid here because of its strong degree of colouration in both acid and base forms, the acid form being a straw yellow colour and the base form a deep purple. This difference is dramatically observed in the visible spectra of the acid (λmax = 440nm) and base forms (λmax = 600nm) (Figure 1). Such spectra allow an estimation of the degree of dissociation, α,

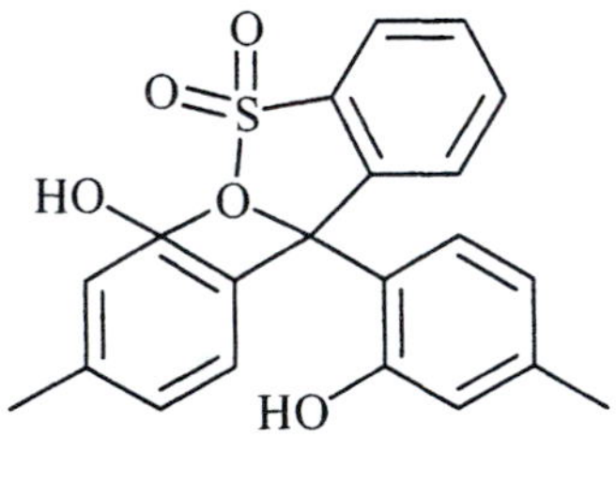

[16] 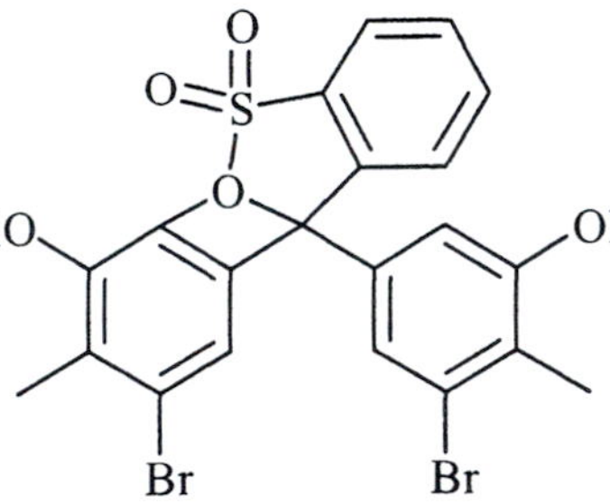 [17]

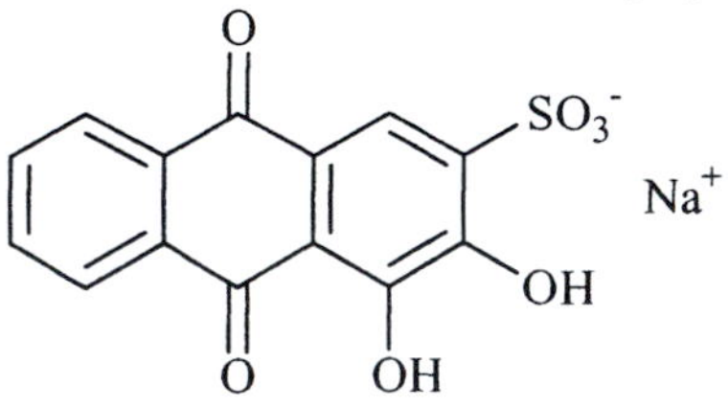

[18]

Scheme 4: Acid/Base indicators; metacresol purple [**16**], bromocresol purple [**17**] and alizarin red S [**18**].

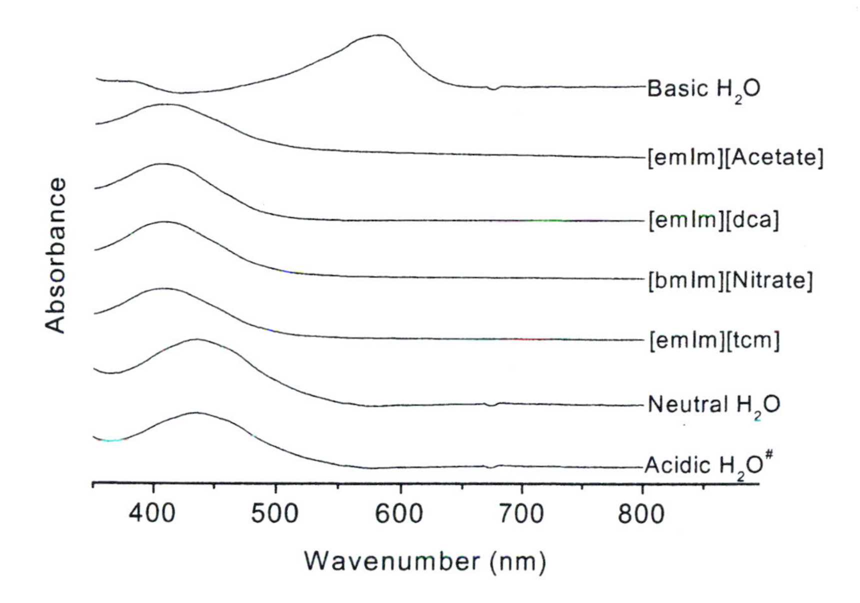

Figure 1: Visible spectroscopy of a weak organic acid, *m*-cresol purple [**16**], in various pure ionic liquids and aqueous solutions. Concentration ~ 1.0 wt% in all cases. #Not shown: [bmIm][PF_6], [emIm][TFSI] and [bmIm][BF_4] also produce spectra equivalent to acidic water.

of the acid. In pure water this acid at the concentration involved is expected to be only approximately 0.02% dissociated into the base form at room temperature. Thus the base form is almost completely absent from the spectra of the acid in pure water.

Equally the base form is not observable in the solutions of the acid in the ionic liquids shown, with the exception of the acetate ionic liquid where a slight lightening in the colour of the solution is observed and an increase in the absorption in the region of the base form absorption maximum (~ 600nm) is seen. Notably, in the dca, tcm, nitrate, PF_6, TFSI and BF_4 ionic liquids the acid appears to be as little dissociated as it is in acidic water. However the comparison between neutral water and the acetate ionic liquid shows that the acid is more dissociated in the latter, as predicted on the basis of the basicity of this anion. Thus we estimate that the degree of dissociation of this acid is very small (<1%) in all but the acetate ionic liquid. Thus all of the ionic liquids studied, except the acetate, act similarly as solvents with respect to this acid.

As a second example of this behaviour we used bromocresol purple [**17**] as the probe acid (Figure 2).

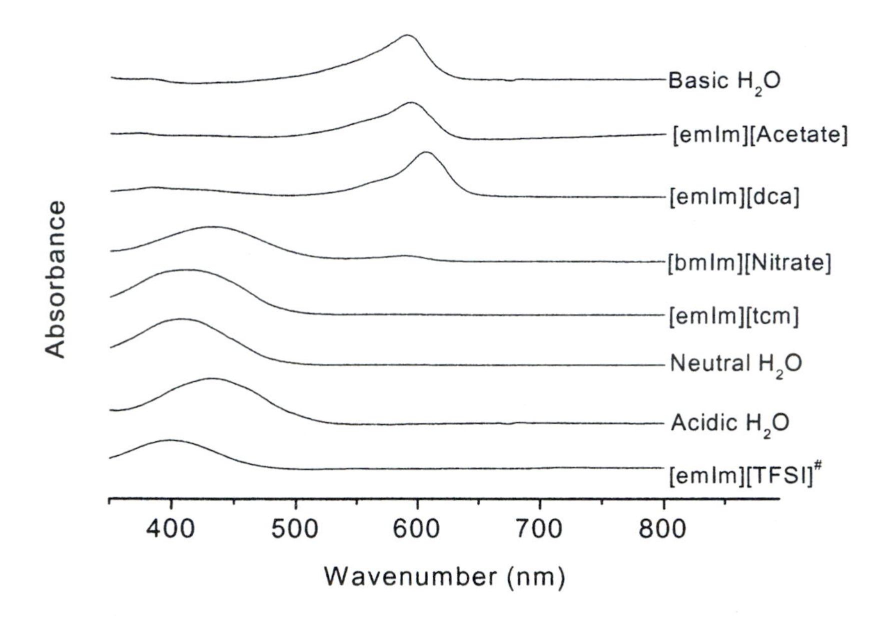

Figure 2: Visible spectroscopy of bromocresol purple [**17**] in various pure ionic liquids and aqueous solutions. Concentration ~ 1.0 wt% in all cases. [#]Not shown: [emIm][PF_6] and [emIm][BF_4] produce spectra equivalent to the TFSI case.

This is a slightly stronger acid than *m*-cresol purple, being about 0.2% dissociated in pure water. Figure 2 shows the visible spectra of the acid in a variety of ionic liquids and aqueous solutions. In this case, the acid remains substantially undissociated in the PF_6, BF_4 (not shown, spectra equivalent to TFSI) and TFSI ionic liquids as well as the nitrate and tricyanomethide ionic liquids. However in the acetate ionic liquid, only the absorption spectrum of the base form is observed, indicating almost complete dissociation in that medium. The spectrum of the acid in the dicyanamide ionic liquid shows strong absorption from the base state as well as a weak absorbance in the region of the acid form. The visual appearance of this sample is also more intermediate in colour than the acetate case. It appears therefore that the acid is not completely dissociated in the dicyanamide case. Notably in both the acetate and dicyanamide cases the acid is more dissociated than it is in pure water.

This acid thus shows varying degrees of dissociation depending on the chemical properties of the anion in the ionic liquid. The PF_6, BF_4, TFSI, tricyanomethide and nitrate ionic liquids are all weaker proton acceptors than the bromocresol purple anion. On the other hand the acetate and dicyanamide anions are sufficiently proton accepting to fully or partially deprotonate, respectively, the bromocresol purple. This observation begins to allow an ordering of anions according to their basicity.

A third example utilizes a still stronger indicator acid, Alizarin red S. This acid is approximately 0.5% dissociated in pure water. Here we see a still greater spread in the colour and spectra of the acid in the various ionic liquids. It is not clear if the difference in the λmax values for the acetate/dicyanamide ionic liquids and basic water is a solvatochromic effect. This spread in behaviours indicates that the proton accepting tendency of the indicator anion now lies intermediate amongst some of the ionic liquids, allowing us to further rank their proton acceptor tendencies. Despite the spreading out of the different ionic liquids, the TFSI ionic liquid remains, not surprisingly, a relatively very poor proton acceptor. We expect that BF_4 and PF_6 ionic liquids will behave similarly.

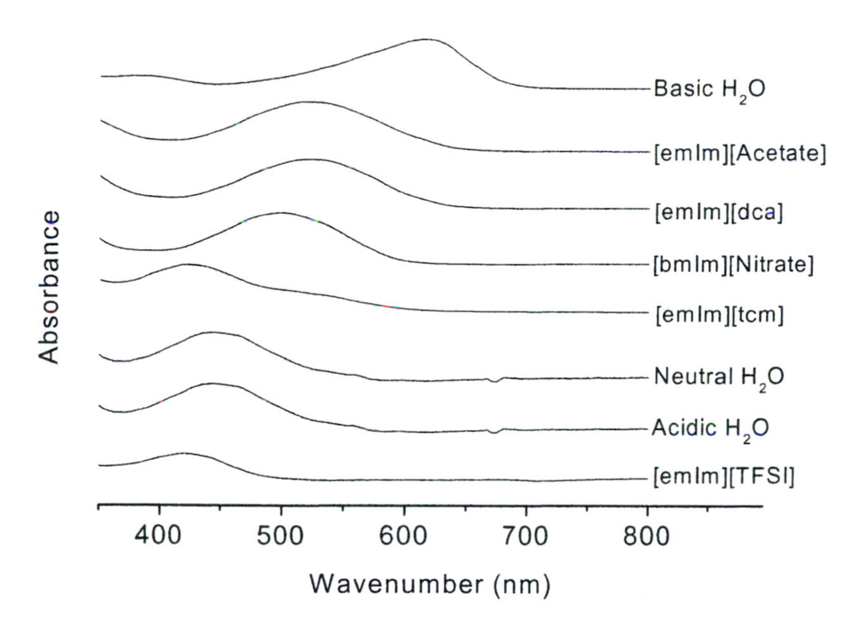

Figure 3: Visible spectroscopy of Alizarin red S [**18**] in various pure ionic liquids and aqueous solutions. Concentration ~ 1.0 wt% in all cases.

Table 1 lists the acid dissociation equilibrium constants, pK_a, for these and a number of more familiar acids in water, as well as the acids of the anions used in the ionic liquids studied here. By making the assumption that at least the relative order of pK_a values is similar in any given ionic liquid as it is in water then we can estimate the α values shown for the ionic liquid media. This assumption allows for the different, as compared with water, nature of the interactions that certainly exist between the dissolved species HA, HS^+, A^- and the solvent ionic liquid species. It assumes that the shift in pK_a caused by this second order effect is approximately the same for all of the acids. The first order effect arises from the relative basicities of the two competing bases, the ionic liquid anion and the conjugate base of the probe acid. There is almost certainly a further strong second order effect arising from the ionic nature of the solvent but this is not easily resolvable from the first order effect. Nonetheless this ionic solvent effect should be relatively the same for all the ionic liquids and therefore not strongly alter the relative order of pK_a values.

We use [emlm][BF_4] and [emlm][acetate] as two archetypal examples of a weakly basic and a basic anion respectively. The basis for the estimations is the first order approximation obtained from the following processes:

Table 1:
Acid Dissociation Constants and Degrees of Dissociation for various Acids in Water and Ionic Liquids.

Acid	Formula	pKa H_2O	α H_2O	α emlm BF4	α emlm acetate
Trifluoromethane sulfonic acid	CF_3SO_3H	*ca.* -13	1	1	1
perchloric acid	$HClO_4$	*ca.* -10	1	1	1
hydrogen iodide	HI	*ca.* -7	1	1	1
hydrogen bromide	HBr	*ca.* -6	1	1	1
tricanomethane	$CH(CN)_3$	*ca.* -5	1	1	1
hydrogen chloride	HCl	*ca.* -3	1	1	1
Methanesulfonic acid	CH_3SO_3H	*ca.* -2	1	1	1
nitric acid	HNO_3	*ca.* -1.5	1	1	1
thiocyanic acid	HSCN	-1.85	1	1	1
hydronium ion	H_3O^+	-1.74	-	1	1
tetrafluoroboric acid	HBF_4	-0.44	0.96	-	1
trifluoroacetic acid	CF_3COOH	0.0	0.92	0.8	1
benzenesulfonic acid	$C_6H_5SO_3H$	0.7	0.7	0.5	1
TFSI acid	$HN(CF_3SO_2)_2$	1.2	0.5	0.3	1
hydrofluoric acid	HF	3.45	0.06	0.04	1
formic acid	HCOOH	3.75	0.04	0.02	1
acetic acid	CH_3COOH	4.75	0.01	0.008	-
alizarin red s	$C_{14}H_7NaO_7S$	5.7	0.004	0.002	0.7
carbonic acid	H_2CO_3	6.4	0.002	0.001	0.4
bromocresol purple	$C_{21}H_{16}Br_2O_5S$	6.4	0.002	0.001	0.4
metacresol purple	$C_{21}H_{18}O_5S$	8.3	<0.001	<0.001	0.05
hydrocyanide	HCN	9.3	<0.001	<0.001	0.02

$$(1)\ HX + H_2O \rightleftharpoons X^- + H_3O^+ \quad pKa(HXaq)$$

$$(2)\ HS + H_2O \rightleftharpoons S^- + H_3O^+ \quad pKa(HSaq)$$

$$(3)\ HX + S^- \rightleftharpoons X^- + HS \quad pKa(HXs)$$

$$(1) - (2) = (3) \Rightarrow pKa(HXs) = pKa(HXaq) / pKa(HSaq)$$

Therefore the relative ordering of pK_a values for acids in an ionic liquid can be approximately calculated in principle from a knowledge of the pK_a value of the acid in water and also the pK_a for the conjugate acid of the ionic liquid anion (HS) in water. These calculations can only be approximate because of the as yet unquantified second order effects discussed above.

We thereby see from Table 1 that acids behave as slightly weaker acids in [emIm][BF_4] than they do in water. Applying the same ideas to ionic liquids of anions from acids further up this list, for example in [emIm][mesylate], we can predict that some acids that we expect to be moderately strong in water, eg trifluoroacetic acid, will in fact behave as moderately weak in the ionic liquid. In other words, an acid will remain protonated in an ionic liquid bearing an anion whose conjugate acid lies above it in this list. If we define Class I (I for "inert") ionic liquids as being those containing the weakly basic anions exemplified in Scheme 2, then we can conclude that many acids will behave as weaker acids in Class I ionic liquids.

However, the ionic liquid [emIm][acetate] is able to act as a stronger acceptor for protons than water or any of the acids above acetic acid in the list. In contrast to the Class I ionic liquids of Figure 1 these ionic liquids can produce strongly dissociated solutions of the probe acid. The anions X^- in these ionic liquids thus act as Lewis bases to promote the dissociation of the dissolved acid:

$$HA + X^- = A^- + HX$$

The sheer concentration of X^- available (activity approaching 1) pushes the position of this equilibrium towards the right. Such Lewis base ionic liquids we describe as Class B (B for basic) ionic liquids. From these observations what we commonly think of as relatively *"weak acids"* can in fact behave as *"strong acids"* in Class B ionic liquids, depending on their relative positions in Table 1.

Base catalysis in Ionic Liquids?

The recognition of Lewis base properties in Class B ionic liquids prompts a question important to the synthetic chemist: Is there a general base catalysis phenomenon possible in these solvents? General base catalysis is a broad group of catalytic reactions involving a basic solvent or additive which acts in a manner to promote a proton abstraction or hydroxide ion attack. In the case of an ionic liquid the catalysis may involve either proton abstraction or a general Lewis base adduct formation with the substrate.

In fact, catalytic effects have been observed in *O*-acetylation reactions of alcohols and sugars in [emIm][dca];[17] in our report on that work we suggested a catalytic action of the dicyanamide anion but were unable to unambiguously describe the action of the catalyst. To investigate the potential for a general base catalysis effect we have now carried out the same reaction in another, chemically quite different, Class B ionic liquid, [emIm][acetate], as described in Table 2. The catalytic effect is clearly present for all examples studied using Class B ionic liquids and absent from the Class I control experiment where the ionic liquid is bmimTFSI. This we believe may indicate a *general base catalysis* effect operating in the Class B ionic liquids. Since the basicity of the ionic liquid is to some extent "tunable" by judicious choice of the anion, a variable base catalysis effect can thus be achieved.

Table 2: *O*-acetylation reactions using no added catalyst.[17]

Substrate	Solvent (eq)	Ac_2O (eq)	Temp (T/°C)	Time (t/h)	Yield %
α-D-Glucose	2 bmIm.dca	5	room temp.	0.2	89
α-D-Glucose	2 bmIm.dca	5	50	0.1	98
β-Me-Glucose	2 bmIm.dca	4.5	room temp.	0.2	92
Sucrose	4 bmIm.dca	8	room temp.	24	93
Raffinose	6 emIm.dca	11	room temp.	24	90
2-Naphthol	2 emIm.dca	1	room temp.	24	85
t-BuOH	2 emIm.dca	1	room temp.	24	91
α-D-Glucose	2 bmIm.tfsi	5	room temp.	24	0
α-D-Glucose	2 emIm.acetate	5	room temp.	1	95

Bases in Ionic Liquids

The state of a base in an ionic liquid is related to the considerations above but is simplified by the absence in most ionic liquids, as the field exists at this point in time, of a proton donor. Thus in both Classes I and B a base such as ammonia or an amino group will remain in its base form and be available, as such, chemically. In principle then, these ionic liquids are solvents that are able to deliver the base in its simplest form to a reaction. This fact has consequences in biochemical applications, since bases such as dopamine, which would exist under physiological conditions in the protonated form, are not so in the ionic liquid. However, recognizing (below) the role of water, presents a partial solution to this problem. This perspective on properties suggests that there may be a further family (Class A) of ionic liquids yet to be widely developed which display distinct proton donor (or, in general, Lewis acid) properties. Work in related fields of solid electrolyte materials has produced boron and boroxine related Lewis acid liquids which may indicate a direction towards Class A ionic liquids.[18]

Role of trace water in ionic liquids

Given the almost ubiquitous presence of water in ionic liquids under normal laboratory conditions, it is of interest to inquire into the role of small amounts of water in the acid/base discussion above. Water contents may be low from a purity point of view (say <0.5%w/w) but may nonetheless be present at molar concentrations close to the concentration of the dissolved acid or base. As an active proton acceptor, water will thus dramatically increase (to the extent of the amount of water present) the degree of dissociation of an acid in a Class I ionic liquid as compared to the pure ionic liquid. Similarly the state of a base will be altered by the presence of water in a Class I ionic liquid. Water will be at least partly dissociated producing OH^- in a Class B ionic liquid. Residual water may also be present in the ionic liquids investigated in this work, but at a level lower than the probe acids studied. The distinctly basic action of the Class B ionic liquids observed here would not be affected by residual water content.

Conclusions

The ever-expanding family of ionic liquids express not a single solvent behaviour but a wide spectrum of behaviours. Using acid/base properties as an indicator we find that we can classify ionic liquids into at least 3 classes, the majority of those described to date belonging to the inert class, Class I, but a growing number are now appearing which belong to the class of bases. The little known Lewis acid class will no doubt receive more attention in the future.

This work represents a first step towards a full understanding of the state of acids and bases in ionic liquids of various types and classes. Quantitative measures of pK_a values in various ionic liquids are clearly required and it is our hope that this discussion will stimulate work in that direction.

References

(1) Walden, P. *Bull. Acad. Sci. St. Petersburg* **1914**, 405-422.

(2) Hitchcock, P. B.; Mohammed, T. J.; Seddon, K. R.; Zora, J. A.; Hussey, C. L.; Ward, E. H. *Inorg. Chim. Acta* **1986**, *113*, L25-L26.

(3) Seddon, K. R. In *Molten Salt Forum*, 1998; Vol. 5-6, pp 53-62.

(4) Welton, T. *Chem. Rev.* **1999**, *99*, 2071-2083.

(5) Lu, W.; Fadeev, A. G.; Qi, B.; Smela, E.; Mattes, B. R.; Ding, J.; Spinks, G. M.; Mazurkiewicz, J.; Zhou, D.; Wallace, G. G.; MacFarlane, D. R.; Forsyth, S. A.; Forsyth, M. *Science,* **2002**, *297*, 983-987.

(6) Sheldon, R. *Chemical Communications* **2001**, 2399-2407.

(7) Wilkes, J. S.; Zaworotko, M. J. *J. Chem. Soc., Chem. Commun.* **1992**, 965-967.

(8) Cooper, E. I.; O'Sullivan, E. J. M. *Proc. - Electrochem. Soc.* **1992**, *16*, 386-396.

(9) Pringle, J. M.; Golding, J.; Forsyth, C. M.; Deacon, G. B.; Forsyth, M.; MacFarlane, D. R. *J.Mater.Chem.* **2002**, *in press.*

(10) Li, T.-S.; Li, A.-X. In *J. Chem. Soc., Perkin Trans. 1*, 1998; pp 1913-1918.

(11) Ngo, H. L.; LeCompte, K.; Hargens, L.; McEwen, A. B. *Thermochimica Acta* **2000**, *357-358*, 97-102.

(12) Golding, J.; Forsyth, S.; MacFarlane, D. R.; Forsyth, M.; Deacon, G. B. *Green.Chem.* **2002**, *4*, 223 - 229.
(13) Huddleston, J. G.; Visser, A. E.; Reichert, W. M.; Willauer, H. D.; Broker, G. A.; Rogers, R. D. *Green Chem.* **2001**, *3*, 156-164.
(14) Bonhote, P.; Dias, A.-P.; Papageorgiou, N.; Kalyanasundaram, K.; Graetzel, M. *Inorg. Chem.* **1996**, *35*, 1168-1178.
(15) Kimizuka, N.; Nakashima, T. *Langmuir* **2001**, *17*, 6759-6761.
(16) MacFarlane, D. R.; Forsyth, S. A.; Golding, J.; Deacon, G. B. *Green Chemistry* **2002**, *4*, 444-448.
(17) Forsyth, S. A.; MacFarlane, D. R.; Thomson, R. J.; von Itzstein, M. *Chem. Commun.* **2002**, 714-715.
(18) Cole, A. C.; Jensen, J. L.; Ntai, I.; Tran, K. L. T.; Weaver, K. J.; Forbes, D. C.; Davis, J. H., Jr. *Journal of the American Chemical Society* **2002**, *124*, 5962-5963.

Chapter 23

Catalytic Olefin Epoxidation and Dihydroxylation with Hydrogen Peroxide in Common Ionic Liquids: Comparative Kinetics and Mechanistic Study

Mahdi M. Abu-Omar, Gregory S. Owens, and Armando Durazo

Department of Chemistry and Biochemistry, University of California, 607 Charles E. Young Drive East, Box 951569, Los Angeles, CA 90095–1569

Methylrhenium trioxide (MTO) catalyzes the epoxidation of alkenes and allylic alcohols with urea hydrogen peroxide (UHP) in room-temperature ionic liquids (RTILs) that are based on *N, N'*-dialkylimidazolium or *N*-alkylpyridinium cations. The ionic liquids must be halide-free because halides catalyze the disproportionation of hydrogen peroxide to molecular oxygen and water. The kinetics and thermodynamics of the reaction of MTO with H_2O_2 in different ionic liquids have been investigated. The rate constant for the formation of the catalytically active diperoxorhenium complex, dpRe, is highly dependent on the concentration of water in the ionic solvent. Also, the rate constants for olefin epoxidation by the peroxorhenium complexes of MTO have been measured by UV-visible and ^2H NMR spectroscopies. ^2H NMR experiments conducted with $[D_3]$dpRe confirmed the speciation of the catalytic system and asserted the validity of the UV-vis kinetics. The dpRe is more reactive in RTILs than its analogous monperoxo species, mpRe. The rate of olefin epoxidation is unaffected by the nature of the ionic liquid's cation; however, a discernable kinetic effect was noted for coordinating anions such as nitrate.

Introduction

A major research goal in green chemistry is the development and study of catalytic reactions that employ benign reagents and solvents. Although the first ionic liquid was reported almost 90 years ago (*1*), their use as alternative reaction media has attracted attention only in the past few years (2). The majority of room-temperature ionic liquids (RTILs) consist of nitrogen-containing organic cations and inorganic anions. Among the most desirable properties of ionic liquids are their (1) negligible vapor pressure, (2) high polarity, (3) weakly coordinating anions, (4) tunable melting temperatures, and (5) ability to dissolve both organic and inorganic compounds. A wide variety of catalytic reactions have been investigated in ionic liquids (*3*). Nevertheless, it is surprising that studies of catalytic oxidations in these media have been limited, particularly, given the importance of such reactions in the laboratory as well as in the chemical industry (*4*).

Hydrogen peroxide is an environmentally friendly oxidant because its only byproduct is water when targeted for oxygen atom transfer reactions. Since reactions of hydrogen peroxide are often slow and nonselective (due to free-radical side reactions), a catalyst is needed for activation and selectivity (*5*). Methylrhenium trioxide (MTO) is one of the best activators for hydrogen peroxide and the most studied organometallic compound (*6*). It reacts with hydrogen peroxide to form η^2-peroxo complexes that are capable of transferring an oxygen atom to olefins and other suitable substrates, Figure 1. The MTO catalytic system in conventional solvents (organic and aqueous) suffers two drawbacks. Catalyst separation from product and catalyst recycling are difficult and cumbersome; for instance, the use of column chromatography is mandatory. Secondly, the peroxorhenium complexes (mpRe and dpRe in Figure 1) require addition of acid for stability. A side effect of the added aqueous acid is facile ring-opening of epoxide to diol.

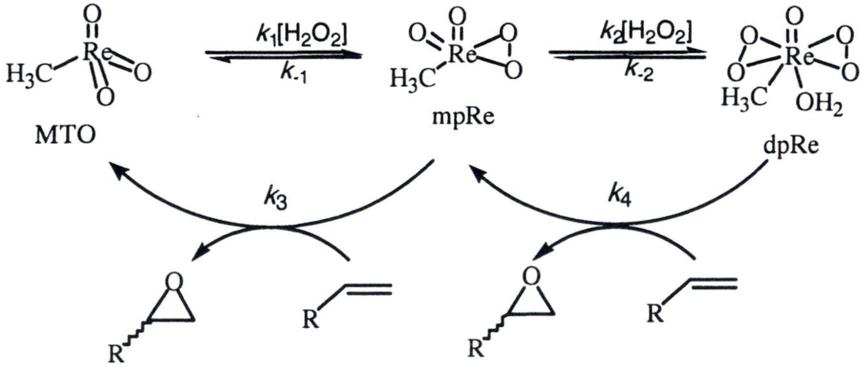

Figure 1. Formation and reactivity of the active catalytic species in the MTO/H_2O_2 System.

We have shown that imidazolium ionic liquids such as [emim]BF_4 can be used with urea hydrogen peroxide (UHP), a water-free source of H_2O_2, and MTO to effect the epoxidation of olefins in high yields (*4b*). Furthermore, the kinetics of dpRe formation (k_2) have been determined in several ionic liquids as well as the kinetics of oxygen atom transfer from mpRe and dpRe to a wide variety of olefinic substrates (*7,8*). It is worth noting that kinetic studies in ionic liquids have been conspicuously scarce in the growing list of physical investigations (*9*). Techniques have been developed in this laboratory to monitor kinetics in ionic media by time-resolved ^{2}H NMR spectroscopy (*10*). In this account, we summarize the chemistry of olefin epoxidation and dihydroxylation as catalyzed by the MTO-peroxide system in RTILs.

MTO-Catalyzed Epoxidations Using UHP

The advantages of ionic liquids and the MTO/peroxide system have been combined to create an oxidation solution that is both environmentally friendly and highly efficient. The ionic liquid *N*-ethyl-*N'*-methylimidazolium tetrafluoroborate ([emim]BF_4) was used with MTO and UHP to oxidize alkenes and allylic alcohols with high conversions to the corresponding epoxides, Table I. This system combines the advantages of heterogeneous and homogeneous catalysis: the epoxide products are easily recovered by simple extraction with diethyl ether, and the reaction times are comparable to those observed with organic solvents.

The importance of a water-free source of hydrogen peroxide is illustrated by equations 1 and 2. When aqueous hydrogen peroxide is employed as oxidant, the 1,2-diol is the major product. Hence, this system is extremely versatile as both epoxides and diols can be easily accessed by simply changing the peroxide from UHP to 30% aqueous H_2O_2. A major difference between the MTO/UHP system in ionic liquids compared to that in organic solvents is the solubility of urea hydrogen peroxide. The UHP complex is sparingly soluble in aqueous and organic ($CHCl_3$ or CH_2Cl_2) media, but UHP is readily soluble in [emim]BF_4, giving homogeneous reaction conditions. This difference accounts for the enhanced reactivity of the MTO/UHP in RTILs. For example, the epoxidation of styrene with MTO/UHP in $CHCl_3$ under the same conditions as in Table I requires 19 h for 45% conversion to styrene oxide. The same reaction is complete in [emim]BF_4 in less than 6 h.

1-phenylcyclohexene + 2 H_2O_2 (aqueous 30%) $\xrightarrow[95\%]{\text{1 mol \% MTO, [emim]BF}_4\text{, 20 °C}}$ 1-phenylcyclohexane-1,2-diol (Ph, OH, OH)

(1)

Table I. Epoxidation of Alkenes with MTO/UHP in [emim]BF_4.[a]

Substrate	*% Epoxide*[b]	*Substrate*	*% Epoxide*[b]
cyclohexene	99	styrene	96
1-methylcyclohexene	95	β-methylstyrene	94
1-phenylcyclohexene (Ph)	98	cyclooctene	95
2-cyclohexen-1-ol (OH)	95[c]	1,5-cyclooctadiene	85[d]
$CH_2=CH(CH_2)_7CH_3$	46		

[a] Conditions: 1.0 mmol substrate, 1.0-2.0 mmol UHP, and 0.02 mmol MTO in 2.0 mL [emim]BF_4 at 20 °C for 4-6 h. [b] % Conversion of substrate into product. [c] 2,3-epoxycyclohexanol (62%) and 2,3-epoxycyclohexanone (32%). [d] This represents the diepoxide; the remainder is the monoepoxide.

1-phenylcyclohexene (Ph) + 1 UHP —(1 mol % MTO; [emim]BF_4, 20 °C; 98%)→ 1-phenylcyclohexene oxide (Ph, O)

(2)

Solvent purity is crucial for successful epoxidations with the MTO-peroxide catalytic system in RTILs. The sensitivity of mpRe and dpRe to pyridine, pyrazole, and imidazole is well documented in the literature (*11*). Hence, ILs must be absolutely free of any residual *N*-alkylimidazole starting materials. The most common procedure for making RTILs involves the metathesis of dialkylimidazolium halide with the desired inorganic anion. Halides (e.g. Br^- and Cl^-) are oxidized by the peroxorhenium complexes of MTO to the hypohalous acids, which in turn catalyze the disproportionation of hydrogen peroxide to molecular oxygen and water (*12*). Halide impurities are detrimental because they are oxidized faster than olefins. For example, k_4 for Br^- is ~2000 times k_4 for styrene. A simple calculation reveals that only 1 mol % residual bromide gives an impurity concentration of 80 mM in the resulting ionic liquid! Hence, care must be taken to remove traces of halide from the ionic liquid. Silver salts are generally most effective reagents for metathesis, but large excess of Ag(I) is also undesirable as its photosensitivity leads to brown ILs (*7*).

Kinetics and Thermodynamics of the MTO/Peroxide Reactions

Kinetics of dpRe Formation

The formation of dpRe (Figure 1) was monitored by observing the absorbance change at 360 nm, the λ_{max} of dpRe (ε = 1100 M^{-1} cm^{-1}). Hydrogen peroxide was used in excess and plots of k_ψ versus [H_2O_2] were analyzed to obtain values of k_2 in a variety of ionic liquids. The rate constants for the formation of two peroxorhenium species from the MTO/H_2O_2 reaction, k_1 and k_2, have been determined in a number of molecular solvents (*6b*). The kinetics of the MTO/H_2O_2 reaction feature biexponential time profiles in organic solvents. However, in ionic liquids, the absorbance change at 360 nm follows a single-exponential curve, since the formation of mpRe occurs extremely rapidly within the mixing time. Because of the viscosity of the ionic liquids ($\geq$ 60 cP), the formation of mpRe was not investigated by stopped-flow techniques. Even though the intercepts from the plots of k_ψ versus [H_2O_2] should, in theory, give values for k_{-2}, the y-intercepts are significantly smaller than the slopes, rendering k_{-2} negligible within experimental precision.

The rate constant k_2 was determined in five different ionic liquids: [emim]BF_4, [bmim]BF_4, [bmim]NO_3, [bmim]OTf, and [bupy]BF_4 (*13*). The rate constant k_2 is the same for all five ionic liquids (k_2 = 0.20 ± 0.02 L mol^{-1} s^{-1}). This result is surprising given the differences in viscosity of the liquids and in the coordination ability of their anions. The rate constant k_2 in ionic liquids is smaller than that in water (5.2 L mol^{-1} s^{-1}) and greater than that in acetonitrile (0.045 L mol^{-1} s^{-1}). In order to gain insight into the effect of water and salt concentrations, values of k_2 were determined in a number of concentrated salt solutions. These results are summarized in Table II. Comparing the rate constants in Table II shows that k_2 in 99% ionic liquid is essentially the same as k_2 in acetonitrile. This is not surprising given that the polarity of ionic liquids has been estimated to be comparable to that of acetonitrile and methanol (*14*). The data in Table II correlated best with the concentration of water. The rate constant k_2 increases as [H_2O] increases. This result provides evidence that in the presence of water RTILs behave like aqueous solutions of high salt concentrations.

Table II. Values of k_2 in Various Media.

Entry	*Solvent*[a]	k_2/ $L\ mol^{-1}\ s^{-1}$	$[H_2O]$/ $mol\ L^{-1}$
1	5.92 M $[etpy]BF_4$	0.20	5.55
2	5.41 M $[etpy]BF_4$	0.38	8.88
3	4.28 M $[etpy]BF_4$	0.65	18.7
4	99% $[bmim]NO_3$	0.053	0.56
5	95% $[bmim]NO_3$	0.12	2.77
6	90% $[bmim]NO_3$	0.22	5.55
7	80% $[bmim]NO_3$	0.36	11.1
8	60% $[bmim]NO_3$	0.79	22.2
9	40% $[bmim]BF_4$	1.17	33.3
10	10% $[bmim]BF_4$	1.90	50.0

[a] Abbreviations: etpy = *N*-ethylpyridinium, and bmim = *N*-butylmethylimidazolium.

Thermodynamics of the MTO-Peroxide Equilibria

The relationship of A_{360} and $[H_2O_2]$, derived using the equilibrium expressions for K_1 and K_2 (Figure 1), is given in equation 3, where $[Re]_T$ = [MTO] + [mpRe] + [dpRe]. A nonlinear least-squares fit of the data using equation 3 provides values for K_1 and K_2, Table III.

$$\frac{A_{360}}{[Re]_T} = \frac{\varepsilon_{mpRe} K_1 [H_2O_2] + \varepsilon_{dpRe} K_1 K_2 [H_2O_2]^2}{1 + K_1 [H_2O_2] + K_1 K_2 [H_2O_2]^2} \tag{3}$$

The results in Table III indicate the same general trends observed for the kinetics studies. However, the influence of water is not quite as dramatic for the equilibrium constants as it is for the rate constant k_2. The equilibrium constants K_1 and K_2 in 90% ionic liquid solution (Table III) fall between water and acetonitrile with $K_2 > K_1$, indicating cooperativity in peroxide binding. However, the values for the reaction in dry $[emim]BF_4$ are K_1 = 110 and K_2 = 160 L mol^{-1}. When less water is present in the ionic liquids, the extent of peroxide binding cooperativity is reduced, as is the case when the reaction is conducted in acetonitrile.

Non-Steady-State Kinetics of Olefin Epoxidation

The goal of this study is to determine the rates of oxygen transfer from the catalytically active peroxorhenium complexes (k_3 and k_4 in Figure 1) to olefinic substrates in ionic liquids, and contrast these values to those obtained in water

Table III. K_1 and K_2 for the MTO-Peroxide Reaction in Different Solutions.[a]

Solvent	K_1/ L mol^{-1}	K_2/ L mol^{-1}
H_2O[b]	16	132
CH_3CN[b]	209	660
[emim]BF_4	49	170
[bupy]BF_4	110	140
[bmim]BF_4	74	130
[bmim]NO_3	28	82
[bmim]OTf	34	120

[a] The equilibrium constants for the ionic liquid solvents were measured in 9:1 (v/v) IL:H_2O media. [b] Reference 6(b).

and organic solvents (*8*). It had become evident early on that two complications needed to be addressed here. First, the effects of water on the kinetics (*7*) and physical properties (*15*) of room temperature ionic liquids have been documented. Hence, a water free source of hydrogen peroxide had to be employed in order to minimize the deleterious effects of water as a co-solvent. Secondly, the UV cutoff of 300 nm imposed by the ionic media necessitated the development of a new and reliable method to monitor these reactions under single turnover conditions. Many solvents were screened for the preparation of essentially water- and peroxide-free dpRe. The best results were obtained with anhydrous THF and UHP, since in this solvent dpRe is stable for prolonged periods of time and UHP is nearly insoluble. Hence, filtration of excess UHP affords a concentrated solution (~ 40 mM) of dpRe in THF.

For the non-steady-state kinetics with alkenes, the THF solution of dpRe was diluted into the desired RTIL to ~ 1 mM such that the final concentration of THF was less than 3% of the ionic solvent. Upon addition of excess alkene (40-150 mM), the reaction progress was followed at 360 nm. The non-steady-state time profiles are biexponential. The absorbance at the end of reaction is essentially zero, which indicates that all of the dpRe and mpRe have reacted with the substrate. The kinetic traces are fitted using the biexponential treatment shown in eq 4, where both α and β are constants. Both k_f and k_s (the fast and slow pseudo-first-order rate constants, respectively) exhibit linear dependences on the alkene concentration.

$$\mathrm{Abs}_t - \mathrm{Abs}_\infty = \alpha \exp(-k_f t) + \beta \exp(-k_s t) \quad (4)$$

The second-order rate constants for the reaction of mpRe (k_3) and for the reaction of dpRe (k_4) with the substrate were obtained from the slopes of the above plots. In order to assign the fast and slow steps, we employed the kinetics simulation program KINSIM (*16*) and determined that the k_4 step, the oxidation of alkene with dpRe, is faster than the k_3 step (*8*). The values of k_3 and k_4 for several olefins have been determined in acetonitrile and in [emim]BF_4, and compared to those previously measured in 1:1 CH_3CN:H_2O (*17*).

k_4 values in [emim]BF_4 for the three styrene substrates shown in Table IV are quite similar to those values in 1:1 CH_3CN/H_2O. The notable exception is the oxidation of cyclohexene for which k_4 in [emim]BF_4 is lower by an order of magnitude than k_4 in 1:1 CH_3CN/H_2O and lower even than k_4 in CH_3CN. This result is anomalous but is reconciled by considering the limited solubility of cyclohexene in [emim]BF_4. The values of k_3 in [emim]BF_4 for the oxidation of the styrenes are only slightly higher than the corresponding k_3 values in acetonitrile. There is, however, a significant difference between the k_3 values obtained in [emim]BF_4 versus those in 1:1 CH_3CN/H_2O. This is in contrast to k_4 for which the values in the ionic liquid and in the CH_3CN/H_2O mixture are virtually the same. The reactivity of the styrenes with the peroxorhenium complexes shows the expected trend for electrophilic oxygen transfer from rhenium: styrene < TBMS < α-methylstyrene (AMS).

Table IV. Values of k_3 and k_4 (L mol^{-1} s^{-1}) in Different Solvents for Several Styrenes.

Substrate	*[emim]BF_4*[a]		*CH_3CN*[a]		*CH_3CN:H_2O*[b]	
	k_3	k_4	k_3	k_4	k_3	k_4
Styrene	0.013	0.13	0.0020	0.015	___	0.11
trans-β-Methylstyrene	0.058	0.26	0.021	0.052	0.51	0.22
α-Methylstyrene	0.10	0.44	0.045	0.045	___	0.47
Cyclohexene	0.031	0.13	0.031	0.22	___	1.06

[a] Reference 8. [b] Reference 17.

The kinetics of α-Methylstyrene epoxidation was investigated in different ionic liquids. The rate constants k_3 and k_4 for the oxidation of α-methylstyrene are summarized in Table V. Entries 3, 4, and 6 show essentially the same rate constants for the ionic liquids [emim]BF_4, [bmim]BF_4, and [bupy]BF_4. This indicates that the cations of these liquids do not have an effect on the reaction of the peroxorhenium complexes with the double bond. However, both k_3 and k_4 are substantially lower when the reaction is done in [bmim]NO_3 (entry 5). This result is rationalized by considering that NO_3^- has much higher coordination ability than BF_4^- and thus hinders approach of substrate to the rhenium peroxo complexes.

Table V. Rate Constants for the Oxidation of α-Methylstyrene in Various Ionic Liquids.[a]

Entry	*Solvent*	k_3/ L mol^{-1} s^{-1}	k_4/ L mol^{-1} s^{-1}
1	CH_3CN	0.045	0.045
2[b]	1:1 CH_3CN/H_2O	—	0.47
3	[emim]BF_4	0.10	0.44
4	[bmim]BF_4	0.074	0.35
5	[bmim]NO_3	0.016	0.12
6	[bupy]BF_4	0.13	0.58

[a] Reference 8. [b] Reference 17.

^{2}H NMR Kinetics

Deuterated alkenes can be used as substrates in proteated ionic liquids to study the MTO-catalyzed oxidation of alkenes under steady-state conditions (*8,10*). The integrals of the ^{2}H resonances for deuterated reactants (alkenes) and products (epoxide or diol) were monitored over time to obtain kinetic information. UHP is soluble in RTILs, a fact that makes the MTO/UHP system homogeneous in ionic liquids and amenable to kinetic investigations. However, in order to avoid complex kinetics and minimize catalyst degradation, one is forced to employ high hydrogen peroxide concentrations (0.5-1.0 M). Under these conditions, the kinetic traces follow a single-exponential decay and are fit with equation 5. The kinetics are first-order in [alkene] and show no dependence on [H_2O_2]. Hence, $k_\psi = k_3$ [mpRe] + k_4 [dpRe] simplifies to $k_\psi \approx k_4$ [dpRe], since at these high [H_2O_2], the major rhenium species is dpRe and (as shown above by UV-vis kinetics) $k_3 < k_4$ in RTILs. The k_4 values obtained by ^{2}H NMR for the epoxidation and dihydroxylation of [D_8]styrene and [D_{10}]cyclohexene in several RTILs are comparable to values obtained by UV-vis kinetics (*8*).

$$I_t - I_\infty = \alpha \exp(-k_\psi t) \tag{5}$$

In an alternative method that more closely resembles the conditions used in the UV-vis kinetic studies, CD_3ReO_3 ([D_3]MTO) was reacted with 2 equivalents UHP in RTILs to generate [D_3]dpRe. The latter was used for single turnover reactions with alkenes, and followed by ^{2}H NMR. The relatively long timescale inherent to NMR kinetic experiments mandated the use of more deactivated styrenes as epoxidation substrates. For this study, styrene, 2-fluorostyrene, and 2,6-difluorostyrene were used (*8*). Over the course of a reaction, the intensity of the [D_3]dpRe resonance (δ = 2.9 ppm) decreases and the intensity of the [D_3]MTO resonance (δ = 2.4 ppm) increases. The ^{2}H resonances due to products of peroxorhenium complex decompositions, namely CD_3OH and CD_3OOH (δ = 3.4 and 3.3 ppm, respectively), increase during the course of reaction. A typical NMR stack plot for the epoxidation of 2-fluorostyrene is shown in Figure 2.

The integration for the [D_3]dpRe resonance is plotted as a function of time. The kinetic traces obtained are exponential, and the integration for the [D_3]dpRe resonance is nearly zero at the end of the reactions, which demonstrates that the reaction between [D_3]dpRe and substrate proceeds to completion. The kinetic traces are treated with a standard exponential decay equation to obtain k_4 from the pseudo-first-order rate constant, $k_\psi = k_4$ [alkene]. The values of k_4 obtained by ^{2}H NMR are comparable to those determined by UV-vis kinetics (*8*).

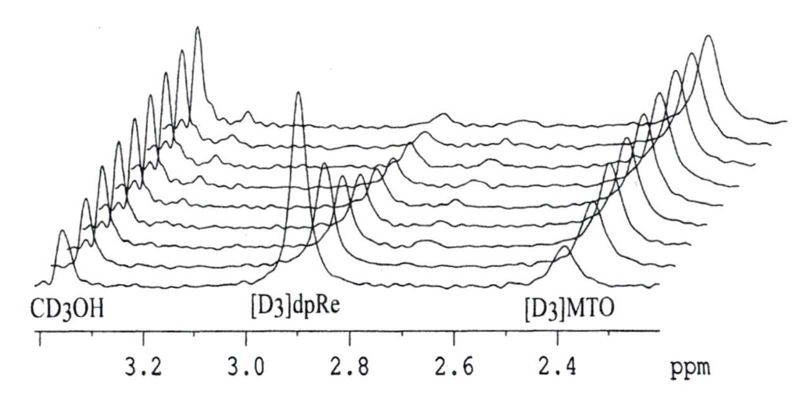

Figure 2. ^{2}H NMR stack plot for the single-turnover epoxidation of 2-fluorostyrene with [D_3]dpRe in [emim]BF_4.

Summary

Little research has been conducted on green catalytic oxidations in ionic liquids. The MTO/UHP is an effective catalytic system for the epoxidation of alkenes and allylic alcohols in different RTILs. When aqueous hydrogen peroxide is used in place of UHP, olefin oxidation proceeds smoothly to give the diol rather than epoxide. Recovery of the organic products is easily achieved by an extraction step in which the catalyst remains in the ionic medium. The purity of the ionic liquid is key to the success of the MTO/UHP system, as small impurities of leftover halide or imidazole result in diminished alkene conversion to epoxide. The kinetics and thermodynamics of the reaction of MTO with hydrogen peroxide in different ionic liquids have been studied, demonstrating the amenability of the system for quantitative measurements. The values of the rate constants are highly dependent on the concentration of water. The rate constants for oxygen atom transfer from mpRe and dpRe to alkenes have been measured by UV-vis and ^{2}H NMR spectroscopies. The diperoxorhenium complex dpRe was found to be about 5 times more reactive than its analogous monoperoxo species mpRe. ^{2}H NMR is an economical and versatile technique for studying the progress of reactions in proteated ionic liquids. In conclusion, we have demonstrated the suitability of ionic liquids for green catalytic oxidations with hydrogen peroxide under ambient and mild conditions. Future directions of interest would include the use of molecular oxygen, possibly the best oxidant, for selective oxidations in ionic liquids, and the employment of task specific ionic media in catalysis.

Acknowledgments

We are grateful to the US National Science Foundation, the Beckman Foundation, and the University of California Toxic Substance Research and Teaching Program for supporting our work on ionic liquids.

References

1. Walden, P. *Bull. Acad. Imper. Sci. (St. Petersburg)* **1914**, 1800.
2. (a) Welton, T. *Chem. Rev.* **1999**, *99*, 2071. (b) Wasserscheid, P.; Keim, W. *Angew. Chem. Int. Ed.* **2000**, *39*, 3772. (c) Earl, M. J.; Seddon, K. R. *Pure Appl. Chem.* **2000**, *72*, 1391. (d) Abu-Omar, M. M. in *2002 McGraw-Hill Yearbook of Science & Technology*, McGraw-Hill, New York, NY, 2001; p. 148.
3. Sheldon, R. *Chem. Commun.* **2001**, 2399.
4. (a) Song, C. E.; Roh, E. J. *Chem. Commun.* **2000**, 837. (b) Owens, G. S.; Abu-Omar, M. M. *Chem. Commun.* **2000**, 1165. (c) Ley, S. V.; Ramarao, C.; Smith, M. D. *Chem. Commun.* **2001**, 2278. (d) Ansari, I. M.; Gree, R. *Org. Lett.* **2002**, *4*, 1507. (e) Farmer, V.; Welton, T. *Green Chem.* **2002**, 97. (f) Seddon, K. R.; Stark, A. *Green Chem.* **2002**, 119. (g) Bortolini, O.; Conte, V.; Chiappe, C.; Fantin, G.; Fogagnolo, M.; Maietti, S. *Green Chem.* **2002**, 94. (h) Nambodiri, V. V.; Varma, R. S.; Sahle-Demessie, E.; Pillai, U. R. *Green Chem.* **2002**, 170.
5. Strukul, G. in *Catalytic Oxidations with Hydrogen Peroxide as Oxidant*, G. Strukul, Editor, Kluwer Academic Publishers, Dordrecht, 1992; p. 1.
6. (a) Romao, C. C.; Kuhn, F. E.; Herrmann, W. A. *Chem. Rev.* **1997**, *97*, 3197. (b) Espenson, J. H. *Chem. Commun.* **1999**, 479. (c) Owens, G. S.; Arias, J.; Abu-Omar, M. M. *Catal. Today* **2000**, *55*, 317.
7. Owens, G. S.; Abu-Omar, M. M. *J. Mol. Catal. A: Chemical* **2002**, 215.
8. Owens, G. S.; Durazo,A.; Abu-Omar, M. M. *Chem. Eur. J.* **2002**, *8*, 3053.
9. (a) Karpinski, Z. J.; Song, S.; Osteryoung, R. A. *Inorg. Chim. Acta* **1994**, *225*, 9. (b) Carter, M. T.; Osteryoung, R. A. *J. Electrochem. Soc.* **1994**, *141*, 1713. (c) Gordon, C. M.; McLean, A. J. *Chem. Commun.* **2000**, 1395. (d) Csihony, S.; Mehdi, H.; Horvath, I. T. *Green Chemistry* **2001**, *3*, 307.

10. Durazo, A.; Abu-Omar, M. M. *Chem. Commun.* **2002**, 66.
11. (a) Herrman, W. A.; Fischer, R. W.; Rauch, M. U.; Scherer, W. *J. Mol. Catal.* **1994**, *86*, 243. (b) Abu-Omar, M. M.; Hansen, P. J.; Espenson, J. H. *J. Am. Chem. Soc.* **1996**, *118*, 4966. (c) Wang, W.; Espenson, J. H. *J. Am. Chem. Soc.* **1998**, *120*, 11335.
12. (a) Espenson, J. H.; Pestovsky, O.; Huston, P.; Staudt, S. *J. Am. Chem. Soc.* **1994**, *116*, 2869. (b) Hansen, P. J.; Espenson, J. H. *Inorg. Chem.* **1995**, *34*, 5839.
13. Abbreviations: emim = *N, N'*-ethylmethylimidazolium, bmim = *N, N'*-butylmethylimidazolium, bupy = *N*-butylpyridinium, and OTf = trifluoromethane sulfonate.
14. Aki, S. N. V. K.; Brennecke, J. F.; Samanta, A. *Chem. Commun.* **2001**, 413.
15. Seddon, K. R.; Stark, A.; Torres, M.-J. *Pure Appl. Chem.* **2000**, *72*, 2275.
16. Barshop, B. A.; Wrenn, C. F.; Frieden, C. *Anal. Biochem.* **1983**, *130*, 134.
17. Al-Ajlouni, A.; Espenson, J. H. *J. Am. Chem. Soc.* **1995**, *117*, 9243.

Chapter 24

Polarity Variation of Room Temperature Ionic Liquids and Its Influence on a Diels–Alder Reaction

Richard A. Bartsch and Sergei V. Dzyuba

Department of Chemistry and Biochemistry, Texas Tech University, Lubbock, TX 79409–1061

The polarity of 1-X-3-methylimidazolium bis(trifluoromethylsulfonyl)imides, as asssessed by $E_T(30)$ values, may be substantially altered by incorporation of functional groups into the X substituent. When such RTILs are employed as solvents for the Diels-Alder reaction of cyclopentadiene and methyl acrylate, increased polarity produces an enhanced *endo*/*exo* ratio in the reaction products.

Introduction

Air- and moisture-stable room-temperature ionic liquids (RTILs) are emerging as important alternatives to conventional molecular organic solvents (*1*). Negligible vapor pressure, as well as ease of recovery and reuse, make RTILs a greener alternative to volatile organic solvents. An intriguing feature of RTILs is the ability to tailor certain bulk properties (*e.g.*, melting point, viscosity, hydrophobicity) by varying the nature of the cation and/or anion (*2*). Hence, RTILs have been deemed 'designer-solvents' (*3*).

Recently, air- and moisture-stable *N,N'*-dialkylimidazolium salts with PF_6^- and BF_4^- anions have been utilized for a wide range of chemical processes (*4*) including alkylation, Baylis-Hillman reactions, Diels-Alder reactions, dimerization, enzymatic catalysis (*5*), Friedel-Crafts reactions, Heck coupling reactions, hydrodimerization, hydrogenation, nucleophilic displacement, polymerization, silica aerogel synthesis (*6*), Suzuki cross-coupling reactions and Wittig reactions. They have also been employed in several separation methodologies as stationary phases for gas chromatography (*7*), diluents for solvent extraction of neutral molecules and ions (*8-13*) and liquid phases in supported liquid membranes (*14, 15*).

Much less attention has been paid to determining the influence of structural variations within the cationic and/or anionic components of RTILs on their physical properties, such as conductivity, density, phase transitions, polarity, refractive index, surface tension, thermal stability and viscosity (*16-32*). For example, only a few studies have been conducted to explore their polarity and how the cationic and anionic components influence solvation on the molecular level (*16, 19, 20, 22-24, 27, 28, 30, 32*).

Probing the Polarity of RTILs

The IUPAC recommended definition of solvent polarity (*33*) is: "Polarity is the sum of all possible, non-specific and specific, intermolecular interactions between the solute ions or molecules and solvent molecules, excluding such interactions leading to definite chemical alterations of the ions or molecules of the solute."

A frequently encountered empirical scale of solvent polarity is $E_T(30)$, which is based the wavelength maximum of the longest intramolecular charge-transfer π-π^* absorption of Reichardt's dye (**1**) (Figure 1). (Historically this is the dye numbered 30 in the initial publication by Dimroth, Reichardt *et al.*) (*34*). This zwitterionic dye exhibits one of the largest observed solvatochromic effects of any known organic molecule (*35*). The charge-transfer absorption wavelength shifts amount to several hundred nanometers in going from a polar solvent ($\lambda_{max} \cong 453$ nm in water) to a non-polar solvent ($\lambda_{max} \cong 925$ nm in hexane). The $E_T(30)$ value is calculated from the wavelength of the absorption maximum by the equation:

$$E_T(30)\ (\text{kcal mol}^{-1}) = 28591/\lambda_{max}\ (\text{nm})$$

In a limited number of investigations, RTIL polarities have been probed using solvatochromic (Figure 1) (*19, 22-24, 28, 30, 32*) and fluorescent (*16, 20, 23, 27*) dyes. The solvatochromic dyes include Reichardt's dye (**1**) (*22-24, 28, 30, 32*), Nile Red (**2**) (*19, 32*) and organometallic complex **3** (*22, 24*).

The measured λ_{max} and calculated $E_T(30)$ values for some 1-alkyl-3-methyl-imidazolium salts (*24*) are presented in Table I.

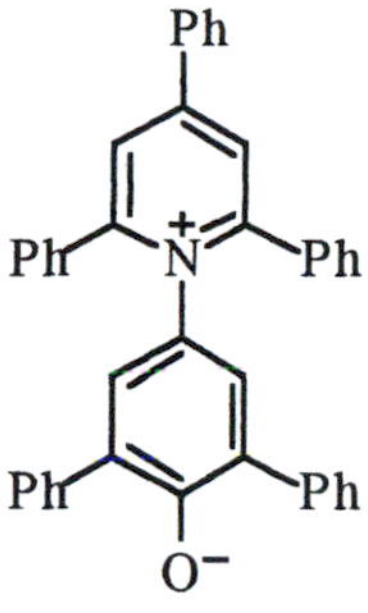

1 (Reichardt's Dye)

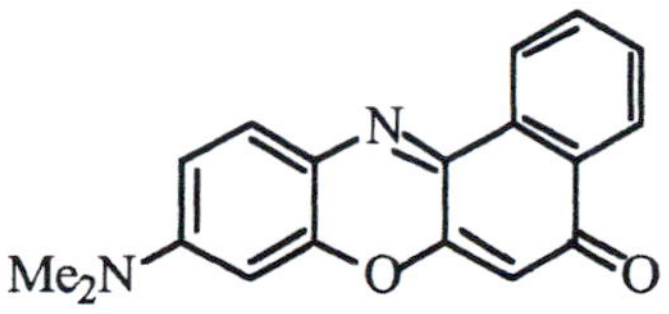

2 (Nile Red)

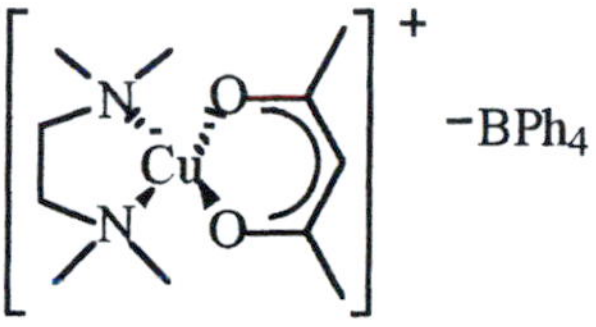

3

Figure 1. Solvatochromic dyes used to probe RTIL polarity.

Table I. Absorption Maxima and $E_T(30)$ Values for Some 1-R-3-methylimidazolium Salts

R	*Anion*	*λ_{max} (nm)*	*$E_T(30)$ (kcal mol^{-1})[1]*
Bu	PF_6^-	546.5	52.3
Bu	BF_4^-	545.0	52.5
Bu	Tf_2N^-	555.5	51.5
Bu	TfO^-	547.0	52.3
Oct	PF_6^-	558.0	51.2
Oct	Tf_2N^-	559.0	51.1

SOURCE: Data from Reference 24, Copyright 2001 Royal Society of Chemistry

For 1-butyl-3-methylimidazolium salts with PF_6^-, BF_4^-, bis(trifluoromethylsulfonyl)imide (Tf_2N^-) and trifluoromethylsulfonate (TfO^-) anions, the $E_T(30)$ values vary only slightly in the range of 51.5-52.5. Changing R from butyl to octyl with PF_6^- and Tf_2N^- anions produces single unit decreases in the $E_T(30)$ values. These observations reveal that neither elongation of the alkyl group nor variation of the anion produces a pronounced change in the $E_T(30)$ value of the RTIL.

The $E_T(30)$ values for several common molecular organic solvents are given in Table II. From comparison of the data in Tables I and II, it is concluded that the 1-R-3-methylimidazolium salts are similar in polarity to ethanol. However, it has been found that the precise positioning of the RTIL polarity on a scale of molecular organic solvent polarities may vary somewhat with the identity of the solvatochromic or fluorescent dye (*23*).

Table II. $E_T(30)$ Values of Molecular Organic Solvents

Solvent	*$E_T(30)$ (kcal mol^{-1})*
Water	63.1
MeOH	55.5
EtOH	51.9
2-PrOH	48.4
DMF	43.2
Acetone	42.2
1,2-Dichloroethane	41.3

SOURCE: Data from Reference 35, Copyright 1994 Amerrican Chemical Society

Expanding the Polarity Range for RTILs

We surmised that incorporation of functional groups into one nitrogen substituent in 1-X-3-methylimidazolium salts might produce a significant alteration in the polarity. (Although others have reported a very limited number of imidazolium salts with functional groups in their substituents, their polarities were not determined (*7, 16, 36-38*). For testing of this proposal, preparation of the series of compounds shown in Figure 2 was undertaken. For this RTIL series, the Tf_2N^- anion was chosen over PF_6^- due to the lower melting points and viscosities of 1,3-disubstituted bistrifylimides (*29*).

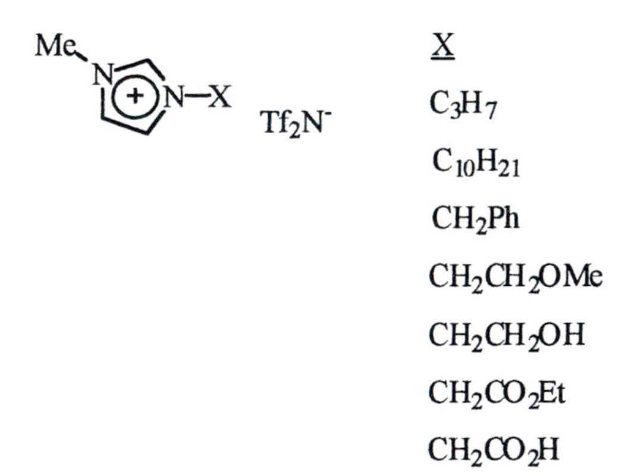

Figure 2. Structures for the proposed 1-X-3-methylimidazolium bistrifylimide series.

Synthesis of members of this series was accomplished in two steps, first by alkylation of 1-methylimidazole under the conditions shown in Figure 3. Then anion metathesis with $LiNTf_2$ in water (*29*) transformed the 1-X-3-methylimidazolium bromide or chloride into the corresponding bistrifylimide salt. (Although the synthesis of 1-(2-methoxyethyl)-3-methylimidazolium bistrifylimide was reported earlier by others (*16*), this new synthetic route is much simpler.) Most members of the series were free-flowing liquids. However, with X = CH_2CO_2Et and CH_2CO_2H, the RTILs were found to be very viscous.

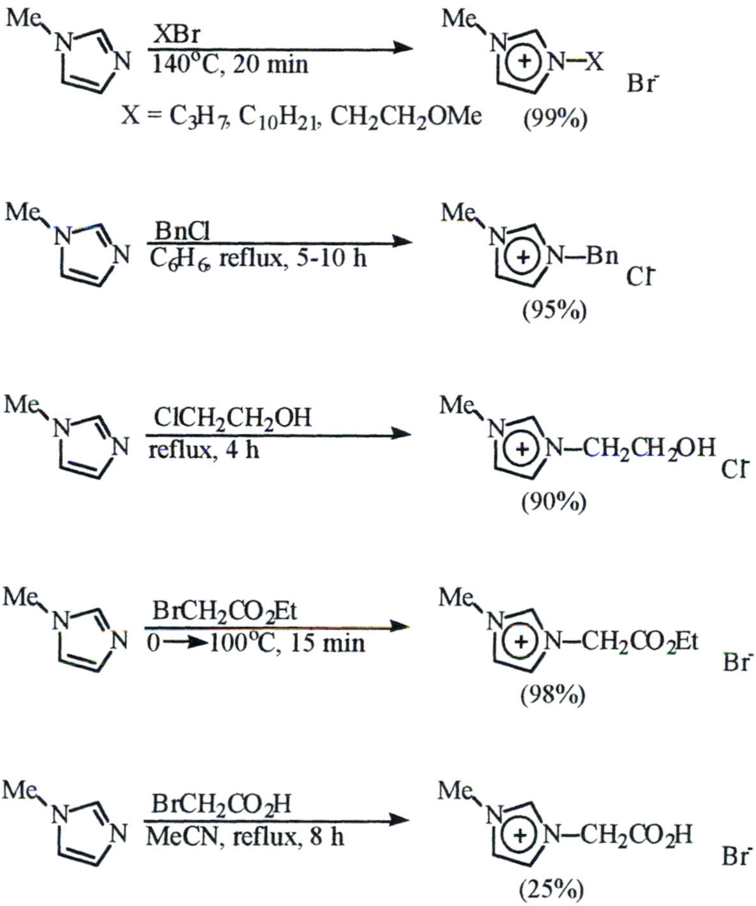

XBr, 140°C, 20 min → Br^-

X = C_3H_7, $C_{10}H_{21}$, CH_2CH_2OMe (99%)

BnCl, C_6H_6, reflux, 5-10 h → N–Bn Cl^- (95%)

$ClCH_2CH_2OH$, reflux, 4 h → N–CH_2CH_2OH Cl^- (90%)

$BrCH_2CO_2Et$, 0→100°C, 15 min → N–CH_2CO_2Et Br^- (98%)

$BrCH_2CO_2H$, MeCN, reflux, 8 h → N–CH_2CO_2H Br^- (25%)

Figure 3. Alkylation of 1-methylimidazole to form 1-X-3-methylimidazolium halides.

Absorption maxima for Reichardt's Dye (**1**) in the remaining five RTILs were measured and the $E_T(30)$ values were calculated. The $E_T(30)$ values and water contents for this series of 1-X-3-methylimidazolium bistrifylimides are recorded in Table III.

Table III. $E_T(30)$ Values for 1-X-3-methylimidazolium bistrifylimides

X	*Water content (ppm)*	*$E_T(30)$ (kcal mol^{-1})*
C_3H_7	3810	52.0
C_3H_7	160	51.9
$C_{10}H_{21}$	1450	52.1
CH_2Ph	2180	52.5
CH_2CH_2OMe	3050	54.1
CH_2CH_2OH	3380	61.4
CH_2CH_2OH	390	60.8

SOURCE: Data from Reference 32, Copyright 2002 Elsevier Science

As expected, the change from X = propyl to decyl did not produce a significant variation in the $E_T(30)$ value. Similarly when X = benzyl, the $E_T(30)$ value was essentially the same as with X = alkyl. On the other hand, when X = 2-methoxyethyl, a significantly larger $E_T(30)$ value shows that the polarity is enhanced. With X = CH_2CH_2OH, a much larger value of $E_T(30)$ was obtained. This reveals a markedly enhanced polarity of the RTIL to one intermediate between water and methanol (Table II).

Since the water content of 1-alkyl-3-methylimidazolium hexafluorophosphates has been shown to influence their $E_T(30)$ values (*30*), two members of the 1-X-3-methylimidazolium bistrifylimide series were subjected to special drying. The $E_T(30)$ values for the dried RTILs with X = C_3H_7 and CH_3CH_2OH were found to be in good agreement with those obtained with the "wet" RTILs.

Thus incorporation of a functional group into the X substituent of 1-X-3-methylimidazolium bistrifylimides is shown to markedly expand the range of RTIL polarities. Further investigation of this phenomena with other functional groups and substituents is underway.

Influence of RTIL Polarity on A Diels-Alder Reaction

Among the most useful synthetic processes for carbon-carbon bond formation is the Diels-Alder reaction (*39*). RTILs are suitable media for Diels-Alder reactions and have been employed as both sovents and catalysts (*40-47*).

For molecular organic solvents, the *endo*/*exo* ratio of Diels-Alder reaction products is related to the polarity of the solvent in which the reaction is performed (*48*). Increasing the solvent polarity enhances the *endo*/*exo* ratio. For comparison, a study of a Diels-Alder reaction (Figure 4) with the five members of the RTIL series was undertaken to probe the influence of ionic liquid solvent polarity on the *endo*/*exo* ratio.

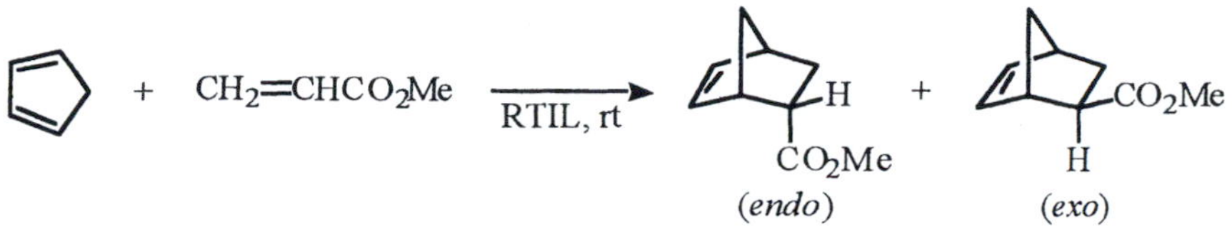

$$\text{cyclopentadiene} + CH_2{=}CHCO_2Me \xrightarrow{\text{RTIL, rt}} \text{(endo)} + \text{(exo)}$$

Figure 4. The Diels-Alder reaction.

The Diels-Alder reaction of cyclopentadiene and methyl acrylate wass reported to be heterogeneous in 1-alkyl-3-methylimidazolium RTILs (*44*) with the *endo*/*exo* ratio depending on both the concentration of reactants and the reaction time. However, we found the reactions to be homogeneous. The influence of concentration and reaction time was evaluated with 1-methyl-3-propylimid-azolium bistrifylimide as solvent. No change in the *endo*/*exo* ratio was observed over a range of concentrations (0.3-1 *M*) and reaction times (2-24 hours).

Reactions of equimolar amounts of cyclopentadiene and methyl acrylate in the RTIL proceeded in high yield (>95%) in 2 hours at room temperature. The *endo*/*exo* ratios obtained for reactions conducted in the RTIL series and $E_T(30)$ values for the solvents are compared in Table IV. Although the variation in the *endo*/*exo* ratio is modest, a correlation with the RTIL polarity is clearly evident. Greater polarity produces a higher *endo*/*exo* ratio.

Table IV. Comparison of the 1-X-3-methylimidazolium bistrifylimide polarity and stereoselectivity of the Diels-Alder reaction of cyclopentadiene with methyl acrylate.

X	*endo/exo*	$E_T(30)$ (kcal mol^{-1})
C_3H_7	4.3	52.0
$C_{10}H_{21}$	4.3	52.1
CH_2Ph	4.9	52.5
CH_2CH_2OMe	5.7	54.1
CH_2CH_2OH	6.1	61.4

SOURCE: Data from Reference 32, Copyright 2002 Elsevier Science Ltd.

Experimental Section

RTIL Synthesis

The 1-methyl-3-propylimidazolium, 1-decyl-3-methylimidazolium and 1-benzyl-3-methylimidazolium bistrifylimides were prepared by the procedures reported in Reference 29.

1-(2-Hydroxyethyl)-3-methylimidazolium chloride

A solution of 1-methylimidazole (10.0 mL, 125 mmol) and 2-chloroethanol (8.4 mL, 125 mmol) was refluxed for 4 hours to give a 90% yield of the product as a tan, oily solid. ^{1}H NMR (300 MHz, DMSO-d_6): δ 3.68 (t, 2H), 3.83-3.90 (m, 3H), 4.23 (t, 2H), 5.09 (br s, 1H), 7.69 (s, 1H), 7.71 (s, 1H). Analysis Calculated (Found) for $C_6H_{11}ClN_2O$: C, 44.32 (44.30); H, 6.82 (6.96); N, 17.23 (17.25).

1-(2-Methoxyethyl)-3-methylimidazolium bromide

Reaction of 1-methylimidazole (5.00 mL, 53 mmol) and 2-bromoethyl methyl ether (4.24 mL, 53 mmol) at 140° C for 20 minutes according to the procedure for the preparation of 1-alkyl-3-methylimidazolium bromides in Reference 49 gave a 99% yield of an oil. ^{1}H NMR (300 MHz, DMSO-d_6): δ 3.27 (s, 3H), 3.61-3.74 (m, 3H), 4.35 (s, 3H), 7.68 (s, 1H), 7.71 (s, 1H), 9.06 (s, 1H). Analysis Calculated (Found) $C_7H_{13}BrN_2O$: C, 38.03 (37.68); H, 5.93 (6.02); N, 12.67 (12.72).

1-(Hydroxycarboxymethyl)-3-methylimidazolium bromide

When 1-methylimidazole (5.00mL, 63 mmol) was added to a solution of bromoacetic acid (8.71 g, 63 mmol) in acetonitrile (25 mL), refluxing commenced. After refluxing for 8 hours and then stirring at room temperature for 8 hours, the white precipitate was filtered, washed with diethyl ether, and dried in vacuo to give a 25% yield of white solid with mp = 165-172°C (dec). ^{1}H NMR (300 MHz, DMSO-d_6): δ 3.96 (s, 3H), 5.27 (s, 2H), 7.81-7.86 (m, 2H), 9.30 (s, 1H). Analysis Calculated (Found) for $C_6H_9BrN_2O_2$: C, 32.60 (32.50); H, 4.10 (4.04), N, 12.67 (12.54).

1-(Ethoxycarboxymethyl)-3-methylimidazolium bromide

To 1-methylimidazole (5.00 mL, 63 mmol) cooled in an ice-bath, ethyl bromoacetate (7.00 mL, 63 mmol) was added dropwise over a 15-minute period forming a gel. The ice bath was removed and the solution was heated at ~100°C for 15 minutes to provide a quantitative yield of a golden oil. ^{1}H NMR (300 MHz, DMSO-d_6): δ 1.25 (s, 3H); 3.84 (s, 3H); 4.21 (q, 2H), 5.31 (s, 2H), 7.79 (s, 2H), 9.19 (s, 1H). Analysis Calculated (Found) for $C_8H_{13}BrN_2O_2$: C, 38.57 (38.67); H, 5.26 (5.39); N, 11.25 (11.11).

1-X-3-methylimidazolium bistrifylimides

Using the procedure given in Reference 29 for conversion of 1-alkyl-3-methylimidazolium bromides into the corresponding 1-alkly-3-methylimidazolium bistrifylimides, the following bistrifylimide salts were prepared in high yields: *X* = *CH₂CH₂OH* as a liquid with T_g = -79°C. ^{1}H NMR (300 MHz, acetone-d_6): δ 3.97 (q, 2H), 3.99 (s, 3H), 4.45 (t, 2H), 7.70-7.76 (m, 2H), 9.00 (s, 1H). Analysis Calculated (Found) for $C_8H_{11}F_6N_3O_5S_2$: C, 23.59 (23.48); H, 2.72 (2.84); N, 10.32 (10.21). *X* = CH_2CH_2OMe as a liquid with T_g = -81°C. ^{1}H

NMR (300 MHz, acetone-d_6): δ 2.06 (t, 3H), 3.35 (s, 3H), 3.82 (t, 2H), 4.10 (s, 3H), 4.55 (t, 2H), 7.72-7.76 (m, 2H), 9.04 (s, 1H). Analysis Calculated (Found) for $C_9H_{13}F_6N_3O_5S_2$: C, 25.66 (25.34); H, 3.11 (3.09); N, 9.97 (9.87). $X = CH_2CO_2H$ as a hygroscopic, viscous oil ^{1}H NMR (300 MHz, acetone-d_6): δ 4.15 (s, 2H), 5.38 (s, 2H), 7.77-7.80 (m, 2H), 9.10 (s, 1H). Analysis Calculated (Found) for $C_8H_9F_6N_3O_6S_2$: C, 22.81 (23.10); H, 2.15 (2.37); N, 9.97 (9.80). $X = CH_2CO_2Et$ as a viscous liquid with T_g = -56°C and T_m = 16°C. ^{1}H NMR (300 MHz, acetone-d_6): δ 1.27 (t, 3H), 4.13 (s, 3H), 4.18-4.30 (m, 2H), 5.34 (s, 2H), 7.70-7.82 (m, 2H), 9.07 (s, 1H). Analysis Calculated (Found) for $C_{10}H_{13}F_6N_3O_6S_2$: C, 26.73 (26.89); H, 2.92 (3.01); N, 9.35 (9.27)

Procedure for Diels-Alder Reactions

A weighed amount of freshly distilled cyclopentadiene was added to 2.0 mL of the RTIL in a vial at room temperature. With magnetic stirring, an equivalent amount of methyl acrylate was added via a syringe and the vial was capped. The contents were stirred for 2 hours at room temperature and then extracted with hexanes and analyzed by gas chromatography on a Carbowax column or in the case of X = $C_{10}H_{21}$ by ^{1}H NMR spectroscopy, since that RTIL was slightly soluble in hexane. Yields in all cases were >95% as determined by ^{1}H NMR spectroscopy.

Acknowledgement

This research was supported by the Texas Higher Education Coordinating Board Advanced Research Program.

References

1. For a recent review see: Sheldon, R. A. *Chem. Commun.* **2001**, 2399.
2. Earle, M. J.; Seddon, K. R. *Pure Appl. Chem.* **2000**, *72*, 1391.
3. Freemantle, M. *Chem. Eng. News* **1998**, *76* (March 30), 32.
4. For reviews see: Welton, T. *Chem. Rev.* **1999**, 2071; Olivier-Bourbigou, H.; Magna, L. *J. Mol. Catal. A; Chemical* **2002**, *182-183*, 419.
5. For a recent review see: Sheldon, R. A.; Lau, R. M.; Sorgedrager, M. J.; van Rantwijk, F.; Seddon, K. R. *Green Chem.* **2002**, *4*, 147.
6. Dai, S.; Ju, H.; Gao, J., Lin, J. S.; Pennycook, S. J.; Barnes, C. E. *Chem. Commun.* **2000**, 243.
7. Armstrong, D. L.; He, L.; Liu, Y. S. *Anal. Chem.* **1999**, *71*, 3873 and references cited therein.
8. Huddleston, J. G.; Willauer, H. D.; Swatloski, R. P.; Visser, A. E.; Rogers, R. D. *Chem. Commun.* **1998**, 1765.

9. Dai, S.; Ju, Y. H.; Barnes, C. E. *J. Chem. Soc., Dalton Trans.* **1999**, 1201.
10. Visser, A. E.; Swatloski, R. P.; Reichert, W. M.; Griffin, S. T.; Rogers, R. D. *Ind. Eng. Chem. Res.* **2000**, *39*, 3596.
11. Chun, S.; Dzyuba, S. V.; Bartsch, R. A. *Anal. Chem.* **2001**, 73, 3737.
12. Deitz, M. L.; Dzielawa, J. A. *Chem. Commun.* **2001**, 2124.
13. Bartsch, R. A.; Chun, S.; Dzyuba, S. V. *Ionic Liquids. Industrial Applications to Ionic Liquids,* ACS Symposium Series 818; Rogers, R. D.; Seddon, K. R., Eds.; American Chemical Society, Washington, DC, 2002; Chapt 5, pp 58-68.
14. Branco, L. C.; Crespo, J. G.; Afonso, C. A. M. *Angew. Chem. Int. Ed.* **2002**, *41*, 2771.
15. Scovazzo, P.; Visser, A. E.; Davis, J. H., Jr.; Rogers, R. D.; Koval, Carl A.; DuBois, D. L.; Noble, R. D. *Ionic Liquids. Industrial Applications to Ionic Liquids,* ACS Symposium Series 818; Rogers, R. D.; Seddon, K. R., Eds.; American Chemical Society, Washington, DC, 2002; Chapt 6, pp 69-87.
16. Bonhôte, P.; Dias, A.-P.; Papageorgiou, N.; Kalyanasundaram, K.; Grätzel, M. *Inorg. Chem.* **1996**, *35*, 1168.
17. Holbrey, J. D.; Seddon, K. R. *J. Chem. Soc., Dalton Trans.* **1999**, 2133.
18. Seddon, K. R.; Stark, A.; Torres, M.-J. *Pure Appl. Chem.* **2000**, *72*, 2275.
19. Carmichael, A. J.; Seddon, K. R. *J. Phys. Org. Chem.* **2000**, *13*, 591.
20. Aki, S. N. V. K. ; Brennecke, J. F.; Samanta, A. *Chem. Commun.* **2001**, 413.
21. Huddleston, J. G.; Visser, A. E.; Reichert, W. M.; Willauer, H. D.; Broker, G. A.; Rogers, R. D. *Green Chem.* **2001**, *3*, 156.
22. Wasserscheid, P.; Dunkin, I. R. *Chem. Commun.* **2001**, 1186.
23. Fletcher, K. A.; Storey, I. A.; Hendricks, A. E.; Pandey, S.; Pandey, S. *Green Chem.* **2001**, *3*, 210.
24. Muldoon, M. J.; Gordon, C. M.; Dunkin, I. R. *J. Chem. Soc., Perkin Trans. 2* **2001**, 433.
25. Noda, A.; Hayamizu, K.; Watanabe, M. *J. Phys. Chem. B* **2001**, *105*, 4603.
26. Lau, G.; Watson, P. R. *Langmuir* **2001**, *17*, 6138.
27. Baker, S. N.; Baker, G. A.; Kane, M. A.; Bright, F. V. *J. Phys. Chem. B* **2001**, *105*, 9663.
28. Park, S.; Kazlauskas, R. J. *J. Org. Chem.* ***2001***,66, *8395*.
29. Dzyuba, S. V.; Bartsch, R. A. *ChemPhysChem* **2002**, *3*, 161.
30. Baker, S. N.; Baker, G. A.; Bright, F. V. *Green Chem.* **2002**, *4*, 165.
31. Quinn, B. M.; Ding, Z.; Moulton, R.; Bard, A. J. *Langmuir*, **2002**, *18*, 1734.
32. Dzyuba, S. V.; Bartsch, R. A. *Tetrahedron Lett.* **2002**, *43*, 4657.
33. Muller, P. *Pure Appl. Chem.* **1994**, *66*, 1077.
34. Dimroth, K.; Reichardt, C.; Siepmann, T.; Bohlmann, F. *Liebigs Ann. Chem.* **1963**, *662*, 1.
35. Reichardt, C. *Chem. Rev.* **1994**, *94*, 2319.
36. Visser, A. E.; Swatloski, R. P.; Reichert, W. M.; Mayton, R.; Sheff, S.; Wierzbicki, A.; Davis, J. H., Jr.; Rogers, R. D. *Chem. Commun.* **2001**, 135.
37. Kimizuki, N.; Nakashima, T. *Langmuir* **2001**, *17*, 6759.

38. Bates, E. D.; Mayton, R. D.; Ntai, I.; Davis, J. H., Jr. *J. Am. Chem. Soc.* **2002**, *124*, 926.
39. For a recent review, see: Kumar, A. *Chem. Rev.* **2001**, *101*, 1.
40. Jaeger, D. A.; Tucker, C. E. *Tetrahedron Lett.* **1989**, *30*, 1785.
41. Howarth, J.; Hanlon, K.; Fayne, D.; McCormac, P. *Tetrahedron Lett.* **1997**, *38*, 3097.
42. Lee, C. W. *Tetrahedtron Lett.* **1999**, *40*, 2461.
43. Earle, M. J.; McCormac, P. B.; Seddon, K. R. *Green Chem.* **1999**, *1*, 23.
44. Fischer, T.; Sethi, A.; Welton, T.; Woolf, J. *Tetrahedron Lett.* **1999**, *40*, 793.
45. Zulfiqar, F.; Kitazume, T. *Green Chem.* **2000**, *2*, 137.
46. Song, C. E.; Shim, W. H.; Roh, E. J.; Lee, S.-g.; Choi, J. H. *Chem. Commun.* **2001**, 1122.
47. Berson, J. A.; Hamlet, Z.; Mueller, W. A. *J. Am. Chem. Soc.* **1962**, *84*, 297.
48. Sethi, A. R.; Wleton, T. *Ionic Liquids. Industrial Applications to Ionic Liquids,* ACS Symposium Series 818; Rogers, R. D.; Seddon, K. R., Eds.; American Chemical Society, Washington, DC, 2002; Chapt 19, pp 241-246.
49. Dzyuba, S. V.; Bartsch, R. A. *J. Heterocyclic Chem.* **2001**, *38*, 265.

Chapter 25

Polar, Non-Coordinating Ionic Liquids as Solvents for Coordination Polymerization of Olefins

Kevin H. Shaughnessy*, Marc A. Klingshirn, Steven J. P'Pool, John D. Holbrey, and Robin D. Rogers*

Department of Chemistry and the Center for Green Manufacturing, The University of Alabama, Tuscaloosa, AL 35487–0336

Increased polymerization yield is observed in the copolymerization of styrene and CO in ionic liquids (ILs) compared to commonly used molecular solvents using palladium catalysts. Conditions for the copolymerizations were optimized and the effect of changes in the cation and anion of the IL solvent were determined. These results suggest that polar, non-coordinating ILs accelerate olefin polymerization catalyzed by electrophilic, charge-separated catalyst species.

Introduction

Ionic liquids (ILs), particularly those that are liquid at room temperature, are an exciting class of neoteric solvents. Although examples of ILs have been known for nearly a century, the past decade has seen a dramatic surge in interest in these materials *(1)*. The development of air- and water-stable ILs based on imidazolium and pyridinium salts of anions, such as PF_6^- and BF_4^- (Figure 1)

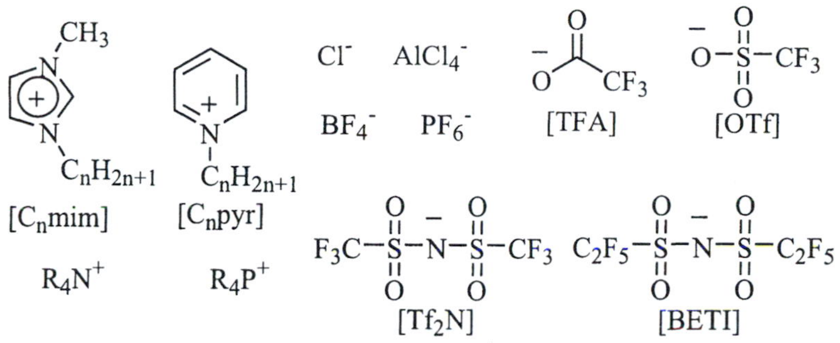

Figure 1. Common ionic liquid cations and anions.

have provided materials that are more convenient to work with than the chloroaluminate ILs that received initial widespread interest.

Since many ILs have negligible vapor pressure, there has been significant interest in their use as environmentally friendly replacements for volatile organic solvents. While the environmental friendliness of ILs remains to be determined, ILs possess a number of other unique properties that have led to their application in a variety of organic synthetic methodologies *(2-4)*. ILs have tunable hydrophobicity that ranges from complete miscibility with water to highly hydrophobic materials. In addition, ILs are immiscible with a variety of nonpolar organic solvents. This tunable miscibility, in combination with the non-volatility of ILs, has lead to their wide application in bi- *(5-13)* and even tri-phasic *(14)* catalytic processes.

ILs are unique among potential reaction media in that they can be polar, yet can also be designed to be non-coordinating. Reported values for the solvent polarity of ILs have ranged from non-polar (hexane) *(15)* to polar (alcohols) *(16,17)* depending on the probes used. Despite the range of reported values, most authors agree that ILs have solvent polarities comparable to aprotic, dipolar solvents, such as acetonitrile and DMF. Although polar, ILs containing non-nucleophilic anions (BF_4^-, PF_6^-, Tf_2N^-) are weakly coordinating, with coordination abilities comparable to methylene chloride *(13)*. Polar, non-coordinating IL solvents may be expected to accelerate certain catalytic processes by stabilizing charge separated catalytic intermediates or transition states. Acceleration of catalytic processes in IL solvents has been attributed to their polar, non-coordinating nature in some cases *(5,18)*, but the true mechanism responsible for the observed rate increases has not been determined.

Catalyst systems for the coordination polymerization of alkenes, such as metallocene-based Ziegler-Natta systems *(19)*, late-transition metal olefin polymerization catalysts *(20)*, and alkene/CO copolymerization catalysts *(21)*, share a common feature. In each case, the active species is a cationic metal complex with a weakly-coordinating anion. These highly electrophilic active species are necessary to allow alkene coordination and insertion. A highly polar, yet non-coordinating solvent may be expected to stabilize these active species and possibly accelerate the propagation steps.

Chloroaluminate ILs have been used as solvents for the cationic *(18,22-26)* and metal-catalyzed *(27,28)* oligomerization or polymerization of olefins. The current generation of stable ILs have been successfully applied to radical polymerization of olefins *(29-31)*. Wasserscheid *(13)* has shown that $[C_n mim][PF_6]$ ILs are effective solvents for ethylene oligomerization using nickel catalysts. Production of high molecular weight polyethylene catalyzed by both zirconocene and palladium catalysts has been claimed in a patent, although supporting data, including polymer yield or catalyst activity, were not reported *(32)*. In this paper we present our initial efforts to apply IL solvents to palladium-catalyzed styrene/CO copolymerization and ethylene homopolymerization.

Results

Alternating Styrene/CO Copolymerization with a Drent-Type Catalyst

To initially test the ability of ILs to be used as solvents in the copolymerization of styrene and CO, we carried out the copolymerizations using a Drent-type catalyst system derived from $LPd(OAc)_2$ (L = 2,2'-bipyridine (bipy)), excess ligand, benzoquinone, and *p*-toluenesulfonic acid (Figure 2) *(33)*. The $[Tf_2N]$ anion was chosen for initial screening due to its stability and low coordinating ability. Initial attempts to copolymerize styrene and CO in $[C_6pyr][Tf_2N]$ gave only styrene homopolymers. Since methanol is known to react with Pd(II) complexes in the presence of CO to give catalytically active $[LPdC(O)OCH_3]^+$ species, we added methanol to the reaction mixture. When the copolymerization was repeated using a 10:1 ratio of $[C_6pyr][Tf_2N]$:MeOH perfectly alternating copolymer was formed with a productivity of 1.11 kgCP/gPd (Trial 2, Table 1) *(34)*. Under these conditions little or no polystyrene is formed as determined by ^1H NMR and FTIR.

The copolymerization conditions were optimized for the $[C_6pyr][Tf_2N]$/methanol solvent system. Doubling the volume of the IL, while holding the methanol volume constant ($[C_6pyr][Tf_2N]$:MeOH = 20:1) resulted in an approximately 50% increase in copolymer yield (Trial 6). If both the IL and methanol volumes were doubled, the productivity of the system increased to 2.7 kgCP/gPd (Trial 7). In addition, the copolymer produced under these conditions had a much higher molecular weight than with other MeOH:IL ratios. Further

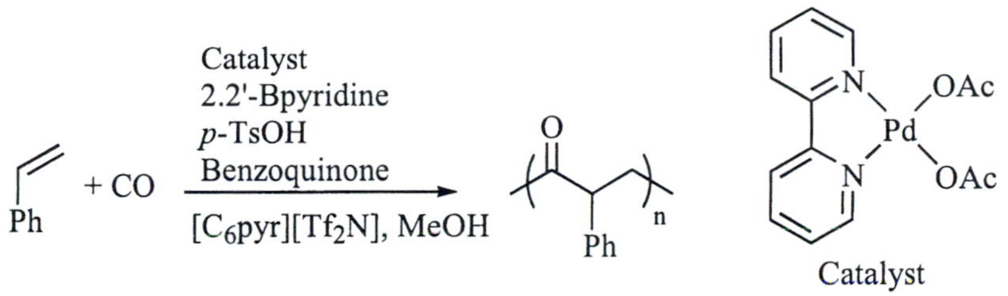

Catalyst
2.2'-Bpyridine
p-TsOH
Benzoquinone
$[C_6pyr][Tf_2N]$, MeOH
Ph + CO
Ph
n
O
N
N
Pd
OAc
OAc
Catalyst

Figure 2. Styrene/CO Copolymerization with a Drent-type Catalyst System.

Table 1. Styrene/CO Copolymerization in $[C_6pyr][Tf_2N]$[a]

Trial	*IL*[b] *(mL)*	*MeOH (mL)*	*T (°C)*	P_{CO} *(bar)*	*TON*[c]	M_w[d]	M_n[d]	*PDI*[d]
1	2	0	70	40	PS[e]			
2	2	0.2	70	40	1.11	9,700	4,700	2.1
3	2	0.2	50	40	0.70	20,800	4,700	2.6
4	2	0.2	90	40	PS[e]			
5	1	0.2	70	40	1.23	6,000	3,100	1.9
6	4	0.2	70	40	1.66	10,000	6,200	1.6
7	4	0.4	70	40	2.73	34,000	25,300	1.3
8	4	0.4	70	20	2.54	12,200	7,200	1.7
9	4	0.4	70	60	1.78	8,400	5,000	1.7

NOTES: [a]See Reference 34 for general procedure. [b]IL = $[C_6pyr][Tf_2N]$ [c]TON = kgCP/gPd [d]Determined by GPC [e]Polymer produced was primarily polystyrene.

SOURCE: (*Chem. Commun.*, **2002**, 1394-1395) Reproduced by permission of the Royal Society of Chemistry

optimization of the reaction temperature and CO pressure resulted in no improvement in the productivity of this system (Table 1).

The copolymers produced in these studies were analyzed by GPC, NMR, and IR to determined their molecular weight and composition. Copolymer molecular weights were generally modest (M_n = 5,000 – 10,000), although the copolymer produced under the optimized conditions had a molecular weight (M_n) of 25,000 (Trial 7, Table 1). The polydispersity values were generally narrow suggesting a single-site catalyst system. The polymer produced under the optimal conditions showed a very narrow PDI value of 1.30. IR and NMR spectroscopic characterization of the copolymer showed it to be perfectly alternating copolymer with a predominately syndiotactic microstructure *(21)*.

Specifically, IR spectra of the copolymers showed a strong absorption at 1709 cm^{-1} for the C=O stretch, ^{13}C NMR showed 8 major resonances, and ^{1}H NMR showed only the expected copolymer resonances with no resonances attributable to polystyrene *(35)*. Analysis of the ipso carbon region of the ^{13}C NMR spectrum showed the copolymer to be approximately 80% syndiotactic.

The effect of variation of the IL cation and anion was determined using the optimized conditions (Table 2). Choice of the IL anion was expected to be important, since the activity of these catalyst systems are known to be strongly anion dependent *(36)*. Addition of small amounts (> 0.1%) of [C_6pyr][Br] significantly inhibits the catalyst system (Trial 2). Thus, it is critical that all residual halide be removed from ILs used in these reactions. Polymerization activity in imidazolium ILs decreased along the series [Tf_2N] ≈ [BETI] > [PF_6] > [TfO] >> [TFA] (Trials 5-8), which seems to correspond to the coordinating ability of the anion. In the [Tf_2N] series, the imidazolium salt gave a slightly lower yield than the pyridinium IL (Trials 1 and 6), although the difference is only about twice the standard deviation (0.112 kg CP/g Pd) obtained in repeated runs under optimal conditions. Copolymer molecular weight and polydispersity were not affected by choice of IL solvent. Concurrent with our own work, Seddon *(37)* reported the copolymerization of styrene and CO with a similar catalyst system and found that [C_6pyr][Tf_2N] gave optimal results.

Styrene/CO Copolymerization with a Cationic Palladium Complex

The Drent-type catalyst system produces a palladium complex with moderately weakly coordinating ligands under the reactions conditions. Since

Table 2. IL Solvent Effects in the Styrene/CO Copolymerization[a]

Trial	*IL*	*TON[b]*	*M_n[c]*	*M_w[c]*	*PDI[c]*
1	[C_6pyr][Tf_2N]	2.73	34,000	25,300	1.3
2	[C_6pyr][Tf_2N][d]	0.01			
3	[C_6pyr][BETI]	2.51	18,200	11,900	1.5
4	[C_6pyr][TFA]	0.00			
5	[C_6mim][BETI]	2.85	18,600	12,300	1.5
6	[C_4mim][Tf_2N]	2.53	20,500	12,600	1.6
7	[C_4mim][PF_6]	2.35	20,100	11,600	1.7
8	[C_4mim][TfO]	1.74	20,300	13,200	1.5
9	[C_4mim][TFA]	0.03			

NOTES: [a]See reference 34 for general procedure. [b]TON = kgCP/gPd [c]Determined by GPC [d][C_6pyr][Br] 0.5% (w/w) in [C_6pyr][Tf_2N]

SOURCE: (*Chem. Commun.*, **2002**, 1394-1395) Reproduced in part by permission of the Royal Society of Chemistry

p-toluenesulfonic acid is used in large excess relative to palladium (ca. 20:1 *p*TsOH:Pd), there is a significant concentration of weakly coordinating anions (TsO^- and AcO^-) in the reaction mixture. We were interested to see what effect IL solvents would have on a catalyst system where there were no coordinating anions present. Brookhart *(38)* has shown that preformed cationic complexes, which are analogous to the presumed active species in the Drent-type system, are active for the copolymerization of styrene and CO.

Copolymerizations were carried out with (bipy)Pd(CH_3)Cl in combination with $NaB(Ar_F)_4$ or $AgSbF_6$ (Figure 3). Initial optimization was carried out in [C_6pyr][Tf_2N]. Our initial results showed that catalysts derived from $AgSbF_6$ gave higher yields of copolymer than those made from $NaB(Ar_F)_4$ (Table 3, Trials 1 and 2 vs. 3 and 4) *(39)*. Further optimization was performed with the $AgSbF_6$ derived catalyst. This catalyst system showed little pressure dependence on catalytic activity. There was a dramatic increase in copolymer yield as the temperature was raised to 50 °C, but the yield decreased at higher temperature (Trials 5-8). An increase in copolymer molecular weight was also observed at 50 °C (M_n = 20,000) compared to lower temperatures ($M_n \approx$ 12,000) Copolymers formed at 50 °C had larger polydispersities than those made at room temperature. IR and NMR spectra of these copolymers were identical to those obtained in the Drent system described previously.*(35)*

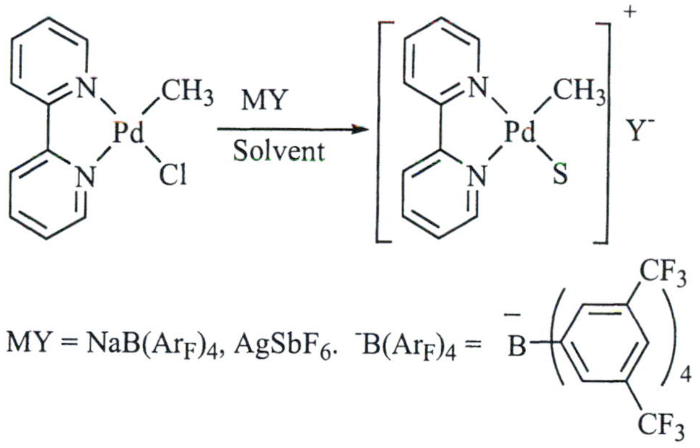

MY = $NaB(Ar_F)_4$, $AgSbF_6$. $^-B(Ar_F)_4$ =

Figure 3. Brookhart-type catalyst system

At 50 °C, copolymerization in [C_6pyr][Tf_2N] gave approximately 3 times more copolymer than was produced in methylene chloride (Trial 9). Molecular weight values and polydispersities were nearly identical in the two solvents. In contrast to the Drent system, higher copolymer yields were obtained in [C_4mim][OTf] than in the [PF_6] IL (Trials 10-11, Table 3). A 1,10-

Table 3. Styrene/CO Copolymerization Catalyzed by (bipy)Pd(Me)Cl/MY[a]

Trial	*Solvent*	*T* (°C)	P_{CO} (bar)	*TON*[b]	M_w[c]	M_n[c]	*PDI*[c]
1	$[C_6pyr][Tf_2N]$[d]	23	3.4	0.07	24,200	9,900	2.45
2	$[C_6pyr][Tf_2N]$[d]	23	6.9	0.11	16,200	10,300	1.57
3	$[C_6pyr][Tf_2N]$	23	3.4	0.18	26,000	14,600	1.78
4	$[C_6pyr][Tf_2N]$	23	6.9	0.14	28,500	15,300	1.87
5	$[C_6pyr][Tf_2N]$	23	13.8	0.13	15,500	12,500	1.24
6	$[C_6pyr][Tf_2N]$	40	6.9	0.27	70,900	24,700	2.87
7	$[C_6pyr][Tf_2N]$	50	6.9	1.14	90,000	20,500	4.39
8	$[C_6pyr][Tf_2N]$	60	6.9	0.59	25,900	11,600	2.24
9	CH_2Cl_2	50	6.9	0.40	57,000	18,700	3.05
10	$[C_4mim][OTf]$	50	6.9	0.97	16,000	10,800	1.47
11	$[C_4mim][PF_6]$	50	6.9	0.56	25,400	9,700	2.63
12	$[C_6pyr][Tf_2N]$	50	6.9	1.22[e]			

NOTES: [a]See reference 39 for procedure. MY = $AgSbF_6$ unless noted. Reaction time = 12 h. [b]TON = kgCP/gPd [c]Determined by GPC [d]MY = $Na[B(Ar_F)_4]$ [e](phen)Pd(Me)Cl

phenanthroline (phen) based catalyst gave about the same yield as was obtained with the bipyridine catalyst (Trial 12).

The crude polymer produced in this system is gray, suggesting that it is contaminated with palladium metal. In contrast, the polymer produced by the Drent-type system was pale yellow in color. Copolymerizations were carried out with different reaction times to probe for catalyst deactivation (Figure 4). A plot of TON vs. time shows that the catalyst was most productive in the first 2 hours, after which it continued to produce polymer at a slower rate for up to 12 hours. After the first two hours, the catalyst produces polymer at a fairly constant rate of approximately 0.08 kgCP/gPd•h. The cumulative TOF decreases slowly throughout the reaction period reflecting the lower activity after the first 2 hours. The apparent increase in Pd precipitation compared with the Drent-type catalyst system may reflect catalyst decomposition upon workup rather than during the reaction period.

Homopolymerization of Ethylene in IL Solvents

Based on the accelerating effect of ILs on the copolymerization of styrene and CO using a cationic palladium complex, we were encouraged to try a class of related ethylene polymerization catalysts (Figure 5). Palladium and nickel *(40-44)*, as well as cobalt and iron *(45-47)*, catalysts supported by bulky diimine ligands have been shown to be active for the polymerization of ethylene and α-olefins. In each system, the active species is thought to be a cationic, monoalkyl metal complex with a non-coordinating anion to provide an open site for alkene coordination. Based on our hypothesis and results with the styrene/CO copolymerization, we would expect ILs to accelerate the polymerization.

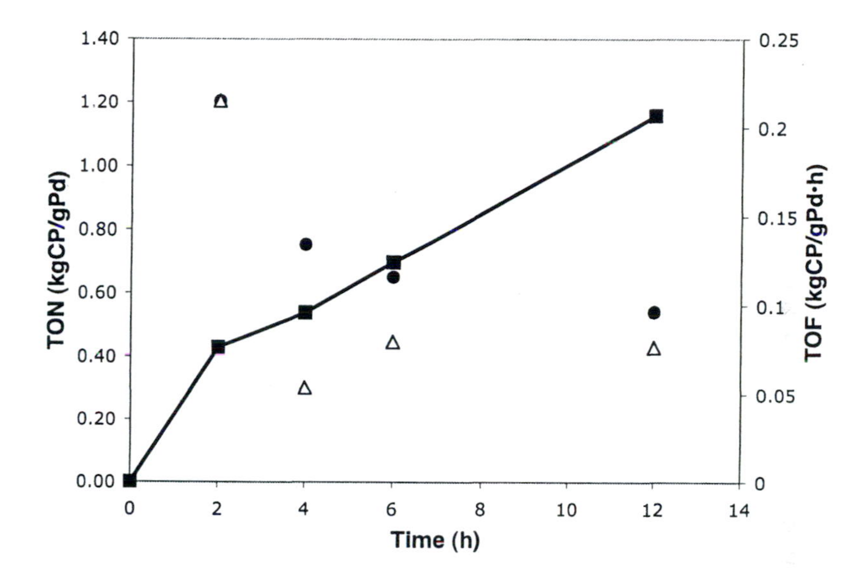

Figure 4. Copolymerization productivity as a function of time. (■ = *TON;* ● = *TOF from start of reaction;* Δ = *incremental TOF.*

Ethylene polymerization using catalyst **1** in combination with $NaB(Ar_F)_4$ and $AgSbF_6$ was carried out in IL solvents and compared to results obtained in methylene chloride (Figure 5, Table 4). Ethylene polymerization in IL solvents gave much lower yields than were obtained in methylene chloride. Of the systems tested, the combination of **1**/$AgSbF_6$ in $[C_6mim][PF_6]$ gave the highest yield of polyethylene (0.19 kgPE/gPd). The best activity obtained in IL solvents was an order of magnitude lower than that observed in methylene chloride. GPC analysis showed the polymer was moderately low molecular weight material ($M_n \approx 10{,}000$). The molecular weight of polyethylene made in ILs was an order of magnitude lower than for polyethylene made in methylene chloride under identical conditions. The branching ratio ranged from 55-89 branches per 1000 C's as determined by ^1H NMR for polyethylene prepared in ILs, which is comparable to that obtained in methylene chloride.

2

Branched Polyethylene

H_3C N(Ar) Pd CH_3 Cl N(Ar) H_3C **1** — MY / Solvent → [H_3C N(Ar) Pd CH_3 S N(Ar) H_3C]$^+$ Y^- **2**

Ar = 2,6-(*i*-Pr)$_2$C$_6$H$_3$

Figure 5. Ethylene polymerization with a cationic Pd-diimine catalyst

Table 4. Ethylene Polymerization in ILs

Trial	*Y*[a]	*Solvent*	*TON*[b]	M_w[c] × 10^{-3}	M_n[c] × 10^{-3}	*PDI*[c]	*Branch Ratio*[d]
1	$^-B(Ar_F)_4$	CH_2Cl_2	1.66	208	153	1.36	81
2	$^-B(Ar_F)_4$	[C_4mim][PF_6]	0.09	23.3	11.7	2.0	89
3	$^-SbF_6$	[C_4mim][PF_6]	0.09	14.8	7.2	2.0	55
4	$^-SbF_6$	[C_6mim][PF_6]	0.19	19.5	12.3	1.6	69
5	$^-B(Ar_F)_4$	[C_6pyr][Tf_2N]	0.01				
6	$^-SbF_6$	[C_6pyr][Tf_2N]	0.003				

NOTES: Catalyst **1**/MY ($^-B(Ar_F)_4$ = Na($^-B(Ar_F)_4$; $^-SbF_6$ = $AgSbF_6$. 10 μmol), IL (2 mL), 20.7 bar, 23 °C. [a]kgPE/gPd [b]Determined by GPC [c]Branches/1000 C's: Determined by ^{1}H NMR

Discussion

Styrene/CO copolymerization is accelerated in [C_6pyr][Tf_2N] relative to commonly used molecular solvents. For the Drent-type system studied, addition of a small volume of methanol proved to be critical to achieving high catalytic activity. For these catalysts, the active species can be generated *in situ* by a number of routes. Under the oxidative conditions used in this study, attack of methanol on palladium-coordinated CO to give a [LPd-C(O)OMe]$^+$ species is a key activation step *(21)*. Based on our results, we believe that this reaction is critical to achieving high catalytic activity. It should be noted that in Seddon's studies of a similar catalyst in [C_6pyr][Tf_2N], methanol was not required to achieve high activity *(37)*.

The concentration effect observed with the Drent system is also interesting. Higher copolymer yields were obtained using larger volumes of [C_6pyr][Tf_2N], while keeping the amount of catalyst constant. Since both styrene and CO are likely at their maximum concentration in [C_6pyr][Tf_2N] in both cases, only the catalyst concentration is likely to change as the IL volume is increased. One possible explanation for the observed trend is that catalyst decomposition is inhibited at lower concentration. Decomposition of the catalyst by precipitation of Pd(0) involves reduction of the Pd(II) catalyst to unstable, soluble Pd(0) species. The Pd(0) species can then combine to form insoluble, catalytically inactive Pd(0) particles. The initial process can be reversed if the Pd(0) species is oxidized by benzoquinone in the presence of excess ligand prior to coalescing with the growing Pd particles.

The importance of a non-coordinating IL solvent can been seen in the anion effect on copolymerization activity. Of the anions tested, [Tf_2N] and [BETI] are expected to be the least coordinating due to delocalization of the negative charge by the electron withdrawing perfluoroalkylsulfonyl groups. Copolymer yield decreases along the series [BETI] $\approx$ [Tf_2N] > [PF_6] > [OTf] >> [TFA]. While each of these are generally considered weakly coordinating ligands, the trend follows the nucleophilicity of the anions. The complete inactivity in [C_4mim][TFA] suggests that CO cannot compete with TFA for the open site on the palladium catalyst, although other deactivation pathways cannot be ruled out. The copolymerization is also strongly inhibited by trace amounts of bromide, which coordinates to Pd(II) more strongly than the other anions used in this study. This result points to the critical importance of ensuring that ILs used for catalytic studies are completely free of residual bromide. There does not appear to be a significant difference in yields obtained in imidazolium and pyridinium ILs with the same anions. Thus the higher acidity of the imidazolium IL does not appear to affect the copolymerization reaction.

There are a number of possible explanations for the accelerating effect of IL solvents on the copolymerization. The very low solubility of CO in pure ILs seems to preclude increased CO concentration as a contributing factor *(48)*, although the CO solubility in styrene/IL biphasic mixtures is unknown. Assuming that the observed increase in productivity is not due to solubility differences, the increased yield is likely due to improved catalyst stability and/or increased propagation rate. Methanol can promote decomposition of the active species by reducing Pd(II) active species to unstable Pd(0) complexes. Reducing the concentration of methanol should inhibit this decomposition pathway leading to improved productivity for the catalyst system. Polar non-coordinating ILs may also accelerate the propagation rate of the catalytic cycle. Both CO and styrene insertion steps are partially dependent on the rate of substitution of neutral ligands (CO and styrene) for weakly coordinating anionic ligands (acetate and tosylate). ILs would be expected to favor formation of solvent separated ionic species, thus accelerating the ligand substitution. Increased activity is also observed in polar, weakly coordinating molecular solvents like 2,2,2-trifluoroethanol *(49)*. Fundamental studies on the effect of IL solvents on organometallic reactivity will be required to better understand the accelerating effects seen with these solvents.

A significant increase in productivity in ILs compared to molecular solvents is also observed with the Brookhart-type catalyst. The increased activity in [C_6pyr][Tf_2N] over CH_2Cl_2 suggests that the cationic catalyst is more stable in the IL solvent, has a higher propagation rate, or both. An increase in propagation rate could be due to decreased interaction between [C_6pyr][Tf_2N] and the active species compared with CH_2Cl_2, which would lead to a greater concentration of active sites. Stabilization of charge separated intermediates may also play a factor in the observed acceleration.

In contrast to the results with olefin/CO copolymerization, catalysts for the homopolymerization of ethylene were much less active in ILs than in CH_2Cl_2. The solubility in ethylene in ILs is much higher than CO, but is still lower than in nonpolar organic solvents *(48)*. Since ethylene is a fairly weakly coordinating ligand, it is possible that the catalyst decomposes when it is starved for ethylene since the IL provides no stabilizing ligands. It is also possible that very minor impurities in the IL coordinate to the active species forming inactive complexes. Since CO is a much better ligand than ethylene, it is possible that even at low concentration it can stabilize the active species in the copolymerization, or that it would be able to displace weakly coordinating solvent impurities that ethylene could not. Further research in this system will be required to answer these questions.

Conclusions

IL solvents have shown promise in the copolymerization of olefins and CO catalyzed by cationic palladium complexes. These results suggest that polar, non-coordinating IL solvents can accelerate this process. The dependence of activity on the coordinating ability of the IL, supports our hypothesis that polar, non-coordinating ILs will prove useful in properly chosen metal-catalyzed reactions. The decreased yields of polyethylene obtained in ILs compared with molecular solvents shows that more work must be done to fully understand these systems. In particular, it will be necessary to carry out studies to address what effect ILs have on the rate and mechanism of fundamental organometallic processes. In this way, ILs can be more systematically paired with catalytic systems to provide highly efficient processes.

Acknowledgements

This research has been funded by the U.S. Environmental Protection Agency's STAR program (R-92925701-0), by the National Science Foundation

(EPS-9977239), and by a grant from the School of Mines and Energy Development at The University of Alabama.

References

1. Wilkes, J. S. *Green Chem.* **2002,** *4*, 73-80.
2. Wasserscheid, P.; Keim, W. *Angew. Chem. Int. Ed.* **2000,** *39*, 3772-3789.
3. Sheldon, R. *Chem. Commun.* **2001**, 2399-2407.
4. Olivier-Bourbigou, H.; Magna, L. *J. Mol. Cat. A: Chem.* **2002,** *182-183*, 419-437.
5. Zimmermann, J.; Wasserscheid, P.; Tkatchenko, I.; Stutzmann, S. *Chem. Commun.* **2002**, 760-761.
6. Dupont, J.; Fonseca, G. S.; Umpierre, A. P.; Fichtner, P. F. P.; Teixeira, S. R. *J. Am. Chem. Soc.* **2002,** *124*, 4228-4229.
7. Dullius, J. E. L.; Suarez, P. A. Z.; Einloft, S.; de Souza, R. F.; Dupont, J.; Fischer, J.; De Cian, A. *Organometallics* **1998,** *17*, 815-819.
8. Dyson, P. J.; Ellis, D. J.; Welton, T.; Parker, D. G. *Chem. Commun.* **1999**, 25-26.
9. Stenzel, O.; Raubenhiemer, H. G.; Esterhuysen, C. *J. Chem. Soc., Dalton Trans.* **2002**, 1132-1138.
10. Brasse, C. C.; Englert, U.; Salzer, A.; Waffenschmidt, H.; Wasserscheid, P. *Organometallics* **2000,** *19*, 3818-3823.
11. Consorti, C. S.; Ebeling, G.; Dupont, J. *Tetrahedron Lett.* **2002,** *43*, 753-755.
12. Kottsieper, K. W.; Stelzer, O.; Wasserscheid, P. *J. Mol. Cat. A: Chem.* **2001,** *175*, 285-288.
13. Wasserscheid, P.; Gordon, C. M.; Hilgers, C.; Muldoon, M. J.; Dunkin, I. R. *Chem. Commun.* **2001**, 1186-1187.
14. Carmichael, A. J.; Earle, M. J.; Holbrey, J. D.; McCormac, P. B.; Seddon, K. R. *Org. Lett.* **1999,** *1*, 997-1000.
15. Bonhôte, P.; Dias, A.-P.; Papageorgiou, N.; Kalyanasundaram, K.; Grätzel, M. *Inorg. Chem.* **1996,** *35*, 1168-1178.
16. Aki, S. N. V. K.; Brennecke, J. F.; Samanta, A. *Chem. Commun.* **2001**, 413-414.
17. Carmichael, A. J.; Seddon, K. R. *J. Phys. Org. Chem.* **2000,** *13*, 591-595.
18. Chauvin, Y.; Einloft, S.; Olivier, H. *Ind. Eng. Chem. Res.* **1995,** *34*, 1149-1155.
19. Brintzinger, H. H.; Fischer, D.; Mülhaupt, R.; Rieger, B.; Waymouth, R. M. *Angew. Chem., Int. Ed.* **1995,** *34*, 1143-1170.
20. Ittel, S. D.; Johnson, L. K.; Brookhart, M. *Chem. Rev.* **2000,** *100*, 1169-1203.

21. Drent, E.; Budzelaar, P. H. M. *Chem. Rev.* **1996,** *96*, 663-681.
22. Ambler, P. W.; Stewart, N. J., Eur. Pat. 0558187, 1993.
23. Abdul-Sada, A. A. K.; Ambler, P. W.; Hodgson, P. K. G.; Seddon, K. R.; Stewart, N. J., PCT Int. Appl. 9521871, 1995.
24. Chauvin, Y.; Olivier, H.; Wyrvalski, C. N.; Simon, L. C.; De Souza, R. F. *J. Catal.* **1997,** *165*, 275-278.
25. Murphy, V., PCT Int. Appl. 0032658, 2000.
26. Wasserscheid, P.; Keim, W., PCT Int. Appl. 9847616, 1998.
27. Carlin, R. T.; Osteryoung, R. A.; Wilkes, J. S.; Rovang, J. *Inorg. Chem.* **1990,** *29*, 3003-3009.
28. Carlin, R. T.; Wilkes, J. S. *J. Mol. Cat.* **1990,** *63*, 125-129.
29. Carmichael, A. J.; Haddleton, D. M.; Bon, S. A. F.; Seddon, K. R. *Chem. Commun.* **2000**, 1237-1238.
30. Biedron, T.; Kubisa, P. *Macromol. Rapid Commun.* **2001,** *22*, 1237-1242.
31. Hong, K.; Zhang, H.; Mays, J. W.; Visser, A. E.; Brazel, C. S.; Holbrey, J. D.; Reichert, W. M.; Rogers, R. D. *Chem. Commun.* **2002**, 1368-1369.
32. Hlatky, G. G., WO Patent 0181436, 2001.
33. Klingshirn, M. A.; Broker, G. A.; Holbrey, J. D.; Shaughnessy, K. H.; Rogers, R. D. *Chem. Commun.* **2002**, 1394-1395.
34. A stainless steel autoclave was charged with a solution of (bipy)Pd$(OAc)_2$ (10 mg, 0.03 mmol), bipy (75 mg, 0.48 mmol) in the desired volume of IL and methanol. *p*-Toluenesulfonic acid (96 mg, 0.50 mmol) and benzoquinone (0.247 g, 2.29 mmol) were added followed by 10 mL of styrene. The reactor was sealed, pressurized with CO to the desired pressure, and heated to the reaction temperature. After 5 hours, the reactor was vented and the reaction mixture poured into methanol to precipitate the polymer.
35. IR (KBr disk): 1708 (s), 1600 (w), 1494 (m), 1453 (m), 751 (m), 698 (s). ^{1}H NMR (1:1 $CDCl_3$:HFIPA-d_2, 360 MHz): ∂ 6.40 - 7.40 (m, 5H), 3.80-4.10 (m, 1H), 2.95-3.10 (m, 1H), 2.65-2.80 (m, 1H). ^{13}C NMR (1:1 $CDCl_3$:HFIPA-d_2, 90.6 MHz): ∂ 209.7, 135.6, 129.2, 128.2, 128.0, 53.9, 42.9.
36. Macchioni, A.; Bellachioma, G.; Cardaci, G.; Travaglia, M.; Zuccaccia, C.; Milani, B.; Corso, G.; Zangrando, E.; Mestroni, G.; Carfagna, C.; Formica, M. *Organometallics* **1999,** *18*, 3061-3069.
37. Hardacre, C.; Holbrey, J. D.; Katadare, S. P.; Seddon, K. R. *Green Chem.* **2002,** *4*, 143-146.
38. Brookhart, M.; Rix, F. C.; DeSimone, J. M.; Barborak, J. C. *J. Am. Chem. Soc.* **1992,** *114*, 5894-5895.
39. In the drybox, a solution of (bipy)Pd(CH_3)Cl (10 mg, 0.03 mmol) and the appropriate salt (0.03 mmol, $AgSbF_6$ or Na$(BAr_F)_4$) were dissolved in the IL (2 mL) and added to a stainless steel autoclave along with 10 mL of styrene. The sealed autoclave was removed from the box and attached to the

pressure system. The reactor was charged with the desired pressure of CO and heated to the reaction temperature. After 12 hours, the reactor was vented and the mixture poured into methanol. The precipitated polymer was recovered and dried *in vacuo*.

40. Johnson, L. K.; Killian, C. M.; Brookhart, M. *J. Am. Chem. Soc.* **1995,** *117*, 6414-6415.
41. Liang, Y.; Yap, G. P. A.; Rheingold, A. L.; Theopold, K. H. *Organometallics* **1996,** *15*, 584-5286.
42. Gibson, V. C.; Tomov, A.; White, A. J. P.; Williams, D. J. *Chem. Commun.* **2001**, 719-720.
43. Killian, C. M.; Tempel, D. J.; Johnson, L. K.; Brookhart, M. *J. Am. Chem. Soc.* **1996,** *118*, 11664-11665.
44. Johnson, L. K.; Mecking, S.; Brookhart, M. *J. Am. Chem. Soc.* **1996,** *118*, 267-268.
45. Small, B. L.; Brookhart, M.; Bennett, A. M. A. *J. Am. Chem. Soc.* **1998,** *120*, 4049.
46. Britovsek, G. J. P.; Bruce, M.; Gibson, V. C.; Kimberley, B. S.; Maddox, P. J.; Mastrianni, S.; McTavish, S. J.; Redshaw, C.; Solan, G. A.; Strömberg, S.; White, A. J. P.; Williams, D. J. *J. Am. Chem. Soc.* **1999,** *121*, 8728-8740.
47. Britovsek, G. J. P.; Gibson, V. C.; Kimberley, B. S.; Maddox, P. J.; McTavish, S. J.; Solan, G. A.; White, A. J. P.; Williams, D. J. *Chem. Commun.* **1998**, 849-850.
48. Anthony, J. L.; Maginn, E. J.; Brennecke, J. F. *J. Phys. Chem. B.* **2002,** *106*, 2002.
49. Milani, B.; Corso, G.; Mestroni, G.; Carfagna, C.; Formica, M.; Seraglia, R. *Organometallics* **2000,** *19*, 3435-3441.

Chapter 26

The Importance of Hydrogen Bonding to Catalysis in Ionic Liquids: Inhibition of Allylic Substitution and Isomerization by [bmim][BF_4]

James Ross and Jianliang Xiao*

Leverhulme Centre for Innovative Catalysis, Department of Chemistry, University of Liverpool, Liverpool L69 7ZD, United Kingdom

Neutral allylic substitution reactions, in which a base is generated in situ and which hence require no external bases, can be significantly retarded when carried out in the ionic liquid 1-butyl-3-methylimidazolium tetrafluoroborate ([bmim][BF_4]). Evidence suggests that this is due to the base or base precursor entering into hydrogen bonding with the imidazolium cation, and interestingly, this intermolecular interaction can be exploited to suppress unwanted allylic isomerization.

Hydrogen bonding has an ubiquitous influence in chemistry and biochemistry (*1*). The effect of hydrogen bonding on reaction chemistry in common molecular solvents is well documented and understood (*2*). Hydrogen bonding in solvents based on room temperature ionic liquids is a relatively new subject. In fact, the perception of hydrogen bonding in imidazolium ionic liquids, the most extensively investigated ionic liquids to date, was still controversial in the mid 1980s (*3, 4*). Thanks to the pioneering studies of several research groups, it is now well established that imidazolium cations and their associated anions form hydrogen bonds both in the solid states and in solution (*3-6*). The three ring protons of the 1,3-dialkylimidazolium cation can act as hydrogen bond donors and interact with counter anions such as Cl^-, OTf^- and BF_4^- acting as hydrogen bond acceptors, which can also enter into hydrogen bonding with external hydrogen bond donors such as H_2O. Of the three imidazolium ring protons, the H^2 proton appears to form the strongest hydrogen bond. However, whilst the concept of hydrogen bonding in ionic liquids has generally been accepted and explosive growth in research on reaction chemistry in these solvents has been witnessed in the past few years (*7-12*), little attention has been paid to the potential effects of hydrogen bonding on catalyzed reactions in ionic liquids (*13, 14*). We report herein that the capability of imidazolium ionic liquids for hydrogen bonding can impose significant effects on neutral allylic alkylation reactions and, interestingly, the effect could be harnessed to suppress allylic isomerization (*15*), a reaction that may deteriorate the stereochemistry of asymmetric allylic substitution (*16*).

Results and Discussion

Following on from our earlier studies on Pd(0)-catalyzed Tsuji-Trost allylic substitution reactions under basic conditions in imidazolium ionic liquids (*17, 18*), we turned our attention to the neutral variants in an attempt to determine the scope of such reactions in ionic liquids. The allylic alkylation of phenylallyl carbonate **1** with dimethyl malonate **2** was first chosen as a model neutral Tsuji-Trost reaction to study in the ionic liquid [bmim][BF_4] (eq 1). The

Ph⌄⌃⌄OCO₂Me (**1**) + MeO₂C⌄CO₂Me (**2**) —[Pd(OAc)₂, 2 mol%; PPh₃, 8 mol%; solvent, 22 °C]→ Ph⌄⌃⌄CH(CO₂Me)₂ (**3**) (1)

reaction requires no external bases, as decarboxylation of $MeOCO_2^-$, which results from the oxidative addition of **1** to Pd(0), generates CO_2 and OMe^- (*15, 16*). Although methanol (*19*) has a higher pK_a than a dialkylimidazolium ion (29 vs ca. 24 in DMSO) (*20, 21*), the methoxide is expected to preferentially deprotonate the malonate **2**, which is about 7 orders of magnitude more acidic than the imidazolium cation (*22*). Furthermore, even if deprotonation of the solvent cations took place, the so generated dialkylimidazol-2-ylidene would readily deprotonate the malonate to give the required nucleophile to attack the palladium-allyl intermediate and complete the catalytic cycle (*20, 21*). After first confirming the reaction to be rapid in THF, reaching complete conversion within 20 min (*23*), we carried out the same reaction in [bmim][BF_4]. Surprisingly, the reaction in [bmim][BF_4] afforded only 38% conversion of **1** after 5 h, and complete conversion giving **3** in 90% isolated yield only after an extended reaction time of 30 h. Even more strikingly, the reaction in THF was significantly inhibited by the addition of a small quantity of [bmim][BF_4] and the rate was progressively decreased by introducing more [bmim][BF_4]. Thus, the reaction in THF reached a conversion of only 63% after 45 min when 4 equiv (relative to the catalyst) of [bmim][BF_4] was added, and the conversion was further decreased to 46% in the presence of 10 equiv [bmim][BF_4] in the same reaction time, indicating that some key intermediate is involved in a pre-equilibrium with the imidazolium additive. A very revealing experiment is the comparison of the reaction performed in [bmim][BF_4] with that in 1-butyl-2,3-dimethylimidazolium tetrafluoroborate ([bdmim][BF_4]), in which the H^2 ring proton is replaced with a methyl group. In [bdmim][BF_4] at 50 °C, the conversion of **1** was 89% in 30 min. By way of contrast, in [bmim][BF_4] at the same temperature a conversion of 48% was reached only after 5 h reaction time, suggesting that the retarding effect of [bmim][BF_4] relates to its H^2 proton. The oxidative addition of **1** to Pd(0)-PPh_3 leads to two ionic species, $MeOCO_2^-$ and a Pd-allyl cation, and as such should not be slowed down on going from THF to an ionic medium. Therefore it is probably the nucleophilic attack step that is affected in the ionic liquid.

The allylic alkylation of phenylallyl acetate **4** with methyl nitroacetate **5** was similarly retarded when conducted in the ionic liquid (eq 2). This is again a neutral reaction, because the OAc^- ion generated in the oxidative addition of **4** to Pd(0) is basic enough to deprotonate **5** (pK_a = 8.0 in DMSO) (*24*), which is much more acidic than **2**. Indeed, the reaction in THF was complete within 0.5 h when catalyzed by Pd(0)-PPh_3 at 75 °C. Repeating this reaction in [bmim][BF_4], only 26% conversion was reached after 0.5 h.

Ph⁄⁄OAc (**4**) + O_2N⁄CO_2Me (**5**) → [$Pd(OAc)_2$, 2 mol%; PPh_3, 8 mol%; solvent, 75 °C] → Ph⁄⁄CH(NO_2)(CO_2Me) (**6**) (2)

The observed inhibition of the neutral Tsuji-Trost reaction by [bmim][BF_4] could be accounted for by assuming that the OAc^- or $MeOCO_2^-$ ion generated in the oxidative addition is strongly solvated or "trapped" by hydrogen bonding with the imidazolium cations and is thus made unavailable to deprotonate the

$$MeOCO_2^- + [\text{Me–Im}^+\text{–Bu (C2–H)}] \rightleftharpoons \text{Me–Im}^+\text{–Bu (C2–H}\cdots{}^-O_2COMe)(\cdots MeOCO_2^-)$$

$$MeOCO_2^- \longrightarrow CO_2 + MeO^-$$

$$HNu + MeO^- \longrightarrow MeOH + Nu^-$$

$$[\text{Ph-allyl–Pd}L_2]^+ + Nu^- \longrightarrow \text{Ph–CH=CH–CH}_2\text{–Nu} + Pd(0)L_2$$

Scheme 1. Solvation of the carbonate by hydrogen bonding and its effect on the nucleophilic attack at the Pd-allyl intermediate.

hydrocarbon acids to give the required nucleophiles for subsequent nucleophilic attack (Scheme 1). OAc^- is known to form a strong hydrogen bond to the H^2 proton of imidazolium ring in both solution and solid states (*25, 26*). A ^1H NMR experiment, in which the concentration of [bmim][BF_4] was kept constant while that of [NBu_4][OAc] was varied in $CDCl_3$, revealed that the H^2 proton chemical shift moved to lower field in almost a linear fashion with increasing concentration of OAc^- until the molar ratio of acetate/bmim reached a value of ca. 2; thereafter the H^2 chemical shift was little affected with additional acetate. Only insignificant changes were observed for the H^4 and H^5 ring protons. These observations suggest that the H^2 proton hydrogen bonds to OAc^- and each H^2 proton is approximately involved with two acetate ions (*27*). Similar observations have been made concerning imidazolium and halide ions (*28*). Amatore, Jutand and co-workers have recently shown that the oxidative addition of allylic carbonates to Pd(0) is reversible and the resulting carbonate anion does not decarboxylate as fast as previously thought (*16*). Hydrogen bonding should certainly enhance its stability. Indeed, it is known that decarboxylation can dramatically be decelerated by using dipolar protic solvents capable of hydrogen bonding (*29*).

The notion that OAc^- or $MeOCO_2^-$ could not function efficiently as a base or base precursor due to hydrogen bonding explains the effects of the added [bmim][BF_4] on the reaction of **1** and **2** in THF. An increasing concentration of imidazolium cation augments the extent of hydrogen bonding and lowers the concentration of free base and nucleophile. The result is slower nucleophilic addition (Scheme 1).

There exists a possibility that the sluggish reaction between **1** and **2** in [bmim][BF_4] could stem from the formation of some inactive dialkylimidazol-2-ylidene complexes of palladium *via* deprotonation of the imidazolium cation by OMe^- (*30-32*). This appears to be unlikely. No Pd-carbene complexes were ever detected by NMR in stoichiometric reactions of Pd(0)-PPh_3 with **1** in the presence of [bmim][BF_4] at room temperature. Furthermore, the aforementioned effect of [bmim][BF_4] on the allylic alkylation in THF casts doubts on this possibility, since the activity of the hypothesised Pd-carbene species would not be affected by additional imidazolium cations. In addition, the reaction of **7** and **2** shown below in eq 3, which involves an in situ generated alkoxide and reaches completion in *both* THF and [bmim][BF_4] within 1 h, suggests that either *N*-heterocyclic carbenes are not formed by alkoxide deprotonation of the imidazolium cation or, if formed, have no effect on the palladium catalysis. The success of this reaction also indicates that strong bases such as the alkoxides do not form stable hydrogen bonds with the acidic imidazolium H^2 ring proton in the presence of much stronger acids.

7 + MeO_2C‿CO_2Me (**2**) —[$Pd(OAc)_2$, 2 mol%; $P(4\text{-}C_6H_4OMe)_3$, 8 mol%; solvent, 22 °C, 1 h]→ HO‿(=)‿CH(CO_2Me)(CO_2Me) (**8**) (3)

Applying the hydrogen bonding proposition between OAc^- or $MeOCO_2^-$ and [bmim][BF_4], it was possible to suppress the Pd(0)-catalyzed isomerization of allylic acetates, which may lead to the loss of regio- and stereo-chemistry in stereospecific allylic substitution (*16*). The isomerization is thought to be due to the key oxidative addition step being reversible, and unequivocal evidence for this has recently been laid out (*16*). As OAc^- can be trapped by hydrogen bonding with [bmim][BF_4], allylic isomerization resulting from reversible attack by the acetate would be expected to be retarded when allyic substitution or isomerization is carried out in an imidazolium ionic liquid. This is indeed the case. Treating **9** with 5 mol% Pd(0)-PPh_3 in CH_2Cl_2 afforded 35% of product **10** after 1 h, comparable with the equilibrium value reported in the literature (Scheme 2) (*33*). However, the same reaction appears to be completely suppressed in [bmim][BF_4]. Thus, **10** was not detected in the crude reaction mixture even after 20 h. The isomerization of **9** in CH_2Cl_2 was even inhibited by a sub-stoichiometric quantity of [bmim][BF_4]. Thus, in the presence of only 0.15

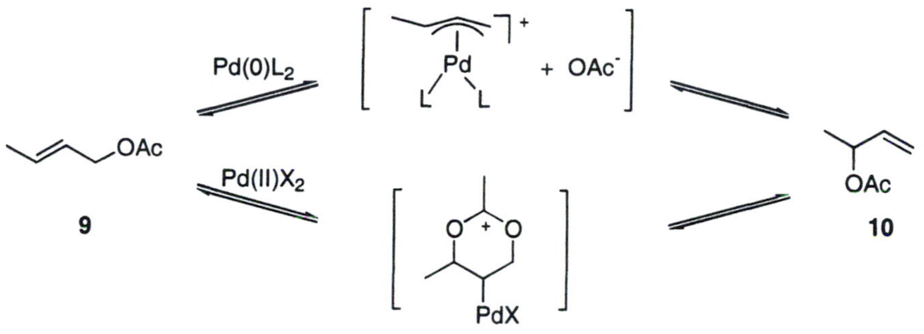

Pd(0)L_2 Pd(II)X_2 Pd + OAc⁻ OAc 9 PdX OAc 10

*Scheme 2. Isomerization of allylic acetate **9** by palladium catalysis.*

equiv [bmim][BF_4] relative to **9**, only 12% of **10** was formed after 3 h. Similar results were obtained starting from **10**. Thus, while 58% of **9** was formed after 1 h in CH_2Cl_2, **10** remained intact in [bmim][BF_4] even after 20 h.

Remarkably, the isomerization can be brought about when a Pd(II) catalyst is employed. Thus, Treatment of **9** with [$PdCl_2(MeCN)_2$] in [bmim][BF_4] afforded 20% of **10** in 1 h reaction time; the same reaction in CH_2Cl_2 gave 33% of **10**. As indicated, this reaction proceeds intra-molecularly via a cyclic intermediate and involves no ionized acetate ions (*34*); therefore it is not expected to be suppressed by hydrogen bonding. Differences between CH_2Cl_2 and [bmim][BF_4] in other solvent properties possibly account for the different extent of isomerization in these solvents. These results show that if the isomerization of an allylic acetate is to be carried out in imidazolium ionic liquids, Pd(II), rather than Pd(0), should be the catalyst of choice.

In summary, neutral Tsuji-Trost reactions can be considerably retarded in dialkylimidazolium ionic liquids and our results suggest that this is due to hydrogen bonding between the H^2 ring proton of [bmim][BF_4] and OAc^- or $MeOCO_2^-$ ions. Being strongly solvated by the ionic liquid via hydrogen bonds, the anions could not function as an effective base to deprotonate a HNu nucleophile, thus rendering slow the nucleophilic attack at the Pd(II)-allyl intermediate. We further demonstrated that this hydrogen bonding could be exploited to suppress unwanted, Pd(0) catalyzed isomerization of allylic acetates. Taken together, these results highlight the potential effect of dialkylimidazolium cations as hydrogen bond donor on any reactions that are performed in ionic liquids containing such cations or in their presence and involve transition and/or ground state with hydrogen bond acceptor characteristics.

Experimental Section

All reactions were carried out in oven-dried glassware under argon, using standard Schlenk and vacuum line techniques. [bmim][BF_4] and [bdmim][BF_4] were prepared according to published procedures and vacuum-dried and stored under argon (*35*). THF was freshly distilled from sodium benzophenone under nitrogen immediately prior to use. CH_2Cl_2 was freshly distilled from calcium hydride under nitrogen immediately prior to use. Phenylallyl carbonate **1** was synthesized according to a literature method (*36*). The synthesis of compound **10** was adapted from a literature method (*37*). Compounds **2**, **4**, **5**, **7** and **9**, [NBu_4][OAc], $Pd(OAc)_2$, $Pd(dba)_2$ and PPh_3 were purchased from commercial suppliers and used as received without further purification.

Typical Tsuji-Trost reaction in [bmim][BF_4] as exemplified for the allylic alkylation of **1** by **2**: $Pd(OAc)_2$ (4.5 mg, 2 mol%) and PPh_3 (21.0 mg, 8 mol%) were stirred in [bmim][BF_4] (2 ml) at 80 °C for 20 min under an atmosphere of argon and then allowed to cool to room temperature. **1** (192.2 mg, 1 mmol) and **2** (198.2 mg, 1.5 mmol) were added and the mixture stirred under argon. The progress of reaction was monitored by ^1H NMR. Upon completion, **3** was isolated in analytically pure form in 90% yield (*38*).

Typical allylic isomerization in [bmim][BF_4] as exemplified for **9**: $Pd(dba)_2$ (28.7 mg, 5 mol%) and PPh_3 (26.2 mg, 10 mol%) were stirred in [bmim][BF_4] (2 ml) at 80 °C for 20 min and then allowed to cool to room temperature. **9** (114.2 mg, 1.0 mmol) was added and the mixture stirred under an atmosphere of argon. Upon completion, **9** and **10** were isolated quantitatively relative to the quantity of starting **9**; the ratio of the two was determined by ^1H NMR (*33*).

Acknowledgement: We are grateful to the EPSRC for a studentship (JR) and the Industrial Partners of the Leverhulme Center for Innovative Catalysis for support. We also thank Johnson Matthey for the loan of palladium.

References

1. Jeffrey, G. A. *An Introduction to Hydrogen Bonding*; Oxford University Press: New York, 1997.
2. Reichardt, C. *Solvents and Solvents Effects in Organic Chemistry*; VCH: Weinheim, 1990.
3. Fannin, A. A.; King, L. A.; Levisky, J. A.; Wilkes, J. S. *J. Phys. Chem.* **1984**, *88*, 2609.
4. Tait, S.; Osteryoung, R. A. *Inorg. Chem.* **1984**, *23*, 4352.

5. Abdul-Sada, A. K.; Greenway, A. M.; Hichcock, P. B.; Mohammed, T. J.; Seddon, K. R.; Zora, J. A. *J. Chem. Soc., Chem. Commun.* **1986**, 1753.
6. Avent, A. G.; Chaloner, P. A.; Day, M. P.; Seddon, K. R.; Welton, T. *J. Chem. Soc., Dalton Trans.* **1994**, 3405 and references therein.
7. Welton, T. *Chem. Rev.* **1999**, *99*, 2071.
8. Wasserscheid, P.; Keim, W. *Angew. Chem. Int. Ed. Engl.* **2000**, *39*, 3772.
9. Earle, M. J.; Seddon, K. R. *Pure Appl. Chem.* **2000**, *72*, 1391.
10. Sheldon, R. *Chem. Commun.* **2001**, 2399.
11. Gordon, C. M. *Appl. Catal. A: General* **2001**, *222*, 101.
12. Olivier-Bourbigou, H.; Magna, L. *J. Mol. Catal. A: Chemicals* **2002**, *182-183*, 419.
13. Fischer, T.; Sethi, A.; Welton, T.; Woolf, J. *Tetrahedron Lett.* **1999**, *40*, 793.
14. Acevedo, O.; Evanseck, J. D. *Abstr. Pap. Amer. Chem. Soc.* **2002**, *224*, 028-IEC.
15. Tsuji, J. *Palladium Reagents and Catalysts-Innovations in Organic Synthesis*; Wiley: Chichester, 1995.
16. C. Amatore, S. Gamez, A. Jutand, G. Meyer, M. Moreno-Manas, L. Morral, R. Pleixats, *Chem. Eur. J.* 2000, *6*, 3372 and references therein.
17. Chen, W.; Xu, L.; Chatterton, C.; Xiao, J. *Chem. Commun.* **1999**, 1247.
18. Ross, J.; Chen, W.; Xu, L.; Xiao, J. *Organometallics* **2001**, *20*, 138.
19. Olmstead, W. N.; Margolin, Z.; Bordwell, F. G. *J. Org. Chem.* **1980**, *45*, 3295.
20. Alder, R. W.; Allen, P. R.; Williams, S. J. *J. Chem. Soc., Chem. Commun.* **1995**, 1267.
21. Kim, Y.-J.; Streitwieser, A. *J. Am. Chem. Soc.* **2002**, *124*, 5757.
22. The pK_a of closely related diethyl malonate in DMSO is 16.4: Bordwell, F. G. *Acc. Chem. Res.* **1988**, *21*, 456.
23. J. Tsuji, *Tetrahedron* 1986, *42*, 4361.
24. Giambastiani, G.; Poli, G. *J. Org. Chem.* **1998**, *63*, 9608.
25. Wilkes, J. S.; Zaworotko, M. J. *J. Chem. Soc., Chem. Commun.* **1992**, 965.
26. Bonhôte, P.; Dias, A.-P.; Papageorgiou, N.; Kalyanasundaram, K.; Grätzel, M. *Inorg. Chem.* **1996**, *35*, 1168.
27. Vinogradov, S. E.; Linnell, R. H. *Hydrogen Bonding*; Van Nostrand Reinhold Company: New York, 1971.
28. Thomas, J.-L.; Howarth, J.; Hanlon, K.; McGuirk, D. *Tetrahedron Lett.* **2000**, *41*, 413.
29. Kemp, D. S.; Cox, D. D.; Paul, K. G. *J. Am. Chem. Soc.* **1975**, *97*, 7312.
30. Xu, L.; Chen, W.; Xiao, J. *Organometallics* **2000**, *19*, 1123.
31. Mathews, C. J.; Smith, P. J.; Welton, T.; White, A. J. P.; Williams, D. J. *Organometallics* **2001**, *20*, 3848.
32. Aggarwal, V. K.; Emme, I.; Mereu, A. *Chem. Commun.* **2002**, 1612.

33. D. C. Braddock, A. J. Wildsmith, *Tetrahedron Lett.* 2001, *42*, 3239.
34. Crilley, M. M. L.; Golding, B. T.; Pierpoint, C. *J. Chem. Soc. Perkin Trans. I* **1998**, 2061.
35. Holbrey, J. D.; Seddon, K. R. *J. Chem. Soc., Dalton Trans.* **1999**, 2133.
36. Lehmann, J.; Lloyd-Jones, G. C. *Tetrahedron* **1995**, *51*, 8863.
37. Hayashi, T.; Yamamoto, A.; Ito, Y.; Nishioka, E.; Miura, H.; Yanagi, K. *J. Am. Chem. Soc.* **1989**, *111*, 6301.
38. Trost, B. M.; Hachiya, I. *J. Am. Chem. Soc.* **1998**, *120*, 1104.

Chapter 27

Recent Developments in the Use of *N*-Heterocyclic Carbenes: Applications in Catalysis

Rebecca M. Kissling, Mihai S. Viciu, Gabriela A. Grasa, Romain F. Germaneau, Tatyana Güveli, Marie-Christiane Pasareanu, Oscar Navarro-Fernandez, and Steven P. Nolan*

Department of Chemistry, University of New Orleans, 2000 Lakeshore Drive, New Orleans, LA 70148

Some recent work where *N*-heterocyclic carbenes (NHC) are used in catalysis is described. A family of $[(NHC)PdCl_2]_2$ complexes has been synthesized and is a storable amination catalyst that is tolerant of both moisture and air. Various NHC add to $[Pd(allyl)(Cl)]_2$, cleaving this dimer and forming a highly active, versatile pre-catalyst for the formation of C-N and C-C bonds. *N*-heterocyclic carbenes are also utilized as highly efficient transesterification catalysts converting vinyl and methyl substrates to more useful or desirable esters.

Introduction

Imidazolium salts have garnered a great deal of attention in recent years as targets for greener chemistry.[1] Many mixed alkyl imidazolium salts are liquids at or near room temperature, and thus these polar materials can serve as solvents for many chemical processes.[2] The highly polar nature and concomitant high vapor pressure of these salts generally leads to simple product recovery by extraction or distillation.[1] These ionic liquids have been successfully employed as "solvents" in a number of transition metal catalyzed reactions.[1] Interestingly, when BmimHX is utilized as a solvent under basic conditions the deprotonated salt, or N-heterocyclic carbene, binds the metal and thus modulates the chemical activity of the catalyst as would a tertiary phosphine.[3] This reaction highlights another area of great interest regarding imidazolium salts, the development of heterocyclic carbenes as phosphine mimics for use in transition metal catalysis.

Early Studies of $[(NHC)PdCl_2]_2$ Mediated Catalysis

In palladium-mediated coupling studies, the (NHC)Pd fragment has been generated *in situ* by action of a base on the imidazolium salt precursor followed by addition of a palladium source.[4] In these studies, the ratio of ligand to metal was optimized to 1:1. These observations led us to explore synthetic avenues leading to palladium complexes bearing a single NHC ligand per palladium; the number of such complexes is still rather limited.[5] Some of these early studies on palladium/NHC systems revealed the Pd-carbene bond to be robust and stable to heat, indicating such a system might not require excess ligand to compensate for ligand metal bond lability.

$$(PhCN)_2PdCl_2 + \textbf{IPr} \xrightarrow[\text{2h, RT}]{\text{THF, Toluene}} [(IPr)PdCl_2]_2 \; \textbf{1} \qquad (1)$$

We have investigated the reaction of 1 equiv IPr with $PdCl_2(PhCN)_2$ in THF/toluene, eq 1. The carbene readily displaces the nitrile ligands generating the tan-orange $[Pd(IPr)Cl_2]_2$ **1**. Single crystal X-ray analysis revealed **1** to be the dimer presented in Figure 1. The geometry at the metal centers is distorted square planar with all Pd and chlorine atoms coplanar, and the aryl groups of the trans-disposed IPr ligands canted normal to each other.

The reaction requires dry solvents and an inert atmosphere, although the synthetic work up can be performed in air with little or no compromise to yield or purity of product. Complex **1** is soluble in polar solvents and sparingly soluble in hexanes and benzene. It is both air- and moisture-stable; and showed no deterioration after storage on the bench for 3 months.

Complex **1** was used as a catalyst precursor in aryl amination. The amination reactions were performed (under optimized conditions) in DME with potassium *tert*-amylate (KO^tAm) as base,[6] as described in Eq 1.[7] The palladium

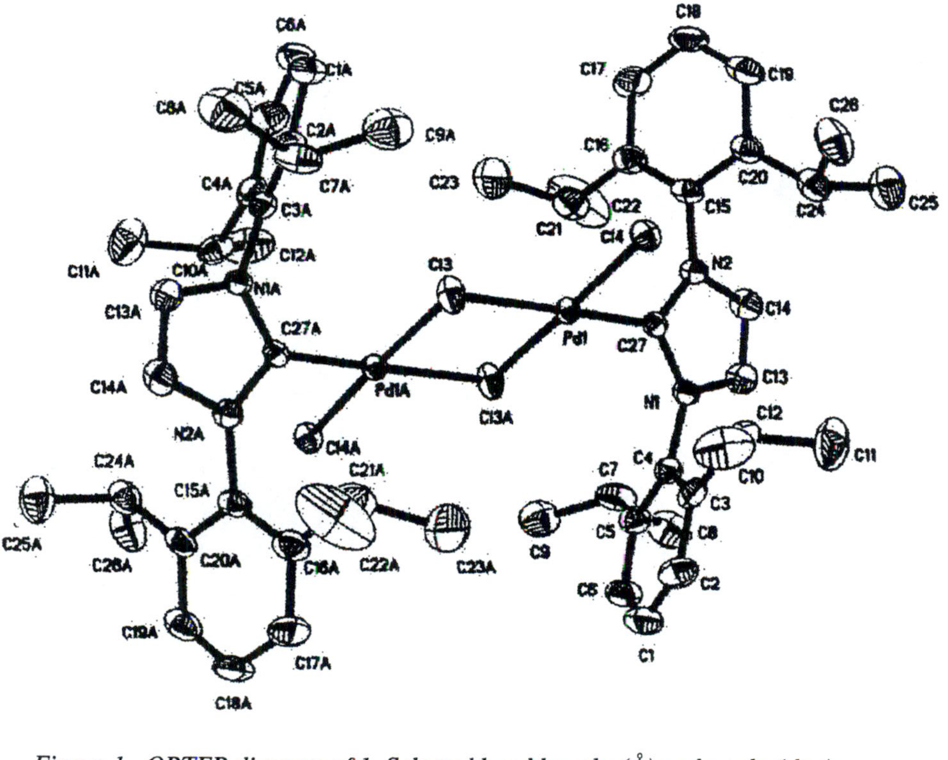

Figure 1. ORTEP diagram of 1. Selected bond lengths (Å) and angles(deg): Pd(1)-C1(3) 2.4029(9); Pd(1)-C1(4) 2.2715(9); Pd(1)-C(27) 1.9553(3); C1(3)-Pd(1)-C1(4) 91.26(3); C1(3)-Pd(1)-C(27) 178.26(9); C1(4)-Pd(1)-C(4) 90.48(9).

loading, 1 mol%(0.5 mol % **1**), is comparable to catalyst loadings in the best Pd/phosphine systems.[8,9 10]

$$\text{Ar-X} \;+\; 1.2\ \text{eq. HNRR'} \xrightarrow[\text{80 °C, DME}]{\text{0.5 mol \% }\mathbf{1}\text{, 1.5 equiv KO}^{t}\text{Am}} \text{Ar-NRR'} \quad (2)$$

The role of the base is two-fold in this aryl amination system. NMR experiments indicate that first the alkoxide reacts with the dimer to form KCl and a monomeric (IPr)Pd(O^{t}Am)(Cl) species transformed via a mechanism recently proposed by Hartwig.[11]

Results of the amination study using aryl and heteroaryl halides are listed in Table 1. We focused on the more difficult to activate aryl chlorides determining some overall steric and reaction condition limitations through reactions with aryl bromides. The reaction of 4-chlorotoluene with aniline proceeds to complete conversion in one hour at 50 °C (entry 1, Table 1); while we found only moderate conversion was achieved at room temperature (75% in 15 h).

Activated aryl bromides and chlorides completely converted to products in just a few minutes at 80 degrees (entries 3 and 4). Electronically deactivated and sterically hindered substrates generally required longer reaction times of up to 3 h (entries 7, 8 and 9). Both primary amines and secondary anilines are capable coupling partners with aryl chlorides, giving high conversions to products in 2h at 80 °C. Chloropyridines are fully compatible with this catalyst system and lead to product in short reaction times (entry 10). Dehalogenation of ArX was only a minor side reaction in all cases (>2%) with the exception of entry 6 where the fluorobenzene/product ratio was 22:78 as determined by GC analysis.

Surprisingly, aryl triflates were not active amine coupling partners in the present system, nor were base sensitive substrates. Substitution of Cs_2CO_3 or K_3PO_4 for KO^{t}Am to address the latter limitation led to rapid formation of palladium black during the course of reactions. Indoles were inert to coupling in the present system; even at elevated temperatures coupling of indoles with aryl bromides did not occur. Current work is being carried out on these substrates.

A remarkable aspect of aryl amination utilizing **1** is the tolerance of these reactions to both air and moisture. As evidence of the robust nature of **1**, we were able to perform the amination reactions loaded on the bench top in reagent grade solvent (stored on the bench without measures to exclude air or water) under air (Table 2).[12] There have only recently been reports of Pd-catalyzed coupling under similar conditions.[13,14] In the present system many substrate pairs reacted on par with those of the air-free system (entries 2 and 3). Although some reactions involving aryl chlorides have markedly diminished activity when conducted in air (e.g. entry 9), moderate activity was found between 4-chlorotoluene and morpholine using as little as 0.05 mol % of **1**.

Complex **1** shows good catalytic activity even under aerobic conditions. Exploration of the reactivity profile of **1** and related palladium complexes in coupling reactions as well as mechanism elucidation are ongoing.

Synthesis and Reactivity of (NHC)Pd(allyl)Cl Complexes

Our recent success in generating Pd(II) complexes of type $[(NHC)PdCl_2]_2$ and the observed catalytic reactivity of these systems led us to explore the synthesis of monomeric palladium(II) complexes. Our synthetic approach makes use of a simple palladium (II) source, $[(allyl)PdCl]_2$ (**2**), and the optimal metal to ligand ratio previously described.[15] Various (NHC)Pd(allyl)Cl complexes, where NHC is SIPr [*N,N'*-bis(2,6-diisopropylphenyl)4,5-dihydroimidazol)-2-ylidene] (**3**), IPr [*N,N'*-bis(2,6-diisopropylphenyl)imidazol)-2-ylidene] (**4**), IMes [*N,N'*-bis(2,4,6-trimethylphenyl) imidazol)-2-ylidene] (**5**) and I^{t}Bu [*N,N'*-bis-*tert*-butyl-imidazol)-2-ylidene] (**6**), were synthesized in excellent yields using the reaction depicted in eq. 3. The reaction of most NHC with **2** can be performed in THF at room temperature in one hour.

The reaction of SIPr and **2** proceeds smoothly at -78°C in Et_2O, affording **3** in 96% yield. Once the complexes are formed, the workup and recrystallization can be performed in air without reduction of yield.

$$[(allyl)Pd(\mu\text{-}Cl)]_2 \ (\mathbf{2}) \xrightarrow[\text{THF, R. T. or ether -78° C}]{\text{2 NHC}} (allyl)Pd(Cl)(NHC) \quad (3)$$

To confirm the structure of this family of complexes, a single crystal X-ray analysis of **3** (formed in Et_2O by slow cooling) was performed (Figure 2). The ORTEP of **3** reveals η^3-coordination of the allyl fragment and distorted square planar coordination around the Pd center.

With complexes **3-6** in hand, activation of this system towards catalysis was examined. In the formation of *in situ* catalysts from Pd(II) sources and imidazolium salts, the exact reaction pathway leading to Pd(0) species was not investigated. With these monomeric compounds, the reduction of Pd(II) to Pd(0) was examined with various bases used under catalytic conditions. The reaction of **6** with NaO^{t}Bu at room temperature led within minutes to a mixture of two different allyl species (60/40) as observed by ^{1}H NMR spectroscopy.

When this solution was warmed to 40°C for 1h, one species was converted to the second species exclusively. The final allylic species is allyl-*tert*-butyl ether, confirmed by comparison with an authentic sample.[16] In palladium-allyl systems, there are precedents for nucleophilic attack by an alkoxide base either at the allyl[17] or at the palladium center.[18] Regardless of the activation route, a NHC-Pd (0) complex is formed. The existence of such a complex was confirmed by a trapping experiment carried out in the presence of PCy_3 (eq. 4).[19]

$$(allyl)Pd(Cl)(I^tBu) \xrightarrow[\text{KO}^t\text{Bu}]{\text{PCy}_3} I^tBu\text{—Pd—PCy}_3 + \text{CH}_2\text{=CHCH}_2\text{O}^t\text{Bu} + \text{KCl} \quad (4)$$

Table I. Air-free aryl amination catalyzed by 1.

1 mmol ArX + 1.2 mol NHRR' $\xrightarrow[\text{1.5 equiv KO}^t\text{Amylate, DME}]{\text{0.5 mol \% of 1}}$ ArNRR'

Entry	ArX	Amine	Product	t, h	T, °C	Yield[a]
1	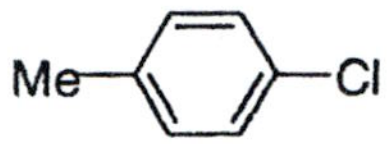 Me, Cl	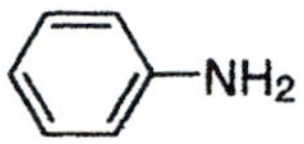 NH_2	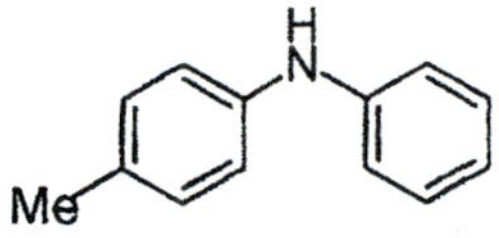 H, N, Me	1	50	100(86)
2	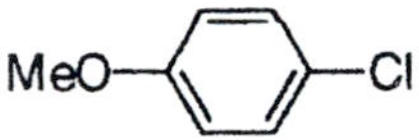 MeO, Cl	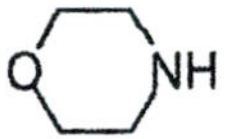 O, NH	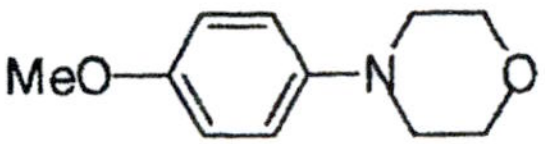 MeO, N, O	1	50	99(79)
3	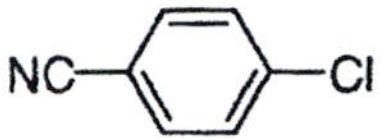 NC, Cl	 O, NH	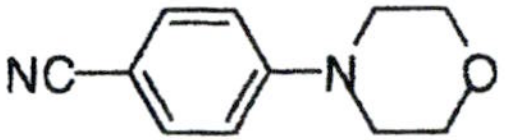 NC, N, O	0.1	80	100(83)
4	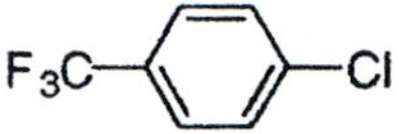 F_3C, Cl	 O, NH	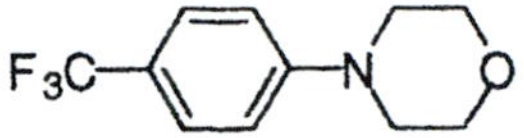 F_3C, N, O	0.25	80	100(97)
5	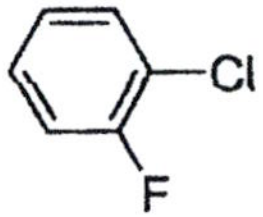 Cl, F	 O, NH	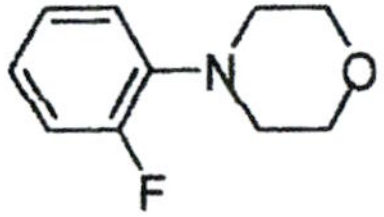 N, O, F	1	80	78(58)

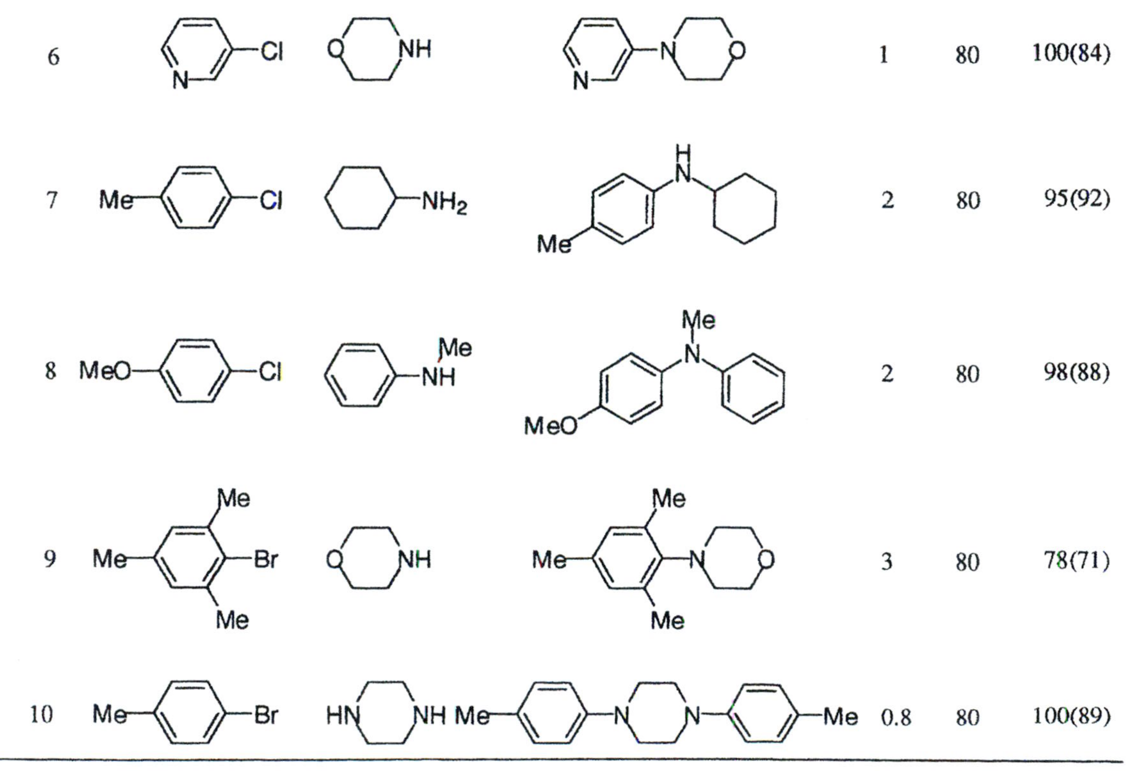

6	1	80	100(84)
7	2	80	95(92)
8	2	80	98(88)
9	3	80	78(71)
10	0.8	80	100(89)

[a]GC conversion(isolated yield); average of 2 runs.

Table 2. Aryl amination reactions catalyzed by 1 under aerobic conditions.

$$\text{1 mmol ArX} + \text{1.2 mol NHRR'} \xrightarrow[\text{1.5 equiv KO}^t\text{Amylate, DME, 80 °C}]{\text{0.5 mol \% of } \mathbf{1}} \text{ArNRR'}$$

Entry	ArX	Amine	Product	Time	Yield[a]
1	 Me … Br	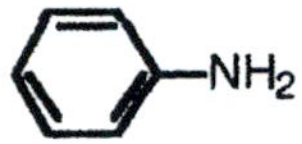 … NH_2	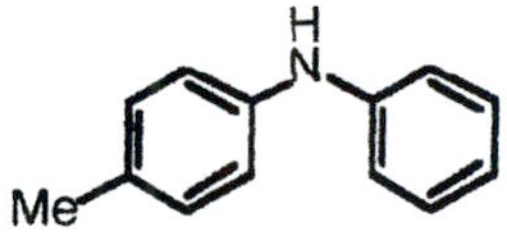 Me … N(H) …	0.3	100(84)
2	 Me … Cl	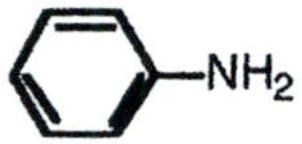 … NH_2	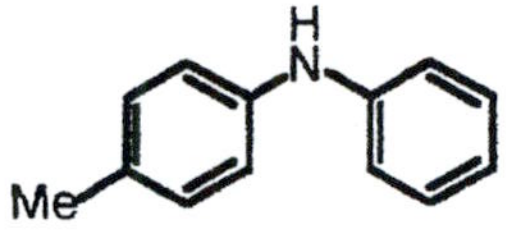 Me … N(H) …	1.3	100(81)
3	 Me … Br	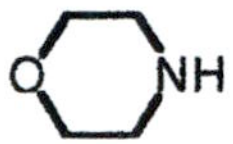 O … NH	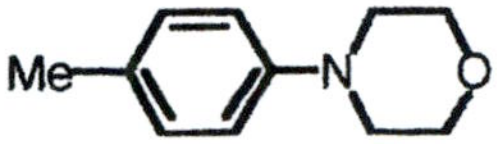 Me … N … O	0.1	100(88)
4	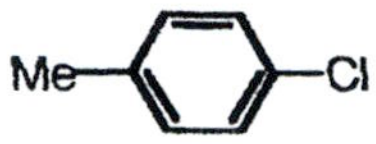 Me … Cl	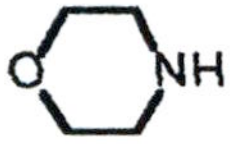 O … NH	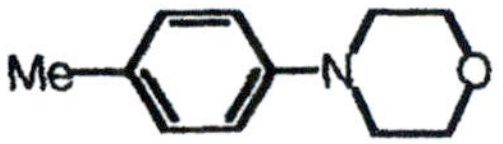 Me … N … O	0.5	99(87)
5	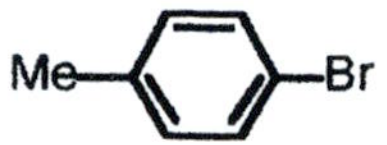 Me … Br	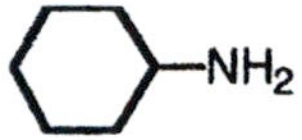 … NH_2	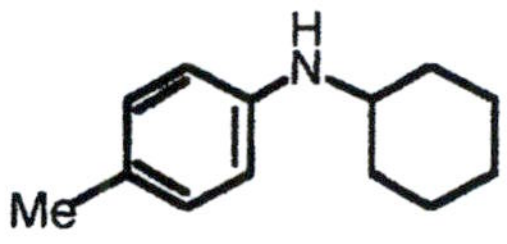 Me … N(H) …	0.6	97(94)

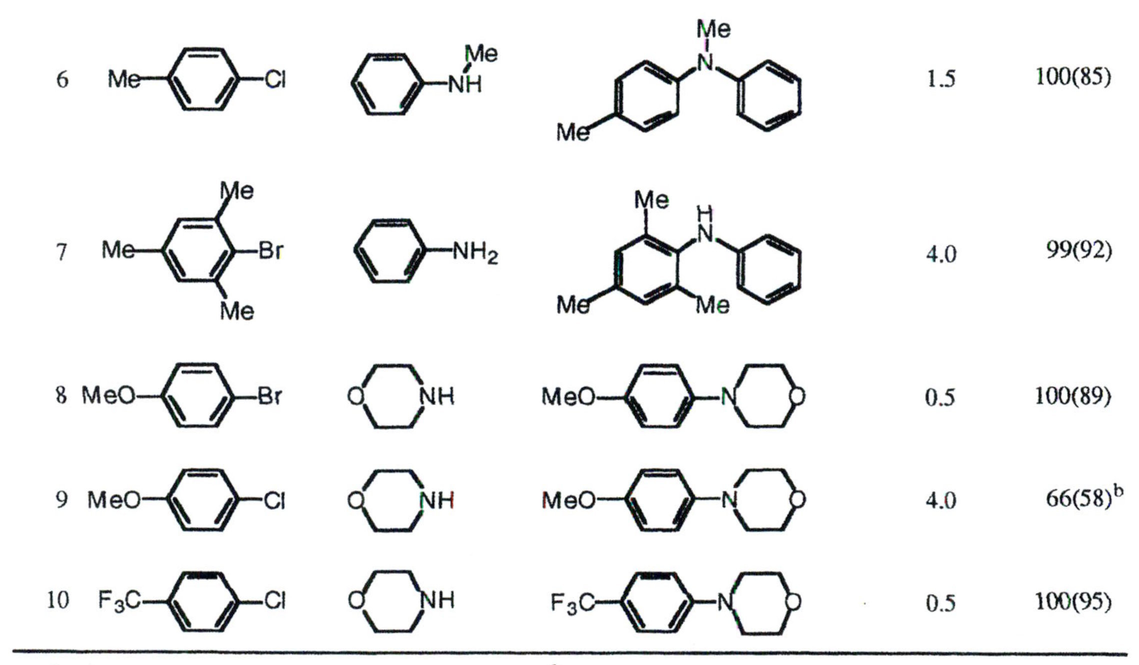

6	Me–C6H4–Cl	PhNHMe	4-Me-C6H4–N(Me)Ph	1.5	100(85)
7	2,4,6-Me3C6H2–Br	PhNH2	2,4,6-Me3C6H2–NHPh	4.0	99(92)
8	MeO–C6H4–Br	morpholine	MeO–C6H4–N(morpholino)	0.5	100(89)
9	MeO–C6H4–Cl	morpholine	MeO–C6H4–N(morpholino)	4.0	66(58)[b]
10	F3C–C6H4–Cl	morpholine	F3C–C6H4–N(morpholino)	0.5	100(95)

[a]GC conversion(isolated yield); average of 2 runs. [b]No Further conversion after 4 h.

We postulate that the new (NHC)Pd(allyl)X species formed is the (NHC)Pd(allyl)(O^tBu) complex. The formation of this species would be the result of a simple metathesis between the (NHC)Pd(allyl)Cl and NaOR. Both alkoxide attack at the allyl position and metathetical alkoxide replacement would lead to the observed NHC-Pd species.

The reductive elimination of ether from a palladium (II) complex would generate a catalytically active (NHC)Pd(0) species. Since a palladium (0) species is formed under basic conditions, the catalytic activity of the pre-catalyst, (NHC)Pd(allyl)Cl, in cross-coupling reactions of aryl chlorides with various substrates was examined. The palladium-mediated aryl amination reaction requires, with few exceptions, the use of a strong base (such as an alkoxide).[20]

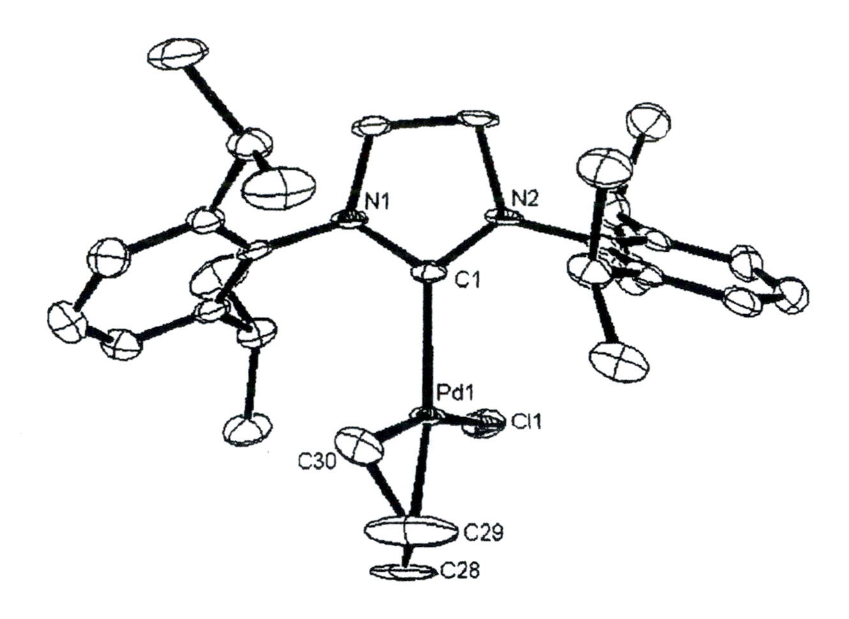

*Figure 2. ORTEP diagram of (SIPr)Pd(allyl)Cl(**3**) . Selected bond lengths (Å) and angles (deg): Pd-C(1), 2.042(5); Pd-C(30), 2.098(6); Pd-C(29), 2.124(7); Pd-C(28), 2.210(6); Pd-Cl(1), 2.3757(14); C(1)-Pd(1)-C(29), 137.4(2); C(1)-Pd(1)-C(28), 169.0(2); C(30)-Pd(1)- C(28), 68.4(2); C(1)-Pd(1)-Cl(1), 92.36(14); C(29)-Pd(1)-Cl(1), 129.1(2).*

The aryl amination reactions were performed in DME with NaO^tBu and **3**. In most cases, reactions were successfully performed at room temperature (Table 3,entries 1-4).

Suzuki-Miyaura reactions have been conducted with a number of NHC-palladium systems.[21] The most commonly used bases are inorganic carbonates, phosphates or fluorides. In our initial attempts at Suzuki-Miyaura coupling with **3**, the use of Cs_2CO_3, CsF, K_3PO_4 or NaOAc alone led to no product conversion involving aryl chlorides and phenyl boronic acid. These bases failed to activate the catalyst. The use of NaO^tBu led to a complete conversion (95% isolated) of the desired product (Table 4 Entry 1). The optimum conversions and reaction times were obtained utilizing a two base system: a catalytic amount of NaO^tBu was used to initiate the Pd(II) complex and Cs_2CO_3 was used as an operating base in Suzuki-Miyaura (Table 4 Entry 2).

The coupling of simple ketones and aryl halides, despite its great synthetic importance, has been less explored.[22] Strong bases are required to generate carbanions from ketones.[23] NaO^tBu is a convenient base in this system since it can activate the catalyst and deprotonate the ketone. Ketone arylation competes with condensation of two ketone molecules due to the presence of substrates with acidic protons and their conjugate bases.[23] This side reaction can be minimized if a rapid oxidative addition of aryl chloride and a fast reductive elimination to the desired product are involved in the catalytic cycle. The size and donating properties of SIPr in **3** were found to be beneficial to this transformation. Aryl chlorides represent good coupling partners in ketone arylation (superior to bromides). In all cases examined (Table 5, Entries 1-4) by-products were formed in less than 5 percent.

Based on the straightforward synthesis of a novel family of (NHC)Pd(allyl)Cl complexes ongoing efforts are aimed at exploring the scope of this facile activation protocol in a number of cross coupling and related reactions.

NHC- Catalyzed Transesterification

The ester moiety is one of the most ubiquitous functional groups in chemistry, playing a paramount role in biology and serving both, as key intermediate or protecting group in organic transformations.[24] As a consequence, highly efficient methods for the synthesis of different esters are potentially very useful. Base[25] or Lewis acid-catalyzed[26] acylation of alcohols by acetic anhydride can suffer from poor selectivity between primary and secondary

Table 3. Palladium-mediated amination of aryl chlorides

$$\text{1 mmol ArCl} + \text{1.2 mol NHRR'} \xrightarrow[\text{1.5 equiv KO}^t\text{Bu, 4 ml DME}]{\text{1 mol \% (SIPr)Pd(Allyl)Cl}} \text{ArNRR'}$$

Entry	ArX	Amine	Product	t, h	T, °C	Yield[a]
1	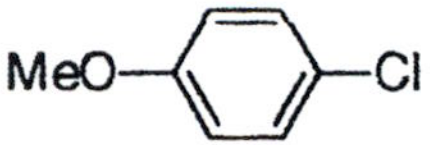 MeO, Cl	 O, NH	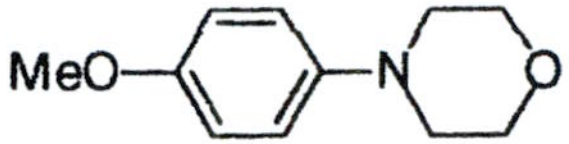 MeO, N, O	1.3	25	93
2	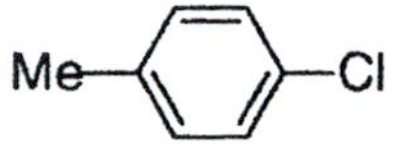 Me, Cl	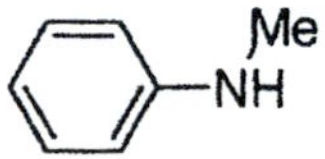 Me, NH	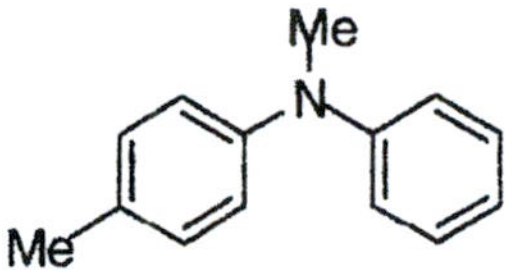 Me, N, Me	14	25	96
3	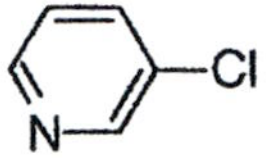 N, Cl	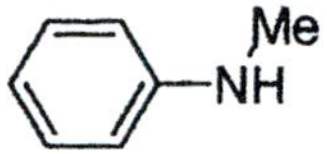 Me, NH	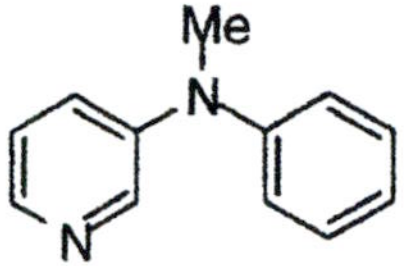 Me, N, N	24	25	88
4	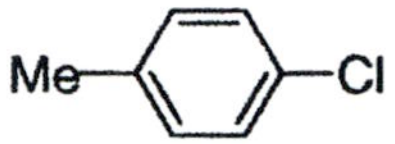 Me, Cl	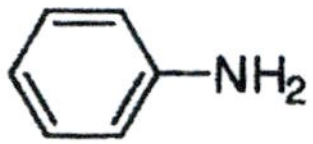 NH_2	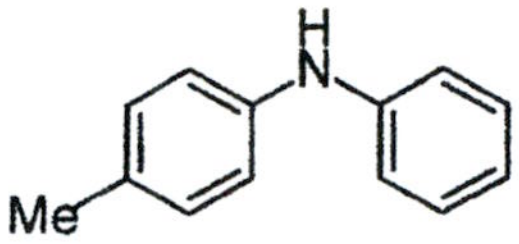 H, N, Me	1	50	95

[a]Isolated yield; average of 2 runs.

Table 4. Palladium-mediated Suzuki-Miyaura reactions of aryl chlorides with phenyl boronic acid.

$$1\ \text{mmol ArX} + 1\text{mmol } PhB(OH)_2 \xrightarrow[\text{1.5 equiv base, 4 mL dioxane}]{\text{0.5 mol \% (SIPr)Pd(allyl)Cl}} \text{Ar-Ph}$$

Entry	ArX	Amine	Product	t, h	T, °C	Yield[a]
1	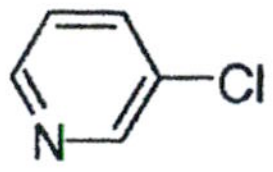 N, Cl	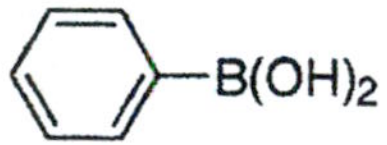 $B(OH)_2$	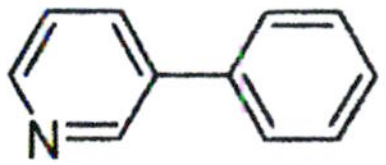 N	1.5	80	95[b]
2	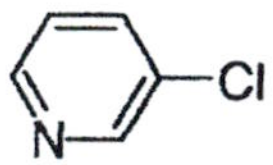 N, Cl	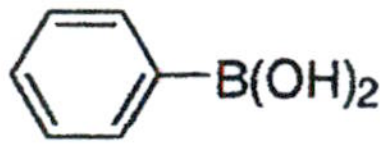 $B(OH)_2$	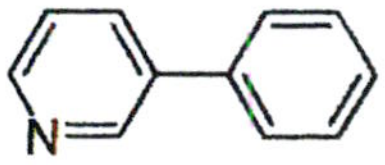 N	1.5	80	97[c]
3	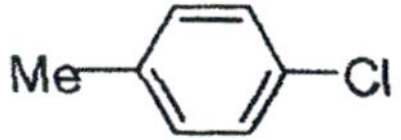 Me, Cl	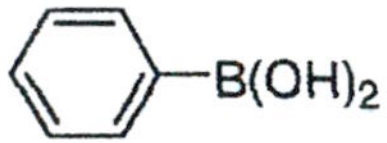 $B(OH)_2$	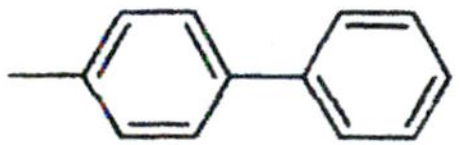	1.5	80	96[c]

[a]Isolated yield; average of 2 runs. [b]Base is Cs_2CO_3. [c]Base is NaO^tBu.

Table 5. Palladium-mediated ketone arylation

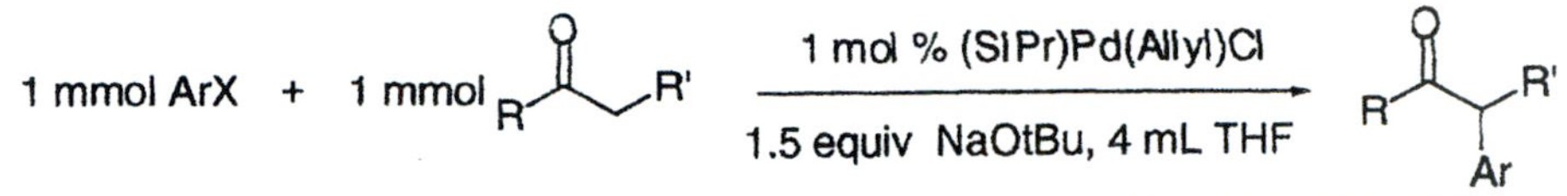

1 mmol ArX + 1 mmol R(C=O)CH$_2$R' → R(C=O)CH(Ar)R'

1 mol % (SIPr)Pd(Allyl)Cl

1.5 equiv NaOtBu, 4 mL THF

Entry	ArX	Amine	Product	t, h	T, °C	Yield[a]
1	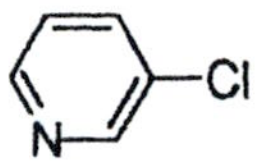	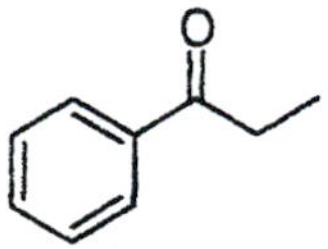	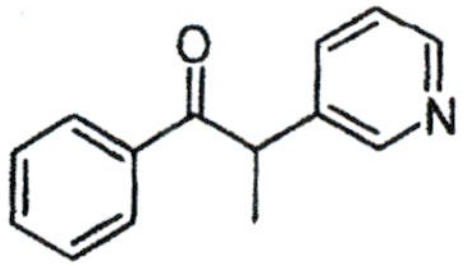	1	80	66
2	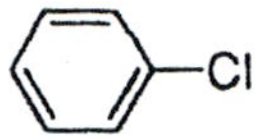	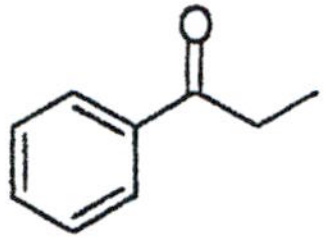	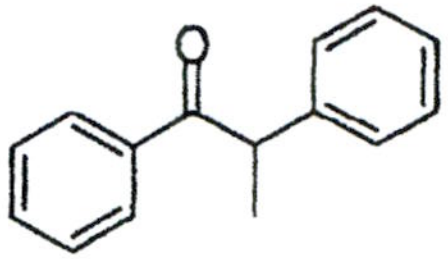	1	50	91
3	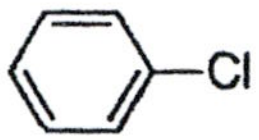	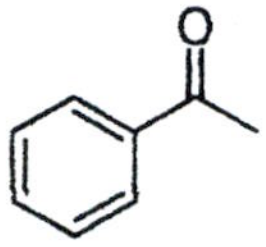	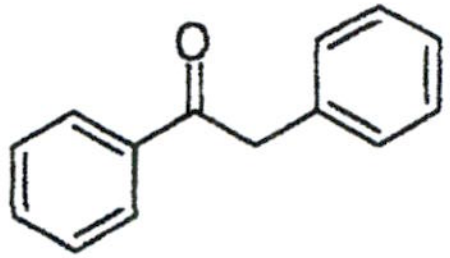	1	80	88
4	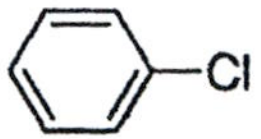		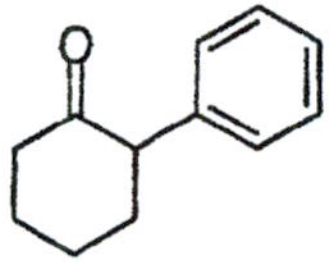	3	80	72

[a]Isolated yield; average of 2 runs.

alcohols or cleavage of acid-sensitive functional groups. Enol esters are convenient acylating agents since the tautomerization of the resulting enolate to the more favored aldehyde or ketone shifts the transesterification equilibrium in the desired direction. Although organometallic catalysts such as $Cp^*_2Sm(thf)_2$[27] and distannoxanes,[28] or the very basic iminophosphoranes[29] have been employed to this end, they either are limited to non acid-sensitive substrates or require high catalyst loading and long reaction times. On the other hand, readily available methyl esters require fairly harsh conditions to enable alcohol deprotection and, at the same time, do not easily undergo transesterification to higher homologues due to the reversibility of the reaction.[30] Utilizing *N*-heterocyclic carbenes (NHC, imidazol-2-ylidenes) (Scheme 1) as nucleophilic catalysts in transesterification reactions leading to the synthesis of various esters.[31]

Scheme 1. *N*-Heterocyclic Carbenes used in this study.

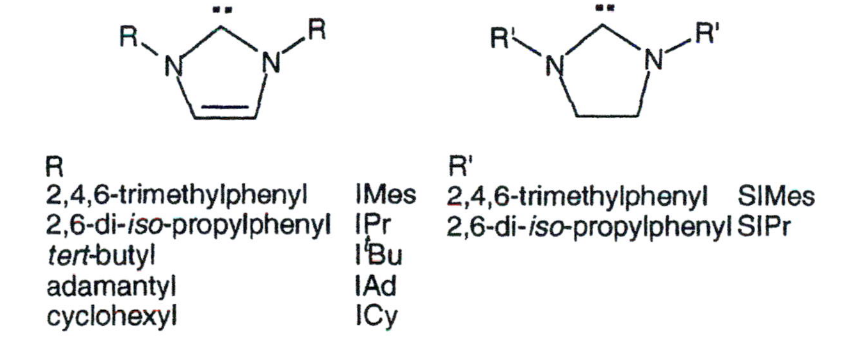

NHC represent a class of ligands with a considerable stabilizing effect in organometallic systems[32] as compared to the widely utilized tertiary phosphines. Our experience with NHC as ligands for transition metal catalyzed processes and the fact that transesterification of enol esters can be affected by basic tertiary phosphines such as PBu_3 or iminophosphoranes, led us to attempt transesterification utilizing NHC. Indeed, as little as 0.5 mol% of the NHC IMes catalyzes the reaction of benzyl alcohol with vinyl acetate in THF, with almost quantitative conversion to benzyl acetate in 5 minutes at room temperature (Scheme 2).

Based on this promising result, the acylation of a commercially available and more challenging substrate, methyl acetate, was tested with different nucleophiles. Two main factors were identified for biasing the reaction in the desired direction. First, 4Å molecular sieves are required to absorb the liberated methanol, leading to quantitative conversion of benzyl alcohol to benzyl acetate in 1 hr with 2.5 mol% of ICy (Table 6, entries 5 and 6). Second, the nature of the nucleophile also influences the efficacy of transesterification.

Scheme 2. IMes-catalyzed acylation of benzyl alcohol with vinyl acetate

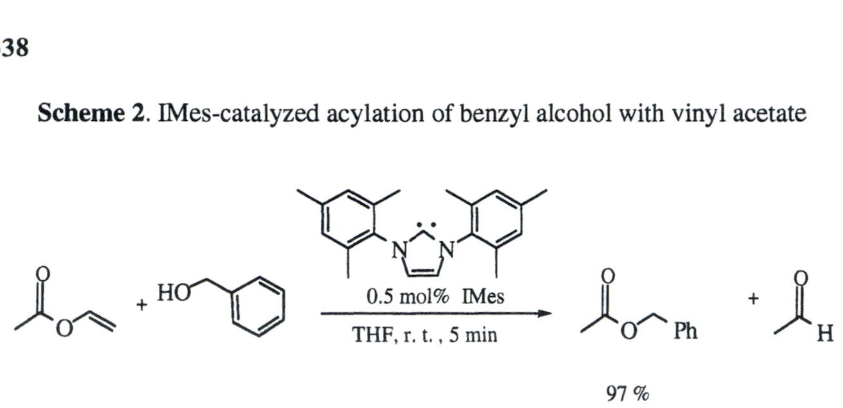

0.5 mol% IMes

THF, r. t., 5 min

97 %

Under similar conditions (2.5 mol% catalyst, 1 ml methyl acetate, 1 hr, molecular sieves) IMes afforded the product in 93% conversion, while IPr led to only a moderate conversion (presumably due to the steric congestion provided by the carbene trending isopropyl groups) (Table 6, entries 1 and 2). The corresponding aryl-substituted imidazolin-2-ylidenes, SIMes and SIPr, afforded the products in only low conversion (entries 3 and 4)[33] while the alkyl-substituted ICy, I^{t}Bu and IAd performed much better in the model reaction affording the product quantitatively (Table 6, entries 5, 7 and 8) presumably owing to their higher nucleophilicity.[34]

Strongly basic species such as DMAP, DABCO[35] and DBU[36] are not effective catalysts for the transesterification of methyl acetate with benzyl alcohol. As expected, the strong inorganic bases NaH and KO^{t}Bu led to high conversions (95%). However, the use of these bases may be problematic for more sensitive substrates.[37]

Having established that NHC are excellent catalysts for transesterification reaction of vinyl acetate/methyl acetate with benzyl alcohol, we investigated various substrates including nitro-containing ones to explore functional group compatibility. In all cases studies thus far excellent conversions are observed.

Efficient selectivity of primary over secondary alcohols is important for acylation catalysts to be useful in organic synthesis. Selective protection of primary alcohols has been achieved using organometallic systems such as distannoxane/enol ester or $Sc(OTf)_3/Ac_2O$.[38] Transesterification selectivity under our standard reaction conditions with respect to primary alcohols was confirmed by the substantially lower activity of 2-butanol to acylation with vinyl acetate. Moreover, benzyl alcohol is almost exclusively acylated by vinyl acetate in the presence of 2-butanol (Scheme 3).

Further explorations into the uses of this NHC catalyst family in organic transformations as well as mechanistic investigations focusing of the mode of action of the NHC catalyst in transesterification and related reactions are ongoing.

Table 6. Transesterification of methyl acetate with benzyl alcohol catalyzed by various nucleophiles

OMe + HO — catalyst (2.5 mol %), 4Å M. S., r. t., 1 hr → O Ph + CH_3OH

Entry	catalyst	yield(%)[a,c]
1	IMes	93
2	IPr	45
3	SIMes	21
4	SIPr	21
5	ICy	100
6	ICy	84[b]
7	IAd	100
8	I^{t}Bu	100

[a]Reaction conditions: 1 mmol benzyl alcohol, 1 mL methyl acetate, 2.5 mol% catalyst, 0.5 g 4 Å molecular sieves, room temperature, 1 hr. [b]No molecular sieves were used. [c]GC yield, average of two runs.

Scheme 3. IMes-catalyzed selective acylation of benzyl alcohol

2 O + HO + HO — 0.5 mol% IMes, THF, r. t., 5 min → O Ph + O + H

20 : 1

Acknowledgements

We gratefully acknowledge partial support of the work described in this contribution by the National Science Foundation, the Petroleum Research Fund administered by the ACS and the Louisiana Board or Regents.

References and Notes

(1) (a)Welton, T. *Chem. Rev.* **1999**, *99*, 2071-2083. (b) Carlin, R.T.; Wilkes, J.S., In *Advances in Nonaqueous Chemistry*; Mamantov, G.; Popov, A. Eds.; VCH Publishing: New York, 1994.

(2) (a)Olivier-Bourbigou, H. In *Aqueous-Phase Organometallic Catalysis: Concepts and Applications*; Cornils, B.; Herrmann, W.A., Eds. Wiley-VCH:Weinheim, 1998.

(3) For example see: Matthews, C.J.; Smith, P.J.; Welton, T.; White, A.J.P.; Williams, D. J. *Organometallics*, **2001**, *20*,3848-3850.

(4) (a) Zhang, C.; Huang, J.; Trudell, M. T.; Nolan, S. P. *J. Org. Chem.* **1999**, *64*, 3804-3805. (b)McGuiness, D.S.; Cavell. K.J.; Skelton, B.W.; White, A.H.; *Organometallics*, **1999**, *18*, 1596-1605. (c) Stauffer, S. R.; Lee, S.; Stambuli, J. P.; Hauck, S. I.; Hartwig, J. F. *Org. Lett,* **2000**, *2*, 1423-1426.

(5) (a)Jackstell, R.; Andreu, G.A.; Frisch, A.; Selvakumar, K.; Zapf, A.; Klein, H.; Spannenberg, A.; Rottger, D.; Briel, O.; Karch, R.; Beller, M. *Angew. Chem., Int. Ed.,* **2002**, *41*, 986-989. (b) Viciu, M. S.; Kissling, R. M.; Stevens, E. D.; Nolan, S. P. *Org. Lett.,* **2002**, *4*, in press.

(6) Other alkoxide bases function well on the test reaction: KO^tBu leads to the desired product in 78% yield in 1.5 hr, as does NaO^tBu (75% in 1.5 hr). The increased solubility of the amylate base in DME may be responsible for the increased reactivity.

(7) For these reactions, a stock solution of **1** and KO^tAm in DME were made and used over the course of several days experiments without apparent loss in activity.

(8) Heck, R. F. *Palladium Reagents in Organic Synthesis*; Academic Press: New York, 1985

(9) See for example: Harris, M. C.; Buchwald, S. L. *J. Org. Chem.* **2000**, *65*, 5327—5333.

(10) Hartwig, J. F.; Kawatsura, M.; Hauk, S. I.; Shaughnessy, K. H.; Alcazar-Roman, L. M. *J. Org. Chem.* **1999**, *64*, 5575—5580

(11) Alcazar-Roman, L. M.; Hartwig, J. F. *J. Am. Chem. Soc.* **2001**, *123*, 12905—12906.

(12) Aerobic conditions: 50 mL vials were charged with **1** and KO^tAm in the glove box. On the bench DME, ArX and amine (used as received) were added to the open vials. The vials were capped to prevent evaporation and suspended in an 80¡ C oil bath for the duration of the reaction.

(13) (a) Li, G. Y.; Zheng, G.; Noonan, A.F. *J. Org. Chem.* **2001**, *66*, 8677—8681. (b) Kwong, F. Y.; Klapars, A.; Buchwald, S. L. *Org. Lett.* **2002**, *4*, 581—584.

(14) Hartwig (reference 4c) was able to load catalyst and excess base in air, evacuate the reaction vessel, and run amination under N_2 with no deleterious effects to substrate conversion to products.

(15) Galardon, E.; Ramdeehul, S.; Brown, J.M.; Cowley, A.; Hii, K.K.; Jutand, A. *Angew. Chem. Int. Ed.*, **2002**, *41*, 1760-1763..

(16) Fellous, R.; Rabine, J.P.; Lizzani-Cuvelier, L.; Luft, R. *Bull. Soc. Chim. Fr.*, **1974**, *11* , 2482-2484.

(17) Tsuji,J.*Transition Metal Reagents and Catalysts*, John Wiley & Sons, **2000**, pp. 109-168 and references therein.

(18) Stanton, S.A.; Felman, S.W.; Parkurst, C.S.; Godleski S.A. *J .Am. Chem. Soc.* **1983**, *105*, 1964-1969.

(19) The $(I^tBu)Pd(PCy_3)$ complex was identified by ^{31}P NMR spectroscopy and compared to the previously isolated complex; Titcomb, L.R.; Caddick, S.; Cloke, F.G.N.; Wilson, D.J.; McKerrecher, D. *Chem. Comm.* **2001**, 1388-1389.

(20) (a) Wolfe, J.P.; Wagaw, S.; Buchwald S.L. *J. Am. Chem. Soc.*,**1996**, *118*, 7215-7216. (b) Driver, M.S.; Hartwig, J.F. *J.Am. Chem. Soc.*, **1996**, *118*, 7217-7218.

(21) See for examples ref 4 a and 4b and B hm, V. P. W.; Gst ttmayr, C. W. K.; Weskamp, T.; Herrmann, W. A. *J. Organomet. Chem.*,**2000**, *595*, 186-190.

(22) (a) Kosugi, M.; Hagiwara, I.; Sumiya, T.; Migita, T. *Bull. Chem. Soc. Jpn.*, **1984**, 242-246. (b) Morgan, J.; Pinhey, J.T.; Rowe, B.A. *J. Chem. Soc., Perkin Trans. 1*, **1997**, 1005-1008. (c) Hamann, B. C.; Hartwig, J. F.; *J. Am. Chem. Soc.* **1997**, *119*, 12382-12383. (d) Palucki, M.; Buchwald, S.L. *J.Am. Chem. Soc.*, **1997**, *119*, 11108-11109. (e) Satoh, T.; Kawamura, Y.; Miura, M., Nomura, M. *Angew. Chem. Int. Ed.*,**1997**, 36, 1740-1742. (f) Culkin, D. A.; Hartwig, J. F. *J. Am. Chem. Soc.*, **2001**, *124*, 5816-5817.

(23) March J. *Advanced Organic Chemistry*, Fourth Edition, John Wiley & Sons, **1992**, pp. 468.

(24) Otera, J. *Chem. Rev.* **1993**, *93*, 1449-1470 and references therein.

(25) (a) Steglich, W.; Hofle, G. *Angew. Chem., Int. Ed. Engl.* **1969**, *8*, 981-983 (b) Shimizu, T.; Kobayashi, R.; Ohmori, H.; Nakata, T. *Synlett.* **1995**, 650-652. (c) D Sa, B.; Verkade, J. G. *J. Org. Chem.* **1996**, *61*, 2963-2966. (d) Vedejs, E.; Diver, S. T. *J. Am Chem Soc.* **1993**, *115*, 3358-3359.

(26) (a) Ishihara, K.; Kubota, M.; Kurihara, H.; Yamamoto, H. *J. Am. Chem. Soc.* **1995**, *117*, 4413-4414. (b) Iqbal, J.; Srivastava, R. R. *J. Org. Chem.* **1992**, *57*, 2001-2007. (c) Miyashita, M.; Shiina, I.; Mukaiyama, T. *Bull. Chem. Soc. Jpn.* **1993**, *66*, 1516-1527. (d) Orita, A.; Tanahashi, C.; Kakuda, A.; Otera, J. *J. Org. Chem.* **2001**, *66*, 8926-8934.

(27) Ishii, Y.; Takeno, M.; Kawasaki, Y.; Muromachi, A.; Nishiyama, Y.; Sakagughi, S. *J. Org. Chem.* **1996**, *61*, 3088-3092.

(28) Orita, A.; Mitsutome, A.; Otera. J. *J. Org. Chem.* **1998**, *63*, 2420-2421.

(29) Ilankumaran, P.; Verkade, J. G. *J. Org. Chem*, **1999**, *64*, 9063-9066.

(30) Ranu, B. C.; Dutta, P.; Sarkar, A. *J. Org. Chem.* **1998**, *63*, 6027-6028.

(31) Recent related reports make use of NHC in polymerization of cyclic esters: Connor, E.F.; Nyce, G.W.; Myers, M.; Mock, A.; Hedrick, J.L. *J. Am. Chem. Soc.* **2002**, *124*, 914-915. and in mediating the asymmetric benzoin condensation: Endeers, D.; Kalfass, U. *Angew. Chem.Int. Ed.* **2002**, *41*, 1743-1745.

(32) (a) Huang, J.; Stevens, E. D.; Nolan, S. P.; Petersen, J. L. *J. Am. Chem. Soc.* **1999**, *121*, 2674-2678. (b) Huang, J.; Schanz, H.-J.; Stevens, E. D.; Nolan, S. P. *Organometallics* **1999**, *18*, 2370-2375.

(33) Arduengo III, A. J.; Calabrese, J. C.; Davidson, F.; Rasika Dias, H. V.; Goerlich, J. R.; Krafczyk, R.; Marshall, W. J.; Tamm, M.; Schmutzler, R. *Helv. Chim. Acta* **1999**, *82*, 2348-2364.

(34) Kim, Y.-J.; Streitwieser,A. *J. Am. Chem. Soc.* **2002**, *124*, 5757-5761.

(35) Aggrawal, V. K.; Dean, D. K.; Mereu, A.; Williams, R. *J. Org. Chem.* **2002**, *67*, 510-514.

(36) Aggrawal, V. K.; Mereu, A. *Chem. Commum.* **1999**, 2311-2312.

(37) Stanton, M. G.; Gagn , M. R. *J. Org. Chem.* **1997**, *62*, 8240-8242.

(38) Procopiou, P. A.; Baugh, S. P. D.; Flack, S. S.; Inglis, G. G. A. *J. Org. Chem.* **1998**, *63*, 2342-2347.

Photochemistry

Chapter 28

An Overview of Photochemistry in Ionic Liquids

Richard M. Pagni

Department of Chemistry, University of Tennessee, Knoxville, TN 37996–1600

Photochemistry in ionic liquids, taken mainly from the authors own work, will be described. It will be seen that ionic liquids provide interesting, unusual and often unique environments in which to carry out photochemistry.

Introduction

Organic, inorganic, organometallic, catalytic, and polymer chemistry is currently being actively studied in ionic liquids.[1-6] Surprisingly, very little of this activity deals with photochemistry. What has been published, however, is very interesting, even unique in some cases. This suggests a rich future for this subject. This article will give an overview of what is presently known about the photophysical and photochemical properties of ionic liquids and solutes therein, with emphasis on work from the author's laboratory. Only photochemistry in imidazolium- and pyridinium-based ionic liquids, the mostly widely used, will be considered.

Results and Discussion

Salt Solutions

A good place to begin is by examining what effect dissolved salts have on the properties of solutions because ionic liquids may be considered "neat" salt solutions. Dissolving a salt in a liquid has the effect of increasing the internal pressure[7], a thermodynamic property, and polarity[8], an empirically derived property, of the resulting solution. Internal pressure, which is known for a few molten salts[9] but not a single imidazolium- or pyridinium-based ionic liquid, does not mimic applied external pressure[10] and has little effect on the rates of reactions.[8] The large internal pressures of lithium perchlorate in diethyl ether,[7] an ionic liquid when the lithium perchlorate concentration is above 4.25 M, has little effect on the lowest B6B* absorption and fluorescence bands of the nonpolar anthracene and the second B6B* absorption and fluorescence bands of the polar azulene.[10] Increases in solution polarity due to added salts does affect the rates of all sorts of reactions including the S_N2 reaction, however.[11]

Photoreactions proceeding through radical ion pairs formed via photoinduced electron transfer[12] are often influenced by added salts.[11] This often occurs by exchange of the anion in the ion pair with the much less basic anion, often ClO_4^-, of the added salt. Many ionic liquids in current use contain weakly basic anions such as PF_6^-, BF_4^-, and $(CF_3SO_2)_2N^-$ (Tf_2N^-).

Relevant Properties of Ionic Liquids

Polarity is a useful, well-known, if poorly defined, property of liquids whose magnitude depends on non-specific electrostatic interactions between charged and dipolar sites on a solute and dipoles on a solvent. It can be defined and thus quantitated in numerous ways using the spectroscopic properties of selected probe solutes. Many of these polarity indices such as E_T, ∀, ∃, B*, and P_y, which involve electronic transitions (absorption, emission), have been applied to ionic liquids and molten salts.[13-25] Several pertinent results arise from these studies. (1) [Emim][Tf_2N] has a very low dielectric constant, a property which cannot be measured directly for ionic liquids.[17] One may presume that other ionic liquids have low dielectric constants as well. (2) By and large, the imidazolium- and pyridinium-based ionic liquids have polarities similar to those of CH_3CN, DMSO, and the lower alcohols.[13,18-21] If an aliphatic side chain is replaced with one containing an hydroxyl group, the polarity of the ionic liquid goes up significantly.[25] (3) Adding small amounts of water to an ionic liquid affects the polarity of the ionic liquid.[19,22,24] (4) The organic cation seems to be the major contributor to the polarity of the ionic liquid.[20] (5) Polarity indices do not reflect polarity alone. Properties such as viscosity and polarizability are also influential.[21]

Viscosity, which is the resistance to flow in a liquid, is known as a function of temperature for all common liquids.[26] The rates of diffusion-controlled bimolecular reactions such as the reaction of NO_2^+ with arenes, quenching of excited states, and photoinduced electron transfer are very much solvent dependent because viscosity and rate of diffusion are interrelated mathematically.[27] Often the internal motions of molecules, excited states, and transients are also viscosity dependent.[28-30]

The viscosities of many imidazolium- and pyridinium-based ionic liquids are known and are much higher than for most common solvents.[31-34] The higher viscosities are undoubtedly due to intermolecular associations between cations and anions by hydrogen bonding, electrostatic attraction and other effects. Ionic liquids thus are excellent media to look for effects where diffusion and molecular motion are important. The effects may be unusual because ionic liquids of necessity are made up of a 1:1 molar ratio of cations and anions. For example, the cations and anions respond differently to rapid changes in structure and polarity of a solute brought about by electronic excitation.[23,34]

Fluorescence quenching of the lowest singlet excited states of six alternant polycyclic aromatic hydrocarbons (PAHs) by nitromethane occurred at the rate of diffusion in [bmim][PF_6], presumably by an electron transfer mechanism.[35] Six non-alternant PAHs, on the other hand, were not so quenched. This lack of quenching is interesting because quenching of the same PAHs did occur under other conditions. MacLean and Gordon have also observed diffusion-controlled quenching in several ionic liquids by transient spectroscopy.[36,37] Several pulse radiolysis studies have been carried out in ionic liquids.[38-41] A wealth of kinetic data on the reactions of transient species has thus been generated. The reaction of CF_3X with pyrene in $(CH_3)(n\text{-Bu})_3N^+(Tf_2N^-)$, for example, occurs at the rate of diffusion.

Organic molecules rarely phosphoresce in fluid solution at room temperature because their long-lived triplet states are easily quenched (self-quenching, trace O_2). 1-Bromonaphthalene, surprisingly, phosphorseces in degassed [emim][Tf_2N] at room temperature.[17] This may be due to the fact that the rates of the biomolecular quenching reactions are retarded in the viscous medium. Added O_2 quenches the phosphorescene of 1-bromonaphthalene, as expected, but not its fluorescence because of the short lifetime of the singlet state.

Virtually no photophysical data are available for any ionic liquid or molten salt. It is generally believed that imidazolium- and pyridinium-based ionic liquids absorb strongly below 300 nm.[42] [Emim][Cl/$AlCl_3$] starts absorbing around 300 nm[43] but [bmim][Tf_2N] doesn't absorb until around 240 nm.[44] Based on the absorption spectrum of the pyridinium ion, pyridinium-based ionic liquids will absorb starting around 300 nm.[45] This is the reason that only substrates having absorption bands above 300 nm have been studied. Owing to the expected large singlet and triplet excitation energies of the ionic liquids,[46,47] quenching of substrates' excited states by ionic liquids is not likely. In fact there is no published evidence that any photoreaction is inhibited by the ionic liquids in any way. Sensitization by the ionic liquids is possible but unlikely because the high concentration of the imidazolium and pyridinium salts in the ionic liquids ensures that self-quenching will dominate.

The ionic liquids may themselves be photoactive. There is one report that photolysis of [bmim][PF_6] results in the formation of unidentified products.[48] Pyridinium and imidazolium salts in solution are known to be photoactive.[49,50]

Photoreactions in Ionic Liquids

One can loosely divide photoreactions in ionic liquids into two categories: those in which the ionic liquid is not directly involved in the chemistry and those in which they are. The discussion will begin with chemistry in the first category.

Gordon and McLean have shown that election transfer from the excited state of the bipyridyl complex, Ru(bypy)$^{+2}$, to methylviologen (MV^{+2}) occurs at the rate of diffusion in butylmethylimidazolium hexafluorophosphate [bmim][PF_6] (Scheme 1, line 1).[36] The energy transfer from the triplet excited state of benzophenone (BP^{3*}) to naphthalene (N), which does not involve electron transfer, occurs at the rate of diffusion in several imidazolium-containing ionic liquids (Scheme 2, line 2).[37]

$$\mathrm{Ru(bpy)_3^{+2*}} + \mathrm{MV^{+2}} \xrightarrow{k_d} \mathrm{Ru(bpy)_3^{+3}} + \mathrm{MV^{+\bullet}}$$

$$\mathrm{BP^{3*}} + \mathrm{N} \xrightarrow{k_d} \mathrm{BP} + \mathrm{N^{3*}}$$

Scheme 1: Electron and energy transfer in ionic liquids

The photochemistry of anthracene (AN) in deoxygenated, basic [emim][Cl/$AlCl_3$] afforded the 4 + 4 dimer,[51] a reaction identical to that seen in more conventional solvents (Scheme 2).[52] In oxygenated, basic [emim]Cl/$AlCl_3$], on the other hand, AN yielded anthraquinone, 9-chloroanthracene, and 9,10-dichloroanthracene.[53] Although the details of this chemistry have not been worked out, it is clear that the chlorinated products arise, in part, by the reaction of the radical cation of AN (AN^{+X}) with Cl^-. The AN^{+X} likely is formed by the electron transfer from the excited state of anthracene to O_2. The ionic liquid thus provides a polar medium to facilitate the electron transfer.

The photochemistry of AN in acidic [emim][Cl/$AlCl_3$] is quite different than seen in the basic melt, yielding an array of monomeric and dimeric

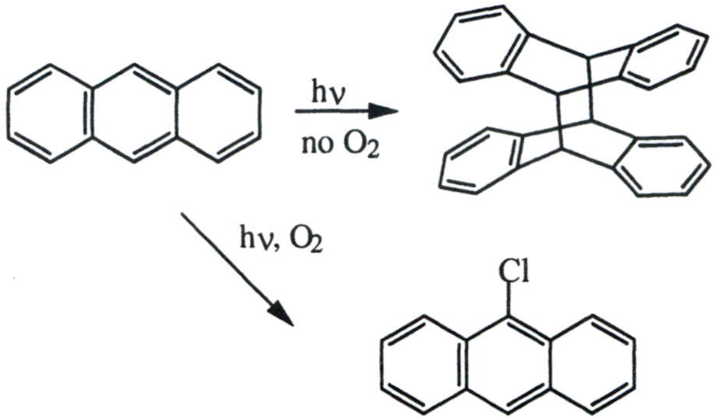

hν
no O_2
hν, O_2
Cl

Scheme 2: Photochemistry of antracene in basic [emim][Cl/AlC.$_3$]

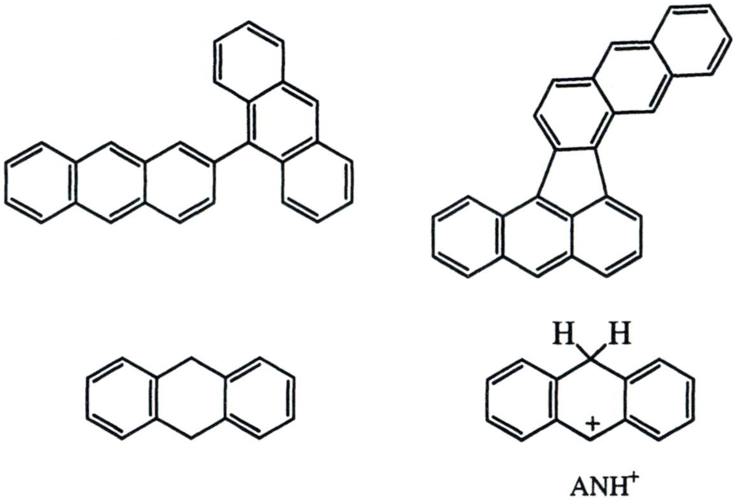

H H
+
ANH^+

Figure 1: Photoproducts from anthracene in acidic [emim][Cl/AlCl$_3$]

reduced, neutral and oxidized products (Figure 1).[51] This is a consequence of the fact that residual HCl is a powerful Brønsted acid[43,54] and protonates AN to yield a small amount of ANH^+, a good electron acceptor from excited states. Furthermore, with the powerful Lewis acid $Al_2Cl_7^-$ present in the ionic liquid, none of the myriad positively charged and neutral molecules generated in the photochemistry will react with the solvent; they will only react with each other. Thus the reaction is inititated by electron transfer from the singlet excited state of AN to ANH^+ followed by a series of bimolecular electron transfer, hydrogen transfer and coupling reactions (Scheme 3). It was possible to mimic this photochemistry using a mixture of the strong acid, trifluoromethanesulfonic acid, and the much weaker acid, trifluoroacetic acid.[55] Clearly the type of chemistry described in this paragraph will not occur in the contemporary one-component ionic liquids, but it might if a Lewis acid were added - [emim][BF_4] + BF_3, for example

$$AN + HCl \rightleftharpoons ANH^+\ Cl^-$$

$$ANH^+ + AN \xrightarrow{h\nu} ANH\bullet + AN^{+\bullet}$$

$$ANH\bullet + AN^{+\bullet} \longrightarrow \longrightarrow \text{Products}$$

Scheme 3: Initial reactions of anthracene in acidic [emim][Cl/AlCl

Jones and coworkers have described a synthetically useful photoreaction in imidazolium-containing ionic liquids, the reduction of benzophenones to benzhydrols by primary amines (Scheme 4).[56] The reaction is initiated by hydrogen abstraction by the ketone to form a radical pair. Electron transfer then yields an ion pair, a reaction facilitated by the polarity of the ionic liquid. Proton transfer within the ion pair completes the reaction. Surprisingly, benzpinacols are formed instead in ionic liquids not containing the imidazolium ring, *sec*-butylammonium trifluoroacetate and *iso*-propylammonium nitrate. Perhaps the imidazolium cation plays a more active role in the chemistry than the mechanistic scheme implies, as is seen in several cases described below.

McLean, Gordon and coworkers have investigated the kinetics of the hydrogen atom abstraction by benzophenone triplet excited state (BP^*) from several imidazolium-containing ionic liquids (Scheme 5).[57] Although no products were isolated, it is reasonable to assume that the hydrogen abstraction occurs on the alkyl substituents of the imidazolium cations as shown below. Interestingly, the energy of activation for the abstraction is considerably higher than seen for abstraction from 1-butanol, cyclohexane and toluene. Why there is a difference is currently unclear.

UCl_6^-, which is stable in acidic [emim][Cl/$AlCl_3$] is reduced when photolyzed into its ligand-to-metal change transfer band, yielding UCl_4, Cl^- and ClX.[58] The chlorine atom apparently reacts with $emim^+$ to form unidentified

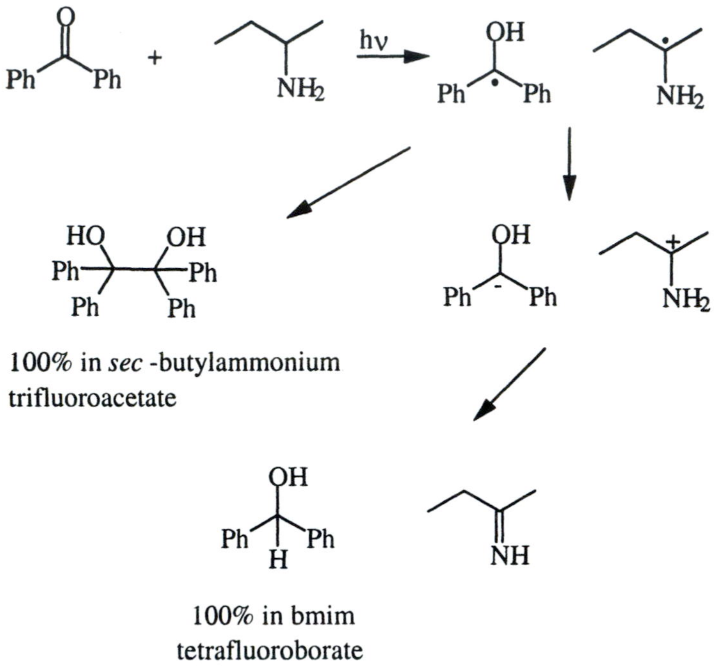

Ph Ph + NH2 hν OH Ph Ph NH2

HO OH Ph Ph Ph Ph

100% in *sec* -butylammonium trifluoroacetate

OH Ph Ph NH2

OH Ph Ph H NH

100% in bmim tetrafluoroborate

Scheme 4: Photoreaction of benzophenone with sec-butylamine in [bmim][BF_4]

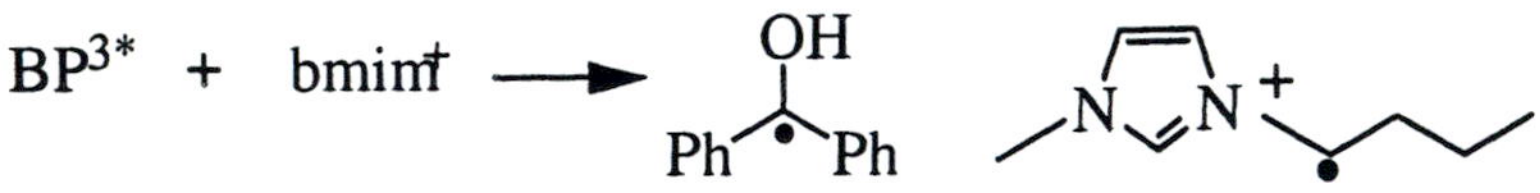

BP^{3*} + $bmim^+$ → OH Ph Ph N N +

Scheme 5: Hydrogen atom abstraction in ionic liquid

products. What makes this hypothesis plausible is the fact that UCl_6^- is photochemically inert in acidic NaCl/$AlCl_3$ where ClX has no chemical outlet.

The first photochemical reaction in an ionic liquid was reported by Osteryoung and coworkers in 1978.[59] Close to a dozen tris Fe(II) complexes including $Fe(PMM)_3^{+3}$ (PMM = 2-pyridinecarboxaldehyde *N*-methylimine) were irradiated in Lewis acidic [*N*-ethylpyridinium][Br/$AlCl_3$]. The Fe(II) complexes were oxidized to the corresponding Fe(III) complexes, with the concomitant reduction of the pyridinium cation to form a dimer (Scheme 6). The reaction occurs by electron transfer from the excited state of the Fe(II) complex to the easily reduced pyridinium cation. For reactions in which the electron transfer was exothermic, the reactions occurred with a quantum yield of 1. Endothermic reactions occurred less efficiently.

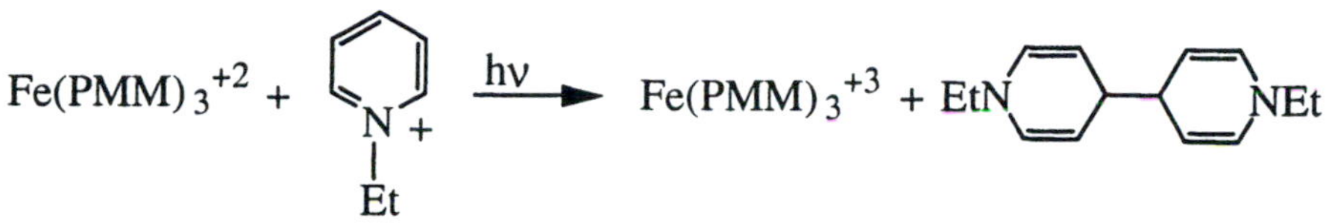

$Fe(PMM)_3^{+2}$ + [N-ethylpyridinium, N+ Et] $\xrightarrow{h\nu}$ $Fe(PMM)_3^{+3}$ + EtN[dimer]NEt

Scheme 6: Photoinduced electron transfer in ionic liquid

Photoinduced electron transfer (PET), as seen above in the examples of Osteryoung, should occur readily in imidazolium- and pyridinium-containing ionic liquids because aromatic cations in general are good election acceptors in such processes.[12] Pyridinium-containing ionic liquids should be especially good in this regard because pyridinium ions are much more easily reduced than imidazolium ions. What is then desirable is a substrate whose PET to the aryl cation is exothermic which is indicative of a fast reaction. Fortunately, it is possible to calculate the free energy of PET using the well-known Rehm-Weller equation.[60]

9-Methylanthracene (9-$ANCH_3$) is an ideal substrate to look for PET in ionic liquids (Figure 2). It is more easily oxidized than AN. The PET reactions to $emim^+$ and the *N*-butylpyridinium cation (bp^+) are exothermic. The resulting radical cation, 9-$ANCH_3^{+X}$, is quite acidic and thus readily deprotonated by Cl^- in basic [emim][Cl/$AlCl_3$] and [bp][Cl/$AlCl_3$], the media used in this study, to form 9-$ANCH_2^X$. In fact five products containing 9-$ANCH_2$ are formed in each ionic liquid. Three of them are shown below (Figure 2).[61]

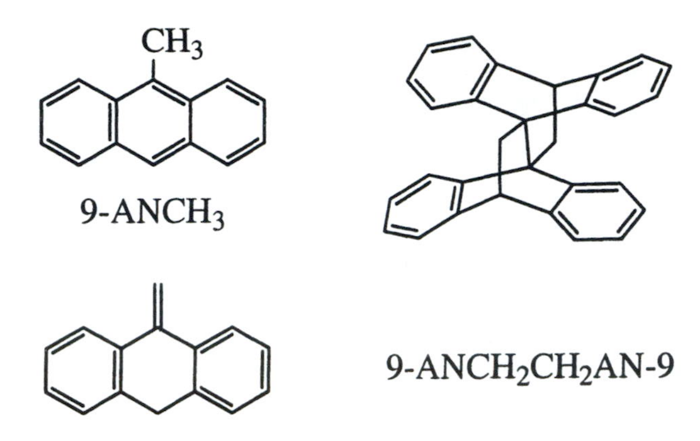

Figure 2: 9-Methylanthracene and three of its photoproducts

Based on a variety of experiments including the use of 9-$ANCD_3$ and 2-deuterio- and 2,4,5-trideuterio [emim][Cl/$AlCl_3$], it is clear that the following transients are formed in the photochemistry (Figure 3).

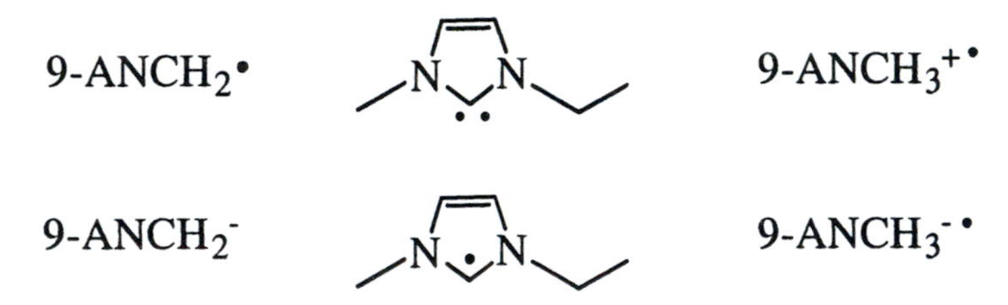

Figure 3: Transients formed in the photochemistry of 9-AnCH$_3$

Of particular note is the formation of 9,10-dihydroanthracene in [emim][Cl /$AlCl_3$], but not [bp][Cl/$AlCl_3$]. This is a consequence of the fact that emimX formed by PET is a potent reducing agent and reduces 9-$ANCH_3$ to its radical anion which eventually yields the reduced hydrocarbon. bp^+ is a better electron acceptor than $emim^+$ but bpX is not powerful enough to reduce 9-$ANCH_3$ to its radical anion. bpX thus reacts with 9-$ANCH_3$ in other ways to give new products such as 9-butyl-10-methylanthracene. One has the very unusual situation in [emim][Cl/$AlCl_3$] where the substrate is oxidized early in the photoreaction and reduced at a later stage.

In the course of the 9-$ANCH_3$ study the photochemistry of 9-(chloromethyl)-anthracene was studied in basic [emim][Cl/$AlCl_3$] (Scheme 7). The initial reaction yields ClX and 9-$ANCH_2X$, which dimerizes and abstracts hydrogen from $emim^+$ to form 9-$ANCH_3$. Interestingly, there is no evidence for the formation of Cl^- and 9-$ANCH_2^+$, as one might expect in a polar solvent.

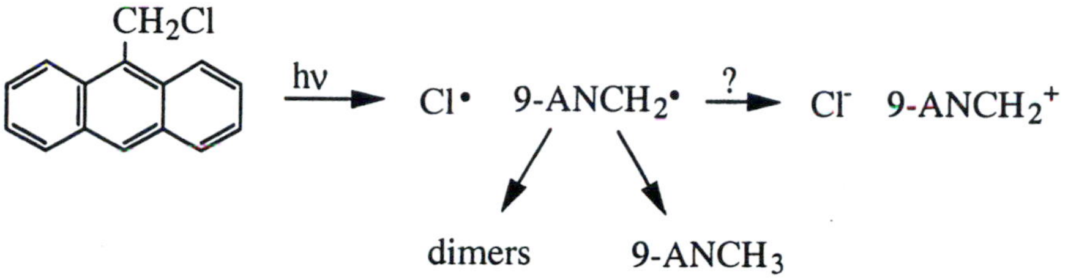

CH_2Cl (9-(chloromethyl)anthracene) $\xrightarrow{h\nu}$ Cl• 9-$ANCH_2$• $\xrightarrow{?}$ Cl^- 9-$ANCH_2^+$

dimers 9-$ANCH_3$

Scheme 7: Photoreaction of 9-chloroanthracene in basic [emim][Cl/Al/Cl_3]

The cis-trans isomerization of *cis*- and *trans*-stilbene is the most studied of all photochemical reactions (Scheme 8).[62] The isomerization can occur by several well-established mechanisms. In the singlet mechanism direct irradiation of the isomers affords singlet excited states that collapse to an excited species in which the two methylene groups are perpendicular (p^{1*}). Radiationless transition of p^{1*} yields p^1 on the ground state energy surface which then decays to *cis*- and *trans*-stilbene. In the triplet mechanism the isomerization is formally similar to the singlet mechanism except that triplet states formed by sensitization are involved. Isomerization will also occur through radical cations formed by PET.

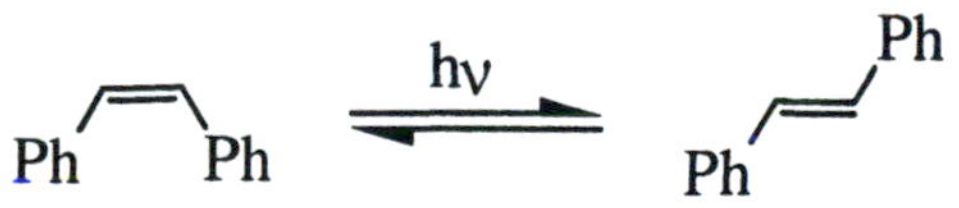

Ph–CH=CH–Ph (cis) $\underset{}{\overset{h\nu}{\rightleftharpoons}}$ Ph–CH=CH–Ph (trans)

Scheme 8: Photointerconversion of cis- and trans-stilbene

Because of the roughly 1 volt difference in the reduction potential of $emim^+$ and bp^+, it should be possible to isomerize the two hydrocarbons in basic [emim][Cl/$AlCl_3$] and [bp][Cl/$AlCl_3$] by different mechanisms. This in fact is observed (Table I).[61] Irradiation of *cis*- and *trans*-stilbene in [emim][Cl/$AlCl_3$] occurs by the singlet mechanism although the distribution of the two hydrocarbons at the photostationary state is a bit unusual. Isomerization in [bp][Cl/$AlCl_3$] occurs via the radical cations of the hydrocarbons formed by PET of the excited states of the hydrocarbons to bp^+.

Table I. Photochemistry of stilbenes in ionic liquids

solvent	photostationary state (%) *cis*-stilbene	*trans*-stilbene	other products (%)	mechanism
[emim][Cl/$AlCl_3$]	41.1	52.2	6.7	singlet
[bp][Cl/$AlCl_3$]	0.6	99.4	0	radical cation

Conclusions

Photochemistry in ionic liquids is now a well-established discipline. It is already clear that there are features in the photophysics and photochemistry of substrates which are distinctive to ionic liquids. Of particular note is that ionic liquids are ideal media to explore photoinduced electron transfer reactions.

References

1. Pagni, R. M. In. *Adv. Molten Salt Chem.*; Mamantov, G., Mamantov, C., Braunstein, J., Eds.; Elsevier: Amsterdam, 1987; Vol. 6, p. 211.
2. Chauvin, Y.; Olivier-Bourbigou, H. *Chemtech.* **1995**, *25*, 26.
3. Seddon, K. R. *J. Chem. Tech. Biotechnol.* **1997**, *68*, 351.
4. Welton, T. *Chem. Rev.* **1999**, *99*, 2071.
5. Wasserscheid, P.; Keim, W. *Angew. Chem. Int. Ed. Engl.* **2000**, *39*, 3772.
6. Gordon, C. M. *Appl. Cat. A* **2001**, 221, 101.
7. Kumar, A. *Pure Appl. Chem. **1998**,* 70, 615.
8. Reichardt, C. *Solvents and Solvent Effects in Organic Chemistry*; 2nd ed.; VCH: Weinheim, 1990.
9. Barton, A. F. M. *Handbook of Solubility Parameters and Other Cohesive Parameters*; CRC Press: Boca Raton, 1983.
10. Springer, G.; Elam, C.; Edwards, A.; Bowe, C.; Boyles, D.; Bartmess, J.; Chandler, M.; West, K.; Williams, J.; Green, J.; Pagni, R. M.; Kabalka, G. W. *J. Org. Chem.* **1999**, *64*, 2202.
11. Loupy, A.; Tchoubar, B. *Salt Effects in Organic and Organometallic Chemistry*; VCH: Weinheim, 1991.
12. Kavarnos, G. J. *Fundamentals of Photoinduced Electron Transfer*; VCH: Weinheim, 1993.

13. Harrod, W. B.; Pienta, N. J. *J. Phys. Org. Chem.* **1990**, *3*, 534.
14. Poole, S. K.; Shetty, D. H.; Poole, C. *Anal. Chim. Acta* **1989**, *218*, 241.
15. Bort, E.; Meltsin, A.; Happert, D. *J. Phys. Chem.* **1994**, *98*, 3295.
16. Reichardt, C.; Harbusch-Görnert, E. *Liebigs. Ann. Chem.* **1993**, 721.
17. Bonhôte, P.; Dias, A.-P.; Papageorgiou, N.; Kalyanasundaram, K.; Grätzel, M. *Inorg. Chem.* **1996**, *35*, 1168.
18. Carmichael, A. J.; Seddon, K. R. *J. Phys. Org. Chem.* **2000**, *13*, 591.
19. Aki, S.N.V.K.; Brennecke, J. F.; Samanta, A. *Chem. Commun.* **2001**, 413.
20. Muldoon, M. J.; Gordon, C. M.; Dunkin, I. R. *J. Chem. Soc., Perkin Trans.2* **2001**, 433.
21. Fletcher, K. A.; Stoney, I. A.; Hendricks, A. E.; Pandey, S.; Pandey, S. *Green Chem.* **2001**, *3*, 210.
22. Fletcher, K. A.; Pandey, S. *Appl. Spectr* **2002**, *56*, 266.
23. Karmakar, R.; Samanta, A. *J. Phys. Chem. A* **2002**, *106*, 4449.
24. Baker, S. N.; Baker, G. A.; Bright, F. V. *Green Chem.* **2002**, *4*, 165.
25. Dzyuba, S. V.; Bartsch, R. A. *Tetrahedron Lett.* **2002**, *43*, 4657.
26. Riddick, J. A.; Bunger, W. B.; Sakano, T. K. *Organic Solvents Physical Properties and Methods of Purification*; Wiley: New York, 1986.
27. Caldin, E. F. *The Mechanisms of Fast Reactions in Solution*; IOS Press: Amsterdam, 2001.
28. Zand, A.; Park, B.-S.; Wagner, P. J. *J. Org. Chem.* **1997**, *62*, 2326.
29. Adam, W.; Diedering, M.; Trofimor, A. *Phys. Chem. Chem. Phys.* **2002**, *4*, 1036.
30. Inoue, Y.; Jiang, P.; Tsukada, E.; Wada, T.; Shimuzu, H.; Tai, A.; Ishikawa, M. *J. Am. Chem. Soc.* **2002**, *124*, 6942.
31. Wilkes, J. S.; Levisky, J. A.; Wilson, R. A.; Hussey, C. L. *Inorg. Chem.* **1982**, *21*, 1263.
32. Fannin, A. A., Jr.; Floreani, D. A.; King, L. A.; Landers, J. S.; Piersma, B. J.; Stech, D. J.; Vaughn, R. L.; Wilkes, J. S.; Williams, J. L. *J. Phys. Chem.* **1984**, *88*, 2614.
33. Dzyuba, S. V.; Bartsch, R. A. *Chem. Phys. Chem.* **2002**, *3*, 161.
34. Karmakar, R.; Sumanta, A. *J. Phys. Chem. A.* **2002**, 106, 6670.
35. Fletcher, K. A.; Pandey, S.; Storey, I. K.; Hendricks, A. E.; Pandey, S. *Anal. Chim. Acta* **2002**, *453*, 89.
36. Gordon, C. M.; McLean, A. J. *Chem. Commun.* **2000**, 1395.
37. McLean, A. J.; Muldoon, M. J.; Gordon, C. M.; Dunkin, I. R. *Chem. Commun.* **2002**, 1880.
38. Behar, D.; Gonzales, C.; Neta, P. *J. Phys. Chem. A.* **2001**, *105*, 7607.
39. Marcinek, A.; Zielonka, J.; Gebicki, J.; Gordon, C. M.; Dunkin, I. R. *J. Phys. Chem. A.* **2001**, *105*, 9305.
40. Behar, D.; Neta, P.; Schultheisz, C. *J. Phys. Chem. A.* **2002**, *106*, 3139.
41. Grodkowski, J.; Neta, P. *J. Phys. Chem. A.* **2002**, *106*, 5468.

42. Owens, G. S.; Durazo, A.; Abu-Omar, M. M. *Chem. Eur. J.* **2002**, *8*, 2053.
43. Smith, G. D.; Dworkin, A. S.; Pagni, R. M.; Zingg, S. P. *J. Am. Chem. Soc.* **1989**, *111*, 525.
44. Lancaster, N. L.; Welton, T.; Young, G. B. *J. Chem. Soc., Perkin Trans. 2* **2001**, 2267.
45. Jaffé, H. H.; Orchin, M. *Theory and Applications of Ultraviolet Spectroscopy*; Wiley: New York, 1962.
46. Ferre, Y.; Vincent, E. J.; Larive, H.; Metzger, J. *J. Chim. Phys. Physicochim. Biol.* **1974**, *71*, 329.
47. Murov, S. L. *Handbook of Photochemistry*; Marcel Dekker: New York, 1973.
48. Ozawa, R.; Hamaguchi, H. *Chem. Lett.* **2001**, 736.
49. Byun, Y.-S.; Jung, S.-H.; Pak, Y.-T. *J. Heterocycl. Chem.* **1995**, *32*, 1835.
50. King, R. A.; Lüthi, H. P.; Schaefer, H. F. III; Glarner, F.; Burger, U. *Chem. Eur. J.* **2001**, *7*, 1734.
51. Hondrogiannis, G.; Lee, C. W.; Pagni, R. M.; Mamantov, G. *J. Am. Chem. Soc.* **1993**, *115*, 9828.
52. Cowan, D. O.; Drisko, R. L. *Elements of Organic Photochemistry*; Plenum: New York, 1976.
53. Pagni, R. M.; Mamantov, G.; Lee, C. W.; Hondrogiannis, G. *Proc.-Electrochem. Soc.* **1994**, *94-13*, 628.
54. Smith, G. P.; Dworkin, A. S.; Pagni, R. M.; Zingg, P. S. *J. Am. Chem. SOC.* **1989**, *111*, 5075.
55. Pagni, R. M.; Mamantov, G.; Hondrogiannis, G.; Unni, A. *J. Chem. Res. (S)* **1999**, 487.
56. Reynolds, J. L.; Erdner, K. R.; Jones, P. B. *Organic Lett.* **2002**, *4,* 917.
57. Muldoon, M. J.; McLean, A. J.; Gordon, C. M.; Dunkin, I. R. *Chem. Commun.* **2001**, 2364.
58. Dai, S.; Toth, L. M. *Proc. Electrochem. Soc.* **1998**, *98-11*, 261.
59. Chum, H. L.; Koran, D.; Osteryoung, R. A. *J. Am. Chem. Soc.* **1978**, *100*, 310.
60. Rehm, D.; Weller, A. *Isr. J. Chem.* **1970**, *8*, 259.
61. Lee, C.; Winston, T.; Unni, A.; Pagni, R. M.; Mamantov, G. *J. Am. Chem. Soc.* **1996**, *118*, 4919.
62. Görner, H.; Kuhn, H. J. *Adv. Photochem.* **1995**, *19*, 1.
63. Lee, C.; Mamantov, G.; Pagni, R. M. *J. Chem. Res. (S)* **2002**, 122.

Chapter 29

Diffusion-Controlled Reactions in Room Temperature Ionic Liquids

Charles M. Gordon[1], Andrew J. McLean[2,*], Mark J. Muldoon[1], and Ian R. Dunkin[1]

[1]Department of Pure and Applied Chemistry, University of Strathclyde, 295 Cathedral Street, Glasgow G1 1XL, Scotland, United Kingdom
[2]Department of Chemistry and Chemical Engineering, University of Paisley, Paisley PA1 2BE, Paisley, Scotland, United Kingdom

Laser flash photolysis has been used to determine bimolecular rate constants for the reaction of triplet excited state benzophenone with naphthalene in a range of room temperature ionic liquids. Comparison of the activation energies for this reaction with the activation energies for viscous flow indicated that the reaction operates under diffusion control in all liquids and at all temperatures studied. Using these data, we have obtained values for the limiting bimolecular rate constants between neutral reactants in these novel solvents. The importance of reliable k_d values in mechanistic studies is illustrated by a photochemical Diels Alder reaction. We also report the experimental procedures necessary for the preparation of colourless "spectroscopic grade" ionic liquids.

1. Introduction

Room temperature ionic liquids (RTILs) are attracting much excitement as useful media in which to carry out organic transformations, as well as finding many other exciting applications (*1*). Recent examples of the range of different possibilities can be found elsewhere in this volume. Despite the wide range of reactions studied to date in room temperature ionic liquids (RTILs), there are very few reports of quantitative reaction rate information for such systems. A pre-requisite of quantitative data of this type is the knowledge of upper (*i.e.* diffusion) limits for bimolecular rate constants in any given solvent. To date, these upper limits have been *estimated* in RTILs using equation 1:

$$k_d = 8000RT/3\eta \qquad (1)$$

where k_d is the diffusion controlled bimolecular rate constant, R is the gas constant in $JK^{-1}mol^{-1}$, T is the absolute temperature, and η the viscosity in cP at that temperature (2). This equation is known to overestimate k_d in many conventional solvents and its application to RTILs must therefore be treated with caution, particularly where fast time scale kinetics are operative such as in most photochemical systems.

We report here our efforts to establish reliable, experimentally derived estimates of k_d in several RTILs together with an example of mixed-regime kinetics that can be interpreted only when knowledge of k_d exits. The reaction systems employed are photochemically based. However, the kinetic treatments used are of general applicability to any reaction system and as such, these results should be of interest to anyone interested in carrying out detailed, quantitative mechanistic studies in RTILs derived from the imidazolium cation.

Our studies have concentrated on using ionic liquids based on the widely used 1-alkyl-3-methylimidazolium cation and the related 2-methyl substituted analogue. RTILs based on these cations remain the most commonly used for most applications, and so the results are likely to be of most general interest. We believe, however, that many of our results and conclusions are likely to be applicable to RTILs based on almost any cation or anion. The anions chosen were $[PF_6]^-$ and $[(CF_3SO_2)_2N]^-$ as they gave RTILs that were immiscible with water, which are relatively easy to prepare in a pure "spectroscopic grade" form. Details of our methodology are given in the experimental section below.

To emphasise the general applicability of our results, we will refer to generic reaction Scheme 1 shown below.

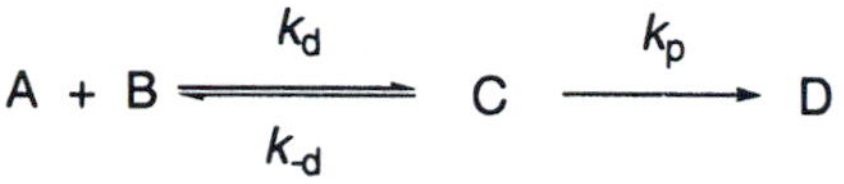

$$A + B \underset{k_{-d}}{\overset{k_d}{\rightleftharpoons}} C \xrightarrow{k_p} D$$

Scheme 1

According to Scheme 1, any bimolecular reaction between two reactants, **A** and **B**, can be broken down into several elementary steps. Initially, an encounter complex, **C**, is formed with a bimolecular rate constant k_d corresponding to diffusion control. This encounter complex can then undergo two competing unimolecular processes: **A** and **B** may be regenerated with a unimolecular rate constant k_{-d}, or an irreversible reaction may occur with a unimolecular rate constant k_p forming products, **D**. The overall bimolecular rate constant for reaction of **A** with **B** is given by:

$$k_q = k_d\, k_p/(k_{-d} + k_p) \qquad (2)$$

Provided $k_p >> k_{-d}$, this expression further reduces to:

$$k_q = k_d \qquad (3)$$

and the reaction is diffusion controlled. Confirmation that k_q values do reflect k_d should come from a comparison of the temperature dependence of k_q with the temperature dependence of viscous flow. Identical activation energies would indicate that both processes are driven by solvent viscous flow, and therefore that k_q does indeed reflect k_d.

We selected the exothermic, irreversible triplet energy transfer process from triplet benzophenone (^{3}**Bp***) to naphthalene (**N**) as our probe of RTIL diffusional properties for the following reasons:

- This system is extremely well characterised and is readily amenable to study by laser flash photolysis (LFP) methods (*3*).
- Electrostatic interaction between charged reactants and ionic liquid components may significantly influence the magnitude of k_d and is difficult to correct for. Therefore we selected neutral reactants and products.
- The molecular size of the reactants is quite representative of that typically employed in studies of reaction kinetics. Although some variation of k_d with different bimolecular reactant systems is anticipated, our values derived from the **Bp/N** system should be quite representative of those expected of any typical reactant system.

2. Results and discussion

Typically, an N_2-saturated 100 mM solution of **Bp** was excited using 355 nm pulsed laser excitation to form ^{3}**Bp***, the absorption of which was monitored at λ = 525 nm. The *pseudo*-first order rate constant for the decay of ^{3}**Bp***, k_2, was measured as a function of the concentration of **N**, and k_q determined over a temperature range of 5–70 °C from plots of k_2 versus [**N**]. An example of the data obtained in [bmim][PF_6] is shown in Figure 1. Arrhenius plots were then constructed (see Figure 2) and the parameters obtained are presented in Table 1 together with those for acetonitrile and toluene - solvents more typically associated with photochemical energy transfer studies. In the absence of naphthalene, ^{3}**Bp*** decays *via* first order kinetics, forming a ketyl radical that is derived from a hydrogen-abstraction process involving the ionic liquid cation. We have recently commented on the unusually large temperature dependence of this process (*4*). The treatment of the data reported here is not influenced by the unusual hydrogen abstraction results.

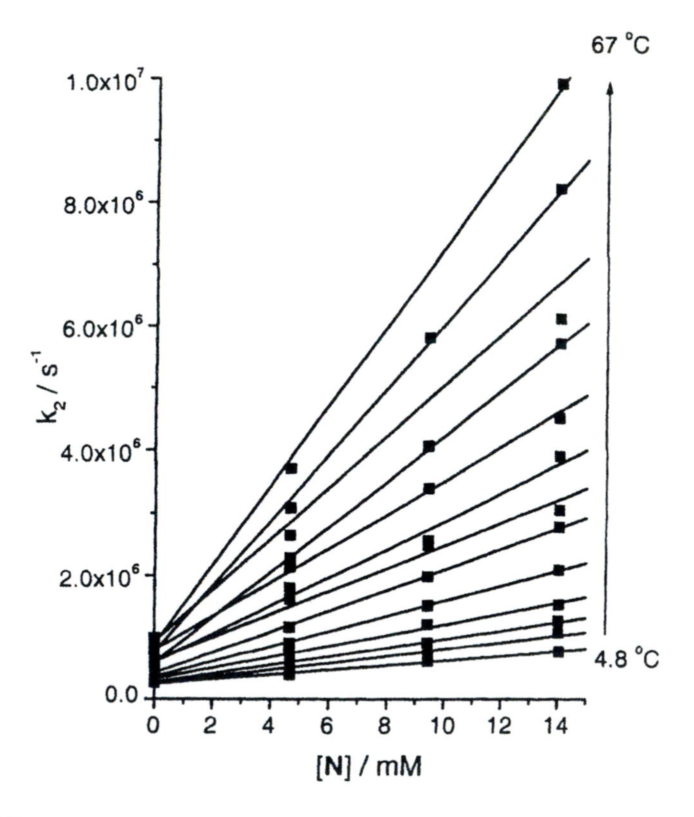

FIGURE 1: Variation of k_2 with temperature and [**N**] for the reaction of ^{3}**Bp*** with **N** in [bmim][PF_6]

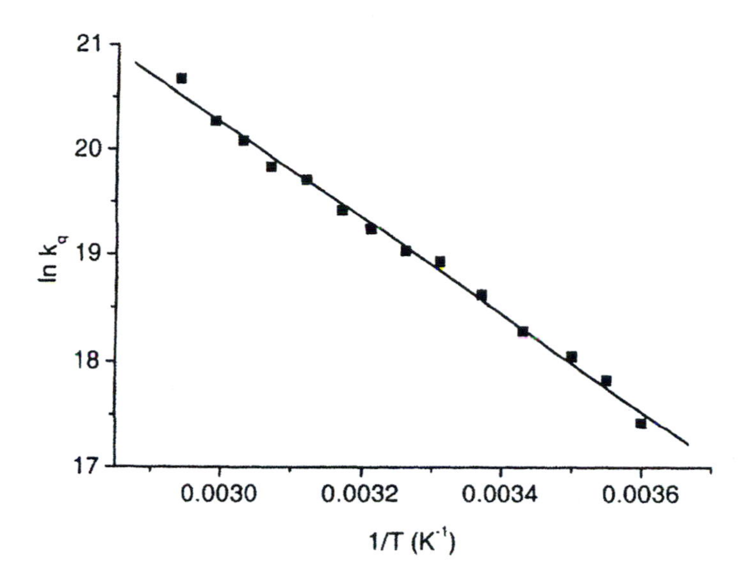

FIGURE 2: Arrhenius plot prepared from the data obtained using Figure 1

Experimentally derived room temperature k_q values in a range of liquids are summarized in Table I together with those values of k_d calculated from equation 1. The calculated k_d values vary over three orders of magnitude whereas the experimental k_q values vary over less than two orders of magnitude. Thus, equation 1 clearly overestimates k_q in less viscous conventional solvents and underestimates it in RTILs.

Table I. Comparison of k_q experimental and calculated

Solvent	k_q (exp.) / $mol^{-1}dm^3 s^{-1}$	k_q (calc.) / $mol^{-1}dm^3 s^{-1}$
[omim][PF_6]	1.10×10^8	9.57×10^6
[bmim][PF_6]	1.26×10^8	2.57×10^7
[omim][Tf_2N]	2.96×10^8	7.13×10^7
[bmmim][Tf_2N]	2.39×10^8	6.80×10^7
[bmim][Tf_2N]	3.83×10^8	1.28×10^8
Acetonitrile	6.83×10^9	1.92×10^{10}
Toluene	8.53×10^9	1.19×10^{10}

The origin of this discrepancy is clearly seen from a glance at Table II which summarises the Arrhenius parameters for both viscosity and k_q. The activation energies for viscous flow and k_q are identical, within experimental error, in all solvent systems. This suggests that k_q is a diffusion controlled rate constant. The breakdown of equation 1 lies in its inability to predict accurate Arrhenius pre-exponential factors, A, for this diffusion controlled process.

The failure of equation 1 to predict k_d values accurately in conventional solvents has previously been examined by other workers (*6, 7*). The chief reason appears to lie in the Stokes/Einstein equation origins of the treatment leading to equation 1. In such treatments the diffusing particle is viewed as moving through a structureless continuum; any equation explaining molecular diffusion must take into account the variation in solute jump probability resulting from its relative size compared to that of the solvent. Larger solvents should create larger holes for solutes to jump into per mole of solvent diffusion which leads to a larger pre-exponential factor for k_d than otherwise may be anticipated. Reasonably good agreement is found between calculated and experimentally derived k_d values when such corrections are made in conventional solvents (*6, 7*). Returning to Table I, it can be seen that equation 1 overestimates k_q in conventional (low molecular weight solvents) and underestimates k_q in RTILs (high molecular weight solvents). This is in broad agreement with what is expected when relative molecular sizes are taken into account.

Table II. Arrhenius parameters for viscosity and k_q

Solvent	η/cP	ln η_0	E_η / kJmol^{-1}	ln A	E_a / kJmol^{-1}
[omim][PF_6]	691	-18.8	45.6	38.1	48.6
[bmim][PF_6]	297	-16.5	37.6	33.9	37.7
[omim][Tf_2N]	92.7	-14.6	30.3	33.2	34.0
[bmmim][Tf_2N]	97.1	-14.9	31.2	31.3	29.7
[bmim][Tf_2N]	51.5	-13.2	25.4	29.4	23.9
Acetonitrile	0.345	-8.42	6.8	26.6	9.8
Toluene	0.556	-8.79	9.1	26.5	9.0

It is clear from Table II that both ln A and E_a depend strongly on the structure of the RTIL. In fact, a striking isokinetic correlation exists between the ln A and E_a values listed in Table II (see Figure 3). This correlation holds for both RTILs and for the two conventional solvents, clearly indicating that these two parameters *cannot* be treated independently of one another.

This correlation suggests that the probability of a diffusional jump of solute into a newly created solvent hole (related to the magnitude of ln A) correlates with amount of energy required to create the hole (related to E_a). However, the magnitude of these parameters does not reflect the molecular mass of the diffusing RTILs; the molecular weight of $[PF_6]^-$ is less than that of $[Tf_2N]^-$. Assuming molecular size correlates directly with molecular mass, this would suggest that a smaller hole should be created in the former case and therefore that there is a lower jump probability. In fact, for RTILs based on equivalent cations, the jump probability appears to be greater in those containing the $[PF_6]^-$ anion. The explanation for this apparent contradiction may lie in the greater interionic interaction available to the $[PF_6]^-$ than $[Tf_2N]^-$, owing to its octahedral

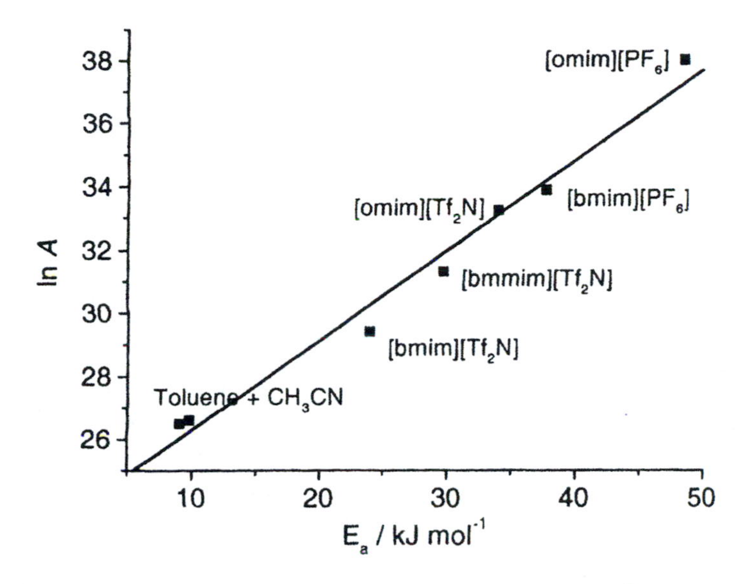

FIGURE 3: Isokinetic relationship obtained for the reaction of ^{3}**Bp*** with **N** in a range of RTILs and conventional solvents

symmetry. Thus $[PF_6]^-$ can potentially interact with a greater number of different cations, which would lead to greater *effective* diffusing mass in the former and therefore greater values of *both* $\ln A$ and E_a.

Finally, there is also a significant dependence of both $\ln A$ and E_a on alkyl chain length and substitution pattern in the cation. This suggests that the breaking of Van der Waals interactions plays a significant role in facilitating diffusion in these systems, even when strong, potentially longer-range, electrostatic interactions are present.

Our conclusions are that the somewhat larger than expected room temperature k_q values observed in RTILs are not as surprising as they may seem at first sight. In RTILs the diffusing masses are very much larger compared with conventional organic solvents, which results in greater solute jump probabilities per mole of diffusing solvent and therefore larger diffusion controlled k_q values than might otherwise be expected in such high viscosity materials. We have also shown that there is a possibly unexpected isokinetic relationship between $\ln A$ and E_a for this process which extends not only to the RTILs, but also to much less viscous organic solvents.

The importance of deriving k_d values in RTILs from experimental k_q values is nicely illustrated by our studies of the bimolecular Diels Alder reaction between diphenylisobenzofuran, **DPBF**, and singlet molecular oxygen, ^{1}**O$_2$*** (the first excited state of molecular oxygen) in [bmim][PF$_6$] (Scheme 2).

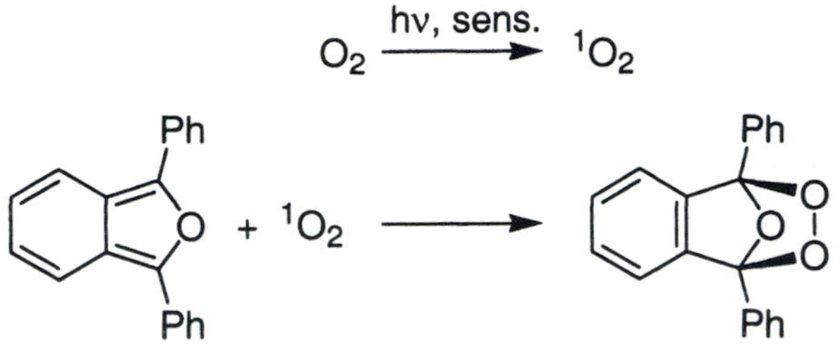

$$O_2 \xrightarrow{h\nu,\ sens.} {}^1O_2$$

Scheme 2

The temperature dependence of k_q for this reaction has been shown by other workers to follow that expected on the basis of Scheme 1 in conventional solvents (*8*). The temperature dependence of k_q for this reaction in [bmim][PF_6] is shown in Figure 4 (square points and solid best fit line). These results may be compared with predicted k_q values calculated assuming purely diffusion limited behaviour (dotted line). Unlike the **Bp/N** system, the Arrhenius plot of this Diels-Alder reaction shows significant curvature over the 70 °C temperature range employed.

The low temperature region is consistent with an ideal diffusion limited situation as indicated by the dotted line through the data (which corresponds to an activation energy of 36.5 $kJmol^{-1}$, *i.e.* identical to that of solvent viscous flow). The room temperature (298 K) experimental k_q value of 2.7×10^8 mol $dm^{-3}\ s^{-1}$ is in good agreement with this estimation, and therefore corresponds to k_d for this system. The importance of knowing that k_q corresponds to k_d at this temperature arises from a superficial comparison of the room temperature value of k_q in [bmim][PF_6] with those in acetonitrile and toluene (9×10^8 and 1.3×10^9 mol $dm^{-3}s^{-1}$ respectively). These results would suggest that the RTIL impedes the reaction with respect to more conventional solvents.

However, it has been shown previously that at 298 K in acetonitrile and toluene $k_p << k_{-d}$; in other words, the second, product forming step, is now rate-determining. This corresponds to a pre-equilibrium rather than diffusion controlled situation, and k_q is given by:

$$k_q = k_d\, k_p \,/\, k_{-d} \tag{4}$$

In this limiting form k_q is a composite rate constant which contains information about the product forming step through k_p. Consequently the room temperature k_q value in [bmim][PF_6] and those in both acetonitrile and toluene correspond to kinetically distinct situations and cannot be directly compared with one another.

Fortunately, at higher temperatures, the experimental k_q values in [bmim][PF_6] clearly deviate away from diffusion limited behaviour. This can be

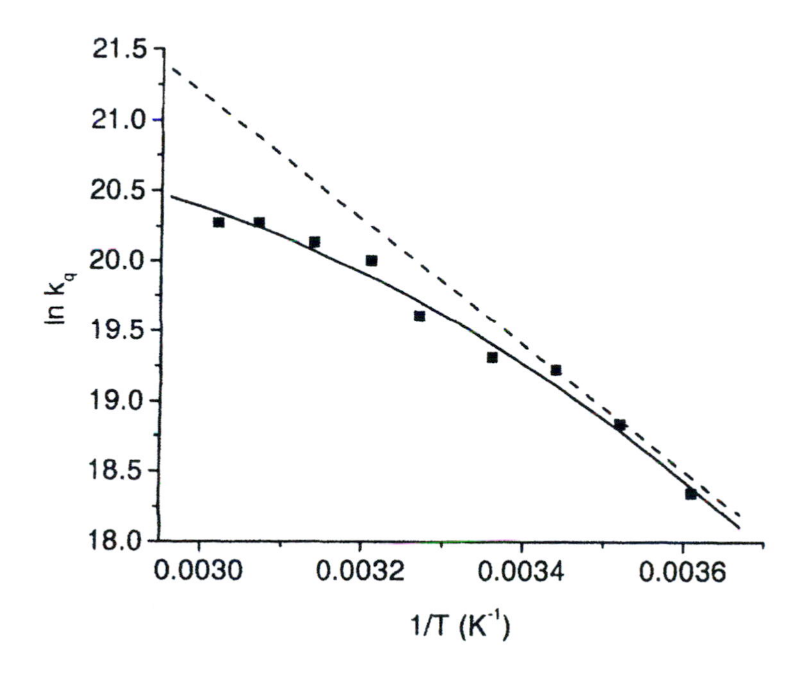

FIGURE 4: Plot showing experimental k_q values (square points) and calculated k_q values (dashed line)

explained using Scheme 1, k_p decreases relative to k_{-d} as temperature is increased. Consequently k_q is no longer diffusion controlled, and is intermediate between diffusion and pre-equilibrium control. The form of k_q is know given by equation 2, which can be re-arranged to give:

$$1/k_q - 1/k_d = 1/k_{act} \qquad (k_{act} = k_d\, k_p \,/\, k_{-d}) \qquad (5)$$

Using equation 5, a pre-equilibrium k_q value of 1.3×10^9 $mol^{-1} dm^3 s^{-1}$ can be estimated at 60°C using the k_d value derived from the Arrhenius line in Figure 4. This pre-equilibrium k_q value can now be compared to those in acetonitrile and toluene (Table III) at the same temperature.

Table III. Comparison of k_{act} for different solvents

Solvent	*k_q / mol^{-1} dm^3 s^{-1} (60°C)*
Toluene	6.5×10^8
Acetonitrile	1.1×10^9
$[bmim][PF_6]$	1.3×10^9

It can now be seen that the situation is reversed compared with that inferred by superficial examination of room temperature k_q values. Under situations of similar kinetic control, the value of k_q in $[bmim][PF_6]$ is greater than those seen at 60 °C in toluene and acetonitrile. This suggests that RTILs can accelerate the rates of Diels Alder reactions even in the absence of Lewis acid catalysis by imidazolium ions. Rate and selectivity enhancements compared with conventional organic solvents have been observed in Diels Alder reactions carried out in RTILs (*9-11*). In general, however, these enhancements have been attributed to specific interactions (particularly Lewis acid catalysis or hydrogen bonding) between the RTIL and substrate molecules. In the reaction of DBPF with $^1O_2$, however, no such interactions are expected to occur. Thus, it appears that RTILs may be able to accelerate the rates of Diels Alder reactions even in the absence of specific solvent effects.

3. Summary and conclusions

In conclusion, the results summarized here show that experimentally derived estimates of k_d in RTILs can be significantly greater than those calculated by commonly accepted methods. The difference between these values is primarily due to very much larger solute jump probabilities in RTILs than anticipated. These large solute jump probabilities are consistent with large diffusing masses in RTILs creating larger holes per mole of solvent diffusion than solvents of low mass.

Finally, we have shown that reliable estimates of k_d in RTILs are critical to accurately determining kinetic control in reactant systems exhibiting large k_q values. In fact, seriously erroneous conclusions may be drawn with respect to detailed reaction mechanisms in the absence of such information.

4. Experimental

Preparation of "spectroscopic grade" ionic liquids

If ionic liquids are to be used for spectroscopic investigations, it is important that they can be prepared in a form that is sufficiently pure. This is particularly true for photophysical and mechanistic studies, where the presence of even small concentrations of impurities can render results meaningless. Furthermore, ionic liquids prepared following literature procedures are often coloured. Quite apart form the impurity problem, this can also mean that the liquid itself will absorb a significant proportion of the photons passing through it. As a result, we have had to develop preparative methods that can be used to generate reliably colourless "spectroscopic grade" ionic liquids. The precautions can be summarized into four sections, which will be expanded on below, notably (i) purification of starting materials, (ii) control of conditions for quaternisation reactions, (iii) anion exchange and (iv) cleanup of end product. The methods described below are discussed for the preparation of 1,3-dialkylimidazolium salts, but there is no apparent reason why they may not be used for any water-stable RTIL. Also, it may not always be necessary to employ all of the procedures, although following them should result in reliably colourless RTILs. Some of these procedures have been adapted from those reported by Welton *et al.* (*12*).

(i) Purification of starting materials. We would generally recommend the purification of organic starting materials following standard literature procedures (*13*). To this end, the alkyl halides are washed with sulfuric acid and then distilled prior to use, while the imidazoles are distilled. Purification of salts or acids used for anion exchange reactions is more problematic, and we have generally found this unnecessary.

(ii) Control of conditions for quaternisation reactions. We have found that keeping the reaction oxygen free, and the temperature as low as possible decreases the chance of discolouration of the resulting halide salt. In order to achieve maximum conversion in a reasonable time, it is generally preferable to use bromoalkanes rather than chloroalkanes. The imidazole is first mixed with an excess of bromoalkane under a flow of dry nitrogen at room temperature. On stirring, the reaction should start spontaneously, and may be identified by the formation of a viscous lower layer and an increase in the temperature of the reaction. If necessary, gentle heating may be applied to initiate the reaction. Being exothermic, care must be taken to prevent the reaction from going out of control (which can result in discoloration of the final product), so at this stage the mixture is placed in an ice/water bath and allowed to stir overnight, maintaining the flow of nitrogen. This generally results in 80-90% conversion, after which the mixture is heated to 60-70 °C for 2-3 hours to take the reaction to completion. On cooling, the excess haloalkane is removed by decantation, and the mixture is then washed with dry ethyl acetate to remove remaining traces of starting material. The resulting halide salt is finally dried by heating at *ca.* 50 °C *in vacuo.* Care must be taken at this stage as overheating can result in breakdown and discolouration of the product. If desired, the halide salt may be further purified by recrystallisation from acetonitrile/ethyl acetate.

(iii) Anion exchange. The anion exchange step is usually less troublesome than the quaternisation reaction, although we have found that some sources of anion, particularly HPF_6, can result in the formation of coloured products. For this reason, we generally now employ ammonium or lithium salts of the desired anion where possible. These have the added advantage of being less exothermic than the neutralization reaction. In the case of the water immiscible salts used in this study, the removal of lithium or ammonium bromide by-products is achieved simply by washing with water until the washing solutions display no bromide on testing with silver nitrate solution. If the free acid is used, testing the washings should be tested with pH paper until neutrality is achieved.

(iv) Cleanup of end product. Despite all our efforts, the liquids often still display some residual colour, so a final cleanup process has been developed. First, the liquid is mixed with a large quantity of activated charcoal and some acidic or neutral alumina. The mixture is then stirred for 24 hours at *ca.* 50 °C under a flow of nitrogen. The neat liquid is then filtered through kieselguhr. This process can be slow, particularly for more viscous RTILs such as those based on the $[PF_6]^-$ anion, but we have found that addition of a co-solvent to reduce the speed up filtration can reintroduce impurities. In this manner we have been able to produce liquids with perfectly clear spectral windows down to λ = 300 nm and below. One disadvantage of the charcoal/alumina cleanup method is that considerable amounts of product are lost, so it is recommended that this approach is used only on relatively large quantities of product.

Once prepared, the liquids are dried by heating *in vacuo* at 70 °C for up to 24 hours. It is also recommended that the liquids are dried immediately prior to use as even water immiscible RTILs may absorb significant quantities of water. It is particularly important that liquids based on the $[PF_6]^-$ anion are carefully dried before storage as breakdown to HF occurs readily in the presence of even small quantities of water.

Acknowledgements

We would like to than Mr. D. Stirling (University of Paisley) for assistance with the laser flash experiments, the EPSRC (Grant No. GR/M56852) for financial support and the award of a studentship (M.J.M.), the Royal Society of Edinburgh for the award of a BP Research Fellowship (C.M.G.).

References

1 Behar, D., Gonzalez, C., Neta, P. *J. Phys. Chem. A*, **2001**, *105*, 7607.
2 J.H. Espenson, *Chemical Kinetics and Reaction Mechanisms*, 2nd ed., McGraw-Hill, New York, 1995, pp 201-202.
3 Gorman, A.A., Hamblett, I., Lambert, C., Prescott, A., Rodgers, M.A.J. *J. Am. Chem. Soc.*, **1987**, *109*, 3091.
4 Muldoon, M.J., McLean, A.J. Gordon, C.M., Dunkin, I.R. *Chem. Commun.*, **2001**, 2364.
5 Muldoon, M.J., Gordon, C.M., Dunkin, I.R. *J. Chem. Soc., Perkin Trans 2*, **2001**, 433.
6 Saltiel, J., Atwater, B.W. *Advances in Photochemistry*, **1988**, *14*, 1.
7 Spernol, A., Wirtz, K. *Z. Naturforsch. A*, **1953**, *8*, 532.
8 Gorman, A.A., Hamblett, I., Lambert, C., Spencer, B., Standen, M.C., *J. Am. Chem. Soc.*, **1988**, *110*, 8053.
9 Fischer T., Sethi A., Welton T., Woolf J. *Tet. Lett.*, **1999**, *40*, 793.
10 Earle M.J., McCormac P.B., Seddon K.R. *Green Chem.*, **1999**, *1*, 23.
11 Lee, C.W. *Tet. Lett.*, **1999**, *40*, 2461.
12 Cammarata L., Kazarian S.G., Salter P.A., Welton T. *Phys. Chem. Chem. Phys.*, **2001**, *3*, 5192.
13 Armarego, W.L.F., Perrin, D.D., *The Purification of Laboratory Chemicals*, 4th ed., Butterworth-Heinemann, Oxford, 1997.

Chapter 30

Amine Mediated Photoreduction of Aryl Ketones in *N*-Heterocyclic Ionic Liquids: Novel Solvent Effects Leading to Altered Product Distribution

Paul B. Jones*, John L. Reynolds, Robert G. Brinson, and Ryan L. Butke

Department of Chemistry, Wake Forest University, Winston-Salem, NC 27109

Amine mediated photoreduction of benzophenones in *N*-heterocyclic ionic liquids produces the corresponding benzhydrols as the major product instead of the expected benzpinacols. The ratio of benzhydrol to benzpinacol is a function of the reduction potential of the intermediate ketyl radical. Results consistent with a mechanism involving a dark electron transfer between the ketyl radical and α-amino radical are presented. Other important factors contributing to the observed benzhydrol/benzpinacol ratio are solvent viscosity and reduction potential of the cation.

Room temperature ionic liquids (RTILs) are salts that remain molten at, or below, 25 °C. Salts of this kind have been known for many years and have been the subject of intensive study.(*1,2*) For much of the history of ionic liquids, the most comprehensive investigations have related to their use in electrochemical applications.(*3*) More recently, the possibility of using RTILs as "green" alternatives to conventional solvents has ignited new interest in these unique compounds.(*1*) Because RTILs are non-volatile, and because many have limited solubility in water and hydrocarbons, solutes may be easily removed via extraction or distillation allowing the RTIL to be recycled. The use of RTILs, therefore, may greatly reduce volatile waste in chemical processes. The RTILs developed in the last fifteen years are stable to air and water and are significantly less corrosive than "first generation" RTILs.(*2*) Ionic liquids show promise as "green" solvents and a number of research groups have already produced clever and exciting results in this regard.(*1-4*) However, to fully realize the green potential of RTILs, much remains to be done.

Recognition of the potential for the use of RTILs in green chemical processes resulted in the investigation of a number of reactions in new, water-stable RTILs. These include Diels-Alder cycloadditions,(*5*) alkene halogenations,(*6*) Stille couplings,(*7*) and even enzymatically catalyzed reactions(*8*) among many others. Furthermore, the need to tailor solvent to reaction has prompted several groups to develop "custom" ionic liquids. A good example is the preparation of an amino-substituted RTIL designed specifically for the removal of carbon dioxide from gas streams.(*9*) Given the environmentally benign nature of photochemistry, we believe that carrying out photochemical reactions in a reusable solvent system could form the basis of a clean method for functional group transformation (*Figure 1*).

Photochemical investigations in chloroaluminate salts showed that the radical cation and radical anion of 9-methylanthracene were produced, in the same experiment, upon irradiation of solutions of anthracene in the RTIL.(*10*)

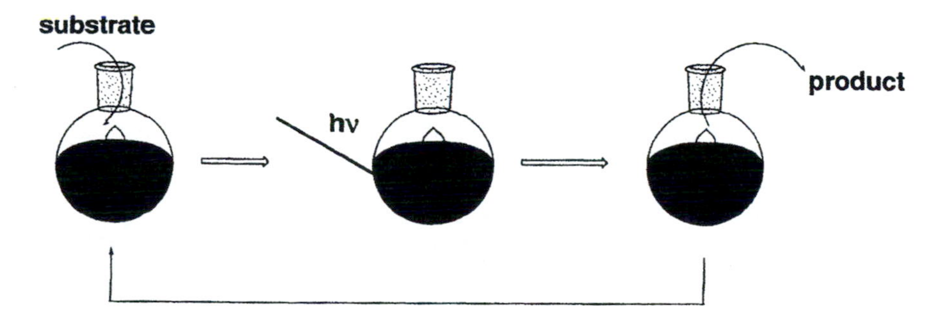

Figure 1. Resusable synthetic photochemistry system.

Later, photoinduced electron transfer in ionic liquids and the abstraction of hydrogen atoms from dialkylimidazolium cations by excited benzophenone were studied by flash photolysis.(*11,12*) Investigations of RTILs under pulse radiolysis conditions confirmed that ionic liquids are ideal media in which to conduct radical ion chemistry.(*13,14*) Both alkylpyridinium and dialkylimidazolium cations were reduced by solvated electrons to yield neutral radicals. These radicals exhibit quite different behavior. The imidazolium radicals were oxidized only by powerful oxidants and at unexpectedly slow rates. In contrast, reduced alkylpyridinium ions were oxidized by a variety of oxidizing agents at a rate near diffusion-controlled.

These observations prompted our consideration of radical pair/ion pair equilibria in RTILs. Radical pairs exist in equilibrium with their corresponding ion pairs and this equilibrium can be affected by external conditions, such as solvent polarity. Given the ionic nature of RTILs, we hypothesized that reactions that normally proceed through a radical pair will instead produce ion pairs when carried out in RTILs (*Figure 2)*. We have since tested this hypothesis and report the results of our investigation to date below.(*15*)

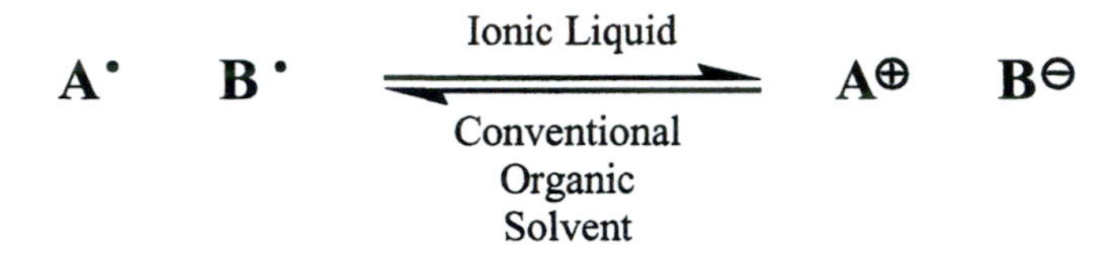

Figure 2. Ion-Pair/Radical Pair Equlibrium Hypothesis.

A simplified mechanism for amine mediated photoreduction is shown in *Figure 3*.(*16*) Charge transfer following excitation of **1** gave a solvent separated radical-ion pair, which diffused to a contact radical-ion pair. Proton transfer produced **2** and **3**, which reduced a second ketone. Coupling of the resulting ketyl radicals gave pinacol (**4**) and imine (**5**). Other evidence suggested an ionic pathway. Photoreduction of 4-benzoylbenzoic acid in aqueous amine gave the benzhydrol.(*17*) Later studies showed that this was general; when benzophenones bearing electron-withdrawing groups were photoreduced, products (**6** and **5**) consistent with the formation of ions (**7** and **8**) were produced as the primary products.(*18*)

Our hypothesis predicted that the reaction pathway of the photoreduction would be altered to favor ion pair **7/8** over radical pair **2/3**. This prediction led to the expectation that photoreduction of benzophenones in RTILs

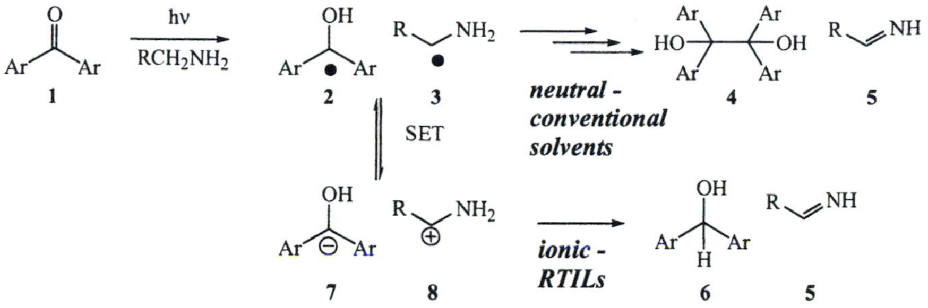

Figure 3. Mechanism of amine mediated aryl ketone photoreduction (reproduced from reference 16. Copyright 2002 American Chemical Society.)

would result in the corresponding benzhydrol rather than the usual benzpinacols. Our results to date are presented in the following section.

Results and Discussion

Product Ratios in Benzophenone Photoreduction

Initial investigation demonstrated that amine-mediated benzophenone photoreduction in RTILs provided products (the corresponding benzhydrol, (**6**) that support the existence of an ionic pathway in the reaction. Control experiments confirmed that reduction did not occur in the absence of either light or amine and that base-catalyzed disproportionation of the pinacol did not account for the formation of benzhydrol. Benzophenone was photoreduced by *sec*-butylamine in 1-butyl-3-methylimidazolium tetrafluoroborate ($BMIBF_4$) to benzhydrol, which was isolated in 90% yield (Table I, entry 3). Benzophenones with electron withdrawing groups on the ring(s) were also photoreduced to the benzhydrol, as expected from the literature (entries 5 and 6).(*18*) Only benzophenones with strong electron donating groups yielded any benzpinacol; 4-methoxybenzophenone was photoreduced under these conditions to give an 75:25 mixture of benzhydrol to benzpinacol (entry 8), while 4,4'-dimethoxybenzophenone gave a 45:55 mixture of the corresponding benzpinacol and 4,4'-dimethoxybenzhydrol (entry 9). Acetophenone was photoreduced to give the pinacol as the sole product. Therefore, the major product is, apparently, dependent on the reduction potential of the intermediate ketyl radical (**2**), which supports the mechanism proposed in *Figure 3*.

Table I. Product Distribution in *sec*-Butylamine Mediated Photoreduction

entry	*compound*	*RTIL*	*benzhydrol*	*benzpinacol*
1	benzophenone	s-butylammonium trifluoroacetate	0	>95
2	benzophenone	EMI(OTf)	>95	0
3	benzophenone	BMI(BF_4)	>95 (90)	0
4	benzophenone	BMMI(BF_4)	>95	0
5	benzophenone	BuPic(BF_4)	---	---
5	methyl 2-benzoyl benzoate	BMI(BF_4)	>95	0
6	methyl 3-benzoyl benzoate	BMI(BF_4)	>95	0
7	4,4'-dichloro benzophenone	BMI(BF_4)	>95 (90)	0
8	4-methoxy benzophenone	BMI(BF_4)	75	25
9	4,4'-dimethoxy benzophenone	BMI(BF_4)	45 (22)	55
10	4,4'-dimethoxy benzophenone	EMI(OTf)	30	70
11	4,4'-dimethoxy benzophenone	BMMI(BF_4)	50	50
12	acetophenone	BMI(BF_4)	0	>95
13	acetophenone	EMI(OTf)	0	>95

NOTE: Data represent ratios determined by ^{1}H NMR. Isolated yields are shown in parentheses.
SOURCE: Some data are taken from reference 16. Copyright 2002 American Chemical Society.

Solvent Effects and Role of the Cation

Because 4,4'-Dimethoxybenzophenone (**9**) provides both benzhydrol (**10**) and benzpinacol (**11**) products when photoreduced in most RTILs, the compound provides a measure of the ionic liquid effect on ketone photoreduction. Photoreduction of (**9**) in a variety of RTILs revealed that the nature of the RTIL affected the product ratio (Table II). There appears to be some correlation of the product ratio to the basicity of the anion in the RTIL, though this is far from an exhaustive list. A similar study involving a larger series of RTILs is underway.

Ketone photoreductions in *N*-butylpicolinyl tetrafluoroborate (BuPic(BF_4)) did not provide either the benzhydrol or benzpinacol (Table II, entry 10). After 24 hours of irradiation, no consumption of ketone was observed. Because the lowest triplet state of pyridinium salts is approximately

25 kcal/mole higher than that for the benzophenone triplet, it is unlikely that the solvent is quenching the reaction through energy transfer.(*19*) Instead, the cation appeared to be reduced by the ketyl radical (and, possibly, the amino radical) resulting in dimer formation (*Figure 4*). Although we had not isolated high yields of the BuPic dimer, we did positively identify it in the ^1H NMR spectrum of the crude photolysis mixture. The dimer appeared to be prone to decomposition under the reaction conditions. This observation was a general trend; all aryl ketones studied failed to undergo photoreduction in BuPic(BF_4).

Table II. Photoreduction of 4,4'-Dimethoxybenzophenone in RTILs.

MeO, O, OMe, **9** → RTIL, *s*-butylamine, hν → MeO, OH, OMe, **10** + MeO, OH, 2, OMe, **11**

entry	*RTIL*	***10***	***11***
1	BMI (OTf)	30	70
2	EMI (OTf)	30	70
3	EMI (BF_4)	35	65
4	BMI (OTs)	40	60
5	BMI (BF_4)	45	55
6	EMI (NO_3)	50	50
7	BMI (PF_6)	55	45
8	BMI (NO_3)	75	25
9	BMMI(BF_4)	50	50
10	BuPic(BF_4)	---	---

NOTE: Data represent ratios. Determined by ^1H NMR.
SOURCE: Some data taken from Reference 16.

Given the observation that benzhydrols form only when the photoreduction was carried out in imidazolium based RTILs, we performed a series of reductions with varying amounts of imidazolium present. These results are shown in Table III.

The results in Table III demonstrate that small amounts of the BMI cation do not result in a shift of product toward benzhydrol. While this observation did not rule out the possibility that the cation played an active role in the reaction, it did indicate that small amounts of cation are not enough to

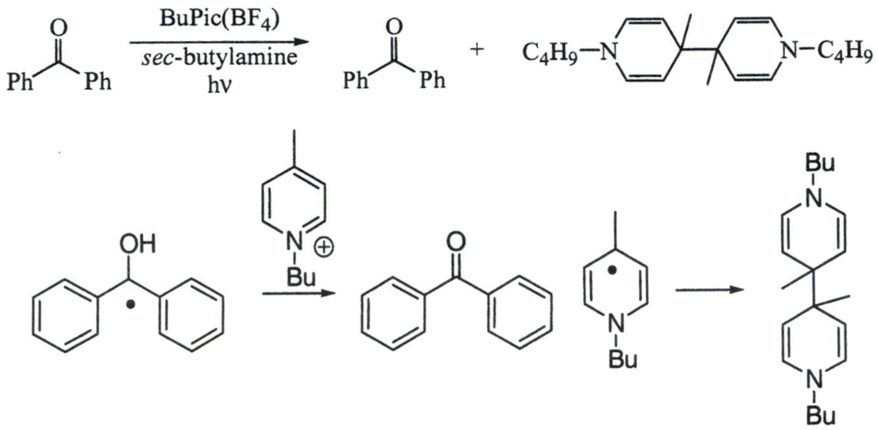

Figure 4. Photoreduction in BuPic(BF_4).

Table III. Photoreduction of benzophenone in the presence of varying amount of BMI cation.

entry	*solvent*	*benzhydrol*	*benzpinacol*
1	sBA-TFA	<5	>95
2	*sec*butylamine	<5	>95
3	secbutylamine + 1 equiv. BMI(BF_4)	<5	>95
4	secbutylamine + 5% (w/w) BMI(BF_4)	<5	>95
5	secbutylamine + 10% (w/w) BMI(BF_4)	<5	>95
6	BMI (BF_4) + 1*M* *sec*butylamine	>95	<5
7	BMMI (BF_4) + 1*M* *sec*butylamine	>95	<5

shift the product ratio. This further supported the theory that the shift of product towards benzhydrol in photoreductions carried out in RTILs is due to a solvent effect and not an indication of more complicated chemistry involving the solvent.

Furthermore, the possibility that the imidazolidene carbene was required for photoreduction to the benzhydrol is ruled out by the observation that the *sec*-butylamine mediated photoreduction of benzophenone was unchanged in going from BMI(BF_4) to BMMI(BF_4). The presence of a large amount of basic amine in BMI and EMI based RTILs likely resulted in the formation of a significant amount of the corresponding carbene.(*20*) However, no such carbene could form in BMMI based RTILs, which had a methyl group blocking the C-2 position. The fact that the photoreduction was not affected by this change demonstrated that the carbene plays no essential role in the photoreduction to benzhydrol.

Role of the C_2-proton

In the photoreduction of benzophenones using primary amines, the source of the bisbenzylic proton is probably the NH_2 of the amine. When tertiary amines are used, no such protons are available so another proton source must be present. In the photoreduction of benzophenone in BMI(BF_4) using triethylamine (TEA), benzhydrol was obtained in over 90% yield. In contrast, when the same photoreduction was carried out in BMMI(BF_4), benzpinacol is the major product (*Figure 5*).

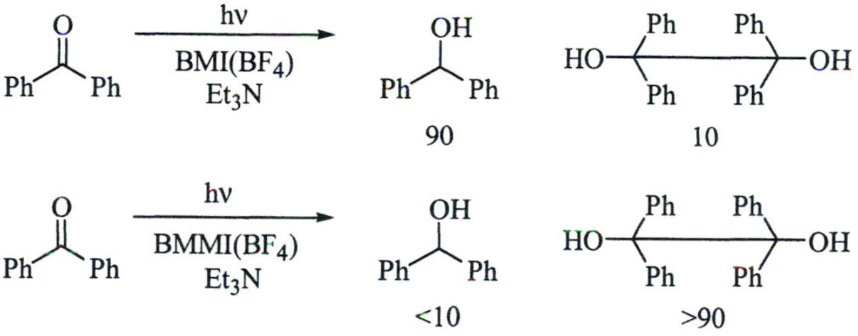

Figure 5. Product Distributions in the photoreduction of benzophenone with triethylamine in BMI(BF_4) and BMMI(BF_4).

As stated above, the production of benzhydrol in TEA mediated photoreduction of benzophenone requires a secondary proton source. We proposed that this proton source is the BMI cation in reactions carried out in BMI(BF_4) (*Figure 6a)*. This suggestion was supported by the fact that benzpinacol was the major product in BMMI(BF_4), which is less acidic than BMI(BF_4) (*Figure 6b*). Assuming an equilibrium between radical pair and ion pair, the absence of a productive decay route for the ion pair would lead to products derived from the radical pair – benzpinacol. This was observed in photoreductions using BMMI(BF_4) and tertiary amines. In this case, the cation

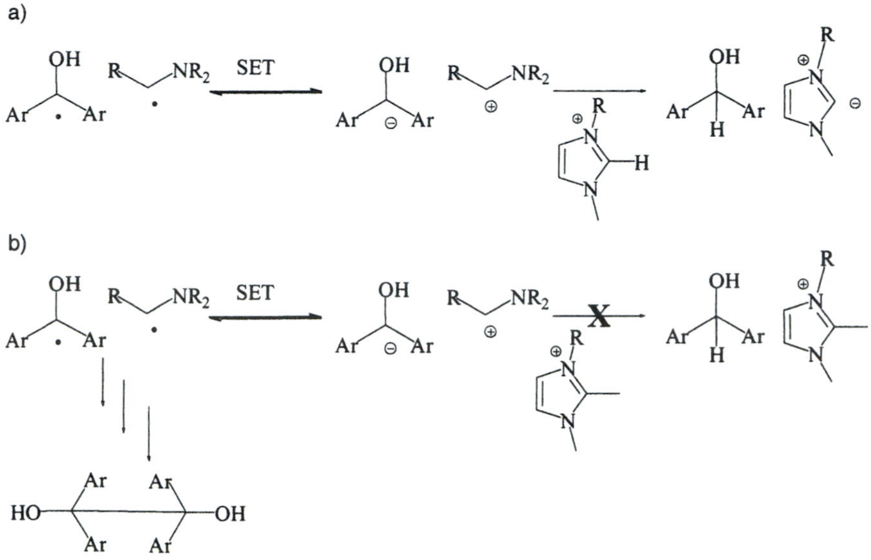

a)

b)

Figure 6. Proposed photoreduction mechanisms in BMI and BMMI RTILs.

was, apparently, playing an active chemical role in determining the major product.

Summary and Conclusions

Benzophenones can be photoreduced cleanly to benzhydrols in RTILs using amines as the source of hydrogen. These reactions are best carried out in imidazolium based ionic liquids as reaction intermediates (*e.g.* ketyl radicals) reduce pyridinium based RTILs and reductions carried out in quaternary ammonium salts lead to mostly benzpinacol. Exisiting data do not indicate whether or not the imidazolium cation plays the role of some sort of "electron shuttle", facilitating the photoreduction. However, it is clear that the carbene corresponding to deprotonation of the imidazolium cation is not required for successful photoreduction. Further experiments are warranted to elucidate the chemical role, if any, of the solvent in these reactions.

The existence of the intermediate bisbenzylic anion has yet to be proven and work continues toward this goal. Our laboratory is also pursuing methods to extend this technique to acetophenones and to understand why quaternary ammonium salts fail to give benzhydrols when used as a solvent in the photoreduction of benzophenones.

General Experimental Notes

Ionic liquids were prepared by literature methods. (*1,15*) RTILs were decolorized by passing a solution of the crude RTIL in dichloromethane through a small length of silica gel. The solvent was then removed *in vacuo* and dried at 80 °C in a vacuum oven. Photoreductions were performed in pyrex vessels, using a medium pressure Hg lamp (Hanovia). Samples were degassed thoroughly prior to irradiation either by exposure to high vacuum or with an argon sparge. Samples were stirred vigorously throughout photolysis. Products of the photoreductions were analyzed by ^{1}H NMR following extraction with diethyl ether of either the neat reaction mixture or after dilution with 0.1 *M* HCl. For detailed experimental conditions, please see the Supporting Information for reference 15.

Acknowledgements

We gratefully acknowledge financial support from the Wake Forest Science Research Fund and the Petroleum Research Fund (Grant #37882-G4). J.L.R. acknowledges the WFU Dept. of Chemistry for a Dean's Fellowship. R.L.B. thanks WFU for a Wake Forest Summer Research Fellowship. The authors appreciate helpful discussions with Dr. Charles Gordon, Dr. Andrew McLean, Dr. Richard Pagni, Dr. Tom Welton, and Dr. Mary Chervenak.

References

1. Wilkes, J.S. *Green Chem.*, **2002**, *4*, 73-80.
2. Welton, T. *Chem. Rev.*, **1999**, *99*, 2071-2084.
3. Lipsztain, M.; Osteryoung, R.A. *Electrochim. Acta*, **1984**, *29*, 1349-1352.
4. Dupont, J. *Quim. Nova*, **2000**, *23*, 825-831.
5. Fischer, T.; Sethi, A.; Welton, T.; Woolf, J. *Tetrahedron Lett.*, **1999**, *40*, 793-796.
6. Chiappe, C.; Capraro, D.; Conte, V.; Pieraccini, D. *Org. Lett.*, **2001**, *3*, 1061-1063.
7. Handy, S.T.; Zhang, X. *Org. Lett.*, **2001**, *3*, 233-236.
8. Madeira Lau, R.; van Rantwijk, F.; Seddon, K.; Sheldon, R.A. *Org. Lett.*, **2000**, *2*, 4189-4191.

9. Bates, E.D.; Mayton, R.D.; Ntai, I.; Davis, J.H., Jr. *J. Am. Chem. Soc.*, **2002**, *124*, 926-927.
10. Lee, C.; Winston, T.; Unni, A.; Pagni, R.M.; Mamantov, G. *J. Am. Chem. Soc.*, **1996**, *118*, 4919-4924.
11. Gordon, C.M.; McLean, A.J. *Chem. Comm.*, **2000**, 1395-1396.
12. Muldoon, M.J.; McLean, A.J.; Gordon, C.M.; Dunkin, I.R. *Chem. Comm.*, **2001**, 2364-2365.
13. Marcinek, A.; Zielonka, J.; Gebicki, J.; Gordon, C.M.; Dunkin, I.R. *J. Phys. Chem. A*, **2001**, *105*, 9305-9309.
14. Behar, D.; Neta, P.; Schultheisz, C. *J. Phys. Chem. A*, **2002**, *106*, 3139-3147.
15. Reynolds, J.L.; Erdner, K.R.; Jones, P.B. *Org. Lett.*, **2002**, *4*, 917-919.
16. Cohen, S.G.; Stein, N.G. *J. Am. Chem. Soc.*, **1971**, *93*, 6542-6554.
17. Cohen, S.G.; Stein, N.G.; Chao, H.M. *J. Am. Chem. Soc.*, **1968**, *90*, 521-522.
18. Jones, P.B.; Pollastri, M.P.; Porter, N.A. *J. Org. Chem.*, **1996**, *61*, 9455-9461.
19. Motten, A.G.; Kwiram, A.L. "Triplet State Properties of Pyridinium." *Chem. Phys. Lett.*, **1977**, *45*, 217-220.
20. Arduengo, A.J. III; Krafczyk, R.; Schmutzler, R.; Craig, H.A.; Goerlich, J.R.; Marshall, W.J.; Unverzagt, M. *Tetrahedron*, **1999**, *55*, 14523-14534.

Chapter 31

Radiation Chemistry of Ionic Liquids: Reactivity of Primary Species

James F. Wishart

Chemistry Department, Brookhaven National Laboratory, Upton, NY 11973

An understanding of the radiation chemistry of ionic liquids is important for development of their applications in radioactive material processing and for the application of pulse radiolysis techniques to the general study of chemical reactivity in ionic liquids. The distribution of primary radiolytic species and their reactivities determine the yields of ultimate products and the radiation stability of a particular ionic liquid. This chapter introduces some principles of radiation chemistry and the techniques used to perform radiolysis experiments. Kinetic studies of primary products and their reactions on short time scales are described and future challenges in ionic liquid radiation chemistry are outlined.

Radiation chemistry has proven to be an invaluable tool for the understanding of chemical reaction mechanisms (*1, 2, 3*). Pulse radiolysis is a particularly useful technique for the measurement of fast redox reactions and reactions of radicals and other energetic transient species. Many of the techniques of radiation chemistry have parallels in photochemistry, but the methods differ significantly in the way the chemical reactions are induced. These differences are complementary and allow many chemical systems to be approached from several directions to provide greater insight into reaction mechanisms. Often, the two techniques can be combined to circumvent their limitations when used in isolation. Despite their versatility and usefulness, many people concerned with understanding chemical reactivity are unaware of radiation chemistry and pulse radiolysis techniques, largely because the necessary equipment is located at a limited number of facilities in the U. S. and abroad. In the hope of making the advantages of radiation chemistry more widely known, this chapter is offered as a general introduction to the field with specific reference to some recent results on radiolysis of ionic liquids and their implications for specific ionic liquid applications.

The rationale for investigating the radiation chemistry of ionic liquids is several fold. First, their solvent properties, non-volatility and combustion resistance make them a very attractive medium for chemical transformations of radionuclides, particularly in the nuclear fuel and waste cycles, as substitutes for volatile organic or aqueous systems. For example, British Nuclear Fuels, Ltd. is exploring the use of electron transfer reactions in ILs to recycle spent nuclear fuel (*4*). Several families of ionic liquids contain good thermal neutron poisons such as boron and chlorine. Calculations from a Los Alamos group indicate that the minimum critical concentrations (above which a solution in a large container would go critical) for plutonium in representative tetrachloroaluminate and tetrafluoroborate ILs are 20 to 100 times greater, respectively, than in water (*5*). Use of ILs could dramatically decrease the risk of criticality accidents such as the one that occurred in Japan in 1999. Successful application of ionic liquids to these problems can only be accomplished if they are sufficiently stable under exposure to high radiation doses. An investigation into the radiochemical stability of certain imidazolium ionic liquids has been conducted using product studies and microsecond-timescale pulse radiolysis (*6*). This approach can be substantially augmented by pico- and nanosecond observations of the primary radiolytic species and the dependence of their yields and reactivities on the composition of the ionic liquid.

Second, applications of ionic liquids as chemical reaction media are expanding at a phenomenal pace (*7, 8, 9, 10*). In order to properly exploit the potential of these new solvents, traditional methods of chemical kinetics studies must be applied, and in some cases adapted, to the study of chemical reactivity in ionic liquids. Pulse radiolysis is a particularly useful technique for the measurement of fast redox reactions and reactions of radicals and other energetic transient species. For water, and to a lesser extent other conventional solvents, elegant methods have been developed to convert primary radiolytic species into specific intermediates for the study of many types of reactions. A similar knowledge base must be assembled for ionic liquids if the versatile methods of pulse radiolysis are to be applied to general questions of chemical reactivity in this exciting new area.

Third, ionic liquids provide a new environment to test the details of theoretical descriptions of charge transfer and other reactions. This can lead to deeper understanding of reactivity in conventional solvents. An example is the reactivity of pre-solvated electrons, which our observations show to feature much more prominently in the radiation chemistry of ionic liquids than in conventional solvents. Pre-solvated electron reactivity is particularly important in concentrated solutions that are exposed to radiation, such as found in nuclear fuel production and waste management.

Radiolytic investigations of ionic liquids have two general goals: understanding the chemistry of ionic liquids under ionizing radiation and learning how pulse radiolysis techniques can be used to study general problems of chemical reactivity in ionic liquids. Some questions which need to be answered are: what primary reactive species are formed when ionic liquids are irradiated, how do their yields depend on the liquid's composition, what reactions do the primary species undergo, and how can their reactivity be exploited to achieve maximum radiation stability, or conversely, provide a high yield of specific products that can be used to study chemical reactions in ionic liquids.

Primary Physical and Chemical Effects of Radiation

Forms of Ionizing Radiation and Modes of Energy Deposition

The term "radiation" can properly be used to refer to emissions in any part of the electromagnetic spectrum, however in common usage it has come to represent photons (X-rays and gamma rays) or energetic particles (electrons, protons, neutrons, alpha particles ($^{4}He^{2+}$), and higher nucleons (e.g., $^{12}C^{6+}$, $^{16}O^{8+}$)) capable of ejecting electrons from molecules. The vernacular usage will be adopted in other parts of this chapter, although the term "ionizing radiation" is more precise. Ionizing radiation is ubiquitous in the universe and represents a significant hazard for long-term space flight. Exposure to high levels of radiation can lead to immediate and long-term illness, yet living organisms cope quite well with chronic exposure to normal background levels from cosmic sources and terrestrial radioactivity. Despite the hazards of ionizing radiation there are many beneficial applications in the diagnosis and therapy of diseases, sterilization of spices, foodstuffs and medical devices, modification of materials, radiography, and (not least) the study of chemical kinetics.

Sources of ionizing radiation for chemical experimentation come in several types (*1, 11*). Most time-resolved, radiation-induced kinetic studies are done with particle accelerators capable of producing short pulses (picoseconds to nanoseconds) of high-energy (1-30 MeV) electrons. Experimental detection systems are similar to those used with laser flash photolysis installations. Product studies are often done through exposure of samples to radioactive sources, such as ^{60}Co which produces gamma rays with an average energy of 1.2 MeV. Radiolysis facilities for heavier ions such as protons and nucleons are far less common, but important for understanding how the physics of radiolysis affects product yields as explained below.

The actions of the various types of ionizing radiation follow similar lines with the exception of neutrons. (*12, 13*) Having no charge, neutrons do not directly ionize molecules but instead act through the charged particles that result from collisions of the neutron with atomic nuclei. X-rays produce energetic electrons through the photoelectric effect and through elastic Compton scattering, processes that occur through interaction of the photon with electrons of the medium. Above energies of 3 MeV the photons may interact with atomic nuclei to produce electron-positron pairs or to excite photonuclear reactions which produce other charged particles. Energetic charged particles ionize molecules by knocking bound electrons loose through coulombic interactions. The result of all these interactions is to convert the primary incident radiation to a cascade of energetic ($\geq$ 5 keV) secondary electrons, which then lose their kinetic energy through ionization and excitation of the medium.

Each primary photon or particle thus produces a "track" of energy deposition events in the irradiated material. The spatial distribution of these events is determined by the velocity and charge of the particle, or the interaction cross section for a photon of a given energy. Interactions transferring large amounts of energy produce electrons capable of generating their own secondary "branch" tracks. Slower and higher-charged particles interact more frequently per unit distance, resulting in high local densities of ionized species. On the picosecond timescale energy deposition is very inhomogeneous, with most of the medium and almost all the solutes completely unperturbed. This is the origin of an important distinction between radiolysis and photolysis. In radiolysis, the energy is deposited in the bulk material, which defines the primary chemistry of reactive intermediates. In photolysis the energy is absorbed by solute chromophores whose photophysics and photochemistry define the subsequent reactivity. Thus, it is clear that the descriptive radiation chemistry of ionic liquids is highly dependent on the composition of the solvent in ways that photochemistry is not. Different classes of ionic liquids will produce different reactive intermediates, which is advantageous considering the wide range of potential applications. Furthermore, with the exception of very high solute concentrations, the primary radiation chemistry is independent of the nature of the solute and initial product yields will remain consistent when different solutes are investigated.

Electron-hole Pairs and Geminate Recombination

Figure 1 is a depiction of the early processes following a single ionization event induced by a high energy electron or other charged particle. Kinetic energy transfers of up to a few tens of electron volts produce energetic electrons which travel tens of nanometers before they have lost all of their excess kinetic energy to the medium. The "thermalized" electrons are spatially delocalized and interact weakly with the medium at first, but the medium rapidly relaxes to accommodate and solvate the excess charge. During this process the electron may pass through one or more phases of partial relaxation, referred to as pre-solvated or "dry" electron states. These phases are typically associated with the characteristic electronic, vibrational, librational and translational responses of the medium. In water or alcohols the entire solvation process may take only a few picoseconds, (*14, 15*) whereas in viscous or glassy materials solvation may not be complete for milliseconds (*16*).

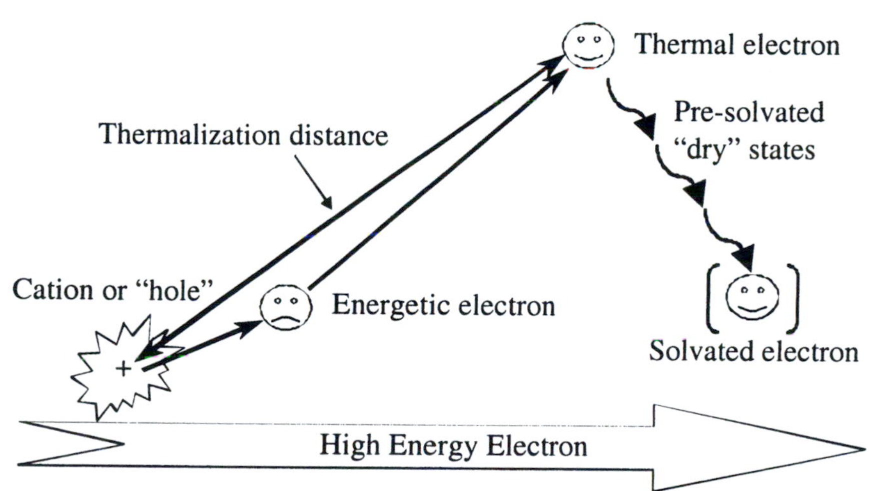

Figure 1. Thermalization and solvation processes following ionization.

The net result of ionization is to produce a pair of reactive intermediates consisting of the solvated electron and the ionized species from which it came, often referred to as the "hole". (In the radiation chemistry of molecular materials it is common to refer to the hole as a cation, but in the context of ionic liquids that term becomes confusing and will be avoided here.) The population of electron-hole pairs are distributed over a range of thermalization distances depending on the frequency of energy transfer events producing ionization and the efficiency of energy loss to the medium from the energetic electron. The efficiency will depend on the strength of coulombic interactions between the electron and the dipoles or ions within the medium, as well as the density of electronic states available to act as energy transfer acceptors. The fact that ionic liquids consist of a disordered lattice of charged species implies that thermalization via coluombic interactions may be extremely efficient and result in a relatively short distribution of electron-hole pair distances.

Electron-hole pair distances are a critical factor in determining the survival probability of the pairs and consequently the ultimate yields of radiolysis products (*17, 18*). Short separation distances increase the probability that motion of the hole and electron will bring them close enough to recombine, resulting in no net chemical change. Recombination can be driven by the Coulomb attraction between the electron and hole if the hole is a cation, or as may obtain in the case of ionic liquids, a neutral radical occupying an anionic lattice site. The coulombic attraction is screened by the dielectric properties of the solvent. Beyond the Onsager radius, defined as the distance at which the screened coulombic potential is equal to the thermal energy, recombination occurs solely through diffusive motion and the survival probability of the "free radicals" (or "free ions" in molecular systems) is high. In molecular solvents with low dielectric constants such as cyclohexane ($\varepsilon = 2.02$), the Onsager radius (28 nm) is large compared to the mean thermalization distance (6-7 nm) (*19, 20*) and the total yield of free ions is low due to loss from coulombic recombination, on the order of 0.15 ion pairs per 100 eV of absorbed radiation. In water, where the dielectric constant is 78 and the Onsager radius is only 0.7 nm, the equivalent free ion yield is 2.7 per 100 eV.

The nature of ionic liquids would seem to make them particularly inappropriate subjects for description by dielectric continuum models. However in the absence of a more detailed theoretical treatment, empirical correlations of their solvatochromic properties with those of molecular solvents allow estimation of their effective polarities and serve as a basis for predicting their effects on free radical yields. Several groups have used solvatochromic dyes such as Reichardt's betaine dye (*21*) or polarity-sensitive fluorophores such as pyrene to estimate that the effective polarities of several imidazolium-containing ionic liquids lie in the range between acetonitrile and ethanol (*22, 23, 24*). The yield of free ions from electron radiolysis in ethanol is 1.8 pairs per 100 eV.

The treatment above has been restricted to the idealized case of isolated electron-hole pairs. As described in the previous section, energy deposition within the radiolysis track is locally dense, and a significant probability exists of neighboring ion pairs occurring within a distance comparable to the thermalization length. The density of ionizations is particularly high at "track ends" when a particle is moving slowly and the frequency of interactions is high, and when the incident particle is a alpha particle or nucleon. The cross reactions that can occur when ion pairs overlap reduce the yield of free radicals but result in a wider variety of products by virtue of radical-radical reactions. Since the frequency of overlapping pairs directly influences the yields of radiolysis products, researchers who are concerned with exposures to multiple types of radiation, as may occur in a fuel or waste processing facility, must determine the yields from each of the different radiation types which apply (*6*).

Chemical Effects of Radiation

Radical pair creation and recombination are very important processes which occur during the first few nanoseconds after radiolysis, but there are several other significant processes to consider as shown in Figure 2.

Energetic particles can directly create excited singlet and triplet electronic states of molecules or ions in the bulk medium through energy transfers too weak for ionization (*18*). The excited states may then undergo subsequent physical and

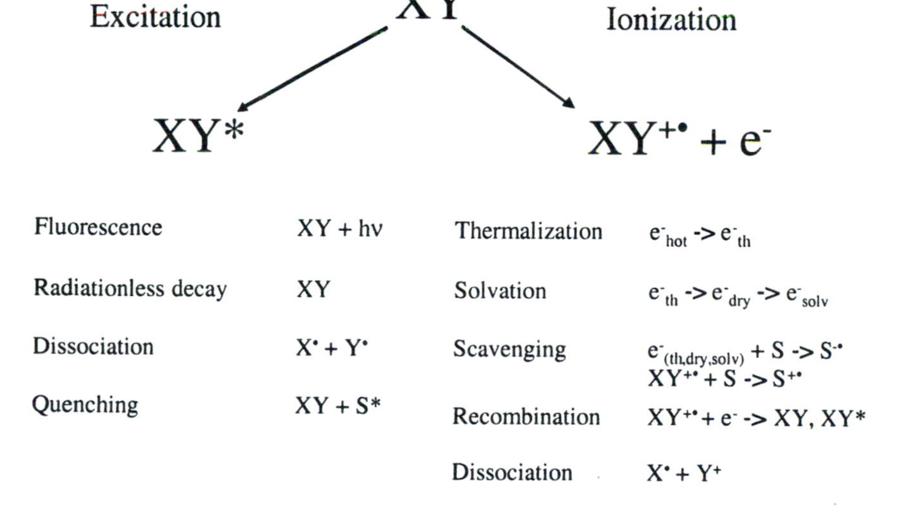

Figure 2. Early Reactions in Radiolysis

chemical processes, such as dissociation or energy transfer to a solute, which may lead to net chemical changes. Excited-state yields are significantly larger than free ion yields in hydrocarbons, particularly aromatic ones. In some cases a large proportion of the excited state yield results from electron-hole recombination to produce excited states (*18, 20*).

Along the ionization pathway, the production of net radiolysis products proceeds through scavenging of primary radicals by solutes, dissociation or fragmentation of holes and excited states, and cross reactions of radicals. In imidazolium and pyridinium ionic liquids the electrons produced by ionization are very rapidly captured by the solvent cations, as observed by the characteristic spectra of imidazoyl and pyridinyl radicals (*25, 26, 27*). Radiation-induced fragmentation of the solvent anion in methyltributylammonium bis(trifluoromethylsulfonyl)imide $[MtBA]^+[NTf_2]^-$, produces $\bullet CF_3$ radicals (*27, 28*). The mechanism may proceed through dissociation of the oxidized-anion "hole" species $[NTf_2]^\bullet$ or dissociation of the excited state $[NTf_2]^{*-}$ formed from geminate recombination of the hole with an electron.

Radiolysis of $[MtBA]^+[NTf_2]^-$

Methyltributylammonium bis(trifluoromethylsulfonyl)imide is an ionic liquid with many desirable properties for the study of fast radiation-induced reactions. It is liquid at room temperature, easy to prepare in relatively pure, colorless form and has a wide electrochemical window: +1.5 to –3 V vs. Ag/Ag^+ (*29*). The very negative cathodic limit permits the solvated electron to exist as a discrete species in solution until it is scavenged by solutes or impurities. The liquid is hydrophobic but hygroscopic and has a viscosity of ~700 cP at room temperature (*27*).

The visible and NIR spectra of the initial transients generated by electron pulse radiolysis of $[MtBA]^+[NTf_2]^-$ are shown in Figure 3 (*30*). The data were obtained at the Brookhaven National Laboratory Laser-Electron Accelerator Facility (LEAF), using the 9 MeV, picosecond electron gun accelerator (*1, 11, 31*). Transients were recorded by digitizer-based transient absorption spectroscopy using silicon and InGaAs photodiodes and fit to single- or double-exponential kinetics as appropriate. The radiolytic dose on the sample was normalized using known values for water in order to express the measured absorbance of the transients as the product of the yield per 100 eV absorbed (G) and the extinction coefficient (ε).

Only two transient species are observed within this wavelength range. Both species appear immediately after the electron pulse. A very broad absorption

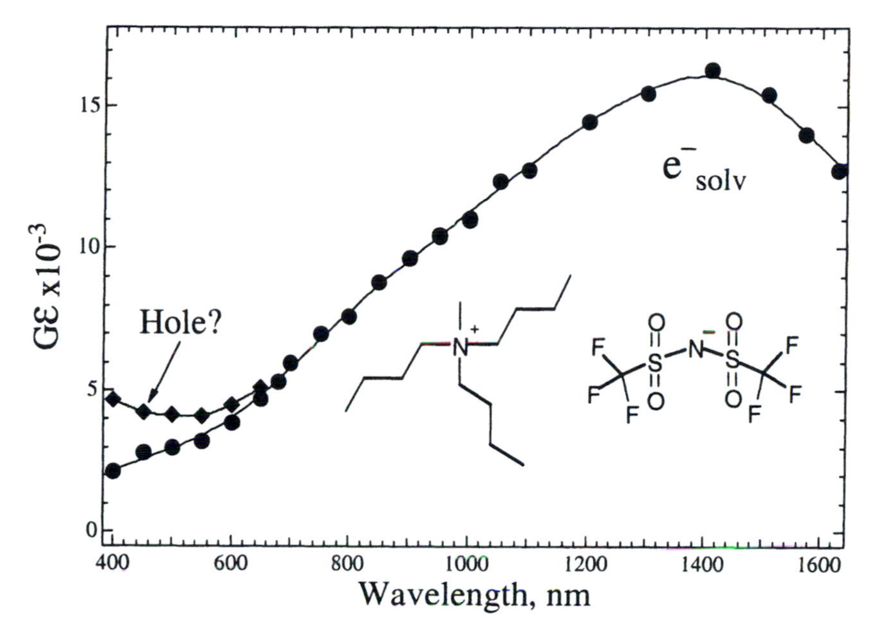

Figure 3. Spectra of the solvated electron and putative hole species produced by electron radiolysis of argon-purged $[MtBA]^+[NTf_2]^-$.

band peaking at 1400 nm is observed to decay by a first order process with a lifetime ($1/k_{obs}$) of 170-300 ns, depending on the batch of ionic liquid. This transient decays faster when electron scavengers are added to the sample and lives slightly longer with addition of the cation scavenger triethylamine. The breadth and shape of the absorption band strongly resembles known spectra of solvated electrons in molecular solvents. The assignment of this species as the solvated electron is therefore difficult to dispute. This is the first observation of the solvated electron in an ionic liquid. The maximum of the solvated electron absorption spectrum in $[MtBA]^+[NTf_2]^-$ (1400 nm) occurs at much longer wavelengths than in water (715 nm) and alcohols (580-820 nm), but not as low energy as in ammonia (1850 nm) and alkylamines (1900-1950 nm) (*32, 33*). The peak in the ionic liquid is practically the same as the peak of the solvated electron in acetonitrile (1450 nm). Another transient absorption band appears on top of the solvated electron spectrum at wavelengths below 700 nm. This species has a lifetime of 50 ns and does not react with electron scavengers. It is most likely a hole species such as $NTf_2^{\bullet}$, however further work is required for identification. One microsecond after irradiation of neat $[MtBA]^+[NTf_2]^-$ there is no significant absorbance anywhere between 400 and 1700 nm.

The reactivity of the solvated electron with pyrene, a typical aromatic electron acceptor, is illustrated in Figure 4. As the concentration of pyrene is increased the decay of the solvated electron observed at 1030 nm gets faster, as does the build-up of pyrene anion absorbance at 490 nm. The second-order rate constant for solvated electron capture obtained from the concentration dependence is 1.7 x 10^8 M^{-1} s^{-1}., about a hundred times slower than observed in alcohols where the reaction is diffusion-limited. Reactions of the solvated electron with other solutes listed in Table 1 show the same general trend, indicating the effect of the substantial viscosity of $[MtBA]^+[NTf_2]^-$.

Dry electron reactivity

It is also evident from inspection of Figure 4 that the initial solvated electron yield, as indicated by the peak absorbance of the transient at 1030 nm, decreases

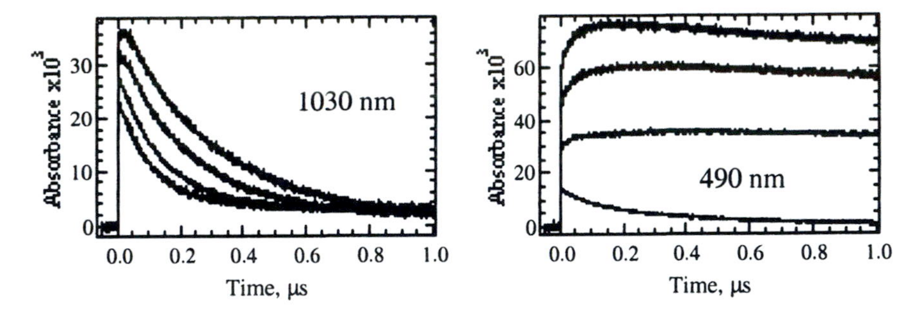

Figure 4. Transient absorption traces showing the reactions of dry and solvated electrons with pyrene in $[MtBA]^+[NTf_2]^-$. The decay of the solvated electron is observed at 1030 nm and the formation of pyrene anion at 490 nm. Pyrene concentrations: 0, 8, 21 and 33 mM from top to bottom on the left and from bottom to top on the right.

Table 1. Rate constants for solvated electron capture and C_{37} values for dry electron capture by several scavengers in $[MtBA]^+[NTf_2]^-$.

Scavenger	$k(e^-_{solv})$, M^{-1} s^{-1}	C_{37}, *M*
Benzophenone	1.6 x 10^8	0.062
Pyrene	1.7 x 10^8	0.063
Phenanthrene	1.3 x 10^8	0.084
Indole	4.3 x 10^7	0.22
H_3O^+ (as 70% $HClO_4$)	6.2 x 10^7	0.20

as the pyrene concentration is increased. This is due to scavenging of the pre-solvated, "dry" electron states depicted in Figure 1, to form pyrene anions at very short times (*34, 35, 36, 37*). The amount of pyrene anion absorption at 490 nm that appears almost instantaneously increases with increasing pyrene concentration. If the pyrene concentration is high enough, some fraction of the electrons produced by solvent ionization will be captured by the pyrene molecules before the electrons can become fully solvated. The fractional yield of solvated electrons remaining after this process can be expressed as a function of the scavenger concentration by the relation $G_c/G_0 = \exp(-C/C_{37})$, where G_c is the yield of solvated electrons at a given scavenger concentration C, G_0 is the yield in the absence of scavenger, and C_{37} is the concentration where only 1/e (37%) of the electrons survive to be solvated. (Some authors prefer to use the reciprocal $Q_{37} = 1/C_{37}$ as a measure of the quenching efficiency.)

The values of C_{37} reported in Table 1 for benzophenone, pyrene and phenanthrene are much lower than those reported (*34, 35, 36, 37*) for efficient dry electron scavengers in water, ethanol and 1-propanol. There is one report of a comparable value, for benzophenone in decanol, ~50 mM (*38*). The higher scavenging efficiency in the ionic liquid is most likely due to less effective competition from the slower electron solvation process, consistent with the higher viscosity of the ionic liquid. Indication of the slower electron solvation process can be seen in the top trace in Figure 4 for the absorbance of the solvated electron at 1030 nm in the absence of pyrene, where the rounded peak reveals a time-dependent Stokes shift with a time constant of 4 ns at room temperature.

Dry electron capture by pyrene provides a convenient method to estimate the initial yield of the solvated electron in $[MtBA]^+[NTf_2]^-$. A plot of the 1030 nm absorbance of the solvated electron extrapolated to time zero versus the extrapolated 490 nm absorbance of the pyrene anion from dry electron capture fits well to a line of slope –0.34. Combined with the reported extinction coefficient of pyrene anion, 5.0×10^4 M^{-1} cm^{-1} (*39*), the observed $G\varepsilon$ value at 1030 nm in Figure 1 results in an estimated solvated electron yield of 0.7 electrons per 100 eV absorbed in $[MtBA]^+[NTf_2]^-$. This a very interesting result which places the yield well below most polar solvents such as alcohols but well above most hydrocarbons (*17, 18*). The solvated electron yield is comparable to that observed in the materials used to process plutonium, indicating that nuclear fuel cycle applications of ionic liquids may be viable.

Dry electron capture has enormous implications for the chemistry of ionic liquids in radiation fields. Quantities of solutes which may seem to be innocent in the context of their solvated electron reactivity may have substantial impact on the radiation chemistry of the ionic liquid because of efficient dry electron capture. This phenomenon must be studied in detail. Because of the broad range of solutes which can be dissolved in ionic liquids, it will be possible for the first time to compare hydrophilic inorganic scavengers and hydrophobic aromatic scavengers in the same medium, which will improve understanding of the process in molecular liquids as well. In addition to its potential impact on radiolytic yields, it also provides a method to rapidly produce reactive intermediates for studies of fast electron transfer and ground- and excited-state reactivity of radicals by avoiding the limitations of diffusion-controlled second-order formation reactions of the intermediates.

Effects of Functionalization

Figure 5 shows how the absorption spectrum and solvation processes of the electron are affected by introduction of an hydroxyl functional group (*40*). The fully-solvated electron spectrum closely resembles the spectra of the solvated electron in alcohols, which tend to peak around 700 nm. The blue shift due to relaxation of the intermediate solvation states that takes about 50 ns to complete in this relatively viscous liquid occurs in about 18 ps in ethanol (*35*) and 300 ps in octanol (*41*). Even at 2 ns a significant amount of electron solvation has occurred since the spectrum is shifted to higher energy than observed for the equilibrated solvated electron in $[MtBA]^+[NTf_2]^-$. The ability to use functionalization to differentiate between various pre-solvated electron states will be a very important tool in the study of dry electron capture. Functionalization also confers a degree of control over the stabilization of the solvated electron, its thermodynamics and reactivity patterns.

Conclusion

The preliminary results reported here demonstrate that there is an enormous potential for development of radiation chemistry in the study of reactivity in ionic liquids and their application to critical technologies of the future. We have measured dry and solvated electron capture by several solutes in one particular ionic liquid. The observation of very efficient dry electron scavenging was a fascinating discovery which we plan to use to test various hypotheses of dry electron scavenging mechanisms, by exploring the competition between electron solvation and electron capture processes. These studies will be very useful also in understanding dry electron capture in molecular solvents where the phenomenon is much harder to observe. The structure of an ionic liquid solvent differs from those of molecular solvents in ways that may significantly influence the solvation process. Neutron diffraction (*42*) and proton NMR (*43*) have verified the intuitive expectation that ionic liquids have a high degree of order typical of a crystalline lattice. The self-imposed regularity of the ionic lattice may reduce the population and depth of pre-formed shallow traps for localization of the thermalized electron relative to molecular solvents. Using the capabilities

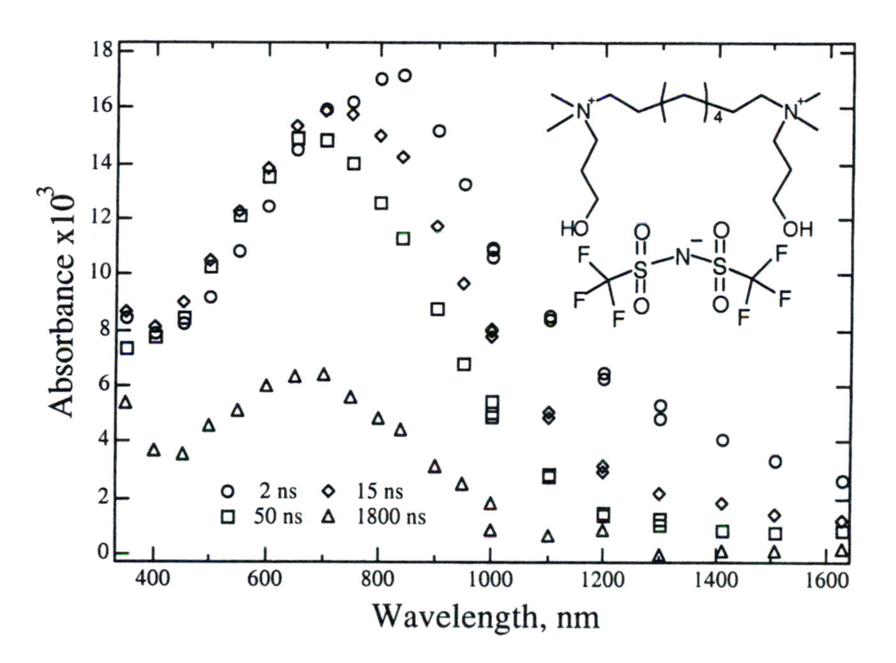

Figure 5. Relaxation of the solvated electron in an ionic liquid with hydroxyl functionalities.

of the LEAF facility we will attempt to observe weakly-localized electrons and follow their evolution. Solvation of the negatively charged electron involves displacement and reorganization of the ionic lattice, which will be influenced by the translational component of the solvent dynamics. In that case, the activation energy for the solvation process should be the same as activation energy for viscous flow. If the solvent dynamics are slow, scavengers can compete effectively for the dry electrons, and it may be possible to use our observations to characterize the reactive dry electron states.

Acknowledgments

The work was performed at Brookhaven National Laboratory under contract DE-AC02-98CH10886 with the U.S. Department of Energy and supported by its Division of Chemical Sciences, Office of Basic Energy Sciences. The author would like to thank Pedatsur Neta, Robert Engel, Sharon Lall, Ravinder Raju and Sherly Bellevue for their collaboration on radiolysis of ionic liquids. The author would also like to thank Richard Holroyd, Sergei Lymar and John Miller for helpful discussions.

References

1. *Photochemistry and Radiation Chemistry: Complementary Methods for the Study of Electron Transfer;* Wishart, J. F. and Nocera, D. G., Eds.; *Adv. Chem Ser.* **254,** American Chemical Society, Washington, DC, 1998.
2. *Radiation Chemistry: Principles and Applications*; Farhataziz; Rodgers M. A. J., Eds.; VCH New York, 1987.
3. *Radiation Chemistry: Present Status and Future Trends; Studies in Physical and Theoretical Chemistry Vol. 87*; Jonah, C. D., Rao, B. S. M., Eds.; Elsevier Science: Amsterdam, 2001.
4. Fields, M., et al.: World Patent, 1999; Vol. WO 99, p 14160.
5. Harmon, C. D.; Smith, W. H.; Costa, D. A. *Radiat. Phys. Chem.* **2001**, *60*, 157-159.
6. Allen, D.; Baston, G.; Bradley, A.; Gorman, T.; Haile, A.; Hamblett, I.; Hatter, J. E.; Healey, M. J. F.; Hodgson, B.; Lewin, R.; Lovell, K. V.; Newton, B.; Pitner, W. R.; Rooney, D. W.; Sanders, D.; Seddon, K. R.; Sims, H. E.; Thied, R. C. *Green Chemistry*, **2002**, *4*, 152-158.
7. Welton, T. *Chem. Rev.* **1999,** *99,* 2071-2083.
8. Sheldon, R. *Chem. Comm.* **2000,** 2399-2407.
9. Zhao, D.; Wu, M.; Kou, Y.; Min, E. *Catal. Today* **2002,** *2654,* 1-33.

10. Dupont, J.; de Souza, R. F.; Suarez, P. A. Z. *Chem. Rev.* **2002,** *102,* 3667-3692.
11. Wishart, J. F. In *Radiation Chemistry: Present Status and Future Trends*; Jonah, C. D., Rao, B. S. M., Eds.; Elsevier Science: Amsterdam, 2001; Vol. 87, pp 21-35.
12. Chatterjee, A. In *Radiation Chemistry: Principles and Applications*; Farhataziz; Rodgers M. A. J., Eds.; VCH New York, 1987, pp 1-28.
13. Mozumder, A. *Fundamentals of Radiation Chemistry*; Academic Press: San Diego, CA, 1999
14. Kambhampati, P.; Son, D. H.; Kee, T. W.; Barbara, P. F. *J. Phys. Chem. A* **2002,** *106,* 2374-2378.
15. Kenney-Wallace, G. A.; Jonah, C. D. *J. Phys. Chem.* **1982,** *86,* 2572-2586.
16. Suwalski, J. P.; Kroh, J. *Radiat. Phys. Chem.* **2002,** *64,* 197-201.
17. Holroyd, R. A. In *Radiation Chemistry: Principles and Applications*; Farhataziz; Rodgers M. A. J., Eds.; VCH New York, 1987, pp 201-235.
18. Swallow, A. J. In *Radiation Chemistry: Principles and Applications*; Farhataziz; Rodgers M. A. J., Eds.; VCH New York, 1987, pp 351-375.
19. Jay-Gerin J.-P., Goulet T.; Billard I., *Can. J. Chem.* **1993,** *71,* 287.
20. Poliakov, P. V.; Cook, A. R.; Wishart, J F.; Miller, J. R., unpublished results.
21. Reichardt, C. *Chem. Rev.* **1994,** *94,* 2319.
22. Aki, S. N. V. K.; Brennecke, J. F.; Samanta, A. *Chem. Comm.* **2001,** 413-414.
23. Muldoon, M.; Gordon, C. M.; Dunkin, I. R. *Perkin Trans. 2* **2001,** 433-435.
24. Fletcher, K. A.; Storey, I. A.; Hendricks, A. E.; Pandey, S.; Pandey, S. *Green Chemistry*, **2002,** *3,* 210-215.
25. Behar, D.; Gonzalez, C.; Neta, P. *J. Phys. Chem. A* **2001**, *105*, 7607-7614
26. Marcinek, A.; Zielonka, J.; Gebicki, J.; Gordon, C. M.; Dunkin, I. R. *J. Phys. Chem. A* **2001**, *105*, 9305-9309.
27. Behar, D.; Neta, P.; Schultheisz, C. *J. Phys. Chem. A* **2002**, *106*, 3139-3147
28. Grodkowski, J.; Neta, P. *J. Phys. Chem. A* **2002**, *106*, 5468-5473
29. Quinn, B. M.; Ding, Z.; Moulton, R.; Bard, A. J. *Langmuir* **2002,** *18,* 1734-1742.
30. Wishart, J. F.; Neta, P., unpublished work.
31. Wishart, J. F. *Houshasenkagaku (Biannual Journal of the Japanese Society of Radiation Chemistry)* **1998,** *66,* 63-64.
32. Belloni, J.; Marignier, J. L. *Radiat. Phys. Chem.* **1989**, *34*, 157-171.
33. Dorfman, L. M.; Galvas, J. F. In *Radiation Research. Biomedical, Chemical and Physical Perspectives;* Nygaard, O. F., Adler, H. J., Sinclair, W. K., Eds.; Academic Press New York, 1975, pp. 326-332.
34. Jonah, C. D.; Miller, J. R.; Matheson, M. S. *J. Phys. Chem.* **1977,** *81,* 1618-1622.

35. Lewis, M. A.; Jonah, C. D. *J. Phys. Chem.* **1986,** *90,* 5367-5372.
36. Jonah, C. D.; Bartels, D. M.; Chernovitz, A. C. *Radiat. Phys. Chem.* **1989,** *34,* 146-156.
37. Glezen, M. M.; Chernovitz, A. C.; Jonah, C. D. *J. Phys. Chem.* **1992,** *96,* 5180-5183.
38. Lin, Y.; Jonah, C. D. *J. Phys. Chem.* **1993,** *97,* 295-302.
39. Gill, D.; Jagur-Grodzinski, J.; Szwarc, M. *Trans. Faraday Soc.* **1964,** *60,* 1424-1431.
40. Lall, S; Engel, R.; Raju, R.; Bellevue, S.; Wishart, J. F., unpublihsed work.
41. Zhang, X.; Lin, Y.; Jonah, C. D. *Radiat. Phys. Chem.* **1999,** *54,* 433-440.
42. Hardacre, C.; Holbrey, J. D.; McMath, S. E. J.; Bowron, D. T.; Soper, A. K. *J. Chem. Phys.* **2002,** *117,* in press.
43. Bonhôte, P.; Dias, A.-P.; Papageorgiou, N.; Kalyanasundaram, K.; Grätzel, M. *Inorg. Chem.* **1996,** *35,* 1168-1178.

Chapter 32

Pulse Radiolysis Studies of Reaction Kinetics in Ionic Liquids

P. Neta, D. Behar, and J. Grodkowski

National Institute of Standards and Technology, 100 Bureau Drive, Gaithersburg, MD 20899–8381

Reaction rate constants were determined by pulse radiolysis in the ionic liquids 1-butyl-3-methylimidazolium hexafluorophosphate and tetrafluoroborate, N-butylpyridinium tetrafluoroborate, and methyltributylammonium bis(trifluoromethylsulfonyl)imide (R_4NNTf_2). Oxidation of chlorpromazine and Trolox by $CCl_3O_2^{\bullet}$ radicals was studied in all the ionic liquids and oxidation of chlorpromazine by $Br_2^{\bullet-}$ radicals was studied in R_4NNTf_2. Reduction of quinones and other compounds was studied both in R_4NNTf_2 and in the pyridinium ionic liquids. The ionic liquids behave as solvent with much lower polarity than water and also inhibit reactions due to their high viscosity. However, electron transfer from the N-butylpyridinyl radical to various acceptors is more rapid than the diffusion-controlled limit, suggesting an electron hopping mechanism. Electron transfer between methyl viologen and quinones takes place several orders of magnitude more slowly in this ionic liquid than in water or 2-PrOH and the direction of the electron transfer is solvent dependent. In contrast, addition and abstraction reactions of $^{\bullet}CF_3$ radicals in R_4NNTf_2 are only slightly slower than those in water and acetonitrile.

Ionic liquids have been shown to enhance the rates of certain reactions and to affect product distribution (*1*). Reaction rate constants may be affected by the solvent viscosity, polarity, hydrogen bonding capacity, and other molecular properties. By comparing the rate constants for elementary reactions in ionic liquids with those in various classical solvents one can draw certain conclusions about the properties of ionic liquids. In addition, since an ionic liquid is composed of two component ions with different properties, and since each reactant may be preferentially solvated by a particular ion, kinetic studies may shed light on the microenvironment in which the reaction takes place and thus on the structure of the solution. To compare reaction rate constants in ionic liquids with those in other solvents we began (*2,3*) with the reactions of $CCl_3O_2^{\bullet}$ radicals with chlorpromazine ($ClPz^+$) (2-chloro-10-(3-dimethylaminopropyl)phenothiazine hydrochloride) and Trolox (6-hydroxy-2,5,7,8-tetramethylchroman-2-carboxylic acid), which have been studied in a wide variety of solvents (*4*). We also studied the oxidation of $ClPz^+$ by $Br_2^{\bullet-}$ radicals (*5*), the reduction of quinones and other compounds (*3*), as well as radical addition and hydrogen abstraction reactions (*6,7*). The ionic liquids used in these studies are 1-butyl-3-methylimidazolium hexafluorophosphate ($BMIPF_6$) and tetrafluoroborate ($BMIBF_4$), N-butylpyridinium tetrafluoroborate ($BuPyBF_4$), N-butyl-4-methylpyridinium hexafluorophosphate ($BuPicPF_6$), and methyltributylammonium bis(trifluoromethylsulfonyl)imide (R_4NNTf_2).

These studies were carried out by the pulse radiolysis technique, using microsecond pulses of 6 MeV electrons to produce radicals from the medium. Formation and decay of transient species were followed by kinetic spectrophotometry. This technique has been widely applied to study radical reactions in various solvents (8) and, recently, to characterize radical ions in ionic liquids (9). Further details on the formation of primary species in the pulse radiolysis of ionic liquids are given by Wishart (10).

Oxidation of Chlorpromazine and Trolox by $CCl_3O_2^{\bullet}$ Radicals

The rate constants for oxidation of chlorpromazine and Trolox by trichloro-

$$CCl_3O_2^{\bullet} + ClPz^+ \rightarrow CCl_3O_2^- + ClPz^{\bullet 2+} \qquad (1)$$

methylperoxyl radicals in several ionic liquids (Table I)(*2,3*) were much slower than those measured in aqueous solutions (*4*) and close to those in alcohols (*4*). Part of the inhibitory effect may be due to the high viscosity of the ionic liquids. However, even after correcting the experimental rate constants for the effect of

Table I. Rate Constants and Activation Energies for Reactions of Chlorpromazine and Trolox with $CCl_3O_2^•$ Radicals

reaction	*solvent*	*T* °C	$k_{exp}{}^a$ L mol^{-1} s^{-1}	$E_a{}^a$ kJ mol^{-1}	*log A*	$k_{diff}{}^b$ L mol^{-1} s^{-1}	k_{act} L mol^{-1} s^{-1}	E_a kJ mol^{-1}	*log A*
$CCl_3O_2^•$ + ClPz	$BMIPF_6$	22	1.2×10^7	30.3	12.4	2.2×10^7	2.6×10^7	22.7	11.4
$CCl_3O_2^•$ + ClPz	$BuPicPF_6$	45	1.9×10^7	30.8	12.3	5.2×10^7	3.0×10^7	25.0	11.6
$CCl_3O_2^•$ + ClPz	$BuPyBF_4$	23	2.4×10^7	25.2	11.9	4.8×10^7	4.8×10^7	18.2	10.9
$CCl_3O_2^•$ + ClPz [c]	R_4NNTf_2	21	4.3×10^6	29	11.8	1.1×10^7	7.2×10^6	22.6	10.8
$RO_2^•$ + ClPz [d]	R_4NNTf_2	21	3.7×10^6	34	12.7	9.2×10^6	6.2×10^6	28.8	11.9
$CCl_3O_2^•$+Trolox	$BMIBF_4$	22	6×10^6			6.7×10^7	6.6×10^6		
$CCl_3O_2^•$+Trolox	$BuPyBF_4$	23	9×10^6	24.2	11.1	4.8×10^7	1.1×10^7	22.8	11.0
$CCl_3O_2^•$ + ClPz	H_2O	25	1.2×10^9	23.7	13.2	7×10^9			
$CCl_3O_2^•$ + ClPz	2-PrOH	22	3.1×10^7	8.7	9.4	3×10^9			

[a]Rate constants and activation energies were determined with an overall estimated standard uncertainties of ± (10 to 15) %. [b]Estimated from the viscosities that were measured with identical solutions. [c]This reaction took place in parallel with reactions of solvent derived peroxyl radicals with ClPz and the processes could not be separated. [d]Reaction involving several solvent radicals but excluding $CCl_3O_2^•$.

the diffusion-controlled limits, estimated from the measured viscosities, the values remained close to those in alcohols. In contrast, the activation energies and pre-exponential factors were closer to those in aqueous solutions than those in alcohols. The results were rationalized in terms of solvation of each reactant by specific solvent ions and the need for the $CCl_3O_2^-$ produced upon electron transfer to accept a proton to form the more stable CCl_3O_2H (*2-4*).

Oxidation of Chlorpromazine by $Br_2^{\bullet -}$ Radicals

$Br_2^{\bullet -}$ radicals have been studied extensively in aqueous solutions (*11*), where they are produced by reaction of $^{\bullet}OH$ radicals with Br^-. In the ionic liquids and in organic solvents, $Br_2^{\bullet -}$ radicals could not be produced by oxidation of Br^- but were produced (*5*) by reduction of $BrCH_2CH_2Br$ via the following sequence of reactions (*12,13*).

$$e_{sol}^- + BrCH_2CH_2Br \rightarrow Br^- + {}^{\bullet}CH_2CH_2Br \quad (2)$$

$${}^{\bullet}CH_2CH_2Br \rightarrow CH_2{=}CH_2 + Br^{\bullet} \quad (3)$$

$$Br^{\bullet} + Br^- \leftrightarrows Br_2^{\bullet -} \quad (4)$$

The rate constants for oxidation of $ClPz^+$ by $Br_2^{\bullet -}$ radicals

$$Br_2^{\bullet -} + ClPz^+ \rightarrow 2Br^- + ClPz^{\bullet 2+} \quad (5)$$

were strongly dependent upon the solvent (Table II) (*5*). The rate constants showed poor correlation with typical solvent polarity parameters but reasonable correlations with hydrogen bond donor acidity (*14*) and with anion- and cation-solvating tendency parameters (*15*) (Figure 1). From the good correlation with the free energy of transfer of Br^- ions from water to the various solvents (*16*) it was suggested (*5*) that the change in the energy of solvation of Br^- in the different solvent is the main factor that affects the rate constant of the reaction through its effect on the reduction potential of $Br_2^{\bullet -}$ and the driving force of the reaction. Therefore, the energy of solvation of bromide ions in this ionic liquid must be much smaller than that in water and alcohols. The energy of solvation of larger, complex ions such $Br_2^{\bullet -}$, may be similar or slightly higher than that in water. As a result, the stability constants of such complex ions are higher in the ionic liquid than in water, which indeed was observed for both $Br_2^{\bullet -}$ and I_3^-.

Table II. Rate Constants for Oxidation of $ClPz^+$ by $Br_2^{\bullet-}$ Radicals

solvent	*k, L mol^{-1} s^{-1}*
H_2O	$(7.7, 5.0) \times 10^9$
MeOH	$(2.8 \pm 0.3) \times 10^9$
EtOH	$(1.2 \pm 0.3) \times 10^9$
i-PrOH	$(1.2 \pm 0.3) \times 10^9$
n-PrOH	$(7.5 \pm 1.1) \times 10^8$
t-BuOH	$(3.0 \pm 0.4) \times 10^8$
MeCN	$(2.0 \pm 0.3) \times 10^7$
DMF	$(5.3 \pm 0.8) \times 10^6$
R_4NNTf_2	$(1.1 \pm 0.1) \times 10^6$
HMPA	$\leq 8 \times 10^4$

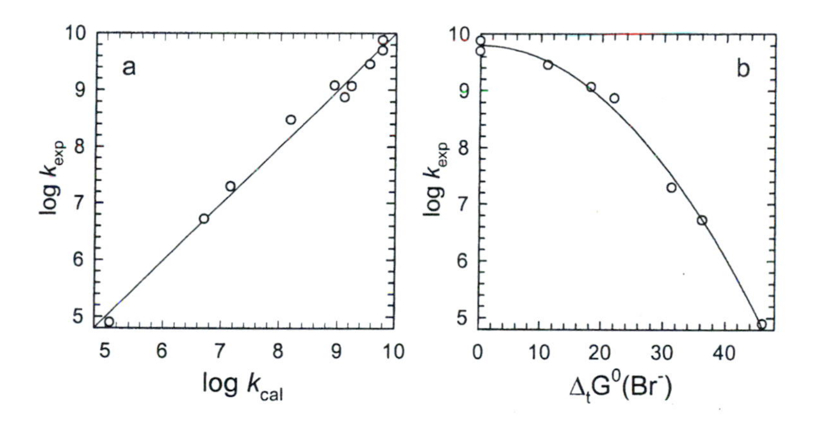

Figure 1. Correlation of the experimental rate constants for oxidation of $ClPz^+$ by $Br_2^{\bullet-}$ radicals (k_{exp}) with solvent parameters. (a) Correlation with A_j and B_j (the anion- and cation-solvating tendency parameters) and (b) correlation with $\Delta_t G^o(Br^-)$ (the standard molar free energy of transfer of the bromide ion from water to the particular solvent)

(Reproduced from reference 5. Copyright 2002 American Chemical Society.)

Reduction of Duroquinone by Benzophenone Ketyl and Pyridinyl Radicals

Radiolysis of the ionic liquids leads to production of solvated electrons (*10*). These species are captured very rapidly by the imidazolium and pyridinium cations (*3*), but in R_4NNTf_2 they are capable of reacting with added solutes. They were reacted with benzophenone (BP) to produce the ketyl radical ($BPH^•$), where the proton is also formed by the radiolysis or supplied by the medium (*3*). The rate constant for reduction of duroquinone (DQ) by $BPH^•$ in R_4NNTf_2 was found to be much lower than that measured in water and 2-propanol but higher than that in glycerol (Table III).

Table III. Rate Constant for Reduction of DQ by $BPH^•$ in Various Solvents

solvent	k_{exp}, $L\ mol^{-1}\ s^{-1}$	k_{diff}, $L\ mol^{-1}\ s^{-1}$
R_4NNTf_2	$(2.0 \pm 0.3) \times 10^7$	1.5×10^7
2-PrOH	$(1.7 \pm 0.2) \times 10^8$	$\sim 3 \times 10^9$
H_2O/2-PrOH (3/1)	$(2.0 \pm 0.4) \times 10^9$	$\sim 3 \times 10^9$
glycerol	$(4 \pm 1) \times 10^6$	$\sim 5 \times 10^6$

By comparing the experimental rate constants with the diffusion-controlled limits estimated from the viscosity, it appears that the rate constants in the ionic liquid, water, and glycerol are close to the diffusion-controlled limits in those solvents and only the value in 2-propanol is considerably lower than the corresponding diffusion limit. This does not necessarily mean that the ionic liquid is behaving here as a polar solvent. If the diffusion did not have a limiting effect, it is possible that the reaction would still be slower in the ionic liquid than in water.

In contrast with the R_4NNTf_2 system discussed above, the solvated electrons produced in the radiolysis of N-butylpyridinium ionic liquids are captured immediately by the solvent cations to produce the corresponding pyridinyl radicals. These radicals cannot reduce benzophenone but can reduce various other solutes which have more positive reduction potentials, such as quinones and nitro compounds.

$$BuPy^• + DQ \rightarrow BuPy^+ + DQ^{•-} \qquad (6)$$

The rate constants for electron transfer from the N-butylpyridinyl radical ($BuPy^•$) and N-butyl-4-methylpyridinyl radical ($BuPic^•$) to duroquinone (DQ), methyl viologen (MV^{2+}), and p-nitrobenzoic acid (p-NBA) were found to be slower than those in water and alcohol (Table IV) (*3*). However, while the rate constants in water were close to the diffusion-controlled limit and those in 2-PrOH were somewhat lower than the diffusion limit, the values in the ionic liquids were higher than the estimated diffusion-controlled limits in these solvents. This finding, which is in contrast with the results in Table III, was rationalized by the suggestion that the specific pyridinyl radical does not have to diffuse to meet its reacting counterpart but rather the electron may be hopping from one pyridinyl radical to a neighboring pyridinium cation until it reached the other reactant. The solvent cations may be somewhat organized in the ionic liquid so as to facilitate the electron hopping.

Table IV. Rate constants for Reduction Reactions in Various Solvents

reaction	*solvent*	*T, °C*	k_{exp}, *L mol^{-1} s^{-1}*	k_{diff}, *L mol^{-1} s^{-1}*
$BuPic^•$ + DQ	$BuPicPF_6$	43	$(4.1 \pm 0.6) \times 10^8$	5×10^7
$BuPy^•$ + DQ	$BuPyBF_4$	24	$(4.4 \pm 0.7) \times 10^8$	5×10^7
$BuPy^•$ + MV^{2+}	$BuPyBF_4$	23	$(2.3 \pm 0.7) \times 10^8$	5×10^7
$BuPy^•$ + p-NBA[a]	$BuPyBF_4$	24	$(4.7 \pm 0.8) \times 10^8$	5×10^7
$BuPic^•$ + DQ	2-PrOH	22	$(1.3 \pm 0.3) \times 10^9$	3×10^9
$BuPic^•$ + DQ	H_2O	22	$(3.3 \pm 0.5) \times 10^9$	7×10^9
$BuPy^•$ + DQ	H_2O	22	$(3.1 \pm 0.5) \times 10^9$	7×10^9
$BuPy^•$ + MV^{2+}	H_2O	22	$(4.0 \pm 0.5) \times 10^9$	7×10^9
$BuPy^•$ + p-NBA[a]	H_2O	22	$(5.4 \pm 0.8) \times 10^9$	7×10^9

[a]The rate constant in the ionic liquid was determined with *p*-nitrobenzoic acid but in water with *p*-nitrobenzoate anion.

Electron Transfer Between Quinones and Methyl Viologen

The rate constant for the equilibrium reaction 7 in $BuPyBF_4$ containing methyl viologen and duroquinone was found to be 1.4×10^5 L mol^{-1} s^{-1}, which is

$$MV^{•+} + DQ \rightleftharpoons MV^{2+} + DQ^{•-} \qquad (7)$$

Table V. Rate constants for Electron Transfer Between Quinones and Methyl Viologen

reaction	*solvent*	*T, °C*	k_{exp}, *L mol⁻¹ s⁻¹*
$MV^{\bullet +}$ + DQ	$BuPyBF_4$	25	$(1.4 \pm 0.3) \times 10^5$
$MV^{\bullet +}$ + DQ	$BuPyBF_4$, 2 v% H_2O	25	$(7.1 \pm 1.1) \times 10^5$
$MV^{\bullet +}$ + DQ	$BuPyBF_4$, 5 v% H_2O	25	$(3.0 \pm 0.5) \times 10^6$
$MV^{\bullet +}$ + DQ	$BuPyBF_4$, 9 v% H_2O	25	$(2.0 \pm 0.3) \times 10^7$
$MV^{\bullet +}$ + DQ	$BuPyBF_4$, 19 v% H_2O	25	$(1.5 \pm 0.3) \times 10^8$
$MV^{\bullet +}$ + DQ	$BuPyBF_4$, 30 v% H_2O	25	$(5.5 \pm 0.3) \times 10^8$
$MV^{\bullet +}$ + DQ	$BuPyBF_4$, 50 v% H_2O	25	$(1.1 \pm 0.1) \times 10^9$
$MV^{\bullet +}$ + DQ	H_2O, 10 v% 2-PrOH	25	$(3.3 \pm 0.5) \times 10^9$
$DQ^{\bullet -}$ + MV^{2+}	2-PrOH, 5 v% TEOA	25	$(1.7 \pm 0.4) \times 10^9$
$AQS^{\bullet 2-}$ + MV^{2+}	$BuPyBF_4$, 2 v% H_2O	22	$(4.0 \pm 0.8) \times 10^6$
$MV^{\bullet +}$ + AQS^-	H_2O, 2 v% 2-PrOH	22	$(6.1 \pm 0.9) \times 10^9$
$AQS^{\bullet 2-}$ + MV^{2+}	2-PrOH, 15 v% water	22	$(1.5 \pm 0.3) \times 10^9$

two orders of magnitude lower than k_{diff}. For comparison, the rate constant for reaction 7 in aqueous solutions was found to be 3.3×10^9 L mol^{-1} s^{-1}, i.e. four orders of magnitude higher (Table V).

The rate constant in $BuPyBF_4$/water mixtures increased sharply upon addition of small amounts of water to the ionic liquid and then slowly approached the aqueous solution value (Figure 2). By comparing with the values of k_{diff} estimated from the viscosities it was clear that in predominantly aqueous solutions the experimental k_7 is about 50% of k_{diff} but as the fraction of ionic liquid approaches 1 the value of k_7 decreases below 1% of k_{diff}. This is due to a change in the equilibrium constant K_7 upon change in solvent properties. In fact, in 2-PrOH solutions reaction 7 was found to proceed to the left (*17*), with k_{-7} = 1.7×10^9 L mol^{-1} s^{-1} (*3*), suggesting that its driving force changes from –0.21 V in water (*18*) to at least +0.2 V in 2-PrOH. Therefore, the driving force in $BuPyBF_4$ appears to be intermediate between these values and thus the reaction occurs much more slowly. It is also possible that reaction 7 in $BuPyBF_4$ is near equilibrium and that it is pulled to the right due to protonation of $DQ^{\bullet -}$ by radiolytically produced protons. Since $DQH^{\bullet}$ has a shorter lifetime than $DQ^{\bullet -}$ and cannot transfer an electron to MV^{2+}, reaction 7 can be driven to the right. In

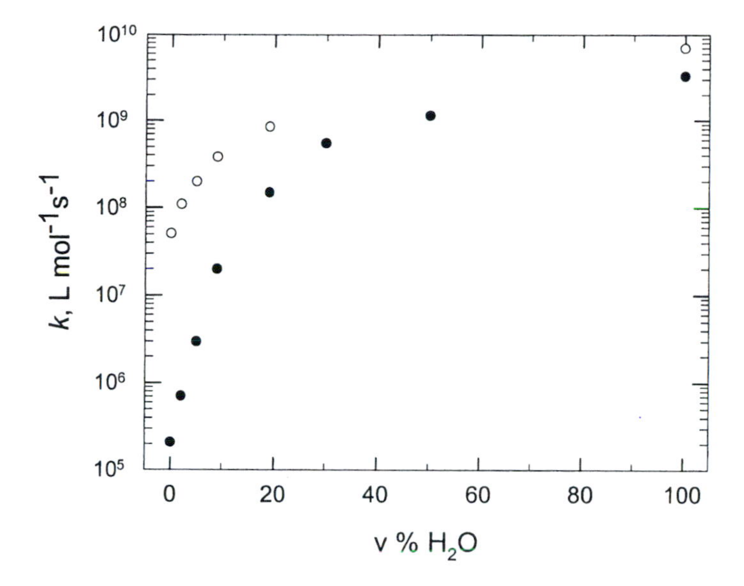

Figure 2. *Rate constant for electron transfer from $MV^{\bullet+}$ to DQ measured in different $BuPyBF_4$/water mixtures (•) and the corresponding values of k_{diff} (o) estimated from the measured viscosities by using equation 12*
(Reproduced with permission from reference 3. Copyright 2002 American Chemical Society.)

confirmation of this suggestion we found that addition of 2 % pyridine to neutralize the protons led to reversal of the direction of reaction 7 and we measured $k_{-7} \approx 10^6$ L mol^{-1} s^{-1}. Thus reaction 7 proceeds in $BuPyBF_4$ in the same direction as in 2-PrOH and in opposite direction to that in water.

This reversal of the reaction driving force was demonstrated more clearly in similar experiments with MV^{2+} and AQS^-. Reaction 8 proceeds to the right with

$$MV^{\bullet+} + AQS^- \rightleftharpoons MV^{2+} + AQS^{\bullet 2-} \quad (8)$$

$k_8 = 6.1 \times 10^9$ L mol^{-1} s^{-1} in aqueous solutions while in 2-PrOH the direction is reversed and $k_{-8} = 1.5 \times 10^9$ L mol^{-1} s^{-1} (Table V). In $BuPyBF_4$ the reaction proceeds to the left, as in 2-PrOH, with $k_{-8} = 4.0 \times 10^6$ L mol^{-1} s^{-1}. Upon addition of 2 % pyridine, to minimize the effect of protonation, k_{-8} increased to 2.7×10^7 L mol^{-1} s^{-1}. These results clearly indicate the reversal of the driving force upon going from water to $BuPyBF_4$ and emphasize the similarity between the ionic liquid and the alcohol.

Hydrogen Abstraction and Addition Reactions

Radiolysis of the ionic liquid $[(C_4H_9)NCH_3]^+[(CF_3SO_2)_2N]^-$ leads to formation of solvated electrons and organic radicals, including $^{\bullet}CF_3$. The solvated electrons were reacted with CF_3Br to produce more $^{\bullet}CF_3$ radicals (*6*). The rate constants for addition of $^{\bullet}CF_3$ radicals to pyrene and phenanthrene were determined to be 1.1×10^7 and 2.6×10^6 L mol^{-1} s^{-1}, respectively (Table VI).

By competition kinetics, the rate constant for reaction of $^{\bullet}CF_3$ radicals with crotonic acid was determined to be 2.7×10^6 L mol^{-1} s^{-1} and the reaction was shown to proceed predominantly via addition to the double bond. Competition kinetics with 2-PrOH in the absence of CF_3Br gave a rate constant of 4×10^4 L mol^{-1} s^{-1} for H-abstraction from 2-PrOH, but in the presence of CF_3Br the rate constant could not be determined because a chain reaction develops through the reduction of CF_3Br by the $(CH_3)_2\dot{C}OH$ radical. The rate constants for reactions of $^{\bullet}CF_3$ radicals in acetonitrile solutions were slightly higher, by a factor of 2.3 for pyrene and phenanthrene and a factor of 1.3 for crotonic acid. The rate constant for pyrene in aqueous acetonitrile solutions was 4 times as high as that

Table VI. Rate Constants for Reactions of $^{\bullet}CF_3$ Radicals

reactant	medium	k, L mol^{-1} s^{-1}
pyrene	R_4NNTf_2, CF_3Br	$(1.1 \pm 0.15) \times 10^7$
	R_4NNTf_2, acidic	$(1.3 \pm 0.25) \times 10^7$
phenanthrene	R_4NNTf_2, CF_3Br	$(2.6 \pm 0.4) \times 10^6$
	R_4NNTf_2, acidic	$(3.2 \pm 0.6) \times 10^6$
crotonic acid	R_4NNTf_2, CF_3Br	$(2.9 \pm 0.5) \times 10^6$ [a]
		$(2.5 \pm 0.5) \times 10^6$ [b]
valeric acid	R_4NNTf_2, CF_3Br	$\leq 1 \times 10^4$ [a]
2-propanol	R_4NNTf_2, acidic	$(4 \pm 1) \times 10^4$ [a]
pyrene	acetonitrile, CF_3Br	$(2.7 \pm 0.4) \times 10^7$
phenanthrene	acetonitrile, CF_3Br	$(7.1 \pm 1.2) \times 10^6$
crotonic acid	acetonitrile, CF_3Br	$(3.4 \pm 0.7) \times 10^6$ [a]
acetonitrile	acetonitrile, CF_3Br	$\leq 5 \times 10^3$
pyrene	acetonitrile/water (7/3), CF_3Br	$(5.0 \pm 0.8) \times 10^7$
2-PrOH	water, CF_3Br	9.2×10^4

[a]From competition with pyrene. [b]From competition with phenanthrene.

in the ionic liquid. Thus rate constants for H-abstraction and addition reactions of $^{\bullet}CF_3$ radicals in the ionic liquid are of the same order of magnitude as in water and acetonitrile, unlike electron transfer reactions which are one or more orders of magnitude slower in ionic liquids than in water or alcohols.

Hydrogen-abstraction reactions of various radicals with 4-mercaptobenzoic acid (MB) to produce the 4-carboxyphenylthiyl radical were studied in aqueous solutions and in the ionic liquid R_4NNTf_2 (*7*). The rate constants in aqueous solutions were in the range of (1 to 3) $\times 10^8$ L mol^{-1} s^{-1} for several alkyl radicals, were higher with reducing radicals (6.4×10^8 L mol^{-1} s^{-1} for $CH_3\dot{C}HOH$ and 1.4×10^9 L mol^{-1} s^{-1} for $(CH_3)_2\dot{C}OH$) and lower with oxidizing radicals ($\leq 10^7$ L mol^{-1} s^{-1} for $^{\bullet}CH_2COCH_3$). Since the bond dissociation energy for the S-H bond is much lower than that for the C-H bonds involved in these reactions, it appears that hydrogen abstraction from mercaptobenzoic acid is not controlled by the relative bond dissociation energies but rather by the electron density at the radical site through a polar transition state. The rate constants for similar reactions in alcohols were slightly lower than those in water, supporting a polar transition state. The rate constants in the ionic liquid were in the range of 10^7 to 10^8 L mol^{-1} s^{-1} and were essentially controlled by the diffusion rate; variations within this range appear to be due mainly to changes in viscosity. The $^{\bullet}CF_3$ radical reacts slightly more slowly (3.6×10^6 L mol^{-1} s^{-1}) with MB in the ionic liquid, in agreement with the low reactivity in water of radicals bearing electron-withdrawing groups.

Conclusions

Ionic liquids generally are more viscous than water and common organic solvents. As a result, the diffusion-controlled rate constants in ionic liquids may be two or three orders of magnitude lower than those in common solvents. Because of this limit, reactions that take place with diffusion-controlled rate constants in all solvents have much lower experimental rate constants in ionic liquids than in common solvents. When the reaction is not limited by the diffusion, the solvent effect on its rate constant depends on the mechanism and the nature of the transition state.

Rate constants for hydrogen-abstraction and addition reactions between radicals and organic molecules are slightly lower in ionic liquids than in aqueous solutions and polar organic solvents. On the other hand, electron transfer reactions are greatly affected by solvent properties. Their rate constants are much lower in ionic liquids than in water and in highly polar organic solvent.

The rate constants of electron transfer reactions depend on the driving force of the reaction, i.e. on the difference between the reduction potentials of the two redox pairs involved in the process. When the driving force is large and the reaction is nearly diffusion-controlled, the rate constant in ionic liquids may be lower than in classical polar solvents because of the high viscosity of ionic liquids. When the driving force is relatively small, however, the rate constant in ionic liquids may be lower than that in water by several orders of magnitude, or the reaction may even proceed in the opposite direction. This large effect is a result of a large change in the driving force of the reaction, which occurs when the ionic liquid affects the reduction potentials of the two redox pairs to different extents. These effects are due mainly to differences in solvation energies. For example, in reaction 8, the products carry higher charges than the reactants and thus are better solvated in water than in organic solvents. This reaction proceeds rapidly to the right in water but to the left in organic solvents and in the ionic liquid. Another example is reaction 5, where both of the products, Br^- and $ClPz^{\bullet 2+}$, are much better solvated by water than by organic solvents whereas both of the reactants, $Br_2^{\bullet -}$ and ClPz+, are equally or better solvated by organic solvents than by water. As a result, the driving force for reaction 5 decreases upon going from water to organic solvents and the rate constant becomes lower. The rate constant in the ionic liquid is very low. By comparison of this rate constant with that in various organic solvents one can estimate the solvation energy in the ionic liquid. When data on solvation energy in ionic liquids become available, certain redox reaction rate constants may become predictable. At this time it can be qualitatively stated that the solvation energies of small anions in ionic liquids are lower than those in water and alcohols and may be comparable to those in DMF or HMPA.

In one type of reaction, i.e., electron transfer from N-butylpyridinyl radical to various acceptors in the ionic liquid N-butylpyridinium tetrafluoroborate, the experimental rate constants appeared to be higher than the diffusion-controlled limit estimated from the viscosity. This finding was explained by a hopping mechanism, whereby the electron is tranferred to the acceptor by hopping through a succession of solvent cations without the need for the specific radical to diffuse to the acceptor. The validity of this mechanism can be acsertained when the diffusion-controlled limit is determined experimentally.

References

1. Welton, T. *Chem. Rev.* **1999**, *99*, 2071, and references therein.
2. Behar, D.; Gonzalez, C.; Neta, P. *J. Phys. Chem. A* **2001**, *105*, 7607.
3. Behar, D.; Neta, P.; Schultheisz, C. *J. Phys. Chem. A*, **2002**, *106*, 3139.
4. Neta, P.; Huie, R. E.; Maruthamuthu, P.; Steenken, S. *J. Phys. Chem.* **1989**, *93*, 7654. Alfassi, Z.B.; Huie, R.E.; Neta, P. *J. Phys. Chem.* **1993**, *97*, 7253.
5. Grodkowski, J.; Neta, P. *J. Phys. Chem. A* **2002**, *106*, 11130.
6. Grodkowski, J.; Neta, P. *J. Phys. Chem. A* **2002**, *106*, 5468.
7. Grodkowski, J.; Neta, P. *J. Phys. Chem. A* **2002**, *106*, 9030.
8. Jonah, C. D.; Rao, B. S. M. Radiation Chemistry, Present Status and Future *Trends*, Elsevier, Amsterdam, 2001.

9. Marcinek, A.; Zielonka, J.; Gebicki, J.; Gordon, C. M.; Dunkin, I. R. *J. Phys. Chem. A* **2001**, *105*, 9305.
10. Wishart, J. F., article in this book.
11. Neta, P.; Huie, R. E.; Ross, A. B. *J. Phys. Chem. Ref. Data*, **1988**, *17*, 1027.
12. Lal, M.; Mönig, J.; Asmus, K.-D. *Free Radical Res. Commun.* **1986**, *1*, 235.
13. Merényi, G.; Lind, J. *J. Am. Chem. Soc.* **1994**, *116*, 7872.
14. Kamlet, M. J.; Abboud, J.-L. M.; Abraham, M. H.; Taft, R. W. *J. Org. Chem.* **1983**, *48*, 2877.
15. Reichardt, C. *Solvents and Solvent Effects in Organic Chemistry*; VCH: Weinheim, 1988.
16. Marcus, Y. *Ion Solvation*, Wiley, New York, 1985, p.168.
17. Pal, H.; Mukherjee, T. *J. Indian Chem. Soc.* **1993**, *70*, 409.
18. Wardman, P. *J. Phys. Chem. Ref. Data* **1989**, *18*, 1637.

Chapter 33

Organic Electrochemistry in Ionic Liquids

Andrew P. Doherty and Claudine A. Brooks

School of Chemistry, David Keir Building, The Queen's University of Belfast, Belfast BT9 5AG, United Kingdom

The reductive electrochemistry of dimethylmaleate and benzaldehyde in room temperature ionic liquids is described. The reduction of the activated olefin involves a one-electron process in a similar fashion to that observed in organic solvent based electrolytes. The reduction of benzaldehyde also exhibits a one-electron reduction process; however, two additional well defined and separated reduction processes are observed at more negative potentials. These have been attributed to the α-hydroxybenzyl radical and the benzaldehyde radical anion. It appears that the remarkable stability of these radical pairs may be due to interaction with the ionic liquid. Also, because of the stability of ionic liquids to electrolysis, and because they tend to stabilise free radical species, they can provide a straightforward means to study radical chemistry.

Introduction

The electrolytic conversion of organic substrates has been pursued extensively since the late nineteenth century. Although many useful conversions and synthetic methodologies have been demonstrated, organic electrochemistry has, for the most part, failed to compete with thermally activated approaches. It appears that the electrochemical route is only considered when all other options have failed to deliver appropriate processing capability. The reasons espoused for the wide-spread reluctance to adopt electrolysis for organic electrosynthetic conversions are many and varied, and include the need for high voltage power supplies, the need of organic solvents/electrolytes in many situations, the sensitivity to processing conditions, the possibility of parasitic reactions, poor selectivity, and frequently low coulombic efficiencies due to O_2 or H_2 evolution under aqueous conditions. Notwithstanding, where superior technical (i.e. economical) advantage can be gained from the electrolytic approach, it has been adopted by industry. The continued use of the adiponitrile process, and the recent introduction of mucic acid electrosynthesis, are excellent examples of successful electrolytic conversions in commercial organic chemistry. Although less well known, electrolysis is frequently used in the pharmaceutical and fine chemicals industry.

Over the last number of years, the potential applications of room temperature ionic liquids in diverse areas such catalysis (*1*), solvents for organic synthesis (*2*), electrodeposition (*3*) and capacitors (*4*) have been demonstrated, with remarkable success. The unique features of ionic liquids are well documented (*5*) and include negligible vapour pressures, differential (selective) solubilities, the ability to "design" solvent systems in terms of function, and their critical role as alternative solvents for "green" chemistry and technology (*6*). In addition to these diverse areas of activity, ionic liquids are potentially useful in many other electrochemical technology fields, including fuel cells and electrosynthesis, because of their electrochemical robustness and inherent electrical conductivity.

In electrosynthesis, ionic liquids have the potential to circumvent the general need for organic solvents as well as negate the requirement for extraneous electrolytes, both of which must be separated, reused or disposed of, post-process. In particular, processes where differential solubility can be achieved to effect efficient product recovery are exceedingly attractive. Also, ionic liquids may be useful alternative solvent/electrolyte systems for traditional aqueous–based electrolytic conversions where O_2 or H_2 evolution reactions can severely limit coulombic efficiencies. Therefore, coupling inherently clean electrolytic conversions with clean ionic liquid solvents potentially offers a new paradigm for electrosynthetic processing. In addition, further coupling ionic liquid and electrochemical processes with, for example, the emerging field of nano-filtration, could provide means for exceptionally clean synthesis and convenient product /ionic liquid recovery/recycle.

Although ionic liquids are eminently attractive for carrying out organic electrosynthesis, the transfer of useful reactions from traditional solvents to the ionic liquid regime may not be straightforward because such conversions are usually very sensitive to operational conditions including solvent/electrolyte classes, substrate concentration, electrode materials, temperature, current density etc. Therefore, useful reactions must be re-examined in terms of selectivity and efficiencies in order to develop useful processes. Furthermore, electrochemical reactor engineering and modelling, require that fundamental issues such as thermodynamics, electrode kinetics and mass transport are carefully understood within the context of this new paradigm.

In this communication, the electrochemical reduction of an activated olefin (dimethylmaleate) and a carbonyl species (benzaldehyde) in ionic liquids will be described. We also speculate on the potential use of ionic liquids in organic electrochemistry.

Experimental

The ionic liquids used here were prepared and characterised according to the literature (*7-9*). The ionic liquids examined were based on the Bmim (where Bmim is 1-methyl-3-butylimidazolium) and Pyrd (where Pyrd is N-methyl-N-butylpyrrolidinium) cations (Fig. 1) with the triflimide (Tfi)) anion. Although the liquids were prepared and dried according to the literature, no exceptional precautions were employed to maintain "anhydrous" conditions i.e. experiments were carried out only under N_2 blanket.

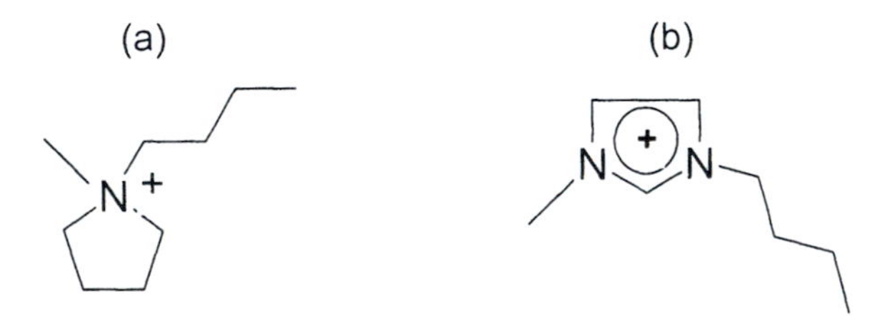

Figure 1 Structure of a) Pyrd and b) Bmim cations.

All voltammetric experiments were carried out using either a Sycopel AEW2, a BAS CW-50, or a Radiometer VoltaLab 40 potentiostat, depending on electrochemical technique being applied. All instruments were operated in the three-electrode configuration using 3 mm glassy carbon disks (BAS) as working electrodes which were prepared, prior to use, by polishing with 0.015 μm alumina (BDH) as an aqueous slurry. The auxiliary electrode, during all experiments, was Pt wire (Goodfellow). In order to minimise "noise" from frits, a solid-state quasi-reference electrode was used throughout which was Au wire (Goodfellow). All experiments were carried out in deoxygenated ionic liquids (O_2-free nitrogen purge overnight) at room temperature.

Results and discussion

Dimethylmaleate reduction

The reduction of dimethylmaleate (DMM), and its structural isomer, dimethylfumarate (DMF) are classical examples of the reductive electrochemistry of activated olefins where dimerisation or cross coupling reactions can frequently be achieved under appropriate conditions. Because substrates such as other olefins, or CO_2, can be involved in cross-coupling reactions, this represents an important class of electrochemically-induced C-C bond forming reactions. The reaction of interest here is practically important because the product, from both their reductions, is tetramethyl butanetetracarboxylate; which may be hydrolysed to 1,2,3,4-butanecarboxylic acid, which finds application as a cross-linking agent in waste paper recycling and the processing of natural fibres such as cotton (*10*). Figure 2(1) shows the reduction of DMM in acetonitrile (with 0.1 mol dm^{-3} tetrabutylammonium chloride electrolyte) where two one-electron reduction processes can be seen. The first process at ca. −1.61 V vs. Au_{SS} is due to the reduction of DMM whereas the second wave, at ca. −1.9 V vs. Au_{SS}, is due to the reduction of DMF impurity. The reduction process results in the corresponding radical anions that subsequently dimerise / protonate to form the final dimer product.

The reduction of DMM in [Bmim][Tfi] ionic liquid is shown in Figure 2(2). The current data in Fig 2(2) has been multiplied by a factor of 10 to allow comparison. The very low current in the ionic liquid is due to the highly restricted diffusion in the viscous ionic liquid. Similar behaviour has been observed for the electrolytic generation of the superoxide ion (*11*). Although the reduction of DMM appears at slightly lower potentials (ca. −1.7 V vs. Au_{SS}), this is probably due to changes in the Au_{SS} reference potential in the ionic liquid compared with that in acetonitrile. The reduction of DMF impurity, however, occurs at slightly higher potentials (ca. −1.8 V vs. Au_{SS}) and is superimposed on the significant diffusional tail of DMF reduction wave. It is therefore evident that

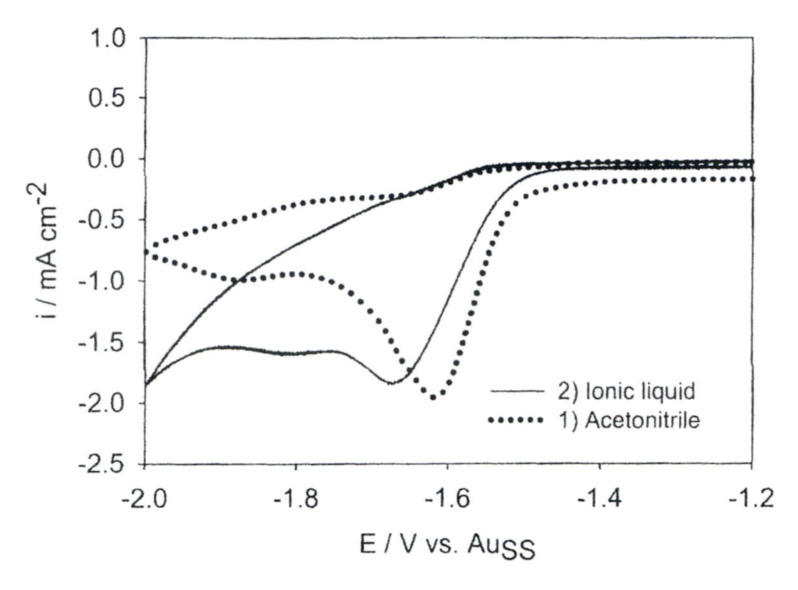

Figure 2 Reduction of DMM in; 1) acetonitrile (0.1 mol dm^{-3} tetrabutyl-ammonium chloride) and 2) (i X 10) [Bmim][Tfi] ionic liquid at 0.1 V s^{-1}.

the reduction, in both environments, yields essentially equivalent electrochemistry.

Square wave voltammetry is a useful technique because it minimises background charging currents which are large in all ionic liquids. Figure 3 (solid line) shows a square wave voltammogram for the reduction of DMM in the ionic liquid obtained using a pulse amplitude of 10 mV. Under such conditions (low pulse amplitude), the peak-width at half height should be 90/n mV where n is the number of electrons involved in the redox process. Indeed, the reduction of DMM exhibits such behaviour as shown in Figure 3. It is known (*12*) that at sufficiently high potential sweep-rates in cyclic voltammetry, the one-electron reduction wave for maleates/fumarates exhibits reversible redox electrochemistry suggesting that the symmetry factor (α) for the electron-transfer reaction is 0.5. With this in mind, simulation of the theoretical square-wave voltammogram, using a diffusion coefficient of 1.0 x 10^{-8} cm^2 s^{-1} and a heterogeneous rate constant of 1.0 x 10^{-3} cm s^{-1}, yields a curve that precisely overlays the experimental curve. The simulated curve is shown as the dotted line in Figure 3 where the overlaying with the experimental curve suggests that the electrochemistry is, indeed, a one-electron transfer process which is favourable in the ionic liquid. This is significant in that it suggests that post-electron transfer

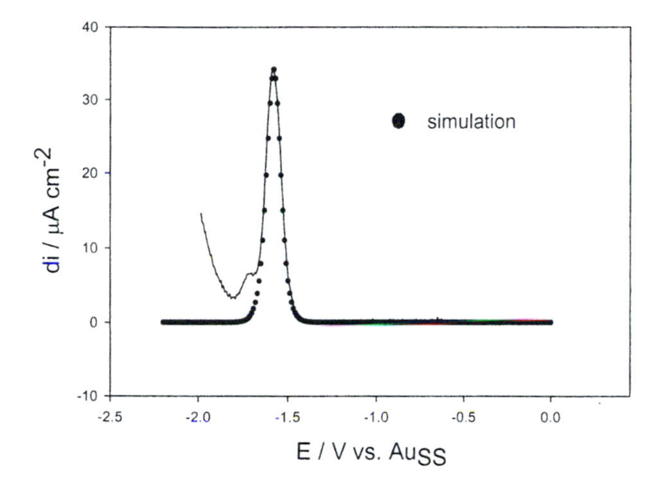

Figure 3 Square-wave voltammogram (solid line) for the reduction of dimethylmaleate in [Bmim][Tfi] ionic liquid and a simulated voltammogram (dotted line).

reactions such as electrocarboxylations, involving electro-generated radical anions of activated olefins, may be possible.

Benzaldehyde reduction

The electrochemical reduction of benzaldehyde in nominally anhydrous ethanol has been described in detail by Saveant et al. (*13*). This substrate is unusual in that radical species generated upon reduction are more difficult to reduce than the initial substrate, which gives rise to further reduction processes that are observable at lower potentials than the initial reduction process. Overall, the reduction results in either the corresponding benzyl alcohol or the dimer product, with the dimerisation reaction being favoured at high substrate concentrations under alkaline conditions.

Figure 4 shows (dotted line) a cyclic voltammogram (reverse sweeo not shown) for the reduction of 3.0 X 10^{-3} mol dm^{-3} benzaldehyde in [Bmim][Tfi] with a lower potential limit of –2.2. V vs. Au_{SS}. This voltammogram exhibits a single irreversible reductive wave with an E_{pc} of –1.98 V vs. Au_{ss}. Decreasing the cathodic limit below the discharge potential of the ionic liquid (<-2.2 V) showed no evidence of the ketyl radical species previously observed in ethanol (*13*); presumably, such processes , if present, are masked by the discharge of the ionic liquid. It can be seen that a small pre-peak is evident; although this could be due to an adsorption process, the feature is also frequently present in background voltammograms and is, therefore, likely to be the reduction of trace water in the ionic liquid.

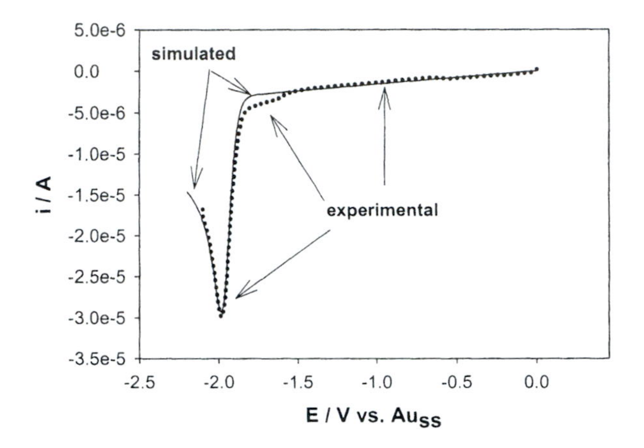

***Figure 4** Cyclic voltammogram for the reduction of benzaldehyde in [Bmim][Tfi] (broken line) at 0.1 V s^{-1} and simulated voltammogram (solid line).*

The value for E_p-$E_{p/2}$ (where $E_{p/2}$ is the potential where $i = i_p/2$) for such waves is typically 65 mV, therefore, using Equation 1, where n_a is the number of electrons in the rate-limiting step;

$$E_p - E_{p/2} = \frac{47.7}{\alpha n_a} \tag{1}$$

α is determined as 0.7, with n_a as 1. Saveant et al. (*13*) have shown the first redox process is a one-electron event. Simulation of linear sweep voltammograms bases on a one-electron transfer process using the following parameters; $E^0 = -1.950$ vs. Au_{ss}, $k^o = 6.0 \times 10^{-3}$ cm s^{-1} and $\alpha = 0.7$, is shown in Fig 4 (solid line) where E_1^0 represents the formal potential for the reduction step;

and k° is the heterogeneous electrochemical rate constant. It is evident that the simulated curve overlays the experimental curve indicating a one-electron process as expected.

In order to detect radical species at lower potentials, the [Pyrd][Tfi] ionic liquid was used because this liquid is electrochemically robust to –3 V vs. Au_{SS}. Cyclic voltammograms for benzaldehyde reduction with progressively decreasing cathodic limits, along with a background voltammogram, are shown in Figure 5.

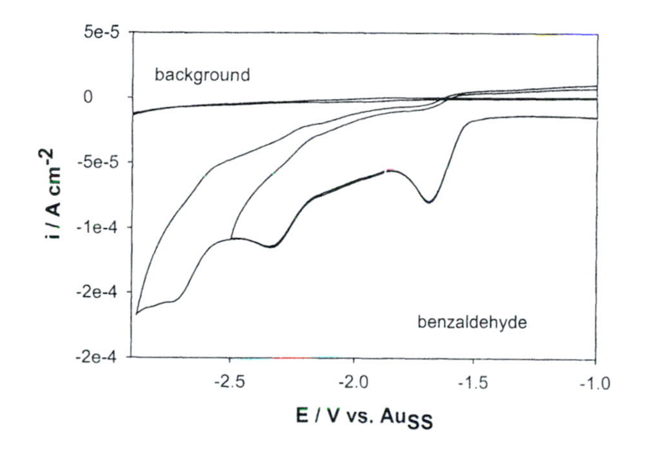

Figure 5 Cyclic voltammograms for the reduction of benzaldehyde in [Pyrd][Tfi] with potential limits of –2.5 and –3.0 V vs. Au_{SS} and a background voltammogram.

It is immediately evident that three irreversible (on the time-scale of our measurements) reduction processes are observed. Either one, or two, additional reduction processes are expected from the evidence presented by Saveant et al. (*13*) with two waves being present at high pH. The observation of two reduction process in the ionic liquid is therefore an unusual observation and obviously suggests the presence of two distinct benzaldehyde ketyl radical species are formed upon reduction.

Figure 6 shows square-wave voltammograms (Osteryoung-type), measured at a pulse frequency of 5 Hz with a pulse amplitude of 5 mV, for benzaldehyde reduction which again demonstrates the three reduction processes with E_1 = -1.75 ,E_2 = -2.38 and E_3 = -2.75 V vs. Au_{SS}.

From both Figures., it is immediately evident that the reduction processes are well separated and appear to be one-electron reduction processes. The first reduction wave (- 1.75 V vs. Au_{SS}) can be attributed to the one-electron reduction of the parent benzaldehyde to form the benzyl radical anion (Equation 2). Now, Saveant et al. (*13*) have attributed the first follow-up electron-transfer process (-2.38 V vs. Au_{SS}) to the reduction of the neutral α-hydroxybenzyl radical (Equation 4), which is formed by the protonation (probably by trace H_2O) of the initial radical anion generated at –1.75 V vs. Au_{SS} (Equation 3), and the 2 second reduction being attributed to the reduction of the benzaldehyde radical anion.

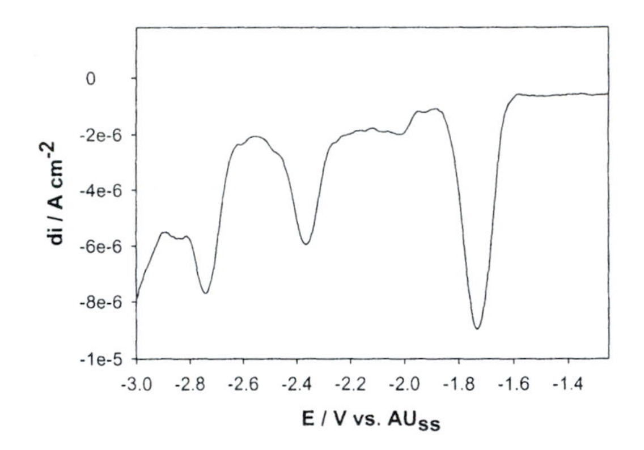

Figure 6 Square-wave voltammogram for the reduction of benzaldehyde in [Pyrd][Tfi] ionic liquid.

$$PhCHO + e^- \rightleftharpoons PhCHO^{\bullet -} \quad (2)$$

$$PhCHO^{\bullet -} + H^+ \rightleftharpoons PhCHOH^{\bullet} \quad (3)$$

$$PhCHO^{\bullet -} + e^- \rightleftharpoons PhCHO^{2-} \quad (4)$$

Comparison of the voltmmograms shown here with Saveant et al.'s (*13*) show remarkably well defined electrochemistry for the radical species in terms of potential separation and the magnitudes of the peak-currents that are attainable in the ionic liquid, relative to that observed ethanolic electrolytes. This suggests that the radical species are stabilised to a significant extent in the ionic liquid. This is of significance since an essential factor governing electrochemical triggering of radical chemistry is the relative reduction potentials values, i.e. electrochemically generated radicals must be stable at the potential of their formation, which is clearly the case for the system studied here. Interaction of the radical anions/neutral radicals with the ionic liquid probably explains the enhanced stability relative to conventional solvent systems. It is therefore evident that room temperature ionic liquids (for [Pyrd][Tfi] at least) may be exceedingly useful in exploring electrochemically-triggered radical chemistry.

Conclusions

The results presented here indicates that the electrochemical behaviour of dimethylmaleate (activated olefin) follows a one-electron reduction process in a similar fashion to that observed in organic solvent based electrolytes, therefore there seems to be the possibility for the (almost) direct transfer of electrochemical reactions to ionic liquid environments. In a similar fashion, the reduction of benzaldehyde is similar the that found previously in ethanolic electrolytes (*13*) in that the reduction of both the radical anion and protonated radical anion are observable. What is of significance is the stability of the radical species, which seems to indicate that the ionic liquid stabilises these species, possibly by charge-transfer complex formation. This open the possibility to easily perform chemistry on such reactive species.

Acknowledgments

APD would like to thank The Royal Society for a University Research Fellowship and CAB thanks the McClay Trust for a studentship. Thanks also to Professor K. R. Seddon and the team at QUILL for their generous support.

References

1. Welton, T. *Chem. Rev.* **1999**, *99*, 2071.
2. Holbrey, J. D.; Seddon, K. R. *Clean Products and Processes*. **1999**, *1*, 23.

3. Endres, F. *Chem. Comm.* **2002**, 892.
4. Nanjundiah, C.; McDevitt, F.; Koch, V. R. *J. Electrochem. Soc.* **1997**, *144*, 3392.
5. Hagiwara R.; Ito, Y. *J. Fluorine Chem.* **2000**, *105,* 221.
6. Earle, M. J; Seddon, K. R. *Pure Appl. Chem.* **2000** *72,* 1391.
7. Bonhote, P.; Dias, A. P.; Papageorgiou, N.; Kuppuswamy, K.; Gratzel, M. *Inorg. Chem.* **1996**, *35*, 1168.
8. Fuller, J.,R.; Carlin, R. T.; De Long, H. C.; Haworth, D. *J. Chem. Soc., Chem. Comm.* **1994**, 299.
9. Wilkes, J. S.; Zaworotko, J. *J. Chem. Soc., Chem Comm.* **1992**, 965.
10. Choi, H. M.; Welch, C. M. *Textile Chemist and Colorist* **1994**, *26*, 23.
11. Nashef, I. M.; Loenard,, M. L.; Kittle, M. C.; Matthews, M. A., Weidner, J.. W. *Electrochemical and Solid-state Letters* **2001**, *4*, D16.
12. Bard, A. J.; Puglisi, V. J.; Kenkel, J. V.; Lomax, A. *Trans. Faraday Soc.* **1973** *56* 353.
13. Andrieux, C. P.; Grzeszczuk, M.; Saveant, J. M. *J. Am. Chem. Soc.*, **1991**, *113*, 880.

Chapter 34

Solvent–Solute Interactions in Ionic Liquid Media: Electrochemical Studies of the Ferricenium–Ferrocene Couple

M. Cristina Lagunas, William R. Pitner, Jan-Albert van den Berg, and Kenneth R. Seddon

The QUILL Research Centre, The Queen's University of Belfast, Belfast, BT9 5AG, Northern Ireland, United Kingdom

Electrochemical studies of the ferricenium/ferrocene ($[Fc]^+/[Fc]$) redox couple have been carried out in ionic liquids containing the 1-butyl-3-methylimidazolium cation $[C_4mim]^+$ and the anions $[CF_3SO_3]^-$, $[N(SO_2CF_3)_2]^-$ and $[PF_6]^-$. The $[Fc]^+/[Fc]$ couple shows electrochemical reversibility in all cases, and controlled-potential electrolysis of ferrocene to the monocation is chemically reversible. The diffusion coefficients and Stokes-Einstein products of both the oxidized and the reduced species have been determined by using cyclic voltammetry, normal pulse voltammetry and rotating disk electrode voltammetry with both platinum and glassy carbon working electrodes. The solvodynamic radii of the diffusing species in each ionic liquid were estimated using the measured diffusion coefficients and a corrected version of the Stokes-Einstein equation. A dependency of the solvodynamic radii upon the anion of the ionic liquid was observed.

Introduction

The use of ionic liquids as green reaction media in industrially relevant processes has increased dramatically in the last years. The haloaluminate(III) ionic liquids, initially developed in the late 1940s by Hurley and Wier [*1*], have found infrequent use as solvent systems due to their moisture and air-sensitivity. The report in 1992 by Wilkes and Zaworotko [*2*] of air- and moisture-stable ionic liquids opened the door for their application in conventional chemical reactions, many of which are metal-mediated processes. Because metal complexes are expected to behave differently in ionic liquids than in conventional solvents, fundamental questions about solvent-solute interactions in these media must be addressed.

Valuable information on solvent-solute interactions can be obtained through electrochemical studies of the transport properties of metal complexes in ionic liquids. Unfortunately, very little fundamental information concerning the electrochemical properties of ionic liquids, especially the non-haloaluminate(III) ionic liquids, has yet been produced. Two recent investigations probed the fundamental nature of mass transport in ionic liquids [*3,4*]. With the aim of developing ionic liquids as components of electrochemical gas phase reactors and gas sensor systems, Schröder *et al.* [*3*] demonstrated the effect of water on the diffusion of ionic and neutral species and proposed a model for the nanoscale structure of water in ionic liquids. Quinn *et al.* [*4*] investigated the partition of water and the ionic liquid in biphasic mixtures. There is a demonstrable need for more investigations into the fundamental nature of mass transport in ionic liquids.

The Ferricenium/Ferrocene Couple

The ferricenium/ferrocene ($[Fc]^+/[Fc]$) redox couple has been thoroughly studied in aqueous and non-aqueous solvents [*5*]. Ferrocene is widely used as an internal standard for reporting electrode potentials, due to the assumption that the standard electrode potential of the $[Fc]^+/[Fc]$ couple remains invariant, irrespective of the nature of the solvent [*6*]. Fundamental studies on metal complexes in chloroaluminate(III) ionic liquids were initiated by Chum *et al.* [*7*] whose investigations included ferrocene. Interest in studying the behaviour of the $[Fc]^+/[Fc]$ model system in ionic liquids has continued [*8-12*]. Cyclic voltammetric studies showed that the couple was electrochemically stable and reversible in acidic (1:2 molar ratio) mixtures of *N*-ethylpyridinium bromide:aluminium(III) chloride and *N*-butylpyridinium chloride:aluminium(III) chloride systems of various compositions (from basic 1:0.75 to acidic 1:2 mixtures) [*9*]. The ferrocene oxidation potential (0.24 V *vs.* the Al(III)/Al reference electrode [*9*]) was reported to be independent of solvent acidity, indicating that no significant interaction occurred between the ionic liquid and ferrocene.

More detailed electrochemical studies in mixtures of *N*-butylpyridinium chloride:aluminium(III) chloride showed that several metallocenes, including ferrocene, exhibited complicated behaviour dependent on solvent acidity [*8b*, *10*].

Both ferrocene and ferricenium were stable in neutral mixtures. In acid mixtures, ferrocene was oxidised by traces of moisture and dioxygen. In basic mixtures, ferricenium reacted with excess chloride. In addition, the electron-transfer rate constant for the couple in basic mixtures of 1-ethyl-3-methylimidazolium chloride:aluminium(III) chloride was found to decrease with increased solvent basicity (*i.e.*, increased viscosity) [*8c*]. This has been related to the slow relaxation of the highly associated basic ionic liquid, rather than to solvation effects, given the constancy of the formal potential of the system over the solvent composition range.

In the first investigation of electrochemical systems in non-haloaluminate(III) ionic liquids, Fuller *et al.* [*11*] examined the behaviour of ferrocene in the 1-ethyl-3-methylimidazolium tetrafluoroborate ionic liquid. The $[Fc]^+/[Fc]$ couple was found to be chemically stable and electrochemically reversible. Hultgren *et al.*, have recently looked at a number of metallocene derivatives, including ferrocene, in the 1-butyl-3-methylimidazolium hexafluorophosphate ionic liquids, in an effort to provide a method for calibrating reference potentials in ionic liquids [*12*].

In this paper the $[Fc]^+/[Fc]$ redox system was characterized electrochemically in ionic liquids containing the same cation (1-butyl-3-methylimidazolium $[C_4mim]^+$) and three different anions (trifluoromethansulfonate $[OTf]^-$, bis(trifluoromethanesulfonyl)imide $[NTf_2]^-$ and hexafluorophosphate $[PF_6]^-$).

Solvent-Solute Interactions

Electrochemistry provides a convenient method for investigating solvent-solute interactions. Many electrochemical techniques, such as cyclic voltammetry, rotating disc electrode voltammetry and normal pulse voltammetry, allow the determination of the diffusion coefficient of electrochemically active species. From the diffusion coefficient, the solvodynamic radius of the diffusion species can be estimated [*13*]. The solvodynamic radius provides dimensional information about the solute and any solvating molecules or ions.

Stokes' law [*14*] (Equation 1)

$$F = \pi\, b\, \eta\, v\, r_s \qquad (1)$$

relates the drag force F opposing a macroscopic sphere moving in an ideal hydrodynamic medium with the viscosity of the medium η and the solvodynamic radius r_s and velocity v of the macroscopic sphere. The numerical constant b usually has a value of 6 or 4, depending on whether 'stick' or 'slip' boundary conditions between the moving sphere and the fluid in contact with it apply, respectively. The 'stick' condition applies for a large spherical particle in a solvent of low relative molecular mass. The 'slip' condition applies when a molecule diffuses through a medium consisting of molecules of comparable size.

The application of Stokes' law to the diffusional movements of particles gives the Stokes-Einstein relation (Equation 2),

$$D = k_B T / \pi \; b \; \eta \, r_s \tag{2}$$

where k_B is the Boltzmann's constant, T is the absolute temperature and D is the diffusion coefficient of the analyte. Equation 2 is sometimes expressed in the form of the Stokes-Einstein product $D \; \eta \,/\, T$ (Equation 2a).

$$D \, \eta / T = k_B / \pi \; b \, r_s \tag{2a}$$

The Stokes-Einstein product should be constant for a given analyte provided that the solvodynamic radius, r_s, of the moving particle does not vary and that Equation 1 is valid. In spite of its approximate nature, Equation 2 holds reasonably well on the molecular level, and it provides a useful starting point for discussing the dimensions of solvated solute molecules.

It has been shown that the Stokes-Einstein products of ferrocene in various chloroaluminate(III) ionic liquids remain almost constant, with values lying in a narrow interval: 6.0-7.7 $\times$ 10^{-10} g cm s^{-2} K^{-1} [*7-10*]. Considering the different methods employed in the determination of the diffusion coefficients and in the measurements of the viscosity of the solvents, the constancy of the Stokes-Einstein product is remarkable. This clearly demonstrates that the same molecular species (with the same r_s) is in motion in all the chloroaluminate(III) ionic liquids previously studied. As no evidence of strong solvent-solute interactions has been found, this diffusion species is assumed to be unsolvated ferrocene.

From Equation 2, if the Stokes-Einstein product for ferrocene lies between 6.0-7.7 $\times$ 10^{-10} g cm s^{-2} K^{-1}, the solvodynamic radius of unsolvated ferrocene would then be in the range 0.98-1.22 Å, for b = 6, or 1.47-1.83 Å, for b = 4. These values are unrealistically low when compared with the known crystallographic radius of ferrocene (3.5 Å) [*15*].

Robinson and Stokes [*16*] demonstrated that while Equation 2 is applicable for particles greater than *ca.* 5 Å in radius, it gives radii that are considerably too small when applied to particles smaller than this. They suggested a correction factor be introduced as the ratio r/r_s, where r represents the radius estimated from molecular volumes or other models, and r_s the radius calculated by Equation 2. Thus, the corrected solvodynamic radius, r_c, would be given by Equation 3:

$$r_c = (k_B T / \pi \; b \; \eta D)(r / r_s) \tag{3}$$

If an analogous approach is to be used in the case of ferrocene in ionic liquids, then when r is substituted by the crystallographic radius of ferrocene (3.5 Å), and r_s by the mean radius calculated from the Stokes-Einstein products (*i.e.*, 1.10 or 1.65 Å for b = 6 or 4, respectively), the solvodynamic radius can be approximated by Equation 3a.

$$r_c = k_B T / 2 \pi \; \eta \, D \tag{3a}$$

Experimental

$[C_4mim]Cl$ was prepared by the reaction of chlorobutane with 1-methylimidazole in ethanenitrile and recrystallised with ethyl ethanoate [*17*]. $[C_4mim]Cl$ was used as a precursor for all the ionic liquids used in this work [*2*]. Its reaction with Na[OTf] in propanone, or with $Li[NTf_2]$ or $H[PF_6]$ in water gave the corresponding ionic liquids $[C_4mim][OTf]$, $[C_4mim][NTf_2]$ and $[C_4mim][PF_6]$, respectively. Residual chloride was removed by dissolving the ionic liquids in dichloromethane and washing the mixture with water until the aqueous washings did not turn cloudy in the presence of silver nitrate. In order to remove residual water and organic solvents, the liquids were first dried under reduced pressure on a rotoevaporator at 60 °C and then dried under high vacuum for 5-12 h at 60 °C.

Ferrocene was sublimed prior to use. $[Fc][PF_6]$ was prepared by oxidation of ferrocene with H_2SO_4 in water in the presence of $K[PF_6]$ [*18*]. Both products were stored under a dry dinitrogen atmosphere.

The ferrocene and ferricenium solutions used for electrochemical experiments were prepared in three different ways. Method 1 involved dissolving a weighed amount (*ca.* 10-15 mg) of solute (ferrocene or $[Fc][PF_6]$) in the corresponding mass (*ca.* 10-15 g) of ionic liquid. Method 2 involved introducing an exact volume (2-3 cm^3) of a 0.02 mol l^{-1} solution of the solute in dry dichloromethane (CH_2Cl_2) into a flask, removing the CH_2Cl_2 by evaporation under vacuum and adding an exact mass of the ionic liquid (*ca.* 10-15 g) to the solute residue. In both Method 1 and 2, the mixtures were stirred under vacuum (2-5 h) until homogeneous orange (ferrocene) or dark blue (ferricenium) solutions were formed. Method 3 involved the bulk electrolysis (oxidation) of a solution of ferrocene in $[C_4mim][PF_6]$ to produce a solution of $[Fc][PF_6]$, and will be discussed in detail below. All ionic liquids and ionic liquid solutions of ferrocene and $[Fc][PF_6]$ were stored either under dinitrogen or under vacuum. Table 1 summarises for the preparation of the solutions used in this investigation.

Table 1. Preparation of solutions for electrochemical investigation.

Solution Label	*Ionic Liquid*	*Solute*	*Preparation Method*	*Concentration / mol l^{-1}*
A	$[C_4mim][PF_6]$	[Fc]	1	0.0048
B	$[C_4mim][PF_6]$	[Fc]	2	0.0043
C	$[C_4mim][PF_6]$	[Fc]	2	0.0039
D	$[C_4mim][PF_6]$	$[Fc]^+$	3[a]	0.0039
E	$[C_4mim][PF_6]$	$[Fc]^+$	1	0.0051
F	$[C_4mim][OTf]$	[Fc]	2	0.0053
G	$[C_4mim][NTf_2]$	[Fc]	2	0.0051

[a]Electrolysis of Solution B.

In all cases, the purity of the ionic liquid was determined by viscosity measurements using an LVDV-II Brookfield cone and plate viscometer. These values were used in the calculation of the diffusion coefficients and solvodynamic radii. Densities of the ionic liquids were measured with an Anton Paar DMA 4500 density meter. The electronic absorption spectra were recorded on a Perkin Elmer Model Lambda2 UV/VIS spectrometer.

All electrochemical experiments were carried out with an EG&G PARC Model 283 potentiostat/galvanostat connected to a PC through an IEEE-488 bus and controlled using EG&G Parc Model 270/250 Research Electrochemistry version 4.23 software. Positive feedback *iR* compensation was employed to eliminate errors due to solution resistance.

Voltammetric experiments were performed at 21 ± 2 °C either inside a controlled-atmosphere box or on a bench under a dry dinitrogen flow, using a three-electrode system. The non-aqueous reference electrode (BAS) was a silver wire immersed in a glass tube containing a 0.1 mol l^{-1} solution of $AgNO_3$ in the $[C_4mim][NO_3]$ ionic liquid separated from the bulk solution by a Vycor plug. All potentials reported are referenced against the Ag(I)/Ag couple. The counter electrode was a platinum coil immersed directly in the bulk solution. The working electrode was a platinum disc (BAS, 1.6 mm diameter or Pine, 5.4 mm diameter) or a glassy carbon disc (BAS, 3.0 mm diameter or Pine, 7.9 mm diameter). Rotating disc electrode experiments were carried out using a Pine Model AFM4RXE Analytical Rotator with MSRX Speed Control.

Results

Spectroscopy

The electronic absorption spectra of ferrocene obtained in the three ionic liquid systems studied were similar to each other and showed no significant differences with the spectra obtained in conventional solvents (*i.e.*, λ_{max} at *ca.* 441 nm [ε = 89 l mol^{-1} cm^{-1}] and 321 nm [ε = 57 l mol^{-1} cm^{-1}] [*19*]). The solutions of ferrocene in these ionic liquids kept under dinitrogen were stable for weeks. Spectra were also recorded for ferricenium in $[C_4mim][PF_6]$ with λ_{max} = 617 nm [ε = 340 l mol^{-1} cm^{-1}]. Typical spectra for ferrocene and ferricenium in $[C_4mim][PF_6]$ are shown in Figure 1.

Cyclic voltammetry

The $[Fc]^+/[Fc]$ couple was found to undergo reversible one-electron transfer in all three ionic liquid systems studied. Cyclic voltammetric experiments were performed at both platinum and glassy carbon (GC) electrodes, over scan rates (ν) ranging from 0.01 to 0.1 V s^{-1}. Typical voltammograms are shown in Figure 2. From data collected at ten different scan rates, the average value of the half-peak potential $E_{1/2}$ and the peak-to-peak separation ΔE_p were calculated for each sample. The ratio of the peak currents $| i_p^c / i_p^a |$ were calculated by using the Nicholson equation [*20*] from sets of data collected at 0.05 V s^{-1} at ten different switching potentials (E_λ = 0.050–0.275 V). Average values for $E_{1/2}$, ΔE_p and $| i_p^c / i_p^a |$ are shown in Table 2.

According to the Randles-Sevcik equation (Equation 4) [*13*]

$$i_p = 2.69 \times 10^5 n^{3/2} A D^{1/2} C \nu^{1/2} \tag{4}$$

for a reversible diffusion controlled system at 25 °C there is a linear relationship between the peak current and the scan rate, where i_p is the peak current in A, ν is the scan rate in V s^{-1}, A is the electrode area in cm^2, D is the diffusion coefficient in cm^2 s^{-1}, and C is the bulk concentration in mol cm^{-3}. Typical plots of i_p *vs.* $\nu^{1/2}$ are given in Figure 3. Using Equation 4, these plots were used to calculate the diffusion coefficients of ferrocene in each of the ionic liquids. These calculated values are collected in Table 3.

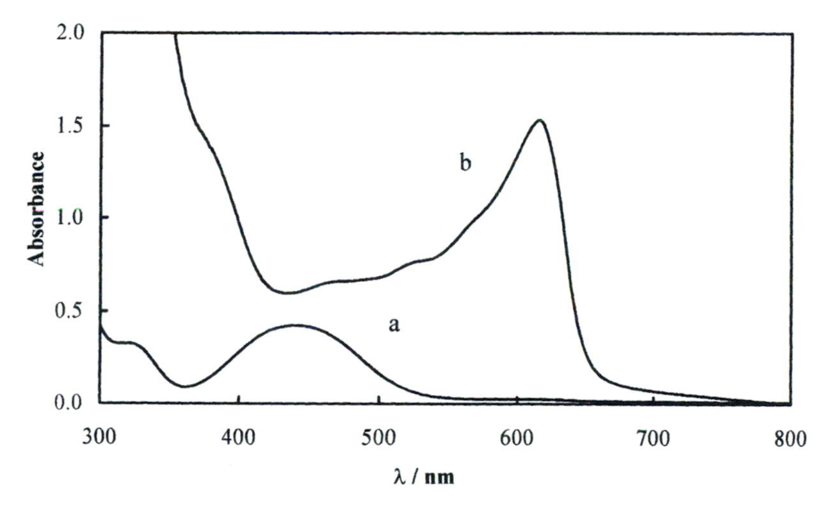

Figure 1. UV/VIS spectra of: (a) a solution of 0.0043 mol l^{-1} [Fc] in $[C_4mim][PF_6]$; (b) the same solution after 92% conversion to $[Fc]^+$ by oxidative electrolysis.

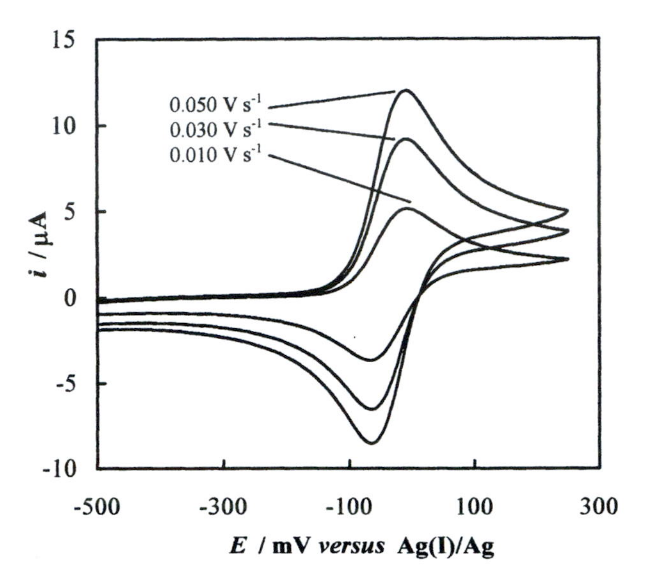

Figure 2. Cyclic voltammograms recorded at a platinum electrode in a 0.0043 mol l^{-1} solution of [Fc] in $[C_4mim][PF_6]$.

Table 2. Cyclic voltammetric data for [Fc] and $[Fc]^+$ in $[C_4mim]X$ ionic liquids.

	Working electrode	$E_{1/2}$ / *mV*	ΔE_p / *mV*	$\lvert i_p^c / i_p^a \rvert$
		[Fc]		
$[PF_6]^-$	Pt[a]	-0.032 ± 0.001	0.050 ± 0.004	0.97 ± 0.02
	GC[a]	-0.034 ± 0.001	0.058 ± 0.004	1.03 ± 0.02
$[OTf]^-$	Pt[a]	-0.011 ± 0.001	0.069 ± 0.005	0.95 ± 0.02
	GC[a]	-0.010 ± 0.001	0.072 ± 0.006	0.97 ± 0.02
$[Tf_2N]^-$	Pt[a]	-0.025 ± 0.001	0.058 ± 0.004	0.98 ± 0.02
		$[Fc]^{+}$[c]		
$[PF_6]^-$	Pt[b]	-0.020 ± 0.001	0.058 ± 0.005	1.11 ± 0.03
	GC[b]	-0.021 ± 0.001	0.053 ± 0.003	1.05 ± 0.02

[a]BAS working electrode. [b]Pine working electrode. [c]$[Fc]^+$ produced by electrolysis (Method 3).

Table 3. Diffusion coefficients of [Fc] in [C_4mim]X calculated from data obtained by cyclic voltammetry (CV), normal pulse voltammetry (NPV) and rotating disk electrode voltammetry (RDE).

Solution label	*Electrochemical technique*	*Working electrode*	*Slope*[c]	R^2 [d]	$D \times 10^7$ / $cm^2\ s^{-1}$
		[C_4mim][PF_6]			
A	CV	Pt[a]	8.01	0.9997	0.95
		GC[a]	30.39	0.9966	1.11
	NPV	Pt[a]	1.70	0.9998	1.05
		GC[a]	6.21	0.9908	1.13
	RDE	Pt[b]	0.76	0.9770	0.76
		GC[b]	1.04	0.9424	0.66
B	CV	Pt[b]	53.71	0.9991	0.80
		GC[b]	82.41	0.9999	0.83
	NPV	Pt[b]	10.81	0.9997	0.79
		GC[b]	15.93	0.9995	0.76
	RDE	Pt[b]	0.68	0.9575	0.77
		GC[b]	1.03	0.9541	0.77
C	CV	Pt[b]	43.94	0.9999	0.65
		GC[b]	68.15	0.9996	0.69
	NPV	Pt[b]	9.82	0.9997	0.79
		GC[b]	13.62	0.9993	0.67
	RDE	Pt[b]	0.56	0.9446	0.67
		GC[b]	0.82	0.9583	0.64
		[C_4mim][OTf]			
F	CV	Pt[a]	7.29	0.9992	0.65
		GC[a]	22.29	0.9999	0.49
	NPV	Pt[a]	1.56	0.9994	0.72
		GC[a]	4.92	0.9985	0.58
		[C_4mim][NTf_2]			
G	CV	Pt[a]	15.21	0.9991	3.04
		GC[a]	50.62	0.9995	2.72
	NPV	Pt[a]	3.19	0.9997	3.27
		GC[a]	10.82	0.9990	3.04

[a]BAS. [b]Pine. [c]Slope from plots of i_p *vs.* $\nu^{1/2}$, $i(t)$ *vs.* $t^{-1/2}$ or and i_L *vs.* $\omega^{1/2}$ from CV, NPV and RDE data, respectively. The least-squares fitted lines were forced to intersect the origin. [d]Correlation coefficients of the least-squares fitted lines.

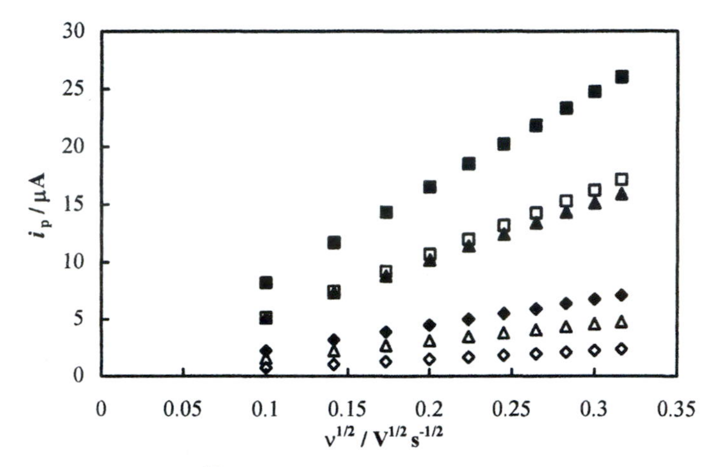

Figure 3. Plots of i_p vs. $v^{1/2}$ from cyclic voltammetric data for a solution of 0.0043 mol l^{-1} [Fc] in [C_4mim]X recorded at platinum (open symbols) and glassy carbon (closed symbols) electrodes: □ , ■ X = $[PF_6]^-$; △ , ▲ X = $[OTf]^-$; ◇ , ◆ X = $[NTf_2]^-$.

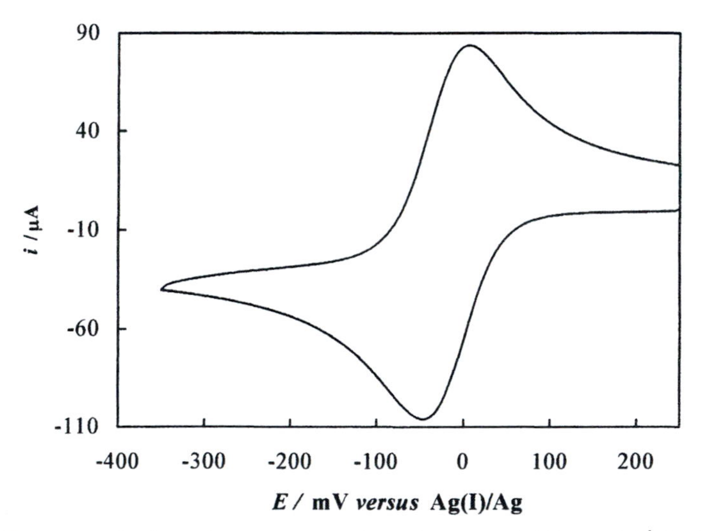

Figure 4. Cyclic voltammogram of a solution of 0.0039 mol l^{-1} $[Fc]^+$ in [C_4mim][PF_6] recorded at a glassy carbon electrode with a scan rate of 0.050 V s^{-1}.

Cyclic voltammograms were also recorded for solutions of ferricenium in the $[C_4mim][PF_6]$ ionic liquid. A typical voltammogram is shown in Figure 4. Cyclic voltammetric data for $E_{1/2}$, ΔE_p and $| i_p^c / i_p^a |$ for the reduction of ferricenium to ferrocene are collected in Table 2. Linear plots of i_p *vs.* $\nu^{1/2}$ for ferricenium (not shown) were used to calculate values for the diffusion coefficient and Stokes-Einstein product of ferricenium (Table 4).

Table 4. Diffusion coefficients of $[Fc]^+$ in the $[C_4mim][PF_6]$ calculated from data obtained by CV, NPV and RDE.

Solution label	*Electrochemical technique*	*Working electrode*	*Slope*[c]	R^2 [d]	$D \times 10^7$ / $cm^2\ s^{-1}$
D	CV	Pt[a]	-29.73	0.9989	0.30
		GC[a]	-47.46	0.9994	0.33
	NPV	Pt[a]	-6.62	0.9948	0.36
		GC[a]	-9.41	0.9915	0.32
	RDE	Pt[a]	-0.35	0.9932	0.32
		GC[a]	-0.53	0.9898	0.34
E	CV	Pt[b]	-4.24	0.9598	0.24
		GC[b]	-16.33	0.9921	0.28
	NPV	Pt[b]	-0.97	0.991	0.30
		GC[b]	-3.25	0.975	0.27

[a]Pine working electrode. [b]BAS working electrode. [c]Slope from plots of i_p *vs.* $\nu^{1/2}$, $i(t)$ *vs.* $t^{-1/2}$ or and i_L *vs.* $\omega^{1/2}$ from CV, NPV and RDE data, respectively. The least squares fitted lines were forced to intersect the origin. [d]Correlation coefficients of the least-squares fitted lines.

Normal pulse voltammetry

When the potential is stepped from a value in which no electrolysis occurs to a value in the mass-transfer-controlled region, the current-time response is given by the Cottrell equation (Equation 5) [*13*].

$$i(t) = n F A C D^{1/2} \pi^{-1/2} t^{-1/2} \qquad (5)$$

To ensure diffusion controlled conditions in our experiments, the potentials at which current-time responses were previously determined by normal pulse voltammetry on each sample. Potential steps between 5 and 400 mV from an initial value (E_1) of –0.25 V were applied to the solutions and the current-time transients measured. A typical normal pulse voltammogram is shown in Figure 5.

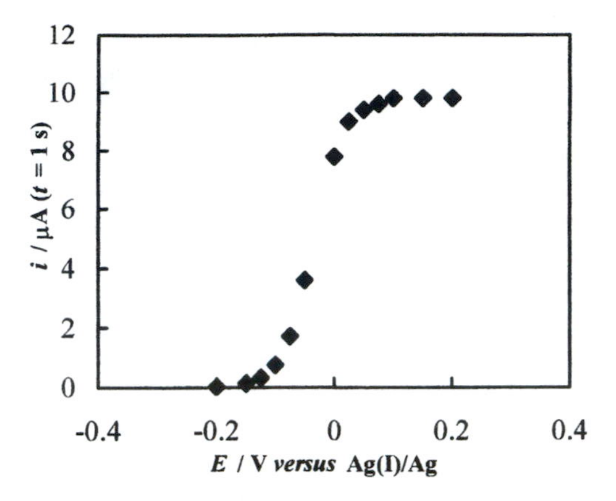

Figure 5. Normal pulse voltammogram of a 0.0039 mol l^{-1} solution of [Fc] in $[C_4mim][PF_6]$ at a platinum disc working electrode.

In each case, three values of potential in which the curve had clearly reached a plateau after $E_{1/2}$ were chosen. The average currents from the chronoamperograms at such potentials *vs.* $t^{-1/2}$ gave linear plots (Figure 6) at times between 1 s and 10 s. Equation 5 could therefore be used to estimate the diffusion coefficients of ferrocene and ferricenium (Table 3) from the slope of the lines passing through the origin.

The half-wave potentials of ferrocene solutions determined by normal pulse voltammetry were –0.034 ± 0.003 V and –0.011 ± 0.002 V, for $[C_4mim][PF_6]$ and $[C_4mim]$[OTf], respectively. These values compared well with those obtained from cyclic voltammetry. Normal pulse voltammetry data were not collected for the $[C_4mim][NTf_2]$ system.

Rotating disk electrode voltammetry

Rotating disc electrode voltammetry was performed in solutions of ferrocene and ferricenium in the ionic liquids rotation speeds between 1000 rpm and 2000 rpm by using both Pt and GC electrodes (scan rate = 0.005 V s^{-1}). Typical voltammograms are shown in Figure 7. In most cases, the curves presented some degree of hysteresis when reverse scans were performed, which may be attributed to adsorption of ferrocene on the electrode surface.

The limiting current (i_L) of a voltammogram obtained at a rotating disk electrode is described by the Levich equation (6) [*13*]

$$i_L = 0.620\, n\, F\, A\, D^{2/3} \gamma^{-1/6} C\, \omega^{1/2} \qquad (6)$$

where ω is the angular rotation rate of the electrode and γ is the kinematic viscosity of the solution. Typical plots of i_L *vs.* $\omega^{1/2}$ are shown in Figure 8 and were used to calculate the diffusion coefficient for ferrocene and ferricenium using Equation 6. These calculated values are collected in Table 3.

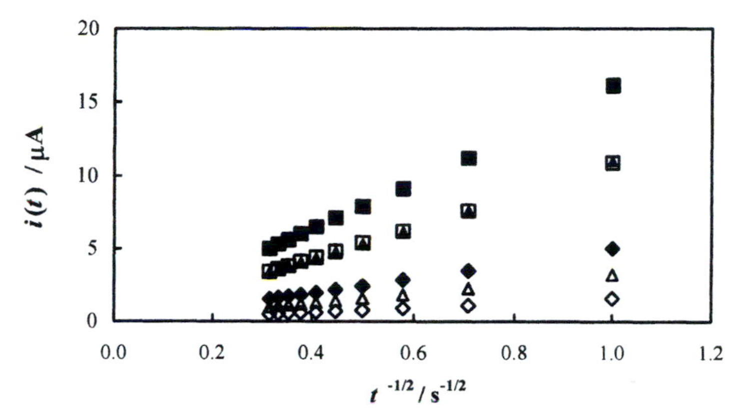

Figure 6. Plots of i(t) vs. $t^{-1/2}$ for solutions of [Fc] in [C_4mim]X ionic liquids recorded at platinum (open symbols) and glassy carbon (closed symbols) electrodes: □, ■ X = [PF_6]$^-$; △, ▲ X = [OTf]$^-$; ◇, ◆ X = [NTf_2]$^-$.

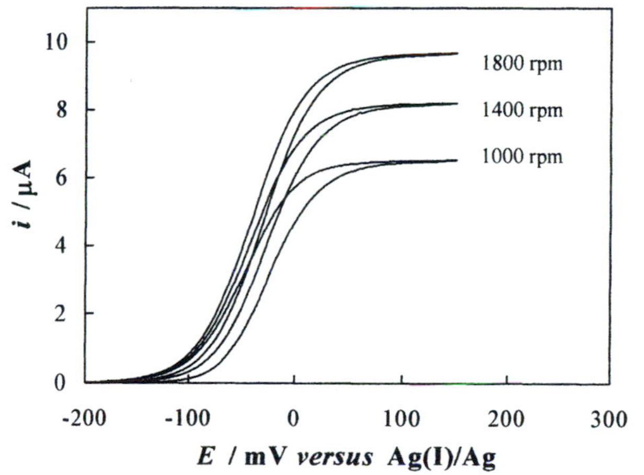

Figure 7. Rotating disc electrode voltammograms of a 0.0043 mol l^{-1} solution of [Fc] in [C_4mim][PF_6] at a platinum electrode.

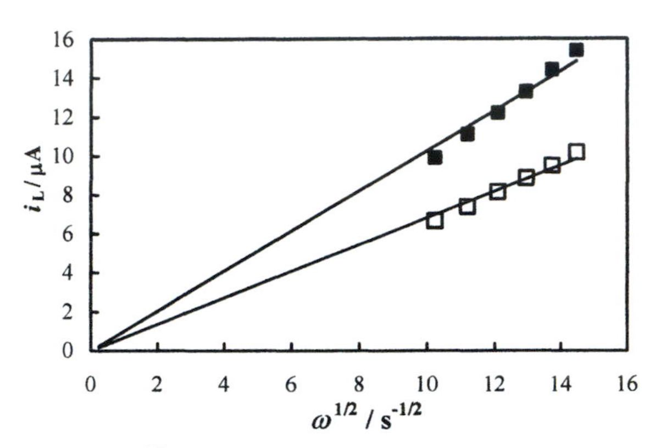

Figure 8. Plots of i_L vs. $\omega^{1/2}$ from the RDE data for [Fc] in [C_4mim][PF_6] at glassy carbon (solid symbol) and platinum (open symbol) electrodes.

Table 5. Mass transport data of the [Fc] and $[Fc]^+$ ionic liquid solution.

Ionic liquid	*Viscosity*[a] */ cp*	*Density*[a] */ g cm^{-3}*	$D_{avg} \times 10^7$ / $cm^2\ s^{-1}$	$D\ \eta / T \times 10^{10}$ / $g\ cm\ s^{-2}\ K^{-1}$	r_c / Å
			[Fc]		
[C_4mim][PF_6]	272	1.363	0.79 ± 0.09	7.2 ± 0.8	3.0 ± 0.4
[C_4mim][NTf_2]	50.2	1.436	3.0 ± 0.2	5.0 ± 0.1	4.3 ± 0.3
[C_4mim][OTf]	79.2	1.295	0.61 ± 0.09	1.6 ± 0.2	13 ± 2
			$[Fc]^+$		
[C_4mim][PF_6]	272	1.363	0.31 ± 0.04	2.8 ± 0.4	7.7 ± 1.0

[a]Viscosity and density measured for the neat ionic liquids.

All the mass transport data of the ferrocene and ferricenium solutions are summarised in Table 5. Average diffusion coefficients were calculated based on the data in Tables 3 and 4. Using these values and the measured viscosity of the ionic liquids, Equation 2a and Equation 3a were used to calculate the Stokes-Einstein products and solvodynamic radii, respectively. The values for viscosity and density measured for these solutions agreed with previously published results [*21*].

Bulk electrolysis of ferrocene to ferricenium

Controlled potential bulk electrolysis (E_{app} = +0.5 V *vs.* Ag(I)/Ag) of ferrocene solutions in $[C_4mim][PF_6]$ resulted in dark blue solutions. Electrolysis of Solution B resulted in the passage of 4.367 C after *ca.* 60 h. Typical cyclic voltammograms (as in Figure 4) showed the expected reversible one-electron wave for ferricenium at both GC and Pt electrodes. Ferricenium solutions produced by electrolysis were stable when kept under dinitrogen. The electronic absorption spectrum (Figure 1) of an electrolysed solutions appeared consistent with that of ferricenium (*i.e.*, λ_{max} = 617 nm). For comparison, a $[C_4mim][PF_6]$ solution of $[Fc][PF_6]$, prepared by chemical oxidation of ferrocene with H_2SO_4, was also studied. Its UV spectra and electrochemical behaviour were similar to the ferricenium solution produced by electrolysis.

Discussion

The data from cyclic voltammetry (Table 2) indicated that the $[Fc]^+/[Fc]$ couple is a reversible one-electron reaction in the ionic liquids studied. The values of $E_{1/2}$ lie between −0.010 and −0.034 V. Significant variations related to the nature of the working electrode, the preparation of the solution, its concentration or its age were not observed. In all cases, except in $[C_4mim][OTf]$, values for ΔE_p were found to be close to the theoretical value of 59 mV for a reversible system [*13*]. A small upward trend on ΔE_p when the scan rate used approached 0.1 V s^{-1} was observed in some sets of data, indicating *quasi*-reversible electron-transfer or, more probably, uncompensated resistance [*13*]. In the case of $[C_4mim][OTf]$, the higher values found for peak-to-peak separation (an average between 69-72 mV) are attributed to uncompensated resistance, since positive feedback *iR* compensation could no be used due to oscillation problems in this particular case. Other criteria for reversibility are well met by all cases. For example, the ratios of $| i_p^c / i_p^a |$ were close to one and the oxidation peak currents (i_p) were proportional to $v^{1/2}$.

When CV or NPV techniques are used, the diffusion coefficients (Tables 3 and 4) can be calculated reliably (*i.e.*, R^2 > 0.99) in all cases with the exception of ferricenium in Solution E (Table 4). When RDE techniques are used, a reliable linear correlation of i_L *vs.* $\omega^{1/2}$ was not obtained and the intercepts were always non-zero. Even so, the values of *D* calculated from the RDE data were in good agreement with those estimated from CV and NPV data. The deviation of the plots of i_L *vs.* $\omega^{1/2}$

from lines intersecting the origin could be related to turbulent flow problems. Accordingly, experiments performed at lower rotation speeds (300-1000 rpm), at scan rate of 1 mV s^{-1}, gave better correlation coefficients for the least-squares i_L *vs.* $\omega^{1/2}$ fitted lines ($R^2 > 0.99$). Alternatively, the non-zero intercept could be due to the high kinematic viscosities of the ionic liquids, which make the hydrodynamic boundary layers too large. Under such conditions, Equation 6 does not apply [*13*].

From our data, the average diffusion coefficients calculated at 25 °C for ferrocene in $[C_4mim][PF_6]$, $[C_4mim][NTf_2]$ and $[C_4mim][OTf]$ (Table 5) were (0.79 ± 0.09), (3.0 ± 0.2) and (0.61 ± 0.09) × 10^{-7} cm^2 s^{-1}, respectively. According to equation (3a), the estimated radii of ferrocene in $[C_4mim][PF_6]$, $[C_4mim][NTf_2]$ and $[C_4mim][OTf]$ would then be of *ca.* 3.0 ± 0.4, 4.3 ± 0.3 and 13 ± 2 Å. Whereas the calculated solvodynamic radii of ferrocene in $[C_4mim][PF_6]$ and $[C_4mim][NTf_2]$ correspond to the crystallographic radius of ferrocene (3.5 Å), the solvodynamic radius of ferrocene in $[C_4mim][OTf]$ is significantly larger (13 ± 2 Å).

The most likely form of interaction between ferrocene and $[OTf]^-$ is through hydrogen bonding. Of the three anions studied, $[OTf]^-$ has the greatest tendency to form hydrogen bonds in the solid state [*22*]. This type of interaction has also been shown to play a part in the liquid structure of ionic liquids which contain $[OTf]^-$ [*23*]. Although the interaction between a neutral molecule, such as ferrocene, and an anion is unexpected, it has been observed in ionic liquids [*24*]. Interaction between hydrogens on the cyclopentadienyl rings and $[PF_6]^-$ or $[NTf_2]^-$ would not be probable. However, the nature of the diffusing species in $[C_4mim][OTf]$ is not clear. Further studies are underway and will be reported in a future paper.

From the calculated diffusion coefficient of ferricenium in $[C_4mim][PF_6]$ (0.31 ± 0.04 × 10^{-7} cm^2 s^{-1}), Equation 3a gives a corresponding solvodynamic radius of *ca.* 7.7 ± 1.0 Å. Since the crystallographic radius of ferricenium is not much different from that of ferrocene, the significantly larger solvodynamic radius of ferricenium in the $[C_4mim][PF_6]$ ionic liquid (3.0 ± 0.4 Å *vs.* 7.7 ± 1.0 Å) may indicate the presence of strong interactions between ferricenium and $[PF_6]^-$, whose crystallographic radius is *ca.* 3.2 Å [*25*]. Coulombic interactions of ferricenium with the tetrachloroaluminate anion have also been suggested, since the diffusion coefficient of ferricenium in neutral *N*-butylpyridinium chloride:aluminium(III) chloride was found to be about 25% smaller than that of ferrocene [*8b*].

This report marks the beginning of the first systematic investigation of the nature of solute-solvent interactions in non-chloroaluminate ionic liquid solutions using electrochemical techniques. As more ionic liquid systems are studied, using a variety of organic, inorganic and organometallic substrates as electrochemical probes, a more certain and detailed picture of these interactions will begin to emerge. Information gathered from crystallographic, spectroscopic and rheological investigations will be invaluable in supporting the theories drawn from such studies.

Acknowledgments

The authors thank the industrial members of the QUILL Research Centre for their financial support. M. C. L. thanks Fundación Flores-Valles, Ministerio de Educación y Ciencia and The Royal Society for their grants.

References

[*1*] (a) Hurley, F. H. U.S. Pat. 2,446,331, Aug. 3, 1948. (b) Wier, T. P.; Hurley, F. H. U.S. Pat. 2,466,349, Aug. 3, 1948. (c) Wier, T. P. U.S. Patent. 2,446,350, Aug. 3, 1948. (d) Hurley F. H.; Wier, T. P. *J. Electrochem. Soc.* **1951**, *98*, 203-206. (e) Hurley, F. H.; Wier, T. P. *J. Electrochem. Soc.* **1951**, *98*, 207-210.

[*2*] Wilkes, J. S.; Zaworotko, M. J. *Chem. Commun.* **1992**, 965-967.

[*3*] Schröder, U.; Wadhawan, J. D.; Compton, R. G.; Marken, F.; Suarez, P. A. Z.; Consorti, C. S.; de Souza, R. F.; Dupont, J. *New J. Chem.* **2000**, *24*, 1009-1015.

[*4*] Quinn, B. M.; Ding, Z.; Moulton, R.; Bard, A. J. *Langmuir* **2002**, *18*, 1734-1742.

[*5*] Kadish, K. M.; Ding, J. Q.; Malinski, T. *Anal. Chem.* **1984**, *56*, 1741-1744.

[*6*] Gritzner, G.; Kuta, J. *Pure Appl. Chem.* **1984**, *56*, 461-466.

[*7*] (a) Chum, H. L.; Koch, J. R.; Miller, L. L.; Osteryoung, R. A. *J. Am. Chem. Soc.* **1975**, *9*, 3264-3265. (b) Chum, H. L.; Koran, D.; Osteryoung, R. A. *J. Organometal. Chem.* **1977**, *140*, 349-359.

[*8*] (a) Karpinski, Z.; Nanjundiah, C.; Osteryoung, R. A. *J. Electrochem. Soc.* **1984**, *131*, C330. (b) Karpinkki, Z. J.; Nanjundiah, C.; Osteryoung, R. A. *Inorg. Chem.* **1984**, *23*, 3358-3364. (c) Karpinski, Z. J.; Song, S.; Osteryoung, R. A. *Inorganica Chimica Acta* **1994**, *225*, 9-14.

[*9*] Robinson, J.; Osteryoung, R. A. *J. Am. Chem. Soc.* **1979**, *101*, 323-327.

[*10*] (a) Gale R. J.; Job, R. *Inorg. Chem.* **1981**, *20*, 40-42. (b) Gale, R. J.; Job, R. *Inorg. Chem.* **1981**, *20*, 42-45. (c) Gale, R. J.; Motyl, K. M.; Job, R. *Inorg. Chem.* **1983**, *22*, 130-133.

[*11*] Fuller, J. Carlin, R. T.; Osteryoung, R. A. *J. Electrochem. Soc.* **1997**, *144*, 3881-3885.

[*12*] Hultgren, V. M.; Mariotti, A. W. A.; Bond, A. M.; Wedd, A. G. *Anal. Chem.* **2002**, *74*, 3151-3156.

[*13*] Bard, A. J.; Faulkner, L. R. *Electrochemical Methods: Fundamentals and Applications*, 2nd ed, Willey & Sons, New York , 2001; pp. 1-369.

[*14*] Bockris, J. O'M.; Reddy, A. K. N. *Modern Electrochemistry*, Plenum Press, New York , 1970; p. 377.

[*15*] Seiler, P.; Dunitz, J. D. *Acta Crystallogr. Sect. B* **1979**, *35*, 2020.

[*16*] Robinson, R. A.; Stokes, R. H. *Electrolyte Solutions*, 2nd ed., Butterworths Scientific Publications, London, 1959; pp. 120-126.

[*17*] Wilkes, J. S.; Levisky, J. A.; Wilson, R. A.; Hussey, C. L. *Inorg. Chem.* **1982** *21*, 1263.

[*18*] Coates, G. E.; Green, M. L. H.; Wade, K. *Organometallic Compounds*, Vol. 2, 3rd ed, Methuen & Co Ltd., London, 1968; p. 104.

[*19*] Rosemblum, M. *Chemistry of the Iron Group Metallocenes: Ferrocene, Ruthenocene, Osmocene*; Willey & Sons, New York, 1965; p. 46-47.

[*20*] Nicholson, R. S. *Anal. Chem.* **1996**, *38*, 1406.

[*21*] (a) Bonhôte P.; Dias, A.-P.; Papageorgiou, N; Kalyanasundaram, K; Michael Grätzel, M. *Inorg. Chem.* **1996**, *35*, 1168-1178. (b) Seddon, K. R.; Stark A.; Torres, M.-J. *Clean Solvents: Alternative Media for Chemical Reactions and Processing*; Abraham, M. A.; Moens, L., Ed.; American Chemical Society, 2002; pp. 34-49. (c) Torres, M.-J. PhD Thesis, The Queen's University of Belfast, 2001.

[*22*] Schaefer, W. P.; Quan, R. W., Bercaw, J. E. *Acta Crystallogr. Sect C [Cr. Str. Comm]* **1992**, *48*, 1610-1612.

[*23*] Bradley, A. E.; Hardacre, C.; Holbrey, J. D.; Johnston, S.; McMath, S. E. J.; Nieuwenhuyzen, M. *Chem. Mater.* **2002**, *14*, 629-635.

[*24*] Holbrey, J. D.; Nieuwenhuyzen, M.; Seddon, K. R., unpublished results (1996-2000).

[*25*] Delaplane, R. G.; Lundgren, J.-O.; Olovsson, I. *Acta Crystallogr. Sect. B* **1975**, *31*, 2208.

Chapter 35

Electrochemical Studies of Ambient Temperature Ionic Liquids Based on Choline Chloride

Andrew P. Abbott, Glen Capper, David L. Davies, Helen Munro, Raymond K. Rasheed, and Vasuki Tambyrajah

Chemistry Department, University of Leicester, Leicester LE1 7RH, United Kingdom

Ionic liquids can be formed between $ZnCl_2$ and a range of substituted quaternary ammonium salts. These ionic liquids are easy to prepare and are insensitive to water and air. They have conductivities comparable to other ionic liquids and as such they are suitable for electrochemical applications. The deposition of zinc is investigated and it is shown that the film formed on an electrode surface following the electroreduction of the melt is non-porous and produces an extremely effective corrosion resistant coating with very high current efficiency. The characteristics of a battery based on a zinc ionic liquid are also demonstrated. Other metals such as chromium are also shown to form ionic liquids with choline chloride and a generic method for depositing metals is demonstrated.

Introduction

Ambient temperature ionic liquids have been used extensively for electrochemical studies. Most of the work has focused on aluminium chloride with imidazolium or pyridinium salts. Work by Osteryoung (*1*) showed that n-butylpyridinium salts gave liquids over a wide range of composition with $AlCl_3$ at ambient temperatures. Similar results were also found using dialkylimidazolium chloroaluminate melts (*2*). The properties of these ionic liquids are controlled by various chloroaluminate anions although their relative concentrations were found to change considerably with melt composition. The density, viscosity and electrical conductivity of these systems have been characterized by Hussey *et al.* (*3*) The electrochemistry of aluminium and a range of transition metal salts have been studied in a variety of chloroaluminate melts and this has found application for batteries and the electrodeposition of aluminium and its alloys. (*4-7*) The chemical and electrochemical properties of $AlCl_3$ containing melts have recently been reviewed in a number of articles. (*8-11*)

The issues encountered using chloroaluminate melts are in their high cost and water sensitivity (hence the need for glove box or Schlenk techniques). It has recently been shown that ionic liquids can be formed using zinc chlorides with pyridinium, (*12*) dimethylethylphenyl ammonium (*13*) and imidazolium salts. (*14-16*) These have higher melting points than the corresponding chloroaluminate melts but are still fluid at ambient temperatures. In recent work (*17*) we have also shown that choline (2-hydroxyethyl trimethyl ammonium) chloride also forms ionic liquids. Not only are these ionic liquids easy to prepare, and water and air insensitive but also their low cost enables their use in large-scale applications such as metal finishing. (*18*) Analogous to $AlCl_3$ containing ionic liquids $ZnCl_2$ forms complex anions with Cl^- and the change in properties with composition and temperature can be predominantly ascribed to the equilibria

$$ZnCl_3^- + ZnCl_2 \leftrightarrows Zn_2Cl_5^-$$

$$Zn_2Cl_5^- + ZnCl_2 \leftrightarrows Zn_3Cl_7^-$$

In the current work we outline the electrochemical properties of zinc chloride/ choline chloride ionic liquids and highlight their applications to metal electrodeposition, and batteries.

EXPERIMENTAL

Anhydrous zinc chloride (Aldrich, 98%) and choline chloride (Aldrich, 99%) were used as received to prepare the ionic liquid. Choline chloride and zinc chloride in a molar ratio of 1:2 were combined and heated at approximately 120 °C until a clear colorless liquid was obtained, (approximately 30 minutes for 0.03 moles of 1:2 choline chloride-zinc chloride). The fluids containing hydrated metal salts were prepared by an analogous method although they were only heated to 70°C.

Voltammetry and chronoamperometry were carried out using an Autolab/PGSTAT12 potentiostat controlled with GPES software. A three-electrode system consisting of a platinum microelectrode ($5x10^{-4}$ cm radius), a platinum counter electrode and a zinc wire reference electrode were used. The working electrode was polished with 0.3μm alumina paste prior to all measurements. For the chromium containing ionic liquids a chromium wire reference electrode was used. All electrochemical measurements were performed at 60 °C and a scan rate of 20 mVs^{-1} was used in voltammetric experiments. The conductivity and its temperature dependence were determined using a Jenway 4071 conductivity meter with temperature and conductivity probes.

The effect of current density on zinc deposit morphology was determined using a Hull Cell. Two nickel plates were gently abraded with glass paper, cleaned with acetone and dried. Bulk electrolysis experiments were performed using a Thurlby Thander power supply with a nickel cathode and a nickel anode. Following electrolysis the cathode was washed with acetone and dried. Surface analysis was carried out using scanning electron microscopy (SEM) and energy dispersive analysis by X-rays (EDAX).

RESULTS AND DISCUSSION

Low melting point ionic compounds were readily prepared from choline chloride and zinc chloride using molar ratios ranging from 1:1.5 to 1:3. The eutectic molar ratio is 1:2 choline chloride-zinc chloride, which has a freezing point of 24 °C. For comparison 1:1.5 and 1:3 choline chloride-zinc chloride were prepared and their melting points were found to be 60 °C and 45 °C respectively. The ionic liquids are thermally stable up to approximately 180 °C; above this temperature the melts begin to discolor becoming pale yellow presumably due to the decomposition of the quaternary ammonium cation. Ionic compounds formed between choline chloride and zinc chloride are insensitive to air and moisture and can be repeatedly frozen and melted with no adverse affects.

Choline chloride-zinc chloride ionic liquids are highly conducting allowing voltammetry to be studied without ohmic distortion. Figure 1 shows the conductivity (σ) of 1:2 choline chloride-zinc chloride as a function of temperature (T). The σ values compare favorably to those for quaternary ammonium triflate salts (*19*) but are lower than those reported for 1:2 *N*-ethylpyridinium bromide-aluminium chloride. (*3*) As recently shown by Noda *et al.* these data can be fitted to a Vogel-Tamman-Fulcher (VTF) equation of the form (*20*) (Figure 1)

$$\sigma = \sigma_0 \exp\left[\frac{-B}{(T - T_0)}\right] \qquad (1)$$

where σ_0, B and T_0 are constants. The best fit parameters were found to be $\sigma_0 = 15.3$ S m^{-1}, $B = 978$ K and $T_0 = 192$ K ($R^2 = 0.999$) which are in very close agreement with those obtained by Noda *et al.* for various imidazolium and pyridinium based ionic liquids. (*20*) The temperature response is similar to a variety of other ionic liquids, which suggests that the overall governing factor in controlling the conductivity of these liquids is the viscosity.

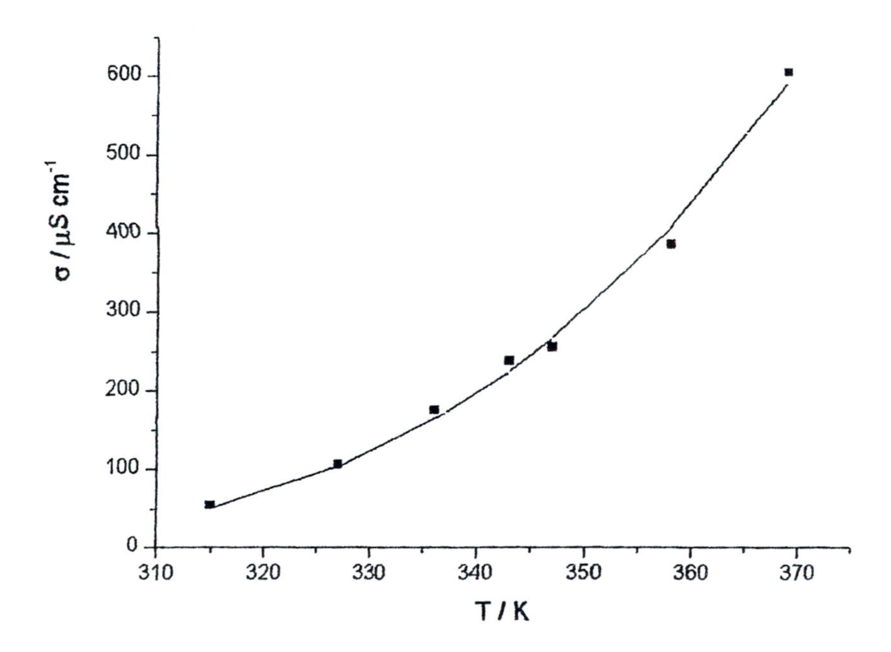

Figure 1: The conductivity of a choline chloride: zinc chloride (1:2) ionic liquid as a function of temperature and fit to eqn. 1.

Metal Deposition

Figure 2 shows a cyclic voltammogram of a Pt electrode in a 1:2 choline chloride-zinc chloride ionic liquid. The liquid has a 2 V potential window, which is limited by zinc deposition and chlorine gas evolution at negative and positive over-potentials respectively.

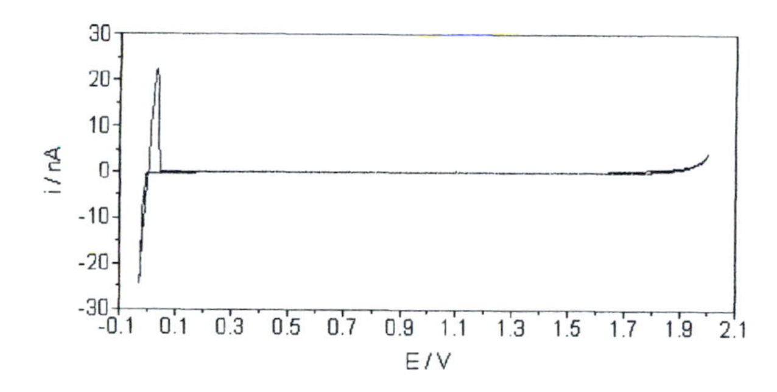

Figure 2: Cyclic voltammogram of a Pt microelectrode in 1:2 choline chloride-zinc chloride.

A more detailed examination of the cyclic voltammogram between +0.055 V and +1.1 V versus zinc reveals the presence of a small cathodic current. This current reaches a maximum value at +0.055 V (Fig. 3) and is attributed to under-potential deposition (u.p.d.) of zinc. The anions present in the melt are chloro-complexes and the most abundant according to mass spectroscopy are $Zn_3Cl_7^-$, $Zn_2Cl_5^-$ and $ZnCl_3^-$. The relative abundances of these three species have been recently determined using potentiometry (*21*) and in a 2:1 $ZnCl_2$: chloline chloride mixture these have been found to be $Zn_3Cl_7^-$ (0.01%), $Zn_2Cl_5^-$ (66%) and $ZnCl_3^-$ (33%). The $ZnCl_3^-$ species can be thought of as a Zn^{2+} cation with an incomplete tetrahedron of Cl^- ions (i.e. it has a vacant co-ordination site) and thus can be likened to an ad-ion. Hence these may allow the complex to adsorb to the electrode surface and produce the observed u.p.d. effect. The chloro-complexes present in the melt exist in equilibrium, the position of which is determined by the Lewis acidity of the melt. In more Lewis acidic melts (zinc chloride: choline chloride > 2) the equilibrium tends towards the larger clusters and vice versa for less Lewis acidic melts, therefore, greater u.p.d. is expected in more basic melts because of the greater profusion of $ZnCl_3^-$. The u.p.d. charge in a 1:1.8 choline chloride-zinc chloride liquid ($ZnCl_3^-$ = 36%) is greater (6.6 nC) than that observed in 1:2 choline chloride-zinc chloride mixture (4.5 nC) confirming this hypothesis. The u.p.d. response is shown in Figure 3 as a function of ionic liquid composition. A similar u. p. d. response was observed for the aluminium chloride / phenyltrimethyl ammonium chloride system by Vandernoot *et al.* (*22*)

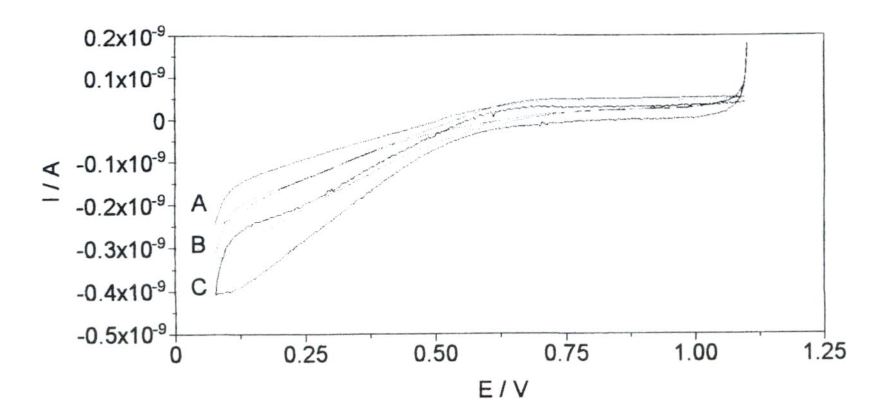

Figure 3: Cyclic voltammogram of zinc u.p.d. at a Pt microelectrode in zinc chloride: choline chloride ionic liquids of various compositions. A = 2.1:1, B = 2:1, C = 1.8:1

Below 0 V versus zinc the electrochemical reduction of Zn^{2+} occurs at the electrode surface (Figure 2). A current loop, indicative of nucleation, is observed on a sweep to -0.04 V and a characteristic stripping peak is observed on the reverse sweep to positive potentials. Figure 4 shows the plot of anodic *vs.* cathodic charge on cycling to different cathodic limits in the 1:2 choline chloride-zinc chloride electrolyte. This shows that the zinc deposition process is almost totally reversible with the small discrepancy being attributable to u.p.d. which is non-reversible. The deposited zinc appears to be stable in the ionic liquid and does not corrode like aluminium deposited under analogous conditions. (*22*) The high degree of reversibility of the zinc deposition/dissolution process suggests that these ionic liquids may have applications for rechargeable batteries (see below).

Hull Cell tests shows bulk zinc deposition can be performed in the 1:2 choline chloride-zinc chloride electrolyte. Analysis of the zinc plated nickel cathode revealed that with current densities between 0.5 mA cm^{-2} and 0.16 mA cm^{-2} a thick white/grey homogenous deposit was obtained, between 0.16 mA cm^{-2} and 0.07 mA cm^{-2} the deposit remained homogenous but became progressively thinner, whilst below 0.07 mA cm^{-2} the zinc deposit was thin and non homogenous. The morphology of zinc deposited onto nickel using a current density of 0.5 mA cm^{-2} is shown in Figure 5 The scanning electron micrograph shows the zinc coverage is very even and non-porous. Closer examination reveals the growth centers vary in size, which is characteristic of progressive nucleation. Because there is no competing electrochemical process with the zinc deposition and the metal formed is stable to corrosion the current efficiency is extremely high (> 99%).

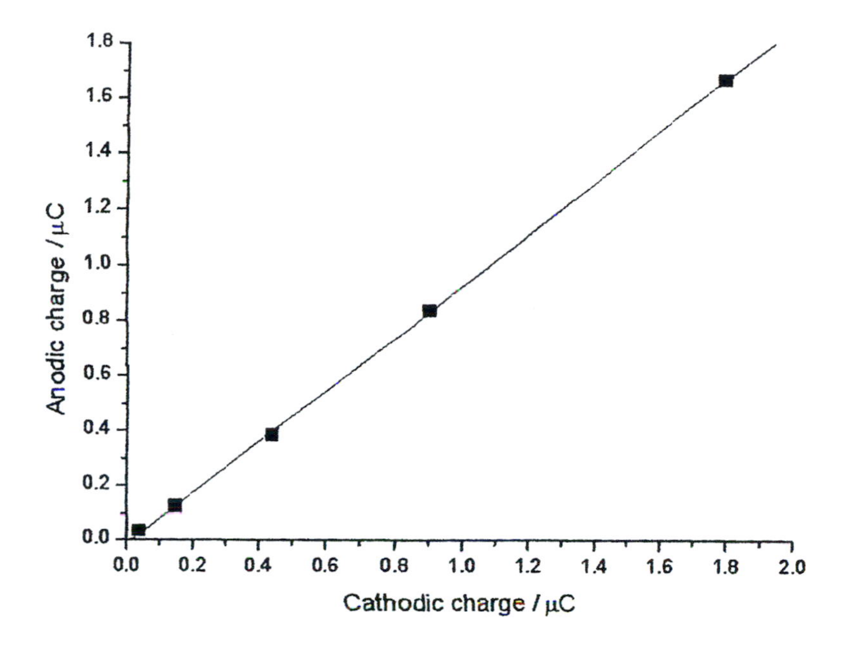

Figure 4: Plot of charge for the deposition and stripping of zinc obtained from the voltammograms shown in Figure 2.

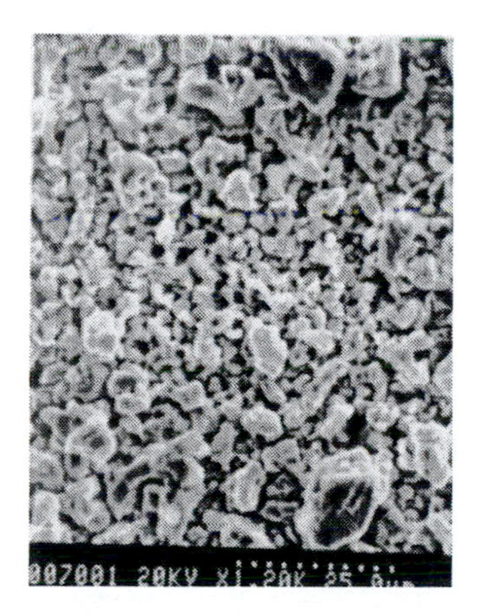

Figure 5: Scanning electron micrograph of zinc deposit on nickel

The porosity and hence the resistance of deposited zinc films to corrosion in aqueous systems was assessed electrochemically using an iron electrode. The electrochemical behavior of a bare iron electrode and a zinc coated iron electrode in 0.1 M potassium nitrate was studied and it was found that zinc plating reduced the anodic current (corrosion) by approximately two orders of magnitude as shown in Figure 6.

Such deposits provide an effective barrier against corrosion and this has been demonstrated on mild steel. Using a current density of 0.5 mA cm^{-2} a 10 μm thick layer of zinc was plated onto mild steel, which was then subjected to a salt spray test for 5 hours. After this time the uncoated regions suffered extensive rusting whilst the zinc plating offered excellent protection with no sign of even pinhole defects.

Reversible Zinc Galvanic Cell

Zinc has been extensively used for battery applications because of it is ready availability and has a high energy density. Its use has largely been limited to primary cells because of the deposition process in only quasi-reversible, which precludes the efficient recharging of the cell. Since the zinc deposition process has already been shown to be reversible these ionic liquids may be useful for secondary cell applications.

To make the anodic and cathodic components compatible, the following cell was constructed

$Zn \mid ZnCl_2(0.643)\ ChCl(0.357) \,\vdots\vdots\, FeCl_3(0.667)\ ChCl(0.33) \mid Fe$

At negligible drain the cell potential was found to be 1.80 V, which is larger than the 1.53 V expected from the aqueous standard electrode potentials. Figure 7 shows that as the drain increases the cell potential decreased markedly, although this was predominantly due to the cell design which had a 4 cm gap between the electrodes and a 2 mm glass frit between the two half cells. This resulted in a significant ohmic resistance in the cell. Following the application of a constant drain of 140 μA cm^{-2} there is an initial decrease in the cell voltage, which may result from a slow dissolution of an oxide film on the zinc electrode, but after 1 hour the cell voltage remained relatively constant at 0.6 V. Clearly the cell is at present un-optimized and has an extremely small power density but these preliminary results show that chlorozincate ionic liquids could be used as an alternatives battery electrolyte.

Hydrated salt mixtures

While the above approach is a novel method of forming ionic liquids it is limited to $ZnCl_2$, $SnCl_2$ and $FeCl_3$. (*17*) We have recently found that mixing choline like salts with a range of hydrated metal salts also form deep eutectics, which are liquid at ambient temperatures. (*23*) Table 1 lists the freezing temperatures of a range of hydrated metal chlorides with choline chloride in a

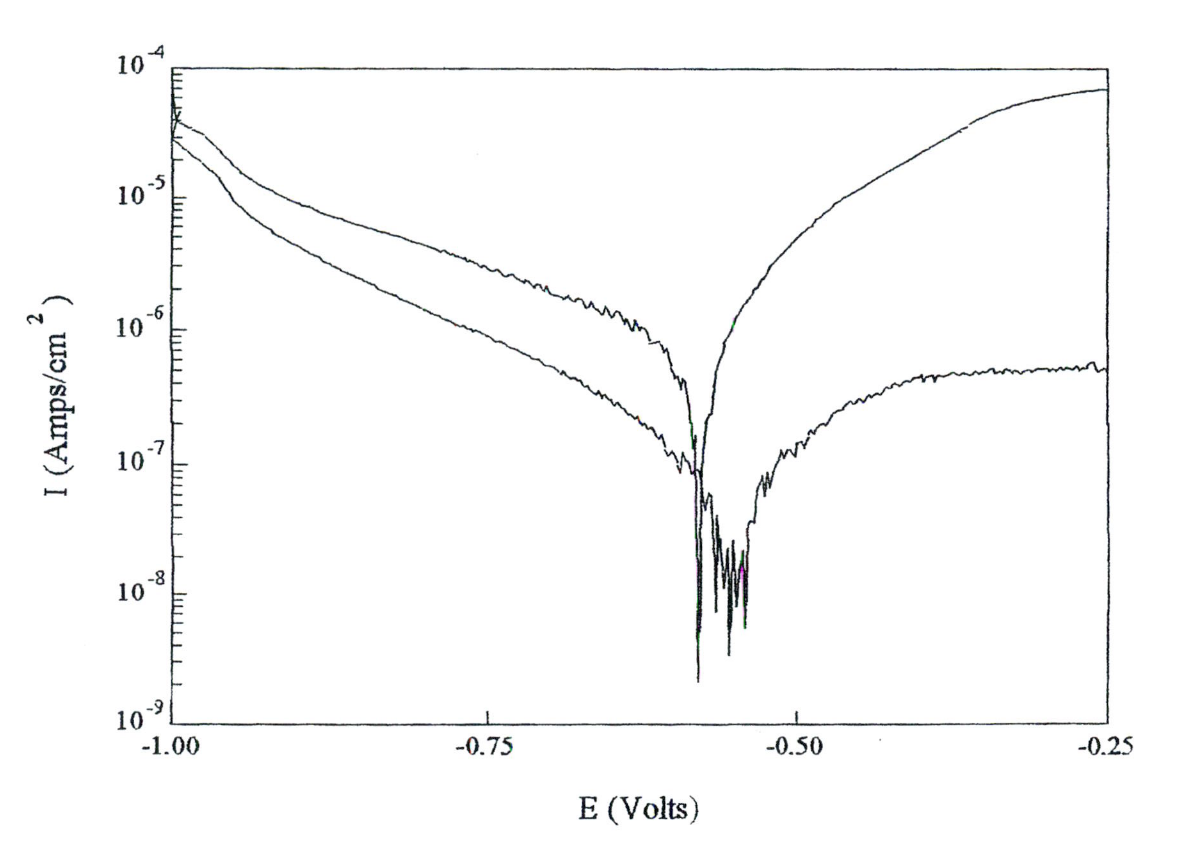

Figure 6: Linear sweep voltammograms of Fe (upper curve) and zinc coated Fe (lower curve) in 0.1 M potassium nitrate.

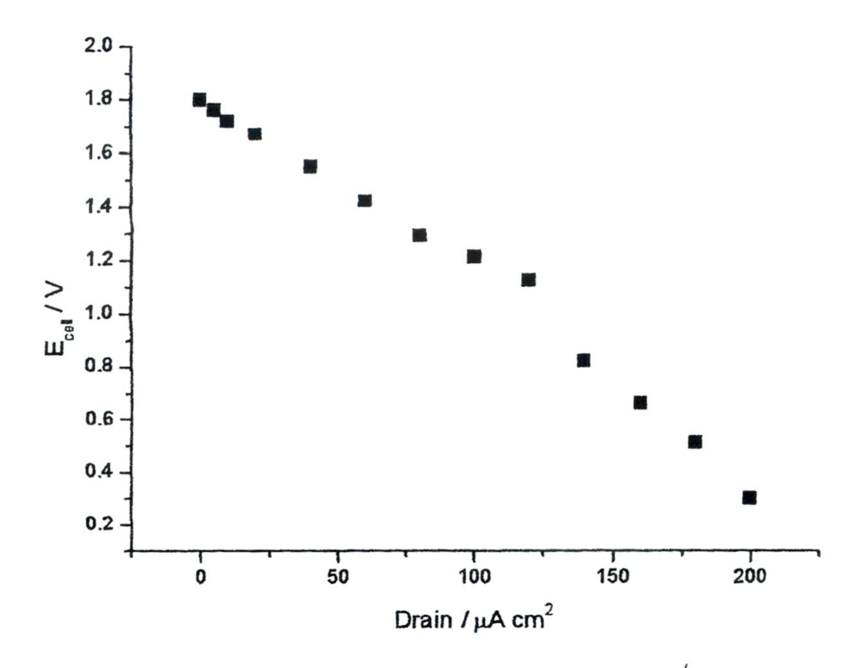

Figure 7: Potential vs. drain characteristics for the cell Zn /$ZnCl_2$(0.643) ChCl(0.357) / /$FeCl_3$(0.667) ChCl(0.33) /Fe at 100°C

Table 1: Freezing point temperatures of choline chloride: hydrated metal salt mixtures in a 2:1 ratio.

Hydrated Salt	T_f / °C
$CrCl_3.6H_2O$	4
$CaCl_2.6H_2O$	5
$MgCl_2.6H_2O$	10
$CoCl_2.6H_2O$	16
$LaCl_3.6H_2O$	6
$CuCl_2.2H_2O$	48

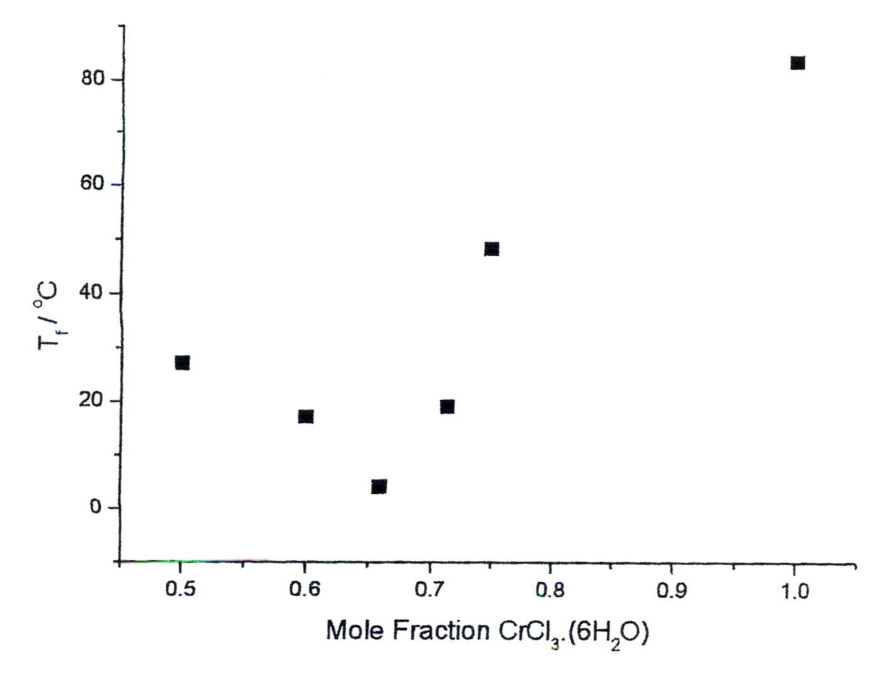

Figure 8: Freezing point temperatures for mixtures of choline chloride and $CrCl_3.6H_2O$ as a function of composition

2:1 ratio mixture. The water molecules appear to be highly coordinated and do not behave like bulk liquid water, hence voltammetry in these liquids does not evolve hydrogen prior to metal ion reduction. Figure 8 shows the freezing point of choline chloride/ $CrCl_3.6H_2O$ as a function of composition and it can be seen that a eutectic is obtained at a 1:2 ratio.

Figure 9 shows that the conductivity and viscosity of a 2:1 mixture of $CrCl_3.6H_2O$ and choline chloride as a function of temperature. The values are considerably higher than those shown in Figure 1 for the choline chloride zinc chloride system showing that they are suitable for electrochemical investigations. Voltammetry of a 2:1 mixture of $CrCl_3.6H_2O$ and choline chloride at a Pt microelectrode is shown in Figure 10. It is noticeable that the chromium deposition process proceeds via a Cr(II) species although the subsequent decrease in the cathodic current may indicate that the Cr(II) species is depositing on the electrode surface. It is also noticeable that the deposition process is irreversible unlike that of zinc but there is no signal for the evolution of hydrogen. The bulk electrolysis of the solution at a constant current density of 3 $mAcm^{-2}$ leads to a thick adherent non-micro-cracked deposit, which contains predominantly chromium with a trace of chlorine arising presumably from the entrapment of salt during the deposition process. The morphology of the deposit is similar to that obtained for zinc in Figure 5. Since there is negligible hydrogen evolution the process has a high current efficiency (>95%), which is considerably better than the current method based on chromic acid (typically 15%). Another advantage of using this type of fluid for chromium deposition is that it uses a Cr(III) salt instead of Cr(VI), which leads to an even higher current efficiency. This method of depositing metals using hydrated salt mixtures can also be applied to a variety of other metals including Mn, Co, Ni and Cu. This technology circumvents the necessity for aggressive acidic or alkaline plating solutions or toxic ligands such as cyanide. The significant increase in current efficiency adds to the overall sustainability of the process.

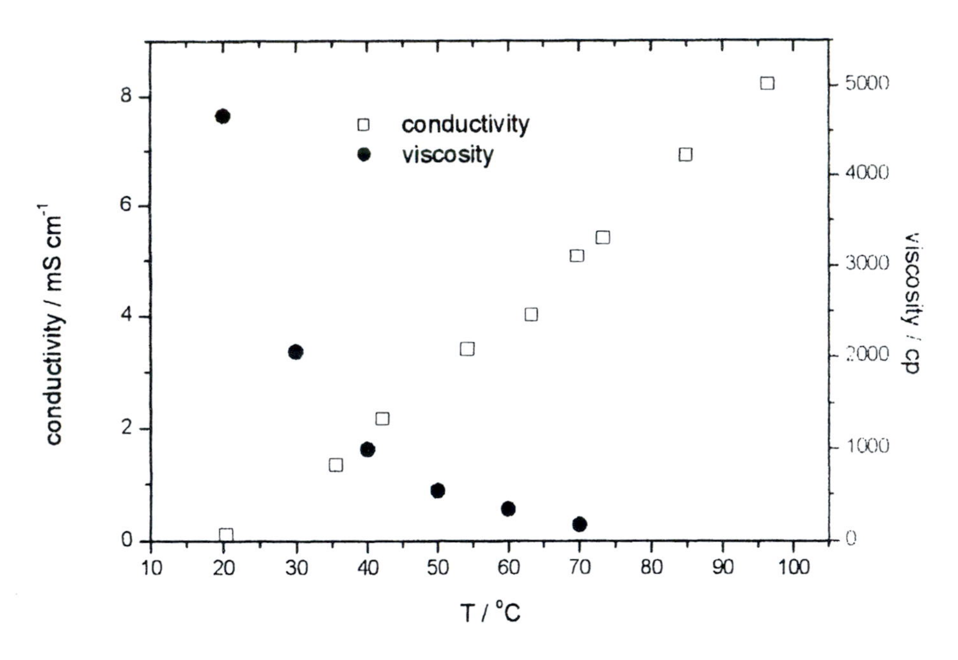

Figure 9: Conductivity and viscosity of a 2:1 mixture of $CrCl_3.6H_2O$ and choline chloride as a function of temperature.

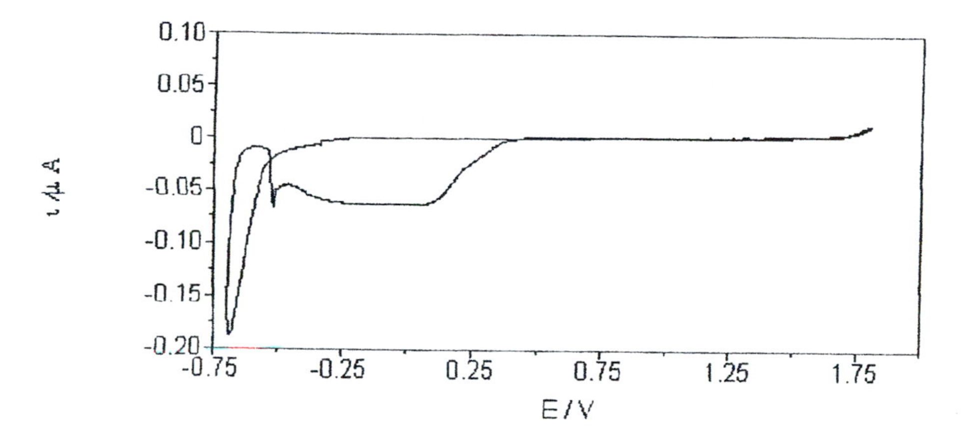

Figure 10: Cyclic voltammogram of a Pt microelectrode in 1:2 choline chloride- $CrCl_3.6H_2O$.

Conclusions

We have established that ionic liquids based on zinc chloride/ choline chloride have analogous electrochemical properties to chloroaluminate melts. Unlike their aluminium counterparts however, they are stable to air and moisture and hence are much easier to synthesize and handle. They are sufficiently conducting to allow voltammetric studies without significant resistance artifacts. These ionic liquids can be used for the deposition of non-porous zinc, which is shown to be an effective anti-corrosion barrier on mild steel. It was also demonstrated that a galvanic cell could be constructed from a chlorozincate ionic liquid and the high reversibility of the deposition process suggested that the cell could be made rechargeable. It is also shown that ionic liquids based on hydrated salt mixtures can be used as an efficient electroplating process for chromium.

Acknowledgements

The authors would like to acknowledge the funding of this work by the EPSRC (GR/R59199) and Scionix Ltd.

References

1. Osteryoung, R. A.; *Molten Salt Chemistry* Mamantov, G.; Marassi, D. Eds.; Reidel Publishing: Holland, 1987.
2. Osteryoung, R. A.; Robinson, J. *J. Am. Chem. Soc.*, **1979**, *101*, 323.

3. Carpio, R. A.; King, L. A.; Lindstrom, R. E.; Nardi, J. C.; Hussey, C. L. *J. Electrochem. Soc.*, **1979**, *126*, 1644.
4. Wier, T. P. *U. S. Patent* 2,446350 **1948**
5. Peled, E.; Gileadi, E. *J. Electrochem. Soc.*, **1976**, *123*, 15.
6. Osteryoung, R. A.; Robinson, J. *J. Electrochem. Soc.*, **1980**, *127*, 122.
7. Dymek, Jr, C. J.; Williams, J. L.; Groeger, D. J.; Auborn, J. J.; *J. Electrochem. Soc.* **1984**, *131*, 2887.
8. Zhao, Y.; Vandernoot, T. J. *Electrochimica Acta*, **1997**, *42*, 3.
9. Welton, T. *Chem. Rev.*, **1999**, *99*, 2071.
10. Holbrey, J. D.; Seddon, K. R. *Clean Products and Processes*, **1999**, *1*, 223.
11. Wasserscheid, P.; Keim, W. *Angew. Chem. Int. Ed.*, **2000**, *39*, 3772.
12. Koura, N.; Endo, T.; Idemoto, Y. *J. Non-Cryst. Solids*, **1996**, *205*, 650.
13. Simanavicius, L.; Stakenas, A.; Starkis, A. *Electrochim. Acta*, **1997**, *42*, 1581.
14. Lin, Y.; Sun, I. *Electrochim. Acta*, **1999**, *44*, 2771.
15. Chen, P.; Lin, M.; Sun, I. *J. Electrochem. Soc.*, **2000**, *147*, 3350.
16. Chen, P.; Sun, I. *Electrochim. Acta*, **2001**, *46*, 1169.
17. Abbott, A.; Capper, G.; Davies, Munro, H.; D. L.; Rasheed, R. K.; Tambyrajah, V. *Chem. Commun*, **2001**, 2010.
18. Abbott, A.; Davies, D. L. *Int. Pat.*, WO 0056700, **2000**.
19. Sun, J.; Forsyth, M.; MacFarlane, D. R.; *J. Phys. Chem. B*, **1998**, *102*, 8858.
20. Noda, A.; K. Hayamizu and M. Watanabe, J. Phys. Chem. **105**, 4603, (2001)
21. Abbott, A.; Capper, G.; Davies, D. L.; Rasheed, R. K.; Tambyrajah, V. paper in preparation.
22. Zhao, Y.; Vandernoot, T. J. *Electrochimica Acta*, **1997**, *42*, 1639.
23. Abbott, A.; Capper, G.; Davies, D. L.; Rasheed, R. K.; Tambyrajah, V. *Int. Pat.* WO 0226701, **2001**.

Chapter 36

Electrodeposition of Nanoscale Metals and Semiconductors from Ionic Liquids

Sherif Zein El Abedin and Frank Endres*

Institute of Metallurgy, Technical University of Clausthal, D–38678 Clausthal-Zellerfeld, Germany

Abstract: In this paper we present a short review on the electrodeposition of metals and semiconductors from ionic liquids, viewed by in situ Scanning Tunnelling Microscopy. Due to their wide electrochemical windows ionic liquids give access to elements, that otherwise cannot be electrodeposited from aqueous or organic solutions, like e.g. light metals and many semiconductors. At the example of Al, Cu and Ge we give a short insight into the nanoscale processes during phase formation. The importance of ionic liquids for electrochemical nanotechnology will be shortly commented.

1. Introduction

Ionic Liquids are – as the name predicts – composed solely of ions. Although the intensively investigated and in technical processes widely used high temperature molten salts fulfill this requirement by nature, it was suggested in the recent years to distinguish from the classical molten salts "artificial" systems with melting points below 100 °C [1, 2].

Historically, ionic liquids can more or less be divided into 3 groups:

a) systems based on $AlCl_3$ and organic salts like the 1-Butylpyridinium chloride (BP^+Cl^-), 1-Ethyl-3-methyl-imidazolium chloride ($[EMIm]^+Cl^-$), 1-Butyl-3-methyl-imidazolium chloride ($[BMIm]^+Cl^-$) and derivatives,

b) systems based on organic cations like in a) and BF_4^-, PF_6^- and SbF_6^- and

c) systems based on the aforementioned organic cations and anions of the type $CF_3SO_3^-$, $(CF_3SO_2)_2N^-$ and similar ones.

Liquids made on the basis of the latter anions are stable under ambient conditions with only low water uptake. The acidity of the first systems can be varied by the relative amounts of organic salt / $AlCl_3$. With a molar excess of

$AlCl_3$ they are acidic, with an excess of the organic salt they are basic. Neutral liquids of this type contain a 50/50 mol-% mixture of both the organic salt and $AlCl_3$. In comparison to aqueous solutions ionic liquids have several advantages: depending on the system they are stable up to temperatures of 200 ° C with low vapour pressures, many organic and inorganic compounds can easily be dissolved in them, they have sufficiently high specific ionic conductivities between 0.001 $(\Omega\ cm)^{-1}$ and 0.01 $(\Omega\ cm)^{-1}$ and they can have wide electrochemical windows of more than 6 Volt. Especially the latter property is pretty interesting for materials electrosynthesis as will also be commented in this article. For an overview on ionic liquids we would like to refer to some recent articles [3-6], a review on ionic liquids as media for the electrodeposition of metals and semiconductors has recently been given by one of us [7].
The aim of the present paper is to summarize recent results on the nanoscale electrodeposition of metals and semiconductors from ionic liquids. Furthermore we will present a few new results on the electrodeposition of thin Ge layers on Au(111), Highly Oriented Pyrolytic Graphite (HOPG) and on hydrogen terminated Si(111). Nanocrystalline metals, i.e. the individual grain sizes are below 100 nm, are interesting for Materials Science, because they are much harder than the microcrystalline ones [8]. The band gap of nanocrystalline semiconductors is a function of the grain size, and the smaller the individual crystal is the wider the band gap becomes [9]. So, in principal, the band gap of semiconductors and thus the optical properties can be tuned over a wide range. Electrochemistry in ionic liquids might give a unique tool here, because semiconductors, that are often made under Ultrahigh Vacuum Techniques with Molecular Beam Epitaxy, could be made easily in one electrochemical step.

2. Experimental

The in situ electrochemical STM experiments were performed with in house designed STM heads, similar to a setup presented recently in [10]. In contrast to commercial systems, due to a special setup the experiments can be performed under inertgas with H_2O and O_2 below 2 ppm, which is a requirement for studies with liquids based on $AlCl_3$ or with hygroscopic halides such as $GeCl_4$ dissolved in otherwise water and oxygen insensitive liquids. STM tips were electrochemically made from Pt-Ir or tungsten wire (Ø 250 μm) and electrophoretically coated by an epoxy resin, in order to reduce the electrochemically active tip area to a minimum. The STM heads were driven either by a Digital Instruments Nanoscope E or a Molecular Imaging PicoScan SPM controller in feedback mode, e.g. a current setpoint was applied which was held constant by the electronics. The tunnelling current is a function of the distance [11], and as the distance tip/sample is held constant by the STM electronics, topographic data can be obtained. The STM pictures are generally given here as 2-dimensional plots where the height differences are given as grey scale. Cyclic Voltammograms were acquired with the electrochemical option of the respective STM controller. For an introduction into the STM technique we would like to refer to [12].

3. Results

Al OPD from acidic $[BMIm]^+Cl^-$ / $AlCl_3$

One of the first systems that was investigated in Karlsruhe by electrochemical in situ STM was the electrodeposition of aluminum from an acidic ionic liquid based on $AlCl_3$ and 1-Butyl-3-methylimidazolium chloride [13]. It is well known that Al can be electrodeposited from such liquids, but here the aim was rather to get an insight into the processes on the nanometer scale. Al is an interesting material for lightweight construction purposes and for surface refining, furthermore it has excellent corrosion protection properties. A disadvantage is, that it is a rather soft material and that the quality of bulk deposits is strongly dependent on the solution composition. Nanocrystalline Al, where an increase in hardness with respect to microcrystalline Al can be expected, would be of great industrial importance, e.g. in automobile industry, where light metals are more and more used for construction. In order to understand the initial processes of phase formation we performed in situ STM studies at the interface electrode / electrolyte in the mentioned ionic liquid. The Cyclic voltammogram of $[BMIm]^+Cl^-$ / $AlCl_3$ (34 : 66 mol-%) is presented in Figure 1.

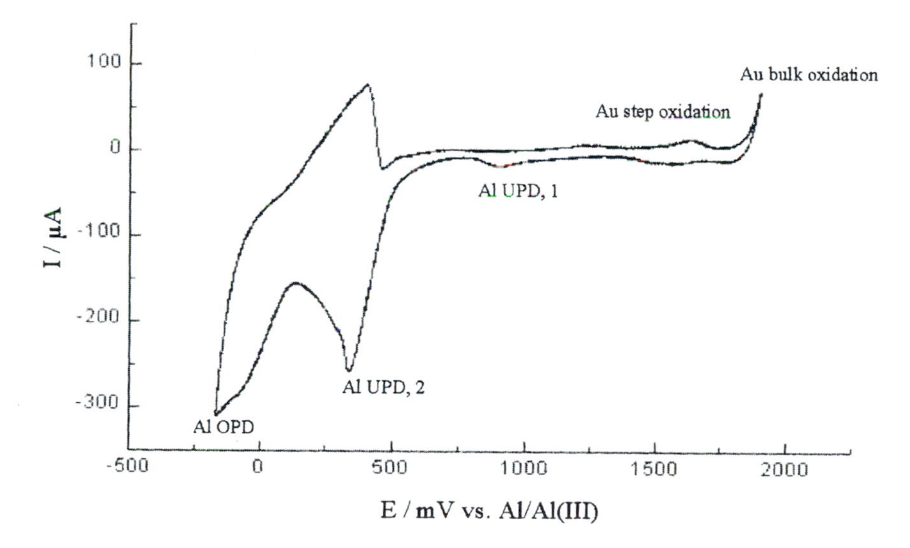

Fig. 1: CV of $[BMIm]^+Cl^-$ / $AlCl_3$ (34 : 66 mol-%) on Au(111). At the anodic limit gold oxidation starts at the steps, at +1000 and about +400 mV underpotential deposition with respect to the bulk phase begins, at electrode potentials below 0 V overpotential deposition of Al sets in (see [13]), v = 100 mV/s

The cyclic voltammogram is characterized by several processes: at about 1.4 V vs. Al/Al(III) gold oxidation starts at the steps, at about 1.7 V pits are rapidly formed that can lead to a complete disintegration of the gold substrate. At about +1000 mV vs. Al/Al(III) 2-dimensional phase formation is clearly observed, 250 pm high islands grow that are most likely the result of a surface alloying / compound formation between Au and Al. At +400 mV nanoclusters with a height of up to 1.5 nm start growing in the UPD regime, and they show alloying with the substrate. At electrode potentials below 0 V the bulk phase of Al begins to grow. A typical picture for the initial growth in the OPD regime is shown in fig. 2 (see also [13]):

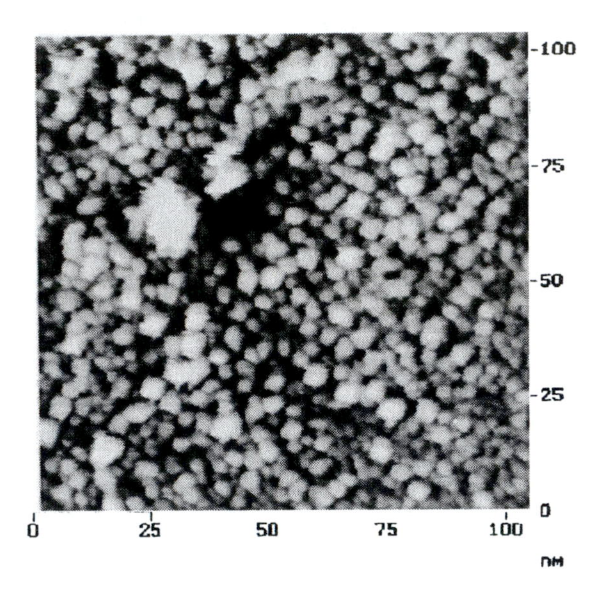

Fig. 2: In situ STM picture during the initial stages of Al OPD from $[BMIm]^+Cl^-$ / $AlCl_3$ (34 : 66 mol-%) on Au(111) at - 100 mV vs. Al/Al(III). The growth starts with the formation of Al nanocrystallites (see also [13])

It is evident, that under these potentiostatic conditions the bulk phase formation starts with the formation of Al nanocrystallites. Unfavorably, thick bulk layers in the millimeter range always consist of microcrystals if they are made potentiostatically. With special electrochemical techniques and special bath compositions, however, it is possible to make millimeter thick, shining and stable deposits of nanocrystalline Al, where the average grain size can be varied electrochemically from 10 nm up to several hundreds of nm [14].

Cu OPD from acidic $[BMIm]^+Cl^-$ / $AlCl_3$ / $CuCl_2$

Copper is an element with extraordinary significance in semiconductor industry. Almost all connections on semiconductor chips are made with copper. Meanwhile also copper electrodeposition plays an important role for such purposes. As a consequence there are numerous studies on the electrodeposition of copper from aqueous solutions (see for example [15] and references therein). Although Cu can be electrodeposited well from aqueous solutions, there is a certain interest to electrodeposit it from ionic liquids. Due to their usually low vapor pressures electrodeposition can also be performed at elevated temperatures. Furthermore, depending on the electrode materials, gas evolution - an unfavored side reaction in closed cells - does usually not occur in ionic liquids due to their wider electrochemical windows. Especially for the electrodeposition of (nanocrystalline) alloys between Cu and less noble elements, ionic liquids are interesting. Cu-Al alloys from $AlCl_3$ based liquids, for example, were recently reported in [16]. Fig. 3 gives an overview on the electrode processes on Au(111) and on glassy carbon in $[BMIm]^+Cl^-$ / $AlCl_3$ / $CuCl_2$ (34/66 mol-%, 5 mmol/l $CuCl_2$):
At 1150 mV vs. Cu/Cu^+ we find the Cu^+/Cu^{2+} peak couple on glassy carbon.

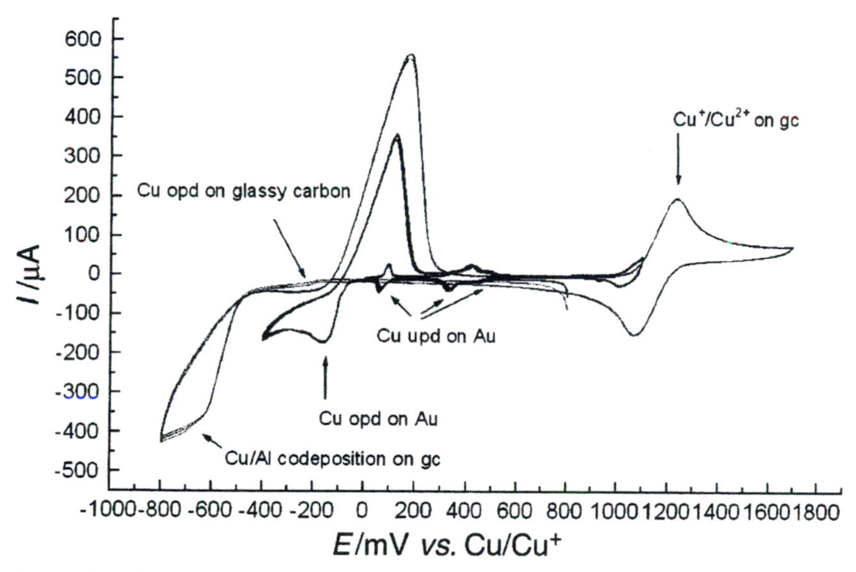

Fig. 3: CV of $[BMIm]^+Cl^-$ / $AlCl_3$ / $CuCl_2$ (34/66 mol-%, 5 mmol/l $CuCl_2$) on glassy carbon (gc) and on Au(111). On gc only OPD of Cu and Cu-Al alloys are observed, whereas on Au(111) several UPD processes can be seen before the growth of the bulk phase sets in ([17] – Reproduced by permission of the Royal Society of Chemistry on behalf of the PCCP owner societies)

This is a remarkable result, because on Au(111) gold step oxidation starts before Cu^+ is oxidized to Cu^{2+}, and - consequently - there is a slight chemical attack of Cu^{2+} on Au(111). On the other hand, Cu^{2+} also attacks elemental Cu to give a stable Cu/Cu^+ reference electrode potential. On glassy carbon there is no visible phase formation in the UPD regime, whereas on Au(111) there are 3 UPD processes before the bulk growth sets in. At about -500 mV vs. Cu/Cu^+ Cu-Al alloys are deposited, both on Au(111) and on glassy carbon (see also [16]). The electrochemical processes in the UPD regime on Au(111) are correlated to the following processes on the gold surface: at +450 mV vs. Cu/Cu^+ Cu nanoclusters deposit at the steps of the terraces. At +400 mV a hexagonal pattern appears on the surface, and at +350 mV 0.7 monolayers of Cu form a well defined 8 x 8 Moiré superstructure with the underlying substrate. This incomplete monolayer finally closes at +200 mV to a complete monolayer, and at +50 mV a second monolayer grows, that may also contain some aluminum. An interesting result is that already in the UPD regime clusters grow, that are mainly composed of Cu but may also contain some aluminum.
Fig. 4 shows an STM picture, where at the top the electrode potential was set from a value in the UPD regime to -10 mV vs. Cu/Cu^+:

Fig. 4: In situ STM picture on Cu OPD on Au(111) in $[BMIm]^+Cl^-$ / $AlCl_3$ / $CuCl_2$ (34/66 mol-%, 5 mmol/l $CuCl_2$) on glassy carbon and on Au(111). The picture shows the slow growth of Cu during 1 hour (picture was acquired from top to bottom), and the initial growth also leads to nanocrystallites ([17] – Reproduced by permission of the Royal Society of Chemistry on behalf of the PCCP owner societies), scan range: 470 x 470 nm^2

The picture was acquired from top to bottom during one hour, and thus it also delivers a certain time resolution. Similar to the beginning OPD of Al the initial stages of Cu bulk deposition lead to Cu nanocrystallites.

Ge UPD and OPD from $GeCl_4$ saturated $[BMIm]^+PF_6^-$

The electrodeposition of germanium has not yet been widely investigated from low temperature electrolytes and only a few studies on the electrodeposition from organic electrolytes, such as glycoles, were performed at around 1950 [18, 19]. Apart from a few attempts to electrodeposit Ge even from aqueous solutions [20], the interest in germanium decreased strongly. This situation changed dramatically when it was found, that nanocrystalline germanium with individual particle diameters of only a few nanometers shows a size dependent photoluminescence. In contrast to the microcrystalline element nanocrystalline Ge rather seems to be a direct semiconductor [21], and it is regarded today as a promising candidate for optical sensors. However, almost all studies on the production or characterization of germanium nanoclusters or quantum dots were performed up to now under Ultrahigh Vacuum conditions. For a technological application this would be a certain shortcoming. So, our motivation was to find a way how to make (nanocrystalline) germanium by electrochemical means. As already mentioned earlier in this article, electrochemistry allows in principal to tailor-make deposits by varying bath composition and electrochemical parameters. In situ STM and in situ tunnelling spectroscopy are valuable tools to analyze the growing structures on a nanometer scale.
In the first paper on Ge electrodeposition the neutral ionic liquid 1-Butyl-3-methylimidazolium hexafluorophosphat ($[BMIm]^+PF_6^-$) saturated with GeI_4 was used [22]. It could be shown, that on Au(111) the first visible phase formation is characterized by the formation of 2-dimensional islands. Later on it was shown, that in this system quasi epitactically grown Ge(111) bilayers can be obtained, into which wormlike nanostructures can be introduced by partial oxidation [23]. This system had some disadvantages: GeI_4 does not dissolve well in the ionic liquid, the maximum thickness of layers was only about 100 nm, and stable Ge nanocrystallites could not be obtained. In the following, experiments were performed with $GeBr_4$ and $GeCl_4$ as germanium source, in both cases the saturation concentration is more than 0.1 mol/l.
Fig. 5a shows the electrochemical window of $[BMIm]^+PF_6^-$ on Au(111), in fig. 5b a typical cyclic voltammogram of $[BMIm]^+PF_6^-$ saturated with $GeCl_4$ is given. For comparison we have calibrated the processes vs. the germanium overpotential deposition, that we observed in this system. In the pure liquid, only capacitive currents are observed between the anodic and the cathodic limit. At the anodic limit gold oxidation starts at the steps, at the cathodic limit the irreversible reduction of the organic cation sets in, likely leading to $[BMIm]_x$ oligomers. If the ionic liquid is saturated with $GeCl_4$, we observe 2 mainly diffusion controlled reduction processes with peaks at about +500 and -500 mV vs. Ge, as well as an oxidation peak at about +1000 mV vs. Ge. We could attribute the first peak to the reduction of Ge(IV) to Ge(II), at potentials below 0 V the bulk deposition of Ge from Ge(II) species sets in. The oxidation peak at +1000 mV is clearly correlated with Ge electrooxidation, whereas the peaks

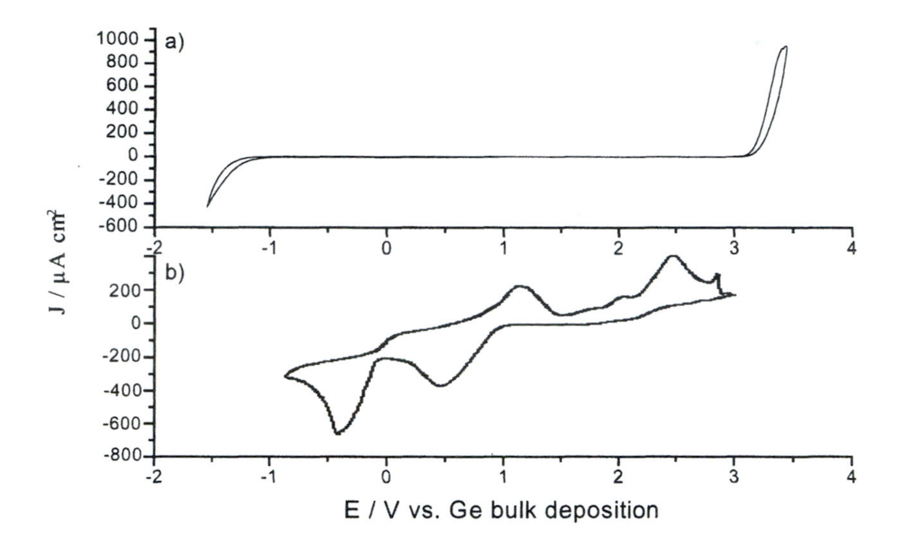

Fig. 5: a) shows the electrochemical window of $[BMIm]^+PF_6^-$ on Au(111): it is a little wider than 4 Volt. Upon saturation with $GeCl_4$ (b) 2 reduction as well as several oxidation peaks are observed (see text and [24]). v = 1 mV/s

above +1500 mV are also observed if the CV is cycled between +1000 and +3000 mV vs. Ge. These redox processes are correlated with the electrooxidation of the gold substrate. On other substrates such as HOPG and Si(111) the Cyclic voltammograms look quite similar. There, electrodeposition of Ge from $GeCl_4$ also proceeds via Ge(II) species. However, at electrode potentials of more than 1.5 Volt vs. Ge there are some differences. HOPG is also strongly attacked, and deep holes can form in the HOPG surface, similar to what was observed in [25]. On Si(111), however, only a very slight increase in current is observed in the CV from 1.5 Volt up to 4 Volt, and hole formation in the substrate, viewed by STM, is remarkably slow. Kinetic effects might play a dominant role here.

Fig. 6 shows a series of typical STM pictures of Ge underpotential deposition on Au(111) from $GeCl_4$ saturated in $[BMIm]^+PF_6^-$. Fig. 6a shows a typical Au(111) surface at +1200 mV vs. Ge. It is characterized by 250 pm high gold terraces and some gold islands. Fig. 6b shows an STM picture, where together with the STM scan (from top to bottom) a cyclic voltammogram was run with a scan rate of 10 mV/s. The electrode potential at the top of the picture is 1000 mV vs. Ge, it is 0 V at the bottom: it is quite evident, that islands grow on the gold surface in the UPD regime. As we have pointed out in more detail in [24], the underpotential deposition of Ge starts at the steps of the terraces at +1000 mV, then 150 pm high Ge islands start to be deposited at +950 mV and at about +750 mV vs. Ge we observe 250 pm high islands, most likely the result of an alloying between Au and Ge. Still in the UPD regime a completely closed monolayer forms on the gold surface, as evidenced in fig. 6b. If this UPD layer is

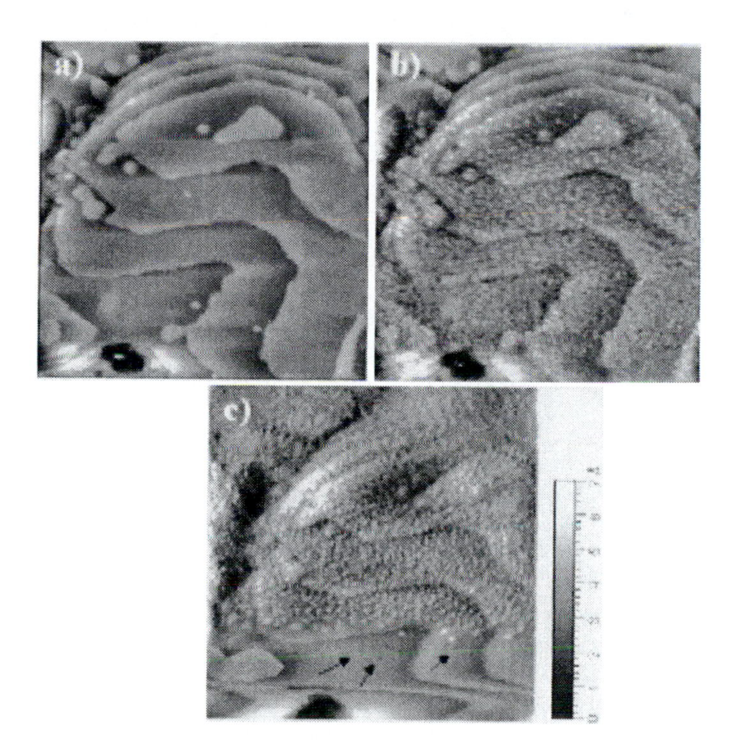

Fig. 6: a) shows a typical Au(111) surface in $[BMIm]^+PF_6^-$ saturated with $GeCl_4$. b) shows the formation of a closed Ge monolayer in the UPD regime, that alloys with the underlying gold surface. c) shows the complete redissolution of the Ge UPD layer, but some holes remain in the surface, typical for surface alloying (see text and [24]). Scan range: 325 x 325 nm^2

redissolved, here by running as Linear Sweep Scan from 0 V back to +1200 mV vs. Ge (STM scan from top to bottom), a closer look to the picture shows that some holes remain in the surface (arrows in fig. 6 c), which is typical for surface alloying between deposit and substrate. For comparison, we have performed the experiment with an another substrate, namely Highly Oriented Pyrolytic Graphite (HOPG). This substrate is routinely used for STM-scanner calibration, as atomic resolution of the graphite surface is a standard procedure. Furthermore, the underpotential deposition of metals has not yet been reported to occur on HOPG due to usually weak interactions between the substrate and metal deposits. On the other hand, in the bulk phase every Ge atom is tetrahedrally coordinated and normally Ge crystallizes in the diamond structure. The question was here, if a UPD of Ge can occur on HOPG. Fig. 7a shows a typical HOPG surface in the UPD regime, here at + 50 mV vs. Ge. The surface exhibits an atomically flat structure, at the lower part of the picture a second

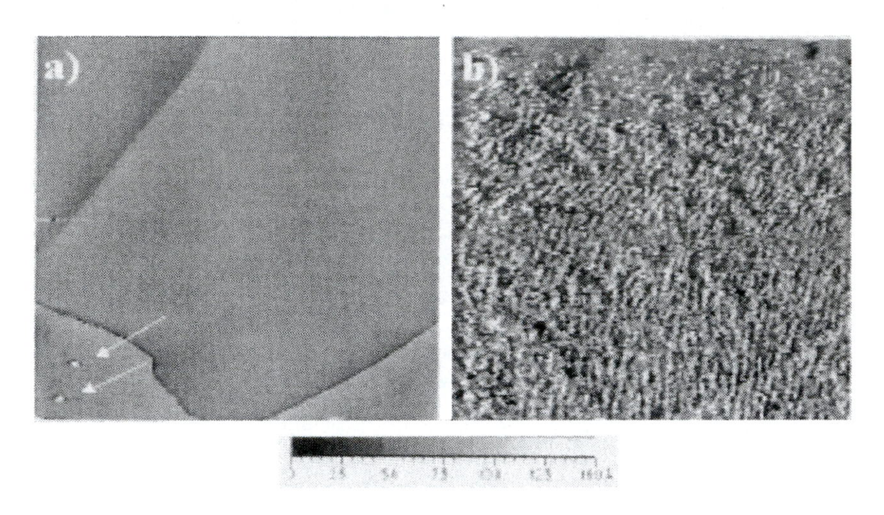

Fig. 7: a) shows a typical HOPG surface in $[BMIm]^+PF_6^-$ saturated with $GeCl_4$: in the UPD regime at +50 mV vs. Ge. b) shows the same site at -450 mV vs. Ge, and it is evident that numerous Ge clusters with a height between 1 and 2 nm and widths up to 20 nm are formed under these conditions (scan from top to bottom, scan range: 2750 x 2750 nm^2)

HOPG layer can be identified. On the flat surface, within the resolution of this experiment, we do not see a phase formation. An interesting observation was, that some nanoclusters with a height of only 1 nm grew at the positions indicated by the arrows. Originally there were 2 defects with a depth of one HOPG monolayer at these positions, and obviously Ge growth set in preferentially at these sites. However, it is not yet clear at the moment, if it is really Ge underpotential deposition on HOPG defects or if local variations in the electrode potential are the reason for this observation. Further studies on this interesting observation are planned in our laboratory. In fig. 7b the same site is shown at an electrode potential of -450 mV vs. Ge. It is evident, that many Ge nanocrystallites grow there, with heights up to 2 nm and widths up to 20 nm. We have to mention here, that these clusters are difficult to probe. It is a known shortcoming for the electrodeposition on HOPG, that the STM tip easily pushes the clusters aside because of the weak interaction between deposits and the HOPG surface. If the electrode potential is set to +500 mV vs. Ge, all clusters including those marked in fig. 7a disappear. Both on Au(111) and on HOPG we do not observe any deposition on the electrode surface in the pure liquid in the respective potential range, so that the deposition of $[BMIm]_x$ oligomers can be excluded.

We would like to mention quite recent results on the electrodeposition of narrowly dispersed Ge nanoclusters. As it has been demonstrated in [24] and [26], from $[BMIm]^+PF_6^-$, that was saturated with $GeCl_4$ or $GeBr_4$, Ge nanocrystallites can be obtained, furthermore micrometer thick nanosized deposits can be made. A disadvantage of these high concentrations is, that the growth in the OPD regime is pretty fast. In [27] we could show, that with $GeCl_4$ concentrations of about 5 x 10^{-3} mol/l narrowly dispersed nanoclusters can be electrodeposited on Au(111). Fig. 8 gives an example:

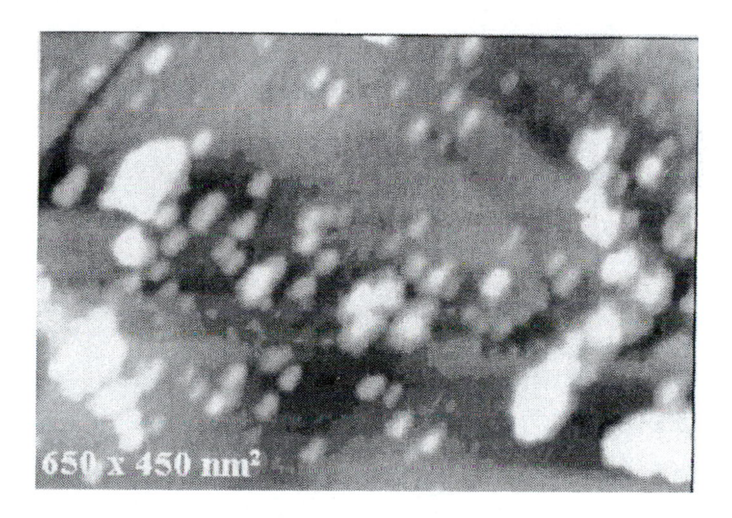

Fig. 8: From $[BMIm]^+PF_6^-$, 5 x 10^{-3} mol/l $GeCl_4$, narrowly dispersed Ge nanoclusters can be electrodeposited in the OPD regime on Au(111). The height of the clusters varies between 1 and 10 nm, with most of them between 1 and 5 nm (see also [27]).

In the OPD regime narrowly dispersed Ge nanoclusters with heights between 1 and 10 nm are deposited, most of them have heights between 1 and 5 nm. In situ current-voltage tunnelling spectroscopy on 10 nm thick clusters has clearly shown a band gap of 0.7 ± 0.1 eV, which is within the limits of error the typical value for microcrystalline germanium.
Both for basic research and for technical applications semiconductors are of great importance as substrates. Furthermore there are numerous studies in literature on the UHV deposition of Ge quantum dots on Si(111), and it was shown, that such quantum dots show a photoluminescence [28]. Our question was, if Ge can be electrodeposited on Si(111) and what the surface processes on the nanometer scale are. For this purpose we used wet etched hydrogen terminated Si(111) substrates. Concerning the etching we would like to refer to recent literature [28]. The electrodeposition was performed as described above

in $[BMIm]^{+}PF_6^{-}$ saturated with $GeCl_4$. Fig. 9 shows an example, where the electrodeposition was performed under potentiostatic conditions, slightly in the OPD regime, on hydrogen terminated Si(111) with a substrate miscut of about 4°.

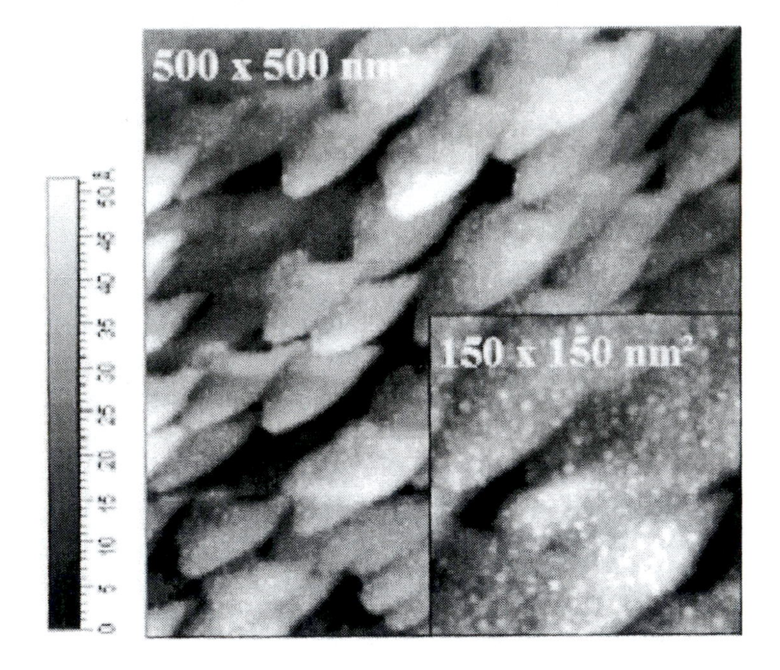

Fig. 9: STM picture of hydrogen terminated Si(111) (4° miscut), covered in the OPD regime with a thin layer of Ge. On the larger picture tongue like islands can be seen on top of which Ge nanoclusters with widths of about 5 nm and heights of about 1 nm grow

The Ge phase formation on Si(111) is a complicated process which is still under investigation in our laboratory: the Ge electrodeposition starts at the edges of the Si substrate and leads to tongue like islands as shown in fig. 9. Upon further reduction, on these islands Ge nanoclusters with a width of 5 nm and a height of 1 nm grow. Layers of more than 10 nm in thickness show clearly a band gap of 0.7 ± 0.1 eV.

These preliminary results show for the first time, that by electrodeposition from ionic liquids semiconductor nanostructures can be obtained, that formerly could only be made under UHV conditions. Further detailed studies on electrosynthesis and characterization are under progress.

4. Summary and Outlook

In this brief review we have summarized some results on the electrodeposition of metals and semiconductors from ionic liquids, probed on the nanometer scale by electrochemical in situ tunnelling microscopy. Due to their wide electrochemical windows ionic liquids give access to light metals, like aluminum, semiconductors (Ge) and rather noble metals, such as copper, for example. The STM is a valuable tool to probe the processes at the interface electrode/electrolyte on the nanometer scale: in $AlCl_3$ based ionic liquids, both Al and Cu show UPD phenomena on Au(111), and the initial stages of bulk growth lead in both cases to nanocrystallites. With the valuable information from the STM electrochemical procedures can be developed that deliver bulk nanocrystalline metals. In comparison to the microcrystalline ones nanocrystalline metals have improved mechanical properties, such as better hardness and often also better corrosion resistance.

During our studies on semiconductor electrodeposition from ionic liquids we could show for the first time, that germanium can be electrodeposited in high quality from the neutral ionic liquid 1-Butyl-3-methylimidazolium hexafluorophosphat. In the OPD regime, micrometer thick nanosized Ge deposits can be obtained on a variety of substrates. Furthermore, Ge nanoclusters with dimensions of only a few nanometers can be made by electrochemical means, for example on hydrogen terminated Si(111). Hitherto, such structures have only been reported to be made under Ultrahigh Vacuum conditions, with Molecular Beam Epitaxy, for example.

In general, electrochemistry is a versatile and powerful tool for materials electrosynthesis. By varying the experimental parameters like overvoltage, current density, temperature and bath composition, the crystal sizes can be adjusted over a wide range. Due to their extraordinary physical properties ionic liquids give now access to nanoscale electrodeposits, that otherwise cannot be made from aqueous or organic solutions.

Nanoscale electrodeposition studies from ionic liquids are still at their infancy. Nevertheless, we are convinced, that ionic liquids will have a strong impact on electrochemical nanotechnology and electrochemistry in general, and we are optimistic that in future they will be used in electrochemistry as water or organic solvents are used today.

5. References

1. K.R. Seddon, J. Chem. Tech. Biotechnol., 68 (1997) 351
2. K.R. Seddon, Kinet. Catal. Engl. Transl., 37 (1996) 693
3. P. Wasserscheid, W. Keim, Angew. Chemie, Int. Ed. 39 (2000) 3772
4. T. Welton, Chem. Rev., 99 (1999) 2071
5. R. Hagiwara, Y. Ito, J. Fluorine Chem., 105 (2000) 221

6. P. Bonhôte, A.-P. Dias, N. Papageorgiou, K. Kalyanasundaram, M. Grätzel, Inorg. Chem., 35 (1996) 1168
7. F. Endres, ChemPhysChem, 3 (2002) 144
8. D. Morris, Trans. Tech., Aedermannsdorf, Switzerland, 1998
9. T. Trindade, P. O Brien, N.L. Pickett, Chem. Mater., 11 (2001) 3843
10. A. Shkurankov, F. Endres and W. Freyland, Rev. Sci. Instrum, 73 (2002) 102
11. J. Tersoff and D.R. Hamann, Phys. Rev. B, 31(2) (1985) 805
12. „Scanning Probe Microscopy and Spectroscopy: Theory, Techniques and Applications", ed. by. D.A. Bonnell, 2nd ed., Wiley-VCH, New York, Chichester, Weinheim, Brisbane, Singapore, Toronto, 2001
13. C.A. Zell, F. Endres, W. Freyland, Phys. Chem. Chem. Phys., 1 (1999) 697
14. M. Bukowski, H. Natter, R. Hempelmann, F. Endres, ChemPhysChem, submitted
15. G. Stafford, T. Moffat, V. Jovic, D. Kelley, J. Bonevich, D. Josell, M. Vaudin, N. Armstrong, W. Huber, A. Stanishevsky, AIP conference proceedings (2001), 550, 402
16. B.J. Tierney, W.R. Pitner, J.A. Mitchell, Ch. L. Hussey, G.R. Stafford, J. Electrochem. Soc., 145 (1998) 3110
17. F. Endres and A. Schweizer, Phys. Chem. Chem. Phys., 2 (2000) 5455
18. C.G. Fink, V.M. Dokras, J. Electrochem. Soc., 95 (1949) 80
19. G. Szekely, J. Electrochem. Soc., 98 (1951) 318
20. M. Schüßler, T. Statzner, C.I. Lin, V. Krozer, J. Horn, H.L. Hartnagel, J. Electrochem. Soc., 143 (1996) L73
21. S. Takeoka, M. Fujii, S. Hayashi, K. Yamamoto, Phys. Rev. B, 58 (1998) 7921
22. F. Endres and C. Schrodt, Phys. Chem. Chem. Phys., 2 (2000) 5517
23. F. Endres, Phys. Chem. Chem. Phys., 3 (2001) 3165
24. F. Endres and S. Zein El Abedin, Phys. Chem. Chem. Phys., 4 (2002) 1649
25. F. Endres, W. Freyland, B. Gilbert, Ber. Bunsenges. Phys. Chem., 101 (1997) 1075
26. F. Endres and S. Zein El Abedin, Phys. Chem. Chem. Phys., 4 (2002) 1640
27. F. Endres and S. Zein El Abedin, Chem. Comm., 8(2002) 892
28. O. Leifeld, A. Beyer, E. Müller, D. Grützmacher, K. Kern, Thin Solid Films, 380 (2000) 176
29. J.C. Ziegler, A. Reitzle, O. Bunk, J. Zegenhagen, D.M. Kolb, Electrochimica Acta, 45 (2000) 4599

Novel Applications

Chapter 37

Plasticizing Effects of Imidazolium Salts in PMMA: High-Temperature Stable Flexible Engineering Materials

Mark P. Scott, Michael G. Benton, Mustafizur Rahman, and Christopher S. Brazel*

Department of Chemical Engineering, University of Alabama, Tuscaloosa, AL 35487

The unique, environmentally-sound applications of ionic liquids (ILs) are well-known, and include use as green solvents for a variety of chemical processes. Because ILs are stable liquids over a wide temperature range, they offer technological advantages over some chemicals used in their liquid phase, such as plasticizers, where polymer flexibility can be enhanced. Common problems with plasticizers include evaporation and leakage from the surface, instability at high temperatures, lack of lubrication at low temperatures, migration within the polymer, and toxicity. We have addressed couple of these issues using poly(methyl methacrylate), PMMA. Systems studied include bulk PMMA, and PMMA plasticized with butyl methylimidazolium hexafluorophosphate $[bmim^+][PF_6^-]$, hexyl methyl imidazolium hexafluorophosphate $[hmim^+][PF_6^-]$, and a traditional plasticizer, dioctyl phthalate (DOP). Experiments indicate that high temperature stability is improved significantly by replacing DOP with an IL. The effect of IL on glass transition temperature and elastic modulus were also determined. Ionic liquids as plasticizers may revolutionize the usage of flexible polymers at high temperatures, without brittleness or loss of mechanical strength.

Background

Plasticizers are polymer additives used to improve the flexibility, processability and workability of plastics. Four million tons of plasticizers are produced annually worldwide[1]. Most of these plasticizers are based on three classes of compounds: phthalates, adipates and trimellitates, with dioctyl phthalate (DOP) accounting for more than 50% of all plasticizers used. Ideally plasticizers should exhibit most of the following characteristics: low volatility, low leachability, high and low temperature stability, thermodynamic compatibility with polymers, low cost and minimal health and safety concerns[2]. However, current plasticizers have relatively small thermal working ranges (due to evaporation or solidification) and have limited lifetimes in products (due to volatility and leaching).

Room temperature ionic liquids (ILs) meet many of the same requirements of plasticizers and offer the potential for improved thermal and mechanical properties. ILs, being salts with melting points of below *ca.* 100 ^{0}C (and reportedly as low as –96 ^{0}C), can be used as solvents under reaction conditions similar to conventional organic liquids. ILs possess a wide liquid temperature range, in some cases in excess of 400 ^{0}C[3]. They have unique properties that have led to their investigation in a wide range of applications, including electrochemical[4] and separation processes[5-7], and as solvents for chemical and biochemical synthesis[8-12]. The principal interests in using ILs is either (1) to utilize their non-volatile behavior in developing environmentally-responsible systems (as replacement solvents for volatile organic compounds, for example), or (2) to give a technological advantage for the creation of novel materials, unique reactions, or more economical processes.

Because of their non-volatile nature and the wide temperature range where ILs can be used as liquids, they provide unique opportunities when used in polymer systems. As plasticizers, ILs have several advantageous properties, as they are nonflammable, have high thermal stability, exhibit no measurable vapor pressure, and can be designed to be highly solvating for specific organic and inorganic compounds. In addition, the ILs may provide environmentally responsible alternatives to traditional plasticizers since the lifetimes of the flexible plastics are increased (reducing landfill waste) and less plasticizer will be released into the environment due to leaching or volatility during a material's use. ILs have been shown to be compatible with certain polymers and free radical polymerizations have been successfully carried out in [$bmim^+$][PF_6^-], although isolation and separation of polymer and IL have proven difficult[13-15]. As plasticizers, ILs offer better mechanical and thermal properties, and may expand the temperature range where flexible polymers can be used.

Objective

The objectives of this study were to determine the effects of ILs as plasticizers on material properties of poly(methyl methacrylate), PMMA. The moduli of elasticity were determined as a function of plasticizer type and concentration, and the effect of plasticizer content on glass transition temperatures and thermal stability of PMMA samples at elevated temperatures was measured.

Experimental

PMMA was synthesized using specific volume ratios of methyl methacrylate (Acros, Fairlawn, NJ) with DOP (Sigma, St. Louis, MO), butyl methylimidazolium hexafluorophosphate $[bmim^+][PF_6^-]$ or hexyl methylimidazolium hexafluorophosphate $[hmim^+][PF_6^-]$ (both ILs were prepared on campus as described elsewhere[16]). Inhibitor was removed from MMA prior to the reactions by passing the monomer through a dehibiting ion exchange column (Aldrich, Milwaukee, WI). Samples were prepared either with 1 mol % ethylene glycol dimethacrylate, EGDMA (Aldrich), or without crosslinking agent. Nitrogen was passed through the monomer solution to remove dissolved oxygen prior to addition of 1 wt % azobisisobutyronitrile, AIBN (Aldrich). The polymers were formed between two siliconized glass sheets separated by Teflon® spacers to provide samples of uniform thickness for mechanical testing. The reactions were carried out at 55 °C for 24 hours. Bulk polymers and samples containing up to 50 vol % of plasticizer were synthesized. Dogbone and rectangular-shaped samples were cut for mechanical analysis and long term thermal exposure at 170 ^{0}C. Small pieces (approximately 10 mg) of polymers were collected for differential scanning calorimetry, DSC, and thermogravimetric analysis, TGA (Models 2920 MDSC and 2950 TGA, both TA instruments, Newcastle, DE). Due to the plasticizer incorporation method, all plasticizer contents are volume percentages based on the feed ratios to the reaction.

Initial mechanical tests were conducted on dogbone-shaped samples at room temperature using an Instron Automated Materials Testing System (Model 4465, Canton, MA). A strain rate of 5 mm/min was applied while stress was measured. Mechanical tests were conducted as a function of temperature using a Rheometrics Solids Analyzer (RSA II, Rheometrics Inc., Piscataway, NJ). Here, elastic moduli were measured using an incremental strain rate of 0.06%/min at temperatures from 25 to 200 ^{0}C. Mass loss was monitored in three ways: (1) by

TGA during a temperature ramp of 10 ^{0}C/min to 350 ^{0}C, (2) by TGA when held isothermally at 250 ^{0}C for 25 minutes, and (3) over an extended period of time held isothermally at 170 ^{0}C in an oven. TGA experiments were conducted using approximately 10 mg polymer samples, while the third method utilized rectangular-shaped samples with approximate dimensions of 40 x 15 x 1.4 mm and 1 g weight. DSC experiments were conducted to determine glass transition temperatures, using a heating/cooling cycle to erase sample thermal history before heating at 10 ^{0}C/min to 150 ^{0}C.

Results & Discussions

Uniform plasticized PMMA samples were successfully polymerized with [bmim$^+$][PF$_6^-$] and [hmim$^+$][PF$_6^-$] containing up to 50 vol % plasticizer. Syntheses of PMMA samples containing up to 50 vol % DOP were also attempted, but samples containing 40 vol % DOP or greater phase separated during polymerization. Therefore, DOP-containing samples had from 10 to 30 vol % of the plasticizer only.

Elastic Moduli

The elastic modulus is the prime consideration in determining the general utility of a polymer. The elastic moduli of PMMA samples (with crosslinking agent) decreased with increasing amounts of plasticizer added, regardless of whether the plasticizer was DOP or an IL (Table 1). This was not unexpected, but the similarity of the curve for the ILs compared to DOP indicates that these ILs are equally good at plasticizing PMMA, with the added benefit that the ILs have lower vapor pressure. Additionally, the ability to incorporate greater amounts of IL into the PMMA samples allows the creation of plastics that have glass transition temperatures at or below room temperature.

The elastic behavior of plasticized PMMA was also determined as a function of temperature, by generating the master curves for each polymer. Figure 1 shows the effect of adding [bmim$^+$][PF$_6^-$] to PMMA formulations in successively increasing quantities. The slope of the leathery region (between the glassy region at low temperatures and the rubbery plateau above the glass transition, T_g, temperatures) is nearly constant, but as the IL content is increased, the T_g drops. The behavior of PMMA plasticized with 30 vol % [hmim$^+$][PF$_6^-$] nearly matched the moduli of DOP-plasticized PMMA (Figure 2). Because ILs were successfully incorporated into PMMA at a higher concentrations than DOP (and 50 vol % is only the highest tested, not necessarily a maximum IL content), ILs allow better control of polymer mechanical properties.

Table 1. Effect of Plasticizer and Plasticizer Content on Elastic Moduli of PMMA Samples as Determined by Instron at Room Temperature

Plasticizer Content (vol %)	Elastic Modulus for PMMA Plasticized with DOP (MPa)	Elastic Modulus for PMMA Plasticized with $[bmim^+][PF_6^-]$(MPa)
0	535	535
10	521	515
20	340	314
30	207	152
40	*	13.8
50	*	0.386

* DOP concentration was above solubility limit to form uniform samples

Dynamic mechanical tests were conducted on samples containing each of the three plasticizing agents to compare the universal curves of the IL-plasticized PMMA with that of the samples containing DOP (Figure 2). PMMA with 30 vol % $[hmim^+][PF_6^-]$ followed the behavior of PMMA with 30 vol % DOP both in magnitude of elastic modulus and the temperature dependence. This indicates that the ILs can do as good a job at plasticizing PMMA as the standard plasticizing agent, while the ILs offer improved high temperature stability.

Thermal Stability

Low volatility is one of the most important characteristics of plasticizers. It must remain in the polymer over for the lifetime of the product, and may be subjected to a range of temperatures. ILs are particularly intriguing in this aspect as they typically have no detectable vapor pressure[17] which enables the formation of flexible materials with significantly extended lifetimes. TGA experiments were used to elucidate the mass loss of bulk and plasticized PMMA. $[bmim^+][PF_6^-]$ and DOP were also tested in bulk. DOP has a boiling point of 380 ^{0}C and is highly volatile at elevated temperatures. In experiments where approximately 10 mg samples of DOP and $[bmim^+][PF_6^-]$ were subjected to a 10 ^{o}C/min ramp, 95 wt% of the DOP evaporated by the time the temperature reached 300 ^{o}C, while only 1.15 wt % of the IL was lost in the same temperature ramp (see data points at 100 % plasticizer content in Figure 3). Plasticized PMMA with 10 to 50 vol % of each plasticizer (DOP and $[bmim^+][PF_6^-]$) were also tested for thermal stability (Figure 3). With increasing concentrations of $[bmim^+][PF_6^-]$, the PMMA samples actually became more thermally stable than PMMA by itself. This is in stark contrast to the loss in mass of the DOP-plasticized PMMA.

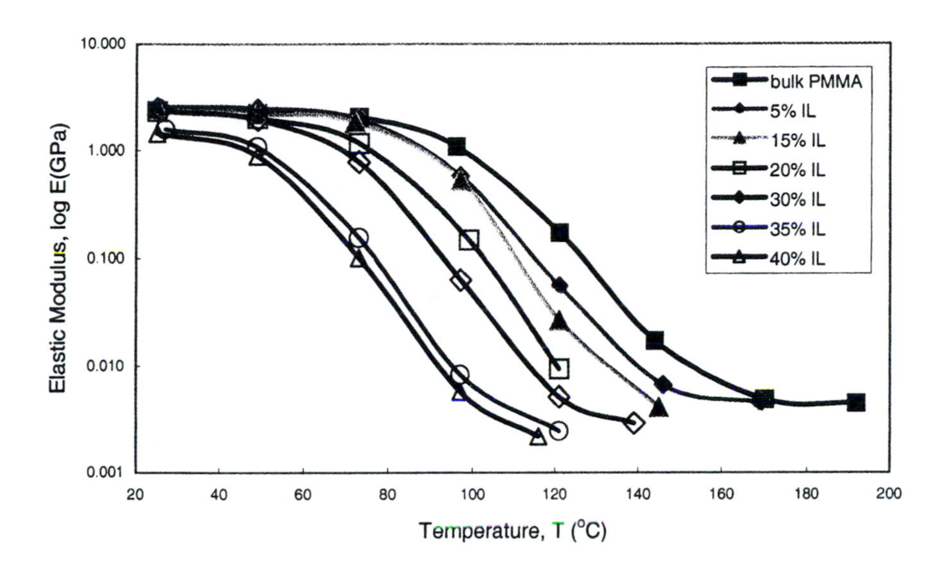

Figure 1. Effect of [bmim⁺][PF₆⁻] Content in Plasticized PMMA on the Master Curve Representing the Mechanical Behavior Over a Range of Temperatures

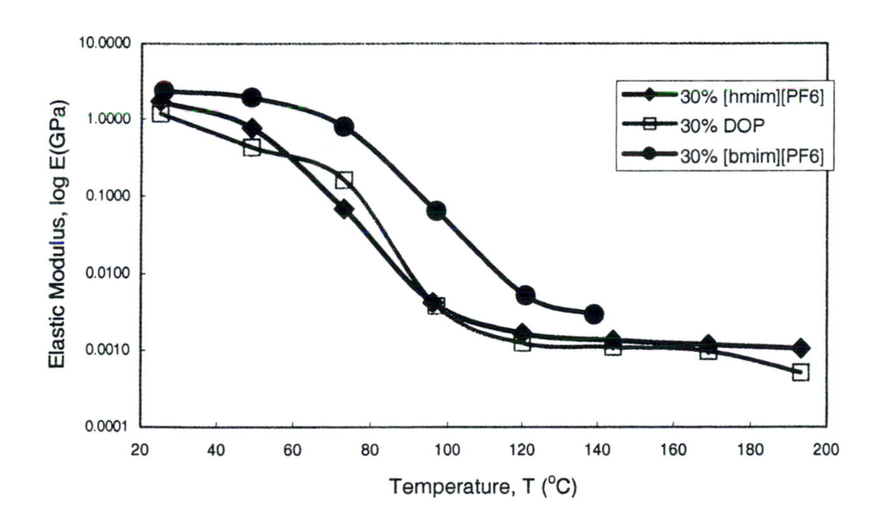

Figure 2. Effect of 30 vol % [hmim⁺][PF₆⁻], DOP and [bmim⁺][PF₆⁻] as Plasticizers for PMMA on the Master Curve

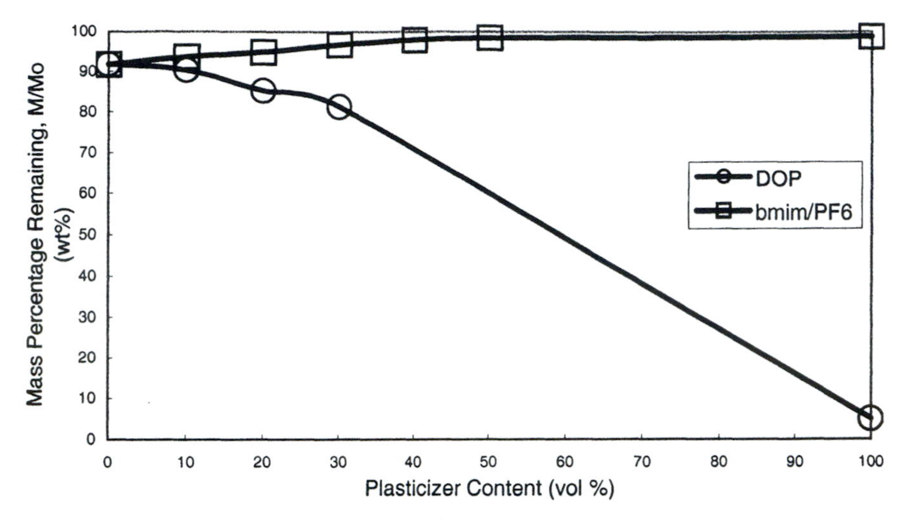

Figure 3. Thermal stability of [bmim$^+$][PF_6^-] and DOP plasticized PMMA samples in TGA experiments

Because the high temperature stabilities tested above are not realistic for the long-term use of flexible plastic, a similar comparison between [bmim$^+$][PF_6^-] and DOP as plasticizers for PMMA was carried out at 170 °C over a period of more than a month (Figure 4). Here, unplasticized PMMA was shown to be the most thermally stable over 42 days, while there was nearly 5 wt % mass loss in the same time period for PMMA plasticized with either 20 or 30 vol % [bmim$^+$][PF_6^-]. This accounts for a significant portion of the IL that was incorporated, as 20 to 25 wt % of all of the IL incorporated had evaporated in this time period. Still, the ILs are much more promising as plasticizers at elevated temperatures than DOP. DOP-plasticized PMMA lost 75 wt % of the plasticizing agent in only 22 days, dropping 15 wt % of total mass for a sample plasticized with 20 vol % [bmim$^+$][PF_6^-], and 24 wt % for PMMA plasticized with 30 vol % [bmim$^+$][PF_6^-]. Further work may focus on other ILs, since the hexafluorophosphate anion in known to decompose to products including HF at high temperatures through hydrolysis.

Glass Transition Temperature

The glass transition temperature (T_g) is important in designing polymeric materials because it draws the distinction between the hard, glassy region and the rubbery plateau in the master curve, thereby determining the temperature range where a polymer will be commercially useful. Plasticizers are often used to lower the glass transition temperature of polymers to make polymers workable

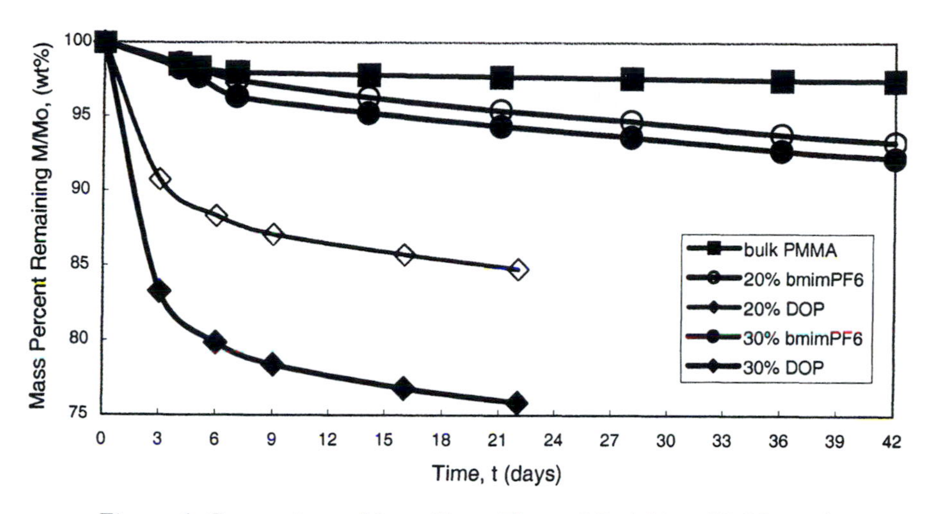

Figure 4. Comparison of Long Term Thermal Stability of DOP- and $[bmim^+][PF_6^-]$-Plasticized PMMA

and flexible at low temperatures. The T_g of PMMA dropped with the addition of DOP, but only up to 52.44 ^{0}C, since DOP was incompatible with formulations containing 40 vol % DOP or greater (Table 2). The two ILs tested, $[bmim^+][PF_6^-]$ and $[hmim^+][PF_6^-]$, on the other hand, lowered the T_g to as low as 10 °C using up to 50 vol% IL in PMMA. Thus, ILs can provide a wider range of material properties, and may be able to drop the glass transition temperature of PMMA significantly below 0 °C with increased concentrations.

Table 2. Effect of Plasticizer Type and Concentration on the Glass Transition Temperature of Bulk and Plasticized PMMA. Errors shown represent the standard deviation for three experiments.

	T_g (^{0}C)		
Plasticizer content (%)	DOP	$[bmim^+][PF_6^-]$	$[hmim^+][PF_6^-]$
0	116.7 (+/- 0.26)	116.7 (+/- 0.26)	116.7 (+/- 0.26)
10	85.09	93.87	--
20	66.68 (+/- 1.88)	71.99 (+/- 1.36)	68.28 (+/- .43)
30	52.44 (+/- 3.37)	48.25 (+/- 2.24)	40.98 (+/- 1.42)
40	Not compatible	28.71	--
50	Not compatible	21.64 (+/- 6.31)	9.49 (+/- 0.48)

Conclusions

From this on-going work, it has been found that $[bmim^+][PF_6^-]$ and $[hmim^+][PF_6^-]$ are able to provide better thermal and mechanical properties to PMMA than does the traditional plasticizer, DOP. Yet PF_6^- anions have their drawback of decomposing at high temperatures to yield HF. Low and moderately high temperature applications are being investigated for them. For additional low temperature applications, ionic liquids such as hexyl and octyl derivatized salts would be advantageous as these ILs have melting points below –70 ^{0}C. Better compatibility of ILs with PMMA than DOP allowed incorporation of over 50 vol % $[bmim^+][PF_6^-]$, which in turn led to a wider range of glass transition temperatures, and may allow PMMA to be used as a flexible material in subzero (Celsius) conditions. This indicates a promising development in the field of flexible polymeric materials that withstand a large range of working temperatures, with minimal evaporation or leaching.

Acknowledgments

This research was supported by National Science Foundation grants NSF-CTS0086874 and a University of Alabama SOMED research grant. The authors also acknowledge Dr. David Nikles and Dr. Mark Weaver for the use of testing equipment and Matthew Reichert for preparation of ILs.

References

1 Cadogan, D. F.; Howick, C. J.,*'Plasticizers'* in Kirk-Othmer *Encyclopedia of Chemical Technology*, ed. Croschwitz, J. I.; Howe-Grant, M., Wiley, New York, **1992;** 258.
2 Scott, M. P.; Brazel, C. S.; Benton, M. G.; Mays, J. W.; Holbrey, J. D.; Rogers, R. D., *Chem. Comm.,* **2002**; p 1370 – 1371.
3 McAuley, B. J.; Seddon, K. R.; Stark, A.; Torres, M. J., *Ind. Eng. Chem.;*American Chemical Society: Washington, DC, **2001**; 221, p 277.
4 Carlin, R. T.; Wilkes, J. S., *Chemistry of nonaqueous solutions*, ed, Mamantov, G.; Popov, A. I., VCH, Weinheim, **1994;** p 227.
5 Huddleston, J. G.; Visser, A. E.; Reichert, W. M.; Willauer, H. D.; Broker G. A.; Rogers, R. D., *Green Chem.*, **2001;** 3, p 156.
6 Visser, A. E.; Swatloski, R. P.; Rogers, R. D., *Green Chem.*, **2000;** 2, 1.

7 Visser, A. E.; Swatloski, R. P.; Reichert, W. M.; Rogers, R. D.; Mayton, R.; Sheff, S.; Wierzbicki, A.; Davis Jr., J. H., *Chem. Commun.*, **2001;** 135.
8 Welton, T., *Chem. Rev.*, **1999;** 99, p 2071.
9 Holbrey, J. D.; Seddon, K. R., *Clean Prod. Proc.*, **1999;** 1, p 223.
10 Wasserscheid, P.; Keim, W., *Angew. Chem., Int. Ed.*, **2000;** 39,p 3772.
11 Sheldon, R., *Chem. Comm.*, **2001**, p 2399.
12 Gordon, C. M., *Appl. Catal. A*, **2002;** 222, p 101.
13 Harrisson, S.; Mackenzie, S.; Haddleton, D. M., *Polym. Prepr.,* **2002**; 43(2), p 883.
14 Benton, M. G.; Brazel, C. S., *Ind. Eng. Chem.;* American Chemical Society: Washington, DC, **2001**; 221, p 165.
15 Hong, K.; Zhang, H.; Mays, J. W.; Visser, A. E.; Brazel, C. S.; Holbrey, J. D.; Reichert, W. M.; Rogers, R. D., *Chem. Commun.*, **2002**; p 1368 – 1369.
16 Huddleston, J.G.; Visser, A.E.; Reichert, W.M.; Willhauer, H.D.; Broker, G.A.; Rogers, R.D. *Green Chem.* **2001**, *3*, 156.
17 Pernak, J.; Czepukowicz, A.; Pozniak, R., *Ind. Eng. Chem. Res.*, **2001;** 40, 2379.

Chapter 38

The Use of Ionic Liquids in Polymer Gel Electrolytes

Hugh C. De Long[1,2], Paul C. Trulove[1,2], and Thomas E. Sutto[2,3,*]

[1]Air Force Office of Scientific Research, 801 Randolph Street, Arlington, VA 22203–1977
[2]Naval Research Laboratory, Chemistry Department Building 207, Department 6170, 4555 Overlook Avenue SW, Washington, DC 20375
[3]U.S. Naval Academy, Chemistry Department, Annapolis, MD 21402

Polymer gel electrolytes composed of 1,2-dimethyl-3-n-alkyl-imidazolium bis-trifluoromethanesulfonylimide (alkyl = propyl or butyl) and polyvinylidenedifluoro-hexafluoropropylene are characterized by ac-impedance and cyclic voltammetry. Two electrode charge-discharge experiments were also performed using graphitic paper or Li metal as the anode, and polymer composites of $LiMn_2O_4$, $LiCoO_2$, or V_2O_5 as cathodes. Results indicated that the polymer composite gel electrolytes were stable for over 50 cycles when used in direct contact with Li metal. High efficiencies and low voltage drop-offs indicate that polymer gel composite electrodes composed of these ionic liquids are a viable alternative to the more common organic solvent electrolytes.

Introduction

By now, any reader of this collection of work should be extremely familiar with the burgeoning field of both ionic liquids as well as their numerous benefits to green chemistry. Ionic liquids themselves exist as a subset of the science of molten salts, which are already extensively used in industry because of their unique physical properties. It would go well beyond the required length of this paper to describe all of the work that has gone into molten salt (and ionic liquid) techniques and technology, as well as the scientists responsible for all of these developments. Suffice it to say that some of these scientists, and their associated research institutes, serve as the heads and homes of several extended families of researchers devoted to molten salts. In order to gain a better understanding of this, the reader is strongly urged to look through the entire thirteen volume series on Molten Salts published by The Electrochemical Society in which the evolution of novel ideas that have led to the recent surge in interest in molten salts can be traced. Additionally, another excellent text would be David Lovering's, "Molten Salt Technology", published in 1982, which summarizes the state of the ecological, economical and technical advantages of molten salts at that time, and which are still true today.

Of all the potential applications for molten salts, one of the least investigated is their use as replacements for some of the more common organic electrolytes in Li-ion batteries. The very physical properties that make them useful as electrolytes for electrochemical deposition techniques, or purification media for plating out even refractory metals, also make them the ideal choice as electrolytes in the rapidly growing area of high energy density power sources. However, in terms of actual attractiveness to the battery industry, it will be important to significantly narrow the scope of molten salts studied. Since most of these systems need to operate in the ambient temperature regime, the molten salt family of choice is clearly the ionic liquids. Furthermore, the question arises as to which ionic liquids to choose.

In terms of the anion, this is a relatively simple matter. Although $AlCl_4^-$, orother similar types of anions have been studied, they are perhaps not the best choice for commercial Li-ion batteries due to their tendency to evolve heat and HCl gas upon contact with moisture.[1] Furthermore, some work has even shown that reactions with the $AlCl_4^-$ anion produce Li-Al alloy coatings on graphitic electrodes rather than allowing for Li intercalation.[2] Thus, for Li-ion battery systems, the best anions to chose from are predominately the air and water stable fluorinated anions, such as BF_4^-, PF_6^-, or the bis-trifluoromethanesulfoylimide ($TFSI^-$) anion. Before we can decide which of these to use, it is important to discuss another significant aspect of ionic liquids as they apply to power sources.

Perhaps the crucial difference between these ionic liquid electrolytes and the more commonly used organic carbonates is the presence of the charged imidazole ring and the anion, both of which move, diffuse, and interact with the electrodes when a current is applied. This can, for electrode materials of a layered structure type such as in the case of graphitic or TiS_2 electrodes, lead to a unique type of battery in which the ionic liquid itself serves as both the electrolyte, and the source of the intercalative guest species. This configuration is termed the dual intercalating molten electrolyte, DIME, batteries.[3-6] Common sense tells us that these DIME batteries will not exhibit high enough energy densities to compete with Li-ion batteries, due to the much larger size of the current carrying species. However, they do represent the simplest form of a battery system, as well as present a very basic test case to explore the electrochemical stability of a variety of cations and anions during extensive series of charge-discharge experiments.

In regards to the anion stability, experiments have shown that the BF_4^- anion is an extremely poor choice due to its electrochemical instability especially when intercalated into graphitic materials.[4] Furthermore, although the PF_6^- anion does behave moderately well, it does create extremely viscous and less ionically conductive ionic liquids.[7-9] However, the $TFSI^-$ anion has performed admirably well in the basic DIME system, and seems a good anion to chose.[10]

In term of the cations to choose from, the substituted imidazoles form ionic liquids with significantly large electrochemical windows,[11-13] and four of the more common types of these cations are shown in figure 1. DIME studies have also produced significant evidence that although the 1-ethyl-3-methylimidazolium, EMI^+, or 1-butyl-3-methylimidazolium, BMI^+, cations are the ones most often studied, they are inherently unstable over prolonged use in DIME systems.[3,9] In fact, both in DIME studies, and in simple Li-ion battery studies, these cations breakdown over time at the high potentials used in high-energy battery sources, most likely due to the reactivity of the lone hydrogen at the 2-position of the imidazole ring.[3,7,9] However, the 1,2-dimethyl-3-n-propylimidazolium, $DMPI^+$, or 1,2-dimethyl-3-n-butylimidazolium, $DMBI^+$, cations have proven to be extremely stable during a number of charge-discharge cycles.[3,7,9] Thus we will concentrate on the two primary ionic liquids, DMPITFSI and DMBITFSI.

The properties of ionic liquids also offer profound advantages when used in polymer gel electrolytes. The discovery of ionically conductive polymers such as polyethylene oxide, PEO, and others led to a wide-ranging amount of research in an attempt to prepare solid state batteries using a pure, solid polymer electrolyte.[14,15] Unfortunately, these solid polymers are very poor ionic

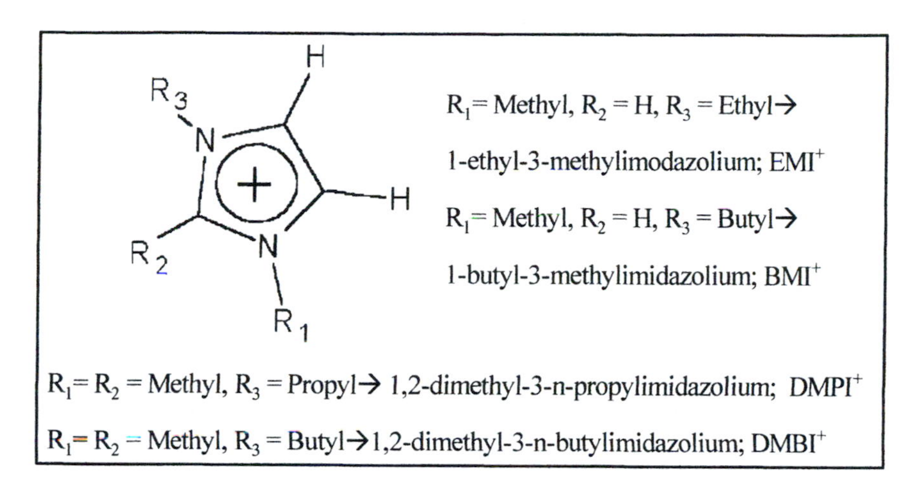

Figure 1. *The basic structure of the four most commonly used imidazolium based ionic liquids.*

conductors at room temperature.[16] In order to overcome this rather significant drawback, investigators have tried to correct this deficiency by creating polymer gel composite electrolytes, in which a liquid electrolyte is added to the polymer in order to improve the ionic conductivity while maintaining the ease of working with a solid polymer. These polymer composite gel electrolytes have been extensively studied, using a variety of polymers, as well as a multitude of electrolytes; the most common of which are the organic carbonates such as propylene carbonate.[17-19] Unfortunately, these liquid components suffer from the same drawbacks as when used as pure liquids. Their volatility leads to the loss of electrolyte over time and the eventually loss of ionic conductivity, and it increases the polymer's flammability. In fact, some formulations have added fire-retardants to the polymer matrix.[20] In order to overcome these shortfalls, ionic liquids can be used in polymer composite gel electrolytes allowing for the formation of much safer electrolytes for Li-ion solid state batteries.

The last decision to be made is which polymer to use as the polymer component in the gels. Initial studies have already pin-pointed one significant problem in choosing PEO. For PEO polymer composite gel electrolytes, the ionic liquid dissolves the PEO somewhat leading to a dimensionally unstable solid like a soft wax, and the resulting gels are fragile and extremely moisture sensitive.[21] However, polyvinylidenedifluoro-hexafluoropropylene, PVdF-HFP, has behaved admirably well in the case of other ionic liquids.[8,9,22,23] Therefore, polymer composite gel electrolytes composed of DMPITFSI or DMBITFSI and PVdF-HFP will be studied and characterized for potential use in Li-ion batteries.

Experimental

Synthesis of the Initial Substituted Imidazolium Salts. The preparation of DM***R***ICl involved the direct reaction of 1,2-dimethylimidazole (Aldrich, 98%) and 1-chloropropane (Aldrich, 99%), or 1-chlorobutane (Aldrich 99.5%) in a 1:1.15 molar ratio. The 1,2-dimethylimidazole (m.p. 38 °C) was vacuum distilled at approximately 10^{-3} torr at 135 °C in order to begin with as pure a product as possible. This was to avoid the formation of the non-volatile yellowish brown tint that often forms in these molten salts formed around 120-130 °C. All subsequent procedures were performed in a dry box (O_2 and H_2O < 1 ppm) unless noted otherwise. The 1,2-dimethylimidazole was melted at 50 °C in the dry box. The starting amount of 1,2-dimethylimidazole, 500 g, and a 15% excess of the corresponding 1-chloro alkane were placed, along with 50 ml of acetonitrile, in a thick-walled, single neck, 2-liter round bottom flask. Once filled, the round bottom flask was removed from the dry box and fitted with a reflux condenser. The solution was degassed several times with dry nitrogen, gradually heated to 60 °C for the DMPICl reaction, and 80 °C for the DMBICl reaction, and allowed to react for 2 days under nitrogen pressure. After this 2-day reaction period, the temperature was increased to 85 °C for the DMPICl reaction, and to 95 °C for the DMBICl reaction. Both reactions refluxed for an additional 5 days. Cooling the solution produced a white precipitate and a very faint, yellow supernatant. 300 ml of ethyl acetate was added to the round bottom flask to precipitate all of the DM***R***ICl. The material was filtered and washed with five 100 ml washings of ethyl acetate in order to remove all of the unreacted 1,2-dimethylimidazole. The solid material was dissolved in a minimum amount of hot acetonitrile and quickly crashed out of solution by the addition of a large excess of ethyl acetate, since it was found that slow recrystallization resulted in the formation of slightly yellowed crystals. The final product was a white crystalline material. Finally, the DM***R***ICl was heated to 100 °C under a dynamic vacuum (10^{-3} torr) for 2 days to remove the volatile contaminants giving a final total yield of approximately 80%. The synthesis of BMICl was performed in a manner identical to that for the DMBICl, except that 1-methylimidazole was distilled and used in place of the 1,2-dimethylimidazole.

Preparation of the ionic liquids. The preparation of DM***R***ITFSI was done by an ion exchange of the respective chloride salt with LiTFSI (Aldrich, 99.98%, dried for 12 hours at 80 °C in a vacuum oven at 10^{-3} torr) in acetonitrile. For a typical reaction, one mole of the DM***R***ICl was placed in a 1-liter reaction flask fitted with a threaded Teflon plug and dissolved in a minimum amount of acetonitrile. To this solution, an equal molar amount of the LiTFSI was added. The flask was sealed and allowed to stir at room temperature for 7 days. After which, the flask was removed from the dry box for the last purification steps.

Preparation of the $BMIPF_6$ was done in a similar manner, except that NH_4PF_6 (Aldrich, 99.5%) was used instead of LiTFSI.

Upon completion of the anion exchange, the solid material (LiCl or NH_4Cl) was removed by vacuum filtration using a glass frit of medium pore size. Subsequently, all of the acetonitrile was removed by rotavaporization, causing much of the remaining LiCl to precipitate out of solution. This impure form of the DM***R***ITFSI or $BMIPF_6$ was washed seven times with 200 mL of distilled water. Since the DM***R***ITFSI and $BMIPF_6$ are hydrophobic, these solutions separated into layers, with the LiCl or NH_4Cl being transferred in the water layer, which was discarded. The solution was then dried for 12 hours at 85 °C under a dynamic vacuum (10^{-3} torr) for 1 day to remove most of the water. 300 ml of acetonitrile, 15 g of decolorizing carbon and 15 g of neutral alumina were added to the solution, which was allowed to stir at room temperature in a sealed reaction flask for 24 hours. The carbon black and alumina were removed by successive filtration. Prior to the final filtration, the solution volume was reduced by half using a rotovap, and re-diluted with acetone. These ionic liquid/acetonitrile/acetone solutions were then filtered through a medium pore size glass frit, a 1 μ filter disc (Whatman, PTFE Membrane), a 0.45 μ filter disc (Whatman, PTFE Membrane), and a 0.2 μ filter disc (Whatman, PTFE Membrane). For the final step in the purification process, the molten salt solutions were heated to 85 °C under a dynamic vacuum (10^{-3} torr) for 2 days to remove the volatile components and any remaining water. The final molten salts were clear and colorless.

Preparation of the lithium containing ionic liquids. LiTFSI (Aldrich, 99.98%, dried for 12 hours at 80 °C in a vacuum oven at 10^{-3} torr) was added to both the DMPITFSI and DMBITFSI to form 1.0 M solutions. For these solutions, intermittent sonication and heating to 50 °C was used to facilitate dissolution of the lithium salt. In every case, the 1.0 M Li ion solutions were slightly more viscous, but still clear and colorless. Preparation of the 1.0M $LiPF_6$ (Aldrich, 99.95%) in $BMIPF_6$ was done in a similar manner.

Preparation of the polymer gel electrolytes. The polymer gels were prepared in an Ar dry box (H_2O and O_2 < 1ppm). The PVdF-HFP (Kynar – 2801-00 from Elf Atochem) was dissolved using 4-methyl-2-pentanone, 4M2P, (98%, Aldrich) as the solvent. Previous work has shown that the nominal composition of these gels should be between 70-85% ionic liquid; therefore, gels that were 75% 1.0 M Li/DM***R***ITFSI or $BMIPF_6$ and 25% PVdF-HFP were prepared. For each sample, 0.625 grams of the polymer was added to 10 mL of the 4M2P solvent. The solvent/polymer mixture was sonicated for 2 minutes at high power, allowed to cool, and then resonicated for 2 additional minutes in order to completely dissolve the polymer in the solvent. 1.875 grams of the ionic liquid were added, and sonication was continued, with intermittent pauses to allow for sample cooling, until there was a marked drop-off in the sound of cavitation produced. The resulting ionic liquid/polymer solution was removed from the sonicator and allowed to stir at 60 °C for 15 to 30 minutes under Ar until the mixture became extremely viscous. It should be noted here that the 4M2P/PVdF-HFP/Li/BMIPF$_6$ solution was white and non-transparent, the 4M2P/PVdF-HFP/Li/DMPITFSI solution was slightly opaque but still transparent, while the solution of 4M2P/PVdF-HFP/Li/DMBITFSI was absolutely clear, indicating a better mixing of the components. The liquid composite material was removed from the heat, poured into a 2.5" diameter Al weighing boat, and cured at room temperature for 24 hours in the dry box. The Al dishes were subsequently placed in a dessicator, and dried in a vacuum oven an additional 24 hours under a dynamic vacuum (approximately 10^{-3} torr) at 75 °C. For all polymer gels prepared, the thickness of the gels was typically between 1-2 mm. The 1.0 M Li/BMIPF$_6$ gel was flexible, but solid white in color. The 1.0 M Li/DMPITFSI gel was flexible and slightly transparent, while the 1.0M Li/DMBITFSI gel was flexible and completely transparent.

Preparation of the polymer composite electrodes. For solid-state test cells, the various electrodes were prepared from polymer solutions of the electrode material. 1.4 grams of the primary electrode materials, $LiMn_2O_4$ (spinel phase, Aldrich, 99.5%), $LiCoO_2$ (Aldrich 99%), or V_2O_5 (Aldrich, 99%), 0.2 grams of graphite (Aldrich, 99.995%, 1-2 micron particle size) and 0.4 grams of PVdF-HFP were mixed with 10 mL 4M2P and sonicated in 2 minute intervals for a total of 14 minutes. This thick solution was carefully brushed on one side of a large rectangle of graphitic paper (Toray Industries, TGP-H-090, 0.01mm thick). The paper was dried in an oven at 110 °C for 12 hours under an active vacuum (10^{-3} torr).

Electrochemical Techniques. Temperature dependent ac-impedance measurements of the gels utilized a Solartron Si 1260 Gain Phase Analyzer at frequencies from 1 Hz up to 3 mHz, and an AC amplitude of 5 mV. The sample holder was a standard T-cell composed of polypropylene and fitted with stainless steel rods in three swage-lock fittings. All measurements started at an initial temperature of 125 °C which was lowered in 10 °C intervals. Between measurements, the cell was thermally equilibrated for one hour. Thickness of the sample was determined by the difference between caliper measurements of the end-to-end length of the stainless steel rods with and without the polymer gel between them. The ionic conductivity was calculated from the measured resistance when the imaginary component at high frequency fell to zero. In order to determine the effects of high voltage charge-discharge cycling, a thick gel composed of 1.0M Li/DMBITFSI, was sandwiched between two pieces of Li metal. Repeated charging and discharging of the cell, in which the Li metal was stripped and deposited over 500 cycles allowed for the monitoring of the change in the ionic conduction of the polymer gel as determined by similar ac-impedance measurements.

Electrochemical experiments were performed using either an EG&G PAR 273A or 263A Potentiostat/Galvanostat with the M270 ver. 4.30 software. A specially designed flat cell was used for all of the electrochemical measurements. In this cell, the working and counter electrodes were simply ½" diameter cutout discs of either graphitic paper, polymer coated graphitic paper, or Li metal. These were separated using two ½" cutout discs of the polymer gel composite electrolyte. For measurements involving a pseudo-reference electrode, a thin, flattened piece of Pt wire was inserted between the two discs of the polymer gel. This thin, solid-state cell was placed between the two Pt current collectors and the entire cell was clamped with a spring-loaded. For the two electrode measurements, the Pt-wire was simply removed from the assembled cell.

Results and Discussion

Ionic Conductivity Results. Figures 2 and 3 illustrate the downward shift in ionic conductivity when 1.0 M LiTFSI is added to the ionic liquid, and then after the lithium containing ionic liquid is entrapped in the polymer gel composite. Additionally, Table 1 lists various details of the measured ionic conductivity data for these ionic liquid systems at room temperature. Physically, these polymer gel composites behave as simple, microporous sponges. Although adding LiTFSI results in a drop of approximately 2 mS/cm in the room temperature ionic conductivity, a drop of only 1 mS/cm results from binding them in the polymer composite. It is possible to prepare gels with an even higher percentage of ionic liquid, which results in an even smaller loss in ionic conductivity. However, these gels tear more easily and are also more subject to weeping, or loss of the ionic liquid, as pressure is applied to the cell by the spring loaded clamps. Therefore, the 75% ionic liquid - 25% polymer composite was chosen as the best compromise between ionic conductivity and physical resilience. This drop in ionic conductivity is relatively inconsequential when compared to the ionic conductivity of 0.0005 mS/cm for pure Li-PEO gels,[16] since all of the observed ionic conductivity values of these gels fall within the 2-5 mS/cm range. Of greater importance will be determination of the Li ion transport properties in these gels, and current NMR experiments are underway to determine to what degree the Li^+ ion, the imidazolium ring, and the $TFSI^-$ are responsible for the observed conductivities.

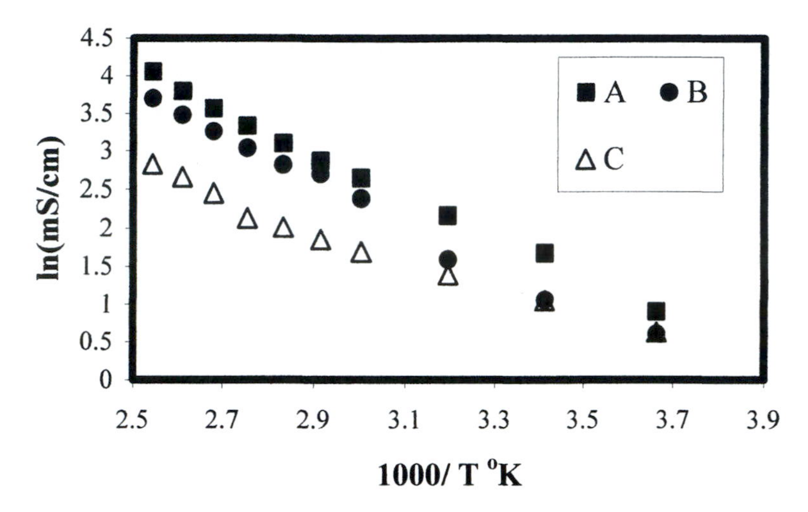

Figure 2. Temperature Dependent Ionic Conductivity of A) Pure DMPITFSI, B) 1.0 M Li/DMPITFSI, C) Polymer Gel of 75% 1.0 M Li/DMPITFSI and 25% PVdF-HFP

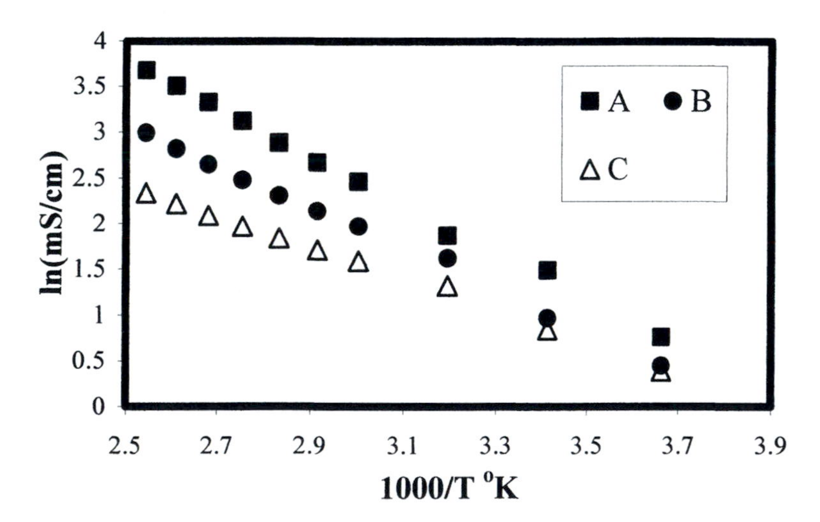

Figure 3 . Temperature Dependent Ionic Conductivity of A) Pure DMBITFSI, B) 1.0 M Li/DMBITFSI, C) Polymer Gel of 75% 1.0 M Li/DMBITFSI and 25%PVdF-HFP

Table 1. Ionic Conductivity Data (± 0.05 mS/cm) at 22 °C of the various DMRITFSI Ionic Liquid Systems

Ionic Conductivity in mS/cm	*Pure Ionic Liquid*		*As a 75% Polymer Gel Composite*	
	Without Li^+	*With Li^+*	*Without Li^+*	*With Li^+*
DMPITFSI	5.31	3.87	3.89	2.88
DMBITFSI	4.44	3.65	3.17	2.86
$BMIPF_6$	5.28	3.47	3.54	2.64

Finally, a detailed analysis of the stability of the polymer gel against metallic Li was studied by sandwiching a gel composed of 1.0M Li/DMBITFSI between two discs of Li metal. The reaction between the polymer gel and Li metal will result in the formation of a solid electrolyte interface at both Li electrodes. This cell was cycled 500 times using a constant current of 250 $\mu A/cm^2$ to strip-off and deposit Li. Caliper measurements gave the gel thickness as 1.65 mm thick and initial ac-impedance measurements of the cell indicated that the ionic conductivity was 2.71 mS/cm (as determined by the point at which the imaginary component of the conductivity fell to zero at high frequency). As the cell was cycled, periodic measurements of the ac-impedance of the cell were performed. As shown in figure 4, after a rapid drop in the ionic conductivity of approximately 20%, a constant value of 2.2 mS/cm was observed. Thus indicating that although the layer between the gel and the Li metal electrode was somewhat less conducting than the gel itself, it did not passivate the Li surface rendering it inaccessible to removal and deposition processes.

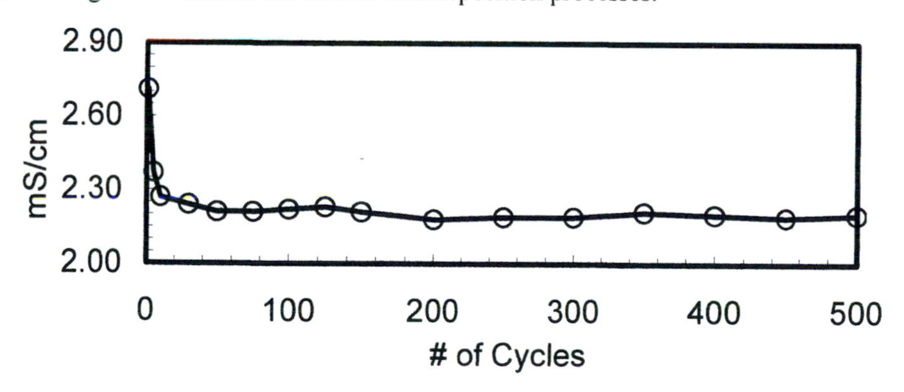

Figure 4. Ionic Conductivity dependency for the 75% 1.0 M Li/DMBITFSI Polymer gel between 2 discs of Li metal. Charging and discharging current was 50 $\mu A/cm^2$

Electrochemical Characterizations of the Polymer Gel Electrolytes. The cyclic voltammogram of the Li/DMPITFSI gel electrolyte, figure 5 curve C, using a disc of graphitic paper as the counter and a disc of Pt as the working electrode indicated an electrochemical window of approximately 4.85 volts (a similar cyclic voltammogram of the Li/DMBITFSI gel was also observed but not shown since they were nearly identical.) Also shown are the cyclic voltammograms of the polymer gel composites between two discs of graphitic carbon. The observed intercalation and deintercalation of the cation and the anion indicates that a type of DIME cell can be formed in which the Li^+, the $DM\underline{R}I^+$ cation, and the $TFSI^-$ anion intercalate into graphitic paper.

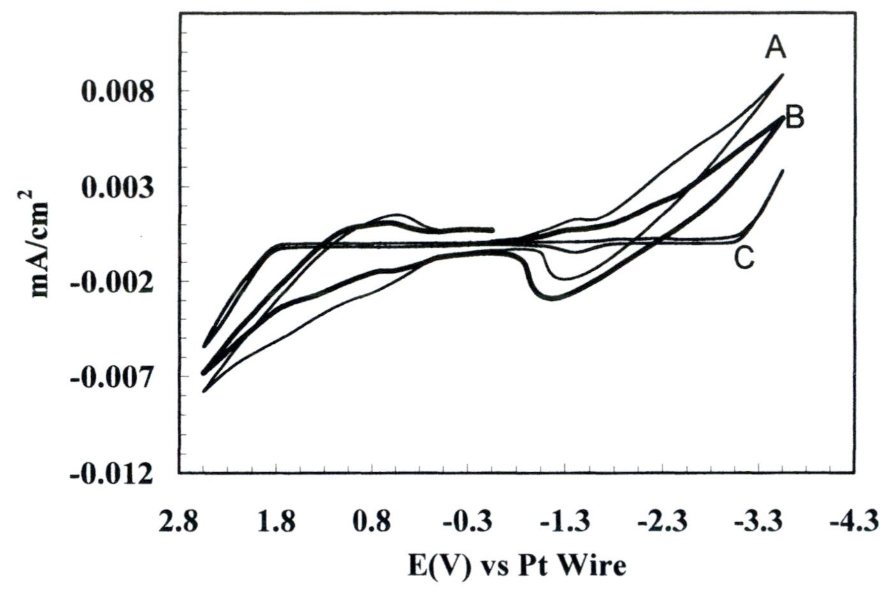

Figure 5. Cyclic Voltammograms of A) the gel containing 1.0 M Li/DMPITFSI, B) the gel containing 1.0 M Li/DMBITFSI, using two discs of Graphitic Paper as the Working and Counter Electrodes and C) a gel containing 1.0 M Li/DMPITFSI using a disc of Pt Foil as the working electrode and a disc of graphitic paper as the counter. The pseudo-reference was Pt wire, and the scan rate was 10 mV/sec.

Thus, by removing the Pt wire, a simple two-electrode battery could be prepared, and the charge-discharge behavior of this simple system is shown in figure 6. The resulting 1.0 volt battery exhibits only 65% efficiency, and does show significant voltage drop-off as the system discharges. Previous work has shown that the cationic intercalation and deintercalation exhibits over 80% efficiencies when measured in a 3-electrode cell, while only 60% efficiency or less is measured for the TFSI$^-$ anion.[21] Thus, the inherent drawback in the DIME system is that the anion charge-discharge efficiencies appear to be the limiting agent.

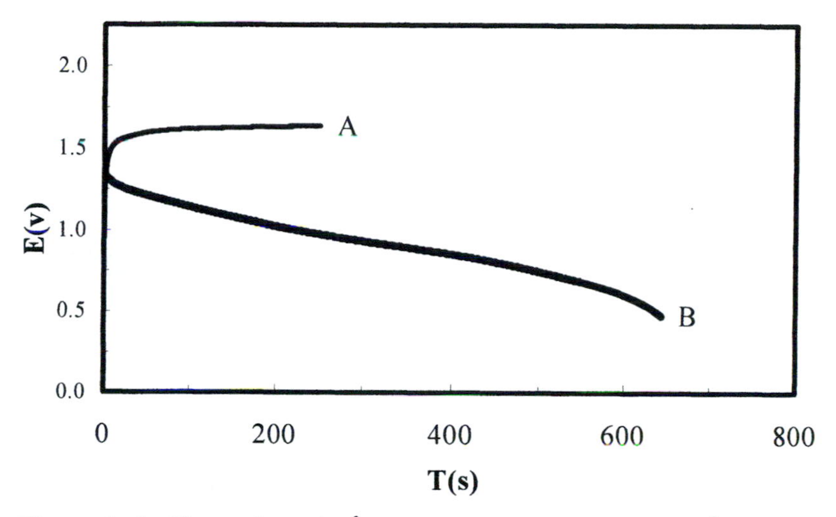

Figure 6. A) Charge (1 mA/cm^2) and B) Discharge (0.25 mA/cm^2) for two discs of graphitic paper and the 75% 1.0M Li/DMPITFSI Polymer Gel Electrolyte.

Li Metal-Metal Oxide Battery Test Cells. For the next series of charge-discharge experiments, polymer composites of $LiMn_2O_4$ and $LiCoO_2$ were used as the cathode, discs of pure Li Metal (Aldrich 99.9%, 0.75mm thick) were the anode material, and single thin discs of the polymer composite gels served as the solid electrolyte. In these experiments, the system was first discharged for a 2 hour period, and then subjected to 50 charge-discharge cycles. As can be seen in Figure 7 and Figure 8, both behaved extremely well. For figure 7, the Li/DMPITFSI gel exhibited very little voltage drop as well as high efficiency for $LiMn_2O_4$. Similarly, for a cell using $LiCoO_2$, Li metal and Li/DMBITFSI, Figure 8, the charging curve was nearly linear, although more of a voltage drop-off over time was observed in the discharge curve. Overall, for both systems, the voltage plateaus and efficiencies did not vary from the 10th cycle up to the 50th cycle, and it is the 50th cycle that is shown. Although a more detailed investigation into the behavior of the metal oxide electrodes is required, it is clear that these ionic liquid gel electrolytes can be used for conventional 4.0V Li-ion batteries. Furthermore, after the charge-discharge

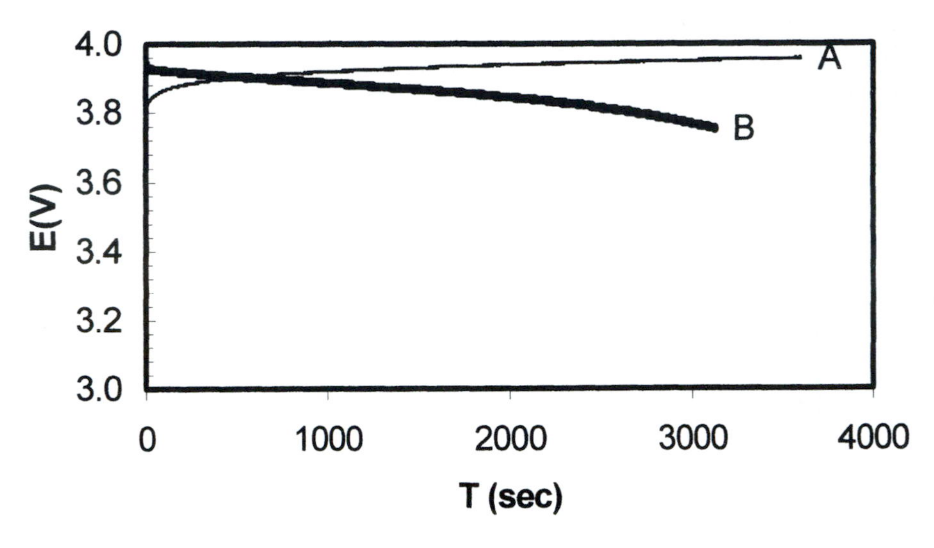

Figure 7. A.) Charging and B)Discharging at 40 μA/cm² For a Li metal, $LiMn_2O_4$ battery. The 50th Cycle is shown

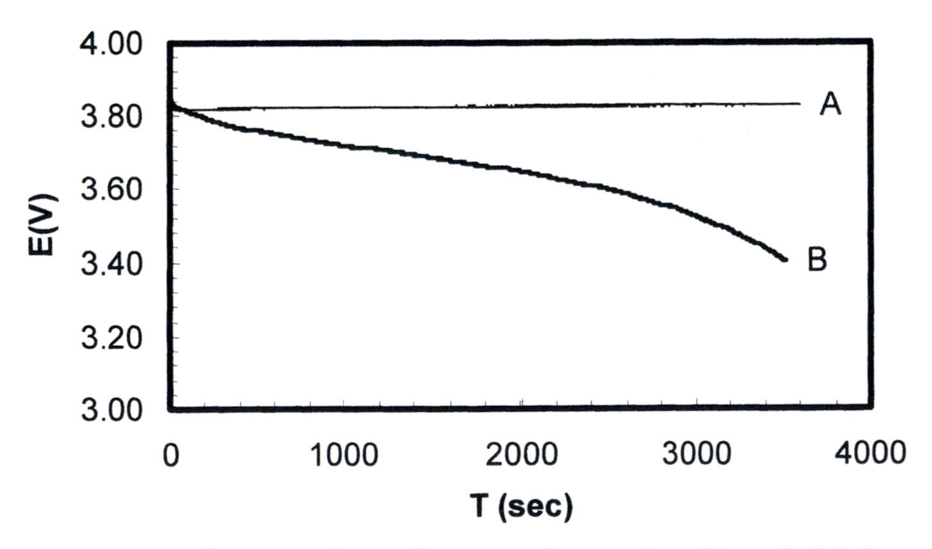

Figure 8. A) Charging and B)Discharging at 40 μA/cm² For a Li metal, $LiCoO_2$ battery. The 50th Cycle is shown.

experiments were complete and the solid-state cells dismantled, the polymer gels were examined and found to still be flexible, transparent discs. Also, the Li metal discs showed what appeared to be a only a light tarnishing. However, as shown in figure 9, a similar experiment using V_2O_5 and Li metal electrodes was performed using a gel composed of 75% 1.0 M Li/$BMIPF_6$ and 25% PVdF-HFP or a gel containing 75% 1.0 M Li/DMBITFSI and 25% PVdF-HFP. The gel with $BMIPF_6$ performed poorly by the 50th cycle, while the cell using the gel electrolyte of Li/DMBITFSI exhibited high efficiency. The degradation of the polymer gel using $BMIPF_6$ can clearly be seen in figure 10, which shows the change in efficiencies over the entire 50 cycles. After these experiments, when the cell was dismantled, the polymer gel composite disc made of 1.0 M Li/$BMIPF_6$ had clearly reacted and turned dark brown and the Li metal surface was covered with a yellow film, indicating that the ionic liquid had likely reacted with the Li metal possibly via the hydrogen atom at the 2-position of the imidazole ring.

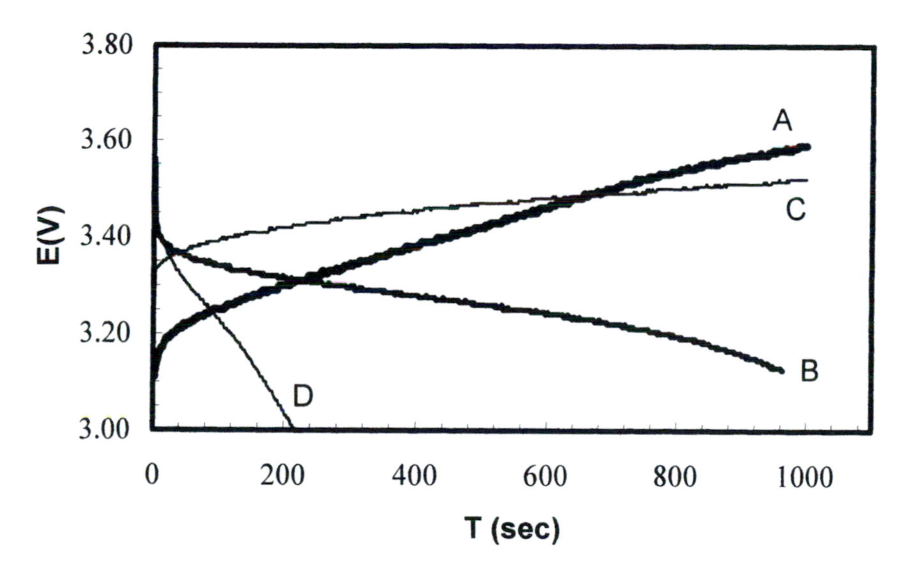

Figure 9. A) Charge and B) Discharge behavior of V_2O_5 vs. Li metal using a gel composed of 1.0 M Li/$DMBIPF_6$, and the C) Charge and D) Discharge behavior of V_2O_5 vs. Li Metal using a gel composed of 1.0 M Li/$BMIPF_6$. Charging and discharging currents were 50 $\mu A/cm^2$. 50th Cycle is shown.

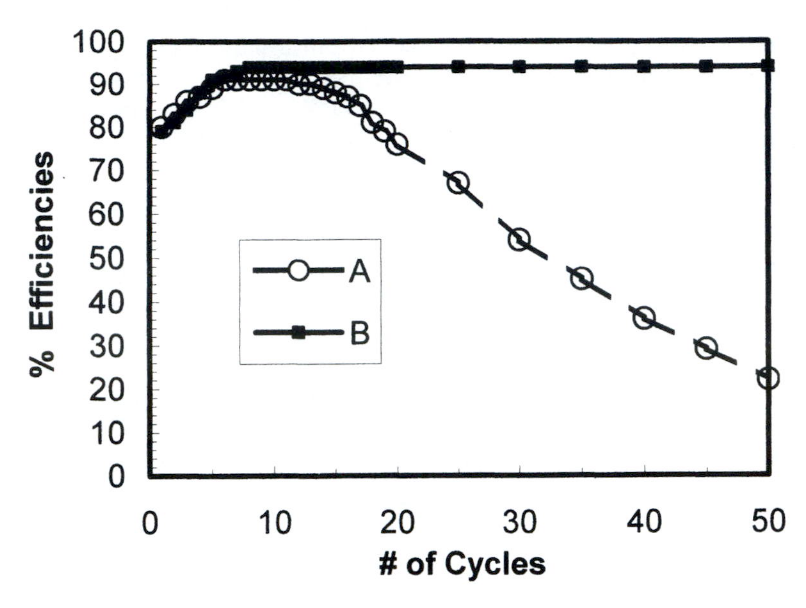

Figure 10. Percent discharge efficiencies vs. # of cycles for cells of Li metal and V_2O_5 using a A.) Polymer gel composite electrolyte of 75% 1.0 M Li/BMIPF$_6$ and 25% PVdF-HFP or B.)Polymer gel composite electrolyte of 75% 1.0 M Li/DMBITFSI and 25% PVdF-HFP.

Conclusions

The above experiments provide strong evidence that it is entirely possible and feasible to use the DM***R***ITFSI ionic liquids as electrolytes in Li ion batteries. Although DIME cells did not perform as well as has been reported for ionic liquids using the $AlCl_4^-$ anion, the Li/DM***R***ITFSI ionic liquids performed as well as any of the more common electrolytes using Li metal and three of the most common metal oxide based cathode materials. Since cycling over 500 times indicated that extensive degradation of the polymer gels did not result in the build up of a non-conductive film on the Li metal, it seems likely a rechargeable Li-metal battery may be possible using these nonflammable polymer gel electrolytes

Further work is now needed to answer several important questions. First and foremost, the primary current carrier needs to be determined, and NMR experiments are currently underway to determine the transport properties of these and other ionic liquid composite gel electrolytes. Secondly, the film that did form over the Li metal during the 500 cycle experiment needs to be well characterized, since this could pinpoint specific weakness of the ionic liquid that could be addressed by further chemical manipulation of either the cation or the anion. Additionally, the actual capacity of the metal oxide cathodes need to be determined to see if any type of secondary limiting interactions occur between the ionic liquid and the metal oxides. However, it seems clear that the use of ionic liquids, with their superior environmental and physical properties are viable alternatives to the other, volatile and flammable organic electrolytes.

Acknowledgments

This work was sponsored by the Air Force Office of Scientific Research, and space was provided by the Naval Research Laboratory. Opinions, interpretations, conclusions and recommendations are those of the authors and are not necessarily endorsed by the United States Air Force.

References

1. F. A. Cotton and G. Wilkinson, *Inorganic Chemistry* 4th ed. p.385, 1980.
2. R. T. Carlin, J. Fuller, W. K. Kuhn, M. J. Lysaght, P. C. Trulove, *J. Applied Electrochem.* **26** (11) 1147, (1996).
3. R. T. Carlin, H. C. DeLong, J. Fuller, and P. C. Trulove, *J. Electrochem . Soc.* **141**, L73, (1994).
4. T. E. Sutto, P. C. Trulove, H. C. DeLong, in the Proceeding of the 12th International Conference on Molten Salts, P. C. Trulove, H. C. De Long, G. R. Stafford, S. Deki, Editors PV 99-41, p. 32, The Electrochemical Society, Inc., Pennington, NJ (2000).
5. T. E. Sutto, P. C. Trulove, H. C. DeLong, in the Proceeding of the 12th International Conference on Molten Salts, P.C. Trulove, H. C. De Long, G. R. Stafford, S. Deki, Editors PV 99-41, p. 43, The Electrochemical Society, Inc., Pennington, NJ (2000).
6. T. E. Sutto, K. D. Sienerth, H. C. DeLong and P. C. Trulove, in the Proceedings of the 12th International Conference on Molten Salts, P.C. Trulove, H. C. De Long, G. R. Stafford, S. Deki, Editors PV 99-41, p. 54, The Electrochemical Society, Inc., Pennington, NJ (2000).
7. T. E. Sutto, P. C. Trulove, H. C. DeLong, in Proceeding of the EUCHEM Conference on Molten Salts, R. W. Berg and H. A. Hjuler, Editors; Elsevier, Paris, p. 511 (2000).

8. T. E. Sutto, H. C. DeLong, and P. C. Trulove, *Zeitschrift für Naturforschung A*, in press.
9. T. E. Sutto, H. C. DeLong, and P. C. Trulove, *Zeitschrift für Naturforschung A*, accepted.
10. T. E. Sutto, H. C. DeLong and P. C. Trulove, in the Proceedings of the 13th International Conference on Molten Salts, The Electrochemical Society, Inc., Pennington, NJ (2002), in press.
11. R. T. Carlin, H. C. De Long, J. Fuller, and P. C. Trulove, *J. Electrochem . Soc.* **141**, L73, (1994).
12. J. Fuller, A. C. Breda, and R. T. Carlin, *J. Electrochem . Soc.* **144**, p. L68, (1997).
13. C. L. Hussey in *Chemistry of Nonaqueous Solvents.* A. Popov and G. Mamantov, Editors. Chap. 4, VCH Publishers, New York (1994).
14. M. Watanbe, M. Kanda, H. Matsuda, K. Tsumemi, E. Tsuchida and I. Shinohara, *Macromol. Chem. Rapid Commun.,* **2**, p. 741 (1981).
15. M. Alamgir and K. M. Abraham, "Room Temperature Polymer Electrolytes", Lithium Batteries, Ed. G. Pistoia, Industrial Chemistry Library, Vol. 5, (1994).
16. O. Bohnke, C. Rousselot, P. A. Gillet and C. Truche, *J. Electrochem. Soc.,* **139**, p. 1862 (1992).
17. F. Lemaitre-Auger and J. Prud'homme, *Electorchem. Acta,* **46**, Issue 9, pg. 1359 (2000).
18. K. Hayamizu, Y. Aihara, S. Arai and C. Garcia-Martinez, *J. Phys. Chem.*, **103,** p. 519 (1999).
19. S. Narang, S. Ventura, B. Dougherty, M. Zhao, S. Smedley and G. Koolpe; SRI International, U. S. Patent 5,830,600, November 3, (1998).
20. S. C. Ventura, S. C. Narang, G. Hum, P. Liu, P. Ranganathan, and L. Sun; SRI International, U. S. Patent 5,731,104, March 24, (1998).
21. T. E. Sutto, H. DeLong and P. C. Trulove, in the 13th International Conference on Molten Salts, The Electrochemical Society, Inc., Pennington, NJ (2002), in press.
22. J. Fuller, A. C. Breda, and R. T. Carlin, *J. Electrochem . Soc.* **144**, L68, (1997).
23. J. Fuller, A. C. Breda, and R. T. Carlin, *J. Electroanalytical Chem.* **459**, 29 (1998).

Chapter 39

Electrochemistry: Ionic Liquid Electroprocessing of Reactive Metals

Jianming Lu and David Dreisinger

Department of Metals and Materials Engineering, University of British Columbia, Vancouver, British Columbia V6T 1Z4, Canada

The ionic liquids were studied as an alternative medium to recover and refine Al, Mg and Ti. The conductivities of 1-butyl-3-methylimidozalium chloride (BMIC), and its mixtures with aluminum, magnesium and titanium chlorides were measured. The electrochemical behaviour of aluminum in BMIC-$AlCl_3$ system was studied. The reduction of $Al_2Cl_7^-$ was quasi-reversible. The deposition of aluminum involved instantaneous three-dimensional nucleation. In acidic liquid, there are two stages for the dissolution of Al. Electrorefining and electrowinning of Al were conducted to evaluate the feasibility of using the room temperature liquids to recover Al.

Introduction

The electrodeposition of reactive metals such as Al, Mg and Ti from aqueous solutions is impossible due to hydrogen evolution at the cathode. Therefore the electrolysis for the reactive metal deposition must be aprotic.

Aluminum is primarily produced by electrowinning (electrolytic recovery) from a molten alumina-cryolite bath according to the following cell reaction [1]:

$$Al_2O_3 + 3/2C \Leftrightarrow 2Al + 3/2CO_2 \quad (1)$$

The energy consumption for this process is around 13-16 kWh/ kg Al [1].

Impure aluminum recovered by recycling is often refined using the Beck process to produce a purified commercial product. This refining process has the disadvantages of high-energy consumption, complicated operation and environmental pollution.

Magnesium is produced by electrowinning from a molten mixture with alkali chloride at 700 - 800 °C. The energy consumption is 10 - 13 kWh/ kg Mg [1]. Titanium is produced by reduction of $TiCl_4$ with magnesium or sodium. Electrowinning from chloride molten salts at very high temperature is also used.

Ionic liquids have the advantages of wide electrochemical windows and low operating temperature. They are interesting solvents for electrodeposition studies of elements such as Al, Ti and Mg that cannot be obtained from aqueous solutions.

Most of the studies on metal deposition have been conducted in $AlCl_3$-based ionic liquid [1-6]. The properties of these ionic liquids are determined by the mole ratio of these components (q/p) [4]:

$$pR^+Cl^- + qAlCl_3 \Leftrightarrow qR^+AlCl_4^- + (p-q)R^+Cl^- \quad (q/p < 1) \quad (2)$$

$$pR^+Cl^- + qAlCl_3 \Leftrightarrow (2p-q)R^+AlCl_4^- + (q-p)R^+Al_2Cl_7^- \; (1 < q/p < 2) \quad (3)$$

The $q/p < 1$ ionic liquids in which Cl^- ion is in excess are termed basic. In the ionic liquids with $q/p > 1$ (acidic ionic liquids), $Al_2Cl_7^-$ ions, which are a chloride ion acceptor, are the principal constituents of the system. Neutral ionic liquids are those with $q/p = 1$.

At $q/p > 2$, in the case of chloride, $Al_3Cl_{10}^-$, $Al_4Cl_{13}^-$ and Al_2Cl_6 are the dominant dissolved species successively with increasing q/p value [7].

At $q/p < 1$, Cl^- is discharged at the anode. At $q/p \geq 1$, $AlCl_4^-$ is discharged on the anode according to the following reaction [5]:

$$4AlCl_4^- = 2Al_2Cl_7^- + Cl_2 + 2e \quad (4)$$

In basic ionic liquids, Al(III) reduction is not observed since the organic cations are reduced at less negative potentials than $AlCl_4^-$. In acidic ionic liquids, Al(III) species are easily reduced to give an Al deposit according to the following reaction.

$$4Al_2Cl_7^- + 3e = Al + 7\,AlCl_4^- \quad (5)$$

Most studies on the electrochemical behaviour of aluminum in the organic molten salts (ionic liquids) have been limited to the use of platinum, glassy carbon and tungsten electrodes [1, 2]. The electrorefining (electrolytic refining) of aluminum in ionic liquid has been conducted at The University of Alabama [8]. There is no report on electrowinning of aluminum from ionic liquids.

Magnesium was not successfully electrodeposited in $MgCl_2$ buffered neutral chloroaluminates possibly due to the passivation [9], but it can be codeposited with aluminum in acidic chloroaluminates [10].

There is little information available for titanium recovery via ionic liquid. Only Takahashi reported that titanium could be plated from $TiCl_3/TiCl_4$ - alkylpyridinium chloride ionic liquid [11].

The objective of this work is to use ionic liquids to develop environmentally clean processes with less energy consumption for production or refining of aluminum, titanium and magnesium.

Experimental

Aluminum chloride (Aldrich 99.99%), magnesium chloride (Aldrich 99.99%) and titanium chloride (Aldrich 99.9%) were used as received. 1-butyl-3-methylimidazolium chloride (BMIC) was received from The University of Alabama. The ionic liquids were prepared by mixing metal salts with BMIC and purified by electrolysis using pure aluminum electrodes for 2 days. The experiments were conducted under an atmosphere of purified nitrogen gas (<1ppm water). The cell containing ionic liquids was placed in an oil-heating bath for controlling the temperature of ionic liquid. Jenway Model 4320 conductivity meter was used to measure the conductivity.

The reference electrodes were made of pure (99.999%) aluminum wire that was immersed in a 1.5:1.0 $AlCl_3$ - BMIC ionic liquid and separated from the working ionic liquid by a fine porous glass frit. The counter electrode was a coiled pure aluminum wire. The working electrodes were either a pure (99.999%) aluminum wire with 2 mm diameter or a pure (99.99%) copper wire with 1 mm diameter. These wires were tightly enclosed in Teflon, leaving an exposed cross section, forming the electrode surface. The electrodes were polished using 1 μm diamond paste. The electrochemical measurements were performed using SOLARTRON 1286 and PAR 273A electrochemical interfaces. IR was compensated using current interruption or positive feedback.

A copper sheet was used to make cathodes for electrorefining and electrowinning. 50 mL of ionic liquid was used for each test. The working surface areas of the anode and the cathode were 1.5 cm^2. The distance between the anode and the cathode was fixed at 1 cm.

Results and Discussions

Conductivity Measurement

BMIC became yellow and further became brown quickly above 160 °C. This change in the color was viewed as the decomposition of BMIC. The conductivities of ionic liquids were measured in the temperature range of 30 – 150 °C. The plots of conductivity vs. temperature are shown in Figure 1. Pure BMIC (supper-cooled liquid below 65 °C) and the $AlCl_3$-BMIC ionic liquids display curved Arrhenius plots indicating that the structures of the ionic liquids may change with temperature. Since the ionic liquids form glass phases, the conductivity dependence on the absolute temperature can be interpreted by the Vogel-Tamman-Fulcher equation: $\kappa = A\ T^{-1/2} \exp\ [-k/(T-T_0)]$ [6], where κ is conductivity, A a scaling factor, T the absolute temperature, T_0 the ideal glass transition temperature and k a constant characteristic of the materials. The excellent fit of the conductivity data to the above equation is shown in Figure 2. The fitted value for T_0 was optimized to maximize the correlation for log $(\kappa T^{1/2})$ against $(T-T_0)^{-1/2}$. The conductivity activation energy (E) can be obtained by the equation: $E= RT^2\ (\partial\kappa/\partial T) = RkT^2/(T-T_0)^2 - T/2$.

The ideal glass transition temperatures are 191, 153, 142, 103, 99, 42 and 79 K for 0:1.0, 0.5:1.0, 0.8:1.0, 1.0:1.0, 1.2:1.0, 1.5:1.0 and 2.0:1.0 $AlCl_3$/BMIC ionic liquids. With increasing temperature from 30 to 150 °C, the activation energy for BMIC decreases from 68.8 to 34 kJ mol^{-1} while the activation energy for 1.5:1.0 $AlCl_3$/BMIC ionic liquid decreases from 18.2 to 16.6 kJ mol^{-1}.

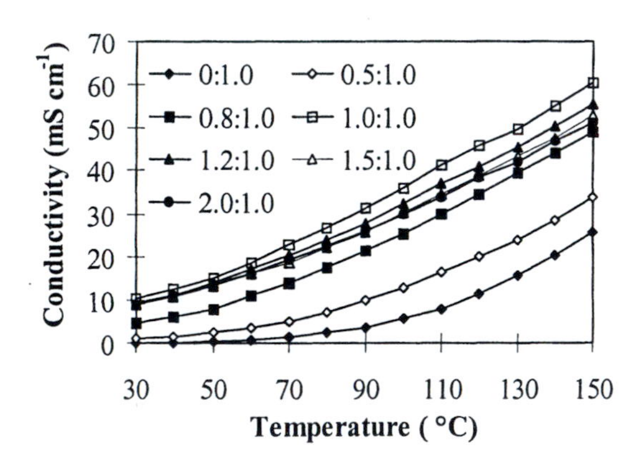

Figure 1. Conductivity vs. temperature for $AlCl_3$ -BMIC ionic liquids.

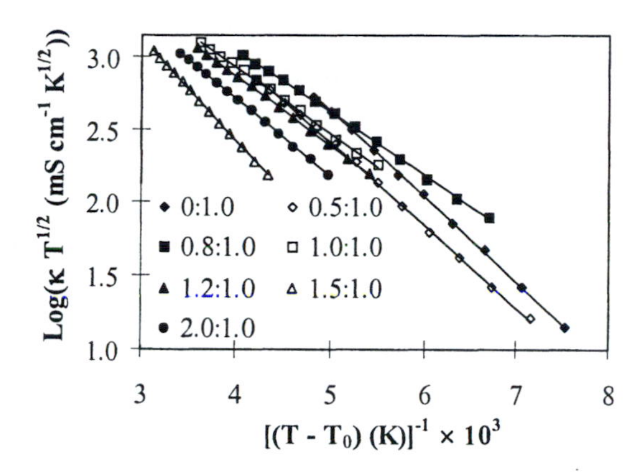

Figure 2. Log (Conductivity $T^{1/2}$) vs. $(T\text{-}T_0)^{-1}$ for $AlCl_3$ - BMIC ionic liquids.

From Figure 3, with increasing mole faction of $AlCl_3$, the conductivity first increases to a peak value and then decreases. At $AlCl_3$/BMIC < 1.0:1.0, with increasing $AlCl_3$, [$AlCl_4^-$] increases, but [Cl^-] decreases. Hence the effect of hydrogen bonds between the hydrogen atoms of the imidazolium cations and the chloride anions [6] decreases resulting in a lower viscosity and a higher conductivity. With further increasing $AlCl_3$, [$AlCl_4^-$] decreases and [$Al_2Cl_7^-$] increases. $Al_2Cl_7^-$ is possibly less conductive due to its large size resulting in decreasing conductivity.

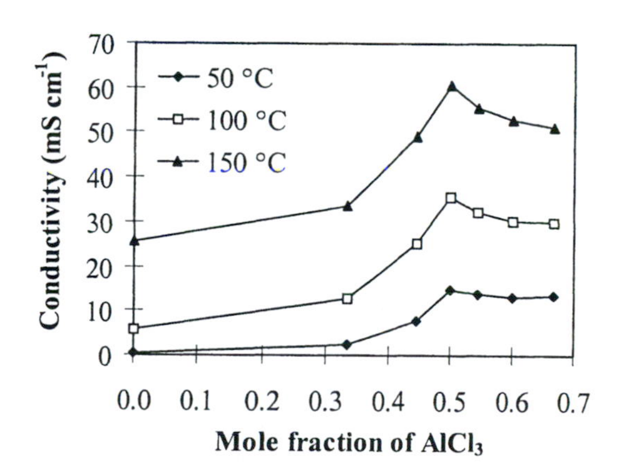

Figure 3. Conductivity vs. aluminum chloride mole fraction.

The dissolution of magnesium chloride in BMIC was very slow. Therefore it was dissolved at 140 °C. When the molar ratio of $MgCl_2$ to BMIC was above 1:1.75, even at 140 °C $MgCl_2$ did not dissolve completely. The plots of conductivity vs. temperature are shown in Figure 4.

Mixing $TiCl_4$ and BMIC generates yellowish precipitate and brownish sticky liquid and it is also rather exothermic. They are requires further analysis and characterization. The conductivity is 1.5 to 22 mS cm^{-1} from 80 - 150 °C.

The conductivity of 2 M H_2SO_4 is about 660 mS cm^{-1} at 25 °C. The ionic liquid conductivity is relatively too small. The low conductivity causes high cell voltage and high-energy consumption during electrowinning and electrorefining

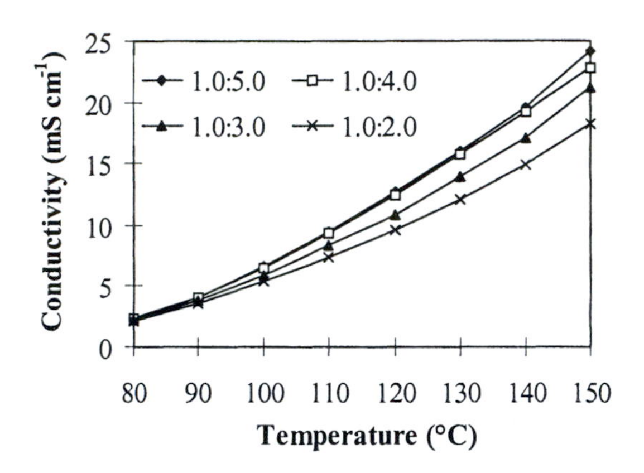

Figure 4. Conductivity vs. temperature for $MgCl_2$ - BMIC ionic liquids.

Electrochemical Behaviour of $AlCl_3$ - BMIC

Cathodic Behaviour of Aluminum

The cyclic voltammograms on a copper electrode in 1.5:1.0 $AlCl_3$/BMIC ionic liquid at 60 °C are shown in Figure 5. The potential was swept in the negative direction to reduce Al(III), then reversed and swept in the positive direction to dissolve deposited aluminum and finally swept back to the starting point.

The analysis of these curves by using the methods given in Ref. 12 indicates that the aluminum reduction is quasi-reversible. Similar results were obtained at 30, 80 and 100 °C. The electrochemical behavior for 1.2:1.0 and 2.0:1.0 $AlCl_3$/BMIC ionic liquids are similar.

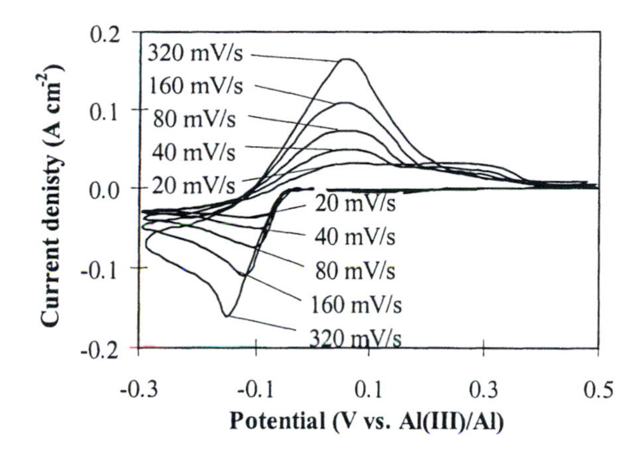

Figure 5. Cyclic voltammograms on a copper electrode in 1.5:1.0 $AlCl_3$/BMIC ionic liquid at 60 °C.

The plots of the peak currents vs. the square root of potential scanning rate for deposition of aluminum give linear lines. For a reversible process, there is the following relation between the peak current (i_p) and the potential scanning rate (v): $i_p = 0.4463nFAC(nF/RT)^{1/2} v^{1/2}D^{1/2}$[12], where n is the number of the electrons transferred, A the surface area of the electrode and D the diffusion coefficient. It is assumed that the peak corresponds to the reversible reduction of $Al_2Cl_7^-$ and so the diffusion coefficients are estimated from the slope of the cathodic peak current vs. potential scanning rate according to Reaction 5. Estimated D values are: 3.4×10^{-7}, 7.7×10^{-7}, 1.4×10^{-6} and 2.0×10^{-6} cm^2 s^{-1} for 30, 60, 80 and 100 °C respectively with a calculated activation energy of 24 $kJmol^{-1}$.

Nucleation of aluminum on the copper

Chronoamperometry was used to examine the deposition of aluminum on copper. A collection of current transients at constant overpotentials (η) are shown in Figure 6. The transients were characterized by an initial regime of current decay due to double-layer charging. This was followed by a rise in the current due to an increase in the electroactive area as independent nuclei grew and/or the number of nuclei increased. During this stage the nuclei developed hemispherical diffusion zones around themselves and as these zones overlapped, hemispherical mass-transfer gave way to linear mass-transfer to what was now effectively a planar diffusion layer [13]. The current passed through a maximum (I_m, t_m) and then decreased with a linear $I - t^{-1/2}$ relationship [13].

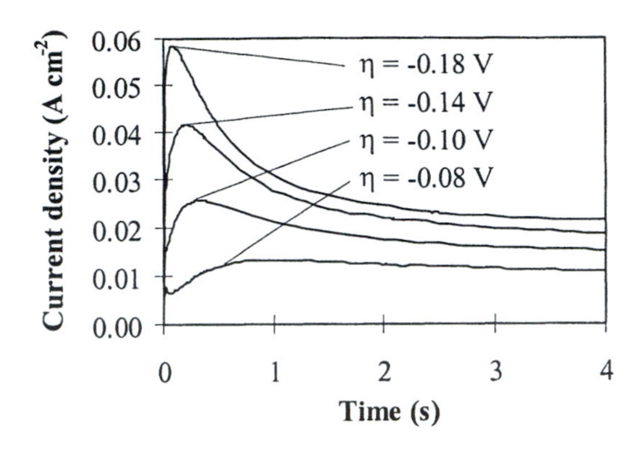

Figure 6. A series of potentiostatic current transients for aluminum deposition on a copper electrode in 1.5:1.0 $AlCl_3$/BMIC ionic liquid at 30 °C.

A definitive test for adherence to the models for progressive nucleation or instantaneous nucleation can be made by comparing data from the entire experimental current-time transient to the appropriate dimensionless theoretical equation. The following two equations relate the dimensionless current to the dimensionless time for diffusion controlled three-dimensional progressive and instantaneous nucleation and growth [13]:

$$(I/I_m)^2 = 1.9542(t/t_m)^{-1}\{1-\exp[-1.2564(t/t_m)]\}^2 \quad \text{(instantaneous)} \qquad (6)$$

$$(I/I_m)^2 = 1.2254(t/t_m)^{-1}\{1-\exp[-2.3367(t/t_m)^2]\}^2 \quad \text{(progressive)} \qquad (7)$$

After deduction of nucleation induction time, the dimensionless plot of the experimental data (I/Im against t/tm) revealed that electrocrystallization occurred by instantaneous nucleation. The diffusion coefficient D of $Al_2Cl_7^-$ can be estimated by the following equation: $(I_m)^2\ t_m = 0.1629(nFC)^2D$ [13]. The estimated diffusion coefficients for $Al_2Cl_7^-$ are $(2.6\pm0.3)\times10^{-7}$, $(7.0\pm0.5)\times10^{-7}$ and $(1.2\pm0.2)\times10^{-6}$ $cm^2\ s^{-1}$ for 30, 60 and 80 °C respectively. These values are in good agreement with those estimated using cyclic voltammetry.

Anodic behaviour of aluminum

The anodic behaviour for aluminum in 1.5:1.0 $AlCl_3$/BMIC at 60 °C is shown in Figure 7. With increasing potential, the current first increased to a peak value, then decreased a little and increased to a second peak value, and finally decreased sharply. The first peak decreased with decreasing potential scanning rate and it disappeared at the potential rate < 20 mV/s. The transient mass transfer of a species shifted to a steady state at slower scanning rate. The second peak is related to the formation of another species resulting in the passivation of the electrode. The dissolution current at a constant potential decreased with increasing mole ratio of $AlCl_3$ to BMIC. At $AlCl_3$/BMIC molar ratios ≥ 2:1, the first peak was not as observed. In those ionic liquids, Cl^- is below 10^{-16} M and its contribution to the dissolution rate is negligible. Based on the ionic liquid density and the aluminum species distribution [7], the concentrations of aluminum species are estimated. The ratio of the $AlCl_4^-$ concentrations in 1.5:1.0 and 2.0:1.0 $AlCl_3$/BMIC ionic liquids are close to the ratio of the limiting currents in the above ionic liquids. This limiting current is probably related to the diffusion of $AlCl_4^-$. So the first stage dissolution reaction is:

$$7AlCl_4^- + Al = 4\ Al_2Cl_7^- + 3\ e \qquad (8)$$

With further increasing potential, Al is dissolved to form $Al_3Cl_{10}^-$, $Al_4Cl_{13}^-$ and Al_2Cl_6. Finally $AlCl_3$ is likely to be precipitated on the anode, resulting in passivation of the anode and hence a decrease in the current.

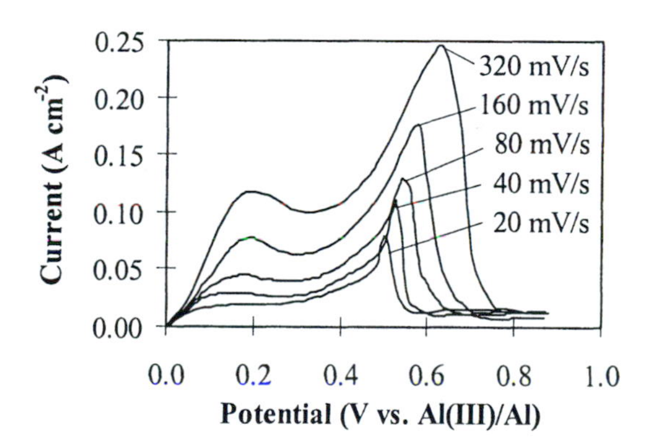

Figure 7. Anodic polarization curves of Al in 1.5:1.0 $AlCl_3$/BMIC ionic liquid at 60 °C.

Electrorefining and Electrowinning of Aluminum

Morphology of Aluminum Deposits

From Figure 8, the cathodic deposit was not smooth and the dendrite grew up to 8 mm at 10 hours. The plane part of the aluminum deposit was < 0.1 mm.

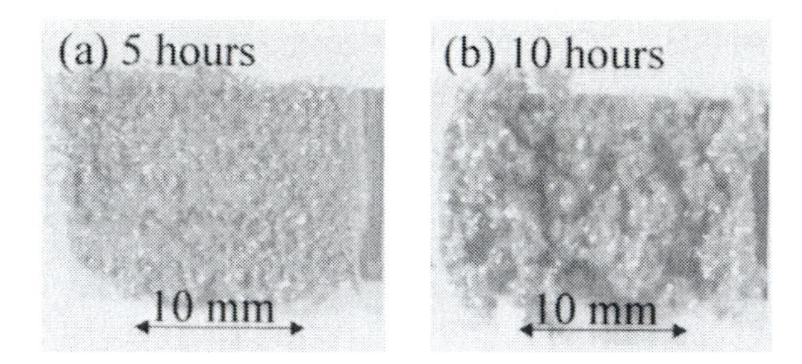

Figure 8. Cathodes after deposition in 1.5:1.0 $AlCl_3$/BMIC liquid at 300 A m^{-2} and 100 °C. Anode materials: pure aluminum.

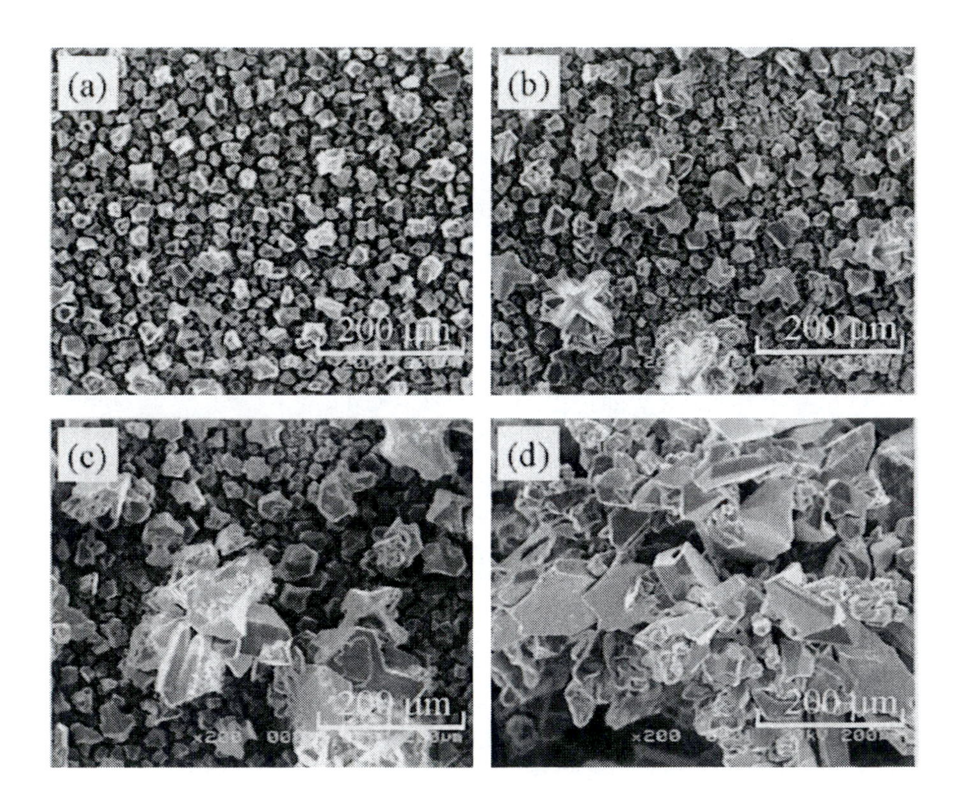

Figure 9. SEM images of aluminum deposits in 1.5:1.0 $AlCl_3$/BMIC liquid at 300 A m^{-2} and 100 °C. Deposition time: (a) 0.5, (d) 1, (c) 2 and (d) 5 hours.

The SEM images of the deposits are shown in Figure 9. The uneven growth of the aluminum crystal was observed at 0.5 hour. At 1 hour, these bigger crystals grew up to 50 μm while the other crystal almost did not grow. At 2 hours, they increased to up to 100 μm. At 5 hours, new crystals were formed on the bigger crystals.

Electrorefining of Impure Aluminum

The electrorefining of impure aluminum was conducted using a constant current. The cathodic reaction is the aluminum deposition process. The anodic reaction is the aluminum dissolution process.

Pure aluminum (99.999%), AA6111 and AA383.1 alloys were used as anodes. The cell voltage against the time of deposition in 1.5:1.0 $AlCl_3$/BMIC liquid is shown in Figure 10. In the case of pure aluminum as an anode, the cell voltage first reached a peak value and then rapidly decreased. This could be due to the nucleation of aluminum on the copper cathode and/or the breakdown of the oxide film on the anode. After several minutes, the cell voltage decreased slowly. Since the cathode dendrite grew continuously, the distance between the anode and cathode decreased. Also the surface area of the deposit became larger indicating that less overpotential for aluminum deposition was needed at a constant current. These factors resulted in a decrease in the cell voltage.

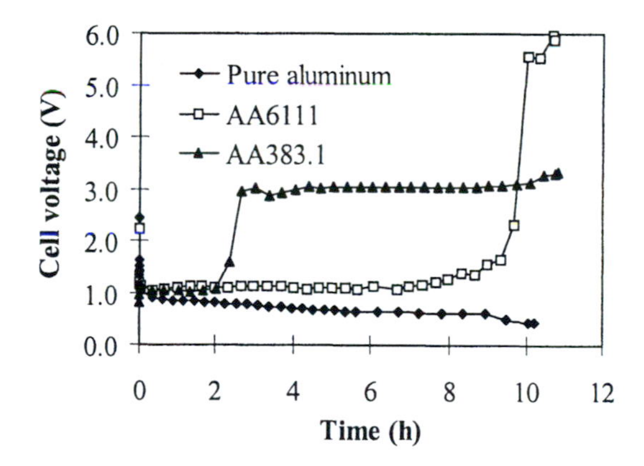

Figure 10. Cell voltage against the electrolysis time in 1.5:1.0 $AlCl_3$/BMIC ionic liquid at 300 A m^{-2} and 100 °C.

In the case of AA6111 alloy as an anode, the initial cell voltage against time was similar to that for pure aluminum as an anode. After 20 minutes, the cell voltage was stabilized. After 8 hours, the cell voltage began to increase, and finally rose up to over 5 V. During the electrolysis, the anode was gradually coated with black precipitate. Therefore the dissolution of the anode was suppressed and the anode potential had to move to a more positive value to maintain a constant current density. This increase in the potential of the anode offset the decrease in the potential due to the cathode deposit resulting in a stable potential. When the anode was completely covered with the precipitate and the cell voltage increased significantly.

When AA383.1 alloy was used as an anode, the cell voltage increased to 3 V after two hours. The impurity contents in AA383.1 are much higher and then more precipitate was formed on the anode and this thicker anode slime caused the anode to passivate in a shorter time. In each case, $AlCl_3$ was precipitated with the anode slime layer, resulting in surface blockage.

At $AlCl_3$:BMIC mole ratio = 1.3:1.0, the passivation was not observed for AA6111 aluminum alloy since a low aluminum chloride concentration favored the dissolution of the aluminum and delayed the passivation of the anode.

After the electrolysis, the anodes were coated with a dark slime. X-ray diffraction patterns show that impure aluminum was purified to pure aluminum.

The main impurity contents of AA6611 and AA383.1 and the cathodic deposits are listed in Table 1. Generally impure aluminum was significantly purified. The purified aluminum may be contaminated by chloride or aluminum oxide due to the reaction of aluminum chloride with water. The current efficiency of aluminum deposition was close to 100%, the cell voltage was 0.6-1 V, the energy consumption was 2 -3 kWh / kg Al.

Table 1 Impurity Contents of Anodes and Cathodic Deposits (wt%)

	Si	*Cu*	*Mg*	*Fe*	*Mn*	*Ni*	*Zn*
AA6111	0.6	0.75	0.77	0.24	0.15	-	-
After refining	0.007	0.008	< 0.01	<0.001	0.0013	-	-
AA383.1	11.24	2.67	0.12	0.97	0.35	0.22	2.65
After refining	0.016	0.02	< 0.01	0.03	< 0.01	< 0.0002	0.07

Electrowinning of Aluminum

During electrowinning, aluminum was deposited at the cathode as expressed by Reaction 5 and Cl_2 is evolved on the anode as expressed by Reaction 4.

Anodes fabricated from the graphite, glassy carbon-coated graphite and pyrolytic graphite disintegrated after 5–10 hours of electrolysis. The dimensionally stable anode was not very stable and some coating was observed to drop off. Only solid glassy carbon and carbon were stable. After 1 hour of electrolysis, the liquid gradually became dark indicating that the chlorine might react with the ionic liquid. The current efficiencies, cell voltage and energy consumption are listed in Table 2.

The current efficiency decreased with increasing time of electrolysis. The concentration of chlorine dissolved in the liquid and the real surface of deposit increased with increasing time resulting in more rapid dissolution of aluminum. The dendrite formation was not significant probably due to the dissolved chlorine, selectively dissolving the aluminum dendrites.

Table 2 Current Efficiency of Aluminum Deposition, Cell Voltage and Energy Consumption in 1.6:1.0 $AlCl_3$/BMIC Ionic Liquid at 300 A m^{-2}

Temp. (°C)	*Electrolysis Time (hour)*	*Current Efficiency (%)*	*Cell Voltage (V)*	*Energy Consumption (kWh/ kg Al)*
100	3	83 ± 2	3.31	12.0
100	5	77 ± 3	3.28	12.8
100	10	60 ± 3	3.21	16.1

Electrolysis of $MgCl_2$-BMIC

The electrolysis in $MgCl_2$-BMIC mixture was conducted using a copper cathode and a magnesium anode. After 5 hours of electrolysis, nothing was deposited on the cathode. The liquid gradually became pink due to the reduction of organic cations. The cell voltage fluctuated from 3 to 14 V possibly due to the formation of the impermeable passive film on the magnesium [1]. The formation and breakdown of the passive film resulted in the fluctuation of the cell voltage.

Conclusions

The ionic liquid conductivities are only 1–61 mS cm^{-1} at 50–150 °C. Such a low conductivity causes a high cell voltage and so high-energy consumption.

The reduction of $Al_2Cl_7^-$ ions was quasi-reversible. The deposition of aluminum on polycrystal copper occurs by instantaneous nucleation. In acidic $AlCl_3$ - BMIC ionic liquid, there are two stages for the dissolution of aluminum.

Impure aluminum can be electrorefined using ionic liquid with 100% of deposition current efficiency and 2-3 kWh/ kg Al of energy consumption. The formation of dendrite on the cathodic deposit is a significant problem for the practical application. Some additives might be added to the refining solution to suppress the dendrite growth. Aluminum can be produced by electrowinning from ionic liquid. However, the dissolved chlorine significantly decreases the current efficiency by reacting with the aluminum deposit. Therefore the energy consumption increases from 12 to 16.1 kWh/ kg Al with increasing time of electrolysis from 2 to 10 hours. For a longer time of electrolysis, the energy consumption may even higher than that (12-16 kWh/ kg Al) for conventional production of aluminum. To prevent the chlorine back reaction, it is best to separate the anode from the cathode possibly through use of a porous diaphragm or an anion membrane cell design.

Acknowledgements

The authors wish to thank P G Research Foundation for supporting this work. 1-butyl-3-methyimidazolium chloride was kindly supplied by the University of Alabama.

References

1. *Nonaqueous Electrochemistry;* Aurbach, D., Eds.; Marcel Dekker Inc., New York, 1999.
2. Zhao, Y.; VanderNoot, T. *J. Electrochim. Acta* **1997**, *42*, 3-13.
3. Robinson, J.; Osteryoung, R. A. *J. Electrochem. Soc.* **1980**,*127,* 122-128.
4. McManis, G. E.; Fletcher, A. N.; Bliss, D. E.; Miles, M. H. *J. Electroanal. Chem.* **1985**, *190,* 171-183.
5. Lipsztajn M.; Osteryoung, R. A. *J. Electrochem. Soc.* **1983**, *130*, 1968-1969.
6. Fannin, A. A., Jr.; Floreani, D. A.; King, L. A.; Landers, J. S.; Piersm, B. J.; Stech, D. J.; Vaughn, R. L.; Wilkes, *J. S.; Williams,* J. *L. J. Phys. Chem.*, **1984**, *88,* 2614-2621.
7. Oye, H. A.; Jagtoyen, M.; Oksefjell, T.; Wilkes, J. S. *Materials Science Forum* **1991**, *73-75,* 183-190.
8. Wu, B.; Reddy, R. G.; Rogers, R. *D. Aluminum Recycling via Near Room Temperature Electrolysis in Ionic Liquids,* Confidential Report of University of Alabama, Tuscaloosa, USA.
9. Fuller, J.; Carlin, R. T.; Osteryong, R. A.; Koranaios, P.; Mantz R. *J. Electrochem. Soc.* **1998**, *145,* 24-28.
10. Xie, J.; Thomas, T. L. *J. Electrochem. Soc.* **2000**, *147,* 424 –4251.
11. Takahashi S. Japanese Patent JP0103199, 1989.
12. Bard, A. J.; Faulkner, *L. R. Electrochemical Methods: Fundamentals and Applications, John Wiley & Sons Inc., New York, 1980, pp.* 213-241.
13. Scharifker, B.; Hills, G. *Electrochim. Acta,* **1983**, *28,* 879-889.

Chapter 40

Electrochemistry: Electrochemically Generated Superoxide Ion in Ionic Liquids: Applications to Green Chemistry

I. M. AlNashef, M. A. Matthews*, and J. W. Weidner

Department of Chemical Engineering, University of South Carolina, Columbia, SC 29208
***Corresponding author: telephone: 803–777–0556; fax: 803–777–8265; email: matthews@engr.sc.edu**

The superoxide ion, which can be electrochemically generated from oxygen in dry room temperature ionic liquids (RTILs), is a safe and mild oxidizing agent that has several potential applications. We have shown that the superoxide ion in RTIL can be used to destroy hexachlorobenzene (HCB), which represents a serious environmental problem. The electrochemically generated superoxide ion was also used to oxidize benzyl alcohol and benzhydrol to benzoic acid and benzophenone, respectively.

Oxidation of alcohols to the corresponding aldehydes, ketones, and carboxylic acids is one of the most important functional group transformations in organic synthesis (1). Several catalytic processes using transition metal complexes have been reported using a stoichiometric amount of terminal oxidant (2). Unfortunately, in most of these methods hazardous or toxic oxidizing reagent is required (3). As an alternative, TEMPO (2,2,6,6-tetramethylpiperidine-1-oxyl-radical) has emerged as the catalyst of choice for mild and selective oxidation of alcohols to the corresponding aldehydes and ketones (4). These transformations employed a catalytic amount of TEMPO and stoichiometric amount of terminal oxidant, but air or molecular oxygen is desirable from environmental and economic point of view to replace the chemical oxidant.

Ionic liquids have been proposed as "green" solvents for organic synthesis due the ease with which they can be recycled whilst offering an inert, dipolar media compatible with much conventional chemistry. These properties have been reflected in their use in a range of reactions (5). Little attention has been paid to performing oxidation in RTILs. Ansari and Gree (6) reported an efficient aerobic oxidation of primary and secondary alcohols to the corresponding aldehydes and ketones by using TEMPO-CuCl catalyst system in [bmim][HFP]. Howarth (7) reported the aerobic oxidation of aromatic aldehydes to the corresponding carboxylic acids using nickel acetylacetonate [$Ni(acac)_2$] and O_2 in [bmim][HFP]. Ley *et al.* (8) showed that the use of tetralkylammonium salts or imidazolium ionic liquids in catalytic oxidation of alcohols with tetra-N-propylammonium perruthenate allowed recovery and reuse of the oxidant. Seddon and Stark (9) used palladium metal as a catalyst for the oxidation of benzyl alcohol to benzaldehyde in ionic liquids. They found that the selectivity to benzaldehyde is strongly dependent on the level of chloride ion, which lead to the formation of dibenzyl ether.

Singh *et al.* (10,11) showed that the electrochemically generated $O_2^{\bullet -}$ in dimethylformamide is able to convert primary and secondary alcohols into the corresponding carboxylic acids and ketones, respectively, according to the following equations:

$$O_2 + e^- \rightarrow O_2^{\bullet -} \quad (1)$$

$$RCH_2OH + O_2^{\bullet -} + O_2 \rightarrow RCOO^- + HO^\bullet + H_2O_2 \quad (2)$$

$$R1CHOHR2 + O_2^{\bullet -} + O_2 \rightarrow R1COR2 + H_2O_2 + O_2^{\bullet -} \quad (3)$$

The net oxidation reaction seems to be a complex multi-step process. Based on earlier studies and known behavior of superoxide ion in aprotic solvents,

Singh *et al.* (10,11) suggested that the initial step of the reaction may be assumed to be the deprotonation of the hydroxyl group by $O_2^{\bullet-}$, dismutation, loss of proton from alcoholic carbon followed by nucleophilic addition of superoxide to carbonyl group leading to the formation of carboxylic acids or ketones.

Casadei *et al.* (12-15) have reported that electrochemically generated $O_2^{\bullet-}$ activates CO_2 which subsequently converts NH protic amines to carbamates (12,13) and primary or secondary alcohols to carbonates (14,15).

$$2\ CO_2 + 2O_2^{\bullet-} \rightarrow C_2O_6^{-2} + O_2 \quad (4)$$

Organic carbamates are valuable synthetic intermediates for chemical and biochemical applications. Commercial synthesis involves ammonolysis of chloroformate or addition of alcohols to isocyanate. Phosgene is frequently used in preparing these reagents, and thus a phosgene-free synthetic route is of great interest from an environmental viewpoint.

The superoxide ion can degrade polychlorinated aromatics and polychlorinated biphenyls, PCBs, that contain three or more chlorine atoms per aromatic ring, to bicarbonates and chlorides (16-19). The mechanism of destruction is thought to be a nucleophilic reaction of $O_2^{\bullet-}$ with the chlorine-carbon bond, as is illustrated for the reaction involving hexachlorobenzene (18):

$$C_6Cl_6 + O_2^{\bullet-} \rightarrow [C_6Cl_6\,O_2^{\bullet-}] \rightarrow C_6Cl_5\,O_2^{\bullet} + Cl^- \quad (5)$$

$$C_6Cl_5\,O_2^{\bullet} + O_2^{\bullet-} \rightarrow C_6Cl_4O_2 + O_2 + Cl^- \quad (6)$$

$$C_6Cl_4O_2 + 10\ O_2^{\bullet-} \rightarrow 3\ C_2O_6^{2-} + 2\ O_2 + 4\ Cl^- \quad (7)$$

$$3\ C_2O_6^{2-} + 3\ H_2O \rightarrow 6\ HCO_3^- + 3/2\ O_2 \quad (8)$$

The nucleophilic addition of superoxide ion in reaction [5], as well as the reactions [6] and [7], must take place in an anhydrous environment. Kalu and White (19) have determined that PCBs were not completely destroyed in a flow-through reactor due to slow kinetics. In any case, however, the present state of knowledge requires use of volatile aprotic organic solvents, which suggests a serious problem with atmospheric solvent emissions and generation of secondary solvent wastes.

AlNashef *et al.* (20) showed that a stable superoxide ion could be generated in RTILs. They also showed that the $O_2^{\bullet-}$ reacts with CO_2 to give what is believed to be a carboxylating reagent (21).

In this paper we show that the electrochemically generated $O_2^{\bullet-}$ in [bmim][HFP] destroys HCB, which is considered as a model compound for poly-halogenated aromatic compounds. In addition, $O_2^{\bullet-}$ can be used in the syntheses of benzoic acid and benzophenone from the corresponding primary and secondary alcohols.

Experimental

Cyclic voltammetry (CV) tests were performed in the ionic liquid 1-butyl-3-methylimidazolium hexafluorophosphate, [bmim][HFP]. [bmim][HFP], (SACHEM) with a stated minimum purity of 97%, was dried overnight in a vacuum oven at 50°C. Impurities, particularly water, will interfere with the generation of the superoxide ion (20-22). We have previously demonstrated that adding 3.2% water to dried, sparged acetonitrile inhibits the reversible generation of O_2 (21). We have also shown that [bmim][HFP], when dried and sparged with nitrogen, will support the reversible generation of the superoxide ion (20). Figure 1 shows cyclic voltammogram, CV, of pure [bmim][HFP] after sparging with N_2 on glassy carbon working electrode. When the RTIL is dried and sparged, an electrochemical window of about 4.5V is obtained and only capacitive currents flow between the cathodic and anodic limits.

Electrochemistry was performed using an EG&G 263A potentiostat/galvanostat controlled by computer and data acquisition software. The electrode configuration was a glassy carbon working (BAS, 3mm dia.) and a platinum mesh counter (Aldrich) using Ag/AgCl reference electrode (Fisher Scientific). All [bmim][HFP] experiments were performed in a dry glove box under an argon atmosphere. The system was sparged prior to electrochemical experiments with UHP nitrogen or oxygen fitted through a Drierite gas purification column (W.A. Hammond). For carboxylic acids and ketones synthesis and for the HCB destruction experiments a glass cell with a glass frit of medium porosity separating the anode and cathode compartments was used (Ace Glass), Figure 2. A reticulated vitreous carbon (BAS) or Pt mesh (Aldrich) was used as a working electrode. The cathode chamber containing RTIL (≈50 mL) was purged with N_2 for 15 min. The catholyte was first pre-electrolyzed at −1.4 vs Ag/AgCl until the background current falls to ≈ 1 mA. For the syntheses experiments about 0.1 g of the alcohol (ACROS), +99% was then added to the cathode chamber. Oxygen is bubbled through the solution during the electrolysis period carried out at a constant potential of −1.1 V vs Ag/AgCl. Agitation of the catholyte was achieved by using a magnet stirrer and

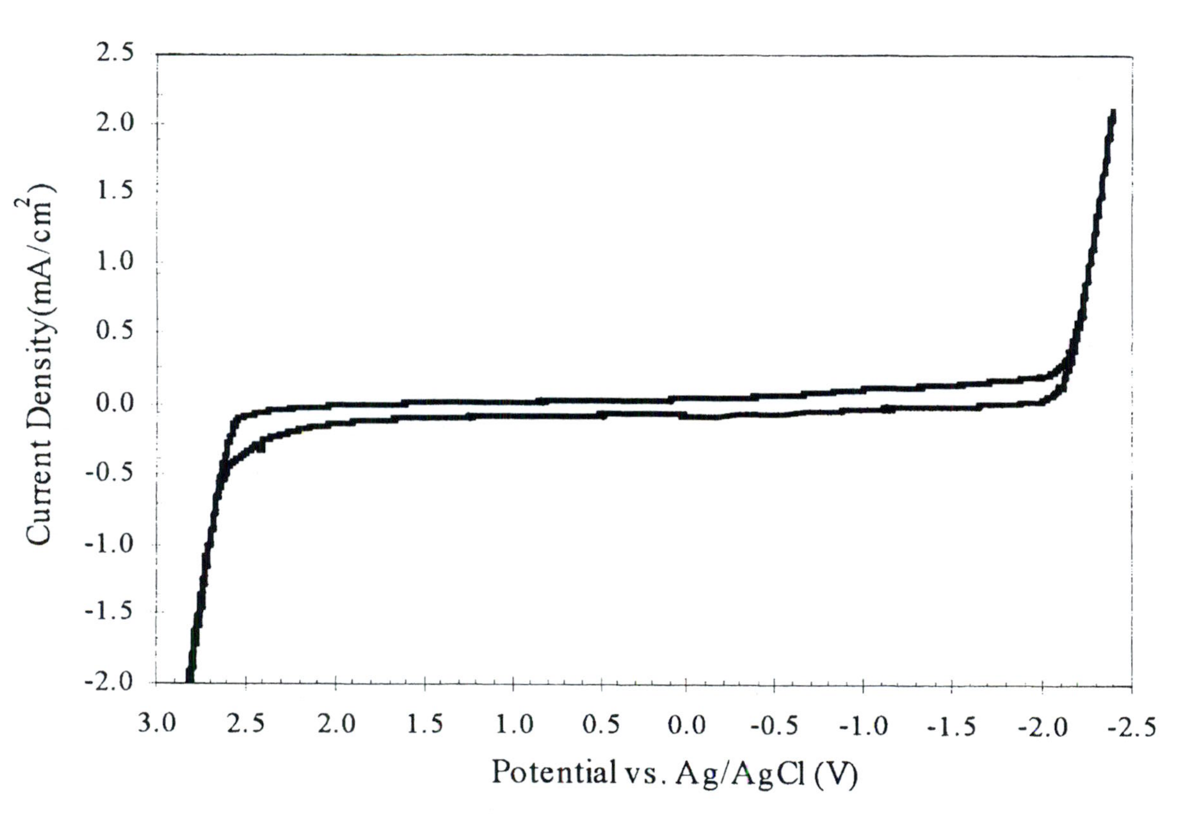

Figure 1 CV showing the electrochemical window for [bmim][HFP]. Working electrode = glassy carbon, sweep rate = 100 mV/s.

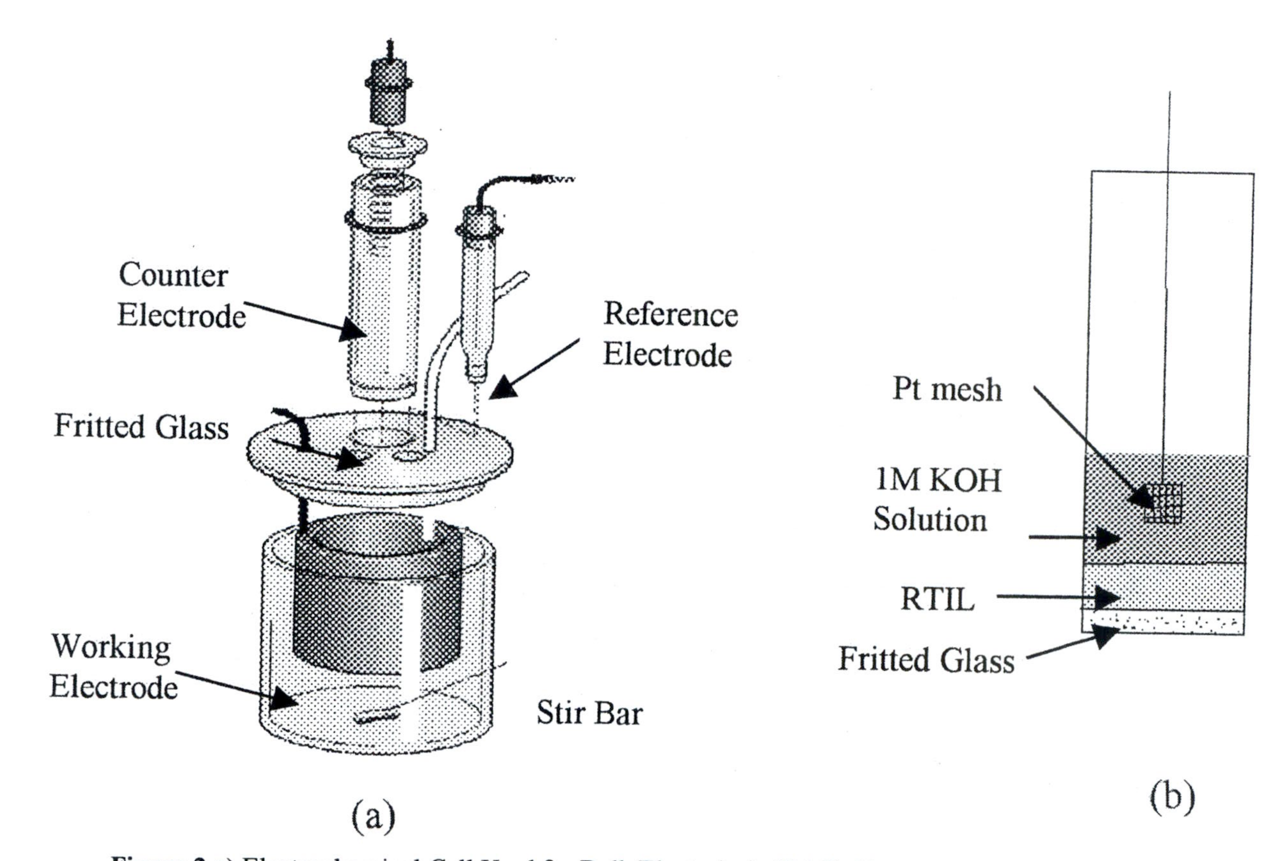

Figure 2 a) Electrochemical Cell Used for Bulk Electrolysis (BAS), b) counter electrode compartment.

through bubbling of oxygen. After electrolysis, diethyl ether was used to extract the products and the remaining reactants from the RTIL. A sample of the extract was then analyzed using GC/MS. For the HCB degradation experiments a weighed amount of HCB was added to the RTIL and the solution was stirred using a magnet stirrer for several hours. The solution was then filtered under vacuum to get rid of the un-dissolved HCB. Oxygen is bubbled through the solution during the electrolysis period carried out at a constant potential of −1.1 V vs Ag/AgCl. Agitation of the catholyte was achieved by using a magnet stirrer and through bubbling of oxygen. After electrolysis, cyclohexane was used to extract the products and the remaining reactant from the RTIL. A sample of the extract was then analyzed using 8610C SRI GC equipped with ECD detector.

Results and Discussion

GC/MS analysis showed that the superoxide ion initiated the oxidation of both primary and secondary alcohols to give the corresponding carboxylic acids and ketones, respectively. In the first case, $O_2^{\bullet -}$ reacted with benzyl alcohol to give benzoic acid with relatively low yield, 13%, Figure 3, compared to 60% yield reported by Singh *et al.* (11). Benzaldehyde, and benzyl benzoate were present as side products with 2% and 3% yield, respectively. Other unidentified products were present at low concentrations. The presence of benzaldehyde was also reported by Singh *et. al.* (11) who used dimethylformamide, DMF, as a solvent in their experiments. When [bmim][HFP] was used as anolyte the products of the oxidation reaction attacked the fritted glass that separates the cathode and anode compartments. Then [bmim][HFP] was replaced by 1M KCl aqueous solution. This worked well for the CV experiments that don't usually take more than a few minutes for completion. In the bulk electrolysis experiments, however, the superoxide ion became unstable after passing a certain amount of charge. During this time it was noted that [bmim][HFP] became more acidic which suggested that the electrolysis of water in the anode compartment, (see Equation 9), produces H^+ that diffused through the fritted glass to the solution in the cathode compartment. H^+ would subsequently react with the superoxide ion.

$$2H_2O \Leftrightarrow 4H^+ + O_2 + 4e \qquad (9)$$

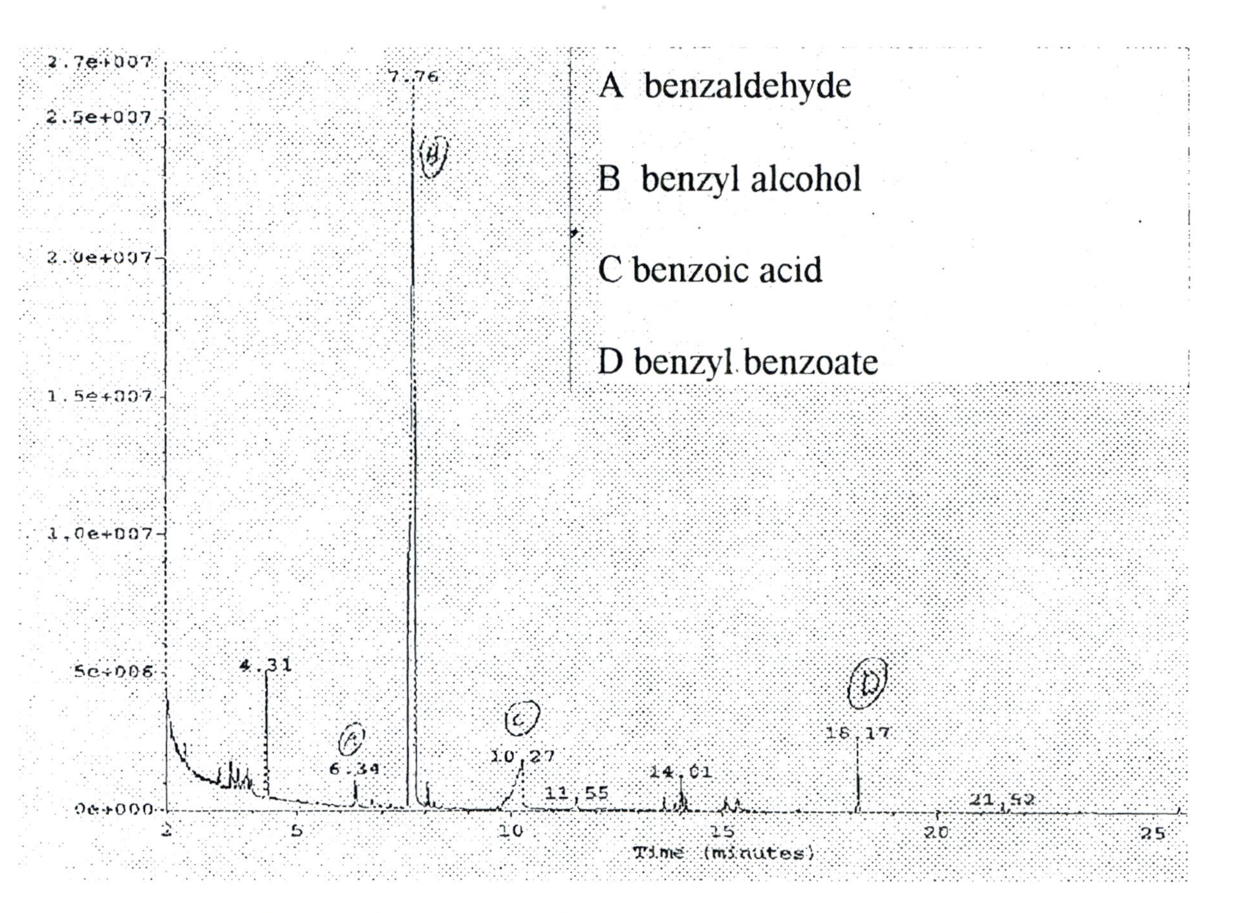

Figure 3 Chromatogram of benzoic acid synthesis products.

In order to solve this problem a layer of [bmim][HFP] was put at the bottom of the anode compartment and above that a layer of 1 M aqueous solution of KOH was used, Figure 2. The RTIL layer serves as a liquid membrane through which ions can move but the water molecules are trapped. Running CVs after the electrolysis indicated that this technique worked well and a well-shaped reverse peak was detected indicating that $O_2^{\bullet -}$ is stable. This new technique increased the yield in the syntheses of benzophenone from benzhydrol from 10% to 35%, Figure 4. Figure 5 shows the mass spectrum for the two overlapped peaks in Figure 4. The reported yield in DMF by Singh *et al.* (10) is 48%. Singh *et al.* (11) stated that the advantages of superoxide-initiated oxidation of alcohols are simple reaction conditions, low cost, easy scale-up, almost no waste problems and for these reasons the process is a good alternative to the known procedures (10,11).

The superoxide ion ($O_2^{\bullet -}$) can degrade polychlorinated aromatics and PCBs that contain three or more chlorine atoms per aromatic ring to bicarbonates and chlorides (16-19). The mechanism of destruction is thought to be a nucleophilic reaction of $O_2^{\bullet -}$ with the chlorine-carbon bond. In this work we show that the superoxide ion ($O_2^{\bullet -}$) generated in [bmim][HFP] can also degrade HCB. Figure 6 shows gas chromatograms for samples of the extract before and after the electrolysis. It is clear from the ratio of the areas of the two peaks that more than 70% of the HCB was degraded. This experiment was performed using the ionic liquid as an anolyte. We suspect that the diffusion of the hydronium ion to the cathode compartment and its reaction with the superoxide ion is one of the main factors for the incomplete destruction of HCB. More work is to be done to study the effect of anolyte, the temperature, working electrode material, and the amount of the electric charge on the efficiency of degradation. Sugimoto *et al.* (18) described the facile complete destruction of HCB in DMF to give bicarbonate and chloride ions as the only products in a batch reactor. Kalu and White (19) studied the effects of current, electrolyte flow, and aprotic media on the extent of degradation of HCB in a flow cell system equipped with a gas fed, porous electrode. They found that complete degradation could not be achieved. They attributed this to the complexity introduced into the system performance due to the flow effects. They found that the type of solvent in use also influenced the degree of degradation. The electrogeneration of $O_2^{-\bullet}$ from dissolved air represents a practical approach for an effective system to destroy HCB and other halogenated aromatic hydrocarbons.

The reported solubility of O_2 in [bmim][HFP] at 25°C is 0.6 mM (23), which is about one order of magnitude less than that in DMF (0.1 M TEAP) that is 4.8 mM (24). The low solubility of O_2 in RTIL may be one of the reasons

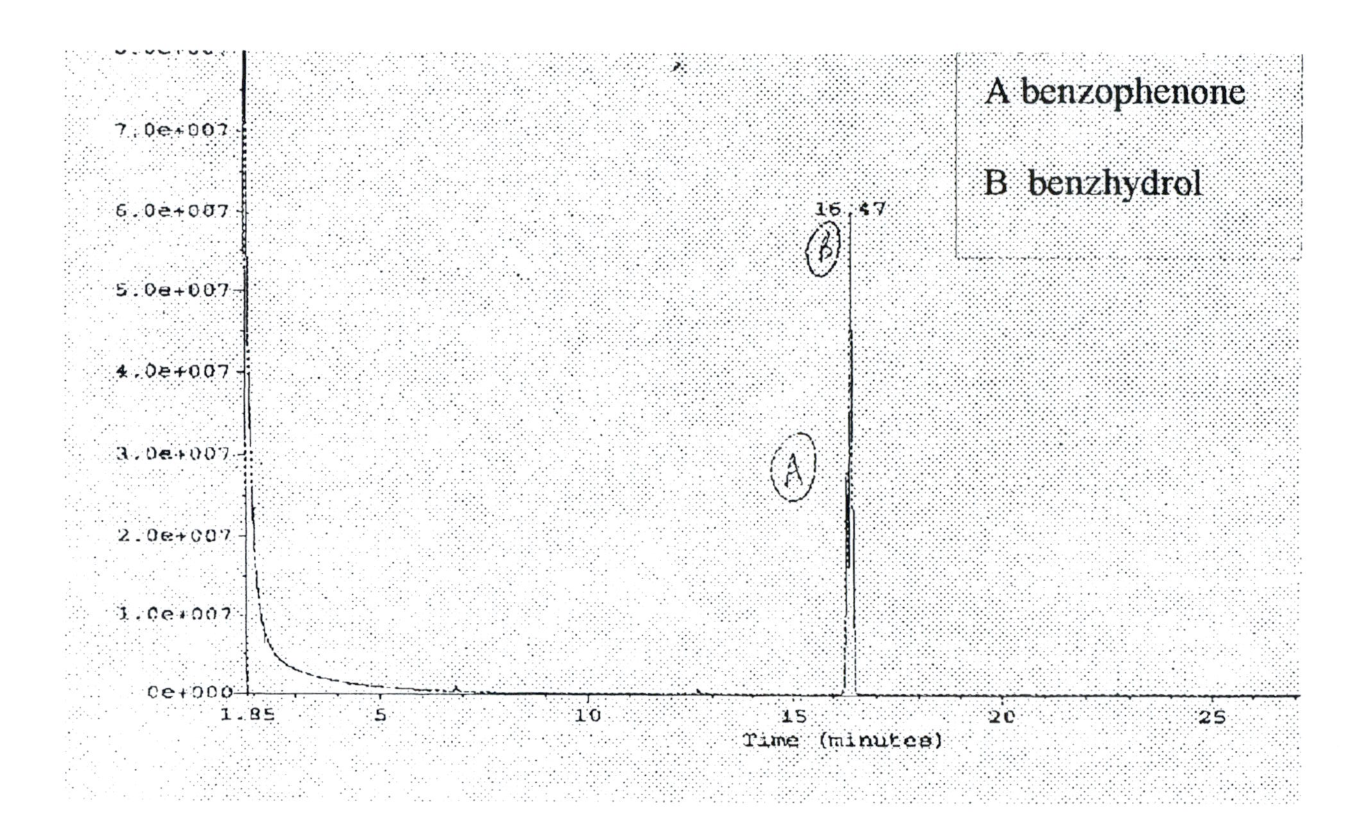

Figure 4 Chromatogram of benzophenone synthesis products.

for the low yield or degradation efficiency. Another factor that may affect the degradation efficiency or products yield is the high viscosity of RTIL at room temperature that affect the diffusion coefficient of O_2 and $O_2^{\bullet-}$ in RTIL. The kinematic viscosity of DMF and [bmim][HFP] are 0.008971 and 2.26 cm^2/s, respectively (25,26). Once the required amount of charge was passed and the current switched off, RTIL was extracted using cyclohexane. We believe that the mixture should be left for several hours while stirring to give $O_2^{\bullet-}$ enough time to diffuse from the electrode surface to react with HCB or the alcohol.

Conclusions

The superoxide ion was used to destroy HCB in [bmim][HFP]. It was also used to synthesize benzoic acid and benzophenone from the corresponding primary and secondary alcohols, respectively. More work is in progress to improve the yields in order to compete with other methods.

Acknowledgments

The authors gratefully acknowledge the financial support from NSF CTS-0086818. The authors would like to thank the Mass Spectrometry Center at the University of South Carolina for the GC/MS analyses.

References

1. Miyata, A.; Murakami, M.; Irie, R.; and Katsuki, T. Chemoselective aerobic oxidation of primary alcohols catalyzed by ruthenium complex. *Tetrahedron Lett.* **2001**, 42, 7067.

2. Ansari, I.A. and Gree R.TEMPO-Catalyzed Aerobic Oxidation of Alcohols to Ketones in Ionic Liquids [bmim][PF6]. *Organic Letters*, **2002**, 4, 1507.

3. Martin, S. E. and Suarez, D. F. Catalytic aerobic oxidation of alcohols by $Fe(NO_3)_3$-$FeBr_3$. *Tetrahedron lett.* **2002**, 43, 4475.

4. de Mico A.; Margarita, R.; Parlanti, L.; Vescovi, A.; Piancatelli, G. A versatile and Highly Selective Hypervalent Iodine (III)/2,2,6,6-Tetramethyl-1-piperidinyloxyl-Mediated Oxidation of Alcohols to Carbonyl Compounds. *J. Org. chem.* **1997**, 62, 6974.

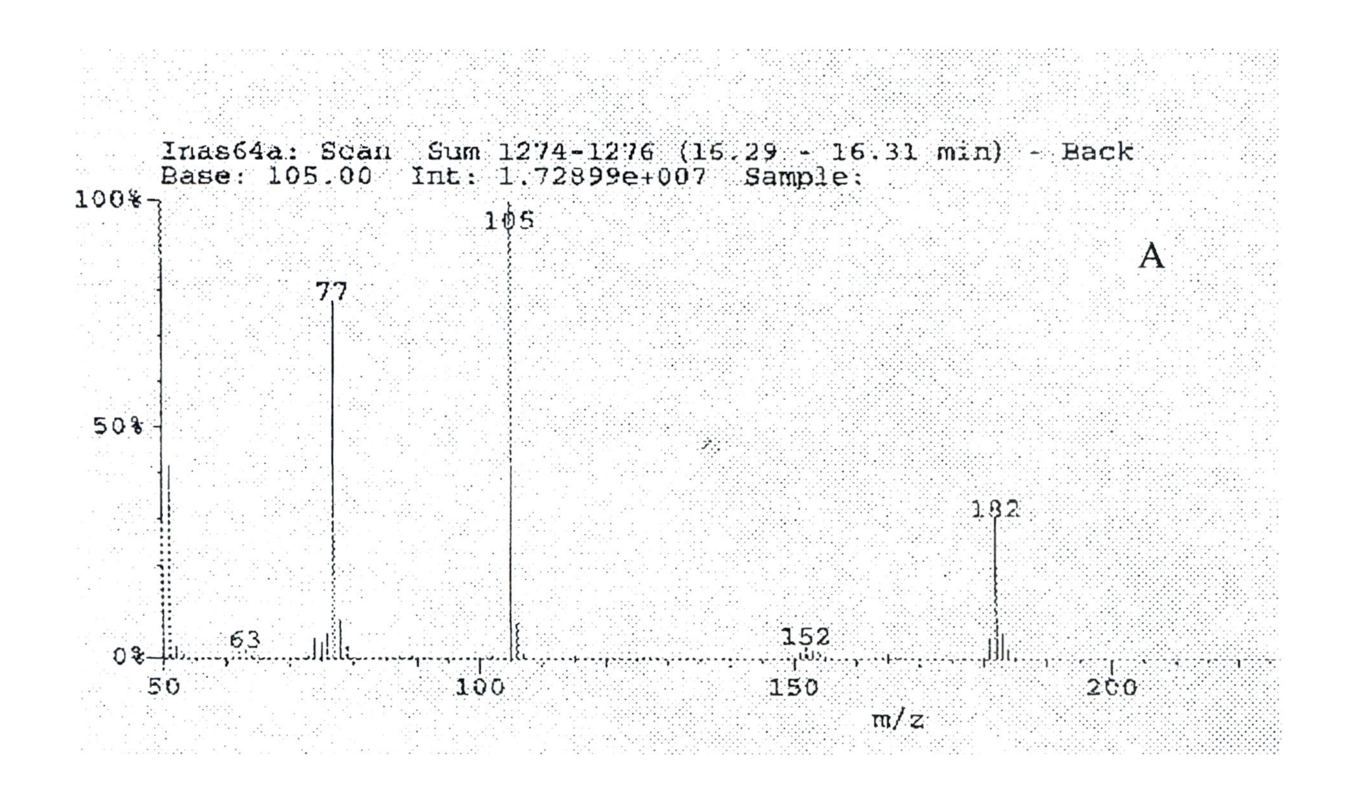

Inas64a: Scan Sum 1274-1276 (15.29 - 16.31 min) - Back
Base: 105.00 Int: 1.72899e+007 Sample:
100%
50%
0%
105
A
77
182
63
152
50
100
150
200
m/z

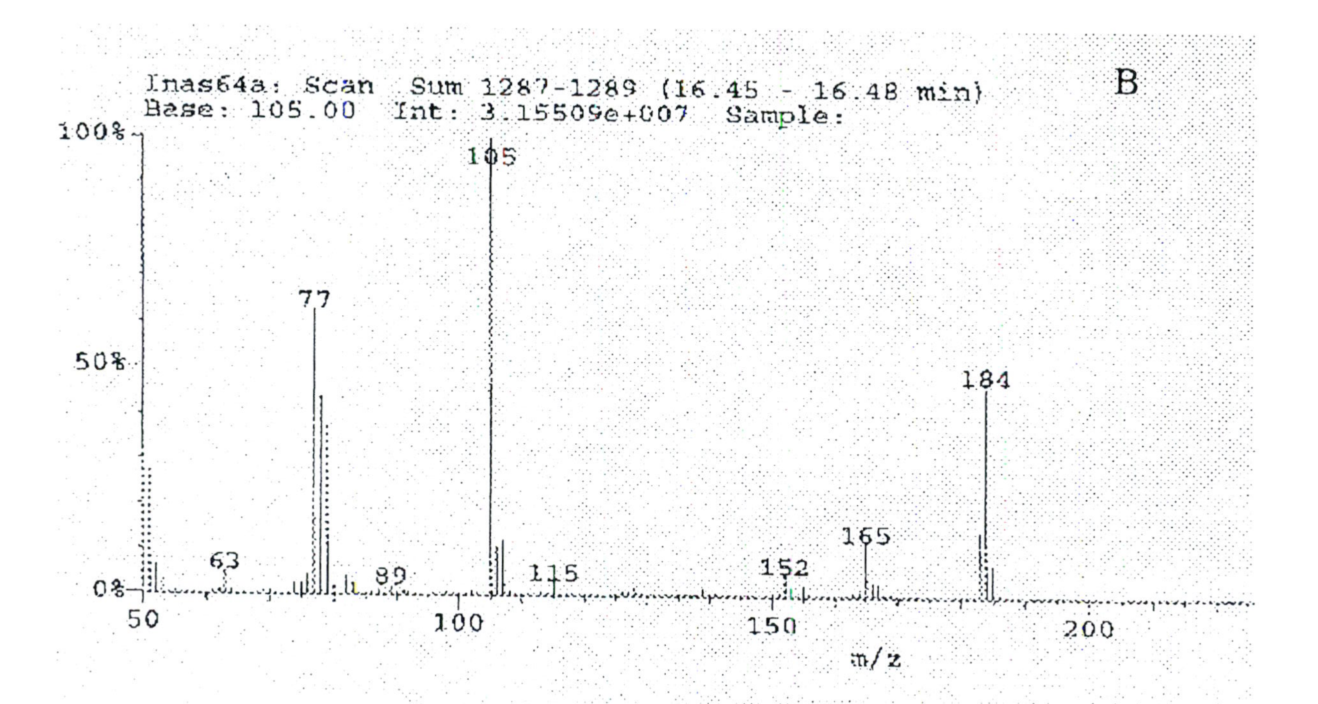

Figure 5 MS spectrum for peaks A and B in Figure 4.

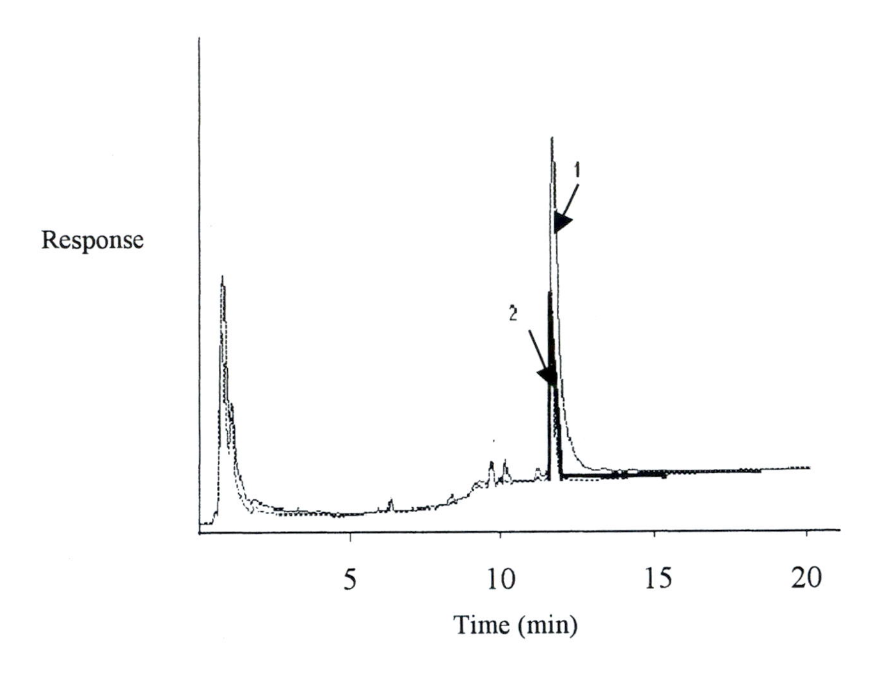

Figure 6 Chromatogram of HCB in cyclohexane before (1) and after (2) electrolysis, at constant potential of -1.1V vs. Ag/AgCl, in [bmim][HFP]. Area 2/Area 1 = 0.222. Working electrode: Pt mesh. After the electrolysis HCB was extracted using cyclohexane and analyzed using GC with ECD detector.

5. Welton, T. Room-Temperature Ionic Liquids. Solvents for Synthesis And Catalysis. *Chem. Rev.* **1999**, 99, 2071.

6. Ansari, I.A and Gree, R. TEMPO- Catalyzed Aerobic Oxidation of Alcohols to Aldehydes and Ketones in Ionic Liquid [bmim][HFP]. *Organic Letters*, **2002**, 4, 1507.

7. Howarth, J. Oxidation of Aromatic Aldehydes in the Ionic Liquid [bmim][HFP]. *Tetrahedron Lett.* **2000**, 41, 6627.

8. Ley, S.V.; Ramarao, C.; and Smith, M.D. Tetra-N-propylammonium perruthenate: a Case Study in Catalyst Recovery and Re-use Involving Tetraalkylammonium Salts. *Chem. Commun.* **2001**, 2278.

9. Seddon, K.R. and Stark, A. Selective catalytic oxidation of benzyl alcohol and alkylbenzene in ionic liquids. *Green Chemistry*, **2002**, 4, 119.

10. Singh, M., Misra, R. A. Electrogenerated Superoxide Initiated Oxidation with Oxygen. A Convenient Method for the Conversion of Secondary Alcohols to Ketones. *Synthesis* **1989**, 403.

11. Singh, K. N., Dwivedi, S., Misra, R. A. Superoxide ($O_2^{\bullet-}$) Initiated Oxidation of Primary Alcohols to Carboxylic Acids. *Synthesis* **1991**, 291.

12. Casadei, M. A., Moracci, F. M., Zappia, G., Inesi, M., Rossi, L. Electrogenerated Superoxide-Activated Carbon Dioxide: A New Mild And Safe Approach To Organic Carbamates *J. Org. Chem.* **1997**, 62, 6754.

13. Casadei, M. A., Cesa, S., Moracci, F. M., Inesi, A., Feroci, M. Activation Of Carbon Dioxide By Electrogenerated Superoxide Ion: A New Carboxylating Reagent *J. Org. Chem.* **1996**, 61, 381.

14. Casadei, M. A., Cesa, S., Feroci, M., Inesi, A., Rossi, L., Moracci, F. M. $O_2^{-\bullet}/CO_2$ System as Mild and Safe Carboxylating Reagent Synthesis of Organic Carbonates. *Tetrahedron* **1997**, 53, 167.

15. Casadei, M. A.; Inesi, A.; Rossi, L. Electrochemcial Activation of Carbon Dioxide: Synthesis of Organic Carbonates. *Tetrahedron Lett.* **1997**, 38, 3565.

16. Sawyer, D.T.; Calderwood, T.S. Degradation and Detoxification of Halogenated Olefinic Hydrocarbons. U.S. Patent 4,468,297 **1984.**

17. Sugimoto, H.; Matsumoto, S.; Sawyer, D. T. Degradation and Dehalogenation of Polychlorbiphenyls And Halogenated- Aromatic Molecules By Superoxide Ion And by Elctrolytic Reduction. *Environ. Sci. Technol.* **1988**, 22, 1182.

18. Sugimoto, H.; Matsumoto, S.; sawyer, D. T. Oxygenation of polychloro aromatic hydrocarbons by superoxide ion in aprotic media. *J. Am. Chem. Soc.* **1987**, 109, 8081.

19. Kalu, E.E.; White, R.E. In situ Degradation Of Polyhalogenated Aromatic-Hydrocarbons By Electrochemically Generated Superoxide Ions. *J Electrochem Soc.* **1991**, 138, 3656.

20. AlNashef, I.M.; Leonard, M.L.; Kittle, L.M.; Matthews, M.A.; and Weidner, J.W. Electrochemical Generation of Superoxide in Room Temperature Ionic Liquids. *Electrochem. Solid-State. Lett.* **2001**, 11, D16.

21. AlNashef, I. M.; Leonard, M. L.; Matthews, M. A.; Weidner, J. W. Superoxide Electrochemistry in an Ionic Liquid. *Ind. Eng. Chem. Res.* **2002**, 41, 4475.

22. Che, Y.; Tsushima, M. T.; Matsumoto, F.; Okajima, T.; Tokuda, K.; and Ohsaka, T. Water Induced Disproportionation of Superoxide ion in Aprotic Solvents. *J. Phys. Chem.*, **1996**, 100, 20134.

23. Anthony, J. L.; Maginn, E. J.; and Brennecke, J. F. Solubilities and thermodynamic properties of gases in ionic liquid 1-n-butyl-3-methylimidazolium hexafluorophosphate. *J. Phys. Chem. B*, **2002**, 106, 7315.

24. Sawyer, D. T.; Chieriticato, G.; Angelis, C. T.; and Nanni, E. J.; and Tsuchiya, T. Effects of media and electrode material on the electrochemical reduction of dioxygen. *Anal. Chem.*, **1982**, 37, 1720.

25. Tsushima, M, Tokuda, K.; and Ohsaka, T. Use of hydrodynamic chronocoulometry for simultaneous determination of the diffusion coefficients and concentrations of dioxygen in various media. *Anal. Chem.* **1994**, 66, 4551.

26. Suarez, P. A. Z.; Einloft, S.; Dullius, J. E. L; de Souza, R. F.; and Dupont, J. Synthesis and physical-chemical properties of ionic liquids based on 1-n-butyl-3-methylimidazolium cation. *J. Chim. Phys.* **1998**, 95, 1626.

Chapter 41

Conventional Aspects of Unconventional Solvents: Room Temperature Ionic Liquids as Ion-Exchangers and Ionic Surfactants

Mark L. Dietz[1], Julie A. Dzielawa[1], Mark P. Jensen[1], and Millicent A. Firestone[2]

Divisions of [1]Chemistry and [2]Materials Science, Argonne National Laboratory, Argonne, IL 60439

With few exceptions, research in the field of room-temperature ionic liquids (RTILs) has sought to understand and exploit the "unconventional" aspects of these solvents, among them their wide electrochemical window, ionicity, and near absence of vapor pressure. The resemblance of certain families of RTILs to various well-known liquid ion-exchangers or ionic surfactants, however, raises a question as to the extent to which the behavior of these neoteric solvents can be understood on the basis of the known properties of these more conventional compounds. In this report, we examine this question as it relates to the development of novel, nanostructured media comprising ionic liquids and RTIL-based methods for metal ion separation and preconcentration. The results presented demonstrate that in certain respects, RTILs can be regarded as conventional chemical reagents.

Introduction

Room-temperature ionic liquids (RTILs) have recently garnered intense interest as potential "green" alternatives to conventional organic solvents in a wide range of synthetic (1-4), catalytic (4-6), and electrochemical (7, 8) applications. Unlike ordinary (*i.e.*, molecular) solvents, and in analogy to classical molten salts, ionic liquids exist in a fully ionized state and as a consequence, typically exhibit little or no vapor pressure (5). In addition, they are charcterized by a wide electrochemical window (4) and an extraordinary degree of tunability, with relatively minor changes in cation and anion structure leading to significant variations in physicochemical properties (4). With few exceptions, research in the field of RTILs has sought to understand and exploit these "unconventional" properties. In certain instances (*e.g.*, the application of ionic liquids in process-scale separations), however, an emphasis on the unconventionality of RTILs may actually serve as an impediment to their adoption as replacements for ordinary solvents. For this reason, our recent work has sought to determine the extent to which RTILs can be regarded as conventional chemical reagents, and the manner in which these more mundane aspects of ionic liquids might be used to advantage in either chemical separations or the design and synthesis of novel materials. While this may not initially appear to be an especially fruitful line of investigation, its potential becomes more evident when one considers the structural similarities between certain families of RTILs and various classical liquid anion-exchangers. The RTILs recently described by MacFarlane *et al.* (9), for example, differ little from the well-known liquid anion-exchanger, Aliquat 336™. The resemblance of the cationic constituent of these RTILs to certain cationic surfactants (*e.g.*, CTAB) is also readily apparent (Figure 1). These similarities raise a question as to the degree to which the behavior of RTILs can be understood on the basis of the known properties of these two classes of compounds. In this chapter, we examine this question as it relates to the development of IL-based methods for the separation of strontium ion for potential application in large-scale nuclear waste processing or chemical analysis and to the preparation of nanostructured media comprising RTILs.

RTILs as Liquid Ion-Exchangers

Strontium Partitioning into 1-Alkyl-3-methylimidazolium-Based Ionic Liquids Containing a Crown Ether. Over the last two decades, work in a number of laboratories has been directed at the development of improved processes for the removal of actinides and fission product radionuclides (particularly Sr-90 and Cs-137) from nuclear waste streams and for their separation and preconcentration from environmental and biological samples for subsequent determination (10-15). These separations are quite demanding; in each case, the matrix is a complex, multi-component mixture, often strongly acidic or alkaline, in which the concentration of the radionuclides present is far exceeded by the levels of any number of other sample constituents. Such separations require chemical systems (*i.e.*, combinations of extractant and organic solvent) exhibiting both high efficiency and extraordinary selectivity. In the case of strontium extraction, the choice of extractant is relatively straightforward. That is, it is well known that certain crown ethers (CEs), in particular, those based upon 18-crown-6 (whose cavity size corresponds closely to that of the diameter of strontium ion), are capable of strong and selective complexation and/or extraction of Sr^{2+} (16-20). When the need to minimize the loss of the crown ether via solubilization in the aqueous phase and to reduce the

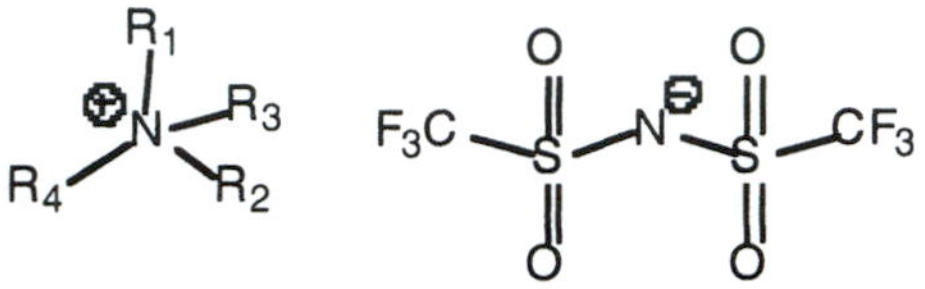

Tetraalkylammonium bis(trifluoromethylsulfonyl)imide salts

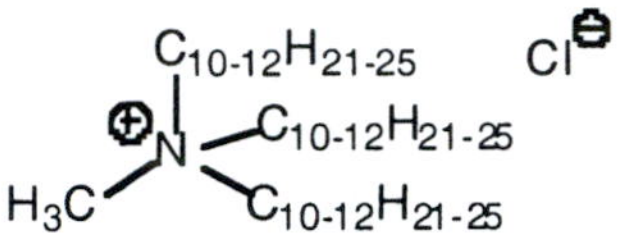

Aliquat 336™

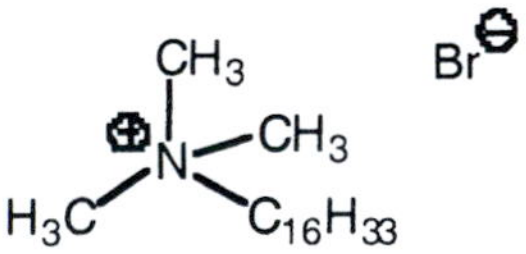

Hexadecyltrimethylammonium bromide (CTAB)

Figure 1. The RTILs of MacFarlane (9), a conventional liquid anion-exchanger ("Aliquat 336"), and a cationic surfactant ("CTAB").

expense and difficulty associated with its preparation are taken into consideration, the most appropriate choices of crown quickly become apparent: dicyclohexano-18-crown-6 (DCH18C6, shown below) or its di-*t*-butyl derivative (DtBuCH18C6) (21, 22).

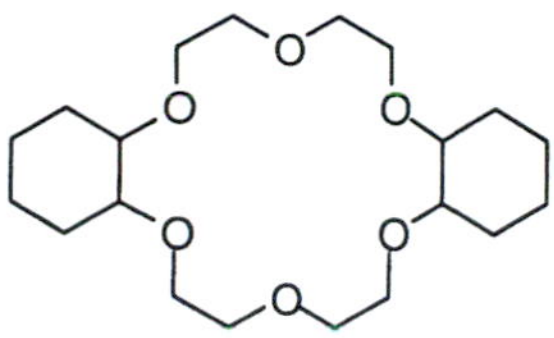

Selection of an appropriate solvent is more problematic. Among classical (*i.e.*, molecular) solvents, chlorinated hydrocarbons would initially appear to offer much promise; solutions of various crown ethers in certain of these solvents have been found to provide relatively large distribution ratios (D_{Sr}, defined as $[Sr]_{org}/[Sr]_{aq}$ at equilibrium) (23-25). As a result of their toxicity and potential for adverse environmental impact, however, chlorinated hydrocarbons have long been regarded as unacceptable for any large-scale application. The lower toxicity of paraffinic hydrocarbons has made them the solvent of choice in many process-scale metal ion separations (11). Solutions of crown ethers alone in these solvents, however, yield extremely poor extraction of strontium ion from acidic media (23).

In the early 1990's, workers at Argonne National Laboratory demonstrated the utility of various oxygenated, aliphatic solvents (in particular, 1-octanol) as diluents for crown ethers in the extraction of strontium (21, 22). Unlike many conventional solvents, 1-octanol exhibits a number of desirable physicochemical properties, most notably, the ability to dissolve significant concentrations of water, which by facilitating the transfer of incompletely dehydrated anions (*e.g.*, nitrate) and cationic metal-crown ether complexes, greatly improves the efficiency of strontium extraction (26). The advantages over previously described approaches to the separation and preconcentration of strontium from aqueous solution afforded by crown ether-octanol systems are numerous (*e.g.*, excellent selectivity, high extraction efficiency, facile recovery of extracted strontium) and have led to their emergence as a benchmark against which the performance of any proposed alternative must be measured (27).

Recently, Dai *et al.* (28) examined the extraction of strontium by DCH18C6 into a series of *N,N'*-dialkylimidazolium-based RTILs, such as are depicted here:

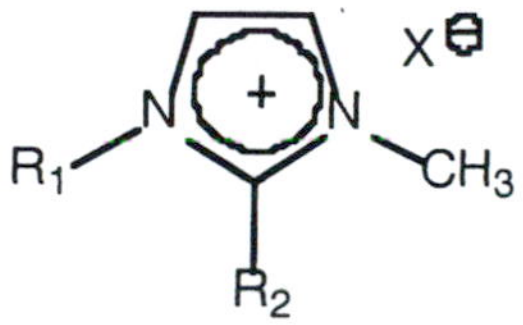

$X = PF_6^-, Tf_2N^-$

In several instances, extraction into 1-ethyl-3-methylimidazolium bis(trifluoromethylsulfonyl)imide (hereafter abbreviated as $C_2mim^+Tf_2N^-$), for example, remarkably large D_{Sr} values were obtained, far exceeding those observed for any conventional solvent. That these results were achieved using an aqueous phase containing only millimolar concentrations of nitrate ion (as $Sr(NO_3)_2$) is all the more remarkable. Interestingly, none of the solvents was found to dissolve any measurable quantity of water, thus indicating that unlike oxygenated aliphatic diluents, the presence of dissolved water is not an important factor in determining the efficiency of strontium partitioning into RTILs. Our own results using acidic, nitrate-containing aqueous phases (systems of greater potential practical value than those employed in Dai's studies) support this conclusion. As shown in Figure 2, which compares the effect of solvent water content on strontium partitioning for solutions of DCH18C6 in 1-octanol and $C_2mim^+Tf_2N^-$, although the water concentration in both solvents rises with increasing acidity, this increase is accompanied by a significant (4-fold) *decrease* in strontium partitioning in the RTIL system, in contrast to 1-octanol, for which a 50-fold increase in D_{Sr} is observed.

This difference in response to changes in water content is not the only way in which strontium partitioning behavior into this solvent (and related RTILs) contrasts with that observed for 1-octanol. As shown in Figure 3, which depicts the dependency of D_{Sr} on DCH18C6 concentration (panel A), the relationship between strontium partitioning and crown ether concentration is decidedly non-linear in 1-octanol, a result of significant aqueous phase Sr-CE complex formation (29). In the RTIL, however, the relationship between D_{Sr} and [DCH18C6] is a line of unit slope, consistent with partitioning of a 1:1 Sr-DCH18C6 complex and the absence of an appreciable equilibrium concentration of this complex in the aqueous phase.

More noteworthy is the difference in the nitric acid dependencies observed for the two solvents (Figure 3, panel B). In 1-octanol, increasing nitric acid concentration is accompanied by an increase in D_{Sr}, as would be expected for extraction of a strontium-nitrato-DCH18C6 complex:

$$Sr^{2+} + 2\ NO_3^- + DCH18C6_{org} \leftrightarrow Sr(NO_3)_2{\cdot}DCH18C6_{org} \quad (1)$$

In the RTIL, however, increasing acidity is accompanied by a decline in strontium partitioning. From a practical perspective, this is a disappointing result, as it indicates that the ionic liquid fails to maintain its dramatic advantage over conventional solvents (28) as the aqueous phase acidity is increased. Moreover, it raises troubling questions as to the ease with which strontium extracted into the RTIL can be recovered, since over the entire range of nitric acid concentrations considered, $D_{Sr} >> 1$. (Higher acidities result in significant dissolution of the RTIL in the aqueous phase.) Despite significant differences in the hydration energies associated with nitrate, chloride, and sulfate anions (26), essentially the same dependency is observed if either hydrochloric or sulfuric

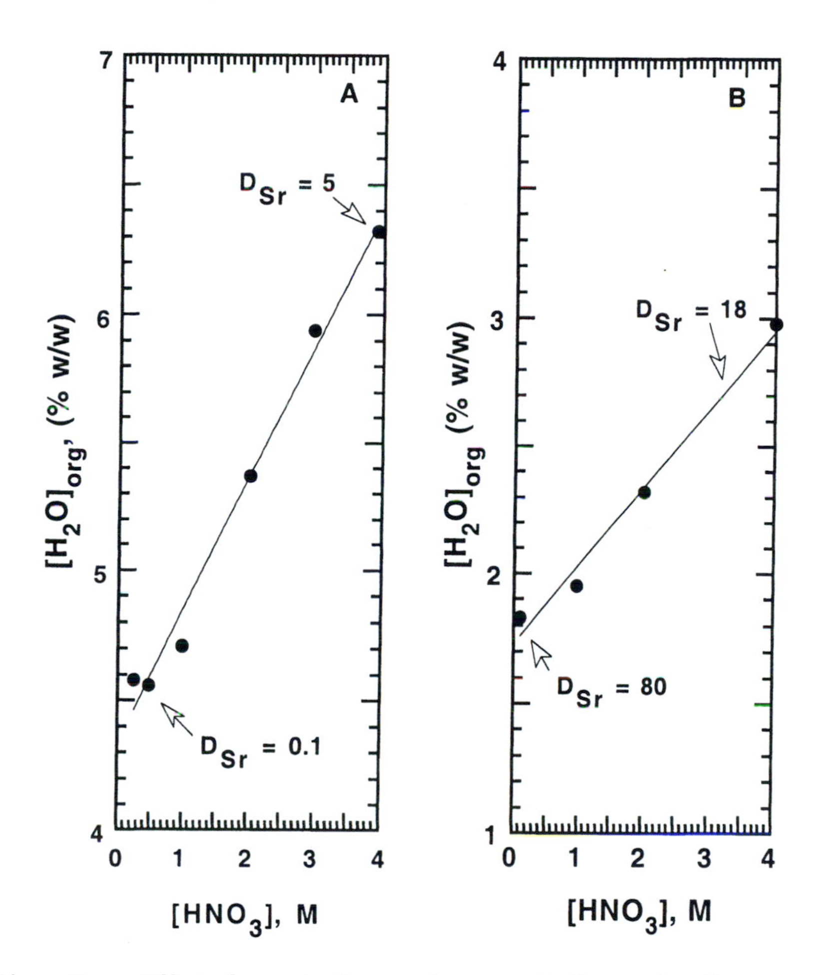

Figure 2. Effect of organic phase water concentration on strontium partitioning between nitric acid and solutions of DCH18C6 in 1-octanol (panel A) or $C_2mim^+Tf_2N^-$ (panel B).

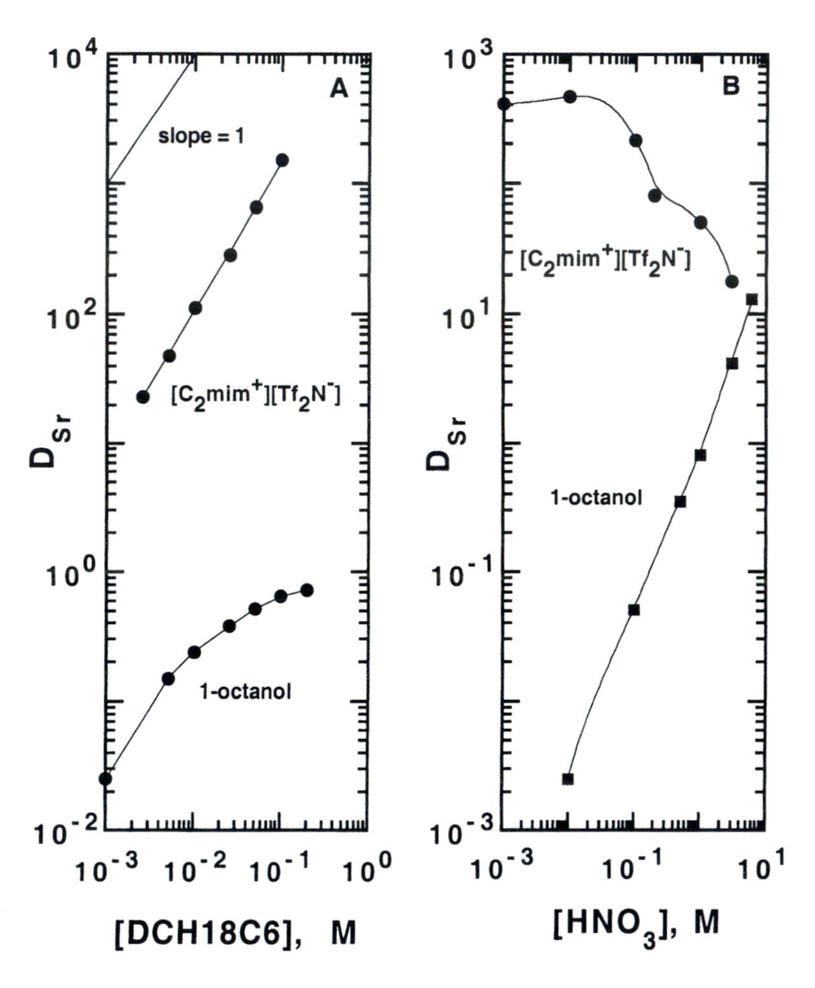

Figure 3. Effect of crown ether (panel A) and aqueous nitric acid concentration (panel B) on the extraction of strontium into $C_2mim^+Tf_2N^-$ and 1-octanol by DCH18C6 (a 1:1 mixture of the *cis-syn-cis* and *cis-anti-cis* isomers). For panel A, $[HNO_3]$=0.1 M and 1 M for the RTIL and 1-octanol, respectively. For panel B, [DCH18C6] = 0.1 M and 0.25 M for the RTIL and 1-octanol, respectively.

acid is employed as the aqueous phase, with the average difference in D_{Sr} values among the three systems amounting to only a factor of 4. In contrast, in 1-octanol, the difference in D_{Sr} values observed for acidic chloride and nitrate aqueous phases is nearly two orders of magnitude (26). Taken together, these results strongly indicate that the mechanism of strontium partitioning between nitric acid and $C_2mim^+Tf_2N^-$ (and, by analogy, related RTILs) is fundamentally different from that observed in 1-octanol, and that this partitioning does not involve extraction of a strontium-nitrato-crown ether complex (as per Eqn. 1). Recently obtained K-edge EXAFS data support this contention.

Figure 4 presents the results of X-ray absorption measurements on the organic phases produced in the partitioning of strontium into either 1-octanol or $C_5mim^+Tf_2N^-$ containing *cis-syn-cis*-DCH18C6 (*csc*-DCH18C6). In this study, solid $Sr(NO_3)_2{\bullet}18C6$, whose crystal structure had been previously determined (30), was used as a standard. Peak assignments and coordination numbers derived from the EXAFS data were based upon this 10-coordinate compound. In $Sr(NO_3)_2{\bullet}18C6$ (shown below, left), strontium sits in the 18C6 ring, where it is coordinated by six ether oxygen atoms and by four oxygen atoms from axially coordinated nitrate anions. Not unexpectedly, three major peaks are observed in the EXAFS, the first (at 2.7 Å) corresponding to 10 oxygen atoms, the next (at 3.5 Å) to 12 carbon atoms, and the last (at 4.3 Å) to a pair of distal (*i.e.*, uncoordinated) oxygen atoms on the axial nitrate anions. When this solid complex is dissolved directly in water-saturated $C_5mim^+Tf_2N^-$, these same three peaks are observed, meaning that the coordination environment of the strontium ion is unaffected by dissolution (*i.e.*, the complex remains intact). These same three peaks are observed when $Sr(NO_3)_2$ is extracted into 1-octanol by *csc*-DCH18C6. Thus, nitrate ion forms only inner-sphere complexes with $Sr{\bullet}CE^{2+}$ cations in either a two-phase water/1-octanol system or a single-phase, water-saturated ionic liquid.

If the strontium complex is prepared by contacting an aqueous solution of $Sr(NO_3)_2$ with a solution of *csc*-DCH18C6 in $C_5mim^+Tf_2N^-$, the peak associated with the distal oxygen atoms nearly disappears. In addition, the coordination number drops from 10 to 8. Both of these observations are consistent with a structure for the Sr-DCH18C6 complex in the RTIL in which a pair of water molecules now occupies the sites that had been occupied by nitrate ion.

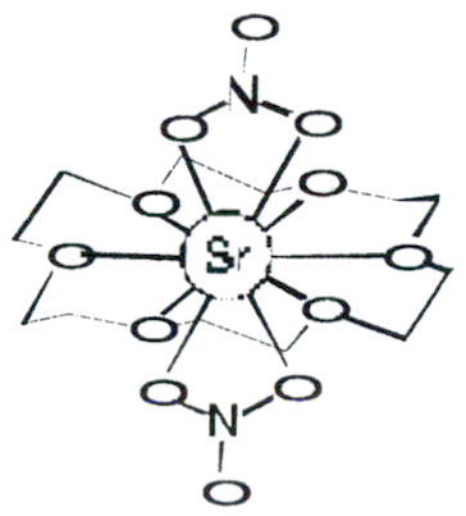

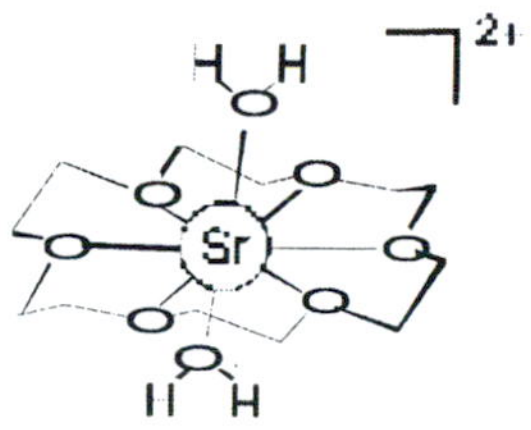

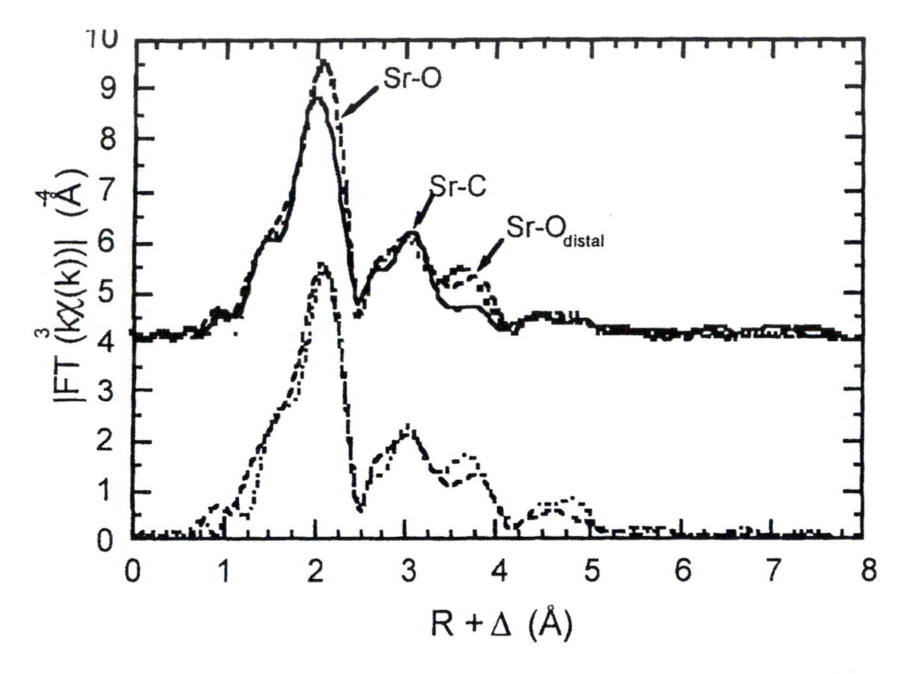

Figure 4. Fourier transform magnitude of the Sr K-edge EXAFS (without phase shift correction) for strontium-crown ether complexes. $Sr(NO_3)_2$(18C6) solid (dotted line), $Sr(NO_3)_2$(18C6) in $C_5mim^+Tf_2N^-$ (dashed line), $Sr(NO_3)_2$(DCH18C6) in 1-octanol (squares), $Sr(DCH18C6)^{2+}$ in $C_5mim^+Tf_2N^-$(solid line).

In principle, nitrate anions could still be present in the $C_5mim^+Tf_2N^-$ sample, but as outer-sphere ligands not discernible in the EXAFS data. Measurements of the nitrate content of the $C_5mim^+Tf_2N^-$ by ^{15}N-NMR following contact of a solution of *csc*-DCH18C6 with an aqueous solution of $Sr(^{15}NO_3)_2$ and ion chromatographic measurements of the change in aqueous phase nitrate concentration caused by contacting a solution of $Sr(NO_3)_2$ with *csc*-DCH18C6 in $C_5mim^+Tf_2N^-$, however, indicate that the amount of nitrate ion extracted is vastly insufficient to produce a neutral complex. Thus, as suggested by the organic phase water content and acid dependency studies (*vide supra*), strontium partitioning in the RTIL systems cannot involve appreciable extraction of a strontium nitrato-crown ether complex. Rather, the data are consistent with a mechanism involving exchange of the strontium-DCH18C6 cation for the cationic constituent of the RTIL:

$$Sr{\bullet}CE^{2+} + 2\ C_nmim^+_{org} \leftrightarrow Sr{\bullet}CE^{2+}_{org} + 2\ C_nmim^+ \quad (2)$$

For such a mechanism, one would expect that for a particular set of conditions, a decrease in strontium partitioning would be observed as the length of the alkyl chain on the *N,N'*-dialkylimidazolium cation is increased, since the transfer of an increasingly hydrophobic species into the aqueous phase would be required, and in fact, this has been observed (31). In addition, increases in strontium "loading" of the crown ether would be expected to lead to increased solubilization of the ionic liquid in the aqueous phase. This too has been reported (31). Thus, despite the obvious resemblance of certain families of RTILs to conventional liquid *anion* exchangers, in the presence of a crown ether, *N,N'*-dialkylimidazolium bis(trifluoromethylsulfonyl)imides function as *cation* exchangers. If this observation proves to be a general characteristic of neutral extractant-RTIL combinations, the implications for the "greenness" of ionic liquids as applied to metal ion separations, at least on the process scale (where high levels of metal loading are not uncommon) are not favorable. Additional work is clearly warranted and is, in fact, now in progress in this laboratory.

RTILs as Cationic Surfactants

Liquid-Crystalline "Ionogels" via Self-Assembly of 1-Alkyl-3-methylimidazolium-Based Ionic Liquids. Among the major areas of current interest in the field of nanoscience is the use of self-assembly to prepare functional materials and devices. Of particular interest has been the development of methods to direct molecular organization and to generate novel supramolecular structures in soft condensed matter phases. These phases, which include such materials as polymers, liquid crystals, and gels (32), exhibit several properties that make them especially useful in the development of functional molecular systems, among them, tunability, ease of processing (*e.g.*, by application of an external field (33)), and a dynamic nature.

Gels constitute one of the most important classes of soft matter, comprising both an essential component of living organisms and the basis of a wide variety of synthetic systems of technological significance in fields ranging from controlled drug delivery to the development of novel chemical sensors (34, 35). Recently, there has been growing interest in the development of gels incorporating ionic liquids, particularly physical gels, which rely upon non-covalent interactions for stabilization. In 1997, for example, Fuller *et al.* (36) described the preparation of novel gel electrolytes consisting of a mixture of a poly(vinylidene fluoride)-hexafluoropropylene copolymer and any of several 1-ethyl-3-methylimidazolium salts (*e.g.*, tetrafluoroborate). More recently, a group of Japanese investigators (37), seeking ways to control the fluidity of ionic liquids for application in electrochemical devices, demonstrated that gelation of various *N,N'*-dialkylimidazolium and *N*-alkylpyridinium-based RTILs could be induced by the addition of small quantities (0.1-0.4% w/w) of a low-molecular weight, cholesterol-based organogelator. At about the same time, Kimizuka and Nakashima (38), in the course of developing means by which to improve the solubility of carbohydrates in ionic liquids, noted that addition of an appropriate carbohydrate to an ether-derivatized RTIL resulted in physical gelation of the ionic liquid. For certain solutes (*i.e.*, amide-enriched glycolipids), gelation was found to be accompanied by the formation of a bilayer structure, thus representing the first example of a self-assembling "ionogel". The formation of molecular assemblies within ionic liquids, like their gelation, has attracted growing recent interest. Work by Yoshio *et al.* (39, 40), for example, directed at the development of anisotropic ion conductors, has shown that the addition of an appropriate mesogen to $C_2mim^+BF_4^-$ results in the formation of a layered structure exhibiting both liquid crystallinity and anisotropic ion conduction.

In conjunction with our ongoing efforts to exploit soft matter in the design and preparation of new materials (41, 42), we have sought a simple means by which to induce both gelation and formation of a supramolecular assembly in an RTIL, one which requires neither organogelators (37) nor exotic mesogens (39) and employs only readily available ionic liquids. The resemblance of certain RTIL cations to conventional cationic surfactants, already noted (Figure 1), the well-known tendency of surfactants to self-assemble in the presence of water (32), and prior reports of the gelation of concentrated aqueous solutions of certain surfactants (32, 43), suggested to us that an appropriate combination of an RTIL and water might produce the desired result. Surprisingly, this seemingly obvious approach to the formation of a liquid-crystalline "ionogel" has not been previously considered, despite a now-extensive literature on the properties and behavior of RTILs.

Figure 5 shows the results of small-angle X-ray scattering (SAXS) studies of the interaction of water with a representative hydrophilic ionic liquid, 1-decyl-3-methylimidazolium bromide, presented as 2-D scattering patterns for a pair of samples containing two different concentrations of water. For $C_{10}mim^+Br^-$ containing only *ca.* 1.6% (w/w) water, the SAXS data (*i.e.*, the corresponding plots of the azimuthally-averaged scattering intensity as a function of scattering vector, which are not shown for brevity) exhibit only a single Bragg peak centered at *ca.* 0.24 $Å^{-1}$, the width of which is consistent with considerable lattice and orientational disorder. The isotropic scattering pattern observed indicates that the sample consists of microdomains in which all spatial orientations are present. By applying the Scherrer equation (44), the average size of these domains can be estimated as 250 Å. Thus, the sample does not exhibit appreciable *long-range* order, an observation consistent with the absence of optical birefringence noted in polarized optical microscopy.

Raising the water concentration 10-fold, to *ca.* 16% w/w, results in the immediate conversion of the sample from a viscous liquid to a stable, homogeneous, optically-birefringent gel that resists flow against gravity for an indefinite period of time. The anisotropic 2-D scattering pattern indicates that the microdomains now exhibit a preferred orientation (*i.e.*, perpendicular to the capillary axis). Also, the average size of these domains has increased to *ca.* 1640 Å, nearly an order of magnitude larger than in the low water content sample. Thus, the increase in water concentration has led to a pronounced increase in the extent of structural ordering in the ionic liquid. The presence of a pair of Bragg peaks at integral order spacing (0.22 and 0.44 $Å^{-1}$) is consistent with the existence of a lamellar structure having a lattice spacing of 29 Å.

It is interesting to note that prior work by a number of authors has clearly established that in the absence of water, only 1-alkyl-3-methylimidazolium salts bearing sufficiently long ($n \geq 12$) alkyl chains will exhibit structural order (*i.e.*, liquid crystallinity) (45-47). Shorter-chain ($n<12$) analogues, such as the $C_{10}mim^+Br^-$ employed here, have been reported to be isotropic liquids (45, 47). It is also noteworthy that concentrations of water insufficient to induce liquid crystallinity are nonetheless sufficient to produce a measurable degree of order in the RTIL.

To determine how the lattice spacing of the lamellar gel phase of $C_{10}mim^+Br^-$ compares with the dimensions of the molecule itself, electronic structure calculations (using *ab initio* molecular orbital theory at the Hartree-Fock level using the 6-31G* basis set (48)) were performed. The results indicate that an isolated, geometry-optimized $C_{10}mim^+Br^-$ molecule will exhibit an overall C-C length of between 12.5 Å (for an all-*trans* conformationally ordered C_{10} chain) and 10.6 Å (for the same molecule containing a pair of *gauche* defects). Thus, regardless of the state of the alkyl chains, the lattice spacing exceeds twice the molecular length, implying the existence of a water channel at least 4 Å in thickness. Given that previous solid-state structural studies of long-chain ($n>12$) C_nmim^+ salts suggest that some degree of alkyl chain interdigitation might reasonably be expected (45), this thickness may well be significantly greater.

In an effort to obtain some insight into the molecular basis for gelation in $C_{10}mim^+Br^-$, infrared spectroscopic studies of its fluid and gel phases have recently been undertaken. While our analysis is not complete at present, initial

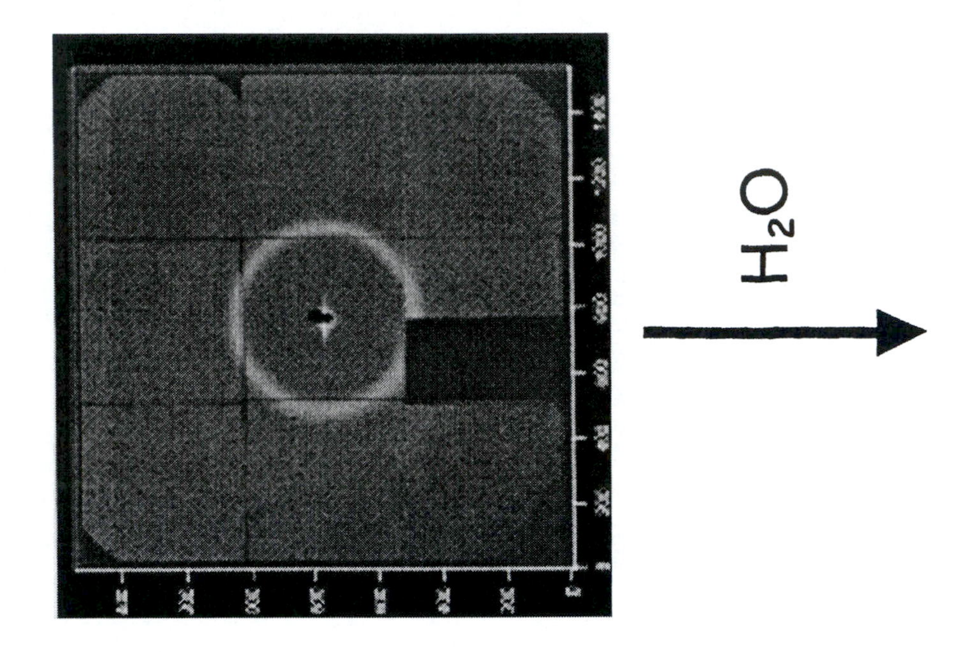
H_2O

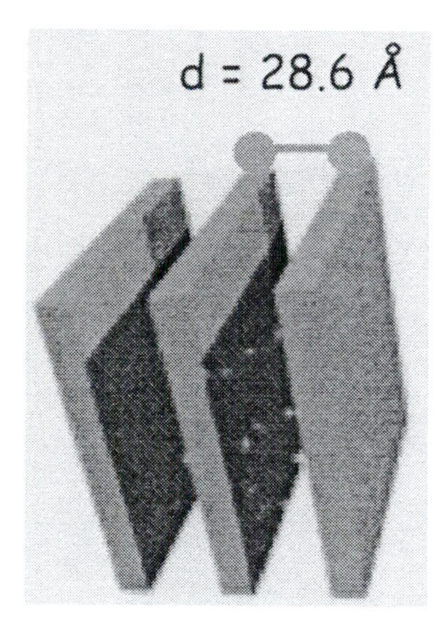

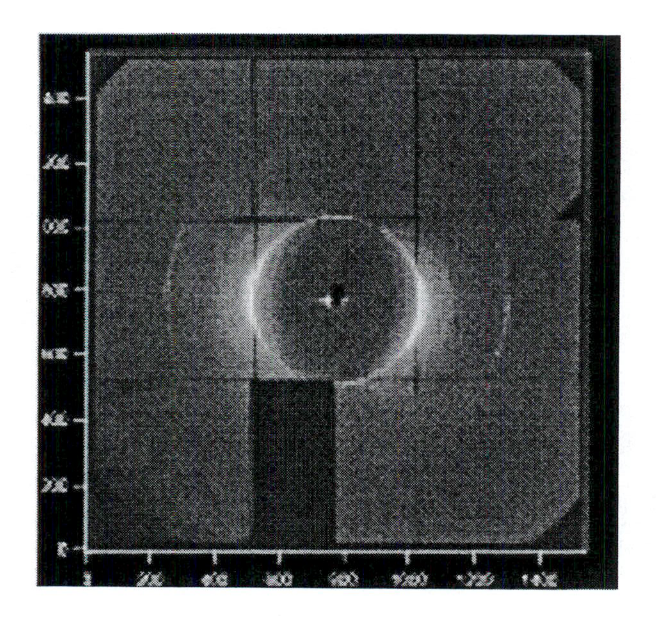

Figure 5. Two-dimensional small-angle X-ray scattering (SAXS) patterns and azimuthally-averaged intensity as a function of scattering vector for the fluid (1.6% w/w water) and gel (16% w/w water) states of 1-decyl-3-methylimidazolium bromide (T = 23 °C).

results indicate that gelation has no significant effect on the shape or position of the symmetric (*i.e.*, υ_s (CH_2)) or asymmetric (υ_{as} (CH_2)) methylene stretching bands for the C_{10} moiety, modes known to be sensitive probes of alkyl chain packing (49). Thus, gelation is not induced by two-dimensional alkyl chain packing or ordering effects. Gelation is, however, accompanied by a diminution in the bands associated with hydrogen bonding between the imidazolium ring (in particular, the proton on C-2) and the bromide ion (*e.g.*, the "Sheppard effect" bands (50)). This observation is consistent with recent results obtained by Cammarata *et al.* (51) in infrared studies of the state of water added to $C_4mim^+Tf_2N^-$ and related RTILs, for which evidence of competition between the imidazolium cation and water for H-bonding with the anion has been observed. This result, taken together with the fact that ionogel formation is not observed for samples containing less than 5% w/w or more than 40% w/w water suggests that gelation relies on a proper balance between $C_{10}mim^+$•••Br^- and Br^- •••H-O-H interactions, and that this balance, which apparently permits formation of a network comprising $C_{10}mim^+$, Br^-, and water molecules, can be achieved only over a specific range of RTIL water contents. Additional work to elucidate the mechanism of gelation of RTILs, to devise means by which to "tune" their gelation (*i.e.*, to adjust the lattice spacing of the ionogel), and to exploit these novel, nanostructured systems in chemical separations and materials synthesis is now underway in this laboratory.

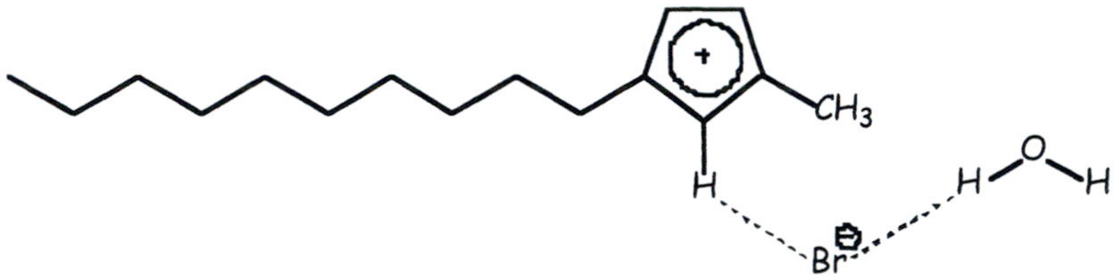

Conclusions

Although room-temperature ionic liquids are normally regarded as "unconventional", the results presented here demonstrate that efforts to understand their behavior by drawing analogies to conventional ion-exchangers and cationic surfactants can yield important and useful insights. In the area of separations, for example, these efforts have led to results that call into question the "greenness" of RTILs as solvents in the extraction of metal ions by neutral ligands. Similarly, in the area of materials chemistry, this approach has led to the development of a simple means by which to induce the formation of a supramolecular assembly in an ionic liquid and thus, to the development of a system expected to provide both a novel medium in which to carry out chemical reactions and a versatile platform for the design and fabrication of new materials. We expect that the utility of this approach will become increasingly evident as investigations of *N,N'*-dialkylimidazolium salts and other families of RTILs progress.

Acknowledgements

The authors thank Larry Curtiss and Peter Zapol (ANL-CHM) for electronic structure calculations on $C_{10}mim^{+}Br^{-}$, Soenke Seifert (ANL-CHM) for assistance with SAXS measurements, Paul G. Rickert (ANL-CHM) for ^{15}N-NMR analyses, and Urs Geiser (ANL-MSD) for X-ray crystallographic analysis of $Sr(NO_3)_2$•18C6. This work was performed under the auspices of the Office of Basic Energy Sciences, Divisions of Chemical (MLD, JAD, MPJ) and Materials (MAF) Sciences, U.S. Department of Energy, under contract number W-31-109-ENG-38.

References

1. Seddon, K. R. *J. Chem. Tech. Biotechnol.* **1997**, *68*, 351-356.
2. Holbrey, J. D.; Seddon, K. R. *Clean Prod. Processes* **1999**, *1*, 223-236.
3. Earle, M. J.; Seddon, K. R. *Pure Appl. Chem.* **2000**, *72*, 1391-1398.
4. Olivier-Bourbigou, H.; Magna, L. *J. Mol. Catal. A* **2002**, *182-3*, 419-437.
5. Seddon, K. R. *Kinet. Catal.* **1996**, *37*, 693-697.
6. Zhao, D.; Wu, M.; Kou, Y.; Min, E. *Catal. Today* **2002**, *74*, 157-189.
7. Fuller, J.; Carlin, R. T.; Osteryoung, R. A. *J. Electrochem. Soc.* **1997**, *144*, 3881-3886.
8. Carlin, R. T.; Fuller, J. In *Proceedings of the 12th Annual Battery Conference on Applications and Advances-1997*; Frank, H. A.; Seo, E. T., Eds.; Institute of Electrical and Electronics Engineers, New York, 1997; pp.261-266.
9. Sun, J.; Forsyth, M.; MacFarlane, D. R. *J. Phys. Chem. B* **1998**, *102*, 8858-8864.
10. Nash, K. L.; Barrans, R. E.; Chiarizia, R.; Dietz, M. L.; Jensen, M. P.; Rickert, P. G. *Solv. Extr. Ion Exch.* **2000**, *18*, 605-631.
11. Horwitz, E. P.; Schulz, W. W. In *Metal-Ion Separation and Preconcentration: Progress and Opportunities*; Bond, A. H.; Dietz, M. L.; Rogers, R. D., Eds.; ACS Symposium Series 716; American Chemical Society: Washington, DC, 1999, pp 20-50.

12. *Science and Technology for Disposal of Radioactive Tank Wastes*; Schulz, W. W.; Lombardo, N. J., Eds.; Plenum: New York, 1998.
13. *Chemical Pretreatment of Nuclear Waste for Disposal*, Schulz, W. W.; Horwitz, E. P., Eds.; Plenum: New York, 1994.
14. Dietz, M. L.; Horwitz, E. P.; Bond, A. H. In *Metal-Ion Separation and Preconcentration: Progress and Opportunities*; Bond, A. H., Dietz, M. L., Rogers, R. D., Eds.; ACS Symposium Series 716; American Chemical Society: Washington, DC, 1999, pp 234-250.
15. Dietz, M. L.; Horwitz, E. P. *LC•GC* **1993**, *11*, 424-436.
16. Gokel, G. *Crown Ethers and Cryptands*; Royal Society of Chemistry: Cambridge, England, 1991.
17. Hiraoka, M. *Crown Compounds: Their Characteristics and Applications*; Elsevier: New York, 1982.
18. Izatt, R. M.; Pawlak, K.; Bradshaw, J. S.; Breuning, R. L. *Chem. Rev.* **1995**, *95*, 2529-2586.
19. Izatt, R. M.; Pawlak, K.; Bradshaw, J. S.; Breuning, R. L. *Chem. Rev.* **1991**, *91*, 1721-2085.
20. Izatt, R. M.; Bradshaw, J. S.; Nielsen, S. A.; Lamb, J. D.; Christensen, J. J. *Chem. Rev.* **1985**, *85*, 271-339.
21. Horwitz, E. P.; Dietz, M. L.; Fisher, D. E. *Solvent Extr. Ion Exch.* **1990**, *8*, 557-572.
22. Horwitz, E. P.; Dietz, M. L.; Fisher, D. E. *Solvent Extr. Ion Exch.* **1991**, *9*, 1-25.
23. Blasius, E.; Klein, W.; Schõn, U. *J. Radioanal. Nucl. Chem.* **1985**, *89*, 389-398.
24. Gloe, K.; Muehl, P.; Kholkin, A. I.; Meerbote, M.; Beger, J. *Isotopenpraxis* **1982**, *18*, 170-175.
25. Filippov, E. A.; Yakshin, V. V.; Abashkin, V. M.; Fomenkov, V. G.; Serebryakov, I. S. *Radiokhimiya* **1982**, *24*, 214-218.
26. Horwitz, E. P.; Dietz, M. L.; Fisher, D. E. *Solvent Extr. Ion Exch.* **1990**, *8*, 199-208.
27. *Standard Test Method for Strontium-90 in Water*, ASTM-D5811-95, American Society for Testing and Materials: West Conshohocken, PA, 1995.
28. Dai, S.; Ju, Y. H.; Barnes, C. E. *J. Chem. Soc., Dalton Trans.* **1999**, 1201-1202.
29. Dietz, M. L.; Bond, A. H.; Clapper, M.; Finch, J. W. *Radiochim. Acta* **1999**, *85*, 119-129.
30. Junk, P. C.; Steed, J. W. *J. Chem. Soc., Dalton Trans.* **1999**, 407-414.
31. Dietz, M. L.; Dzielawa, J. A. *Chem. Commun.* **2001**, 2124-2125.
32. Hamley, I. W. *Introduction to Soft Matter: Polymers, Colloids, Amphiphiles, and Liquid Crystals*, John Wiley: New York, 2000.
33. Firestone, M. A.; Tiede, D. M.; Seifert, S. *J. Phys. Chem. B* **2000**, 104, 2433-2438.
34. Rethwisch, D. G.; Chen, X.; Martin, B. D.; Dordick, J. S. In *Biomolecular Materials by Design*; Alper, M.; Bayley, H.; Kaplan, D.; Navia, M., Eds.; Materials Research Society: Pittsburgh, PA, 1994, pp 225-230.
35. Tess, M. E.; Cox, J. A. *J. Pharm. Biomed. Anal.* **1999**, *19*, 55-68.

36. Fuller, J.; Breda, A. C.; Carlin, R. T. *J. Electrochem. Soc.* **1997**, *144*, L67-L70.
37. Ikeda, A.; Sonoda, K.; Ayabe, M.; Tamaru, S.; Nakashima, T.; Kimizuka, N.; Shinkai, S. *Chem. Lett.* **2001**, 1154-1155.
38. Kimizuka, N.; Nakashima, T. *Langmuir* **2001**, *17*, 6759-6761.
39. Yoshio, M.; Mukai, T.; Kanie, K.; Yoshizawa, M.; Ohno, H.; Kato, T. *Adv. Mater.* **2002**, *14*, 351-354.
40. Yoshio, M.; Mukai, T.; Kanie, K.; Yoshizawa, M.; Ohno, H.; Kato, T. *Chem. Lett.* **2002**, 320-321.
41. Firestone, M. A.; Thiyagarajan, P.; Tiede, D. M. *Langmuir* **1998**, *14*, 4688-4698.
42. Firestone, M. A.; Williams, D. E.; Seifert, S.; Csencsits, R. *Nano Lett.* **2001**, *1*, 129-135.
43. Nagamine, S.; Kurumada, K.; Tanigaki, M. *Adv. Powd. Technol.* **2001**, *12*, 145-156.
44. Guinier, A. *Crystals, Imperfect Crystals, and Amorphous Bodies*, Dover: New York, 1994.
45. Bowlas, C. J.; Bruce, D. W.; Seddon, K. R. *Chem. Commun.* **1996**, 1625-1626.
46. Gordon, C. M.; Holbrey, J. D.; Kennedy, A. R.; Seddon, K. R. J. *Mater. Chem.* **1998**, *8*, 2627-2636.
47. Holbrey, J. D.; Seddon, K. R. *J. Chem. Soc., Dalton Trans.* **1999**, 2133-2139.
48. Hehre, W. J.; Radom, L.; Schleyer, P. V. R.; Pople, J. A. *Ab Initio Molecular Orbital Theory*, John Wiley: New York, 1986.
49. Firestone, M. A.; Shank, M. L.; Sligar, S. G. Bohn, P. W. *J. Am. Chem. Soc.* **1996**, *118*, 9033-9041.
50. Elaiwi, A.; Hitchcock, P. B.; Seddon, K. R.; Srinivasan, N.; Tan, Y.; Welton, T., Zora, J. A. *J. Chem. Soc., Dalton Trans.* **1995**, 3467-3472.
51. Cammarata, L.; Kazarian, S. G.; Salter, P. A.; Welton, T. *Phys. Chem. Chem. Phys.* **2001**, *3*, 5192-5200.

Chapter 42

Extraction of Chlorophenols from Water Using Room Temperature Ionic Liquids

Evangelia Bekou[1], Dionysios D. Dionysiou[1,*], Ru-Ying Qian[2], and Gregory D. Botsaris[2,*]

[1]Department of Civil and Environmental Engineering, University of Cincinnati, Cincinnati, OH 45221–0071
[2]Department of Chemical Engineering, Tufts University, Medford, MA 02155
[*]Correspond author: phone: 513–556–0724; fax 513–556–2599; email: dionysios.d.dionysiou@uc.edu

ABSTRACT. This study deals with extraction of chlorinated phenols from aqueous solutions using two room temperature water immiscible ionic liquids, 1-ethyl-3-methylimidazolium bis(perfluoroethylsulfonyl)imide, [emim]Beti, and 1-butyl-3-methylimidazolium hexafluorophosphate, [bmim]PF_6. Partitioning of phenol, 2-chlorophenol, 2,4-dichlorophenol, 2,4,6-trichlorophenol, 2,3,4,5-tetrachlorophenol, and pentachlorophenol were measured in both aqueous and ionic liquid phases using HPLC. Extraction efficiency was found to be greater when [bmim]PF_6 was used and when the pH of the aqueous solution was at least one unit below the value of the dissociation constant (pK_a). Partitioning, for both ionic liquids, was increased as the number of chlorine atoms in the chlorophenol increased, illustrating the same behavior as 1-octanol-water partition coefficient. The ionic liquid-water distribution ratios for the extraction efficiency of chlorinated phenols using these two ionic liquids was one order of magnitude lower than the corresponding 1-octanol-water partition coefficients. The ionic strength of the aqueous phase had no significant effect on the ionic liquid-water distribution ratios of chlorinated phenols but had a dramatic effect on the solubility of ionic liquids in water. In addition, the ionic liquid-water distribution ratio of chlorinated phenols was not influenced significantly by the number of chlorophenols present in the aqueous phase.

1. INTRODUCTION

Chlorophenols are detected in natural water and impose health risk to humans and hazard to biota. Heavy and acute exposure to chlorophenols can cause death, and small amounts or long-term exposure can cause liver and kidney problems to animals (*1*). The side effects of chlorophenols are perilous and the presence of such compounds in natural water should be nullified. Chlorophenols have been reported as toxic substances under emergency planning and community right-to-know act (EPCRA), according to US EPA (*2*). Herein, chlorophenols were separated from aqueous solutions using ionic liquids as extraction media. The goals of the study were to examine the feasibility of such process using two different room temperature ionic liquids and investigate the dependence of extraction efficiency on pH, ionic strength, and composition of aqueous solution.

The use of ionic liquids in environmental applications is new. Nevertheless, the unique properties of ionic liquids attract the interest for exploring their use as solvents for the removal of contaminants from aqueous streams. Ionic liquids are characterized as neoteric solvents (*3*) and are considered as environmentally friendly compounds (*4*, *5*), mainly because they are non-flammable and non-volatile. These two properties prompt the interest for replacing the well-known volatile organic solvents with ionic liquids (*6*). Other important properties are their wide liquid phase temperature range (around 300 °C), their high thermal stability (up to 400 °C), their miscibility or immiscibility with water depending on their chemical composition, and their good solvent characteristics for many organic, inorganic and polymeric materials (*7*). Some of the applications of ionic liquids as solvents and catalysts are related with Friedel-Craft reactions (*8*, *9*), Diels-Alder reactions (*10*, *11*), and Heck reactions (*12-16*). More detailed information relating their properties and applications can be found in published reviews dealing with ionic liquids (*17-20*).

Ionic liquids have also been investigated as extraction media (*21-30*). Among the ionic liquids studied is 1-ethyl-3-methylimidazolium chloride/aluminum trichloride, first-generation ionic liquid (*18*), which was tested for the extraction of zicronium cluster $[Zr_6(B)Cl_{18}]^{5-}$ from a solid precursor (*21*). This ionic liquid can dissolve many cluster compounds and is redox stable over a wide electrochemical window, but its air and water sensitivity makes the extraction process very delicate, since it has to take place in inert atmosphere or under vacuum. The development of second-generation ionic liquids (*31*) made the extraction process easier and possible even for water-ionic liquid systems, because these ionic liquids are air and water stable.

1-butyl-3-methylimidazolium hexafluorophosphate is a widely used second-generation ionic liquid and has been investigated for liquid-liquid (ionic liquid-water) extractions. Extraction of various substituted benzenes is one of the applications of this ionic liquid (*22*). The distribution ratio of the benzene compounds depends on the pH value of the aqueous solution and is high for neutral or uncharged species, and low for the charged species. pH dependence provides a flexibility in the extraction process, since by changing the pH value of the solution it is possible to extract a compound from water and then back extract it to a different solvent. Another example of partitioning dependence on pH is the preference of thymol blue on aqueous or ionic liquid phase in acidic, neutral, or basic solutions (*23*). Thymol blue is a dye and shows a preference for ionic liquid phase at low pH, when it is present in its neutral form, and for aqueous phase at higher pH, where it is present in its ionized form. In addition, the distribution ratio is an order of magnitude greater when 1-octyl-1-methylimidazolium hexafluorophosphate is used compared with 1-butyl-3-methylimidazolium hexafluorophosphate, indicating the importance of the alkyl chain length of the imidazolium ring to the extraction process. Comparison of 1-octanol-water partition coefficient and ionic liquid-water distribution ratio demonstrates higher values for the 1-octanol-water system by one order of magnitude. This difference in magnitude is ascribed either to the less hydrophobic character of ionic liquid phase or to the fact that ionic liquids have strong interactions, since they are composed from ions (*22*).

Extraction of aromatic compounds from mixtures of aromatics and paraffins to ionic liquids is also promising (*24*). The application of this process can lead to less complex and less expensive separation processes than existing aromatic/paraffin separations. The separation factor indicates that toluene can be extracted efficiently from heptane when toluene is present at low concentration.

Various ionic liquids were examined for their ability to extract metal ions from aqueous solutions using crown ethers, 1-(2-pyridylazo)-2-naphthol (PAN), or 1-(2-thiazolylazo)-2-naphthol (TAN) as extractants (*25-27*). The presence of crown ethers proved necessary, because the distribution ratio is very small without their presence (*25*). The organic extractants PAN and TAN display pH dependence in their extraction efficiency for metal ions (*26*). In general, partitioning depends on the hydrophobicity of the extractant and on the species that are present in the aqueous phase. It has been shown that the configuration of ionic liquids has an effect on the distribution ratio since ionic liquids with anion bis[(trifluoromethyl)sulfonyl]imide show higher distribution ratio than those with hexafluorophosphate anion. In addition, the distribution ratio is larger when the imidazolium cation is not substituted in the second atom of the ring, since the hydrogen bonding between crown ethers and ionic liquid may be higher in this case resulting to higher partitioning (*25*). The problem is that highly acidic aqueous solution can decompose the hexafluorophosphate anion and can significantly increases the solubility of ionic liquid in the aqueous phase as well as the water content in the ionic liquid phase. On the other hand, low acid content in the aqueous phase decreases metal ion distribution ratio and water content in the ionic liquid phase. In addition, highly hydrated salts, such as $Al(NO_3)_3$ and $LiNO_3$, salt out the ionic liquid ions and the crown ethers (*27*). The increase of metal ion transfer in ionic liquid phase has been reported to cause increase in the solubility of ionic liquid in water (*28*). The mechanism of metal transfer into ionic liquids was studied with extended X-ray absorption fine structure (EXAFS) measurements (*29*).

Ionic liquids with various modifications of the imidazolium cations have been investigated for extraction of metal ions (*30*). The scope of these task specific ionic liquids was to examine simpler and more efficient metal ion extraction systems, since the conventional extractants (i.e., crown ethers) produced a complicated system for chemical analysis. The imidazolium cation modified by the addition of thiourea, thioether and urea into the alkyl chain of the cation, gives to the ionic liquid a double role of a hydrophobic medium and an extractant (30). The more efficient additions were those with thiourea or urea derivative with long chain; they were inserting sulfur or oxygen atoms, respectively, into the molecule.

The anion of [emim]Beti contains both sulfur and oxygen atoms in contrast to [bmim]PF_6 which has neither. The different molecular structure of these two ionic liquids, and in combination with the pH and the ionic strength of the aqueous solution, may affect the ionic liquid-water distribution ratio, as it is illustrated in this study.

2. EXPERIMENTAL SECTION

Materials: 1-ethyl-3-methylimidazolium bis(perfluoroethylsulfonyl)imide, [emim] Beti (98% purity), was purchased from Covalent Associates, Inc. (Woburn, Massachusetts), while 1-butyl-3-methylimidazolium hexafluorophosphate, [bmim]PF_6 (97% purity) was obtained from Sachem, Inc. (Austin, Texas). Both ionic liquids were used as received. All chlorophenols were obtained from Aldrich and their purities were between 97 and 99%. Sodium nitrate was purchased from Fisher and used as received. The aqueous solutions were prepared with double distilled deionized water (12 MΩ) and the pH was adjusted with HNO_3 or NaOH whenever it was necessary.

Extraction procedure: Two ml of ionic liquid ([bmim]PF_6 or [emim]Beti) were placed in a 50 ml polypropylene tube and 10 ml of aqueous chlorophenol solution were added. The ionic liquid and water form two different phases, with the ionic liquid at the bottom of the tube and the aqueous solution at the top. The density of [emim]Beti is 1.57 g/cm^3 (*32*) while that of [bmim]PF_6 is 1.37 g/cm^3 (*33*).

The two phases were mixed vigorously for ten minutes in order to achieve equilibrium partitioning of chlorophenol between the two phases. This time was determined in preliminary experiments (*34*). After mixing, two distinct phases were formed, with ionic liquid being the bottom phase due to its higher density. However, small liquid droplets of the other phase could be seen in each phase. These liquid droplets were subsequently removed by centrifugation of the sample at 3578 g for 30 min. Clear solutions with no visible droplets of ionic liquid in aqueous phase or water in ionic liquid phase were obtained after centrifugation.

The aqueous phase was analyzed using High Pressure Liquid Chromatography (HPLC), for the quantification of chlorophenols and ionic liquid, and by Total Organic Carbon (TOC) analyzer for the measurement of organic carbon in the sample. It has been previously published that the solubility of [bmim]PF_6 is approximately 19 g/l (*35*), which is in agreement with the results obtained using the HPLC method developed in this work. The upper phase was diluted with double distilled deionized water (12 MΩ) (1ml of the aqueous phase was placed into a 100 ml volumetric flask and deionized water was added), before analyzing a sample for the ionic liquid concentration using HPLC or TOC. The dilution contributes to better results, because the calibration curve for the ionic liquid in the HPLC is linear up to 0.5 g/l and the amount of organic carbon in the sample must be below 0.5 g/l for accuracy in TOC analysis. The ionic liquid phase was analyzed for its chlorophenol content using HPLC. The ionic liquid phase was also diluted (0.5 ml of the ionic liquid phase was diluted with 5 ml of acetonitrile), because ionic liquids are very viscous and may block the flow if injected undiluted into HPLC. All experiments and analyses were performed in triplicate. The average error in the analysis was 2 %.

Analysis: An Agilent 1100 series HPLC equipped with a reverse phase amide column was used for the analysis. The mobile phase was acetonitrile and 0.01 N H_2SO_4 aqueous solution (the ratio of the solvents was changing from 60:40 to 10:90 depending on the IL and the chlorophenol used). The absorbance wavelength was set at 212 nm for ionic liquids and at 197 nm or 212 nm for chlorophenols. The correlation coefficient of the calibration curves was 0.996 for ILs, 0.999 for chlorophenols, and 0.995 for phenol. Dilution of ionic liquid phase with acetonitrile was necessary because it reduces the viscosity of ionic liquid phase and enhances HPLC performance. Using these new HPLC methods, chlorophenol partitioning was quantified in both phases.

3. RESULTS AND DISCUSSION

3.1 Single solute extraction using [emim]Beti

Phenol, 2-chlorophenol, 2,4-dichlorophenol, 2,4,6-trichlorophenol, 2,3,4,5-tetra-chlorophenol and pentachlorophenol were extracted from their aqueous solution using [emim]Beti. Their distribution ratios are compared with 1-octanol-water partition coefficient (*36*) and are illustrated in Figure 1.

The pH of all the initial aqueous chlorophenol solutions was at least one unit lower than the pK_a values of the compounds (*37*), and the dominant species in the solution were the neutral species. Most of the initial solutions had a pH value close to 5.5, which is at least one unit lower than the pK_a value of the phenol (9.99), 2-chlorophenol (8.55), and 2,4-dichlorophenol (7.85). The pH of the aqueous solution containing 2,4,6-trichlorophenol, or pentachlorophenol was adjusted at 3.0 with H_2SO_4, because the pK_a values of these two chlorophenols are 6.23, and 4.75, respectively. All D_{bw} values were calculated based on a volume ratio 1:5; 2 ml of ionic liquid phase and 10 ml of aqueous chlorophenol solution. All chlorophenols were used at initial concentration in water of 50 mg/l, except for pentachlorophenol, which was used at an initial concentration of 4 mg/l due to its lower solubility in water. After the extraction process, the detection of pentachlorophenol in the aqueous phase was not possible using this HPLC method, when a volume ratio 1:5 was used. The volume ratio was altered to 1:10, 1:20, 1:30 and 1:40 of [emim]Beti to aqueous phase, and the distribution ratio for these conditions was calculated at 1468 (std. dev. 308).

The trends of chlorophenol partitioning are similar for water-1-octanol and water-[emim]Beti extraction systems. Specifically, the [emim]Beti-water distribution ratio, D_{bw}, increases with the increase of chlorine atoms in the solute, like 1-octanol-water partition coefficient, P_{ow}. However, P_{ow} is an order of magnitude higher than D_{bw}, indicating that [emim]Beti is not as hydrophobic as 1-octanol as also discussed in previous studies (22).

3.2 Single solute extraction using [bmim]PF_6

Phenol, 2-chlorophenol, 2,4-dichlorophenol, 2,4,6-trichlorophenol, and 2,3,4,5-tetrachlorophenol were extracted as single solutes from aqueous solution using [bmim]PF_6. The [bmim]PF_6 distribution ratio, D_{pw}, of these experiments was compared with the P_{ow} and the results are illustrated in Figure 2. It is evident that the [bmim]PF_6-water extraction system demonstrates similar behavior as the 1-octanol-water and [emim]Beti-water extraction systems; increasing D_{pw} for increasing number of chlorine atoms of the chlorophenols. Furthermore, P_{ow} values are higher than D_{pw} values by one order of magnitude.

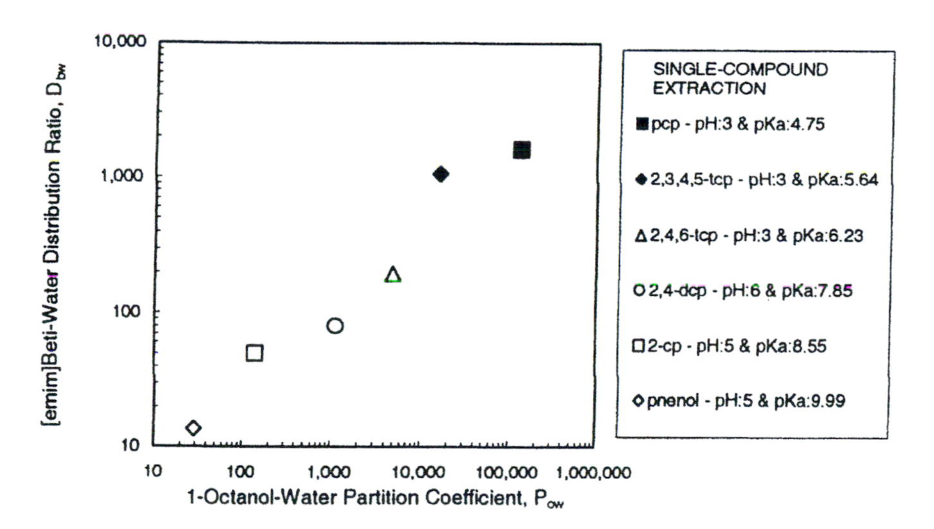

Figure 1: Extraction of chlorophenols using [emim]Beti

The detection of pentachlorophenol in the aqueous phase after the extraction process was again not possible with HPLC methods used in this study for volume ratio 1:5; indicating that pentachlorophenol may be completely removed from the aqueous phase with this process by both ionic liquids. It should be noted that the detection limit of the method was 0.001 mg/L. It was noticed that pH adjustment of the initial aqueous chlorophenol solution was not necessary for the extraction process with [bmim]PF_6. It was observed that after extraction with [bmim]PF_6, the aqueous phase had a pH around 3.0. On the other hand, the pH of the aqueous phase after extraction with [emim]Beti was around 5.0. This means that the pH of the aqueous phase after the extraction can be affected by the presence of ionic liquid and perhaps any impurities they contain. The solubilities of [bmim]PF_6 and [emim]Beti were measured with HPLC and were close to 20 g/l and 5 g/l, respectively. The solubility of [bmim]PF_6 as determined using the HPLC method described in this study is close to a published value of 18.8 g/L (*35*).

3.3 Extraction using solution with multiple chlorophenols

The same extraction process was repeated with an initial aqueous solution containing 50 mg/l each of 2-chlorophenol, 2,4-dichlorophenol, and 2,4,6-trichlorophenol using [emim]Beti or [bmim]PF_6. The objective of these experiments was to examine the influence of the aqueous composition of multi-component mixture on the extraction efficiency. The latter is defined here as the ratio of the amount of chlorophenol transferred into ionic liquid phase versus the maximum theoretical amount that can be transferred. The extraction efficiencies of the multi-compound solution were

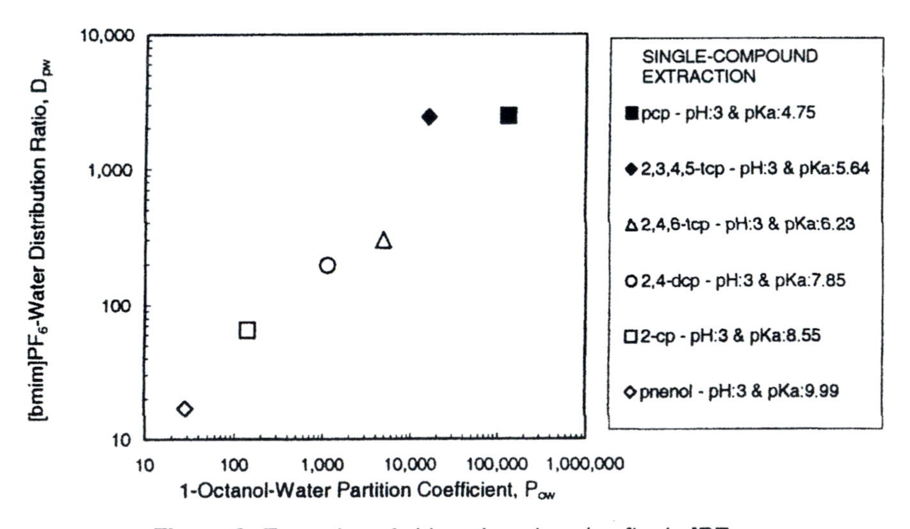

Figure 2: Extraction of chlorophenols using [bmim]PF_6

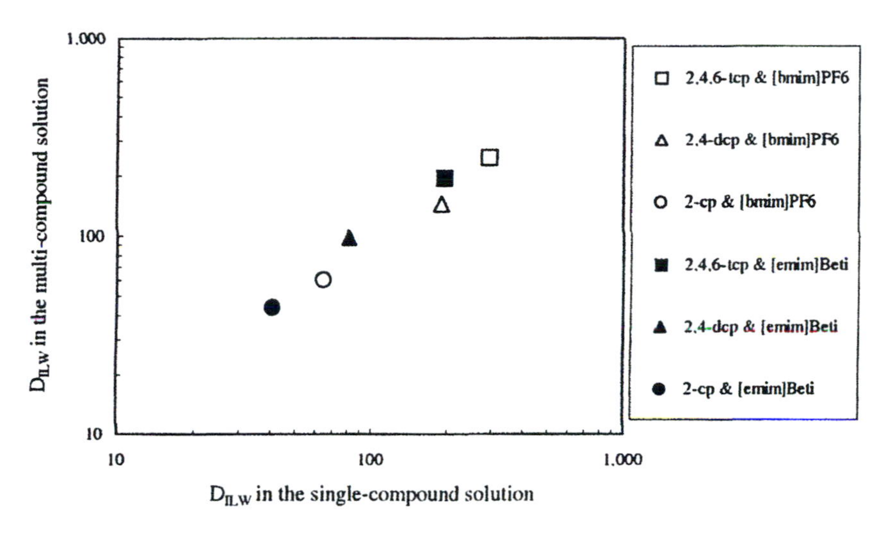

Figure 3: Extraction of aqueous solution with three chlorophenols

slightly lower than those of the single solute systems: 87% vs. 91% for 2-cp, 89% vs. 94% for 2,4-dcp, and 90% vs. 98% for 2,4,6-tcp. As shown in Figure 3, the decrease of the extraction efficiency is not significant on the ionic liquid-water distribution ratio, D_{ILW}.

A similar experiment was performed with a different initial solution containing 4-chlorophenol, 2,4-dichlorophenol, and 2,4,6-trichlorophenol. The extraction efficiencies of the multi-component mixture and single-solute solutions were the same. The corresponding values were 88, 94 and 98 % for 4-cp, 2,4-dcp, and 2,4,6-tcp, respectively. The D_{ILW} was not changed in the multi-component and single-component solutions.

3.4 Effect of acidic and basic solution

The influence of the pH on the extraction was best demonstrated by the extraction of a solution with five different chlorophenols with varying pK_a values. Specifically, a solution with 2-chlorophenol, 4-chlorophenol, 2,4-dichlorophenol, 2,4,6-trichlorophenol, and pentachlorophenol was used for the extraction process using [emim]Beti or [bmim]PF_6. Eight experiments were performed with this solution. The following parameters were varied: (a) the pH of the initial solution (3.0 or 5.0), and (b) the volume ratio of the two phases (1:5 or 1:30). The pH value affects the presence of the neutral and ionized species, while the volume ratio affects the final concentration of pentachlorophenol in the aqueous phase and thus its detection with HPLC. The initial pH solution was adjusted at 3.0 with H_2SO_4 whenever it was necessary. The initial concentration for each of the chlorophenols was 50 mg/l, except for pentachlorophenol, which was 4 mg/l. The results of this series of experiments are presented in Figure 4. It is apparent from Figure 4 that the highest effect of the pH is on pentachlorophenol, which has the lowest pK_a value of these compounds (4.75). The distribution ratio remained intact for 2-chlorophenol, 4-chlorophenol and 2,4-dichlorophenol with the change of pH from 5.0 to 3.0. The distribution ratio was affected for 2,4,6-trichlorophenol and pentachlorophenol, since their pK_a values are 6.23 and 4.75, respectively. These results corroborate that the transferred species to ionic liquids is the neutral and not the ionized.

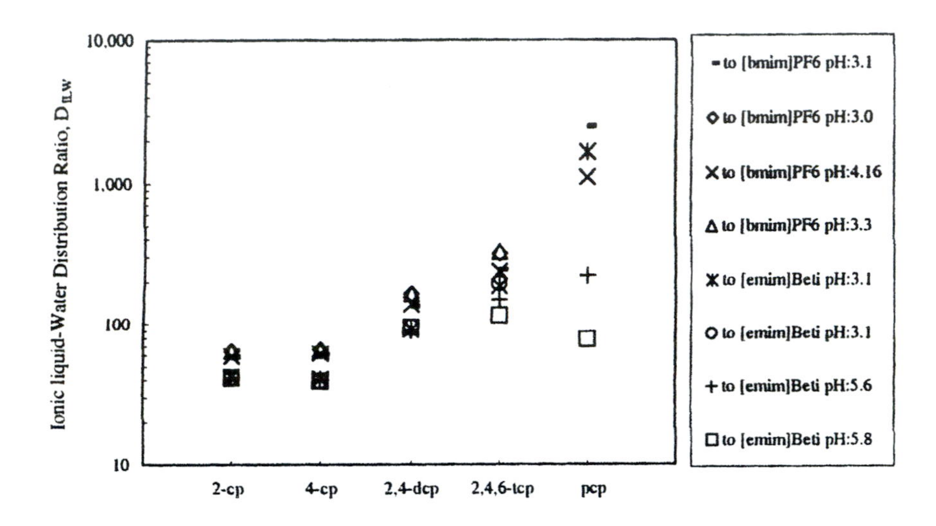

Figure 4: Extraction of aqueous solution with five chlorophenols

The influence of high pH was demonstrated with the extraction of a solution that contained 50 mg/l phenol, 2-chlorophenol, or 2,4-dichlorophenol. The results are presented in Table-I. The pH of the initial solution was adjusted at 11.0 with NaOH and extracted using [bmim]PF_6 or [emim]Beti. The D_{pw} was calculated to be 0.7 and 0.2 for phenol and 2,4-dichlorophenol, respectively. The D_{bw} was calculated at 1.0, 0.1, and 0.05 for phenol, 2-chlorophenol, and 2,4-dichlorophenol, respectively.

Further work on the pH effect of initial aqueous solution on the distribution ratio was performed for the system 2-chlorophenol/[bmim]PF_6. Five ml of [bmim]PF_6 and 25 ml of aqueous solution with high 2-cp concentration of 400 mg/l were mixed well in a 60 ml bottle by a magnetic stirrer at 800 rpm for 4 minutes. A stirring time longer than 4 minutes was found to have no effect on extraction in the range of 4-150 minutes for the case of "neutral" 2-cp solution. The acidity of the solution was adjusted as above by adding dilute H_2SO_4 solution while the alkalinity was adjusted by dilute NaOH solution. The chlorophenol concentration in the aqueous solution was determined by the well-established 4-aminoantipyrine colorimetric method (*37*). HP8452A diode array UV/Vis spectrometer was used. The results are listed in Table II.

Table-I: Influence of pH on the distribution ratio of ionic liquids

Characteristics	Extraction with [emim]Beti			Extraction with [bmim]PF_6		
	ph	2-cp	2,4-dcp	ph	2-cp	2,4-dcp
D_{ILW}: Neutral	14	46	105	17	70	180
pH: 3	13	46	106	17	68	172
pH:11	**1.0**	**0.1**	**0.05**	**0.7**	-	**0.2**
Efficiency: Neutral	73	90	95	76	92	98
pH: 3	68	92	94	76	92	97
pH:11	**17**	**2.0**	**0.9**	**11**	-	**3**

Table-II: The effect of acid and alkaline in aqueous 2-cp solution on the distribution ratio of [bmim]PF_6

Concentration of H_2SO_4 or NaOH	H_2SO_4 0.0114N	H_2SO_4 0.00114N	None	NaOH 0.0032N	NaOH 0.0096N	NaOH 0.048N	NaOH 0.48N
Distribution Ratio	46.1	49.7	50.7	1.16	0.061	0.015*	0.038*

* 10 ml [bmim]PF_6 and 5 ml aqueous 2-cp solution were used.

The pH-dependence of thymol blue in ionic liquids was reported *(23)*. In this work the dramatic decrease of distribution ratio to a very low value in alkaline solution is very significant since it could provide an effective technology for back extraction of chlorophenol to an aqueous phase and regeneration of the ionic liquid. In such a technology, the volume of the alkaline back-extract phase should be less than that of ionic liquid in order to achieve concentration of chlorophenol. Thus, 10 ml [bmim]PF_6 and 5 ml aqueous solution were used for the determination of distribution ratio at NaOH concentration of 0.048 or 0.48 N.

3.5 Effect of ionic strength

The effect of the ionic strength was examined for three different solutions that contained 50 mg/l phenol, 2-chlorophenol or 2,4-dichlorophenol. The extraction was performed using [bmim]PF_6 or [emim]Beti and the ionic strength was varied form 0 to 500 mM $NaNO_3$. The results are illustrated in Figure 5. It is obvious from Figure 5, that the ionic strength does not affect the distribution ratio. It only seems to be a small decrease in the distribution ratio for ionic strength 500 mM, when 2,4-dichlorophenol is extracted using [bmim]PF_6. Even though the distribution ratio is not affected by the ionic strength, the solubility of ionic liquids in water increases significantly, which is in agreement with previously published results (*28*). The solubility of [bmim]PF_6 and [emim]Beti in water increases with an increase in the ionic strength. The solubilities of [bmim]PF_6 and [emim]Beti increase from 20 g/l and 4.2 g/l at zero ionic strength to 25 g/l and 5.3 g/l at 500 mM $NaNO_3$, respectively.

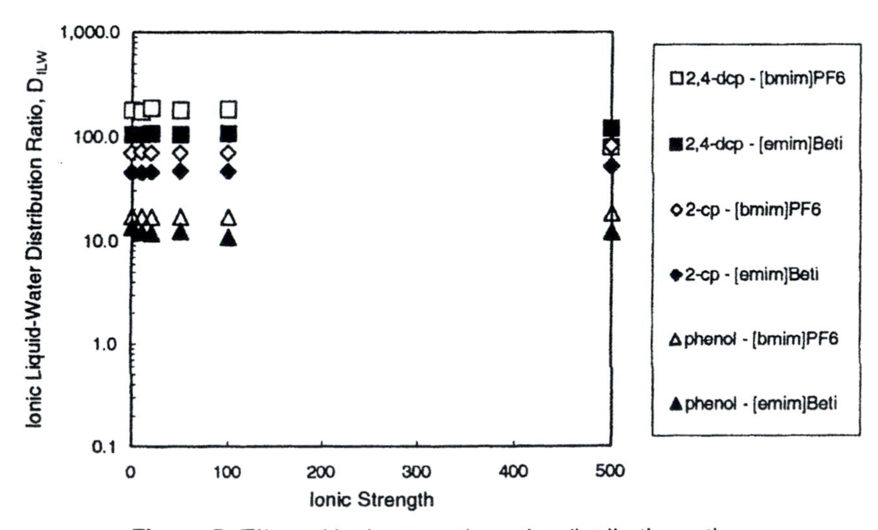

Figure 5: Effect of ionic strength on the distribution ratio

3.6 Comparison of ionic liquid-water distribution ratios

Comparison of D_{pw} with D_{bw} shows that D_{pw} is slightly higher than D_{bw}, (Figure 6) and the correlation is adequately high, 0.983. This is apparent for individual extractions with single solute systems as well as for multi-component systems of chlorophenols (Figure 4). The distribution ratio D_{pw} is higher for all the chlorophenols comparing to that of D_{bw}. The reason for this difference in values is not attributed to hydrophobicity because [emim]Beti is more hydrophobic than

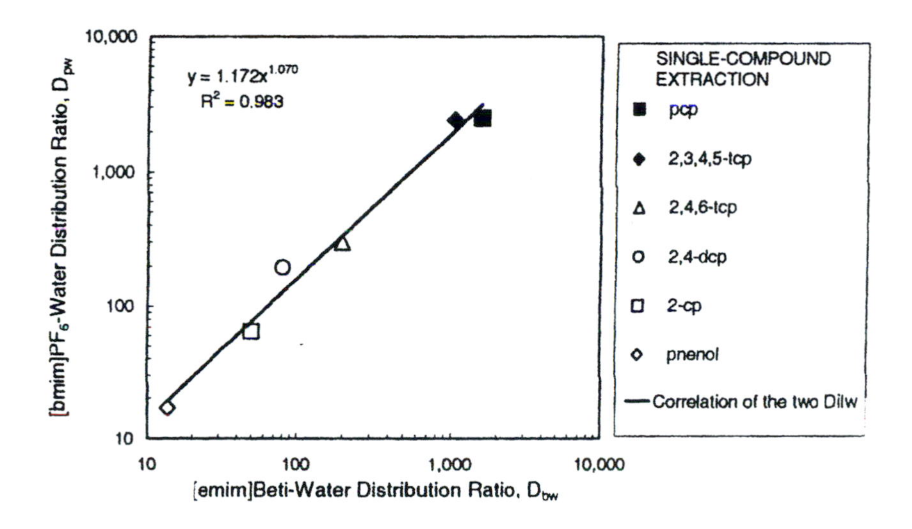

Figure 6: Correlation of ionic liquids distribution ratios

[bmim]PF_6; their solubility in water is 5 and 19 g/l, respectively. While further fundamental studies are necessary to examine the mechanism of chlorophenol partitioning in these two ionic liquids, it is believed that chlorophenol chemical properties, ionic liquid properties, and aqueous solution matrix characteristics affect the distribution ratio of different chlorophenols. Such properties of ionic liquids include the molecular structure, size of ions, water content, and tendency to form hydrogen bonding (*38-40*). Further studies are necessary to provide more insights on the detailed mechanism of chlorophenol partitioning in ionic liquids.

Comparison of ionic liquid-water distribution ratio with 1-octanol-water partition coefficient is illustrated in Figure 7. Correlation for both systems is equitably high for both ionic liquids, 0.950 for P_{ow}-D_{bw} and 0.938 for P_{ow}-D_{pw}.

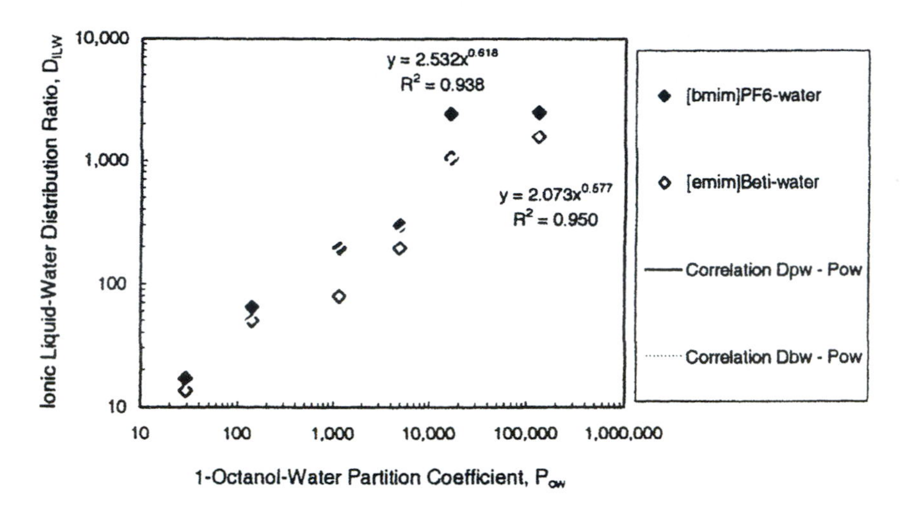

Figure 7: Correlation of D_{bw} - P_{ow} and D_{pw} - P_{ow}

4. CONCLUSIONS

The objective of this study was to explore the potential of ionic liquids as extractant systems for the purification of water contaminated with chlorinated aromatics. Several conventional organic solvents are currently used and considered good choices. For example, 1-octanol is considered as a good solvent for extraction for many compounds; it is immiscible with water (540 mg/l solubility) and it has low viscosity (10 cP at 10 °C) (*41*). In addition, the hydroxyl group can act as donor as well as acceptor for hydrogen bonding. However, 1-octanol is volatile and has low resistance to oxidation. The latter is important in the second step of this process, which is applied to treat the chlorinated aromatics *in situ* in the ionic liquids (*42*). The use of ionic liquids has the potential to eliminate these problems, because they are non-volatile and are resistant to oxidation/reduction (i.e., based on their wide electrochemical window). In fact, chemical oxidation, UV radiation, and Advanced Oxidation Technologies (AOTs) are currently employed in our laboratories after the extraction process for the *in-situ* destruction of organic contaminants in the ionic liquid extractant phase. This approach can simultaneously achieve regeneration of the ionic liquid phase, which is also a subject of current investigation in our group. In this second step, the chlorinated phenols and other contaminants are treated at higher concentrations, and thus at higher photonic efficiency (i.e., assuming first order photodegradation reaction), and using significantly lower volume of solvent (i.e., compared to the initial contaminated aqueous solution). This may have potential to enhance process efficiency, considering regeneration of ionic liquid is feasible.

In this study, we showed that chlorophenols could be extracted to water immiscible room temperature ionic liquids. The ionic liquid-water distribution ratios of various chlorophenols using [bmim]PF_6 and [emim]beti are approximately an order of magnitude lower than those of 1-octanol-water partition coefficients but are high enough from an engineering perspective. This could give further motivation to investigate the extraction of contaminants such as pesticides or polycyclic aromatic hydrocarbons (PAHs) that have low solubility in water, either from water or soil. However, of major environmental and engineering concern, is the extremely high solubility of ionic liquids in water. This could be disastrous if ionic liquids are proved to have toxicity concerns. In addition, ionic liquids are currently much more expensive, and in many cases more viscous, than conventional organic solvents. For example, the viscosity of 1-octanol is only 10 cP at 10 °C (*32*) and is much lower than those of [emim]Beti (61 cP at 26 °C) (*32*) and [bmim]PF_6 (312 cP at 30 °C) (*43*). Nevertheless, ionic liquid research is currently very intense and the preparation of much more water immiscible ionic liquids could be a matter of time. If the toxicity of ionic liquids proves to be comparable to common organic solvents, then this liquid-liquid extraction process may be an option from an engineering point of view. In addition, engineering designs can be constructed that confine the ionic liquids and prevent their release into the aqueous phase. Certainly, these areas will be the subjects of future investigations.

Acknowledgments

The authors are grateful to National Science Foundation for funding this project (Grant CTS-0086725). D. D. Dionysiou is also grateful to the NOAA/UNH Cooperative Institute for Coastal and Estuarine Environmental Technology (CICEET) for providing partial financial support for the experiments dealing with pentachlorophenol.

Symbols used:

K_a	Dissociation Constant
P_{ow}	1-Octanol-Water partition coefficient
D_{ILW}	Ionic Liquid-Water distribution ratio
D_{bw}	[emim]Beti-Water distribution ratio
D_{pw}	[bmim]PF_6-Water distribution ratio

Literature Cited

(1) a) http://www.atsdr.cdc.gov/toxprofiles/phs107.html, b) Davi, M. L., and Gnudi, F. 1999. "TECHNICAL NOTE: Phenolic compounds in surface water". *Wat. Res.*, 33(14) 3213-3219.

(2) a) http://www.epa.gov/tri;
b) http://www.epa.gov/region08/toxics_poisons/epcra/epcra.html

(3) Seddon, K. R. 1996. "Room-Temperature Ionic Liquids: Neoteric Solvents for Synthesis and for Clean Catalysis". *Kinet. Catal.*, 37, 693-697.

(4) Earle, M. J., and Seddon, K. R. 2000. "Ionic liquids. Green solvents for the future". *Pure Appl. Chem.*, 72 (7), 1391-1398.

(5) Gordon, C. M., and McCluskey, A. 1999. "Ionic liquids: a convenient solvent for environmentally friendly allylation reactions with tetraallylstannane". *Chem. Commun.*, 1431-2.

(6) Blanchard, L. A., and Brennecke, J. F. 2001. "Recovery of Organic Products from Ionic Liquids Using Supercritical Carbon Dioxide". *Ind. Eng. Chem. Res.*, 40(11), 287-292.

(7) Seddon, K. R. 1997. "Ionic Liquids for Clean Technology". *J. Chem. Tech. Biotechnol.*, 68, 351-356.

(8) Adams, C. J., Earle, M. J., Roberts, G., and Seddon, K. R. 1998. "Friedel-Crafts reactions in room temperature ionic liquids". *Chem. Commun.*, 2097-2098.

(9) Valkenberg, M. H., DeCastro, C., and Holderich, W. F. 2001. "Friedel-Crafts acylation of aromatics catalyzed by supported ionic liquids". *Applied Catalysis A: General*, 215(1-2), 185-190.

(10) Fischer, T., Sethi, A., Welton, T., and Woolf, J. 1999. "Diels-Alder Reactions in Room-Temperature Ionic Liquids". *Tetrahedron Letters*, 40, 793-796.

(11) Earle, M. J., McCormac, P. B., and Seddon, K. R. 1998. "Diels-Alder reactions in ionic liquids. A safe recyclable alternative to lithium perchlorate-diethyl ether mixtures". *Green Chemistry*,

(12) Carmichael, A. J., Earle, M. J., Holbrey, J. D., McCormac, P. B., and Seddon, K. R. 1999. "The Heck Reaction in Ionic Liquids: A Multiphasic Catalyst System". *Organic Letters*, 1, (7), 997-1000.

(13) Xu, L., Chen, W., and Xiao, J. 2000. "Heck Reaction in Ionic Liquids and the in Situ Identification of *N*-Heterocyclic Carbene Complexes of Palladium". *Organometallics,* 19, 1123-1127.

(14) Calo, V., Nacci, A., Monopoli, A., Lopez, L., and DiCosmo, A. 2001. "Heck reaction of β-substituted acrylates in ionic liquids catalyzed by a Pd-benzothiazole carbene complex". *Tetrahedron*, 57, 6071-6077.

(15) Hagiwara, H., Shimizu, Y., Hoshi, T., Suzuki, T., Ando, M., Ohkubo, K., and Yokoyama, C. 2001. "Heterogeneous Heck reaction catalyzed by Pd/C in ionic liquid". *Tetrahedron Letters*, 42, 4349-4351.

(16) Nair, D., Scarpello, J. T., White, L. S., DosSantos, L. M. F., Vankelecom, I. F. J., and Livingston, A. G. 2001. "Semi-continuous nanofiltration-coupled Heck reactions as a new approach to improve productivity of homogeneous catalysts". *Tetrahedron Letters*, 42, 8219-8222.

(17) Hussey, C. L. 1983. "Room Temperature Molten Salt Systems". *Advances in Molten Salt Chemistry,* 5, 185-230.

(18) Holbrey, J. D., and Seddon, K. R. 1999."Ionic Liquids". *Clean Products and Processes*, 1, 223-236.

(19) Welton, T. 1999. "Room-Temperature Ionic Liquids. Solvents for Synthesis and Catalysis". *Chemical Review*, 99, 2071-2083.

(20) Wasserscheid, P., and Keim, W. 2000. "Ionic Liquids-New "Solutions" for Transition Metal Catalysis". *Angew. Chem. Int. Ed.*, 39, 3772-3789.

(21) Tian, Y., and Hughbanks, T. 1995. "Extraction and Isolation of the $[(Zr_6B)Cl_{18}]^{5-}$ Cluster from a Solid State Precursor". *Inorg. Chem.*, 34, 6250-6254.
(22) Huddleston, J. G., Willauer, H. D., Swatloski, R. P., Visser, A. E., and Rogers, R. D. 1998. "Room temperature ionic liquids as novel media for 'clean' liquid-liquid extraction". *Chem. Commun.*, 1765-1766.
(23) Visser, A. E., Swatloski, R. P., and Rogers, R. D. 1999. "pH-Dependent partitioning in room temperature ionic liquids provides a link to traditional solvent extraction behavior". *Green Chemistry*, 1-4.
(24) Selvan, M. S., McKinley, M. D., Dubois, R. H., and Atwood, J. L., 2000. "Liquid-Liquid Equilibria for Toluene+Heptane+1-Ethyl-3-methylimidazolium Triiodide and Toluene+Heptane+1-Butyl-3-methylimidazolium Triiodide". *J. Chem. Eng. Data*, 45, 841-845.
(25) Dai, S., Ju, Y. H., and Barnes, C. E. 1999. "Solvent extraction of strontium nitrate by a crown ether using room-temperature ionic liquids". *J. Chem. Soc., Dalton Trans.*, 1201-1202.
(26) Visser, A. E., Swatloski, R. P, Griffin, S. T., Hartman, D. H., and Rogers, R. D. 2001. "Liquid/liquid extraction of metal ions in room temperature ionic liquids". *Separation Science and Technology*, 36(5&6) 785-804.
(27) Visser, A. E., Swatloski, R. P., Reichert, W. M., Griffin, S. T., and Rogers, R. D. 2000. "Traditional Extractants in Nontraditional Solvents: Groups 1 and 2 Extraction by Crown Ethers in Room-Temperature Ionic Liquids". *Ind. Eng. Chem. Res.*, 39, 3596-3604.
(28) Dietz, M.L., and Dzielawa, J.A. 2001. "Ion-exchange as a mode of cation transfer into room-temperature ionic liquids containing crown ethers: implications for the 'greenness' of ionic liquids as diluents in liquid-liquid extraction," *Chem. Commun.*, 2124-25.
(29) Jensen, M.P., Dzielawa, J.A., Rickert, P., and Dietz, M.L. 2002. "EXAFS Investigations of the Mechanism of Facilitated Ion Transfer into a Room-Temperature Ionic Liquid," *J. Am. Chem. Soc.*, 124, 10664-5.
(30) Visser, A. E., Swatloski, R. P, Reichert, W. M., Sheff, S., Wierzbicki, A, Davis, J. H. D, and Rogers, R. D. 2001. "Task-specific ionic liquids for the extraction of metal ions from aqueous solutions". *Chem. Commun.*, 135-136.
(31) Wilkes J. S., and Zaworotko M. J., 1992, "Air and Water Stable 1-ethyl-3methylimidazolium Based Ionic Liquids". *J. Chem. Soc., Chem. Commun.*, 965-967.
(32) McEwen, A., Ngo, H. L., Le Compte, K., and Goldman, J. L. 1999. "Electrochemical Properties of Imidazolium Salt Electrolytes for Electrochemical Capacitor Applications". *Journal of The Electrochemical Society*, 146, (5), 1687-1695.
(33) Dullius, J. E. L., Suarez, P. A. Z., Einloft, S., DeSouza, R. F., Dupont, J., Fischer, J., and DeCian, A. 1998. "Selective Catalytic Hydrodimerization of 1,3-Butadiene by Palladium Compounds Dissolved in Ionic Liquids". *Organometallics*, 17, 815-819.

(34) Bekou, E. "*Extraction of Chlorinated Phenols from Water Using Water Immiscible Room Temperature Ionic Liquids*", 2003, M.S. Thesis, University of Cincinnati, Cincinnati, Ohio.
(35) Chun, S., Dzyuba, S. V., and Bartsch, R. A. 2001. "Influence of Structural Variation in Room-Temperature Ionic Liquids on the Selectivity and Efficiency of Competitive Alkali Metal Salt Extraction by a Crown Ether". *Analytical Chemistry*, 73(15), 3737-3741.
(36) Hansch C, Leo A., and Hoekman D., 1995, "Exploring QSAR: Hydrophobic, Electronic, and Steric Constants", ACS Professional Reference Book, American Chemical Society, Washington, DC, USA.
(37) Eaton A. D., Clesceri L. S., Greenberg A. E. (eds.), "Standard Method for the Examination of Water and Wastewater" 19th ed., American Public Health Association: Washington D. C., 1995, pp. 5-36 to 5-39.
(38) Law, G., Watson, P. R., Carmichael, A. J., and Seddon, K. R. 2001. "Molecular composition and orientation at the surface of room-temperature ionic liquids: Effect of molecular structure". *Physical Chemistry Chemical Physics, a Journal of European Physical*, 3(14), 2879-2885.
(39) Seddon, K. R., Stark, A, and Torres, M. J. 2000. "Influence of chloride, water, and organic solvents on the physical properties of ionic liquids". *Pure Appl. Chem.*, 72(12), 2275-2287.
(40) Huang, J. F., Chen, P. Y., Sun, I. W., and Wang, S. P. 2001. "NMR evidence of hydrogen bonding in 1-ethyl-3-methylimidazolium-tetrafluoroborate room temperature ionic liquid". *Inorganica Chimica Acta*, 320, 7-11.
(41) Dean J. A., 1992, "Lange's Handbook of Chemistry. 14th edition", McGraw Hill, Inc., New York, USA.
(42) Yang, Q. and Dionysiou, D. D., 2002, "Use of Water Immiscible Room Temperature Ionic Liquids for Liquid-Liquid Extraction and *In Situ* Photodegradation of Chlorinated Aromatics". *Journal of Chemical Society Chemical Communications* - submitted for publication.
(43) Wu B., Reddy R. G., and Rogers R. D., 2001, "Novel Ionic Liquid Thermal Storage for Solar Thermal Electric Power Systems". *Proceedings of Solar Forum 2001. Solar Energy: The Power to Choose.* April 21-25, 2001, Washington, DC.

Indices

Author Index

Abbott, Andrew P., 439
Abu-Omar, Mahdi M., 277
Acevedo, Orlando, 174
Aki, Sudhir N. V. K., 110
AlNashef, I. M., 509
Anthony, Jennifer L., 110
Aresta, M., 93
Baker, Gary A., 212
Baker, Sheila N., 212
Bartsch, Richard A., 289
Behar, D., 397
Bekou, Evangelia, 544
Benton, Michael G., 468
Bösmann, Andreas, 57
Botsaris, Gregory D., 544
Bowron, D. T., 151
Bradaric, Christine J., 41
Brazel, Christopher S., 468
Brennecke, Joan F., 110
Brinson, Robert G., 370
Brooks, Claudine A., 410
Butke, Ryan L., 370
Capper, Glen, 439
Carmichael, Adrian J., 14
Carrié, Daniel, 239
Crosthwaite, Jacob M., 110
Davies, David L., 439
Davis, James H. Jr., 100
De Diego, Teresa, 239
De Long, Hugh C., 478
Deetlefs, Maggel, 14
Dietz, Mark L., 526
Dionysiou, Dionysios D., 544
Doherty, Andrew P., 410
Downard, Andrew, 41
Dreisinger, David, 495
Dunkin, Ian R., 357
Durazo, Armando, 277
Dzielawa, Julie A., 526
Dzyuba, Sergei V., 289
Earle, Martyn J., 14
El Abedin, Sherif Zein, 453
Endres, Frank, 453
Eßer, Jochen, 57
Evanseck, Jeffrey D., 174
Firestone, Millicent A., 526
Forsyth, Stewart A., 264
Fox, Phillip A., 100
Fröhlich, Ute, 14
Germaneau, Romain F., 323
Gordon, Charles M., 357
Grasa, Gabriela A., 323
Grodkowski, J., 397
Güveli, Tatyana, 323
Hardacre, C., 151
Heintz, Andreas, 134
Hert, Daniel G., 110
Holbrey, John D., 2, 121, 151, 300
Hult, Karl, 225
Iborra, José L., 239
Itoh, Toshiyuki, 251
Jensen, Mark P., 526
Jess, Andreas, 57
Jones, Paul B., 370
Kaftzik, Nicole, 206
Kashiwagi, Masaya, 251
Kazlauskas, Romas J., 225
Kennedy, Christine, 41

Kissling, Rebecca M., 323
Klingshirn, Marc A., 300
Kragl, Udo, 206
Kula, Maria-Regina, 206
Lagunas, M. Cristina, 421
Lau, R. Madeira, 192
Lehmann, Jochen K., 134
Lozano, Pedro, 239
Lu, Jianming, 495
MacFarlane, Douglas R., 264
Maginn, Edward J., 110, 162
Matthews, M. A., 509
McCleskey, T. Mark, 212
McLean, Andrew J., 357
McMath, S. E. J., 151
Morrow, Timothy I., 162
Muldoon, Mark J., 357
Munro, Helen, 439
Navarro-Fernandez, Oscar, 323
Neta, P., 397
Neumann, Sebastian, 206
Nieuwenhuyzen, M., 151
Nishimura, Yoshihito, 251
Nolan, Steven P., 323
Onaka, Makoto, 251
Owens, Gregory S., 277
P'Pool, Steven J., 300
Pagni, Richard M., 344
Park, Seongsoon, 225
Pasareanu, Marie-Christiane, 323
Pitner, William R., 421
Qian, Ru-Ying, 544
Rahman, Mustafizur, 468
Rakita, Philip E., 32
Rasheed, Raymond K., 439
Reddy, Ramana G., 121
Reichert, W. Matthew, 121
Ren, Rex X., 70
Reynolds, John L., 370
Robertson, Allan J., 41
Rogers, Robin D., 2, 121, 300
Ross James, 314
Scott, Mark P., 468
Seddon, Kenneth R., 14, 421
Shaughnessy, Kevin H., 300
Sheldon, Roger A., 192
Soper, A. K., 151
Sutto, Thomas E., 478
Tambyrajah, Vasuki, 439
Tkatchenko, I., 93
Tommasi, I., 93
Trulove, Paul C., 478
Turner, Megan B., 2
van den Berg, Jan-Albert, 421
van Hal, Roy, 57
van Rantwijk, F., 192
Varma, Rajender S., 82
Vaultier, Michel, 239
Verevkin, Sergey P., 134
Viciu, Mihai S., 323
Viklund, Fredrik, 225
Wasserscheid, Peter, 57
Weidner, J. W., 509
Werner, James H., 212
Wishart, James F., 381
Xiao, Jianliang, 314
Zhou, Yuehui, 41

Subject Index

A

Ab initio calculations, Diels–Alder, 17
Ab initio molecular structures. *See* Molecular structures
Abraham's generalized solvent equation, multiple linear regression, 6
Absorption, ionic liquids, 346–347
Acetylation of glucose, enzyme-catalyzed in ionic liquids, 202
N-Acetyllactosamine, synthesis in ionic liquids, 200, 201
Acid-base neutralization, ionic liquid synthesis, 34–35
Acidic melt, chloroaluminates, 175
Acids and bases in ionic liquids
 O-acetylation reactions using no added catalyst, 274*t*
 acid/base indicators, 268
 acid dissociation constants, pKa, 267
 acid dissociation constants and degrees of dissociation for various acids in water and ionic liquids, 272*t*
 Alizaris red S indicator acid, 271
 anions X^- in ionic liquids acting as Lewis bases, 273
 base catalysis, 273–274
 bases, 274
 Brönsted–Lowry acidity, 267
 general base catalysis effect, 274
 ionic liquids for study, 268
 Lewis and Brönsted–Lowry bases, 267
 pKa values in water and ionic liquids, 271–272
 relative ordering of pKa values, 273
 role of trace water, 275
 visible spectroscopy of Alizarin red S, 271*f*
 visible spectroscopy of bromocresol purple, 270*f*
 visible spectroscopy of *m*-cresol purple, 269*f*
Activated C–H bond, transfer of CO_2-moiety, 98
Activation, hydrogen peroxide, 278
Activation energy
 calculating for transition structures, 184, 187
 chlorpromazine and Trolox by CCl_3O_2• radicals, 399*t*
 See also Diels–Alder reactions
Activity
 enzymes in ionic liquids, 198–199
 lipases in ionic liquids, 230
Activity coefficients
 solutes in ionic liquids at infinite dilution, 135–137
 solutes in ionic liquids over whole concentration range, 137, 142
 vapor pressures and, of diols in ionic liquid, 143*f*
Acylations
 benzyl alcohol, 338, 339
 carbohydrates in ionic liquids, 202
 procedures for ascorbic acid and glucose, 236–237
 regioselective, of ascorbic acid in ionic liquids, 232–234, 235
 regioselective, of glucose in ionic liquids, 230–232, 234–235
 See also Enantioselective acylation; *N*-Heterocyclic carbenes (NHC)
Addition reactions, radiolysis, 406–407

Air-stable ionic liquids, anion metathesis route, 73–74
L-Alanine ethyl ester hydrochloride, availability, 103
Alcohol conversion
green synthesis of ionic liquids, 78*t*
synthesis of non-halide ionic liquids, 75–76
waste-free process, 76*f*
Alcohols
activity coefficients and vapor pressure of diols in ionic liquid, 143*f*
heat of solution at infinite dilution in ionic liquid, 137, 141*t*
limiting activity coefficients in ionic liquids, 139*f*
See also Superoxide ion; Transesterifications
Alizarin red S
probe indicator, 271
structure, 268
visible spectroscopy, 271*f*
n-Alkanes
heat of solution at infinite dilution in ionic liquid, 137, 141*t*
limiting activity coefficients in ionic liquids, 138*f*
Alkylbenzenes
heat of solution at infinite dilution in ionic liquid, 137, 141*t*
limiting activity coefficients in ionic liquids, 139*f*
Alkyl imidazolium halides, synthesis in microwave oven, 86*t*
1-Alkyl-3-methylimidazolium halides
specific heat capacity, 124*t*
ultrasound-assisted preparation, 89*t*
Alkyloligoethersulfate ionic liquids
synthesis, 62–63
thermal properties, 63–64
viscosity, 64–65
Alkylsulfate ionic liquids
thermal properties, 63–64
viscosity, 64–65
Alkyl sulfonates, halide-free ionic liquids from, 23–24
Allylic acetates, isomerization by palladium catalysis, 318–319
Allylic alkylation
phenylallyl acetate, 316
phenylallyl carbonate, 315–316
See also Hydrogen bonding
Aluminum
Al(III) reduction, 496
anodic behavior, 502–503
cathodic behavior, 500–501
electrodeposition from 1-butyl-3-methylimidazolium (bmim)Cl/$AlCl_3$, 455–456
electrorefining of impure, 505–506
electrowinning, 495–496, 506–507
morphology of deposits, 503, 504*f*, 505
nucleation of, on copper, 501–502
studies of electrochemical behavior, 497
See also Electrodeposition; Electroprocessing of reactive metals
Aluminum chloride. *See* Electroprocessing of reactive metals
Amidation, peptide amidase catalyzed, 209–210
Amination. *See* Aryl amination
Amine mediated photoreduction mechanism, 372–373
See also Photochemistry
Ammoniolysis, octanoic acid, 194
Anions
common for ionic liquids (ILs), 3
controlling reactivity, coordination, and hydrophobicity, 3–4
very weakly basic anions, 265–266
See also Acids and bases in ionic liquids
Anions and cations, selection for phosphonium ionic liquids, 44–45
Anodic behavior, aluminum, 502–503
Anthracenes

acidic [1-ethyl-3-methylimidazolium (emim)][Cl/$AlCl_3$], 347–349
basic [emim][Cl/$AlCl_3$], 347, 348
9-chloroanthracene in basic [emim][Cl/$AlCl_3$], 353
9,10-dihydroanthracene in [emim][Cl/$AlCl_3$], 352
initial reactions in acidic [emim][Cl/$AlCl_3$], 349
9-methylanthracene and photoproducts, 351–353
See also Photochemistry
Aryl amination
air-free, 328*t*, 329*t*
palladium complexes containing *N*-heterocyclic carbene ligands, 324, 326
reactions under aerobic conditions, 330*t*, 331*t*
See also *N*-Heterocyclic carbenes (NHC)
Ascorbic acid
acylation procedure, 236–237
Candida antarctica (CAL-B) catalyzed acylation, 226–227
regioselective acylation in ionic liquids, 232–234

B

Bases. *See* Acids and bases in ionic liquids
Basic anions, ionic liquids, 265–266
Battery
charging and discharging, 490*f*
dual intercalating molten electrolyte (DIME), 480
Li metal-metal oxide test cells, 489, 491
potential for molten salts in Li-ion, 479
reversible zinc galvanic cell, 446, 448*f*
See also Polymer gel electrolytes
Benzaldehyde
cyclic voltammograms for reduction in 1-methyl-3-butylimidazolium triflimide [Bmim][Tfi], 416*f*
cyclic voltammograms for reduction in *N*-methyl-*N*-butylpyrrolidinium triflimide ([Pyrd][Tfi]), 417*f*
electrochemical reduction, 415–419
square-wave voltammogram for reduction in [Pyrd][Tfi], 418*f*
See also Organic electrochemistry
Benzene, solubility in [bmim][BF_4], 116, 117*f*
Benzophenone
experimental, 379
hydrogen atom abstraction, 349, 350
photoreaction with *sec*-butylamine, 349, 350
product ratios in photoreduction, 373, 374*t*
reduction of duroquinone by, ketyl and pyridinyl radicals, 402–403
role of C_2-proton, 377–378
solvent effects and role of cation, 374–377
triplet energy transfer to naphthalene, 359
See also Diffusion controlled reactions; Photochemistry
Binding energy, melt model, 178
Binding enthalpies, ionic liquid complexes, 178*t*
Biocatalysis. *See* Nonaqueous biocatalysis with polar substrates
Biocatalyst, recycling, 203–204
Biosolvation. *See* Solvent properties
Biotransformations
N-acetyllactosamine synthesis, 200, 201
activity and stability of enzymes in ionic liquids, 198–199
acylations of carbohydrates in ionic liquids, 202
added value in ionic liquids, 195–196
ammoniolysis of octanoic acid, 194
Candida antarctica lipase B (CaL B), 194

enantioselectivity, 196–198
first example in ionic liquid, 193–194
glycosidases in ionic liquids, 200, 201
in situ generation of peroxycarboxylic acid via CaL B, 194–195
ionic liquids, 193–195
kinetic resolution of 1-phenylethanol, 196–197
oxidoreductases in ionic liquids, 203
product recovery and catalyst recycling, 203–204
proteases in ionic liquids, 200
scope in ionic liquids, 195–196
thermal stability of CaL B preparations, 199
thermolysin, 200
transesterification, 194
transesterification of chiral allylic alcohol, 197
transesterification of chiral primary alcohol, 197–198
turnover frequencies (TOF), 198
turnover numbers (TON), 198
use of isolated enzyme, 194
Bis(trifyl)imide $[NTf_2]^-$
distribution ratios of organic solutes between, and $[C_4mim][NTf_2]$, 8*f*
ionic liquid anion, 3
Bromine radicals
oxidation of chlorpromazine, 400, 401*f*
See also Pulse radiolysis
Bromocresol purple
probe indicator, 269–270
structure, 268
visible spectroscopy, 270*f*
Brookhart-type catalyst
productivity in ionic liquids, 310
schematic of system, 305*f*
See also Coordination polymerization of olefins
Brönsted–Lowry acidity, solvents, 267
Brönsted–Lowry bases, ionic liquids, 267
Bulk ionic liquids
characterization, 60
definition, 59
industrial use on larger scale, 59
1-Butyl-3-methylimidazolium bis(trifyl)amide ($[bmim][(CF_3SO_2)_2N]$)
ab initio structure, 168, 169*f*
dipole moment, 169
1-Butyl-3-methylimidazolium butanoate
ab initio structure, 170, 172*f*
dipole moment, 170
1-Butyl-3-methylimidazolium chloride (BMIC). *See* Electro-processing of reactive metals
1-Butyl-3-methylimidazolium hexafluorophosphate ($[C_4mim][PF_6]$)
ab initio structure, 164–165
calculated volumetric heat capacity, 130*t*
case against, 4–6
cost, 37–38
dipole moment, 165
effect as plasticizer for PMMA, 473*f*
effect of microwave power level on formation, 87
hazards, 36–37
Henry's law constants for gases, 115*f*
hydrolysis of anion, 58
interaction parameters for partitioning of organic solutes, 8*t*
liquid clathrate formation, 8–10
liquid-liquid extraction, 545
measuring translational diffusion within, 215, 217–218
most widely used, 101
organic synthesis and transition metal catalysis, 73
process parameters, 36*t*
properties, 35*t*
purity, 38–39
raw materials for synthesis, 35–36

single solute extraction using, 549–550
solubility isotherms at 25°C for CH_4, 115*f*
solubility isotherms at 25°C for CO_2, 114*f*
solubility of gases, 110–111
specific heat capacity, 124*t*, 126*f*
survey of reactions, 101–102
synthesis and separation, 5*f*
thermal stability of, and dioctyl phthalate plasticized poly(methyl methacrylate), 474*f*, 475*f*
toxicology, 37
typical analysis, 39*t*
See also Chlorophenols; Electrodeposition; Plasticizers; Pulse radiolysis; Superoxide ion
1-Butyl-3-methylimidazolium nitrate ([bmim][NO_3]
ab initio structure, 166, 167*f*
dipole moment, 167
1-Butyl-3-methylimidazolium octylsulfate, potential impurities, 61
1-Butyl-3-methylimidazolium pivalate
ab initio structure, 170, 171*f*
dipole moment, 170
1-Butyl-3-methylimidazolium *p*-toluenesulfonate ([bmim][*p*-TSA])
ab initio structure, 169, 171*f*
dipole moment, 170
1-Butyl-3-methylimidazolium tetrafluoroborate ([bmim][BF_4])
activity of *Candida antarctica* B (CALB) for butyl butyrate synthesis, 245*f*
apparatus for determining liquid-liquid solubility, 112*f*
calculated volumetric heat capacity, 130*t*
densities, 111
density and temperature, 114, 116*f*
density determination method, 112
experimental, 112–113
gas solubility measurement method, 112
Henry's law constant, 113–114
Henry's law constants for gases, 115*f*
liquid-liquid equilibrium (LLE), 111–112
LLE determination method, 113
LLE of 1-propanol and, as function of mol% IL, 118*f*
LLE of 1-propanol and, as function of wt% IL, 117*f*
materials, 113
solubility isotherms at 25°C for CH_4, 115*f*
solubility isotherms at 25°C for CO_2, 114*f*
solubility of benzene, 116, 117*f*
solubility of CO_2, CH_4, N_2, and CO, 113
solubility of gases, 111
stabilizing effect for α-chymotrypsin, 243
temperature dependence of density, 114, 116*f*
See also Pulse radiolysis
1-Butyl-3-methylimidazolium tetraphenylborate ([C_4mim][BPh_4])
ab initio structure, 166
dipole moment, 166
specific heat capacity vs. temperature, 125*f*
1-Butyl-3-methylimidazolium trifluoromethaneacetate ([bmim][CF_3CO_2])
ab initio structure, 167
dipole moment, 168
1-Butyl-3-methylimidazolium trifluoromethanesulfate ([bmim][CF_3SO_3])
ab initio structure, 168
dipole moment, 168
1-Butyl-3-methylimidazolium (trifluoromethanesulfonyl)imide ([C_4mim][NTf_2])
calculated volumetric heat capacity, 130*t*
specific heat capacity, 124*t*, 126*f*
Butyl butyrate

Candida antarctica B (CALB)
catalyzed synthesis in ILs, 244–246
synthesis method in ionic liquids, 241
N-1-Butylpyridinium chloride (BPC)
infrared study, 179–180
structure, 175
See also Diels–Alder reactions
N-Butylpyridinium tetrafluoroborate ($BuPyBF_4$), electron transfer from methyl viologen to duroquinone in $BuPyBF_4$/water mixtures, 405*f*

C

Candida antarctica lipase B (CaL B)
activity and stability in ionic liquids, 198–199
ammoniolysis of octanoic acid, 194
butyl butyrate synthesis in ionic liquids-supercritical CO_2, 245–246
butyrate synthesis in ionic liquids, 244–245
kinetic resolution of *rac*-1-phenylethanol in ILs-$scCO_2$, 247–249
regioselective acylation of ascorbic acid, 232–234
regioselective acylation of glucose, 230–232
stability and activity in ionic liquids, 243–244
stability in liquid media, 241
transesterification, 194
transesterification of 5-phenyl-1-penten-3-ol, 253–254
use in ionic liquid, 194–195
See also Enzymatic catalysis; Nonaqueous biocatalysis with polar substrates
Carbohydrates, acylations in ionic liquids, 202
Carbon dioxide
solubility in [bmim][BF_4] and [bmim][PF_6], 113, 114*f*
See also Supercritical carbon dioxide ($scCO_2$)
Carbon monoxide
solubility in [bmim][BF_4], 113
See also Coordination polymerization of olefins
Carboxylates. *See* 1,3-Dialkylimidazolium-2-carboxylates
Catalysis
hydroformylation of 1-octene, 65–67
See also Hydrogen bonding
Catalyst, recycling, 203–204
Catalyst systems, coordination polymerization of alkenes, 301–302
Cathodic behavior, aluminum, 500–501
Cationic surfactants
absence of water, 537
gels, 536
hexadecyltrimethylammonium bromide (CTAB), 527, 528*f*
lattice spacing of lamellar gel phase of [C_{18}mim]Br, 537
liquid-crystalline ionogels, 535–541
mechanism of gelation, 540
molecular basis for gelation in [C_{10}mim]Br, 537, 540
self-assembly, 535
small-angle X-ray scattering (SAXS) studies, 536–537, 538*f*, 539*f*
water concentration, 537
Cations
ether functionality, 266
quaternized nitrogen cations, 265
See also Acids and bases in ionic liquids
Chemical effects, radiation, 387–388
Chemical reactions
effects of ionic liquids, 175–176
See also Diels–Alder reactions
Chemical recognition. *See* Solvent properties
Chemical technology, greening, 2–3
Chiral ionic liquids

cations and anions, 17*f*
physical properties of chiral $[dpeim]^+$ salts, 27*t*
structure of S-[dpeim]bromide, 28*f*
synthesis, 16–17, 26–28
Chloride, ionic liquid anion, 3
Chloroacetamidine hydrochloride, starting material, 102
Chloroaluminates
acidic melt, 175
basic melt, 175
electrochemistry, 440
photochemical investigations, 371–372
polymerization of olefins, 302
preparation, 72
See also Diels–Alder reactions
Chlorophenols
analysis methods, 548
1-butyl-3-methylimidazolium hexafluorophosphate [bmim][PF_6] for liquid-liquid extraction, 545–546
comparison of ionic liquid-water distribution ratios, 555, 556*f*
effect of acid and alkaline in aqueous 2-chlorophenol solution on distribution ratio of [bmim][PF_6], 553*t*
effect of acidic and basic solution, 551–554
effect of ionic strength, 554
environmental applications of ionic liquids, 545
experimental, 547–548
extraction of aqueous solution with five chlorophenols, 552*f*
extraction procedure, 547–548
extraction using solution with multiple, 550–551
influence of high pH, 552
influence of pH on distribution ratio of ionic liquids, 553*t*
materials, 547
metal ion extraction, 546
partitioning dependence on pH, 545–546
pH dependence of thymol blue, 554
side effects, 545
single solute extraction using 1-ethyl-3-methylimidazolium bis(perfluoroethylsulfonyl)imide ([emim]Beti), 548–549
single solute extraction using [bmim][PF_6], 549–550
thymol blue dye, 546
trends of partitioning, 549
Chlorpromazine
oxidation by Br_2. radicals, 400, 401*f*
oxidation by CCl_3O_2• radicals, 398, 400
rate constants and activation energies for reactions with CCl_3O_2• radicals, 399*t*
See also Pulse radiolysis
Choline chloride based ionic liquids
conductivity and viscosity of 2:1 mixture of $CrCl_3$•$6H_2O$ and choline chloride, 449–450
conductivity of 1:2 choline chloride:zinc chloride vs. temperature, 441–442
cyclic voltammogram of Pt electrode in 1:2 choline chloride:zinc chloride, 443*f*
cyclic voltammogram of Pt microelectrode in 1:2 choline chloride: $CrCl_3$•$6H_2O$, 451*f*
effect of current density on zinc deposit morphology, 441
experimental, 440–441
freezing point temperatures, 448*t*
freezing point temperatures of mixtures vs. temperature, 449*f*
Hull Cell tests showing bulk zinc deposition, 444
hydrated salt mixtures, 446, 449–450
linear sweep voltammograms of Fe and zinc coated Fe, 447*f*
metal deposition, 443–446

morphology of zinc deposited onto nickel, 445*f*
plot of anodic vs. cathodic charge on cyclic, 445*f*
porosity and resistance of deposited zinc films to corrosion, 446
potential vs. drain characteristics for cell, 448*f*
preparation, 441
reversible zinc galvanic cell, 446, 448*f*
underpotential deposition (UPD) of zinc, 443
UPD response vs. ionic liquid composition, 444*f*
voltammetry and chronoamperometry methods, 441
See also Electrochemistry
Chromium chloride. *See* Choline chloride based ionic liquids
α-Chymotrypsin
stability and activity in ionic liquids, 243–244
stability in liquid media, 241
See also Enzymatic catalysis
CO_2-moiety transfer, activated C–H bond, 98
Commercial battery, Li-ion, 479
Commercially available, low melting salts, 104*t*, 105*t*, 106*t*
Commercial production
costs, 37–38
hazards, 36–37
importance of ionic liquids, 34
purity, 38–39
raw material issues, 35–36
synthesis routes to ILs, 34–35
Condensation reactions. *See* Enzymatic condensation reactions
Conductivity
aluminum chloride-BMIC (1-butyl-3-methylimidazolium chloride) ionic liquids, 498–500
choline chloride:zinc chloride ionic liquid, 441–442
$CrCl_3 \cdot 6H_2O$ and choline chloride mixture, 449–450
See also Electro-processing of reactive metals; Polymer gel electrolytes
Coordination polymerization of olefins
acceleration of styrene/CO copolymerization, 308
alternating styrene/CO copolymerization with Drent-type catalyst, 302–304
Brookhart-type catalyst system, 305*f*
catalyst systems, 301–302
chloroaluminate ionic liquids (ILs), 302
common IL cations and anions, 301*f*
concentration effect with Drent system, 308–309
copolymerization productivity vs. time, 307*f*
effect of variation of IL cation and anion, 304
ethylene polymerization in ILs, 308*t*
explanations for accelerating effect of IL solvents, 309–310
homopolymerization of ethylene in ILs, 306–307, 310
IL solvent effects in styrene/CO copolymerization, 304*t*
importance of non-coordinating IL solvent, 309
polar, non-coordinating ionic liquids, 301
polar ionic liquids containing non-nucleophilic anions, 301
productivity with Brookhart-type catalyst, 310
styrene/CO copolymerization with cationic palladium complex, 304–306
styrene/CO copolymerization by (bipy)Pd(Me)Cl/MY, 306*t*
styrene/CO copolymerization in $[C_6pyr][Tf_2N]$, 303*t*

Copolymerization. *See* Coordination polymerization of olefins
Copper
electrodeposition from $[bmim]Cl/AlCl_3/CuCl_2$, 457–458
See also Electrodeposition
Corrosion, deposited zinc films, 446
Costs, commercial use, 37–38
Coulomb parameters, 1,3-dimethylimidazolium hexafluorophosphate $[dmim][PF_6]$ and [dmim]Cl, 154*t*
Coupling reactions. *See* *N*-Heterocyclic carbenes (NHC)
Cross-linked enzyme aggregates (CLEA), enyzmes in ionic liquids, 198–199
Cross-linked enzyme crystals (CLEC), enyzmes in ionic liquids, 198–199
Crown ethers
extracting metal ions from aqueous solution, 546
extraction of strontium by dicyclohexano-18-crown-6 (DCH18C6), 529–530
relationship between strontium partitioning and, 530, 532*f*
See also Ion exchangers
Crystal structures
comparison with liquid structures, 160
1,3-dimethylimidazolium hexafluorophosphate $[dmim][PF_6]$ and [dmim]Cl, 155
See also Liquid structures
Current density, zinc deposit morphology, 441
Cyclic voltammetry
1-butyl-3-methylimidazolium hexafluorophosphate [bmim][HFP], 512, 513*f*
diffusion coefficients of ferricenium/ferrocene, 429*t*, 431*t*
ferricenium/ferrocene, 426–427, 431
See also Choline chloride based ionic liquids; Electrodeposition; Ferricenium/ferrocene; Organic electrochemistry; Superoxide ion
Cyclopentadiene. *See* Diels–Alder reactions

D

Decomposition, phosphonium ionic liquids, 46
1-Decyl-3-methylimidazolium bromide $[C_{10}mim]Br$
interaction of water with, 536–537, 538*f*, 539*f*
molecular basis for gelation in, 537, 540
See also Cationic surfactants
Degradation efficiency, superoxide ion, 517, 519
Density
$[bmim][BF_4]$, 114, 116*f*
accuracy for analyzing gas solubility, 111
determination method, 112
function of temperature for phosphonium ionic liquids, 48*f*
heat capacity and, 129–130
mixtures containing ionic liquids, 145, 147
phosphonium ionic liquids, 47, 49
temperature dependence, 116
thermal storage, 129–130
See also 1-Butyl-3-methylimidazolium tetrafluoroborate $[bmim][BF_4]$
Density functional theory (DFT) calculations, 177
See also Diels–Alder reactions
1,3-Dialkylimidazolium-2-carboxylates
reactivity with $[HOEt_2]BF_4$, 96–98
synthesis, 94–96
transfer of CO_2-moiety to organic substrate with activated C–H bond, 98

1,3-Dialkylimidazolium tetrafluoroborates, microwave-assisted, 88*t*
Diels–Alder reactions
ab initio and density functional theory (DFT) calculations, 177
acidic melt model, 178
activation barrier, 184
approaches for calculating activation energies for transition structure, 184, 187
basic melt model, 178
binding energy for $AlCl_4^-$...1-ethyl-3-methylimidazolium (EMI^+) and $Al_2Cl_7^-$...EMI^+ complexes, 178
binding enthalpies of ionic liquid complexes, 178*t*
common ionic liquids EMI chloride (EMIC) and *N*-1-butylpyridinium chloride (BPC), 175
comparison of computed thermodynamic activation parameters, 187*t*
comparison of peak frequency changes with melt acidity, 180*t*
computed activation energies, enthalpies, and Gibbs energies, 181–182
computed activation energies for EMI^+ and four stereospecific transition structures, 185*t*
computing isolated ions forming in melt, 177–178
cyclopentadiene and methyl acrylate, 180, 295*f*
diphenylisobenzofuran (DPBF) and singlet oxygen, 363–364
effects of ionic liquids, 175–176
geometry of anions in molten chloroaluminate salts, 178
hydrogen bonding, 176–177
hydrophobic effects, 176–177
influence of ionic liquid polarity on, 294–295
influence of solvent on endo/exo selectivity and reaction rate, 180–181
infrared study of EMIC and BPC, 179–180
methods, 177
NC (*endo*, *s-cis* methyl acrylate) transition structures, 182, 183*f*
NC transition structures for cyclopentadiene and methyl acrylate with basic and acidic environment, 186*f*
possible reaction pathways for cyclopentadiene and methyl acrylate, 181
procedure, 297
rate accelerations, 176
rate and endo selectivity, 182
rate and selectivity enhancements in ILs, 366
X-ray and theoretical data for EMI^+, 179*t*
X-ray crystal structure of melt involving EMIC and $AlCl_3$, 178–179
Diffusion, measuring translational, within ionic liquid, 215, 217–218
Diffusion coefficients
ferricenium in $[C_4mim][PF_6]$, 431*t*
ferrocene in $[C_4mim]X$, 429*t*
See also Ferricenium/ferrocene
Diffusion controlled reactions
anion exchange, 368
Arrhenius parameters for viscosity and k_q, 362*t*
Arrhenius plot, 361*f*
cleanup of end product, 368–369
comparison of k_{act} for different solvents, 366*t*
comparison of k_q experimental and calculated, 361*t*
controlling conditions for quaternization reactions, 367–368
dependence of ln A and E_a on alkyl chain length and substitution pattern, 363

deriving k_d values in room temperature ionic liquids (RTILs) from experimental k_q values, 363–364
Diels–Alder reaction between diphenylisobenzofuran (DPBF) and singlet oxygen, 363–364
diffusion controlled bimolecular rate constant, 358
energy transfer from triplet benzophenone ($^3Bp^*$) to naphthalene (N), 359
experimental, 367–369
failure in predicting k_d values, 362
higher temperatures, 364, 366
isokinetic relationship for reaction of $^3Bp^*$ with N, 363*f*
k_q values in RTILs, 363
low temperature region, 364
overall bimolecular rate constant, k_q, 359
preparation of spectroscopic grade ionic liquids, 367–369
purification of starting materials, 367
rate and selectivity enhancements for Diels–Alder reactions, 366
reaction rate information, 358–359
room temperature, 364
solute jump probability, 362–363
temperature dependence of k_q for DPBF and $^1O_2^*$ reaction in [bmim][PF_6], 364, 365*f*
variation of k_2 with temperature and [N] for reaction of $^3Bp^*$ with N, 360*f*
Dimethylcarbonate, alkylating agent, 95
1,2-Dimethyl-3-ethyl-imidazolium bis(trifluoromethyl-sulfonyl) amide ([emmim][NTf_2])
limiting activity coefficients of alcohols in, 139*f*
limiting activity coefficients of alkylbenzenes in, 139*f*
limiting activity coefficients of *n*-alkanes in, 138*f*
structures, 138*f*
1,3-Dimethylimidazolium [dmim] salts
atom connectivity and numbering scheme of [dmim][PF_6], 153*f*
cation-cation distributions, 159
crystal structures of [dmim]Cl and [dmim][PF_6], 155
empirical potential structure refinement process (EPSR), 155
experimental, 152–155
experimental and ESPR fitted cross sections vs. function of Q for deuteriated and protiated [dmim]Cl, 156*f*
Lennard–Jones and coulomb parameters for simulation of chloride and hexafluorophosphate salts, 154*t*
liquid structures of [dmim]Cl and [dmim][PF_6], 155, 157–160
liquid structures of salts using molecular dynamics simulations, 159–160
liquid vs. crystal structures, 160
neutron diffraction method, 153–155
partial radial distribution functions for anions and cations surrounding central cation, 157*f*
single crystal X-ray diffraction, 152–153
three dimensional probability distributions, 158–159
Dimethylmaleate reduction, electrochemistry, 413–415
Dioctyl phthalate (DOP)
effect as plasticizer for poly(methyl methacrylate) (PMMA), 473*f*
elastic moduli of PMMA plasticized with DOP, 472*f*
glass transition temperature of plasticized PMMA, 474–475
plasticizer, 469
thermal stability of plasticized PMMA, 472, 474
See also Plasticizers

Diphenylisobenzofuran (DPBF), Diels–Alder reaction with singlet oxygen, 363–364
Dipole moment
 1-butyl-3-methylimidazolium [bmim][BF_4], 166
 [bmim][CF_3CO_2], 167, 168
 [bmim][$(CF_3SO_2)_2N$], 169
 [bmim][CF_3SO_3], 168
 [bmim][NO_3], 167
 [bmim][PF_6], 165
 [bmim] butanoate, 170
 [bmim] pivalate, 170
 [bmim] *p*-toluenesulfonate, 170
Direct combination, ionic liquid synthesis, 34–35
Disaccharides, β-galactosidase catalyzed synthesis, 207–208
Dissociation constants
 acids, 267
 relative ordering, 273
 See also Acids and bases in ionic liquids
Drent-type catalyst, palladium complex, 304–305
Dry electron states
 radiation chemistry, 385
 reactivity, 390–392
Dual intercalating molten electrolyte (DIME), batteries, 480
Duroquinone (DQ)
 electron transfer between, and methyl viologen, 403–405
 reduction by benzophenone ketyl and pyridinyl radicals, 402–403

E

Economics, commercial use, 37–38
Elastic moduli, plasticized poly(methyl methacrylate) (PMMA), 471–472, 473*f*
Electrochemistry
 aluminum and transition metal salts, 440
 method for superoxide ion, 512, 515
 solvent-solute interactions, 423–424
 superoxide ion generation, 510–511
 See also Choline chloride based ionic liquids; Electro-processing of reactive metals; Ferricenium/ferrocene; Organic electrochemistry; Superoxide ion
Electrodeposition
 Al overpotential deposition (OPD) from acidic [bmim]Cl/$AlCl_3$, 455–456
 Cu OPD from acidic [bmim]Cl/$AlCl_3$/$CuCl_2$, 457–458
 cyclic voltammogram of [bmim]Cl/$AlCl_3$/$CuCl_2$ on glassy carbon and Au(111), 457
 cyclic voltammogram of [bmim]Cl/$AlCl_3$ on Au(111), 455*f*
 electrochemical window of [bmim][PF_6] on Au(111), 460*f*
 experimental, 454
 Ge underpotential deposition (UPD) and OPD from $GeCl_4$ saturated [bmim][PF_6], 459–464
 highly oriented pyrolytic graphite (HOPG), 461–462
 narrowly dispersed Ge nanoclusters, 463
 outlook, 465
 scanning tunneling microscopy (STM) picture during initial stages of Al OPD on Au(111), 456*f*
 STM of Cu OPD on Au(111), 458*f*
 STM picture of hydrogen terminated Si(111), 464*f*
 STM pictures of Ge UPD on Au(111), 461*f*
 typical HOPG surface in [bmim][PF_6] saturated with $GeCl_4$, 462*f*
 ultrahigh vacuum (UHV), of Ge quantum dots on Si(111), 463–464
 See also Choline chloride based ionic liquids
Electrolysis

ferrocene to ferricenium, 435
$MgCl_2$–BMIC (1-butyl-3-methylimidazolium chloride), 507
Electrolytes. *See* Polymer gel electrolytes
Electron-hole pairs, radiation chemistry, 385–387
Electron transfer
quinones and methyl viologen, 403–405
See also Photochemistry; Pulse radiolysis
Electro-processing of reactive metals
anodic behavior of aluminum, 502–503
cathodes after deposition in $AlCl_3$–BMIC (1-butyl-3-methylimidazolium chloride), 504*f*
cathodic behavior of aluminum, 500–501
cell voltage vs. electrolysis time in $AlCl_3$/BMIC, 505*f*
conductivity measurement, 498–500
conductivity vs. aluminum chloride mole fraction, 499*f*
conductivity vs. temperature for $AlCl_3$–BMIC ionic liquids, 498*f*
conductivity vs. temperature for $MgCl_2$–BMIC ionic liquids, 500*f*
current efficiency of aluminum deposition, cell voltage, and energy consumption, 507*t*
cyclic voltammograms on copper in $AlCl_3$/BMIC, 501*f*
definitive test for adherence to nucleation models, 502
electrochemical behavior of $AlCl_3$–BMIC, 500–503
electrolysis of $MgCl_2$–BMIC, 507
electrorefining of impure aluminum, 505–506
electrowinning of aluminum, 506–507
experimental, 497
impurity contents of anodes and cathodic deposits, 506*t*
morphology of aluminum deposits, 503, 505
nucleation of aluminum on copper, 501–502
scanning electron microscopy (SEM) images of Al deposits in $AlCl_3$/BMIC, 504*f*
Electrorefining, impure aluminum, 505–506
Electrowinning
aluminum, 495–496, 506–507
magnesium, 496
Enantioselective acylation
acylation of mandelic acid methyl ester, 255
anchoring lipase by ionic liquid, 256
evaluation of supporting materials for lipase-catalyzed transesterification, 255*t*
lipase-catalyzed reaction system anchored in ionic liquid solvent, 253–256
lipase-catalyzed reaction under reduced pressure, 257–258
lipase-catalyzed transesterification in ILs, 254*t*
normal pressure conditions, 252
reaction system anchored to solvent, 256*f*
reduced pressure conditions, 252–253
repeated use of lipase in [bmim][PF_6] solvent, 257*f*
repeated use of lipase in [bmim][PF_6] solvent under reduced pressure, 260*f*
transesterification of 5-phenyl-1-penten-3-ol, 253–254
transesterification of 5-phenyl-1-penten-3-ol using methyl pentanoate, 258, 259*t*
Enantioselectivity, reactions in ionic liquids, 196–198
Enforced, term, 176
Enthalpy. *See* Diels–Alder reactions

Environmental applications, ionic liquids, 545
Environmentally benign solvents, ionic liquids, 163
Enzymatic catalysis
activity, selectivity, and enantioselectivity profiles of *Candida antarctica* B (CALB)–(1-ethyl-3-methylimidazolium triflimide) (CALB–[emim][Tf_2N]) system, 247*f*
activity of CALB for butyl butyrate synthesis in different ionic liquids (Ils), 245*f*
butyl butyrate synthesis in ILs, 241
CALB-catalyzed butyl butyrate synthesis in ILs-supercritical CO_2 ($scCO_2$), 245–246
CALB-catalyzed butyrate synthesis in ILs, 244–245
CALB-catalyzed kinetic resolution of *rac*-1-phenylethanol in ILs-$scCO_2$, 247, 249
effect of $scCO_2$ density on activity and half-life of CALB–[bmim][Tf_2N], 246*f*
effect of temperature on synthetic activity and half-life time of CALB-ILs, 248*f*, 249
experimental setup of bioreactor with ILs-$scCO_2$, 242*f*
GC analysis method, 243
materials and methods, 240–243
protective effect of different media on α-chymotrypsin and CALB stability, 244*f*
reactions in ILs-$scCO_2$, 242
stability and activity of α-chymotrypsin and CALB in ILs, 243–244
stability of α-chymotrypsin in liquid media, 241
stability of CALB in liquid media, 241
synthesis of ionic liquids, 241
Enzymatic condensation reactions
β-galactosidase catalyzed synthesis of disaccharides, 207–208
peptide amidase catalyzed amidation, 209–210
reverse hydrolytic activity of β-galactosidase, 208*f*
yield of lactose as function of ionic liquid amount, 208*f*
Enzymes
activity and stability in ionic liquids, 198–199
See also Biotransformations
Epoxidations
alkenes with methylrhenium trioxide/urea hydrogen peroxide (MTO/UHP), 280*t*
^2H NMR kinetics, 285–286
MTO catalyzed, using UHP, 279–281
non-steady-state kinetics of olefin, 282–284
solvent purity, 280–281
Ethanol, alcohol conversion route, 77
Ether functionality, cations, 266
1-Ethyl-3-methylimidazolium chloride (EMIC)
binding enthalpies of complexes, 178*t*
computed activation energies for EMI^+ and transition structures, 185*t*
configurations of transition structures, 182, 183*f*
infrared study, 179–180
solvent for Diels–Alder reaction, 180–181
structure, 175
X-ray and theoretical data for EMI^+, 179*t*
X-ray crystal structure of melt of, and $AlCl_3$, 178–179
See also Diels–Alder reactions
1-Ethyl-3-methylimidazolium bis(perfluoroethylsulfonyl)imide

[emim][Beti], single solute extraction using, 548–549
1-Ethyl-3-methylimidazolium bis(trifluoromethane sulfonyl)imide ([emim][Tf_2N])
activity of *Candida antarctica* B (CALB) for butyl butyrate synthesis, 245*f*
porphyrin self-assembly, 219, 220*f*
1-Ethyl-3-methylimidazolium tetrafluoroborate ([emim][BF_4])
activity of *Candida antarctica* B (CALB) for butyl butyrate synthesis, 245*f*
melting points, 14–15
Ethylene
homopolymerization in ionic liquids, 306–307, 310
See also Coordination polymerization of olefins
Excess volume
4-methyl-*N*-butyl-pyridinium tetrafluoroborate ([4-M-nBP][BF_4]) + methanol, 148*f*
mixtures containing ionic liquids, 145, 147
Extraction
comparison of ionic liquid-water distribution ratios, 555, 556*f*
effect of acidic and basic solution, 551–554
effect of ionic strength, 554
ionic liquids, 545
procedure, 547–548
single solute extraction using 1-ethyl-methylimidazolium bis(perfluoroethylsulfonyl)imide [emim][Beti], 548–549
single solute extraction using [bmim][PF_6], 549–550
solution with multiple chlorophenols, 550–551
See also Chlorophenols; Ion exchangers

F

Ferricenium/ferrocene
bulk electrolysis, 435
Cottrell equation for current-time response, 431
current-time response for [Fc] in [C_4mim]X ionic liquids, 433*f*
cyclic voltammetric data for, in [C_4mim]X ionic liquids, 428
cyclic voltammetry, 426–427, 430*f*, 431
diffusion coefficients, 435–436
diffusion coefficients of $[Fc]^+$ in [C_4mim][PF_6], 431*t*
diffusion coefficients of [Fc] in [C_4mim]X, 429*t*
experimental, 425–426
Levich equation for limiting current, 432
mass transport data of [Fc] and $[Fc]^+$ ionic liquid solution, 434*t*
normal pulse voltammetry (NPV), 431–432
preparation of solutions, 425
RDE (rotating disk electrode voltammetry), 432, 435
RDE voltammograms of [Fc] in [C_4mim][PF_6], 433*f*
redox couple, 422–423
solvent-solute interactions, 423–424
solvodynamic radius, 424
spectroscopy, 426, 427*f*
Stokes–Einstein relation, 423–424
Stokes' law, 423
voltammetric experiments, 426
Fluorescence
ionic liquids, 346
probe as thermodynamic stability of protein, 219–221
Fluorescence correlation spectroscopy (FCS)
autocorrelation curves for rhodamine 6G, 218*f*
measuring translational diffusion within ionic liquid, 215, 217–218

method, 214–215
schematic of confocal FCS system, 216*f*
temporal autocorrelation function, 215, 217
Fluorescent probes, solvatochromatic, 7
Fluorinated esters, halide-free ionic liquids from, 23–24
Formate dehydrogenase (FDH), activity in ionic liquids, 203
Freezing point, hydrated salt mixtures, 448*t*, 449*f*
Functional groups, cations, 266
Functionalization, hydroxyl group, 392, 393*f*

G

β-Galactosidase, disaccharide synthesis, 207–208
Galvanic cell, reversible zinc, 446, 448*f*
Gas phase. *See* Molecular structures
Gelation, molecular basis in [C_{10}mim]Br, 537, 540
Geminate recombination, radiation chemistry, 385–387
Generally regarded as safe (GRAS), pharmaceutical and food additive industries, 5–6
Geometry, anions in chloroaluminate salts, 178
Germanium
disadvantages of GeI_4, 459
electrochemical window of [bmim][PF_6] on Au(111), 460*f*
electrodeposition from $GeCl_4$ saturated [bmim][PF_6], 459–464
highly oriented pyrolytic graphite (HOPG) surface, 461–462
narrowly dispersed Ge nanoclusters, 463
scanning tunneling microscopy (STM) pictures of Ge underpotential deposition (UPD) on Au(111), 461*f*
ultrahigh vacuum (UHV) deposition of Ge quantum dots on Si(111), 463–464
See also Electrodeposition
Gibbs energy
solvation for Diels–Alder reaction, 176
See also Diels–Alder reactions
Glass transition temperature, plasticized poly(methyl methacrylate) (PMMA), 474–475
Glucose
acetylation procedure, 236
Candida antarctica (CAL-B) catalyzed acetylation, 226
enzyme-catalyzed acetylation in ionic liquids, 202
regioselective acylation in ionic liquids, 230–232
Glycosidases, use in ionic liquids, 200, 201
Gold surfaces. *See* Electrodeposition
Graphite paper, charge and discharge behavior, 489*f*
Green chemistry
aims, 71
ionic liquids (ILs), 4, 192, 479
Green paradigm, use of $[PF_6]^-$ anion, 5
Green solvents
end-of-life factors, 5
ionic liquids for organic synthesis, 510
Green synthesis of ionic liquids
alcohol conversion results, 78*t*
anion metathesis route to air- and water-stable ionic liquids (ILs), 73–74
bromide-based ILs as choice, 76–77
challenges for research and development, 72–74
chloroaluminate-based ILs, 72
ethanol use, 77
future work, 79

halide-based ILs as precursors to non-halide, 74–75
mild conversion of 1-butanol to 1-bromobutane, 74*f*
starting material and product purity, 77–78
synthesis of non-halide based ILs via alcohol conversion route, 75–76
synthetic organic chemistry, 71
waste-free process via alcohol conversion route, 76*f*
waste minimization and pollution prevention, 71

H

Halide-free ionic liquids
fluorinated esters and alkyl sulfonates to, 23–24
halide-based ionic liquids as precursors, 74
imidazolium-based carbenes for, 24–26
preparation, 22–26
routes to phosphonium ionic liquids, 52
synthesis of tri-*iso*-butyl(methyl)phosphonium tosylate, 52
See also Green synthesis of ionic liquids
Haloaluminate(III) ionic liquids, solvent systems, 422
Halogen-free ionic liquids
cause for development, 58–60
literature, 60
Hazards, industrial issues, 36–37
Heat capacity
$[C_nmim][PF_6]$ ionic liquids (Ils), 123, 124*t*
anions with high hydrogen bond acceptor characteristics, 127
calculated volumetric, 130*t*
composition and operating temperature ranges for commercial thermal fluids, 128*t*
density and, of ILs, 129–130
experimental, 122–123
heat transfer fluid applications, 128
hydrogen-bonding interactions, 127–128
importance, 122
linear equation, 123
linear increase with temperature, 123–124
modulated differential scanning calorimetry (MDSC) method, 123
molar, vs. temperature for hexafluorophosphate $[PF_6]$-containing ILs, 127*f*
properties and structure changes, 130
series of $[C_4mim]X$ ILs, 123, 124*t*
specific, vs. temperature for $[C_4mim][BPh_4]$, 125*f*
specific, vs. temperature for $[C_4mim]Cl$, 126*f*
specific, vs. temperature for $[PF_6]$-containing ILs, 126*f*
thermal storage, 129
volumetric, vs. temperature for ILs and commercial heat transfer fluids, 131*f*
Heat of solution, liquid solute in ionic liquid at infinite dilution, 137, 141*t*
Heat transfer fluid
applications, 128
composition and operating temperature ranges, 128*t*
Henry's coefficient, definition, 135–136
Henry's law constant
definition, 113–114
gases in $[bmim][PF_6]$ and $[bmim][BF_4]$, 114, 115*f*
N-Heterocyclic carbenes (NHC)
acylation of benzyl alcohol with vinyl acetate, 337–338

air-free aryl amination catalyzed by complex 1, 328*t*, 329*t*
aryl amination by complex 1 under aerobic conditions, 330*t*, 331*t*
complex 1 catalyzing aryl amination, 324, 326
complex 1 formation, 324
coupling simple ketones and aryl halides, 333
early studies of $[(NHC)PdCl_2]_2$ mediated catalysis, 324–326
palladium-mediated amination of aryl chlorides, 332, 334*t*
palladium-mediated ketone arylation, 333, 336*t*
selective acylation of benzyl alcohol, 338, 339
single crystal X-ray analysis, 325*f*
single crystal X-ray analysis of complex 3, 327, 332*f*
structures, 337
Suzuki–Miyaura reactions, 333, 335*t*
synthesis and reactivity of (NHC)Pd(allyl)Cl complexes, 327, 332–333
transesterification, 333, 337–338
transesterification of methyl acetate with benzyl alcohol, 339*t*
Hexachlorobenzene, degradation by superoxide ion, 517, 522*f*
Hexafluorophosphate $[PF_6]^-$
exchange of halide for, 102
green paradigm, 5
ionic liquid (IL) anion, 3, 193*f*
molar heat capacity for PF_6-containing ILs, 127*f*
specific heat capacity for PF_6-containing ILs, 126*f*
Hexyl methylimidazolium hexafluorophosphate $[hmim][PF_6]$
effect as plasticizer for poly(methyl methacrylate) (PMMA), 473*f*
polymerizing PMMA, 470–471
See also Plasticizers
Homopolymerization
ethylene in ionic liquids, 306–307, 310
See also Coordination polymerization of olefins
Hull cell tests
method, 441
zinc deposition, 444, 445*f*
Hydrated salt mixtures
choline chloride based ionic liquids, 446, 449–450
freezing point temperatures, 448*t*, 449*f*
Hydroformylation reactions, ionic liquids, 65–67
Hydrogen abstraction, radiolysis, 406–407
Hydrogen bonding
allylic alkylation of phenylallyl acetate with methyl nitroacetate, 316
allylic alkylation of phenylallyl carbonate with dimethyl malonate, 315–316
anions and cations interactions, 127–128
anions with imidazolium cation, 170, 172*f*
concept in ionic liquids, 315
Diels–Alder reaction, 176, 176–177
effect on reaction chemistry, 314
experimental, 320
formation of inactive dialkylimidazol-2-ylidene complexes of palladium, 318
gelation of $[C_{10}mim]Br$, 540
increasing concentration of imidazolium cation, 317
inhibition of neutral Tsuji–Trost reaction, 316–317
isomerization of allylic acetate by palladium catalysis, 318–319
Pd(0)-catalyzed isomerization of allylic acetates, 318–319
solvation of carbonate by, and its effect on nucleophilic attack, 317
Tsuji–Trost reaction, 315, 319

Hydrogen peroxide
activation and selectivity, 278
importance of water-free source, 279
urea hydrogen peroxide (UHP), 279
See also Methylrhenium trioxide (MTO)
Hydrogen terminated Si(111), ultrahigh vacuum (UHV) deposition of Ge quantum dots on, 463–464
Hydrophobic effects, Diels–Alder reaction, 176–177
6-Hydroxy-2,5,7,8-tetramethylchroman-2-carboxylic acid (Trolox)
oxidation by CCl_3O_2• radicals, 398, 400
rate constants and activation energies for reactions with CCl_3O_2• radicals, 399*t*
See also Pulse radiolysis
Hydroxyl group, functionalization, 392, 393*f*

I

Ideal solvent, requirements, 3
Imidazolium-based carbenes
halide-free ionic liquids from, 24–26
See also *N*-Heterocyclic carbenes (NHC)
Imidazolium halide salts
green chemistry, 323–324
large scale, 22
preparation, 19–20
Imidazolium salts. *See* Plasticizers; Polymer gel electrolytes
Industrial chemistry, green chemistry, 83
Infrared study, 1-ethyl-3-methylimidazolium chloride and *N*-1-butylpyridinium chloride, 179–180
Ion exchangers
conventional Aliquat 336, 527, 528*f*
dicyclohexano-18-crown-6 ether (DCH18C6), 529
effect of organic phase water concentration on strontium partitioning, 530, 531*f*
extraction of strontium by DCH18C6 into *N,N*-dialkylimidazolium-based series, 529–530
mechanism involving exchange of Sr–DCH18C6 cation for ionic liquid cation, 535
nitrate content, 535
nitric acid dependencies in solvents, 530, 532*f*, 533
1-octanol as diluents for crown ethers in Sr extraction, 529
relationship between strontium partitioning and crown ether concentration, 530, 532*f*
selection of solvent, 529
Sr partitioning into 1-alkyl-3-methylimidazolium-based ionic liquids containing crown ether, 527, 529–535
structure for Sr–DCH18C6 complex, 533
X-ray absorption measurements, 533, 534*f*
Ionic conductivity. *See* Polymer gel electrolytes
Ionic liquids (ILs)
Abraham's generalized solvent equation, 6
activity and stability of enzymes in, 198–199
advantage of modern scientific microwave equipment, 17–18
anions containing fluorinated alkyl groups, 58–59
anions controlling solvent reactivity, 4
anions with tetraalkylphosphonium cations, 43*f*
case against 1-butyl-3-methylimidazolium

hexafluorophosphate ([C_4mim][PF_6]), 4–6
catalogs and chemical supply, 102–103
challenges of current research, 72–74
characterization, 71
characterization using polarity scales, 7–8
chemical reaction media, 382–383
chiral cations and anions, 17*f*
chiral ILs, 16–17, 26–28
commercial availability and applications, 103
commercially available, 104*t*, 105*t*, 106*t*
common anions, 3, 301*f*
common cations, 301*f*
definition, 3, 100–101
1,3-dialkylimidazolium-2-carboxylates, 94–96
distribution ratios for solutes between toluene and [C_4mim][PF_6], 8*f*
environmentally friendly volatile organic solvent replacements, 301
1-ethyl-3-methylimidazolium tetrafluoroborate [C_2mim][BF_4], 14–15
examples, 71*f*
first generation, 58
general syntheses, 15–17
green chemistry, 2–3, 4
halide-free ILs from fluorinated esters and alkyl sulfonates, 23–24
halide-free ILs from imidazolium-based carbenes, 24–26
high purity, 93
historical development, 58
history of chemistry, 42
hydroformylation of 1-octene, 65–67
ideal solvent requirements, 3
imidazolium and pyridinium halide salts, 19
imidazolium carbene routes to halide-free ILs, 25*f*
interaction parameters from linear solvent energy relationship (LSER) model for partitioning organic solutes, 9*t*
large scale [Rmim]X and [Rpy]X salt preparations, 22
liquid clathrate formation, 8–10
LSER and property contributions, 6–8
methanesulfonic acid and recycling, 24
microwave irradiation for preparation, 17–22
molecular structures, 163
partitioning of organic molecules in IL/water, 7–8
physical properties of chiral [dpeim]$^+$ salts, 27*t*
plasticizers, 469
possibilities, 3–4
preparation of 1-alkyl-3-methylimidazolium ILs, 16*f*
preparation of [Rmim]X salts on medium scale, 20*t*
preparation of [Rpy]X salts on medium scale, 21*t*
preparation of halide-free ILs, 22–26
preparation of imidazolium and pyridinium halides in open vessels, 19–20
preparation of lithium containing, 483
previous [Rmim]$^+$ syntheses using domestic microwave heating, 18*f*
properties, 345–347
quaternized nitrogen cations, 265
ratio of aromatic to IL in lower phase of liquid clathrate biphasic systems, 9*f*
reaction emphasis, 14
reaction media for organic synthesis, 42, 44
ready synthesis and separation of [C_4mim][PF_6], 5*f*
recycling of trifluoroethanoic acid, 24*f*

[Rmim][CF_3CO_2] and [Rmim][CH_3SO_3] preparations, 23*f*
[Rmim]X and [Rpy]X preparations using controlled microwave (mw) heating, 19*f*
salt solutions, 345
second generation, 58
solvatochromatic probes, 7
sonochemical synthesis, 88–90
structure of *S*-[dpeim] bromide, 28*f*
synthesis of 1-alkyl-3-methylimidazol-2-ylidenes using K[$OCMe_3$], 26*t*
synthesis of alkyloligoethersulfate, 62–63
synthesis of octylsulfate, 60–62
synthesis of *S*-[dpeim][CH_3CO_2], 27*f*
task specific, 59, 102
very weakly basic anions, 265–266
volatile organic solvent replacements, 301
See also Coordination polymerization of olefins; Green synthesis of ionic liquids; Microwave irradiation; Molecular structures; Phosphonium ionic liquids; Polymer gel electrolytes; Room temperature ionic liquids (RTILs)
Ionic strength, chlorophenol extraction, 555, 556*f*
Isomerization, allylic acetates by palladium catalysis, 318–319

K

Ketone arylation, palladium-mediated, 333, 336*t*
Kinetics
diffusion controlled bimolecular rate constant, 358
See also Diffusion controlled reactions; Pulse radiolysis

L

Lennard–Jones parameters, 1,3-dimethylimidazolium hexafluorophosphate [dmim][PF_6] and [dmim]Cl, 154*t*
Levich equation, limiting current, 432
Lewis bases, ionic liquids, 267, 273
Limiting current, Levich equation, 432
Linear solvent energy relationship (LSER)
interaction parameters for partitioning, 8*t*
modeling solvent properties of ionic liquids (ILs), 6–7
Lipases
activity in ionic liquids, 230
See also Candida antarctica lipase B (CaL B); Enantioselective acylation
Liquid clathrate formation, ionic liquid/aromatic mixtures, 8–10
Liquid-liquid equilibrium (LLE)
coexistence curves for 1-methyl-3-ethyl-imidazolium bis(trifluoromethyl-sulfonyl) amide ([emim][NTf_2]) and alcohols, 146*f*
determination method, 113
experimental setup, 144*f*
LLE of 1-propanol and [bmim][BF_4] as function of mol% IL, 118*f*
LLE of 1-propanol and [bmim][BF_4] as function of wt% IL, 117*f*
mixtures containing ionic liquids, 142, 145
solubility of reactants and products in ionic liquids, 111–112
See also 1-Butyl-3-methylimidazolium tetrafluoroborate [bmim][BF_4]
Liquid structures
cation-cation distributions, 159
comparison with crystal structures, 160

1,3-dimethylimidazolium hexafluorophosphate [dmim][PF_6] and [dmim]Cl, 155, 157–160
experimental and modeled differential cross sections vs. function of Q, 156*f*
molecular dynamics simulations, 159–160
partial radial distribution functions, 157*f*
three dimensional probability distributions, 158–159
See also Crystal structures
Lithium. *See* Polymer gel electrolytes
Low melting salts, commercially available, 104*t*, 105*t*, 106*t*

M

Magnesium
electrolysis of $MgCl_2$–BMIC (1-butyl-3-methylimidazolium chloride), 507
electrowinning, 496
See also Electro-processing of reactive metals
Mechanism, superoxide ion degrading polychlorinated aromatics, 511
Melting points
1-ethyl-3-methylimidazolium tetrafluoroborate [C_2mim][BF_4], 14–15
limits of ionic liquids, 100–101
Metacresol purple
probe indicator, 268
structure, 268
visible spectroscopy, 269*f*
Metal deposition
$AlCl_3$-based ionic liquid, 496
See also Choline chloride based ionic liquids
Metathesis
anion route to air- and water-stable ionic liquids, 73–74
ionic liquid synthesis, 34–35
phosphonium halide conversion by, 45
process parameters, 36*t*
routes to phosphonium ionic liquids, 49–50
Methane, solubility in [bmim][BF_4] and [bmim][PF_6], 113, 115*f*
Methanesulfonic acid, recycling, 24
Methyl acetate, transesterification, 338, 339*t*
Methyl acrylate (MA). *See* Diels–Alder reactions
1-Methyl-3-butylimidazolium triflimide ([Bmim][Tfi])
benzaldehyde reduction, 416*f*
structure of Bmim cation, 412*f*
See also Benzaldehyde
Methyl-*N*-butyl-pyridinium tetrafluoroborate [4-M-nBP][BF_4]
limiting activity coefficients of alcohols in, 139*f*
limiting activity coefficients of alkylbenzenes in, 139*f*
limiting activity coefficients of *n*-alkanes in, 138*f*
molar excess volume of, with methanol, 148*f*
structure, 138*f*
N-Methyl-*N*-butylpyrrolidinium triflimide [Pyrd][Tfi]
cyclic voltammograms for benzaldehyde reduction, 417*f*
square-wave voltammograms for benzaldehyde reduction, 418*f*
See also Benzaldehyde
1-Methyl-3-ethyl-imidazolium bis(trifluoromethyl-sulfonyl) amide ([emim][NTf_2])
activity coefficients and vapor pressures of diols in, 143*f*
limiting activity coefficients of alcohols in, 139*f*
limiting activity coefficients of alkylbenzenes in, 139*f*
limiting activity coefficients of *n*-alkanes in, 138*f*

liquid-liquid coexistence curves for, and alcohols, 146*f*
structure, 138*f*
Methylrhenium trioxide (MTO)
activator for hydrogen peroxide, 278
epoxidation of styrene with MTO/UHP, 279, 280*t*
epoxidations using urea hydrogen peroxide (UHP), 279–281
formation and reactivity of peroxorhenium complexes mpRe and dpRe, 278*f*
^{2}H NMR kinetics, 285–286
kinetics and thermodynamics of MTO/peroxide reactions, 281–282
kinetics of dpRe formation, 281
non-steady-state kinetics of olefin epoxidation, 282–284
solvent purity, 280–281
thermodynamics of MTO-peroxide equilibria, 282
Methyltributylammonium bis(trifluoromethylsulfonyl)imide [MtBA][NTf_2]
radiolysis, 388–392
See also Pulse radiolysis; Radiation chemistry
Methyl viologen (MV), electron transfer between quinones and, 403–405
Microwave irradiation
advantage of modern scientific equipment, 17–18
advantages of high yield method, 88
effect of power level on formation of 1-butyl-3-methylimidazolium tetrafluoroborate [C_4mim][BF_4], 87
effect of power on reactions using alkyl halides and 1-methyl imidazole (mim), 84
emerging tool, 83
general schematic, 85*f*
imidazolium and pyridinium halide salt preparation, 19–20
ionic liquid halide salt preparation, 18
ionic liquids (ILs) preparation, 17–22
preparation of 1,3-dialkyl imidazolium tetrafluoroborates, 88*t*
preparation of [Rmim]X salts on medium scale, 20*t*
preparation of [Rpy]X salts on medium scale, 21*t*
preparation of ionic liquids with longer alkyl chains or higher boiling points, 84–85
protocol, 87–88
purity of ionic salts, 85
[Rmim]$^+$ syntheses using domestic, 18*f*
solvent-free conditions, 84
synthesis of alkyl imidazolium halides, 86*t*
synthesis of ionic liquids with [BF_4] anions, 87
water- and air-stable ionic liquids, 85, 87
Molar excess volume
4-methyl-*N*-butyl-pyridinium tetrafluoroborate ([4-M-nBP][BF_4]) + methanol, 148*f*
mixtures containing ionic liquids, 145, 147
Molar heat capacity. *See* Heat capacity
Molecular structures
atom numbering convention for 1-butyl-3-methyl imidazolium [bmim] cation, 163*f*
[bmim][$(CF_3SO_2)_2N$], 168–169
[bmim] butanoate, 170, 172*f*
[bmim][PF_6], 164–165
[bmim][NO_3], 166–167
[bmim] pivalate, 170, 171*f*
[bmim] *p*-toluenesulfonate [bmim][*p*-TSA], 169–170, 171*f*
[bmim][BF_4], 166
[bmim] trifluoromethaneacetate [bmim][CF_3CO_2], 167–168

[bmim] trifluoromethanesulfate [bmim][CF_3SO_3], 168
methodology, 164
vibrational frequencies, 170, 172*f*
Molten salt technology, publications, 479
Monellin
thermal denaturation curves in water and *N,N*-butylmethylpyrrolidinium bis(trifluoromethane sulfonyl)imide ([bmp][Tf_2N]), 221*f*
thermal stability, 221
Morphology
aluminum deposits, 503, 504*f*, 505
method using Hull cell, 441
zinc deposition, 444, 445*f*

N

Na[*n*-C_8H_{17}-O-SO_3], synthesis, 60–61
Naphthalene
triplet energy transfer from benzophenone, 359
See also Diffusion controlled reactions
Neutralization reactions
ionic liquid synthesis, 34–35
process parameters, 36*t*
Neutron diffraction, 1,3-dimethylimidazolium hexafluorophosphate [dmim][PF_6], 153–155
NHC. *See N*-Heterocyclic carbenes (NHC)
Nile Red, solvatochromatic probe, 7
Nitrate. *See* Ion exchangers
Nitrogen, solubility in [bmim][BF_4], 113
Nonaqueous biocatalysis with polar substrates
acetylation of glucose, 226, 234–235
acetylation of glucose procedure, 236
activity of lipases in ionic liquids, 230
acylation of ascorbic acid procedure, 236–237
Candida antarctica (CAL-B) catalyzed acylation of L-ascorbic acid, 226–227, 233*t*, 234
conversion of *Pseudomonas-cepacia*-lipase catalyzed acetylation of racemic 1-phenylethanol, 229*f*
experimental, 235–237
ionic liquids expanding solvent polarity range, 234
polarity of ionic liquids vs. polar organic solvents, 228–230
preparation and purification of 3-alkyl-1-methylimidazolium tetrafluoroborate, 228
purification methods, 236
purification of ionic liquids, 227–228
regioselective acylation of ascorbic acid in ionic liquids, 232–234
regioselective acylation of glucose in ionic liquids, 230–232
regioselective CAL-B-catalyzed acetylation of glucose, 231*t*
solubility of glucose, 232
solvent purification, 234
synthesis of ionic liquids, 227, 228, 235–236
transesterification of *sec*-phenethyl alcohol, 236
Non-coordinating ionic liquids. *See* Coordination polymerization of olefins
Non-toxic pharmaceutically acceptable ions, pharmaceutical and food additive industries, 5–6
Normal pulse voltammetry (NPV)
diffusion coefficients of ferricenium/ferrocene, 429*t*, 431*t*
ferricenium/ferrocene, 431–432
Nucleation
aluminum on copper, 501–502
definitive test for adherence to models, 502
Nucleophilic addition

phosphonium ionic liquids, 44
superoxide ion, 511

O

Octanoic acid, ammoniolysis, 194
1-Octanol, diluent for crown ethers, 529
1-Octene, Rh-catalyzed hydroformylation, 65–67
Octylsulfate ionic liquids, synthesis, 60–62
Olefins. *See* Coordination polymerization of olefins; Epoxidations
Onsager radius, definition, 386
Organic electrochemistry
benzaldehyde reduction, 415–419
cyclic voltammogram for benzaldehyde reduction in 1-methyl-3-butylimidazolium triflimide ([Bmim][Tfi]), 416*f*
cyclic voltammograms for benzaldehyde reduction in *N*-methyl-*N*-butylpyrrolidinium triflimide ([Pyrd][Tfi]), 417*f*
dimethylmaleate reduction, 413–415
electrolytic conversion of organic substrates, 410–411
experimental, 412–413
potential of ionic liquids in electrosynthesis, 411
sensitivity to operational conditions, 411–412
square wave voltammetry technique, 414
square wave voltammogram for benzaldehyde reduction in [Pyrd][Tfi], 418*f*
square wave voltammogram for dimethylmaleate reduction, 415*f*
structure of Pyrd and Bmim cations, 412*f*
unique features of ionic liquids, 411
Organic molecules, partitioning in ionic liquids/water systems, 7–8, 9*t*
Organic solvents, polarity values, 292*t*
Overpotential deposition (OPD). *See* Electrodeposition
Oxidation reactions
chlorpromazine by $Br_2\bullet$ radicals, 400, 401*t*
chlorpromazine by $CCl_3O_2\bullet$ radicals, 398, 400
Trolox by $CCl_3O_2\bullet$ radicals, 398, 400
See also Pulse radiolysis; Superoxide ion
Oxidoreductases, activity in ionic liquids, 203
Oxygen, singlet, Diels–Alder reaction with diphenylisobenzofuran, 363–364

P

Palladium catalysis, isomerization of allylic acetate, 318–319
Palladium complexes
styrene/CO copolymerization, 304–306
See also *N*-Heterocyclic carbenes (NHC)
Partitioning, organic molecules in ionic liquids/water systems, 7, 8*t*
Partitioning, strontium. *See* Ion exchangers
Patent activity, phosphonium salts, 53
Peptide amidase, amidation, 209–210
Phenols. *See* Chlorophenols
1-Phenylethanol
acetylation of racemic, 229*f*
kinetic resolution, 196–197
kinetic resolution of racemic, in ILs-supercritical CO_2, 247–249
5-Phenyl-1-penten-3-ol
transesterification, 253–254
transesterification under reduced pressure, 258, 259*f*, 260
See also Enantioselective acylation

Phosphonium ionic liquids
  addition of metal chlorides to phosphonium chlorides, 45
  decomposition, 46
  densities, 47, 49
  density vs. temperature, 48*f*
  features and advantages, 44–49
  general formula, 45–46
  halide free routes to, 52
  interest, 53
  metathesis methods, 45
  metathesis routes to, 49–50
  patent activity, 53
  phosphonium phosphinates, 49
  potential, 44
  preparation by nucleophilic addition, 44
  quaternization reactions, 44–45
  selection of anion and cation, 44–45
  stability, 46
  stability with respect to temperature, 47*f*
  synthesis of trihexyl(tetradecyl)phosphonium bis(2,4,4-trimethylpentyl) phosphinate, 50, 51
  synthesis of trihexyl(tetradecyl)phosphonium chloride, 50
  synthesis of tri-*iso*-butyl(methyl)phosphonium tosylate, 52
  viscosity, 46–47
  viscosity with respect to temperature, 47*f*
Phosphorescence, ionic liquids, 346
Photochemistry
  AN (anthracene) in acidic 1-ethyl-3-methylimidazolium [emim][Cl/$AlCl_3$], 347–349
  AN in deoxygenated, basic [emim][Cl/$AlCl_3$], 347, 348
  electron and energy transfer in ionic liquids, 347
  electron transfer from excited bipyridyl complex $Ru(bypy)^{+2}$ to methylviologen (MV^{+2}), 347
  experimental, 379
  formation of 9,10-dihydroanthracene in [emim][Cl/$AlCl_3$], 352
  hydrogen atom abstraction by benzophenone triplet excited state, 349, 350
  initial reactions of anthracene in acidic [emim][Cl/$AlCl_3$], 349
  investigations in chloroaluminate salts, 371–372
  irradiation of tris Fe(II) complexes in Lewis acid [*N*-ethylpyridinium][Br/$AlCl_3$], 351
  ketone photoreduction of *N*-butylpicolinyl tetrafluoroborate [BuPic(BF_4)], 374–375, 376*f*
  9-methylanthracene (9-$ANCH_3$) ideal for photoinduced electron transfer (PET), 351, 352*f*
  parallels with radiation chemistry, 382
  photoinduced electron transfer in ionic liquid (IL), 351
  photointerconversion of *cis*- and *trans*-stilbene, 353
  photoreaction of 9-chloroanthracene in basic [emim][Cl/$AlCl_3$], 353
  photoreduction of 4,4'-dimethoxybenzophenone in room temperature ionic liquids (RTILs), 375*t*
  photoreduction of benzophenone in presence of various amount of 1-butyl-3-methylimidazolium (BMI) cation, 376*t*
  product distributions in photoreduction of benzophenone with triethylamine in BMI(BF_4) and BMMI(BF_4), 377*f*
  product ratios in benzophenone photoreduction, 373, 374*t*
  proposed mechanisms in BMI and BMMI RTILs, 378*f*

radical pair/ion pair equilibria in RTILs, 372
reduction of benzophenones to benzhydrols by primary amines, 349, 350
reduction of UCl_6^-, 349, 351
reusable synthetic system, 371*f*
role of C_2-proton, 377–378
simplified mechanism for amine mediated photoreduction, 372–373
solvent effects and role of cation, 374–377
stilbenes in ionic liquids, 354*t*
transients in photochemistry of 9-$ANCH_3$, 352*f*
Photoinduced electron transfer (PET)
ionic liquids, 351
See also Photochemistry
Photoreduction. *See* Photochemistry
Plasticizers
additives, 469
comparison of long term thermal stability of diocyl phthalate (DOP)- and [1-butyl-3-methylimidazolium][PF_6] ([bmim][PF_6])-plasticized poly(methyl methacrylate) (PMMA), 475*f*
diocyl phthalate (DOP), 469
effect of, and content on elastic moduli of PMMA samples, 472*f*
effect of 30 vol% [hmim][PF_6], DOP, and [bmim][PF_6] as, in PMMA, 473*f*
effect of [bmim][PF_6] content in plasticized PMMA, 473*f*
elastic moduli, 471–472
experimental, 470–471
glass transition temperature, 474–475
ionic liquids, 469
mass loss monitoring, 470–471
mechanical tests, 470–471
phthalates, adipates, and trimellitates, 469
PMMA synthesis, 470
thermal stability, 472, 474
Polarity
absorption maxima and, of 1-R-3-methylimidazolium salts, 291*t*
activity of lipases in ionic liquids, 230
expanding, range for ionic liquids, 292–294
ionic liquids, 345
organic solvents, 292*t*
probing, of ionic liquids, 290–291
solvatochromic dyes for probing, 290, 291*f*
Polar organic solvents, polarity of ionic liquids vs., 228–230
Polar substrates. *See* Nonaqueous biocatalysis with polar substrates
Pollution prevention, strategy, 71
Polychlorinated aromatics, degradation by superoxide ion, 511, 517, 522*f*
Polymer gel electrolytes
advantages of ionic liquids, 480–481
charge and discharge behavior of V_2O_5 vs. Li metal, 491*f*
charging and discharging for Li metal, $LiCoO_2$ battery, 490*f*
charging and discharging for Li metal, $LiMn_2O_4$ battery, 490*f*
dual intercalating molten electrolyte (DIME), 480
electrochemical characterizations of, 488–489
electrochemical techniques, 484–485
experimental, 482–485
ionic conductivity results, 485, 487
Li metal-metal oxide battery test cells, 489, 491
percent discharge efficiencies vs. number of cycles for cells of Li metal and V_2O_5, 492*f*
preparation, 483–484
preparation of ionic liquids, 482–483
preparation of lithium containing ionic liquids, 483

preparation of polymer composite electrodes, 484
structure of common imidazolium based ionic liquids, 481*f*
synthesis of initial substituted imidazolium salts, 482
temperature dependent ionic conductivity, 486*f*
Polymerization. *See* Coordination polymerization of olefins
Poly(methyl methacrylate) (PMMA). *See* Plasticizers
Porphryin, self-assembly in ionic liquid, 219, 220*f*
Pre-solvated electron states, radiation chemistry, 385
Pressure. *See* Enantioselective acylation
Prices, commercial use, 37–38
1-Propanol, liquid-liquid equilibrium of, and [bmim][BF_4], 117*f*, 118*f*
Proteases, enzyme in ionic liquids, 200
Publications, 1-butyl-3-methylimidazolium hexafluorophosphate [bmim][PF_6], 101
Pulse radiolysis
electron transfer between quinones and methyl viologen (MV), 403–405
hydrogen abstraction and addition reactions, 406–407
oxidation of chlorpromazine and Trolox by CCl_3O_2. radicals, 398, 400
oxidation of chlorpromazine by Br_2. radicals, 400, 401*f*
radiation chemistry, 382
radical reactions in various solvents, 398
rate constant for reduction of duroquinone (DQ) by BPH• in various solvents, 402*t*
rate constant for reduction reactions in various solvents, 403*t*
rate constants and activation energy for reactions of chlorpromazine and Trolox, 399*t*
rate constants for electron transfer between quinones and MV, 404*t*
rate constants for electron transfer from $MV^{\bullet +}$ to DQ in different *N*-butylpyridinium tetrafluoroborate ($BuPyBF_4$)/water mixtures, 405*f*
rate constants for oxidation of chlorpromazine by Br_2• radicals, 401*t*
rate constants for reactions of •CF_3 radicals, 406*t*
reduction of DQ by benzophenone ketyl and pyridinyl radicals, 402–403
Purity
commercial use, 38–39
high, products by decarboxylation, 94
ionic salts by microwave heating, 85
issue for ionic liquids, 93
raw materials in synthesis, 77–78
solvent, for epoxidations, 280–281
Pyridinium halide salts
large scale, 22
preparation, 19–20, 21*t*
1-(2-Pyridylazo)-2-naphthol (PAN), extracting metal ions, 546

Q

Quaternary phosphonium salts
ionic liquids, 42
See also Phosphonium ionic liquids
Quaternized nitrogen cations, ionic liquids, 265
Quinones, electron transfer between, and methyl viologen, 403–405

R

Radiation chemistry

actions of various types of ionizing radiation, 384
chemical effects of radiation, 387–388
concentration where 37% of electrons survive to be solvated, C_{37}, 391
density of ionizations, 387
dry electron capture, 391–392
dry electron reactivity, 390–392
early reactions in radiolysis, 387*f*
effective polarities, 386
effects of functionalization, 392, 393*f*
electron-hole pair distances, 386
electron-hole pairs and geminate recombination, 385–387
forms of ionizing radiation and modes of energy deposition, 383–385
goals of radiolytic investigations of ionic liquids, 383
hole, 386
idealized case of isolated electron-hole pairs, 387
ionic liquids as chemical reaction media, 382–383
Onsager radius, 386
parallels in photochemistry, 382
polarity-sensitive fluorophores, 386
pre-solvated or dry electron states, 385
production of net radiolysis products, 388
radiolysis of methyltributylammonium bis(trifluoromethylsulfonyl)imide [MtBA][NTf_2], 388–392
rate constants for solvated electron capture and C_{37} values for dry electron capture by scavengers, 390*t*
rationale for investigating, 382–383
relaxation of solvated electron in ionic liquid with hydroxyl functionality, 393*f*
solvatochromatic dyes, 386
sources of ionizing radiation, 384
spectra of solvated electron and putative hole species by electron radiolysis, 389*f*
term, 383
thermalization and solvation processes following ionization, 385*f*
thermalized electrons, 385
track of energy deposition events, 384
transient absorption traces showing reactions of dry and solvated electrons with pyrene in [MtBA][NTf_2], 390*f*
Radicals. *See* Pulse radiolysis
Radiolysis
methyltributylammonium bis(trifluoromethylsulfonyl)imide [MtBA][NTf_2], 388–392
reactions, 387–388
See also Pulse radiolysis; Radiation chemistry
Rate constants
chlorpromazine and Trolox by CCl_3O_2• radicals, 399*t*
electron transfer between quinones and methyl viologen (MV), 404*t*
electron transfer from MV to duroquinone (DQ), 405*f*
oxidation of chlorpromazine by Br_2• radicals, 401*t*
reactions of •CF_3 radicals, 406*t*
reduction of DQ by BPH• in various solvents, 402*t*
reduction reactions in various solvents, 403*t*
See also Pulse radiolysis
Raw materials, industrial issues, 35–36
Reaction kinetics. *See* Pulse radiolysis
Reactive metals. *See* Electro-processing of reactive metals
Recovery, product, 203–204
Recycling
catalyst, 203–204

methanesulfonic acid, 24
trifluoroethanoic acid, 24
Reduction reactions
benzaldehyde, 415–419
dimethylmaleate, 413–415
duroquinone by benzophenone ketyl and pyridinyl radicals, 402–403
rate constants for, in various solvents, 403*t*
See also Organic electrochemistry; Pulse radiolysis
Regioselective acylation
ascorbic acid in ionic liquids, 232–234
glucose in ionic liquids, 230–232
Reichardt's dye, solvatochromatic probe, 7
Retention volume, equation, 136
Reversible zinc galvanic cell, battery, 446, 448*f*
Rh catalysis, hydroformylation of 1-octene, 65–67
Rhenium. *See* Methylrhenium trioxide (MTO)
Room temperature ionic liquids (RTILs)
absorption maxima and ET(30) solvent polarity values for some 1-R-3-methylimidazolium salts, 291*t*
acid-base neutralization reactions, 34–35
alkylation of 1-methylimidazole to form 1-X-3-methylimidazolium halides, 293*f*
1-butyl-3-methylimidazolium hexafluorophosphate ($BMIPF_6$), 35
commercial importance, 34
comparative prices of various solvents, 38*t*
cost for commercial use of $BMIPF_6$, 37–38
definition, 100–101
description, 33, 371
direct combination, 34–35
expanding polarity range, 292–294
experimental, 295–297
field and phenomenon, 42
green alternatives, 192, 289, 371
hazard classifications of materials for $BMIPF_6$ synthesis, 37*t*
hazards, 36–37
influence of RTIL polarity on Diels–Alder reaction, 294–295
interest, 33–34
means of acquisition, 34–35
metathesis route, 34–35
novel class of solvents, 33
probing polarity of, 290–291
procedure for Diels–Alder reactions, 297
process parameters for $BMIPF_6$ synthesis, 36*t*
properties of $BMIPF_6$, 35*t*
purity of $BMIPF_6$, 38–39
quaternary phosphonium salts, 42
raw material issues, 35–36
reaction rate information, 358
selecting synthetic route, 35–39
solvatochromic dyes, 290, 291*f*
structures of proposed 1-X-3-methylimidazolium bistrifylimide series, 292*f*
synthesis method, 295–297
synthetic organic chemistry, 71
toxicology, 37
typical analysis of $BMIPF_6$, 39*t*
unconventional properties, 526–527
unique properties, 469
See also Ionic liquids (ILs)
Rotating disk electrode voltammetry (RDE)
diffusion coefficients of ferricenium/ferrocene, 429*t*, 431*t*
ferricenium/ferrocene, 432, 435

S

Salts, low melting
commercially available, 104*t*, 105*t*, 106*t*

intrinsic utility, 103
Salt solutions, properties, 345
Scanning tunneling microscopy (STM). *See* Electrodeposition
Scope, biotransformations in ionic liquids, 195–196
Selection. *See* Ionic liquids (ILs)
Selectivity
enhancements in Diels–Alder reactions in ILs, 366
hydrogen peroxide, 278
solvent influence on endo/exo, 180–181
See also Enantioselective acylation
Self-assembly, porphyrin in ionic liquid, 219, 220*f*
L-Serine methyl ester hydrochloride, availability, 103
Silicon, ultrahigh vacuum (UHV) deposition of Ge quantum dots on, 463–464
Simulations, 1,3-dimethylimidazolium hexafluorophosphate [dmim][PF_6] and [dmim]Cl, 154*t*
Single crystal X-ray diffraction, 1,3-dimethylimidazolium hexafluorophosphate [dmim][PF_6], 152–153
Singlet oxygen, Diels–Alder reaction with diphenylisobenzofuran, 363–364
Small-angle X-ray scattering (SAXS), interaction of water with 1-decyl-3-methylimidazolium bromide, 536–537, 538*f*, 539*f*
Solubility
apparatus for determining liquid-liquid, 112*f*
benzene in [bmim][BF_4], 116, 117*f*
density accuracy, 111
gas measurement method, 112
Henry's law constant, 113–114, 115*f*
motivation for measuring gas, 111
solutes in ionic liquids at infinite dilution, 135–137
See also 1-Butyl-3-methylimidazolium tetrafluoroborate [bmim][BF_4]
Solute jump
probability, 362–363
See also Diffusion controlled reactions
Solvatochromic dyes, probing polarity, 290–291
Solvent properties
autocorrelation curves, 218*f*
diffusion coefficients of *N,N,N,N*-tetramethyl-*p*-phenylenediamine (TMPD) in [bmim][PF_6], 218
effects of controlled wetting, 217–218
experimental, 214–215
fluorescence correlation spectroscopy (FCS) method, 214–215
instrumentation, 214–215
ionic liquids, 213
kinetics of acid-triggered *meso*-tetrakis(4-sulfonatophenyl)porphine (TSPP), 220*f*
materials, 214
measuring translational diffusion within ionic liquids, 215, 217–218
native fluorescence as probe of thermodynamic stability of single tryptophan (Trp) protein within ionic liquid, 219–221
porphyrin self-assembly within ionic liquids, 219, 220*f*
radiation chemistry, 382
schematic of confocal FCS system, 216*f*
steady state fluorescence measurements, 215
temporal autocorrelation function, 215, 217
thermal denaturation for monellin in water and *N,N*-butylmethylpyrrolidinium bis(trifluoromethane

sulfonyl)imide ([bmp][Tf_2N]), 221*f*
Solvent-solute interactions. *See* Ferricenium/ferrocene
Solvodynamic radius, equation, 424
Sonochemical synthesis
ionic liquids, 88–90
See also Ultrasound
Specific heat capacity. *See* Heat capacity
Spectroscopy, ferrocene and ferricenium in [C_4mim][PF_6], 426, 427*f*
Square-wave voltammetry
benzaldehyde reduction, 418*f*
dimethylmaleate reduction, 415*f*
technique, 414
See also Organic electrochemistry
Stability
enzymes in ionic liquids, 198–199
function of temperature for phosphonium ionic liquids, 47*f*
phosphonium ionic liquids, 46
See also Thermal stability
Stilbenes
photochemistry, 353, 354*t*
See also Photochemistry
Stokes–Einstein, diffusional movements, 423–424
Stokes' law, equation, 423
Strontium partitioning. *See* Ion exchangers
Structure-property, ionic liquids, 130
Styrene
epoxidation with methylrhenium trioxide/urea hydrogen peroxide (MTO/UHP), 279, 280*t*
kinetics in different solvents of several styrenes, 284*t*
rate constants for oxidation of α-methylstyrene, 285*t*
See also Coordination polymerization of olefins
Supercritical carbon dioxide ($scCO_2$)
butyl butyrate synthesis in ILs-$scCO_2$, 245–246
enzymatic reactions in ILs-$scCO_2$, 242
kinetic resolution of *rac*-1-phenylethanol, 247–249
reaction media for biocatalysis, 240
use with ionic liquids, 203–204
See also Enzymatic catalysis
Superoxide ion
chromatogram of benzoic acid synthesis products, 516*f*
chromatogram of benzophenone synthesis products, 518*f*
chromatogram of hexachlorobenzene (HCB) in cyclohexane before and after electrolysis, 522*f*
counter electrode compartment, 514*f*
cyclic voltammogram of pure 1-butyl-3-methylimidazolium hexafluorophosphate [bmim][HFP], 513*f*
degradation efficiency, 517, 519
degrading polychlorinated aromatics and biphenyls, 511, 517, 522*f*
electrochemical cell for bulk electrolysis, 514*f*
electrochemically generated, 510
experimental, 512, 515
generation in room temperature ionic liquids (RTILs), 511–512
mass spectra for overlapping peaks in benzophenone synthesis, 520*f*, 521*f*
mechanism of destruction, 511
net oxidation reaction, 510–511
nucleophilic addition, 511
oxidation of alcohols, 510
oxidation of primary and secondary alcohols, 515, 517
Surfactants. *See* Cationic surfactants
Suzuki–Miyaura reactions, palladium-mediated, 333, 335*t*
Synthesis
1,3-dialkylimidazolium-2-carboxylates, 94–96
ionic liquids, 15–17
microwave irradiation, 17–22, 83–88
ultrasound, 88–90

T

Task-specific ionic liquids, definition, 59, 102
Temperature
dependence for Diels–Alder reaction, 364–366
See also Diffusion controlled reactions
Temporal autocorrelation function, equation, 215, 217
Tetrafluoroborate $[BF_4]^-$
exchange of halide for, 102
ionic liquid anion, 3, 193*f*
microwave synthesis of ionic liquids with, 87
meso-Tetrakis(4-sulfonatophenyl)porphine (TSPP), self-assembly in ionic liquid, 219, 220*f*
Thermalized electrons, radiation chemistry, 385
Thermal properties, alkylsulfate and alkyloligoethersulfate ionic liquids, 63–64
Thermal stability
Candida antartica B (CaL B) preparations, 199
fluorescence as probe, 219–221
monellin in water and *N,N*-butylmethylpyrrolidinium bis(trifluoromethane sulfonyl)imide ([bmp][Tf_2N]), 221*f*
plasticized poly(methyl methacrylate) (PMMA), 472, 474, 475*f*
See also Stability
Thermal storage, ionic liquids, 129
Thermodynamic properties
activity coefficients and solubilities of solutes at infinite dilution, 135–137
activity coefficients and vapor pressures of diols in diol + 1-methyl-3-ethylimidazolium bis(trifluoromethyl-sulfonyl) amide ([emim][NTf_2]) mixture, 143*f*
activity coefficients of solutes in whole concentration range, 137, 142
densities and excess volumes of mixtures containing ionic liquids, 145, 147
experimental setup for liquid-liquid equilibrium, 144*f*
heat of solution of liquid solute at infinite dilution in ionic liquid, 137, 141*t*
Henry's coefficient, 135–136
limiting activity coefficients of alcohols in ionic liquids, 139*f*
limiting activity coefficients of *n*-alkanes in ionic liquids, 138*f*
limiting activity coefficients of alkylbenzenes in ionic liquids, 139*f*
liquid-liquid coexistence curves for [emim][NTf_2] + alcohols, 146*f*
liquid-liquid equilibria containing ionic liquids, 142, 145
molar excess volume, V^E, 145
retention volume, 136
schematic of transpiration apparatus, 140*f*
structures of [4-M-nBP][BF_4], [emim][NTf_2], and [emmim][NTf_2], 138*f*
V^E of 4-methyl-*N*-butyl-pyridinum tetrafluoroborate [4-M-nBP][BF_4] + methanol, 148*f*
Thermolysin, enzyme in ionic liquids, 200
1-(2-Thiazolylazo)-2-naphthol (TAN), extracting metal ions, 546
Thymol blue
dye, 546
pH dependence in ionic liquids, 554
See also Chlorophenols
Titanium
electrodeposition, 495

See also Electro-processing of reactive metals
Toluene, distribution ratios of organic solutes between, and $[C_4mim][NTf_2]$, 8*f*
Toxicology, reagents for synthesis, 36–37
Transesterification
chiral allylic alcohols, 197
chiral primary alcohols, 197–198
chiral secondary alcohols, 197
enantioselectivity, 196–198
enzyme-catalyzed, in ionic liquids, 194
N-heterocyclic carbene (NHC)-catalyzed, 333, 337–338
sec-phenethyl alcohol, 236
See also *N*-Heterocyclic carbenes (NHC)
Transition structures
calculating activation energy for, 184, 187
See also Diels–Alder reactions
Translational diffusion, measuring within ionic liquid, 215, 217–218
Trichloromethylperoxyl radicals, oxidation of chlorpromazine and Trolox, 398, 400
Trifluoroethanoic acid,recycling, 24*f*
Trolox (6-hydroxy-2,5,7,8-tetramethylchroman-2-carboxylic acid)
oxidation by CCl_3O_2• radicals, 398, 400
rate constants and activation energies for reactions with CCl_3O_2• radicals, 399*t*
See also Pulse radiolysis
Tryptophan, fluorescence as probe of thermodynamic stability, 219–221
Tsuji–Trost reaction
hydrogen bonding, 315, 319
inhibition of neutral, 316–317
Turnover frequencies (TOF), enzyme stability and activity, 198
Turnover numbers (TON), enzyme stability and activity, 198

U

Ultrasound
effect on reactions of alkyl halides and 1-methylimidazole (mim), 88
generality of reaction, 90
in-situ generation of ionic liquids and utilization, 90
preparation of 1-alkyl-3-methylimidazolium halides, 89*t*
schematic, 89*f*
visual monitoring ionic liquid formation, 90
Underpotential deposition (UPD)
zinc, 443, 444*f*
See also Electrodeposition
Urea hydrogen peroxide (UHP). *See* Methylrhenium trioxide (MTO)

V

Value, biotransformations in ionic liquids, 195–196
Vanadium pentoxide, charge and discharge behavior, 491*f*
Vibrational frequencies, C_2–H bond stretch in imidazolium ring vs. anion, 170, 172*f*
Viscosity
alkylsulfate and alkyloligoethersulfate ionic liquids, 64–65
$CrCl_3$•$6H_2O$ and choline chloride mixture, 449–450
function of temperature for phosphonium ionic liquids, 47*f*
ionic liquids, 346
phosphonium ionic liquids, 46–47
Volatile organic compounds (VOCs)
advantages of ionic liquids over, 4
green chemistry redesigning, 2–3

Volumetric heat capacity
ionic liquids, 130*t*
performance vs. temperature, 131*f*
See also Heat capacity

W

Waste minimization, strategy, 71
Water. *See* Cationic surfactants
Water-stable ionic liquids, anion metathesis route, 73–74
Weakly basic anions, ionic liquids, 265–266

X

X-ray crystal structure
melt involving 1-ethyl-3-methylimidazolium chloride (EMIC) and $AlCl_3$, 178–179
(*N*-heterocyclic carbene (NHC))Pd(allyl)Cl complex, 332*f*
palladium complex containing NHC ligands, 325*f*
X-ray diffraction, 1,3-dimethylimidazolium hexafluorophosphate [dmim][PF_6], 152–153

Y

Yeast alcohol dehydrogenase (YADH), activity in ionic liquids, 203

Z

Zinc chloride. *See* Choline chloride based ionic liquids
Zinc galvanic cell, reversible, 446, 448*f*